International Conference on Science and Applied Science 2017

Journal of Physics: Conference Series Volume 909

Solo, Indonesia
29 July 2017

Editors:

Dra. Suparmi
Fuad Anwar

ISBN: 978-1-5108-5339-3
ISSN: 1742-6588

Printed from e-media with permission by:

Curran Associates, Inc.
57 Morehouse Lane
Red Hook, NY 12571

CURRAN ASSOCIATES INC. proceedings.com

Some format issues inherent in the e-media version may also appear in this print version.

Printed by Curran Associates, Inc. (2018)

For permission requests, please contact the Institute of Physics at the address below.

Institute of Physics
Dirac House, Temple Back
Bristol BS1 6BE UK

Phone: 44 1 17 929 7481
Fax: 44 1 17 920 0979

techtracking@iop.org

Additional copies of this publication are available from:

Curran Associates, Inc.
57 Morehouse Lane
Red Hook, NY 12571 USA
Phone: 845-758-0400
Fax: 845-758-2633
Email: curran@proceedings.com
Web: www.proceedings.com

Table of contents

Volume 909

International Conference on Science and Applied Science 2017
29 July 2017, Solo, Indonesia

Accepted papers received: 23 October 2017
Published online: 27 November 2017

Preface

Keynote Speaker

Physics

Educational Physics

Educational Mathematics

Educational Biology

Civil Engineering

Chemistry

Chemical Engineering

Biology

International Conference on Science and Applied Science 2017 (ICSAS 2017)

Lorin Solo Hotel
Surakarta, Indonesia
29 July 2017

Preface

International Conference on Science and Applied Science 2017 (ICSAS 2017) was held at the Lorin Solo Hotel, Surakarta, Indonesia on 29 July 2017. The ICSAS 2017 conference is aimed to bring together scholars, leading researchers and experts from diverse backgrounds and applications areas in Science. Special emphasis is placed on promoting interaction between the science theoretical, experimental, and education sciences, engineering so that a high level exchange in new and emerging areas within Mathematics, Chemistry, Physics and Biology, all areas of sciences and applied mathematics and sciences is achieved.

In ICSAS 2017, there are eight parallel sessions and four keynote speakers. It is an honour to present this volume of Journal of Physics: Conference Series (JPCS) and we deeply thank the authors for their enthusiastic and high-grade contribution. From the review results, there are 93 papers which will be published in JPCS. We would like to express our sincere gratitude to all in the Programming Committee who have reviewed the papers and developed a very interesting Conference Program, as well as thanking the invited and plenary speakers. Finally, we would like to thank the conference chairman, the members of the steering committee, the organizing committee, the organizing secretariat and the financial support from the Sebelas Maret University that allowed ICSAS 2017 to be a success.

The Editor of the ICSAS 2017 Proceedings:

Prof. Dra. Suparmi, M.A.,Ph.D.
Dr. Fuad Anwar, S.Si., M.Si.

Organizer of International Conference on Science and Applied Science 2017 (ICSAS 2017)

Organizer

Graduate Program, Physics Department, Sebelas Maret University, Indonesia
Jl. Ir. Sutami 36A Kentingan Jebres Surakarta 57126, Indonesia
Phone/fax : (0271) 632450 psw 308
Email : icsas@mail.uns.ac.id

Chairman

1. Prof. Suparmi, M.A., Ph.D, Sebelas Maret University, Indonesia
2. Dr. Fuad Anwar, S.Si., M.Si, Sebelas Maret University, Indonesia

Organizing Committee

1. Prof. Cari, M.A., M.Sc., PhD, Sebelas Maret University, Indonesia
2. Ahmad Marzuki, S.Si., Ph.D., Sebelas Maret University, Indonesia
3. Dr. Eng Budi Purnama, S.Si, M.Si., Sebelas Maret University, Indonesia
4. Dr. Fahru Nurosyid, S.Si., M.Si., Sebelas Maret University, Indonesia
5. Drs. Harjana, M.Si. M.Sc., Ph.D, Sebelas Maret University, Indonesia
6. Dr. Agus Supriyanto, S.Si, M.Si. Sebelas Maret University, Indonesia
7. Dr. Yofentina Iriani, S.Si., M.Si., Sebelas Maret University, Indonesia
8. Dr.Eng. Risa Suryana, S.Si, M.Si., Sebelas Maret University, Indonesia
9. Khairuddin, S.Si., M.Phil, Ph.D., Sebelas Maret University, Indonesia
10. Drs. Iwan Yahya, M.Si., Sebelas Maret University, Indonesia
11. Mohtar Yunianto, S.Si, M.Si., Sebelas Maret University, Indonesia
12. Nuryani,S.Si,M.Si, Ph.D., Sebelas Maret University, Indonesia
13. Beta Nur Pratiwi, S.Si., M.Si., Sebelas Maret University, Indonesia
14. Dewanta Arya Nugraha, S.Pd., M.Pd., M.Si., Sebelas Maret University, Indonesia

ICSAS 2017

International Conference on Science and Applied Science 2017

UNS
SEBELAS MARET
UNIVERSITY

IOP Conf. Series: Journal of Physics: Conf. Series **909** (2017) 011003 doi:10.1088/1742-6596/909/1/011003

Photographs of International Conference on Science and Applied Science 2017 (ICSAS 2017)

Lorin Solo Hotel
Surakarta, Indonesia
29 July 2017

ICSAS
UNS

IOP Conf. Series: Journal of Physics: Conf. Series **909** (2017) 011003 doi:10.1088/1742-6596/909/1/011003

ICSAS
UNS

ICSAS

IOP Conf. Series: Journal of Physics: Conf. Series **909** (2017) 011003 doi:10.1088/1742-6596/909/1/011003

IOP Conf. Series: Journal of Physics: Conf. Series **909** (2017) 011004 doi:10.1088/1742-6596/909/1/011004

Peer review statement

All papers published in this volume of *Journal of Physics: Conference Series* have been peer reviewed through processes administered by the proceedings Editors. Reviews were conducted by expert referees to the professional and scientific standards expected of a proceedings journal published by IOP Publishing.

IOP Conf. Series: Journal of Physics: Conf. Series **909** (2017) 012001 doi:10.1088/1742-6596/909/1/012001

Examining live cell cultures during apoptosis by digital holographic phase imaging and Raman spectroscopy

Alexander Khmaladze

SUNY University at Albany, Physics Department, 1400 Washington Avenue, Albany, NY 12222

akhmaladze@albany.edu

Abstract. Cellular apoptosis is a unique, organized series of events, leading to programmed cell death. In this work, we present a combined digital holography/Raman spectroscopy technique to study live cell cultures during apoptosis. Digital holographic microscopy measurements of live cell cultures yield information about cell shape and volume, changes to which are indicative of alterations in cell cycle and initiation of cell death mechanisms. Raman spectroscopic measurements provide complementary information about cells, such as protein, lipid and nucleic acid content, and the spectral signatures associated with structural changes in molecules. Our work indicates that the chemical changes in proteins, which were detected by Raman measurements, preceded morphological changes, which were seen with digital holographic microscopy.

1. Introduction

Digital Holographic Microscopy (DHM) is an imaging technique which is becoming a routine method for three dimension real time imaging and inspection of microstructures and biological systems on a nanometer scale. When the light wave is reflecting from (or propagating through) a microscopic object, it slows down. The resulting wave front changes reveal variations in the optical thickness of the object (see figure 1(a)). This optical path length can be obtained for each point of the object from the phase of light for each pixel of the image $\varphi(x,y)$. The optical path length can be then converted to physical thickness (assuming that the index of refraction is known), providing the sample height information:

$$t(x,y)=\frac{\lambda}{2\pi}\frac{\varphi(x,y)}{n-n_0} \quad (1),$$

where λ is the wavelength of light, n is the index of refraction of the sample, and n_o is the index of refraction of the surrounding medium.

IOP Conf. Series: Journal of Physics: Conf. Series **909** (2017) 012001 doi:10.1088/1742-6596/909/1/012001

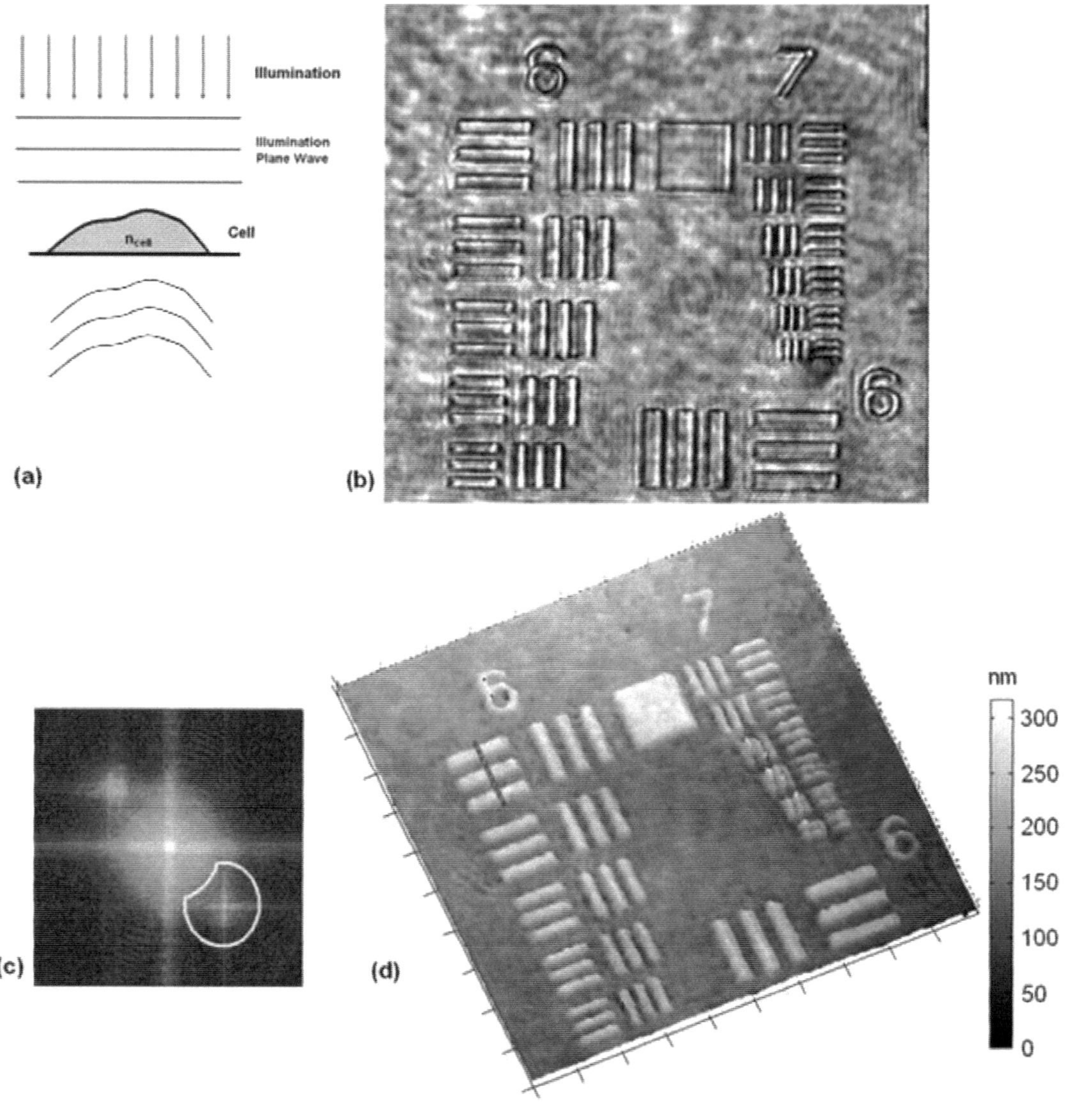

Figure 1. The principles of Digital Holographic Microscopy: (a) light wave propagating through a cell, showing the wavefront distortion, which resembles the shape of the cell, (b) the hologram of a reflective resolution target, (c) Fourier transform of (b), showing the first maximum, selected for subsequent reconstruction and (d) 3D reconstruction of the resolution target, showing the height of the steps of 100 nm.

In order to obtain the phase of light propagating through or reflecting from and object, DHM records the interference pattern between an object and a reference waves. The difference between DHM and "regular" holographic setup is that the interference pattern is recorded by a digital camera (see figure 1(b)), which can then be reconstructed numerically. Angular spectrum method, which is used here for numerical reconstruction, is based on performing a Fourier transform of the original hologram, with the subsequent suppression of a DC term (see figure 1(c)). The resulting plane waves are propagated in space numerically by multiplying them by a phase factor. Finally, after performing the inverse Fourier transform, the phase of light for each pixel is obtained as a phase of a complex number. Figure 1(d) shows the height map of the object in figure 1(b), obtained from the phase values

IOP Conf. Series: Journal of Physics: Conf. Series **909** (2017) 012001 doi:10.1088/1742-6596/909/1/012001

for each pixel, which is then converted to height values using equation (1). Thus, DHM is a technique that can produce nanometer-scale measurements of height and can be used to image microstructures and biological systems in three-dimensions[1-5].

Furthermore, DHM provides three-dimensional cell imaging in real-time, enabling monitoring of cell morphological changes during apoptosis[6]. Primary signatures of apoptosis include nuclear condensation, fragmentation in internucleosomal DNA, and formation of apoptotic bodies. In addition to that, one of the fundamental apoptosis characteristic is cell volume loss, termed apoptotic volume decrease (AVD), which can be monitored using DHM. Thus, since cell volume is associated with cellular apoptosis, DHM can be used as a great tool to provide information about volume change during this process.

Additionally, Raman spectroscopic studies of cellular cultures have demonstrated significant changes in Raman spectra as a result of cells undergoing apoptosis[7-10]. In this paper, we have monitored apoptotic volume changes of live central nervous system (CNS) cells (C6), while also acquiring its Raman spectral information during several stages of apoptosis to non-invasively monitor the chemical changes within these cells in real time.

2. Experimental

2.1. Digital Holographic Microscopic Measurements

The C6 glial cells were obtained from ATCC (Cat #ATCC® CCL-107™). They were grown to confluency in DMEM/F12 (50:50) media supplemented with 5% FBS and antibiotics. TO induce apoptosis, the cells were treated with 1μM by Doxorubicin (DOX).

Our transmission DHM setup is shown in figure 2. The 633 nm He-Ne laser beam is reflected by plane mirrors M1 and M2 into a microscope objective L1 and a plano-convex lens L2 to obtain a collimated beam with a large cross-section. Beamsplitter B1 then splits the laser beam into the reference and object arms. Lens L3 focuses the beam on the back focal plane of the microscope objectives OBJ1 and OBJ2. The reference beam is reflected by a plane mirror M3, passes through the microscope objective OBJ1, and into the beamsplitter B2, where it recombines with the object beam. Plane mirror M4 directs the object beam through the cell culture, which is kept in the environmental chamber (IBIDI Nanolive Heating System/Temperature Controller/Gas Incubation System), mounted on a three-dimensional translation stage. OBJ2 is also used to forms an image on the CCD camera. Numerical compensation is used to correct the curvature between the reference and object wavefronts[6].

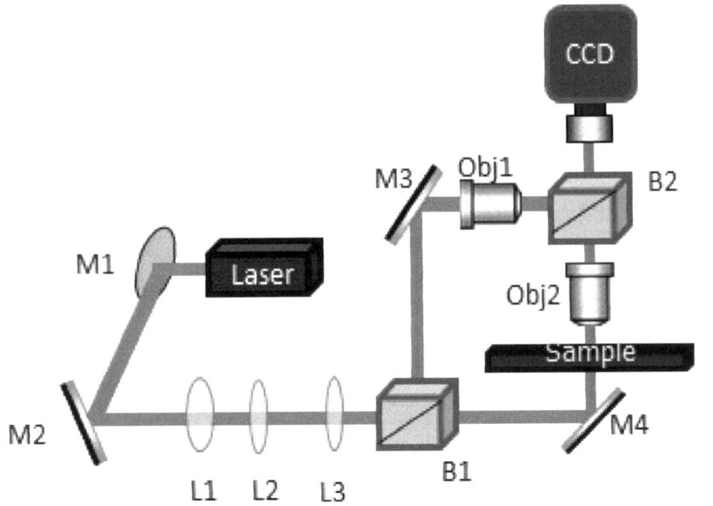

Figure 2. Transmission DHM setup (see text for details).

IOP Conf. Series: Journal of Physics: Conf. Series **909** (2017) 012001 doi:10.1088/1742-6596/909/1/012001

2.2. Raman Measurements
Raman spectra were acquired by a commercial micro-spectroscope (HORIBA XploRA PLUS). This Raman spectrometer is equipped with a 1024×256 TE air cooled CCD chip (pixel size 26 micron, temperature -60°C). HORIBA's LabSpec6 software was used for data acquisition, fluorescent background removal, baseline correction, and peak fitting. Excitation wavelength was set to 785 nm. Entrance slit of 50 μm produced an approximate spectral resolution of 4 cm^{-1}.

Cells were kept in the incubator and removed for measurements at several time points. For each of the time points, three measurements of three accumulations of 60 seconds each were acquired. To minimize the effects of fluorescence, cells were grown and imaged on quartz substrates. Quartz substrates also provide a known Raman signature that can be used for normalization of Raman signal from different cell cultures.

3. Results and Discussion
Our DHM experiments were done on the cultured cells (~5000 cells) that were placed in an environmental chamber, where apoptosis was induced by adding 1μM DOX. The changes in cell volume were monitored in real time over a 3 hour period. Figure 3 shows the percent volume changes in C6 cells as a result of AVD. A substantial decrease in cell volume was observed within 40 mins after DOX treatment. Cellular volume continued to decrease significantly with time. After 150 minutes, the cells were detaching from the bottom of the petri dish, making further volume measurements unreliable. By then, the cell culture volume has decreased to 40% of the original volume[11].

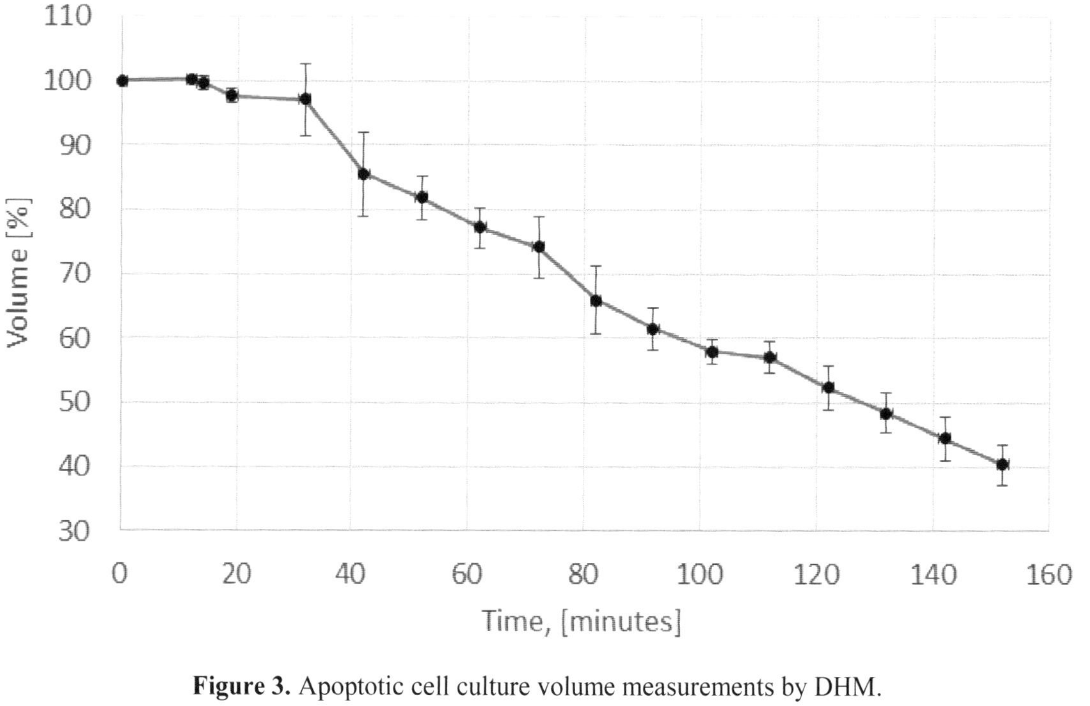

Figure 3. Apoptotic cell culture volume measurements by DHM.

For Raman measurements, the cells were removed from the incubator, placed on the sample stage of the Raman microscope, and their spectra were collected at several time points (*t=0 min, t=50 min, t =100 min* and *t=150 min*) after the induction of apoptosis with 1μM DOX. Control cell cultures spectra that were taken 200 minutes apart were also acquired.

The spectra were background subtracted, and the following peaks (which are indicative of molecular vibrations in lipids, nucleic acid and proteins[8]), were fitted: 953, 1003, 1092, 1165, 1305,

1338, 1447, 1544, 1611, and 1661 cm^{-1}. In addition to that, we have observed peaks at 1054 cm^{-1} and 1220 cm^{-1}. These peaks were identified as substrate peaks from our prior measurements of Raman signal from the substrate. Of these peaks, the height of 1220 cm^{-1} peak was used for normalization.

The heights of normalized 1003 cm^{-1}, 1165 cm^{-1}, 1305 cm^{-1}, 1447 cm^{-1} and 1661 cm^{-1} peaks are shown in figure 4. The intensity of phenylalanine peak at 1003 cm^{-1} remained approximately the same through the experiment. The peak at 1165 cm^{-1} (C-C and C-N stretch of proteins) increased in intensity, presumably because of accumulation of proteins and progression of nuclear condensation[9]. Then, at the last time point at $t=150$ *min*, when the cells were dying, the intensity decreased, which may indicate the breakdown of proteins. The peak at 1447 cm^{-1} CH_2 (bending mode of proteins) and 1661 cm^{-1} Amide I peak also showed similar behaviour. The peak at 1305 cm^{-1} (CH_2 deformation in lipids) reduced in intensity, which likely indicates the decreased contribution of lipids to Raman signal obtained from these cells.

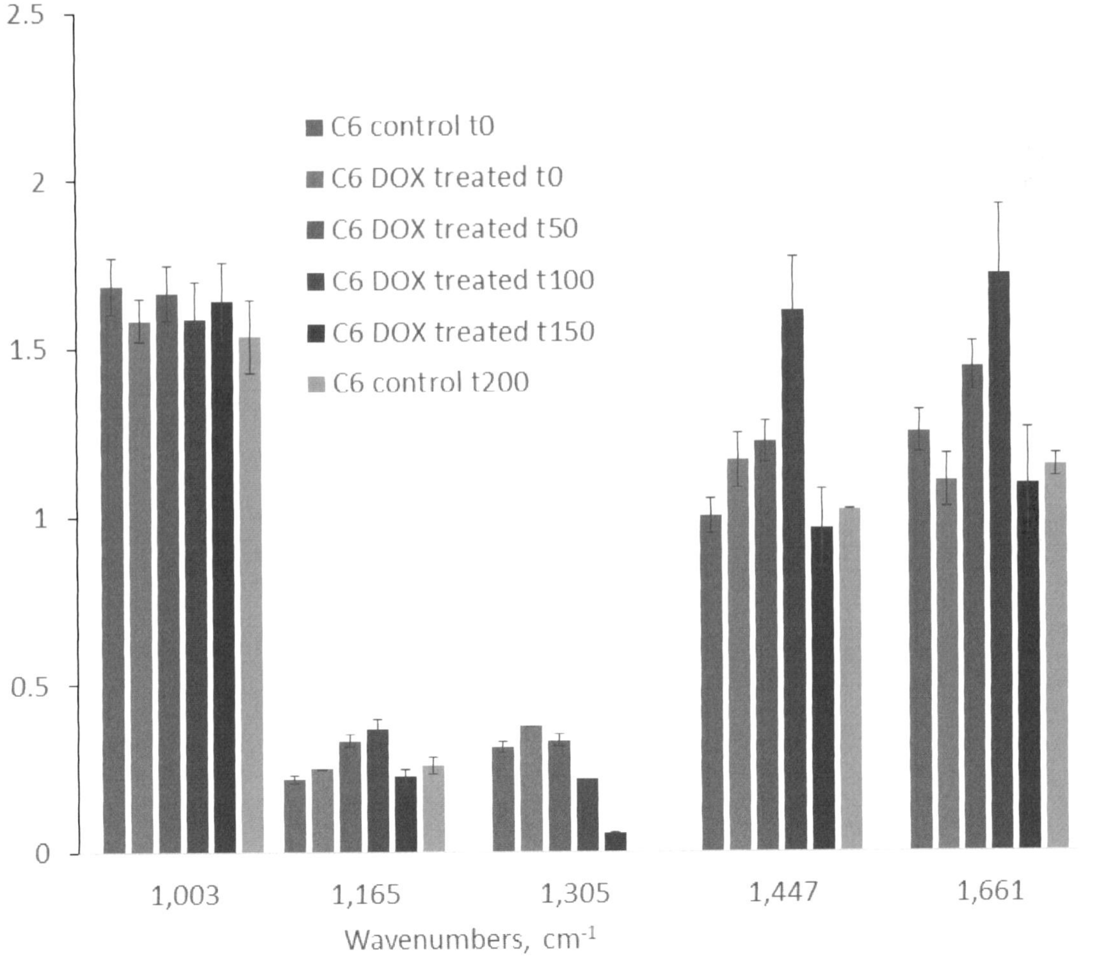

Figure 4. Heights of characteristic Raman peaks for apoptotic and control cell cultures (see text for details).

4. Conclusion

In this work, the changes in cell volume of C6 glial cells undergoing apoptosis were monitored by digital holographic microscopy, while the chemical changes within these cells were studied by Raman spectroscopy. Glial cells exposed to 1μM DOX exhibited a significant decrease in cell volume, a primary characteristic of apoptosis, which was clearly seen within 40 mins post treatment. A digital holographic microscope, with the attached environmental chamber, permitted accurate measurements of cell volume for several hundred cells in real time over an extended time period.

Our Raman data indicates that the chemical changes in proteins preceded morphological changes, which were seen with DHM. Based on the results of this study, it is apparent that apoptosis is a unique process, which consists of organized series of events, leading to programmed cell death. This study also emphasizes that DHM and Raman based imaging tools can be developed for non-invasive and simultaneous monitoring of both morphological and chemical changes in cells during apoptosis, which can also be used to monitor other dynamic cell processes.

References

[1] Haddad, W.S., Cullen, D., Solem, J. C., Longworth, J. W., McPherson, A., Boyer, K., and Rhodes, C. K., "Fourier-transform holographic microscope," Appl. Opt. 31, 4973-4978 (1992).

[2] Xu, W., Jericho, M. H., Meinertzhagen, I. A., and Kreuzer, H. J., "Digital in-line holography for biological applications," Proc. Natl. Acad. Sci. USA 98, 11301-10305 (2001).

[3] Cuche, E., Bevilacqua, F., and Depeursinge, C., "Digital holography for quantitative phase-contrast imaging," Opt. Lett. 24, 291-293 (1999).

[4] Mann, C. J., Yu, L., Lo, C.-M., and Kim, M. K., "High-resolution quantitative phase-contrast microscopy by digital holography," Opt. Express 13, 8693-8698 (2005).

[5] Khmaladze, A., Matz, R. L., Epstein, T., Jasensky, J., Banaszak Holl, M. M., and Chen, Z., "Cell volume changes during apoptosis monitored in real time using digital holographic microscopy," J. Struct. Biol. 178, 270-278 (2012).

[6] Khmaladze, A., Kim, M.K., Lo, C-M., "Phase imaging of cells by simultaneous dual-wavelength reflection digital holography" Opt. Express 16, 10900-10911 (2008).

[7] Brauchle, E., Thude, S., Brucker, S. Y. & Schenke-Layland, K. "Cell death stages in single apoptotic and necrotic cells monitored by Raman microspectroscopy," Sci. Rep. 4, 1-4 (2014).

[8] Yao, H., Tao, Z., Ai, M., Peng, L., Wang, G., He, B., and Li, Y., "Raman spectroscopic analysis of apoptosis of single human gastric cancer cells," Vibrat. Spectrosc. 50, 193–197 (2009).

[9] Zoladek, A., Pascut, F. C., Patel, P., and Notingher, I., "Non-invasive time-course imaging of apoptotic cells by confocal Raman micro-spectroscopy," J. Raman Spectrosc. 42, 251–258 (2011).

[10] Kunapareddy, N., Freyer, J. P., and Mourant, J. R., "Raman spectroscopic characterization of necrotic cell death," J. Biomed. Opt. 13, 054002-1-054002-9 (2008).

[11] Park, J. Y., Maloney, M. C., Desta, H., Mahajan, S., Sharikova, A., Khmaladze, A., "Phase Imaging of Live Central Nervous System Cells during Apoptosis by Digital Holographic Microscopy," OSA Digital Holography and Three-Dimensional Imaging, Paper# JW4A.11 (2016).

Approximate solution for the minimal length case of Klein Gordon equation for trigonometric cotangent potential using Asymptotic Iteration Method

C Cari, A Suparmi, and Isnaini Lilis Elviyanti

Physics Department, Graduate Program, Sebelas Maret University

Email: cari@staff.uns.ac.id

Abstract. The approximate solution of Klein Gordon equation within minimal length formalism for trigonometric cotangent potential was solved using Asymptotic Iteration Method. Asymptotic Iteration Method was used to obtain relativistic energy and wave function of alternative solution Klein Gordon equation within minimal length formalism. The relativistic energy was calculated numerically using Matlab software, and unnormalized wave function was expressed in hypergeometric terms. The minimal length parameter which was used in this research was very small value.The results show that the relativistic energy within minimal length formalism increased by the increase of minimal length parameter.

1. Introduction

The dynamic of spin-zero particles in relativistic quantum mechanics was described using the Klein Gordon equation [1,2]. The vector potential (*V*) and scalar potential (*S*) of Klein Gordon equation had an exact solution in the case of theexact symmetric spin condition and symmetric pseudospin condition. In these two cases, Klein Gordon equation was reduced to Schrodinger like equation thus a state solution was obtained using various methods in non-relativistic quantum mechanics [3]. The Klein Gordon equation has been solved by Nugraha et al.[4,5],Ikot et al.[6], and Poszwa A [7].

In quantum mechanics, the minimal length idea was based on the following deformed commutation relations between position and momentum operators in Generalized Uncertainty Principle (GUP) [8]. The minimal length may be viewed as an intrinsic scale characterizing the system [9]. The Klein Gordon equation within minimal length can be solved using Heun's function for Coulomb potential [10] and approach Algebraic for trigonometric potential [11]. The approximate solution of minimal length in Bohr-Mottelson equation has been investigated by Chabab et al. [12].

In this paper, we solved a radial part of Klein Gordon equation within minimal length for symmetric spin condition using approximate solution for trigonometric cotangent function potential. The trigonometric cotangent potential was used to explain nucleon excitation [13]. The Asymptotic Iteration Method was used to obtain the relativistic energy and radial wave functions in an approximate solution of minimal length. The work was organized as follows. In Section 2, the approximate solution of Klein Gordon equation with minimal length formalism was introduced. In Section 3, trigonometric cotangent potential was presented. We described the Asymptotic Iteration Method which was used to solve the approximate solution of Klein Gordon equation was presented in section 4. In Section 5, the results and discussion were presented. The conclusion was presented in Section 6.

2. Klein Gordon Equation with Minimal Length Formalism

The uncertainty of particle position was explained using commutation relations between position and momentum operators as expressed in Heisenberg's uncertainty principle. The modified Heisenberg's uncertainty principle was given as [14,15],

$$[X,P]\geq i\hbar\left(1+\alpha_{ML}\left(\Delta P\right)^{2}\right) \tag{1}$$

where X was a position, P was a corresponding momentum and α_{ML} was a minimal length parameter that has very small positive values. By equation (1) the position and momentum operators [14-16] were defined as,

$$\hat{X}_i=\hat{x}_i \tag{2}$$

$$\hat{P}_i=\left(1+\alpha_{ML}\hat{p}^{2}\right)\hat{p}_i \tag{3}$$

Klein Gordon equation was given by [10,16]

$$\left(E-V\left(r\right)\right)^{2}\psi\left(r\right)=\left[P^{2}c^{2}+\left(M_{o}c^{2}+S\left(r\right)\right)^{2}\right]\psi\left(r\right) \tag{4}$$

Where $V\left(r\right)$ and $S\left(r\right)$ were vector and scalar potentials. E and M_o were energy and rest mass. By setting $S\left(r\right)=V\left(r\right)$ for case symmetric spin condition in equation (4) and substituting equation (3) into equation (4) with $\hat{p}=-i\hbar\nabla$ and $c=\hbar=1$. By setting $2V\left(r\right)\rightarrow V\left(r\right)$, we had

$$-\left(\Delta-2\alpha_{ML}\Delta^{2}\right)\psi\left(r\right)-\left(E^{2}-M_{o}^{2}-\left(E+M_{o}\right)V\left(r\right)\right)\psi\left(r\right)=0 \tag{5}$$

Equation (5) was Klein Gordon equation within minimal length which can be solved using an approximate solution by substituting a new wave function [12], was given by

$$\psi\left(r\right)=\left(1+2\alpha_{ML}\right)\phi\left(r\right) \tag{6}$$

So we got,

$$\left[-\Delta-\frac{\left(E^{2}-M_{o}^{2}-\left(E+M_{o}\right)V\left(r\right)\right)}{\left(1+2\alpha_{ML}\Delta\left(E^{2}-M_{o}^{2}-\left(E+M_{o}\right)V\left(r\right)\right)\right)}\right]\phi\left(r\right)=0 \tag{7}$$

Equation (7) was an approximate solution of Klein Gordon equation within the minimal length. It can be solved using Binomial expansion, so equation (7) becomes

$$\Delta\phi\left(r\right)+\left(\left(E^{2}-M_{o}^{2}\right)-\left(E+M_{o}\right)V\left(r\right)-2\alpha_{ML}\left(E^{2}-M_{o}^{2}-\left(E+M_{o}\right)V\left(r\right)\right)^{2}\right)\phi\left(r\right)=0 \tag{8}$$

From equation (8) was obtained α_{ML}^{2} that had very small value for the approximate solution, so α_{ML}^{2} was ignored. By applying spherical Laplacian operator, was given as,

$$\Delta=\frac{1}{r^{2}}\frac{\partial}{\partial r}r^{2}\frac{\partial}{\partial r}+\frac{1}{r^{2}}\left(\frac{1}{\sin\theta}\frac{\partial}{\partial\theta}\sin\theta\frac{\partial}{\partial\theta}+\frac{1}{\sin^{2}\theta}\frac{\partial^{2}}{\partial\varphi^{2}}\right) \tag{9}$$

And by setting $\phi\left(r\right)=Q\left(r\right)\mathrm{Y}_{LM}(\theta,\phi)$, where we have used

$$\left(\frac{1}{\sin\theta}\frac{\partial}{\partial\theta}\sin\theta\frac{\partial}{\partial\theta}+\frac{1}{\sin^{2}\theta}\frac{\partial^{2}}{\partial\varphi^{2}}\right)Y_{LM}(\theta,\phi)=-L(L+1)Y_{LM}(\theta,\phi) \tag{10}$$

Then by inserting equations (9) and (10) into equation (8) we get

$$\left[\begin{array}{l}\left(\frac{1}{r^2}\frac{\partial}{\partial r}r^2\frac{\partial}{\partial r}-\frac{L(L+1)}{r^2}\right)+\left(E^2-M_o^{\ 2}-(E+M_o)V(r)\right)\\-2\alpha_{ML}\left(\left(E^2-M_o^{\ 2}\right)^2-2\left(E^2-M_o^{\ 2}\right)(E+M_o)V(r)+(E+M_o)^2V^2(r)\right)\end{array}\right]Q(r)=0 \qquad (11)$$

Equation (11) was an approximate solution for the radial part of Klein Gordon equation with the minimal length for the case symmetric spin condition.

3. Asymptotic Iteration Method

Asymptotic Iteration Method is used to solve the second order differential equation which is expressed as [17-19],

$$y_n''(x)=\lambda_0(x)y_n'(x)+s_0(x)y_n(x) \qquad (12)$$

where n is a quantum number, $\lambda_0(x)\neq 0$ and $s_0(x)$ are the coefficient of a differential equation. By writing the $(k+1)^{th}$ derivative of y which is obtained from equation (12), we got

$$y_n^{k+1}(x)=\lambda_{k-1}(x)y_n'(x)+s_{k-1}(x)y_n(x) \qquad (13)$$

where,

$$\lambda_k(x)=\lambda'_{k-1}(x)+s_{k-1}(x)+\lambda_0(x)\lambda_{k-1}(x) \qquad (14)$$

$$s_k(x)=s'_{k-1}(x)+s_0(x)\lambda_{k-1}(x) \qquad (15)$$

$$k=1,2,3... \qquad (16)$$

The eigenvalue was obtained from the quantization condition, is given by

$$\Delta_k(x)=\lambda_k(x)s_{k-1}(x)-\lambda_{k-1}(x)s_k(x)=0 \qquad (17)$$

The one-dimensional Schrodinger like equation was reduced to a hypergeometric type differential equation which is given as,

$$y_n''(x)=2\left(\frac{ax^{N+1}}{1-bx^{N+2}}-\frac{t+1}{x}\right)y_n'(x)-\frac{wx^N}{1-bx^{N+2}}y_n(x) \qquad (18)$$

The eigenfunction was obtained from equation (9), is given as,

$$y_n(x)=(-1)^n C'(N+2)^n(\sigma)_n\ {}_2F_1\left(-n,\rho+n;\sigma;bx^{N+2}\right) \qquad (19)$$

where

$$(\sigma)_n=\frac{\Gamma(\sigma+n)}{\Gamma(\sigma)} \quad \sigma=\frac{2t+N+3}{N+2} \quad \rho=\frac{(2t+1)b+2a}{(N+2)b} \qquad (20)$$

C' is normalization constant and ${}_2F_1$ is a hypergeometric function. The wave function of Klein Gordon equation can be obtained by using equations (18-20) [17-19].

4. Trigonometric Cotangent Potential

The equation of trigonometric cotangent potential [12] was given as,

$$V(r)=V_o\cot(\mu r)+V_1 \qquad (21)$$

where V_o and V_1 were potentialconstants. μ was a range of potential. By setting $V_0=0.1, V_1=0.01$ and $\mu=0.1$, the trigonometric cotangent potential was given by

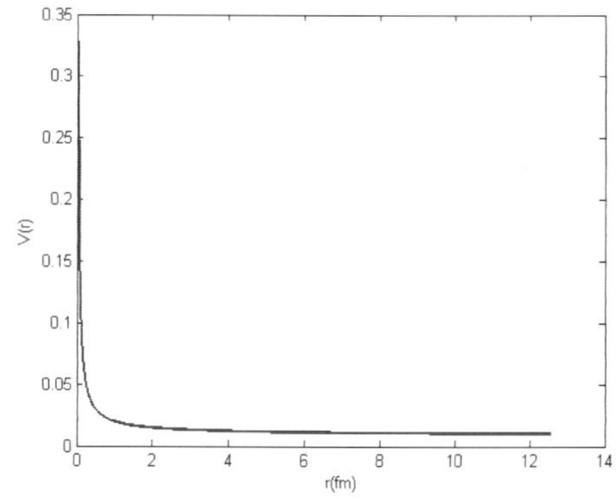

Figure 1. The trigonometric cotangent potential

Figure 1 was a visualization of trigonometric cotangent potential in *r* function. In various value of *r*, we can see in Figure 1, a value of *r* was approximately from 0 *fm* until 14 *fm,* the trigonometric cotangent potential had the different value. The trigonometric cotangent potential comes to infinity value in a very small value of *r*, while for the higher value of *r*, the trigonometric cotangent potential incline to be constant.

5. Result and Discussion

By applying equation (21) into equation (11), we got

$$\begin{bmatrix} \left(\frac{1}{r^2}\frac{\partial}{\partial r}r^2\frac{\partial}{\partial r}-\frac{L(L+1)}{r^2}\right)+\left(E^2-M_o^{\,2}-(E+M_o)\left(V_o\cot(\mu r)+V_1\right)\right) \\ -2\alpha_{ML}\begin{pmatrix}\left(E^2-M_o^{\,2}\right)^2-2\left(E^2-M_o^{\,2}\right)(E+M_o)\left(V_o\cot(\mu r)+V_1\right) \\ +(E+M_o)^2\left(V_o\cot(\mu r)+V_1\right)^2\end{pmatrix}\end{bmatrix}Q(r)=0 \tag{22}$$

Equation (22) was Klein Gordon equation with the minimal length for trigonometric cotangent potential. By setting $Q(r)=\frac{1}{r}F(r)$ and applying the centrifugal approximate [20],

$$\frac{1}{r^2}=\frac{\mu}{\sin^2(\mu r)} \tag{23}$$

Then equation (22) becomes

$$\begin{bmatrix} \frac{d^2F(r)}{dr^2}-\left(\frac{\mu L(L+1)+2\alpha_{ML}(E+M_o)^2V_o^{\,2}}{\sin^2\mu r}\right)F(r) \\ +\left(\left(4\alpha_{ML}\left(E^2-M_o^{\,2}\right)(E+M_o)-(E+M_o)\right)V_o-2\alpha_{ML}(E+M_o)^2V_2\right)\cot\mu rF(r) \\ +\begin{pmatrix}\left(4\alpha_{ML}\left(E^2-M_o^{\,2}\right)(E+M_o)-(E+M_o)\right)V_1+\left(E^2-M_o^{\,2}-2\alpha_{ML}\left(E^2-M_o^{\,2}\right)^2\right) \\ +2\alpha_{ML}(E+M_o)^2V_o^{\,2}-2\alpha_{ML}(E+M_o)^2V_1^2\end{pmatrix}F(r)\end{bmatrix}=0 \tag{24}$$

Then equation (24) was reduced becomes,

$$\frac{d^2F(r)}{dr^2}-\left[\frac{v(v+1)}{\sin^2(\mu r)}-2q\cot(\mu r)+K^2\right]F(r)=0 \tag{25}$$

By setting,

$$v(v+1)=\left(\mu L(L+1)+2\alpha_{ML}(E+M_o)^2V_o^2\right) \tag{26}$$

$$2q=\left(\left(4\alpha_{ML}(E^2-M_o^2)(E+M_o)-(E+M_o)\right)V_o-2\alpha_{ML}(E+M_o)^2V_2\right) \tag{27}$$

$$-K^2=\begin{pmatrix}\left(4\alpha_{ML}(E^2-M_o^2)(E+M_o)-(E+M_o)\right)V_1+\left(E^2-M_o^2-2\alpha_{ML}(E^2-M_o^2)^2\right)\\ +2\alpha_{ML}(E+M_o)^2V_o^2-2\alpha_{ML}(E+M_o)^2V_1^2\end{pmatrix} \tag{28}$$

Equation (24) was a differential equation which had to be reduced to hypergeometric differential equation type. By using the suitable variable change $\cot(\mu r)=i(1-2u)$, we got,

$$u(1-u)\frac{d^2F(r)}{dy^2}+(1-2u)\frac{dF(r)}{dy}+\left[v'(v'+1)-\frac{4\alpha^2}{4u}-\frac{4\beta^2}{4(1-u)}\right]F(r)=0 \tag{29}$$

with,

$$\frac{2qi-K^2}{\mu^2}=4\alpha^2\ ,\ \frac{-2qi-K^2}{\mu^2}=4\beta^2\ ,\ v'(v'+1)=\frac{v(v+1)}{\mu^2} \tag{30}$$

Intermediate of the hypergeometric differential equation was shown in equation (29). Equation (29) was reduced by the new wavefunction as

$$F(r)=u^{\alpha}(1-u)^{\beta}g(u) \tag{31}$$

and we got

$$u(1-u)\frac{g^2(u)}{du^2}+\left[(2\alpha+1)-(2\alpha+2\beta+2)u\right]\frac{g(u)}{du}+\left[v'(v'+1)-(\alpha+\beta)(\alpha+\beta+1)\right]g(u)=0 \tag{32}$$

The equation (32) must be reduced to AIM type equation, by dividing it with $u(1-u)$, then we got

$$g''(u)+\left[\frac{(2\alpha+1)-(2\alpha+2\beta+2)u}{u(1-u)}\right]g'(u)+\left[\frac{v'(v'+1)-(\alpha+\beta)(\alpha+\beta+1)}{u(1-u)}\right]g(u)=0 \tag{33}$$

By comparing equation (12) and equation (32), we had

$$\lambda_0=\frac{-(2\alpha+1)}{u}+\frac{(2\beta+1)}{1-u} \tag{34}$$

$$s_o=\frac{(\alpha+\beta)(\alpha+\beta+1)-v'(v'+1)}{u}+\frac{(\alpha+\beta)(\alpha+\beta+1)-v'(v'+1)}{(1-u)} \tag{35}$$

Using equations (18-21) and equation (33), we obtained the eigenvalue, was given was

$$v'(v'+1)=(\alpha+\beta+n)(\alpha+\beta+(n+1)) \tag{36}$$

From the eigenvalue in equation (36), we obtained the relativistic energy equation by using equations (30) and (36).The relativistic energy equation of Klein Gordon equation for the minimal length was given as

$$\left(E^2-M_o^{\;2}\right)=\varpi^2\left[\left(\chi_{\alpha_{ML}}\right)^2-\frac{\left(2\delta_{\alpha_{ML}}\right)^2}{4\left(\mu^2\chi_{\alpha_{ML}}\right)^2}\right]\left(4\alpha_{ML}\left(E^2-M_o^{\;2}\right)\left(E+M_o\right)-\left(E+M_o\right)\right)V_1+\gamma_{\alpha_{ML}} \tag{37}$$

where,

$$\chi_{\alpha_{ML}}=\sqrt{\frac{\mu L(L+1)+2\alpha_{ML}\left(E_0+M_o\right)^2V_o^{\;2}}{\mu^2}+\frac{1}{4}}-n-\frac{1}{2} \tag{38}$$

$$\delta_{\alpha_{ML}}=\left(\left(4\alpha_{ML}\left(E^2-M_o^{\;2}\right)\left(E+M_o\right)-\left(E+M_o\right)\right)V_o-2\alpha_{ML}\left(E+M_o\right)^2V_2\right) \tag{39}$$

$$\gamma_{\alpha_{ML}}=2\alpha_{ML}\left(E^2-M_o^{\;2}\right)^2+2\alpha_{ML}\left(E+M_o\right)^2V_1^2-2\alpha_{ML}\left(E+M_o\right)^2V_o^{\;2} \tag{40}$$

and n was a quantum number which was expressed in equation (38). Equation (37) was the relativistic energy of Klein Gordon equation with the minimal length using an approximate solution. Equation (36) was calculated numerically by using the Matlab software. The result was shown in Table 1.

Table 1.The relativistic energy of Klein Gordon equation for M_o=1, V_o=0.01, V_1=0.1, V_2=0.001, $\mu=0.1$ and L=0

α_{ML}	E(eV)		
	n=1	*n=2*	*n=3*
0	0.0100	0.6277	0.8527
0.001	0.6282	0.8533	0.9688
0.003	0.6293	0.8546	0.9700
0.005	0.6303	0.8558	0.9712
0.007	0.6313	0.8570	0.9724
0.009	0.6322	0.8582	0.9735
0.010	0.6327	0.8588	0.9740

Table 1 shown, that the relativistic energy of an approximate solution Klein Gordon equation with the value of minimal length parameter was very small. The relativistic energy within the minimal length was higher than relativistic energy without the minimal length. The increasing of minimal length parameter caused the increase of the relativistic energy value. The relativistic energy increased by the increase of quantum number n.

The general un-normalized wave function of Klein Gordon equation was obtained by inserting equations (30-31,36) and $\cot(\mu r)=i(1-2u)$,so we got wave functions that were shown in Table 2.

Table 2. The wave function of Klein Gordon equation within the minimal length

n	Wavefunction
0	$F_0(r)=C'\left(\frac{1+i\cot\mu r}{2}\right)^{\alpha}\left(\frac{1-i\cot\mu r}{2}\right)^{\beta}$
1	$F_1(r)=-C'\left(\frac{1+i\cot\mu r}{2}\right)^{\alpha}\left(\frac{1-i\cot\mu r}{2}\right)^{\beta}(2\alpha+1)\left[1+\frac{(-1)(2\alpha+2\beta+2)\left(\frac{1+i\cot\mu r}{2}\right)}{(2\alpha+1)}\right]$

2	$F_2(r)=C'\left(\frac{1+i\cot\mu r}{2}\right)^{\alpha}\left(\frac{1-i\cot\mu r}{2}\right)^{\beta}(2\alpha+1)(2\alpha+2)\left[1+\frac{(-2)_1(2\alpha+2\beta+3)_1\left(\frac{1+i\cot\mu r}{2}\right)^1}{(2\alpha+1)_1 1!}+\frac{(-2)(-1)(2\alpha+2\beta+3)(2\alpha+2\beta+4)\left(\frac{1+i\cot\mu r}{2}\right)^2}{(2\alpha+1)(2\alpha+1)2}\right]$

6. Conclusion

We investigated the approximate solution of Klein Gordon equation within the minimal length for trigonometric cotangent potential using Asymptotic Iteration method. Asymptotic Iteration method was used to obtain the relativistic energy and radial wave functions for Klein-Gordon equation within the minimal length. The approximate solution was used to minimal length parameter that had very small value. The increasing of minimal length parameter caused the increase of the relativistic energy value. The relativistic energy increased by the increase of quantum number *n*.

Acknowledgement
This research was partly supported by Sebelas Maret University Higher Education Project Grant 623/UN27.21/PP/2017.

References

[1] IkhdairSM and Sever R 2008 *arXiv:0808.1002v1 [quant-ph]* 14.
[2] Momtazi E, Rajabi A A, and Yazarloo B H 2014 *Turkish J. Phys* **38**181–85.
[3] Greiner W 2000 Springer-Verlag Berlin Heidelberg, New York.
[4] Nugraha D A, Suparmi A, Cari C and Pratiwi B N 2017*J. Phys. Conf. Ser* **795**.
[5] Nugraha D A, Suparmi A, Cari C and Pratiwi B N 2017*J. Phys. Conf. Ser* **820**.
[6] Poszwa A, 2014 *Acta Physica Polonica A* **126** 6.
[7] A N Ikot, L E Akpabio and E J Uwah 2011 *Electron. J. Theor. Phys* **25** 25 225-232.
[8] Hassanabadi H, Maghsoodi E, Ikot N A, and Zarrinkamar S 2013 *Hindawi Publishing Corporation.*
[9] Jana K T and Roy P 2009*Physics Letters A***373**1239–41.
[10] Bouaziz D 2013 arXiv:1311.7405v1 [quant-ph].
[11] Merad M, Zeroual F and Benzair H 2010 *Electronic Journal of Theoretical Physics* 723 41-56.
[12] Chabab M, Batoul A, Lahbas A, and Oulne M 2016 *Physics Letter B* **758**, 212-218.
[13] Cari, Suparmi, Deta A U and Werdiningsih I S 2013 *Makara J. Sci.*
[14] Sprenger M, Nicolini P and Bleicher M 2012 *European Journal Of Physics* **33** 853-62.
[15] Garay J L 1994 *International Journal of Modern Physics A* **10**2 145-65.
[16] Chabab M, ElBatoul A, Lahbas A and Oulne M 2016 *Physics Letters B* 212-16.
[17] H Ciftci, R L Hall and N Saad 2003 *J. Phys. A. Math. Gen* **36** 47 11807-16.
[18] Pratiwi N B, Suparmi A, Cari C and Husein S A 2017 *Pramana. J. Phys.*
[19] Pramono S, Suparmi A, and Cari C 2016 *Adv. High Energy Phys* **2016**.
[20] Naderia L and Hassanabadi H 2016 *The European Physical Journal Plus* 131:133.

Solution of Klein Gordon equation for trigonometric cotangent potential in the presence of a minimal length using Asymptotic Iteration Method

A Suparmi, C Cari, and Isnaini Lilis Elviyanti

Physics Department, Graduate Program, Sebelas Maret University

Email: soeparmi@staff.uns.ac.id

Abstract. Analysis of energy eigenvalue and wave function for Klein Gordon equation within minimal length formalism which was influenced by trigonometric cotangent potential has done. Asymptotic iteration method was used to solve the modified Klein Gordon equation to obtain energy eigenvalueand wave function. The energy eigenvalue was calculated numerically using Matlab software, and the general unnormalized wave function was expressed in hypergeometric terms. The results showed that the energy eigenvalue within the minimal length formalism was higher than energy eigenvalue without the minimal length formalism. The energy eigenvalue increased by the increase of minimal length parameters. The minimal length parameter caused the shape changein wave functions.

1. Introduction

The Klein Gordon equation was an equation to describe dynamic of spin-zero particles in relativistic quantum mechanics [1,2]. The vector potential (*V*) and scalar potential (*S*) of Klein Gordon equation have an exact solution in the case of exact symmetric spin condition when $S=V$and $S = -V$ for exact symmetric pseudospin condition. In these two cases, Klein Gordon equation was reduced to Schrodinger like equation thus a state solution which was obtained using various methods in non-relativistic quantum mechanics [3].

The minimal length idea in quantum mechanics was based on the following deformed commutation relations between position and momentum operators in Generalized Uncertainty Principle (GUP) [4]. The minimal length may be viewed as an intrinsic scale characterizing the system [5]. The Klein Gordon equation in the presence of minimal length has been solved using Heun's functionfor Coulomb potential [6], approach Algebraic for trigonometric potential [5] and approach Feynman for linear potential [7]. On the other than, the Klein Gordon equation without the minimal length has been solved using Asymptotic Iteration Method (AIM) for Kratzer potential [8] and Trigonometric Poschl-Teller potential [9], approach Algebraic in Harmonic potential [10] and Nikiforov Uvarof method in Hulten Potential [11].The solution of minimal length has been investigated by Alimohammadi and Hassanabadi using Bohr-Mottelson equation [12].

In this paper, we solved symmetric spin condition of Klein Gordon equation in the presence of a minimal length. The radial part solution of Klein-Gordon equation for trigonometric cotangent function potentialin the presence of a minimal length has been studied[13]. The trigonometric cotangent potential was used to explain nucleon excitation [13]. The Asymptotic Iteration Method was applied to obtain the energy eigenvalues and radial wave functions in the presence of minimal length. AIM was used to solve equation because of its easy method [14] and theresult was accurate

[9].Thework was organized as follows. In Section 2, Klein Gordon equation with minimal length formalismwas introduced. In Section 3, we described the Asymptotic Iteration Method which was used to solve Klein Gordon equation. In Section 4, the results and discussion were presented. The conclusion was presented in Section 5.

2. Klein Gordon Equation with Minimal Length Formalism

The uncertainty of particle position was explained using commutation relations between position and momentum operators as expressed in Heisenberg's uncertainty principle. In the case of minimal length was occurring in position uncertainty and momentum uncertainty was influenced by a field. The general equation was given as [15,16],

$$[X,P] \geq i\hbar\left(1+\alpha_{ML}(\Delta P)^2\right) \tag{1}$$

where X was a position, P was a corresponding momentum and α_{ML} was a minimal length parameter that hasvery small positive values. By equation (1) the position and momentum operators [15-17] were defined as,

$$\hat{X}_i = \hat{x}_i \tag{2}$$

$$\hat{P}_i = \left(1+\alpha_{ML}\hat{p}^2\right)\hat{p}_i \tag{3}$$

The general equation of Klein Gordon equation was given by [10,17]

$$\left(E-V(r)\right)^2\psi(r) = \left[P^2c^2 + \left(M_o c^2 + S(r)\right)^2\right]\psi(r) \tag{4}$$

where $V(r)$ was vector potential and $S(r)$ was scalar potential. E and M_o were energy and rest mass, respectively. By setting $S(r)=V(r)$ in equation (4) and substituting equation (3) into equation (4) with $\hat{p}=-i\hbar\nabla$ and $c=\hbar=1$.We must set $2V(r)\rightarrow V(r)$ and we had,

$$\left(E^2 - M_o^{\ 2} - (E+M_o)V(r)\right)\psi(r) = -\left(\Delta - 2\alpha_{ML}\Delta_0^{\ 2}\right)\psi(r) \tag{5}$$

From equation (5) was obtained $\alpha_{ML}^{\ 2}$ that had very small value, so $\alpha_{ML}^{\ 2}$ was ignored. In the case of Klein Gordon without the minimal length formalism with $\alpha_{ML}=0$ [12], so square term of equation (5) was given as,

$$\Delta_0^{\ 2} = \left(E_0^{\ 2} - M_o^{\ 2} - (E_0+M_o)V(r)\right)^2 \tag{6}$$

E_0 was energy eigenvalue at any level without the presence of minimal length. In equation (6) the expression of $\Delta_0^{\ 2}$ depends only on the potential $V(r)$ and it was not considered to be Laplacian.

By inserting equation (6) into equation (5) we got

$$\left(E^2 - M_o^{\ 2} - (E+M_o)V(r)\right)\psi(r) = -\left(\Delta - 2\alpha_{ML}\left(E_0^{\ 2} - M_o^{\ 2} - (E_0+M_o)V(r)\right)^2\right)\psi(r) \tag{7}$$

and the spherical Laplacian operator was given as,

$$\Delta = \frac{1}{r^2}\frac{\partial}{\partial r}r^2\frac{\partial}{\partial r} + \frac{1}{r^2}\left(\frac{1}{\sin\theta}\frac{\partial}{\partial\theta}\sin\theta\frac{\partial}{\partial\theta} + \frac{1}{\sin^2\theta}\frac{\partial^2}{\partial\varphi^2}\right) \tag{8}$$

And by setting $\psi(r)=U(r)\mathrm{Y}_{LM}(\theta,\phi)$,yields

$$\left(\frac{1}{\sin\theta}\frac{\partial}{\partial\theta}\sin\theta\frac{\partial}{\partial\theta} + \frac{1}{\sin^2\theta}\frac{\partial^2}{\partial\varphi^2}\right)Y_{LM}(\theta,\phi) = -L(L+1)Y_{LM}(\theta,\phi) \tag{9}$$

where L was angular momentum. Then equation (8) was substituted into equation (7), so we got

$$\left[\begin{array}{l}\left(\frac{1}{r^2}\frac{\partial}{\partial r}r^2\frac{\partial}{\partial r}-\frac{L(L+1)}{r^2}\right)+\left(E^2-M_o^{\,2}-\left(E+M_o\right)V(r)\right)\\ -2\alpha_{ML}\left(\left(E_0^{\,2}-M_o^{\,2}\right)^2-2\left(E_0^{\,2}-M_o^{\,2}\right)\left(E_0+M_o\right)V(r)+\left(E_0+M_o\right)^2V^2\right)\end{array}\right]U(r)=0 \qquad (10)$$

Klein Gordon equation in the presence of a minimal length for the case symmetric spin condition was shown by equation (10).

3. Asymptotic Iteration Method

Asymptotic Iteration Method is used to solve the second order differential equation which is expressed as[18-20],

$$y_n''(x)=\lambda_0(x)y_n'(x)+s_0(x)y_n(x) \qquad (11)$$

where n is a quantum number, $\lambda_0(x)\neq 0$ and $s_0(x)$ are the coefficient of a differential equation. By writing the $(k+1)^{th}$ derivative of y which is obtained from equation (11), is given as,

$$y_n^{k+1}(x)=\lambda_{k-1}(x)y_n'(x)+s_{k-1}(x)y_n(x) \qquad (12)$$

where,

$$\lambda_k(x)=\lambda'_{k-1}(x)+s_{k-1}(x)+\lambda_0(x)\lambda_{k-1}(x) \qquad (13)$$

$$s_k(x)=s'_{k-1}(x)+s_0(x)\lambda_{k-1}(x) \qquad (14)$$

$$k=1,2,3... \qquad (15)$$

The eigenvalue can be obtained from the quantization condition which is given by

$$\Delta_k(x)=\lambda_k(x)s_{k-1}(x)-\lambda_{k-1}(x)s_k(x)=0 \qquad (16)$$

The one-dimensional Schrodinger like equation was reduced to a hypergeometric-type differential equation which is given as,

$$y_n''(x)=2\left(\frac{ax^{N+1}}{1-bx^{N+2}}-\frac{t+1}{x}\right)y_n'(x)-\frac{wx^N}{1-bx^{N+2}}y_n(x) \qquad (17)$$

The eigenfunction which is obtained from equation (9) is given as,

$$y_n(x)=(-1)^n C'(N+2)^n(\sigma)_n\ {}_2F_1\left(-n,\rho+n;\sigma;bx^{N+2}\right) \qquad (18)$$

where

$$(\sigma)_n=\frac{\Gamma(\sigma+n)}{\Gamma(\sigma)} \qquad (19)$$

$$\sigma=\frac{2t+N+3}{N+2} \qquad (20)$$

$$\rho=\frac{(2t+1)b+2a}{(N+2)b} \qquad (21)$$

C' is normalization constant and ${}_2F_1$ is a hypergeometric function. The wave function of Klein Gordon equation can be obtained by using equations(18-21) [18-20].

4. Result and Discussion

The function $1/r^2$ of equation (10) was approximated using the centrifugal approximation [21], was given as,

$$\frac{1}{r^2} = \frac{\varpi}{\sin^2(\varpi r)} = f(r) \tag{22}$$

Equation (22) was used for the small value of variable r and the visual approximate was given in Figure 1.

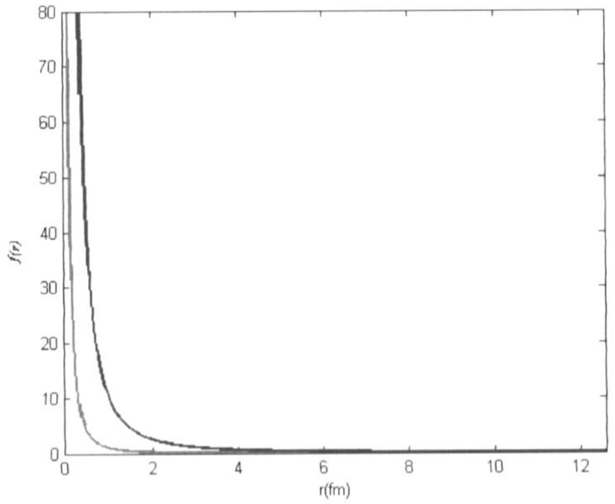

Figure 1. The function $1/r^2$ and $\varpi/\sin^2(\varpi r)$

Figure 1 shows that line blue was function $1/r^2$ and line red was a function $\varpi/\sin^2(\varpi r)$. So, the function $1/r^2$ of equation (10) can be approximated using the centrifugal approximation.

The trigonometric cotangent potential [13] was given as,

$$V(r) = V_o \cot(\varpi r) + V_1 \tag{23}$$

where V_o and V_1 were potential constants, ω was a range of potential. By applying equation (23) into equation (10) and by setting $U(r) = \frac{1}{r}\phi(r)$ and the centrifugal approximation, then equation (10) becomes,

$$\begin{bmatrix} \frac{d^2\phi(r)}{dr^2} - \left(\frac{\varpi L(L+1) + 2\alpha_{ML}\delta^2 V_o^2}{\sin^2 \varpi r}\right)\phi(r) + \left(\left(4\alpha_{ML}\chi - (E+M_o)\right)V_o - 2\alpha_{ML}\delta^2 V_2\right)\cot \varpi r \phi(r) \\ + \left(\left(4\alpha_{ML}\chi - (E+M_o)\right)V_1 + \left(E^2 - M_o^2 - 2\alpha_{ML}\eta^2\right) + 2\alpha_{ML}\delta^2 V_o^2 - 2\alpha_{ML}\delta^2 V_1^2\right)\phi(r) \end{bmatrix} = 0 \tag{24}$$

where,

$$\delta = (E_0 + M_o), \quad \chi = (E_0^2 - M_o^2)(E_0 + M_o), \quad \eta = (E_0^2 - M_o^2) \tag{25}$$

Then equation (24) was reduced to,

$$\frac{d^2\phi(r)}{dr^2} - \left[\frac{v(v+1)}{\sin^2(\varpi r)} - 2q\cot(\varpi r) + \kappa^2\right]\phi(r) = 0 \tag{26}$$

with,

$$v(v+1) = \left(\varpi L(L+1) + 2\alpha_{ML}\delta^2 V_o^2\right) \tag{27}$$

$$2q = \left(\left(4\alpha_{ML}\chi - (E+M_o)\right)V_o - 2\alpha_{ML}\delta^2 V_2\right) \tag{28}$$

$$-\kappa^2 = \left(\left(4\alpha_{ML}\chi - \left(E + M_o\right)\right)V_1 + \left(E^2 - M_o^2 - 2\alpha_{ML}\eta^2\right) + 2\alpha_{ML}\delta^2 V_o^2 - 2\alpha_{ML}\delta^2 V_1^2\right) \tag{29}$$

Equation (26) was a differential equation which must be reduced to hypergeometric differential equation type. By using the suitable variable change $\cot\left(\varpi r\right) = i\left(1-2y\right)$, so we got,

$$y(1-y)\frac{d^2\phi(r)}{dy^2} + (1-2y)\frac{d\phi(r)}{dy} + \left[v^{'}\left(v^{'}+1\right) - \frac{4\alpha^2}{4y} - \frac{4\beta^2}{4(1-y)}\right]\phi(r) = 0 \tag{30}$$

with,

$$\frac{2qi-\kappa^2}{\varpi^2} = 4\alpha^2, \quad \frac{-2qi-\kappa^2}{\varpi^2} = 4\beta^2, \quad v^{'}\left(v^{'}+1\right) = \frac{v\left(v+1\right)}{\varpi^2} \tag{31}$$

Equation (30) was an intermediate of the hypergeometric differential equation. Equation (30) was reduced by using new wave function,

$$\phi\left(r\right) = y^{\alpha}\left(1-y\right)^{\beta} f\left(y\right) \tag{32}$$

So we get

$$y(1-y)\frac{f^2(y)}{dy^2} + \left[\left(2\alpha+1\right) - \left(2\alpha+2\beta+2\right)y\right]\frac{f(y)}{dy} + \left[v^{'}\left(v^{'}+1\right) - \left(\alpha+\beta\right)\left(\alpha+\beta+1\right)\right]f\left(y\right) = 0 \tag{33}$$

The equation (33) must be reduced to AIM type equation, by dividing it with $y(1-y)$, then we got

$$f^{''}\left(y\right) + \left[\frac{\left(2\alpha+1\right) - \left(2\alpha+2\beta+2\right)y}{y\left(1-y\right)}\right]f^{'}\left(y\right) + \left[\frac{v^{'}\left(v^{'}+1\right) - \left(\alpha+\beta\right)\left(\alpha+\beta+1\right)}{y\left(1-y\right)}\right]f\left(y\right) = 0 \tag{34}$$

By comparing equation (11) and equation (34), we had

$$\lambda_0 = \frac{-\left(2\alpha+1\right)}{y} + \frac{\left(2\beta+1\right)}{1-y} \tag{35}$$

$$s_o = \frac{\left(\alpha+\beta\right)\left(\alpha+\beta+1\right) - v^{'}\left(v^{'}+1\right)}{y} + \frac{\left(\alpha+\beta\right)\left(\alpha+\beta+1\right) - v^{'}\left(v^{'}+1\right)}{\left(1-y\right)} \tag{36}$$

Using equations (18-21) and equation (34), we can obtain the eigenvalue, was given as

$$v^{'}\left(v^{'}+1\right) = \left(\alpha+\beta+n\right)\left(\alpha+\beta+\left(n+1\right)\right) \tag{37}$$

From the eigenvalue in equation (37), we obtained the energy eigenvalue equation by using equations (31) and (37). The energy eigenvalue equation of Klein Gordon equation for minimal length, was given by,

$$\left(E^2 - M_o^2\right) = \varpi^2\left[\left(L_{\delta\alpha_{ML}}\right)^2 - \frac{\left(\frac{2\left(\left(4\alpha_{ML}\chi - \left(E+M_o\right)\right)V_o - 2\alpha_{ML}\delta^2 V_2\right)}{\varpi^2}\right)^2}{4\left(L_{\delta\alpha_{ML}}\right)^2}\right] - \left(4\alpha_{ML}\chi - \left(E+M_o\right)\right)V_1 + \gamma_{\alpha_{ML}} \tag{38}$$

where,

$$\gamma_{\alpha_{ML}} = 2\alpha_{ML}\eta^2 + 2\alpha_{ML}\delta^2 V_1^2 - 2\alpha_{ML}\delta^2 V_o^2 \tag{39}$$

$$L_{\delta\alpha_{ML}} = \sqrt{\frac{\varpi L\left(L+1\right) + 2\alpha_{ML}\delta^2 V_o^2}{\varpi^2} + \frac{1}{4}} - n - \frac{1}{2} \tag{40}$$

and n was a quantum number. Equation (38) was energy eigenvalue of Klein Gordon equation in the presence of a minimal length for trigonometric cotangent potential. The energy eigenvalue of equation (38) was calculated numerically by using the Matlab software. The results were shown in Table 1.

Table 1. The relativistic energy for M_o=1, V_o=0.01, V_1=0.1, V_2=0.001, $\varpi = 0.1$ and L=0 with various minimal length parameter and quantum number n.

n	$En\ (eV)$			
	$\alpha_{ML}=0$	$\alpha_{ML}=0.02$	$\alpha_{ML}=0.04$	$\alpha_{ML}=0.06$
1	1.0840	1.0841	1.0843	1.0844
2	1.1136	1.1137	1.1138	1.1139
3	1.1397	1.1398	1.1400	1.1401
4	1.1723	1.1727	1.1731	1.1735
5	1.2121	1.2131	1.2141	1.2152

Table1 shows that the energy eigenvalue within the minimal length formalism was higher than energy eigenvalue without the minimal length formalism. The increasing of the minimal length parameter causes the increase of the energy eigenvalue. The energy eigenvalue increased by the increase of quantum number n.

We got the wave functions by inserting equations (26-35) into equation (37).The wave functions of Klein Gordon equation in the presence of minimal length for n=0, n=1 and n=2, were given as

$$\phi_0(r)=C'\left(\frac{1+i\cot\varpi r}{2}\right)^{\alpha}\left(\frac{1-i\cot\varpi r}{2}\right)^{\beta} \tag{41}$$

$$\phi_1(r)=-C'\left(\frac{1+i\cot\varpi r}{2}\right)^{\alpha}\left(\frac{1-i\cot\varpi r}{2}\right)^{\beta}(2\alpha+1)\left[1+\frac{(-1)(2\alpha+2\beta+2)\left(\frac{1+i\cot\varpi r}{2}\right)}{(2\alpha+1)}\right] \tag{42}$$

$$\phi_2(r)=C'\left(\frac{1+i\cot\varpi r}{2}\right)^{\alpha}\left(\frac{1-i\cot\varpi r}{2}\right)^{\beta}(2\alpha+1)(2\alpha+2)\left[1+\frac{(-2)_1(2\alpha+2\beta+3)_1\left(\frac{1+i\cot\varpi r}{2}\right)^1}{(2\alpha+1)_1 1!}+\frac{(-2)(-1)(2\alpha+2\beta+3)(2\alpha+2\beta+4)\left(\frac{1+i\cot\varpi r}{2}\right)^2}{(2\alpha+1)(2\alpha+1)2}\right] \tag{43}$$

5. Conclusion

We investigated Klein Gordon equation in the presence of a minimal length for trigonometric cotangent potential using Asymptotic Iteration methodto obtain the energy eigenvalues and radial wave functions. The energy eigenvalue within the minimal length formalism was higher than energy eigenvalue without the minimal length formalism. The increase of minimal length parameter causes the increase of the energy eigenvalue. The minimal length effect gives influence in increasing binding energy value. The binding energy eigenvalue increased by the increase of quantum number n.

Acknowledgement
This research was partly supported by Sebelas Maret University Higher Education Project Grant 623/UN27.21/PP/2017.

References

[1] Ikhdair S M and Sever R 2008 *arXiv:0808.1002v1 [quant-ph]* 14.
[2] Momtazi E, Rajabi A A, and Yazarloo B H 2014 *Turkish J. Phys* **38** 1 81–85.
[3] Greiner W 2000 Springer-Verlag Berlin Heidelberg, New York.
[4] Hassanabadi H, Maghsoodi E, Ikot N A, and Zarrinkamar S 2013 *Hindawi Publishing Corporation.*
[5] Jana K T and Roy P 2009 *Physics Letters A* **373** 1239–41.
[6] Bouaziz D 2013 *arXiv:1311.7405v1 [quant-ph].*
[7] Merad M, Zeroual F and Benzair H 2010 *Electronic Journal of Theoretical Physics***7** 23 41-56.
[8] Nugraha D A, Suparmi A, Cari C, and Pratiwi B N 2017 *J. Phys. Conf. Ser***795**.
[9] Nugraha D A, Suparmi A, Cari C, and Pratiwi B N 2017 *J. Phys. Conf. Ser* **820**.
[10] Poszwa A, 2014 *Acta Physica Polonica A* **126** 6.
[11] A N Ikot, L E Akpabio and E J Uwah 2011 *Electron. J. Theor. Phys* **25** 25 225-232.
[12] Alimohammadi M and Hassanabadi H 2017 *Nuclear Physics A* **957** 439-449.
[13] Cari, Suparmi, Deta A U and Werdiningsih I S 2013 *Makara J. Sci.*
[14] Farisky N M, Suparmi A, Cari C and Yunianto M 2016 *J. Phys. Conf. Ser.*
[15] Sprenger M, Nicolini P and Bleicher M 2012 *European Journal Of Physics***33** 853-62.
[16] Garay J L 1994 *International Journal of Modern Physics A* **10** 2 145-65.
[17] Chabab M, ElBatoul A, Lahbas A and Oulne M 2016 *Physics Letters B* 212-16.
[18] H Ciftci, R L Hall and N Saad 2003 *J. Phys. A. Math. Gen* **36** 47 11807-16.
[19] Pratiwi N B, Suparmi A, Cari C and Husein S A 2017 *Pramana. J. Phys.*
[20] Pramono S, Suparmi A, and Cari C 2016 *Adv. High Energy Phys***2** **016**.
[21] Naderia L and Hassanabadi H 2016 *The European Physical Journal Plus* 131:133.

Structural characterization of hydrothermally synthesized MnO_2 nanorods

D Q A'yuni[1], I Alkian[1], F K Sya'diyah[1], Kadarisman[1], A Darari[1], V Gunawan[1] and A Subagio[1]

[1]Department of Physics, Faculty of Science and Mathematics, Diponegoro University Jl. Prof. Soedarto, Tembalang, Semarang City, 50275, Indonesia

E-mail: dewiqurrota@st.fisika.undip.ac.id

Abstract. We prepared the hydrothermal method to synthesize MnO_2 nanorods with controlled structure. $KMnO_4$ and HCl with the various molar ratio (1:2,1:6,1:8) reacted at 160°C for three hours to form MnO_2 nanorods. The study found that changing the molar ratio can control the structure and morphology of MnO_2. The result revealed that MnO_2 formed in nanorod microstructures with different crystallographic structure and phase composition of each molar ratio. The diffraction peaks observed at 2θ values of 28.9°, 37.8°, 40.9°, 49.7° and 60.5° respectively indexed to (110), (101), (200), (411) and (521) plane reflections of a tetragonal phase of β-MnO_2 and α-MnO_2. The characterization of the morphology showed that the diameters of nanorod microstructures of MnO_2 ranging from 30 to 145 nm with length ranging from 0.5 to 3 μm. These MnO_2 nanorods product would be potentially used in energy storage devices.

1. Introduction

MnO_2 was one of the most attractive materials because of its applications, such as catalysts, lithium-ion batteries and Mg batteries, electrochemical supercapacitors, ionic or molecular sieves, for its advantages of low cost, earth abundance, environmentally friendly, and superior performance in energy capacity [1,2]. The physical properties of MnO_2 relied on the crystalline phase and morphology of MnO_2 nanostructures [2,3]. The previous studies investigated the morphology and crystal structure of MnO_2 in the different morphology, including nanorods [1,3,4], nanoflower [5,6], nano urchin [3,7], nanowires [8,9], nanoneedles [10]. Nanorods were one of the most interesting morphologies because it can control self-aggregation effectually [11]. The various method has been developed to synthesize MnO_2 with controlled morphologies, including thermal, refluxing, hydrothermal, sol-gel, electrochemical, solid-state reaction [12]. The hydrothermal method was attractive because it is a cheap, environmentally friendly method to prepare materials in different nanostructures [13]. For example, Li et al. [3] studied the electrochemical properties on supercapacitor with the different morphology of MnO_2 including nanorod, hollow urchin, and smooth ball, and nanorod structure displayed the best electrochemical capacity.

In this study, we demonstrate a hydrothermal method to synthesize MnO_2 nanorods with controlled structure. We changed the molar ratio of Mn precursor solution to HCl to control the structure of MnO_2. To calculating the crystallite sizes of MnO_2, we used the Scherrer equation given below [14]:

$$D = \frac{K\lambda}{\beta \sin \theta} \quad (1)$$

where λ is X-Ray wavelength (nm), β is the peak width of the diffraction peak profile at half maximum height (rad) and K is a constant related to crystallite shape. Furthermore, the products of MnO_2 nanorods potentially would be used in energy storage and other devices which are need storage of electron.

2. Experimental

2.1. Synthesis MnO_2 Nanorods

MnO_2 were synthesized by using $KMnO_4$ and HCl with the different molar ratio (1:2, 1:6, 1:8). All chemicals were analytical grade reagents from Merck, Germany. 0.395 g $KMnO_4$ was dissolved completely in deionized water + HCl at room temperature. The solution was kept continuous stirring to form a clear solution. The mixture was transferred into a 20 mL Teflon-lined autoclave. The autoclave was heated at 160 °C for three hours (**Figure 1.**) in an oven and then cooled down to room temperature naturally.

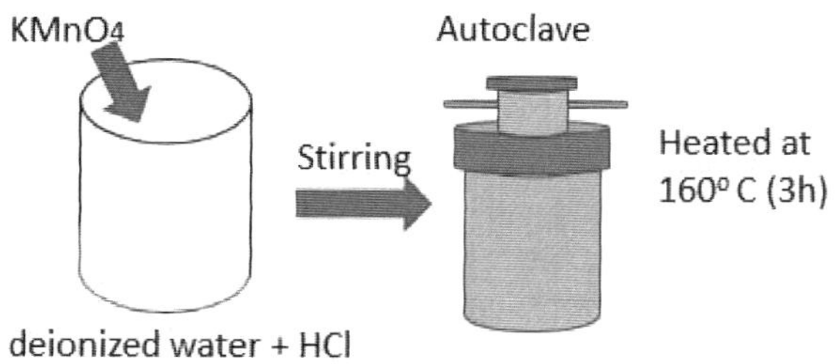

Figure 1. Schematic of hydrothermally synthesized MnO_2 nanorods

2.2. Characterization

The samples were characterized by X-ray diffraction spectroscopy (XRD, Philip Analytical X-Ray B. V) with Cu Kα radiation (λ=1.5418 Å) at 40 kV, and Scanning Electron Microscopy (SEM) to study the morphology of MnO_2 nanorods.

3. Results and Discussion

Analysis role of the molar ratio of $KMnO_4$ to HCl, we made three different samples with the molar ratio of 1:2, 1:6, and 1:8. The reaction was carried out at the temperature of 160 °C for 3 h. The morphology of samples was characterized by SEM. **Figure 2.** shows the morphology of MnO_2 nanorods at the different molar ratio. The results show nanorod microstructures with aggregation with length ranging from 0.5-3 μm. However, the product of molar ratio 1:2 shows more spherical nanostructure than nanorod structure (**Figure 2.** (a)). **Figure 2.** (b) and (c) shows nanorod and nanowire structures with more nanorods in the molar ratio of 1:8. The products of nanorod structures consist with the diameter ranging from 30-145 nm. It is likely that a larger amount of HCl will produce more nanorod structures with the same reaction time. Based on the reaction process, the reaction on the formation of MnO_2 using $KMnO_4$ and HCl is according to the following reaction [1]:

$$KMnO_4 + H_2O + HCl \rightarrow MnO_2.H_2O + KCl + H_2O$$

Figure 2.SEM images of MnO_2 nanorods at the different molar ratio: (a) 1:2, (b) 1:6, (c) 1:8

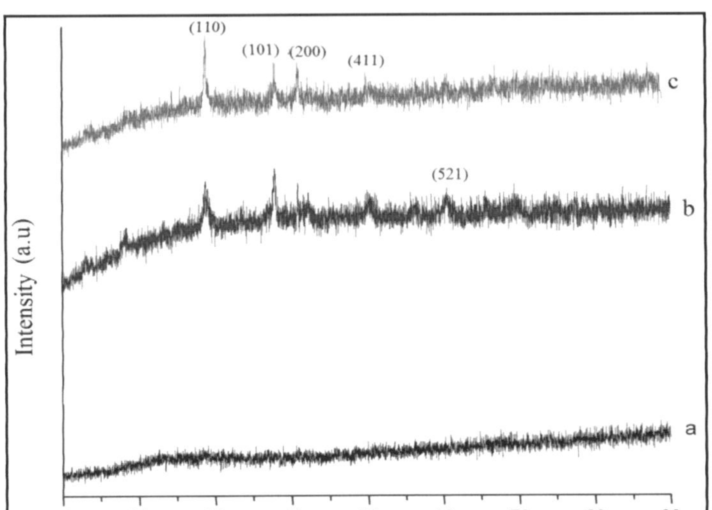

Figure 3. XRD patterns of MnO_2 nanorods at the different molar ratio: (a) 1:2, (b)1:6, (c) 1:8

The powder X-Ray diffraction (XRD) pattern was shown in **Figure 2.** for the samples synthesized with the three different molar ratio. The diffraction peaks observed at 2θ values of 28.9°, 37.8°, 40.9°,

49.7°, and 60.5° respectively indexed to (110), (101), (200), (411), and (521) plane reflections of a tetragonal phase of β-MnO_2 and α-MnO [1,3]. However, the crystallinity of the molar ratio of 1:2 has small rods and irregularity (**Figure 2.** (a)), indicating that the product presents an amorphous structure. The diffraction patterns show that, of the three samples, the molar ratio of 1:8 has the best crystallinity. However, the peaks of intensity are broad and low, indicating the presence of low degree crystallinity of MnO_2 nanorods.

Table 1. shows the crystallite size of MnO_2 nanorods as calculated according to the equation (1). The crystallite sizes increase with increasing the molar ratio of samples. However, because of the amorphous structure of MnO_2 with molar ratio 1:2, the crystallite size of the sample is below the detection limit for XRD.

Table 1. Crystallite sizes of MnO_2 nanorods

Molar Ratio	Crystallite Size (nm)
1:2	-
1:6	10.09
1:8	20.61

The mechanism of the crystallographic structure transformation has been illustrated in **Figure 4.** At the molar ratio of 1:2, the high concentration of precursors lead to rapid formation of numerous nuclei, which these nuclei self-assemble to form nearly amorphous spheres [15,16]. Increasing molar ratio from 1:2 to 1:6 and 1:8, the spherical morphology with amorphous structures are changed to nanorod morphology with the crystal structure. This is attributed that more HCl could lead to a decrease in the nucleation rate, thus the structures becoming to nanorods [5].

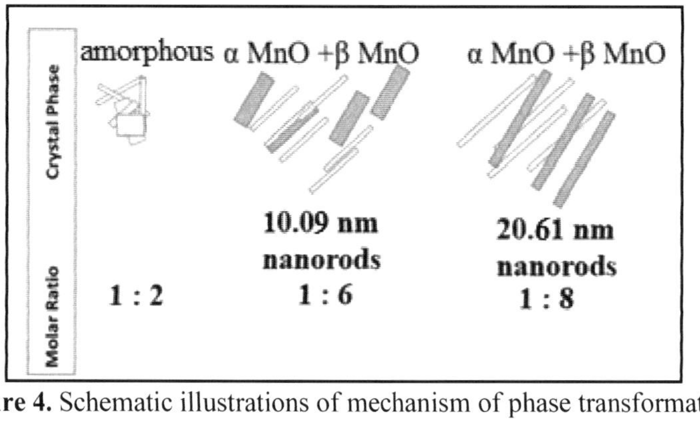

Figure 4. Schematic illustrations of mechanism of phase transformation

4. Conclusions

This work shows that MnO_2 nanorods have been successfully prepared by hydrothermal method. We have synthesized β-MnO_2 and α-MnO_2 nanorods with the combination of sphere and wire nanostructures by tuning the molar ratio of Mn precursor solution to HCl. The peaks of intensity are still broad and low, indicating the presence of low degree crystallinity of MnO_2 nanorods. The uniform morphology can be controlled by increasing the molar ratio. Synthesis process studies reveal that the molar ratio is one of the parameters to get the nanorod structures. These products would be potentially used in energy storage devices.

5. Acknowledgments

We would like thank to Diponegoro University, Ministry of Research Technology and the Higher Education Republic of Indonesia for funding our research from Program KreativitasMahasiswa (PKM). We are also indebted to Mr. Chasbullah Hindratmo for his valuable assistance in Laboratory Nanotechnology, Diponegoro University.

References

[1] Wu J, Huang H, Yu L and Hu J 2013 *Advances in Materials Physics and Chemistry* **3** 201
[2] Darari A, Ardiansah H R, Arifin, Rismaningsih N, Ningrum A N and Subagio A 2016 *AIP Conference Proceedings* **1725** 020012
[3] Li N, Zhu X, Zhang C, Lai L, Jiang R and Zhu J 2017 *Journal of Alloys and Compound* **692** 26
[4] Kumar A, Sanger A, Singh A K, Kumar A, Kumar M and Chandra R 2017 *Journal of Magnetism and Magnetic Materials* **433** 227
[5] Wang G, Tang B, Zhuo L, Ge J and Xue M 2006 *Eur. J. Inorg. Chem* 2313
[6] Chu J, Lu D, Ma J, Wang M, Wang X and Xiong S 2017 *Material Letters***193** 263
[7] Li W et al. 2012 *Journal of Materials Chemistry* **22** 14864
[8] Devaraj S and Mundichairah N 2008 *J. Phys. Chem* **112** 4406
[9] Guan H, Chen G, Zhang S and Wang Y 2010 *Material Chemistry and Physics* **124** 639
[10] Deng S, Sun D, Wu C, Wang H, Liu J, Sun Y and Yan H 2013 *Electrochimica Acta* **111** 707
[11] Chen Y, Liu C, Li F and Cheng H M 2005 *Journal of Alloys and Compounds* **397** 282
[12] Liu X, Fu S and Huang C 2005 *Powder Technology* **154** 120
[13] Subramanian V, Zhu H and Wei B 2006 *Journal of Power Sources* **159** 361
[14] Feng L, Xuan Z, Zhao H, Bai Y, Guo J, Su C and Chen X 2014 *Nanoscale Research Letter* **9**:290
[15] Xiao W, Wang D and Lou X W 2010 *J. Phys. Chem. C* **114** 1694
[16] Zhu J, Tang S, Xie H, Dai Y and Meng X 2014 *ACS Appl. Mater. Interfaces* 1

IOP Conf. Series: Journal of Physics: Conf. Series **909** (2017) 012005 doi:10.1088/1742-6596/909/1/012005

Lagrangian for RLC circuits using analogy with the classical mechanics concepts

Albertus Hariwangsa Panuluh and Asan Damanik

Department of Physics Education, Sanata Dharma University
Kampus III USD Paingan, Maguwoharjo, Sleman, Yogyakarta, Indonesia

Email: panuluh@usd.ac.id

Abstract. We study and formulate the Lagrangian for the LC, RC, RL, and RLC circuits by using the analogy concept with the mechanical problem in classical mechanics formulations. We found that the Lagrangian for the LC and RLC circuits are governed by two terms i. e. kinetic energy-like and potential energy-like terms. The Lagrangian for the RC circuit is only a contribution from the potential energy-like term and the Lagrangian for the RL circuit is only from the kinetic energy-like term.

1. Introduction

Lagrangian formalism is a powerful way to obtain the equation of motion of a physical system. The Lagrangian formalism is turned up to solve problems that are not simple by using Newtonian Mechanics [1]. In Newtonian mechanics, we usually formulate the mechanical problem (physical system) in the form of force or vector. Meanwhile, in Lagrangian mechanics, we approach the mechanical problem in the form of energy which is a scalar. If we know the Lagrangian of a physical system, then we can obtain the equation of motion of a physical system by using the Euler-Lagrange equation [2].

In the electrical subject, the RLC (Resistance-Inductance-Capacitance) circuit is commonly discussed in student physics textbooks as an example of the application of electronics principles to build an electronic device [3]. In the mathematical physics subject, the RLC circuit is commonly used to show the application of differential equation in the physical system [4]. The knowledge of RLC circuit is certainly of great physical interest both from experimental (applied) and theoretical sides. Recently, the RLC circuit and its physical properties to be an interesting research subject due to the development of mesoscopic physics and nanophysics that need a quantum mechanics because the mesoscopic system and nanophysics are not anymore a macroscopic system.

The analysis of RLC circuit as a mesoscopic system by using quantum mechanics based on Cardirola-Kanai Hamiltonian and quantum invariant method to solve the Schrödinger equation for the RLC circuit and to obtain the corresponding wave functions in term of a particular solution of Milne-Penney equation was reported by Pedrosa and Pinheiro [5]. The quantum mechanics treatment of the RLC circuit with discrete charge and semiclassical consideration can be found in Utreras-Diaz [6] that obtained the approximation of energy eigenvalues in term of a dimensionless parameter $\sqrt{L/C}\, e^2/h$ (e is the electron charge, and h is the Planck constant). In literature, we found most analysis of the RC, RL, LC, and RLC circuits, as a mesoscopic system or nanophysics, are commonly discussed and formulated by using the concept of quantum mechanics [7-10]. As we have already known that the energy operator in quantum mechanics is the Hamiltonian which is defined as the sum of energy kinetic and energy potential of the physical system. In classical mechanics, we also know that we can formulate the physical system in form of Lagrangian defined as the kinetic energy minus potential energy of the physical system. Thus, it is also an interesting problem to know the explicit form of the Lagrangian for RC, LC, RL, and RLC circuits and its phenomenological implications.

In this paper, we formulate the Lagrangian for LC, RC, RL, and RLC circuits by using the analogy with the classical mechanics formulation for a physical system. We also discuss the advantages of the lagrangian formulation of LC, RC, RL, and RLC circuits compare to the o electrical formulation as one can find in the most of the student textbooks. We organized this paper as follow: in section 2 we derive the Lagrangian for LC circuit; in section 3 we derive the Lagrangian formulation for the RC circuit; in section 4, we formulate the lagrangian for RL circuit; in section 5 we derive the Lagrangian for RLC circuit. Finally, in section 6 we write the conclusions.

2. Lagrangian for the LC Circuit

In this section, we formulate the Lagrangian for the LC circuit. We begin with by writing the differential equation of Kirchhoff's rule for LC circuit without source as follow

$$\frac{\mathrm{d}^2 q}{\mathrm{d}t^2}+\frac{q}{LC}=0\,. \tag{1}$$

By analogy with the harmonic oscillator, we can put the angular frequency regarding L and C as follow

$$\omega=\sqrt{\frac{1}{LC}}\,. \tag{2}$$

As we know from the mathematical physics, the general solution of Eq. (1) is given by

$$q(t)=k_1\mathrm{e}^{\mathrm{i}\omega t}+k_2\mathrm{e}^{-\mathrm{i}\omega t} \tag{3}$$

where we can determine the constants k_1 and k_2 from the appropriate boundary values.

Mathematically, there is two possible initial conditions for the capacitor C. First, at time t = 0 the capacitor has zero charges: $q(0)$ = 0 and at time $t=\infty$ the charge of the capacitor reach a maximum value that is Q and this process is known as charging process. Second, at time t = 0 it is possible that the capacitor C has initial charge Q_0 and at time $t=\infty$ the capacitor has no charge which is known as discharging a capacitor. If we put the the initial condition $q(0)$ = 0 into Eq. (3) , then we have

$$q(0)=0=k_1\mathrm{e}^0+k_2\mathrm{e}^0 \tag{4}$$

that implies

$$k_1=-k_2 \tag{5}$$

By inserting Eq. (5) into Eq. (3), we obtain

$$q(t)=k_1(\mathrm{e}^{\mathrm{i}\omega t}-\mathrm{e}^{-\mathrm{i}\omega t})\,. \tag{6}$$

If we use Euler's equation, then Eq. (6) reads

$$q(t)=2\mathrm{i}k_1\sin(\omega t)\,. \tag{7}$$

It is clear from Eq. (7) that the charge as a sine function. The solution of the differential equation of Eq. (1) with the initial condition $q(0)=0$ is not relevant to the physical reality although it is possible from the mathematics point of view. Therefore we discard the charging process with the solution in the form of sine function.

If we put the initial condition $q(0)=Q_0=k_1+k_2$ for discharging process into Eq. (3), then we have

$$k_2=Q_0-k_1\,. \tag{8}$$

By inserting the Eq. (8) into the Eq. (3) and after doing a little algebra, the Eq. (3) become

$$q(t)=Q_0\mathrm{e}^{-\mathrm{i}\omega t}+2\mathrm{i}k_1\sin(\omega t)\,. \tag{9}$$

The second term of Eq. (9) does not satisfy the initial condition that $t=\infty\rightarrow q$ = 0, so we drop the second term of Eq. (9) and finally, we have the solution for discharging capacitor in the LC circuit as follow

$$q(t)=Q_0\mathrm{e}^{-\mathrm{i}\omega t}. \tag{10}$$

Now, we are in the position to formulate the Lagrangian for LC circuit. To formulate the lagrangian of a physical system, we must know the potential energy and kinetic energy of the system. The potential energy for the LC circuit is given by

$$V(q)=\frac{1}{2C}q^2. \tag{11}$$

By analogy with the potential energy of harmonic oscillator, we can put the corresponding parameters as follow

$$m\leftrightarrow L, \qquad x\leftrightarrow q, \qquad k\leftrightarrow\frac{1}{C}. \tag{12}$$

By inserting Eq. (10) into Eq. (11), then we find the potential energy of the LC circuit as follow

$$V_{LC}=\frac{1}{2C}Q_0^2\mathrm{e}^{-2\mathrm{i}\omega t}, \tag{13}$$

and the kinetic energy of oscillator harmonic system is

$$T_{ho}=\frac{1}{2}m\dot{x}^2. \tag{14}$$

By using the analogy in Eq. (12) and take derivative of Eq. (10), we have the kinetic energy of the LC circuit as follow

$$T_{LC}=-\frac{1}{2}L\omega^2Q_0^2\mathrm{e}^{-2\mathrm{i}\omega t}. \tag{15}$$

From Eqs. (13) and (15), we finally can write the Lagrangian for the LC circuit as follow

$$L_{LC}=-\frac{1}{2}L\omega^2Q_0^2\mathrm{e}^{2\mathrm{i}\omega t}-\frac{1}{2C}Q_0^2\mathrm{e}^{-2\mathrm{i}\omega t}, \tag{16}$$

which is composed of two terms that can be put as kinetic energy-like and potential energy-like.

3. Lagrangian for the RC Circuit

In this section, we formulate the Lagrangian for the RC circuit. The differential equation of Kirchhoff's rule for RC circuit without a source is given by

$$R\frac{\mathrm{d}q}{\mathrm{d}t}+\frac{q}{C}=0. \tag{17}$$

It has already known that there are also two processes that can be assigned to the RC circuit i.e. charging or discharging of the capacitor C.

By integrating Eq. (17), and using the values of q = 0 at time $t=0$ and q = q at the time t = t, we then have

$$q(t)=\mathrm{e}^{-t/RC}. \tag{18}$$

From Eq. (18) it is apparent that the solution is unacceptable because if we put $t\rightarrow\infty$, then the charge becomes zero which is contrary to the physical process as a charging process. By integrating Eq. (18), and using the values of q = Q_0 at time t = 0 and q = q at time t = t, then we obtain

$$q(t)=Q_0\mathrm{e}^{-t/RC}. \tag{19}$$

which is relevant to the physical process as previously stated. It is clear that there is no kinetic energy term for the RC circuit. The potential energy in the RC circuit as follows

$$V_{RC}=\frac{1}{2C}Q_0^2\mathrm{e}^{-2t/RC}. \tag{20}$$

Because there is no kinetic energy in the RC circuit, then the Lagrangian of RC circuit as follows

$$L_{RC}=\frac{1}{2C}Q_0^2\mathrm{e}^{-2t/RC}. \tag{21}$$

IOP Conf. Series: Journal of Physics: Conf. Series **909** (2017) 012005 doi:10.1088/1742-6596/909/1/012005

4. Lagrangian for the RL Circuit

The differential equation for the RL circuit is given by

$$L\frac{\mathrm{d}I}{\mathrm{d}t}+RI=0. \tag{22}$$

The solution of the Eq. (22) read

$$I(t)=I_0\mathrm{e}^{-Rt/L}. \tag{23}$$

For the RL circuit, we can make an analogy with the equation of motion of a particle with the force is velocity dependent as follow

$$m\frac{\mathrm{d}v}{\mathrm{d}t}+kv=0, \tag{24}$$

By putting L correspond to mass m and the resistance R correspond to friction constant k, then we can put the current I in Eq. (23) as a corresponding velocity in Eq. (24).

To formulate the kinetic energy of the RL circuit, we use the analogy concept by using the usual mechanical kinetic energy ½ mv^2 and electric current I correspond with velocity v, and then we have

$$T_{RL}=\frac{1}{2}LI_0\mathrm{e}^{-2Rt/L} \tag{25}$$

as the kinetic energy term for RL circuit. For the RL circuit, there is no potential energy term. Thus, the lagrangian for the RL circuit read

$$L_{RL}=\frac{1}{2}LI_0\mathrm{e}^{-2Rt/L}. \tag{26}$$

5. Lagrangian for the RLC Circuit

The differential equation for the RLC circuit as follows

$$L\frac{\mathrm{d}^2q}{\mathrm{d}t^2}+R\frac{\mathrm{d}q}{\mathrm{d}t}+\frac{q}{C}=0$$

or

$$\frac{\mathrm{d}^2q}{\mathrm{d}t^2}+\frac{R}{L}\frac{\mathrm{d}q}{\mathrm{d}t}+\frac{q}{LC}=0 \tag{27}$$

By using analogy with a damped harmonic oscillator, we have

$$\gamma=\frac{R}{2L}, \qquad \omega_0=\sqrt{\frac{1}{LC}} \tag{28}$$

where γ and ω_0 as damping factor and natural frequency respectively.

The general solution of Eq. (27) as follows

$$q(t)=C_1\mathrm{e}^{-(\gamma-\lambda)t}+C_2\mathrm{e}^{-(\gamma+\lambda)t} \tag{29}$$

where

$$\lambda=\sqrt{\gamma^2-\omega_0^2}. \tag{30}$$

There are three possible cases for the parameter λ i.e. overdamping, critical damping and underdamping.

5.1 Overdamping

Overdamping will occur when $\lambda > 0$ so the exponents in the first and second term in Eq. (29) are real. If we insert the discharging capacitor boundary condition into Eq. (29), then we have the solution as follow

$$q(t)=Q_0\mathrm{e}^{-(\gamma+\lambda)t}. \tag{31}$$

By using the same procedure to formulate the Lagrangian for LC circuit, we then have the Lagrangian for overdamping RLC circuit as follow

$$L_{RLCover} = \frac{1}{2}L(\gamma+\lambda)^2 {Q_0}^2 \mathrm{e}^{-2(\gamma+\lambda)t} - \frac{{Q_0}^2\mathrm{e}^{-2(\gamma+\lambda)t}}{2C}. \tag{32}$$

After doing a little algebra that is by expanding and simplifying λ and γ, the Eq. (32) become

$$L_{RLCover} = \frac{1}{2}\left(\frac{R^2}{2L} - \frac{1}{C} + \frac{R}{2}\sqrt{\frac{R^2}{L^2} - \frac{4}{LC}}\right){Q_0}^2\mathrm{e}^{-\left(\frac{R}{L}+\sqrt{\frac{R^2}{L^2}+\frac{4}{LC}}\right)t} - \frac{{Q_0}^2\mathrm{e}^{-\left(\frac{R}{L}+\sqrt{\frac{R^2}{L^2}+\frac{4}{LC}}\right)t}}{2C}. \tag{33}$$

One can see that the Lagrangian for overdamping RLC circuit as shown in Eq. (33) is composed of two terms i. e. kinetic energy-like and potential energy-like.

5.2 Critical Damping

Critical damping will occur when $\lambda = 0$. As shown in [6], the Eq (29) becomes

$$q(t) = C_1 t\mathrm{e}^{-\gamma t} + C_2\mathrm{e}^{-\gamma t}. \tag{34}$$

By inserting the discharging capacitor boundary condition then Eq (34) becomes

$$q(t) = Q_0\mathrm{e}^{-\gamma t} \tag{35}$$

and by using the same procedure as previously, we find the Lagrangian for the critical damping RLC circuit as follow

$$L_{RLCcrit} = -\frac{1}{8}\frac{R^2}{L}Q_0^2\mathrm{e}^{-Rt/L} - \frac{{Q_0}^2\mathrm{e}^{-Rt/L}}{2C}, \tag{36}$$

which is composed of two terms i. e. energy kinetic like and potential energy-like.

5.3 Underdamping

The RLC circuit will undergo the underdamping when $\lambda < 0$, and the general solution for the case of underdamping RLC circuit can be read in [6]

$$q(t) = \mathrm{e}^{-\gamma t}\left(A\cos(\omega_d t - \theta_0)\right) \tag{37}$$

where

$$\beta = \sqrt{\omega_0^2 - \gamma^2} = \sqrt{\frac{1}{LC} - \frac{R^2}{4L^2}}. \tag{38}$$

By inserting the discharging capacitor boundary condition and we set the initial phase angle θ_0 by 0°, then Eq (37) become

$$q(t) = \mathrm{e}^{-\gamma t}Q_0\cos(\beta t) \tag{39}$$

and we obtain the lagrangian for the underdamping RLC circuit as follow

$$L_{RLCund} = -\frac{1}{2}L\left(\frac{R}{2L}Q_0 e^{-\gamma t}\cos(\beta t) + \sqrt{\frac{1}{LC} - \frac{R^2}{4L^2}}Q_0 e^{-\gamma t}\sin(\beta t)\right)^2 - \frac{{Q_0}^2\mathrm{e}^{-Rt/2L}\cos^2(\beta t)}{2C} \tag{40}$$

By inspecting Eq. (4), we can see that the first term is kinetic energy-like and the second term is potential energy-like.

It is clear from Eqs. (33), (36), and (40) that the Lagrangians for the RLC circuit are the sum of the kinetic energy-like and potential energy-like as well as one can find in classical mechanics. It can be stated that the Lagrangian terms of the RLC circuit independent of the value of λ qualitatively.

6. Conclusions

We have studied and evaluated the Lagrangian for LC, RC, LR, and RLC circuits by using the analogy with the classical mechanics concept to formulate the lagrangian of a physical system. We find that the lagrangian for LC and RLC circuits are composed of terms that can be assigned as kinetic energy and potential energy terms in corresponding with the Lagrangian of a physical system in classical mechanics. Meanwhile, the Lagrangian for the RC circuit is only from potential energy-like contribution and the and the lagrangian for the LR circuit is only kinetic energy-like contribution. We

IOP Conf. Series: Journal of Physics: Conf. Series **909** (2017) 012005 doi:10.1088/1742-6596/909/1/012005

also find that the contribution terms to the Lagrangian of the LRC circuit independent of the value of λ qualitatively.

References

[1] Arya A P 1998 *Introduction to Classical Mechanics* 2nd Edition (New Jersey: Prentice-Hall)
[2] Fowles R and Cassiday G L 2005 *Analytical Mechanics* 7th Edition (Connecticut: Thomson Learning Inc.)
[3] Serway R A and Jewett J W 2008 *Physics for Scientists and Engineers with Modern Physics* 7th Edition (Connecticut: Thomson Learning Inc.)
[4] Boas M L 2006 *Mathematical Methods in the Physical Sciences* 3rd Edition (New Jersey: John Wiley&Sons)
[5] Pedrosa I A and Pinheiro A P 2011 On the quantization of the dissipative systems *Prog. Theor. Phys.* **125** 6 1133-1141.
[6] Utreras-Diaz C A 2008 Discrete charge quantum circuits and electrical resitance *Phys. Lett. A* **372** 30 5059-5063.
[7] Buot F A 1993 Mesoscopic physics and nanoelectronics: nanoscience and nanotechnology *Phys. Rep.* **234** 2 73-174.
[8] Li Y Q and Chen B 1996 Quantum theory for mesoscopic electric circuits *Phys. Rev. B.* **53** 7 4027
[9] Lu T and Li Y Q 2002 Mesoscopic circuit with linear dissipation *Mod. Phys. Lett. B* **16** 26 975
[10] Flores J C and Utreraz-Diaz C A 2002 Mesoscopic circuits with charge discreteness: Quantum current magnification for mutual inductances *Phys. Rev. B* **66** 15 153410

IOP Conf. Series: Journal of Physics: Conf. Series **909** (2017) 012006 doi:10.1088/1742-6596/909/1/012006

QRS peak detection for heart rate monitoring on Android smartphone

Trio Pambudi Utomo, Nuryani Nuryani, and Darmanto

Physics Department of Universitas Sebelas Maret,
Jl. Ir. Sutami 36A Kentingan Jebres Surakarta 57126, INDONESIA

E-mail: triopambudiutomo@gmail.com

Abstract. In this study, Android smartphone is used for heart rate monitoring and displaying electrocardiogram (ECG) graph. Heart rate determination is based on QRS peak detection. Two methods are studied to detect the QRS complex peak; they are Peak Threshold and Peak Filter. The acquisition of ECG data is utilized by AD8232 module from Analog Devices, three electrodes, and Microcontroller Arduino UNO R3. To record the ECG data from a patient, three electrodes are attached to particular body's surface of a patient. Patient's heart activity which is recorded by AD8232 module is decoded by Arduino UNO R3 into analog data. Then, the analog data is converted into a voltage value (mV) and is processed to get the QRS complex peak. Heart rate value is calculated by Microcontroller Arduino UNO R3 uses the QRS complex peak. Voltage, heart rate, and the QRS complex peak are sent to Android smartphone by Bluetooth HC-05. ECG data is displayed as the graph by Android smartphone. To evaluate the performance of QRS complex peak detection method, three parameters are used; they are positive predictive, accuracy and sensitivity. Positive predictive, accuracy, and sensitivity of Peak Threshold method is 92.39%, 70.30%, 74.62% and for Peak Filter method are 98.38%, 82.47%, 83.61%, respectively.

1. Introduction

Today, heart disease is one of the dominant causes of death in the world [1]. At least 6 million deaths were caused by heart disease in 2002, and increasing to 7.4 million deaths in 2012 [2]. To prevent more death from this disease, special attention is needed, one of them is early heart disease diagnosis. One of the diagnostic techniques for heart disease is Electrocardiogram (ECG) [3-5].

Electrocardiogram is a representation of heart's electrical activity [6]. This electrical activity can be recorded by placing some electrodes on the particular body's surface [7]. The abnormalities of the heart can be seen on this recorded graph of ECG [8].

Lately, the advances in electronics technology and the development of digital signal processing techniques; many ways have been developed to recognize the heart abnormalities automatically through ECG signals [9]. Lou et al. (2013) have developed a wireless health monitoring system based on the smartphone operating system. The system is capable of measuring, analyzing, storing various physiological parameters of the patient [10]. Compared to monitors on Personal Computer (PC) or Personal Digital Assistant (PDA), smartphones are smaller and easier to carry, have the ability to communicate remotely and other advance functions [11].

Another ECG function is the measurement of heart rate. Heart rate is a number of complete pulsation of the heart when cardiac muscles contract and relax in one minute. In the ECG, one pulsation is referred as two R peaks or QRS complex peak of ECG waveform [12]. To detect QRS complex peak in the ECG waveform, many algorithms have been developed [13-15]. This work contributes to developing QRS complex peak detection and heart rate monitoring system based on Android smartphone.

2. Methods

2.1. ECG Monitoring System

This ECG monitoring system consists of the AD8232 ECG Kit for recording cardiac activity, Arduino UNO R3 to process data that is obtained from AD8232 Kit, Bluetooth HC-05 is used for transmitting data from Arduino to Android, SDCard module to store data read by Arduino, and Android smartphone to display the ECG graph and heart rate.

Small bio potential signals such as EKG can be read by the AD8232 module because this module has 1100 times amplification [16]. Besides that, AD8232 module can filter the signal when there are disturbance and interference due to body movement [17]. Recorded signal form AD8232 module is decoded by Arduino UNO R3 into analog data and processed to find the QRS complex peak. Then, the data is converted into voltage value (mV) and is stored into memory card using SDCard module. Furthermore, the data will be sent to Android smartphone using Bluetooth HC-05. Heart rate is calculated in the Arduino UNO R3. The function of Android smartphone is displaying ECG data into graph and represent the heart rate. Android Studio is used for building the application [18]. Figure 1 shows block diagram of this ECG monitoring system.

Figure 1. Block diagram of ECG monitoring system

2.2. QRS Complex Peak Detection

Figure 2 (a) shows algorithm of QRS complex detection using Peak Filter method. This method consists of three main processes; Linear High Pass Filter, Non-Linear Low Pass Filter, and Decision Making. The high pass filter can be written as equation (1) which is using M-point moving average filter and an ideal delayed system by group delay of (M+1)/2 samples. M value can be 5 or 7 [14].

Low pass filter can be written into equation (2), which is constructed by point to point simple squaring operation and moving window integration. According to Pan and Tompkins (1985), the value of K is 30 for sampling rate 200 Hz [13]. In the decision-making process, adaptive threshold is utilized. Thresholding scheme can be formulated as equation (3) to (5).

$$y[n] = x[n - \frac{M-1}{2}] - \frac{1}{M}\sum_{m=1}^{M} x[n-m] \tag{1}$$

$$z[n] = \sum_{k=0}^{K-1} y^2[n-k] \tag{2}$$

$$if\ z[n] \geq Threshold \rightarrow QRS\ complex \tag{3}$$

$$if\ z[n] < Threshold \rightarrow Not\ QRS\ complex \tag{4}$$

$$Threshold = \alpha\gamma PEAK - (1-\alpha)Threshold \tag{5}$$

PEAK is new local maximum detected in the wave feature, α is referred as "forgotten factor", where $0 \leq \alpha \leq 1$. γ is a weighting factor; used for the peak value contribution to the adjustment of the threshold determination, empirically its values are ranged from 0.15 to 0.2 γ [14].

IOP Conf. Series: Journal of Physics: Conf. Series **909** (2017) 012006 doi:10.1088/1742-6596/909/1/012006

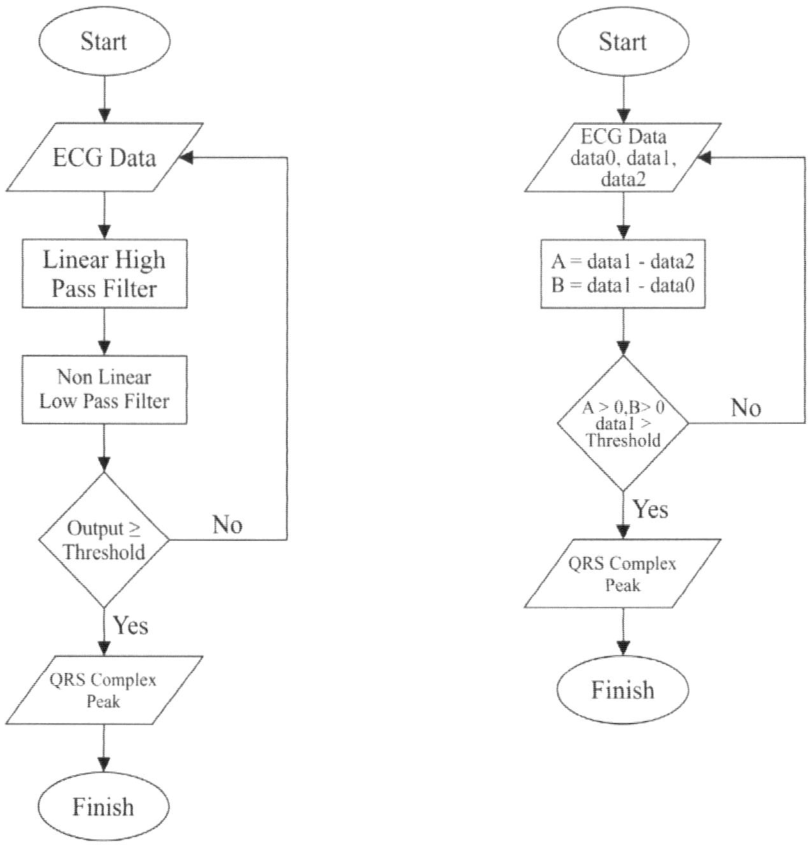

(a) (b)

Figure 2. (a) Peak Filter method (b) Peak Threshold method algorithm

Figure 2 (b) shows algorithm of QRS complex detection using Peak Threshold method. A data point (x_n) will be considered as QRS complex peak if the subtraction result of the current data point (x_n) and previous data point (x_{n-1}); then current data point (x_n) and its next data point (x_{n+1}), is positive. Besides that, the current data point's value is more than threshold value. The determination of the QRS complex in this method can be illustrated in Figure 3 and can be written as,

$$x_n - x_{n-1} > 0 \tag{6}$$

$$x_n - x_{n+1} > 0 \tag{7}$$

$$x_n > Threshold \tag{8}$$

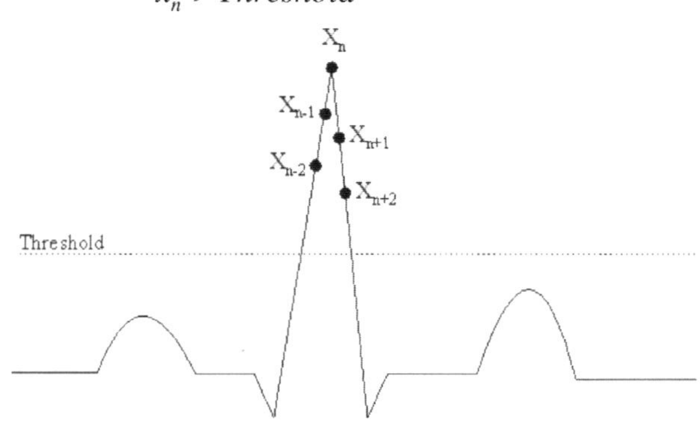

Figure 3. Determination of QRS complex peak with Peak Threshold method

2.3. Performance of QRS complex peak detection method

To calculate the performance of QRS complex detection algorithm, the following parameters can be used [19],

$$Acc = \left(1 - \frac{F_p + F_n}{Totalbeat}\right) x100\% \quad (9)$$

$$Se = \left(\frac{T_p}{T_p + F_n}\right) x100\% \quad (10)$$

$$P = \left(\frac{T_p}{T_p + F_p}\right) x100\% \quad (11)$$

Where Acc is accuracy, Se is sensitivity, P is positive predictive, T_p is the number of QRS complexes detected correctly, F_p is the number of non-QRS complexes detected as QRS complexes, and F_n is the number of undetectable QRS complexes.

2.4. Heart rate

The number of QRS complex peak (R wave) that occur in one minute is used for heart rate (HR) calculation which the unit is beat per minute (bpm). To calculate HR in real time, following equation is utilized,

$$HR = \frac{60,000}{Interval\_R - R(ms)} \quad (12)$$

For example, if one of the beats (from R to R or from QRS complex peak to another peak) takes 750 ms, then in one minute (60*1000 = 60000ms), there is (800000/750 = 80bpm) 80 beats. This 80bpm is the heart rate against time [20].

3. Result and Discussion

The Android application has several main functions; they are displaying data in graphical form, displaying heart rate, and recording data that has been received from Arduino. Figure 4 shows the application interface of Android smartphone while receiving ECG data and heart rate from Microcontroller Arduino UNO R3.

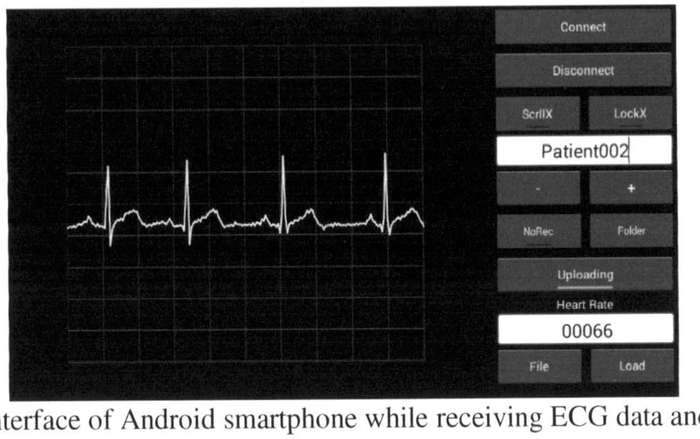

Figure 4. Interface of Android smartphone while receiving ECG data and heart rate

In this study, the determination of QRS complex use two methods; Peak Threshold method and Peak Filter method. The determination of this QRS complex is used for calculating the patient's heart rate in one minute or bpm. In this monitoring system, the determination of QRS complex is done in

Arduino. Then, Arduino will send a parameter to Android when the program in Arduino detects the QRS complex via Bluetooth HC-05. To calculate the heart rate in real time, equation (12) is utilized. Where the R-R interval represents the time from R peak (QRS complex peak) to the next R peak.

The QRS complex detection algorithm for the Peak Filter method is shown in Figure 2 (a). In this QRS complexes detection method, ECG data is continuously processed by Linear High Pass Filtering (HPF), Non-Linear Low Pass Filtering (LPF), and Decision Making. The function of the HPF is to assert the QRS complex and suppress the source of low-frequency noise on ECG signals, e.g. P and T waves. Therefore, the selection of the variable M value of the equation (1) in this HPF must be carefully considered. Different M value will result in different frequency of HPF responses, and lead to different QRS detection performance [14]. According to Chen (2003), the exact M value is 5 or 7 [14]. In this study, the value of M is 5.

In LPF, the number of samples (N) in the moving window is very important. The window width should be approximately equal to the possible area of the QRS complex. According to Pan, for a sampling rate of 200 Hz, the value of N is 30 [13]. In Decision Making, adaptive thresholding schemes on the wave features that is generated from the non-linear LPF stage (z [n]) is utilized for detecting the QRS Complex. Where, when z [n] is greater than Threshold; z [n] is the peak of the QRS complex, and vice versa. The adaptive threshold value will always be updated according to equation (5). The forgetting factor, α; is a correction factor in the determination of the adaptive threshold and γ is the weighting factor used for determining the contribution of the peak value to the threshold adjustment [14]. Figure 5 shows QRS complex peak detection with Peak Filter method from a patient.

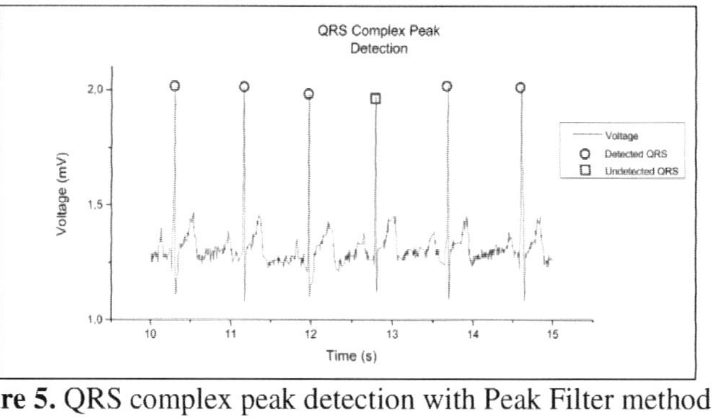

Figure 5. QRS complex peak detection with Peak Filter method

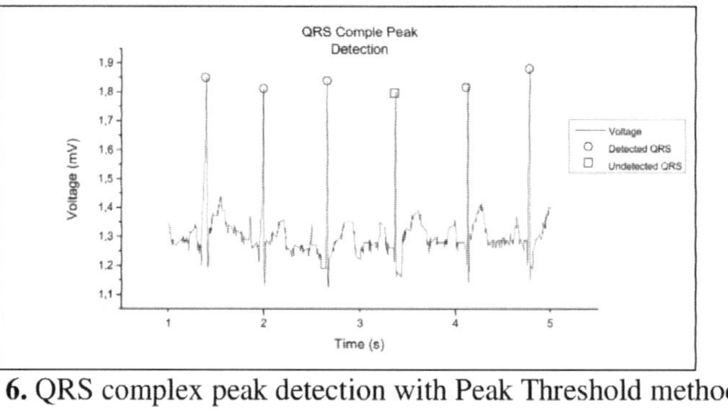

Figure 6. QRS complex peak detection with Peak Threshold method

The QRS complex detection algorithm with Peak Threshold method is shown in Figure 2 (b). In this method, Arduino will store three analog ECG data in the form of arrays; x [n], x [n-1], and x [n + 1], then with equations (6), (7), and (8) The peak of the QRS complex can be determined. If the QRS complex is not found or have been found in those three arrays, the data will be shifted. So that, x [n + 1] = x [n], x [n] = x [n-1], and x [n- 1] is the new analog data. After that, the program will do the determination of QRS complex again. Figure 6 shows QRS complex peak detection with Peak Filter method from a patient.

To evaluate the performance of QRS complex peak detection method, three parameters are used; they are positive predictive, accuracy and sensitivity. Accuracy shows the number of correctly detected QRS complexes compared with the number of detected pulses. In this system, the accuracy value is influenced by false negative and false positive where false negative is the number of undetectable QRS complexes and the false positive is the number of non-QRS complexes detected as QRS complexes. Sensitivity shows the sensitivity of the system in detecting QRS complex correctly. The sensitivity value is affected by true positive and false negative where true positive indicates the number of QRS complexes detected correctly. Positive predictive shows correctly detected QRS complexes compared to non-QRS complexes but are detected as QRS complexes. Table 1 shows positive predictive, accuracy and sensitivity of the Peak Threshold method and Peak Filter method.

Table 1. performance of QRS complex peak detection method

Method	Positive Predictive (%)	Accuracy (%)	Sensitivity (%)
Peak Threshold	92.39	70.30	74.62
Peak Filter	98.38	82.47	83.61

The low value of the QRS detection performance in the Peak Threshold method is caused by the peaks of the QRS complex's value is less than the threshold value that is specified in the program. So, the peak of the QRS complex cannot be detected, as seen in figure 6. This is because of equation (8) is not fulfilled.

4. Conclusion

In this research, we have made remote monitoring system of electrocardiogram and heart rate using Android smartphone. The ECG monitoring system is composed of AD8232 EKG Kit, Arduino UNO R3, Bluetooth HC-05, SDCard module, and an Android smartphone. Heart rate is based on QRS peak detection. Two methods are studied to determine the peak of QRS complex; they are Peak Threshold method and Peak Filter method.

To evaluate the performance of QRS complex peak detection method, three parameters are used; they are positive predictive, accuracy and sensitivity. Positive predictive, accuracy, and sensitivity of Peak Threshold method is 92.39%, 70.30%, and 74.62% and for Peak Filter method are 98.38%, 82.47%, 83.61%, respectively.

References

[1] Agung M. A and Basari 2017 *AIP Conf. Proc.* **1817** 040015
[2] World Health Organization 2000 *The World Health Report 2000: Health Systems: Improving Performance* (Jenewa: World Health Organization)
[3] Meijs L P B, Galeotti L, Pueyo E P and others 2014 *Americ. J. of Phys.-Heart and Circulatory Phys.* 00419
[4] Shaw L J, Xie J X, Phillipsand L M and others 2016 *Heart Asia* **8** 1-7
[5] Grond M, Jauss M, Hamann G and others 2013 *Stroke* 113
[6] Nair H K 2012 *Basic Concepts of EKG: A Simplified Approach* (USA: APRN World)
[7] NCHS 1991 *National Health and Nutrition Examination Survey III: Electrocardiogram.* (Rockville: Westat, Inc.)
[8] Gonna H, Gallagher M M, Guo X H and others 2014 *Annals of Noninvasive Electrocardiology* **19** 57 – 62
[9] Walker J S 1999 *Wavelet and Their Scientific* (USA: CRC Press)
[10] Lou D, Chena X, Zhaoa Z and others 2013 *IERI Procedia* **4** 208 – 215
[11] Li H-B, Takizawa I and others 2007 *Mobile and Wireless Com. Summit* 2007 16 IST 1-5

IOP Conf. Series: Journal of Physics: Conf. Series **909** (2017) 012006 doi:10.1088/1742-6596/909/1/012006

[12] Trivedi P and Ayub S 2014 *International Journal of Computer Applications* **97** 10-13
[13] Pan J and Tompkins W J 1985 *IEEE Transactions on Biomedical Engineering* **32** 230-236
[14] Chen H and Chen S 2003 *IEEE Computers in Cardiology* **30** 585-588
[15] Hamilton P S and Tompkins W J 1986 *IEEE Transactions on Biomedical Engineering* **12** 1157-65
[16] Nguyen L, Perry R., Seneres L and Seyedmahmoud N 2015 *Analog Integrated Circuit Applications* (Massachusetts: Worcester Polytechnic Institute)
[17] Analog Devices I 2013 *Data Sheet Single-Lead, Heart Rate Monitor Front End (AD8232)* (Cambridge: Analog Devices, Inc.)
[18] Wolfson M 2013 *Android Developer Tools Essentials* (USA: O'Reilly Media, Inc.)
[19] Nugroho A A, Nuryani N and Yahya I 2015 *ICEVT & IMECE 2015 Joint International Conference* 212-214
[20] Ahmed M U, Begum S and Islam M S 2010 *MRTC Report*

IOP Conf. Series: Journal of Physics: Conf. Series **909** (2017) 012007 doi:10.1088/1742-6596/909/1/012007

dLOGIS: Disaster Logistics Information System

Sorja Koesuma[1,2,*], Rio Riantana[2], Budi Siswanto[1], Fendi Aji Purnomo[3] and Sarjoko Lelono[1]

[1]Disaster Research Center, Sebelas Maret University, INDONESIA

[2]Department of Physics, Sebelas Maret University, INDONESIA

[3]Department of Informatics, Sebelas Maret University, INDONESIA

[*]E-mail: sorja@uns.ac.id

Abstract. There are three timing of disaster mitigation which is pre-disaster, emergency response and post-disaster. All of those is important in disaster mitigation, but emergency response is important when we are talking about time. Emergency response has limited time when we should give help. Rapid assessment of kind of logistic, the number of survivors, number children and old people, their gender and also for difable person. It should be done in emergency response time. Therefore we make a mobile application for logistics management system. The name of application is dLOGIS, i.e. Disaster Logistics Information System. The application is based on Android system for mobile phone. Otherwise there is also website version. The website version is for maintenance, data input and registration. So the people or government can use it directly when there is a disaster. After login in dLOGIS, there is five main menus. The first main menu shows disaster information, refugees conditions, logistics needed, available logistics stock and already accepted logistics. In the second menu is used for entering survivors data. The field coordinator can enter survivors data based on the rapid assessment in disaster location. The third menu is used for entering kind of logistic. Number and kind of logistics are based on the BNPB needed standard for the survivor. The fourth menu displays the logistics stock available in field coordinator. And the last menu displays the logistics help that already accepted and sent by donation. By using this application when a disaster happened, field coordinator or local government can use maintenance distribution of logistics base on their needs. Also for donor people who will give help to survivor, they can give logistics with the corresponding of survivor needs.

1. Introduction

Nowdays, governments and humanitarian organization have coordinated the aids from the donor to victims of the disaster. The assistance provided should match the needs of disaster victims. The kind of assistance is in the form of food, clothing, home or special needs for the disabled. Some researcher has made a disaster management system [1]–[3], web-based system for disaster logistics [4] and also logistics system during or post emergency response [5]–[9]. dLOGIS (Disaster Logistics Information System) is made base on the experience so far, that the assistance provided by the community or community organizations does not meet the needs of the refugees. For instance, at the time of landslide disaster in Banjarnegara, the community gave aid in the form of used-clothing, whereas many other needs, such as blankets and baby diapers. Also at the time of the eruption of Merapi volcano, the number of fast food in Boyolali was abundant, but at that time the need was clothing. Therefore, a disaster logistics management system is needed that can regulate the need and willingness of assistance for the refugees. With this dLOGIS system can facilitate government and community organizations in providing appropriate assistance for the refugees.

2. Method

We have created an Android-based dLOGIS application. Hopefully, this application can be used by people who coordinate the aid and also the community who will distribute the aid. dLOGIS is a mobile application, which will make it easier for coordinators and aid providers to work together. This application is temporarily available in Bahasa Indonesia only. There is five main menus in the main display Mobile version (Disaster Information, Refugees, Data Needs, Logistics Stock, Logistics Accepted). Here are the menus contained in the application dLOGIS:

Disaster Information, contains information on the types of disaster events. Disaster information consists of disaster names, disaster locations, current conditions and field coordinators for aid disbursement. This disaster information is input by the admin or by the field coordinator and up to date of any changes in disaster conditions. In this early version only one disaster was shown, but for the next will be developed and displayed several types of disasters that occur together so that people who will provide assistance can choose the location or type of disaster which will be helped.

Refugees, on the refugee's menu displayed the current condition of the refugees. Which includes the number of refugees, the number of families, the number of refugees by age and gender, the number of refugees pregnant and the number of displaced persons with disabilities. This refugee data is updated daily based on the dynamics of the number of refugees in shelters or refugee posts based on data collection, because usually there are refugees who have returned to their homes or stay in their family. The grouping of the types of refugees above is based on the regulation of the head of BNPB number 18/2009 [10]. The following is captured of Disaster Information and Refugees menus, in figure 1.a and 1.b.

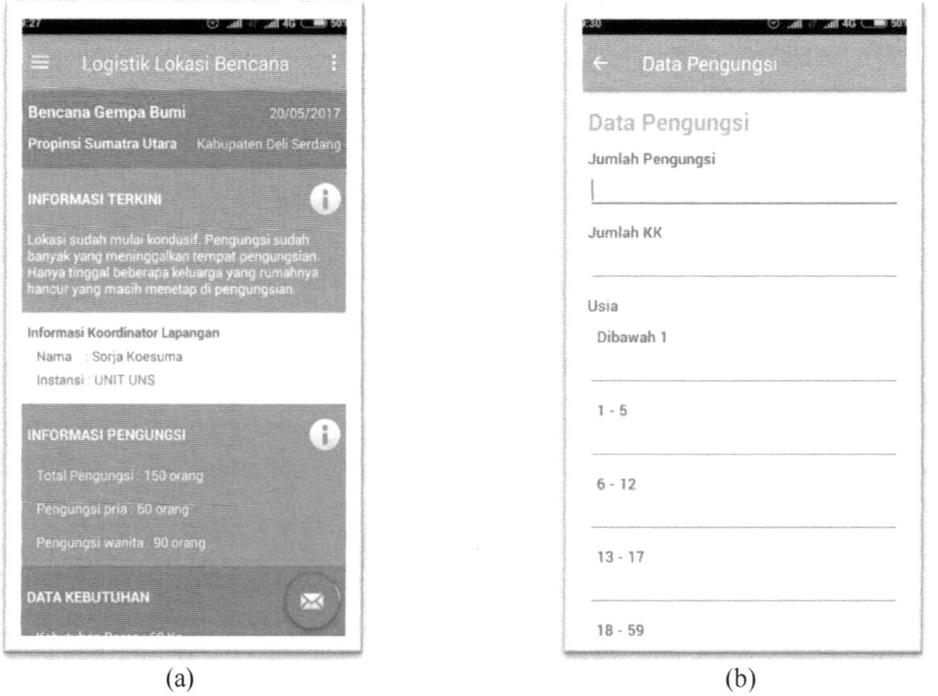

(a) (b)

Figure 1. Captured examples Disaster Information and Refugees menus

In the menu of **Logistics Needs**, displays the number of refugee needs for daily life. Needs include food, drinks, clothing, toiletries, and blankets. This requirement is based on based on the regulation of the head of BNPB number 18/2009 [10] which is the requirement of each person for each day.

Available Logistics Stock menu, shows the available logistics in refugee's camp or shelter. The last menu is **Accepted Logistics** menu that displays the logistics that deliver by the donor or humanitarian organization. There are two statuses for accepted logistics, delivery process and accepted. At the end of the menu, there is a contact address for coordinate and delivering the aids for refugees. Figure 2 shows the captured examples of Logistics Needs, Available Logistics Stock and Accepted Logistics menus.

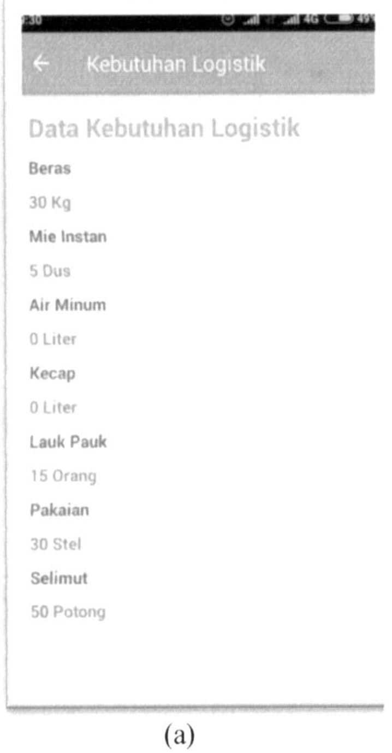

(a)

(b)

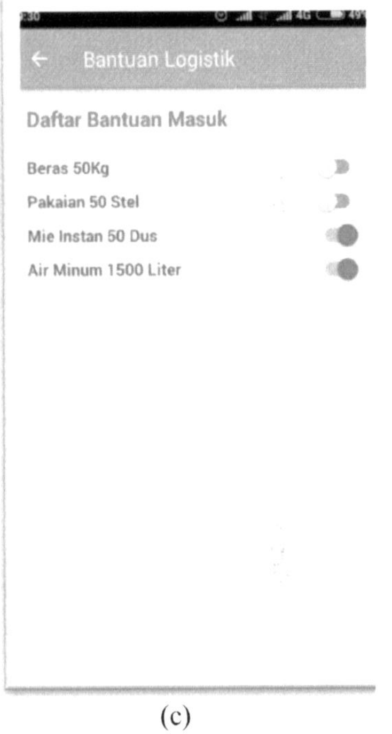

(c)

Figure 2. Captured examples of Logistics Needs, Available Logistics Stock and Accepted Logistics menus.

3. Application

dLOGIS is an information system for disaster logistics. This application can be used anywhere in disaster location as long as mobile phone signal or the internet available. Although this application has not been released yet and still in trial stage, the potential use of this application quite a lot. The main function of dLOGIS is to manage and coordinate the required logistics during the emergency response period. However, it can also be used as information on the availability of logistics stock in BPBD (Badan Penanggulangan Bencana Daerah). Also it can be used as training material for disaster logistics management for government, volunteers or humanitarian community. After some improvements, this application will soon be released and uploaded on the Google Play Store so that it can be used by government or community agencies for disaster logistics management during the emergency response period.

4. Conclusion

A disaster logistics information system (dLOGIS) based on Android has been developed that can be used mobile in disaster areas during emergency response. This application can be downloaded and shared in a community as information to logistic contributors of disaster.

References

[1] B. Kusumasari and Q. Alam, "Bridging the gaps: The role of local government capability and the management of a natural disaster in Bantul, Indonesia," *Nat. Hazards*, vol. 60, no. 2, pp. 761–779, 2012.
[2] A. D'Uffizi, M. Simonetti, G. Stecca, and G. Confessore, "A simulation study of logistics for disaster relief operations," *Procedia CIRP*, vol. 33, pp. 157–162, 2015.
[3] M. R. S. Scarpin and R. De Oliveira Silva, "Humanitarian logistics: Empirical evidences from a natural disaster," *Procedia Eng.*, vol. 78, pp. 102–111, 2014.
[4] R. A. Hadiguna, I. Kamil, A. Delati, and R. Reed, "Implementing a web-based decision support system for disaster logistics: A case study of an evacuation location assessment for Indonesia," *Int. J. Disaster Risk Reduct.*, vol. 9, pp. 38–47, 2014.
[5] D. Alem, A. Clark, and A. Moreno, "Stochastic network models for logistics planning in disaster relief," *Eur. J. Oper. Res.*, vol. 255, no. 1, pp. 187–206, 2015.
[6] J. Holgu??n-Veras, E. Taniguchi, M. Jaller, F. Aros-Vera, F. Ferreira, and R. G. Thompson, "The Tohoku disasters: Chief lessons concerning the post disaster humanitarian logistics response and policy implications," *Transp. Res. Part A Policy Pract.*, vol. 69, pp. 86–104, 2014.
[7] R. Kaynak and A. T. Tuğer, "Coordination and Collaboration Functions of Disaster Coordination Centers for Humanitarian Logistics," *Procedia - Soc. Behav. Sci.*, vol. 109, pp. 432–437, 2014.
[8] M. A. G. Bastos, V. B. G. Campos, and R. A. de M. Bandeira, "Logistic Processes in a Post-disaster Relief Operation," *Procedia - Soc. Behav. Sci.*, vol. 111, pp. 1175–1184, 2014.
[9] J. B. Sheu, "Post-disaster relief-service centralized logistics distribution with survivor resilience maximization," *Transp. Res. Part B Methodol.*, vol. 68, pp. 288–314, 2014.
[10] BNPB, "Pedoman Standarisasi Logistik," pp. 1–14, 2009 (Text in Bahasa Indonesia).

IOP Conf. Series: Journal of Physics: Conf. Series **909** (2017) 012008 doi:10.1088/1742-6596/909/1/012008

Optimization of active cell area on the dye-sensitized solar cell efficiency

A W Putri, F Nurosyid*, and Agus Supriyanto

Department of Physics, Faculty of Mathematic and Natural Sciences, Universitas Sebelas Maret, Ir. Sutami Street 36 A Kentingan, Surakarta 57126, Indonesia

*Email : nurosyid@yahoo.com

Abstract. This study is aimed to obtain optimal active area producing high efficiency of DSSC module. The DSSC structure is constructed of TiO_2 as working electrode, *dye* as photosensitizer, platinum as counter electrode, and electrolyte as electron transfer media. TiO_2 paste was deposited on Fluorine-doped Tin Oxide (FTO) by screen printing method. Meanwhile, platinum was also coated on FTO via brush painting method. Keithley *I-V* meter was performed to characterize DSSC electrical property. The active area of each cell was varied of 4.5 cm^2, 9 cm^2, and 13.5 cm^2. Each cell was assembled into a module using an external series connection of Z type. The module was consisted of 12 cells, 6 cells, and 4 cells with module active area of 54 cm^2. The optimal active area of DSSC cell is 4.5 cm^2 resulting 0.4149% efficiency. In addition, the highest efficiency of DSSC module is 0.2234% acquired by 6 cells assembling.

1. Introduction

DSSC is the third generation of solar cell which converts light into electrical energy. It exploits a dye as a photon absorber in the converting process [1-3]. The DSSC structure constructs of FTO glass that serves as the working electrode and the counter electrode. The working electrode consists of deposited TiO2 and dye, while the other one is coated Platinum. Between the two electrodes is given electrolyte [4-7].

Initially, DSSC was fabricated with a small active area size of about 1 cm^2 [8]. As the research grows rapidly, the researchers have developed DSSC cells into larger sizes. Various studies have combined several cells into a single module with varying circuit connections. The module can be applied as solar window [8-10].

Ramasamy *et al.*, 2007 has studied the design of monolithic type of DSSC module with efficiency of 1.02%, while the parallel connection type with 5 cells combination yielded power conversion efficiency of 5.45% [11]. Lee *et al.*, 2008 compared the two types of DSSC modules with various counter electrodes. The module using Pt electrode had efficiency of 5.26%, while using carbon electrode generated 4.23% [12]. Pranoto has manufactured DSSC module varying the connection type. It was monolithic, Z type and W type connection. The research yielded that the highest efficiency was 0.77% from Z type connection [13]. Hastuti *et al.*, 2016 examined circuit types such as monolithic, external Z type series, internal Z type series, and parallel external connections. The best DSSC module efficiency that on an external series connection of Z type was 0.078% [14].

In order to be applied as a solar window, the DSSC module must be built into a large active area. However, it is difficult on the fabrication process. Alternatively, the DSSC module can be created by combining several DSSC cells into a module. The active area has effect on the efficiency of DSSC. Therefore, it is necessary to study the optimal active area of the cell and the DSSC module. The size of the cells active area is varied by 4.5 cm^2, 9 cm^2, 13.5 cm^2. Each cell is arranged using series external circuit to be a module.

2. Methods

In this research, the materials involved FTO glass substrate with resistance of 7 Ω / sq. 18NR-T Transparent TiO_2 Paste dyesol was functioned as the working electrode. Dye Ruthenium N719 dyesol was worked as photosensitizer. Platinum dyesol used as the counter electrode. EL-HPE dyesol electrolyte solution was performed as electron transfer medium. Silver paste was played as contact between cells. The FTO glass substrate was cut into various size of active area. They were 4.5 cm^2, 9 cm^2, and 13.5 cm^2 symbolized by A, B, and C, respectively. Because the module size must be 54 cm^2, there must be 12, 6, 4 combined cells to be a module for A, B, and C cell active area respectively. The substrate was washed via ultrasonic cleaner. The TiO_2 paste was deposited on FTO glass by screen printing method. It was dried by hotplate at temperature of 130°C for 10 min and then annealed in furnace at temperature of 500°C for 10 min. The working electrode was immersed in Ruthenium dye N-719 for 24 h. The platinum was coated by brush painting method on the counter electrode, and heated at temperature of 250°C for 10 min. The working electrode and the counter electrode were arranged like sandwiches. The electrolyte was injected into DSSC cells. The silver paste was applied as the intercellular contact. The DSSC sandwiches or cells were tested by Keithley I-V Meter. The cells then were assembled to be a module with external series circuit. The module was also characterized via Keithley I-V Meter. The characterization using Keithley I-V Meter was proposed to observe the DSSC electrical property.

3. Results and Discussion

The DSSC prototypes of each cell are shown by Figure 1. Each cell is connected to a unitary module by using the series external connection as in Figure 2. The DSSC module is made by connecting 12 cells (Figure 2a), 6 cells (Figure 2b), 4 cells (Figure 2c). All three modules have the same total active area of 54 cm^2.

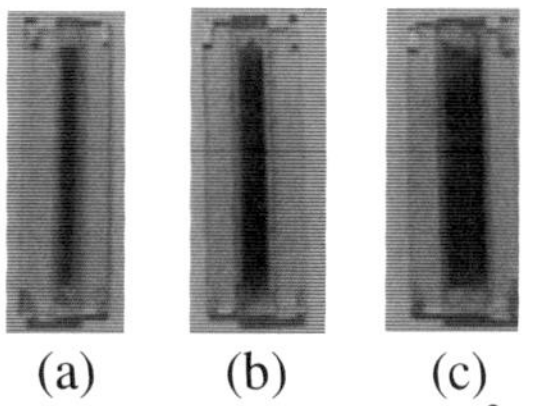

(a) (b) (c)

Figure 1. Single cells DSSC at area of (a) 4.5 cm^2 (b) 9 cm^2 and (c) 13.5 cm^2.

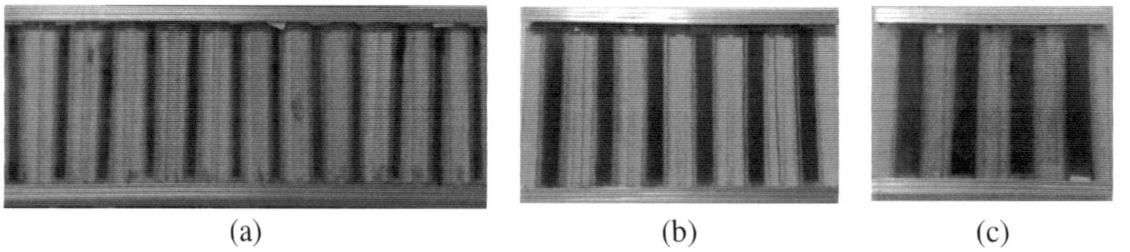

(a) (b) (c)

Figure 2. DSSC module consist of (a) 12 cells (b) 6 cells (c) 4 cells

Several parameters that affect the efficiency of DSSC include short circuit current (I_{sc}), open circuit voltage (V_{oc}), and fill factor (FF). I_{sc} is a current when the output voltage is zero. V_{oc} is the voltage in the open circuit which has zero current because there is no current flowing [15]. *FF* is the ratio of the output power of the solar cell to the actual power. It can be calculated by Eq. 1 [11]

$$Fill\ Factor = \frac{V_m I_m}{V_{oc} I_{sc}} \quad (1)$$

V_m and I_m are the voltage and the current at maximum power, respectively. The efficiency of DSSC is calculated using Eq. 2. P_{in} is the incident solar radiation, expressed in Watt [11].

$$Efficiency = \frac{V_{oc}\, I_{sc}}{P_{in}} \times FF \times 100\ \% \quad (2)$$

Figure 3 presents that the voltage of each cell has output voltage and currents. Table 1 demonstrates *I-V* characteristic of each cell. The increasing of I_{sc} occurs when the active area is larger. The larger active area leads the more photon absorbed in the dye which means more excited electrons produced. It makes the more current generated [16]. However, V_{oc} has the similar voltage. Efficiency decreases as the larger active area. The optimal active area of DSSC cell is 4.5 cm^2 resulting efficiency of 0.4149%.

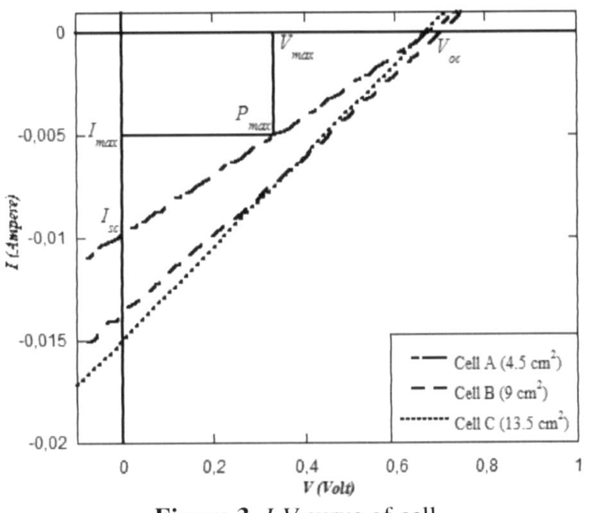

Figure 3. *I-V* curve of cell

Table 1. *I-V* Characteristics of cell

Samples	I_{sc} (A)	I_{max} (A)	V_{oc} (V)	V_{max}(V)	*FF*	P_{max}(W)	P_{in}(W)	η (%)
Cell A (4.5 cm^2)	0.0086	0.0050	0.6556	0.3702	0.3320	0.0019	0.45	0.4149
Cell B (9 cm^2)	0.0124	0.0078	0.6846	0.3401	0.3139	0.0027	0.9	0.2955
Cell C (13.5 cm^2)	0.0133	0.0076	0.6498	0.3501	0.3089	0.0027	1.35	0.1973

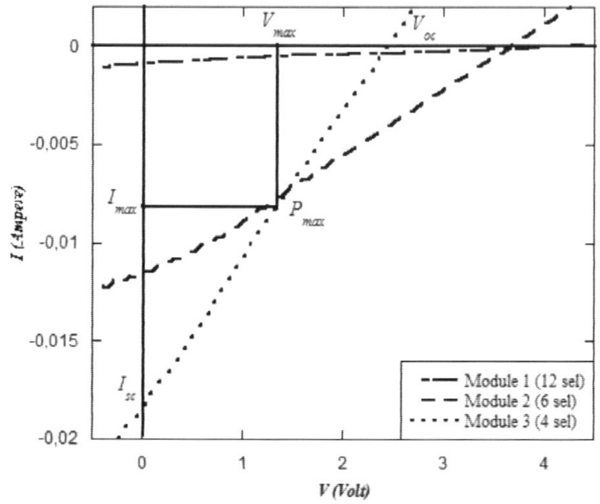

Figure 4. *I-V* curve of module

Table 2 *I-V* Characteristics of module

Samples	I_{sc} (A)	I_{max} (A)	V_{oc} (V)	V_{max}(V)	*FF*	P_{max}(W)	P_{in}(W)	η (%)
Module 1(12 cells)	0.0009	0.0004	4.0114	2.7465	0.2870	0.0010	5.4	0.0193
Module 2(6 cells)	0.0119	0.0066	3.6407	1.8401	0.2772	0.0121	5.4	0.2234
Module 3(4 cells)	0.0181	0.0089	2.3105	1.3204	0.2818	0.0117	5.4	0.2176

Table 3. Resistance of module

Samples	Resistance (kΩ)
Module 1 (12 cells)	258.67
Module 2 (6 cells)	120.88
Module 3 (4 cells)	114.97

In the series circuit produces a greater voltage as a result of the more cell assembled presented by Figure 4 [14]. The DSSC module has a series resistance which effect the efficiency of DSSC. Series resistance (R_s) is caused by the resistance between the metal contact and the cells [17]. It is denoted that module 1 has the lowest efficiency because of the highest R_s. However, in the module 2 and module 3, it does not influence significantly. This is due to the larger active area the higher the resistance. The highest efficiency of DSSC module is 0.2234% acquired by 6 cells assembling.

4. Conclusion

Based on the cell variation of the active area, it is obtained that the optimal active area of DSSC cell is 4.5 cm^2 resulting 0.4149% efficiency. The highest efficiency of DSSC module is 0.2234% acquired by 6 cells assembling.

Acknowledgment
Ministry of Research, Technology and Higher Education Directorate General of Learning and Student Affairs which has funded PKM-PE

References

[1] Mathew S A, Yella P, Gao R., Baker H, Ashari A., Tavernelli I & M.Grätzel. 2014 *J.Nat.Chem* **6** 242–247.
[2] M.G. Brian O'Regan 1991 *J. Nature.***353** 737–740.

[3] Kundu, S., Sarojinijeeva, P., Karthick, R., Anantharaj, G., Saritha, G., Bera, R., ... & Jeyakumar, D. 2017 *J. Electrochimica Acta*. S0013-4686(**17**)30996-9
[4] Goldstein, J., Yakupov, I., & Breen, B. (2010). *Solar Energy Materials and Solar Cells*, **94**(4), 638-641.
[5] Nosheen, E., Shah, S. M., Hussain, H., & Murtaza, G. 2016 *Journal of Photochemistry and Photobiology B: Biology*, **162**, 583-591.
[6] Son, M. K., Seo, H., Kim, S. K., Hong, N. Y., Kim, B. M., Park, S., ... & Kim, H. J 2013 *Journal of Power Sources*, **222**, 333-339.
[7] L. Wang, X. Fang, Z. Zhang. 2010 *J. Renew. Sust. Energ.* Rev. **14** 3178e3184
[8] Chen CY, Wang MK, Li JY, Pootrakulchote N,Alibabaei L, Ngocle CH & Grätzel M 2009 ***ACS Nano***2009;**3**: 3103–3109.
[9] Rong, Y., Li, X., Ku, Z., Liu, G., Wang, H., Xu, M., ... & Shu, T. 2012*Solar Energy Materials and Solar Cells*, **105**, 148-152.
[10] S. Dai, J. Weng, Y. Sui, S. Chen, A. Xiao, Y. Huang, F. Kong, X. Pan, L. Hu, C. Zhang, K. Wang, Inorg. Chem. Acta **361** 786–791.
[11] Ramasamy E, LeeWJ, Lee DY, & Song, JS 2007 *Journal of Power Sources* **165**: 446–449.
[12] Lee WJ, Ramasamy E, Lee YD, & Song JS 2008 *Journal of Photochemistry and Photobiology A: Chemistry***194** 27–30.
[13] Pranoto, L.M. 2013. *Laporan Tematik* PPET LIPI.
[14] Hastuti, SD, Nurosyid F, Supriyanto, A& Suryana 2016*Journal Of Physics: Conference Series* ***776*** *012003*
[15] Hagfeldt, A., Boschloo, G., Sun, L., Kloo, L., & Pettersson, H 2010 *Chemical reviews*, *110*(11), 6595-6663
[16] Nurosyid F, Furqoni L, Supriyanto A & SuryanaR 2016 *Journal of Physics: Conference Series* **776** *012011*
[17] Veurman W, Brandt H, Aguirre RL, Bialecka K, & Hinsch A, 2010 *Proceedings of 25th European Photovoltaic Solar Energy Conference and Exhibition/5th World Conference on Photovoltaic Energy Conversion*

IOP Conf. Series: Journal of Physics: Conf. Series **909** (2017) 012009 doi:10.1088/1742-6596/909/1/012009

Numerical simulation of dimension effect on critical field of rectangular superconductor

R Rosyida, Fuad Anwar, and Darmanto*

*Physics Department, Faculty of Mathematics and Natural Science, Universitas Sebelas Maret, Surakarta

E-mail: rezarosyida@student.uns.ac.id

Abstract. Numerical simulation of dimension variations was observed to determine the critical field of rectangular superconductors. The object was superconductor type-II and considered in the vacuum. The external magnetic field is applied to the *XY*-plane superconductors perpendicularly. The TDGL equations and the boundary condition were applied to the assumptions and solved using ψU method numerically. Afterward, it yielded numerical simulation that can be used to study the lower critical field (H_{c1}) and the higher critical field (H_{c3}). The input is dimension variations N_xxN_y and the value of Ginzburg-Landau parameter κ=2.00 for all dimension variations. The result of the numerical simulation showed that when the ratio between circumference and area of the rectangular superconductor decreased, the value of H_{c1} and H_{c3} would also decrease.

1. Introduction

The Time Dependent Ginzburg-Landau (TDGL) equations can be used to analyze irreversible phenomena in a superconductor. The phenomena are related to the dynamical properties of superconducting magnetic vortices. In other words, the TDGL equations can be used to study superconducting properties [1-3]. The TDGL equations have high nonlinearity, so numerical solution is a possible tool to do [4]. Numerical analysis of TDGL equations for studying superconducting properties had been widely practiced in earlier studies [5-7]. The ψU method is the most commonly used numerical method to solve TDGL equations. The approach using ψU-method is to preserves the gauge invariance of TDGL equations after they are discretized and to obtain numerical convergence at high magnetic field [8].

The numerical analysis of the TDGL equations by the ψU-method is interesting to study. Dynamical superconductor and its properties have been studied in various geometric shapes and superconducting states by applying TDGL equations which were solved by ψU method [5,9].

The TDGL equations can be written as [5,7],

$$\frac{\partial \psi(\mathbf{r},t)}{\partial t} = (\nabla - i\mathbf{A}(\mathbf{r},t))^2 \psi(\mathbf{r},t) + \psi(\mathbf{r},t) - |\psi(\mathbf{r},t)|^2 \psi(\mathbf{r},t) \tag{1}$$

$$\sigma \frac{\partial \mathbf{A}(\mathbf{r},t)}{\partial t} = \frac{1}{2i}\left(\overline{\psi}(\mathbf{r},t)\nabla \psi(\mathbf{r},t) - \psi(\mathbf{r},t)\nabla \overline{\psi}(\mathbf{r},t) - 2i|\psi(\mathbf{r},t)|^2 \mathbf{A}(\mathbf{r},t)\right) - \kappa^2(T)\nabla \times (\nabla \times \mathbf{A}(\mathbf{r},t) - \mathbf{H}_{ext}(\mathbf{r},t)) \tag{2}$$

If the superconductor in vacuum, the boundary condition for ψ and $\mathbf{A}$ as follows

$$\hat{\mathbf{n}}\cdot[\nabla - i\mathbf{A}]\psi = 0 \quad (3)$$

and the magnetic field boundary condition is

$$\mathbf{H}_{\mathbf{ext}} = \nabla\times\mathbf{A} \quad (4)$$

where n is the normal vector of the superconducting boundary, ψ is (the complex-valued) *parameter order*, $\mathbf{A}$ and $\mathbf{H_{ext}}$ are the vector potential and the external magnetic field. In the above equations, $\mathbf{r}$ as a length is expressed in unit scale of the coherence length $\xi(T)$, t as a time is expressed in unit scale $\tau(T)= \xi(T)^2/D$, ψ in $\psi_0=(|\alpha(T)|/\beta)^{1/2}$, $\mathbf{A}$ in $A_0=\mu_0 H_{c2}(T)\xi(T)$, H_{ext} is scaled in $H_{c2}(T)$, σ in $\sigma_0=1/(\mu_0\kappa(T)^2 D)$, $H_{c2}(T)$ is the upper critical field, σ is the electrical conductivity, $\alpha(T)$ and β are expansion coefficients, $\kappa(T)$ is Ginzburg-Landau parameter, D is phenomenological diffusion constant [5,7].

In this paper, we study about the critical field that appeared due to the influence of dimension variations by applying the Time Dependent Ginzburg-Landau equations and the boundary conditions that were solved by ψU method numerically. It would yield numerical simulation that can be used to study the lower critical field (H_{c1}) and the higher critical field (H_{c3}).

2. Numerical Method

The object studied was superconductor type-II and considered in the vacuum. The superconductor is represented as the rectangular shape in the xy-direction with dimension N_xxN_y. External magnetic field (H_{ext}) uniformly was applied to the XY-plane superconductor perpendicularly as shown in Figure 1. The Time Dependent Ginzburg-Landau equations and the boundary condition were applied to the assumptions and solved using the ψU method numerically [1,4-7]. The ψU method involves the magnitude of ψ and U where the superconductor consists of size cells Δ_xxΔ_y. H_{ext} was applied to the superconductor as time increases slowly, so that we could obtain the output ψ as a function of H_{ext}. Because of external magnetic field (H_{ext}) was applied to the superconductor, we could obtain the output $\psi(r,H_{ext})$ where r is a function of space (x,y).

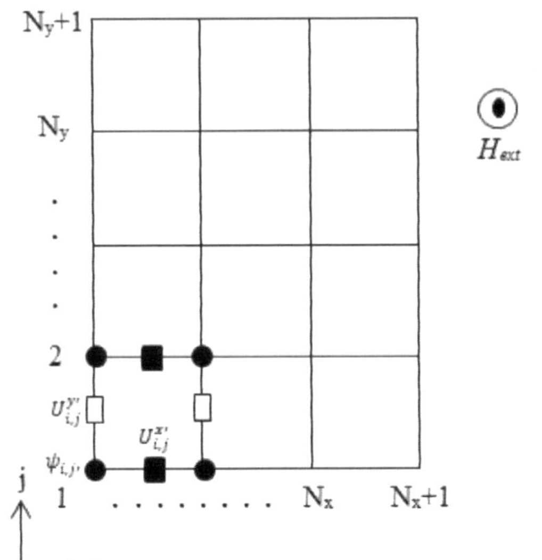

Figure 1. The rectangular superconductor is applied external magnetic field perpendicularly

3. Result and Discussion

The input-dimensional for simulation program is given in Table 1 with a value of κ = 2.00 for all dimension variations, where each grid are composed of cells with size Δ_xxΔ_y = 0,5ξ(T)x0,5ξ (T). Based on the input, we could obtain $\psi(r,H_{ext})$. Afterward, we could acquire figure $|\psi|^2$ of H_{ext} function. H_{c1} and H_{c3} can be determined from the first local maximum point and zero points of the $|\psi|^2$-H_{ext} graph

[10] as shown in Fig. 2, Fig. 3 and Fig.4. From the $|\psi|^2$-H_{ext} figure, it appeared that the dimension influences the graph pattern. Since H_{c1} and H_{c3} were determined by the $|\psi|^2$-H_{ext} graph, the values of H_{c1} and H_{c3} were affected by the dimension as well.

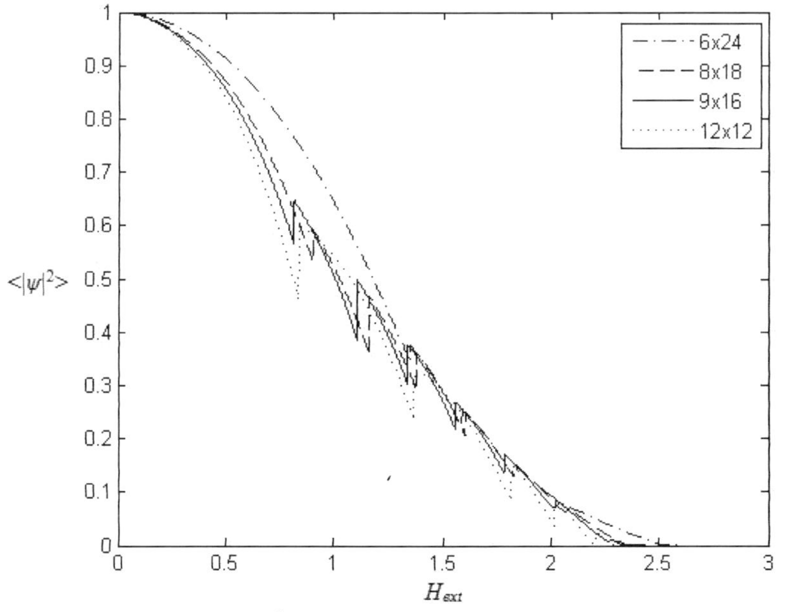

Figure 2. $|\psi|^2$-H_{ext} graph for basis N_xxN_y=12x12

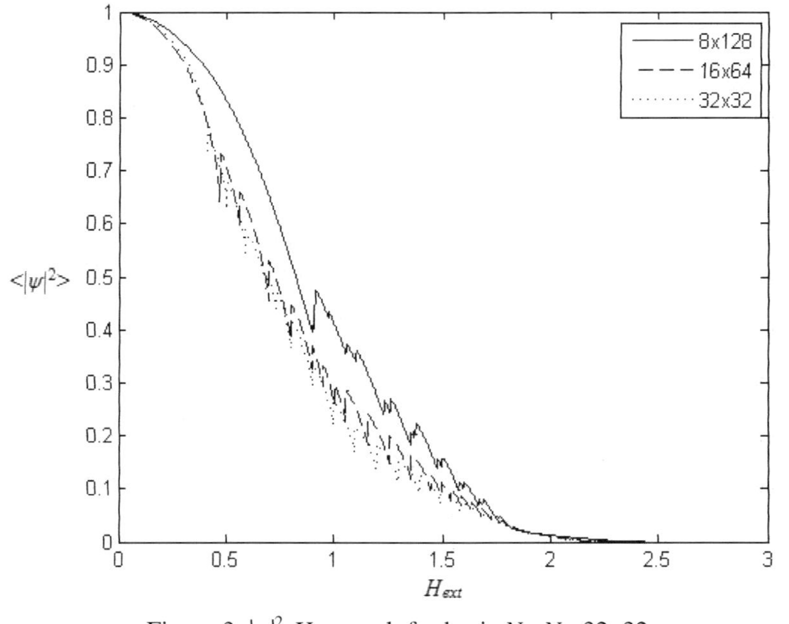

Figure 3. $|\psi|^2$-H_{ext} graph for basis N_xxN_y=32x32

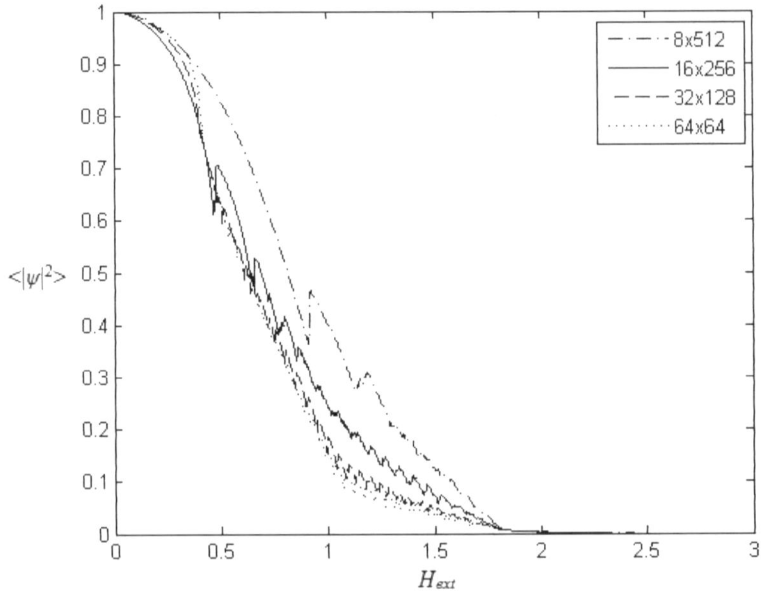

Figure 4. $|\psi|^2$-H_{ext} graph for basis N_xxN_y=64x64

The values of H_{c1} and H_{c3} for each dimension variations were shown in Fig. 5 and Fig. 6. Based on the figures, it can be seen that the input of the dimensions affected the values of H_{c1} and H_{c3} due to simulation calculation $\psi(r,H_{ext})$ used the TDGL equations and boundary condition. The TDGL equations can be used to determine the states of ψ in the middle superconductor. Meanwhile, the boundary condition can be used to determine the states of ψ in the edge superconductor. If the dimension of superconductor is greater, the effect of the TDGL equations on the numerical calculation is more dominant than the boundary conditions. If the dimension of the superconductor is smaller, the effect of boundary condition on the numerical calculation becomes more dominant.

Based on the numerical simulations, we observed the values of H_{c1} and H_{c3} on each basis N_xxN_y. The circumference of the rectangular superconductor was compared with the area of the rectangular superconductor on each basis N_xxN_y. On the basis N_xxN_y, the values of H_{c1} and H_{c3} decreased for each dimension variations. It showed that when the ratio between circumference and area of the rectangular superconductor decreased, the value of H_{c1} and H_{c3} would also decrease. When the result of the ratio differ considerably, the difference in H_{c1} and H_{c3} values would be significant.

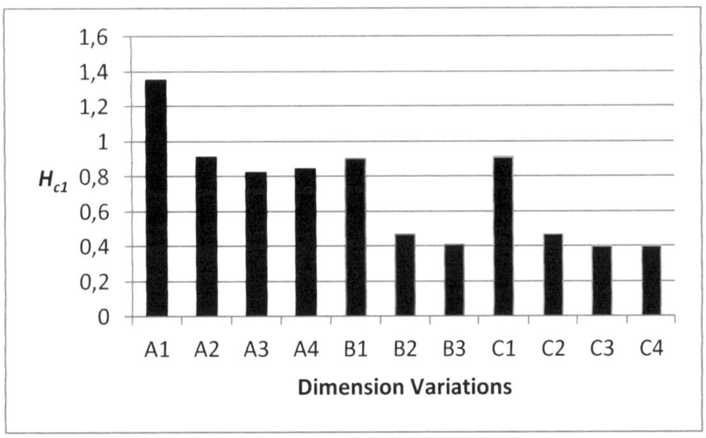

Figure 5. Value of the lower critical field (H_{c1}) for each dimension of the superconductor

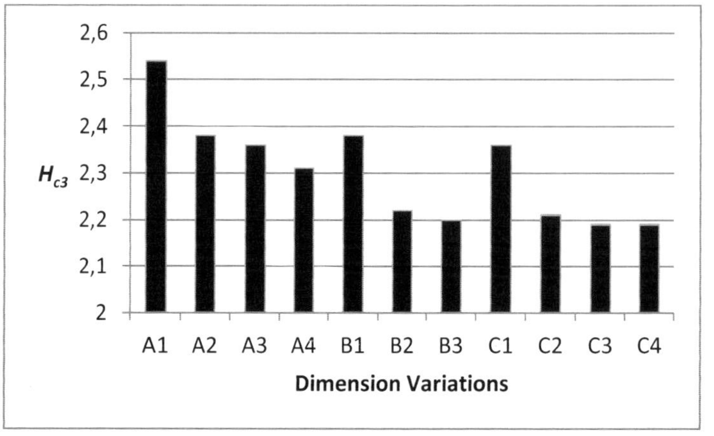

Figure 6. Value of the higher critical field (H_{c3}) for each dimension of the superconductor

Table 1. The input of superconductor dimension

Basis N_x x N_y	Input Name	N_x x N_y	Circumference	Area	Circumference/Area
12 x 12	A1	6 x 24	60	144	0.42
	A2	8x 18	52	144	0.36
	A3	9 x 16	50	144	0.35
	A4	12 x 12	48	144	0.33
32 x 32	B1	8 x 128	272	1024	0.26
	B2	16 x 64	160	1024	0.16
	B3	32 x 32	128	1024	0.12
64 x 64	C1	8 x 512	1040	4096	0.25
	C2	16 x 256	544	4096	0.26
	C3	32 x 128	320	4096	0.08
	C4	64 x 64	256	4096	0.06

4. Conclusion

We have made the numerical simulation to determine the critical field (H_{c1} and H_{c3}) of the rectangular superconductor. We applied the TDGL equations and the boundary condition to the studied superconductor and then it solved by using the ψU method. In this simulation, we varied the dimension of the superconductor to observe its influence on the value of H_{c1} and H_{c3}. The result of the numerical simulation showed that when the ratio between circumference and area of the rectangular superconductor decreased, the value of H_{c1} and H_{c3} would also decrease.

References

[1] Barba-Ortega J, Gonzalez J D and Joya M R 2013 *Journal of Physics* **410** 1
[2] Kato R, Enomoto Y and Maekawa S 1993 *Physical Review B* **47** 8016
[3] Barba-Ortega, J., Becerra, A., & Aguiar, J. A 2010 *Physica C* **470** 225
[4] Du Q 2005 *Journal of Mathematical Physics* **46** 095109-1
[5] Bolech C, Buscaglia G C and Lopez A 1995 *Physical Review B* **52** R15719
[6] Gropp W D, Kaper H G, Leaf G K, Levine D M, Palumbo M and Vinokur V M 1996 *Journal of Computational Physics* **123** 254
[7] Winiecki T and Adams C S 2002 *Journal of Computational Physics* **179** 127
[8] Pascolati M C V, Sardella E and Lisboa-Filho P N 2010 *Physica C* **470** 206
[9] Barba-Ortega J, Sardella E, Aguiar J A and Brandt E H 2012 *Physica C* **479** 49
[10] Anwar T, Nurwantoro P and Hermanto A 2013 *Journal of Natural Sciences Research* **3** 99

IOP Conf. Series: Journal of Physics: Conf. Series **909** (2017) 012010 doi:10.1088/1742-6596/909/1/012010

Comparison XRD pattern of $CoFe_2O_4$ thin films and nanoparticles

E Fitriyanti, Utari and B. Purnama[1]

Department of Physics Faculty of Mathematics and Natural Sciences Universitas Sebelas Maret,
Jl. Ir. Sutami 36A Kentingan Jebres Surakarta 57126, INDONESIA

[1] E-mail: bpurnama@mipa.uns.ac.id

Abstract. In this study, preparation of $CoFe_2O_4$ coated film on glass using CSD method with spin coating technique. The films deposition were carried out at room temperature. The angular velocity on the spin coating process of $CoFe_2O_4$ films was selected for 1000 and 3000 rpm. The $CoFe_2O_4$ thin film was deposited using $CoFe_2O_4$ nanoparticles powder. The X-ray Diffraction (XRD) was utilized to characterizing of crystal structures. The XRD analysis of $CoFe_2O_4$ thin films was compared with the $CoFe_2O_4$ nanoparticles produced and resulting the exact same spectral. The result of XRD spectral corresponds to peak data from ICDD number 221086 that indicating the sample formed was $CoFe_2O_4$ with inverse spinel face center cubic (FCC) structure. Furthermore, the intensity of $CoFe_2O_4$ nanoparticles was greater than $CoFe_2O_4$ thin film. The crystallite size of the $CoFe_2O_4$ thin films sample with 1000 rpm obtained 70.6347 nm with the strain 0.00160, whereas for 3000 rpm it was obtained 60.5354 nm with the strain 0.00187.

1. Introduction

The rapid development of nanotechnology causing a drastic reduction in magnetic device size for dimensions below one micron and thus an increase in the ratio between the surface and volume of the system. It is clear that the surface effects in such low-dimensional systems will be very important and will affect the overall magnetic behavior in various ways. In the magnetic nanoscale system, nanoparticles and thin films are examples of systems where surface effects play an important role in magnetic properties [1].

Spinel ferrite with face-centered cubic structure has the formula MFe_2O_4 where M represents divalent metal ions such as zinc, nickel, manganese or cobalt. Among these, $CoFe_2O_4$ is a spherical inverse ferrite with all or part of the Co^{2+} ion occupying the octahedral sites (B sites) and Fe^{3+} tetrahedral ions in both (A sites) and B sites [2]. Cobalt ferrite is a material suitable for practical applications such as biomedicine in human pancreas and ovarian cancer cells [3], catalysts for alkene oxidation [4], adsorbents for efficient arsenic removal [5], catalysts for methanol decomposition [6], and microwave absorption media [7]. The size of nanocrystalline spinel ferrite has been the center of attention for researchers in recent years due to excellent electrical and magnetic properties and capable of alignment (as needed). Among the ferrite spinel families, cobalt ferrite has attracted extensive research among the research community because of its high coercivity and moderate saturation magnetization with chemical stability and promising mechanical hardness suggests it to be a candidate for good magnetic storage device. The magnetic properties of cobalt ferrite (CFO) are heavily dependent on morphology and are therefore strongly influenced by the particle size [8]. These properties make $CoFe_2O_4$ material the right candidate for magnetic recording media, high-performance electromagnetic spintronics, cancer treatments, high-frequency applications, and materials ideal for future magnetostrictive applications [9, 10].

The magnetic behavior of the nanoparticles depends heavily on the shape, size, and purity of the nanoparticles [11]. Bulk saturation magnetization, Curie temperature, crystalline anisotropy, magnetostriction, etc. of $CoFe_2O_4$ depend on cation distribution [12]. The magnetic thin film is very important in technology because most electronic devices are exploiting magnetic behavior using thin film architecture [13]. The magnetic thin film is widely used for technological applications such as magnetic recording or micro-electromechanical applications [14].

There are several conventional methods for synthesizing cobalt nano ferrites such as sol-gel method, ball milling process, coprecipitation, hydrothermal, and much more [15-20]. Appropriate selection of processes and nano ferrite synthesis conditions will affect the crystal structure, size, morphology, and distribution affecting the magnetic properties of cobalt nano ferrite [12]. The coating of $CoFe_2O_4$ has many methods such as pulsed laser deposition [21, 22], electrophoresis [23], sputtering [24], molecular beam epitaxy [25, 26], chemical vapor deposition [27], chemical solution deposition [28-37]. Due to its wide application, it is expected to grow a coercive layer of $CoFe_2O_4$ coated thin film and nanoparticles using low cost, high quality, and time-consuming methods.

The creation of lattice strain on CFO is a center of attraction by different researchers. One way to create a lattice strain is to create a nanopattern substrate and deposition CFO on it [38]. This study emphasizes the structural enhancement of thin films and cobalt ferrite nanoparticles that comparable lattice strain values and their crystallite size.

2. Experimental

Preparation of cobalt ferrite nanoparticles by coprecipitation method

Cobalt nitrate ($Co(NO_3)_2.6H_2O$) and iron nitrate ($Fe(NO_3)_2.9H_2O$) as raw material, distilled water were used as solvents. The conditions and procedure routes for nano particles syntesis refer to previous experiment [39]. Cobalt nitrate and iron nitrate were dissolved in distilled water with stirring for 20 minutes (solution 1). NaOH 4.8 M (solution 2) was heated to 85°C and then titrated with solution 1 to exhaust and mixed homogeneously. The solution was washed with ethanol and then double distilled deionized water by allowed to form a precipitate. After completion of precipitate, hydrolysis process with temperature 100°C for 12 hours, and pounded clockwise so that the form of powder. Then powder on annealing 800°C for 5 hours, and pounded again clockwise.

Preparation of cobalt ferrite thin films coated on glass by CSD method

Cobalt ferrite thin films prepared by a spin coating technique using cobalt ferrite nanoparticle powder. Spin coating rotation speed varied 1000 rpm and 3000 rpm. The area of the cobalt ferrite-coated glass was 1 cm^2. The cobalt ferrite-coated glass was air dried and sintered at 350°C in air to remove residual solvents and any organic compounds to increase contact between the film and the substrate.

Characterization of cobalt ferrite thin films and nanoparticles

Cobalt ferrite thin films and nanoparticles were characterized by X-ray diffractometer using Cu Kα radiation. The crystallite size D of the sample was estimated using the Scherer equation, $(0.9\lambda) / (\beta\cos\theta)$, by measuring the width of the main intensity peak, where λ is the wavelength of Cu Kα radiation, β is full width at half maximum, and θ is the brag angle. Glass substrate coated using spin coating method (Chemat Technology, Spin Coater KW-4A).

3. Results and discussion

The XRD pattern of the $CoFe_2O_4$ thin films and nanoparticles is shown in Figure 2. The shape of the ferrite structure with the diffraction peak corresponds to the peak data of ICDD number 221086 that indicating the sample formed was $CoFe_2O_4$ with the inverse spinel face center cubic (FCC) structure. The XRD patterns of nanoparticles and thin films are identical, indicating that no other oxides are formed. The result of this comparison shows that the intensity of nanoparticles is larger than thin film. The peaks can be indexed by (220), (311), (222), (400), (422), (511), and (440) lattice planes. Average crystal size for the $CoFe_2O_4$ thin film was determined from the XRD line broadening of the most intense (311) diffraction peak using the Scherrer formula [40-42]:

$$D = \frac{0.9\,\lambda}{\beta \cos\theta} \qquad (1)$$

Where λ is the wavelength of Cu Kα, β is the full width at half maximum, and θ is the Bragg angle. From Scherrer formula, it is calculated that the size of the thin film crystallites is 70.6347 nm, while the nanoparticle crystallite size is 42.3837 nm.

The strain in $CoFe_2O_4$ thin films are calculated by using Hall-Williamson's plot using the formula

$$\varepsilon = \frac{\beta_{hkl}}{4\tan\theta} \qquad (2)$$

Where ε is a strain and β_{hkl} is full width at half maximum. In addition, it can also compare the size of crystallites and strains, where these results indicate that the thin film crystallite size is larger than that of nanoparticles. While the strain of thin film is smaller than nanoparticles, i.e., for thin film 0.00160 and nanoparticles 0.00267.

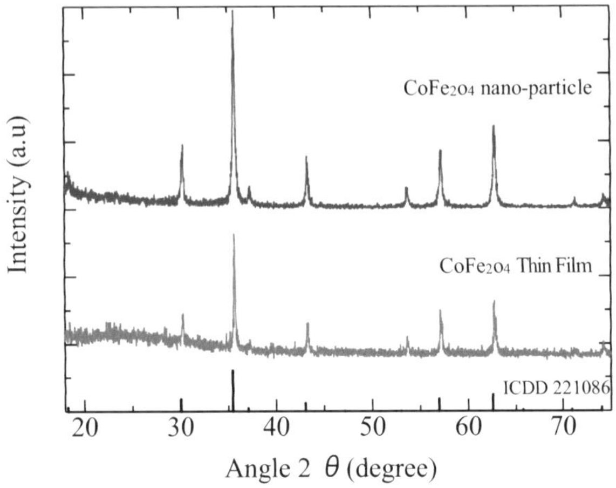

Fig. 1. XRD patterns of $CoFe_2O_4$ thin films and nanoparticles

Figure 2 shows the XRD pattern of the $CoFe_2O_4$ thin films with different spin coating speeds. All of the $CoFe_2O_4$ layers were spin coated for five times. From Scherrer formula, it is calculated that crystallite size decreases from 70.6347 nm to 60.5354 nm by increasing spin coating speed from 1000 rpm to 3000 rpm. While the lattice strain is found to increase from 0.00160 to 0.00187 by increasing spin coating speed from 1000 rpm and 3000 rpm.

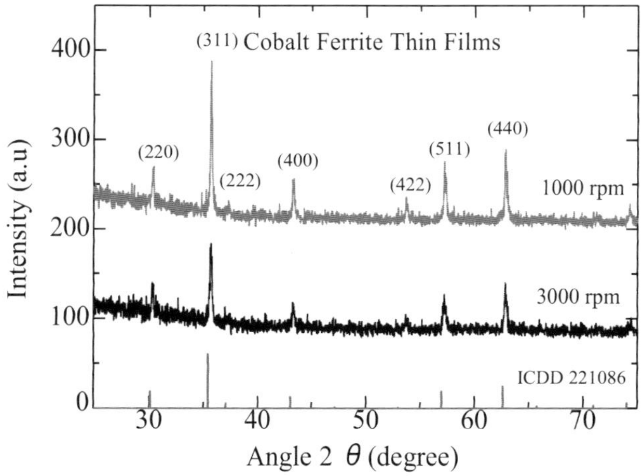

Fig. 2. XRD patterns of $CoFe_2O_4$ thin films deposition for 1000 rpm and 3000 rpm.

4. Conclusions

The $CoFe_2O_4$ thin films have been successfully prepared using CSD method with a spin coating technique. The results were compared with $CoFe_2O_4$ nanoparticles. The XRD characterization is performed to ensure the crystal structure of $CoFe_2O_4$ which produces peak data according to ICDD number 221086 that the sample formed is $CoFe_2O_4$ with inverse spinel face-centered cubic (FCC) structure. The XRD patterns of nanoparticles and thin films are identical, indicating that no other oxides are formed. The average of the $CoFe_2O_4$ thin films crystallite size was found in the range of 60-71 nm, so it can be concluded that the crystallite size of thin films is greater than that of nanoparticles. While inversely proportional to the strain, where the strain of thin films is smaller than nanoparticles. The enhancement of the structural properties of the ferrite $CoFe_2O_4$ at the nano level makes this material a suitable candidate for different applications.

References

[1] Díaz R Y 2011 *Doctoral dissertation Ph.D. thesis* (Universidad Autónoma de Madrid)

[2] Swatsitang E, Phokha S, Hunpratub S, Usher B, Bootchanont A, Maensiri S, and Chindaprasirt P 2016 *J. Alloys Compd.* **664** 792-797

[3] Pašukoniene V, Mlynska A, Steponkiene S, Poderys V, Matulionyte M, Karabanovas V, Statkute U, Purviniene R, Kras′ko J A, Jagminas A, Kurtinaitiene M, Strioga M, and Rotomskis R 2014 *Medicina* **50** 237–244

[4] Kooti M, and Afshari M 2012 *Scientia Iranica* **19** 1991-1995

[5] Dey A, Singh R, and Purkait M K 2014 *J. Water Process Eng.* **3** 1-9

[6] Genova I, Tsoncheva T, Dimitrov M, Paneva D, Tsyntsarski B, Ivanova R, Cherkezova-Zheleva Z, Budinova T, Kovacheva D, Mitov I, and Petrov N 2014 *Catalysis. Commun.* **55** 43-48

[7] Sodaee T, Ghasemi A, Paimozd E, Paesano Jr A, and Morisako A 2013 *J. Magn. Magn. Mater.* **330** 169–173

[8] Chiu W S, Radiman S, Abd-Shukor R, Abdullah M H, and Khiew P S 2008 *J. Alloys. Compd.* **459** 291-297

[9] Mohaideen K K, and Joy P A 2014 *J. Eur. Ceram. Soc.* **34** 677-686
[10] Maaz K, Mumtaz A, Hasanain S K, and Ceylan A 2007 *J. Magn. Magn. Mater.* **308** 289-295
[11] Song Q, and Zhang Z J 2004 *J. American. Chem. Soc.* **126** 6164-6168
[12] Chinnasamy C N, Narayanasamy A, Ponpandian N, Joseyphus R J, Jeyadevan B, Tohji K, and Chattopadhyay K 2002 *J. Magn. Magn. Mater.* **238** 281-287
[13] Spaldin N 2010 *Magnetic Materials Fundamental and Applications* (New York: Cambridge University Press)
[14] Garcia-Sanchez F, Chubykalo-Fesenko O, Mryasov O, Asselin P, and Chantrell R W 2008 *Appl. Phys. Lett.* **103** 07F505
[15] Chen Q, Du P, Jin L, Weng W, and Han G 2007 *Appl. Phys. Lett.* **91** 022912
[16] Kodama R H 1999 *J. Magn. Magn. Mater.* **200** 359-372
[17] Thakur A, Chevalier A, Mattei J L, and Queffelec P 2010 *J. Appl. Phys.* **108** 014301
[18] Thakur A, Thakur P, and Hsu J H 2012 *J. Appl. Phys.* **111** 07A305
[19] Sajjia M, Oubaha M, Hasanuzzaman M, and Olabi A G 2014 *Ceram. Int.* **40** 1147-1154
[20] Gul I H, Abbasi A Z, Amin F, Anis-ur-Rehman M, and Maqsood A 2007 *J. Magn. Magn. Mater.* **311** 494-499
[21] Suzuki Y, Van Dover R B, Gyorgy E M, Phillips J M, Korenivski V, Werder D J, ... and Do K B 1996 *Appl. Phys. Lett.* **68** 714-716
[22] Rigato F, Geshev J, Skumryev V, and Fontcuberta J 2009 *J. Appl. Phys.* **106** 113924
[23] Barbosa J G, Pereira M R, Mendes J A, Proença M P, Araújo J P, and Almeida B G 2010 *J. Phys. Conf. Series* (Vol. 200, No. 7, p. 072009) IOP Publishing
[24] Lee J G, Chae K P, and Sur J C 2003 *J. Magn. Magn. Mater.* **267** 161-167
[25] Xie S, Cheng J, Wessels B W, and Dravid V P 2008 *Appl. Phys. Lett.* **93** 181901
[26] Huang W, Zhou L X, Zeng H Z, Wei X H, Zhu J, Zhang Y, and Li Y R 2007 *J. Cryst. Growth* **300** 426-430
[27] Pan M, Liu Y, Bai G, Hong S, Dravid V P, and Petford-Long A K 2011 *J. Appl. Phys.* **110** 034103
[28] Mustaqima M, Lee M Y, Kim D H, Lee B W, and Liu C 2014 *J. Magn.* **19** 227-31
[29] Sathaye S D, Patil K R, Kulkarni S D, Bakre P P, Pradhan S D, Sarwade B D, and Shintre S N 2003 *J. Mater. Sci.* **38** 29-33
[30] Sun J, Wang Z, Wang Y, Zhu Y, Shen T, Pang L, ... and Li F 2012 *Mater. Sci. Eng.* B **177** 269-273
[31] Lee J G, Park J Y, Oh Y J, and Kim C S 1998 *J. Appl. Phys.* **84** 2801-2804
[32] Lee J G, Lee H M, Kim C S, and Young-Jei O 1998 *J. Magn. Magn. Mater.* **177** 900-902
[33] Wang X W, Zhang Y Q, Meng H, Wang Z J, and Zhang Z D 2011 *J. Alloys Compd.* **509** 7803-7807
[34] Cheng F, Peng Z, Liao C, Xu Z, Gao S, Yan C, ... and Wang J 1998 *Solid State* Commun. **107** 471-476
[35] Inui C, Tsuge Y, Kura H, Fujihara S, Shiratori S, and Sato T 2008 *Thin Solid Films* **516** 2454-2459
[36] Cheng F, Liao C, Kuang J, Xu Z, Yan C, Chen L, ... and Liu Z 1999 *J. Appl. Phys.* **85** 2782-2786
[37] Quickel T E, Le V H, Brezesinski T, and Tolbert S H 2010 *Nano Lett.* **10** 2982-2988
[38] Kumar R, and Kar M 2016 *J. Magn. Magn. Mater.* **416** 335-341
[39] Purnama B, Rahmawati R, Wijayanta AT, and Suharyana 2015 *J.Magn.* 20 207
[40] Erdem D, Bingham N S, Heiligtag F J, Pilet N, Warnicke P, Heyderman L J, and Niederberger M 2016 *Advanced Functional Mater.* ***26*** 1954-1963
[41] Habibi M H, and Parhizkar J 2015 Spectrochimica acta part A: Molecular and biomolecular spectroscopy **150** 879-885
[42] Rana K, Thakur P, Sharma P, Tomar M, Gupta V, and Thakur A 2015 *Ceram. Int.* ***41*** 4492-4497

Parallel-line number dependence of magneto-impedance effect in multilayer permalloy $[Ni_{80}Fe_{20}/Cu]_N$ films

R H Yohanasari, Utari and B Purnama[1]

Department of Physics, Faculty of Mathematics and Natural Sciences, Universitas Sebelas Maret, Jl. Ir. Sutami 36A Kentingan Surakarta 57126, Indonesia

E-mail: bpurnama@mipa.uns.ac.id

Abstract. In this paper, we studied the magneto-impedance effect in multilayered $[Ni_{80}Fe_{20}/Cu]_N$ with variation in the number of parallel-line on Cu PCB substrate. The method used in this research is the electrodeposition at a room temperature with Pt as an electrode. The results show that the magneto-impedance ratio increases with the increase in the number of parallel-line on Cu PCB. The maximum magneto-impedance ratio obtained in Cu PCB substrate which four parallel lines were 4.5%. Likewise, frequency variation, the magneto-impedance ratio increases with increasing frequency.

1. Introduction

Magneto-Impedance (MI) is a phenomenon of the change in impedance caused by changes in the external magnetic field [1-8]. Interestingly, the magnetic sensor based on magneto-impedance operating at room temperature, with sensitivity ultra high, high thermal stability, high spatial resolution, small size, light weight and low power consumption [9].

The magneto-impedance effect depends on different factors, such as the alloy composition, temperature, frequency and geometry of the sample. In this study, the sample was made using electrodeposition method [10-16]. It caused metal layers can deposit on the substrate instead of conduction. Additionally, the deposited layer typically has a microstructure and strong corrosion resistance [17].

Determination of the magneto-impedance ratio using Equation (1),

$$MI\% = \frac{Z(H) - Z(H_{max})}{Z(H_{max})} \times 100 \tag{1}$$

with $Z(H_{max})$ is applied to the impedance maximum magnetic field and $Z(H)$ is a rated impedance of the magnetic field [4,18-20].

Skin effect can influence the magneto-impedance at low frequencies less than 100 kHz. In thin films, values depth δ_m expressed by the transverse permeability (μ_T), to obtain Equation (2).

$$\delta_m = \frac{c}{\sqrt{4\pi^2 f \sigma \mu_T}} \tag{2}$$

where c is the speed of light, σ is the electrical conductivity and f is the frequency of AC current flowing in the sample [6].

IOP Conf. Series: Journal of Physics: Conf. Series **909** (2017) 012011 doi:10.1088/1742-6596/909/1/012011

In this paper, we will examine the phenomenon of magneto-impedance on Cu PCB substrate electrodeposition results $[NiFe/Cu]_N$ at low frequencies. Based on the results obtained in this study, it will be known parallel-line number dependence of the magneto-impedance on Cu PCB substrate.

2. Experimental Methods

The method used in this research is electrodeposition at room temperature with Pt as an electrode. Pt was used as an electrode, while the sample used was Cu PCB. The area of the deposited sample of 1 cm^2. Modification of patterned samples the parallel-line number on Cu PCB substrates was 1, 2 and 4 (see Figure 1). Electrolyte solution used was $Ni_{80}Fe_{20}$ and Cu. The materials utilized for the manufacture of the electrolyte as in Table 1. The ingredients in Table 1 then added H_2SO_4 0.1 M as producing acidic conditions.

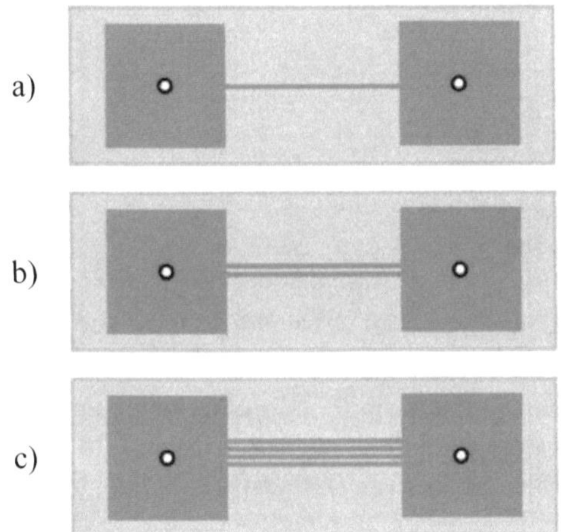

Figure 1. Modification of parallel-line number on the Cu PCB

Cu PCB was deposited with $Ni_{80}Fe_{20}$ using 2 nm/s deposition rate. As for the Cu layer using 6 nm/s deposition rate. The deposition process is carried out as in Figure 2. Each sample was deposited to form 5 layers. No treatment furthermore of the deposition.

Table 1. Material for making the electrolyte solution

Electrolyte	Material	Concentration
NiFe	$NiSO_4.6H_2O$	0.099 M
	$FeSO_4.7H_2O$	0.012 M
	H_3BO_3	0.149 M
	$C_6H_8O_3$	0.002 M
Cu	$CuSO_4.5H_2O$	0.065 M
	$C_6H_8O_3$	0.002 M

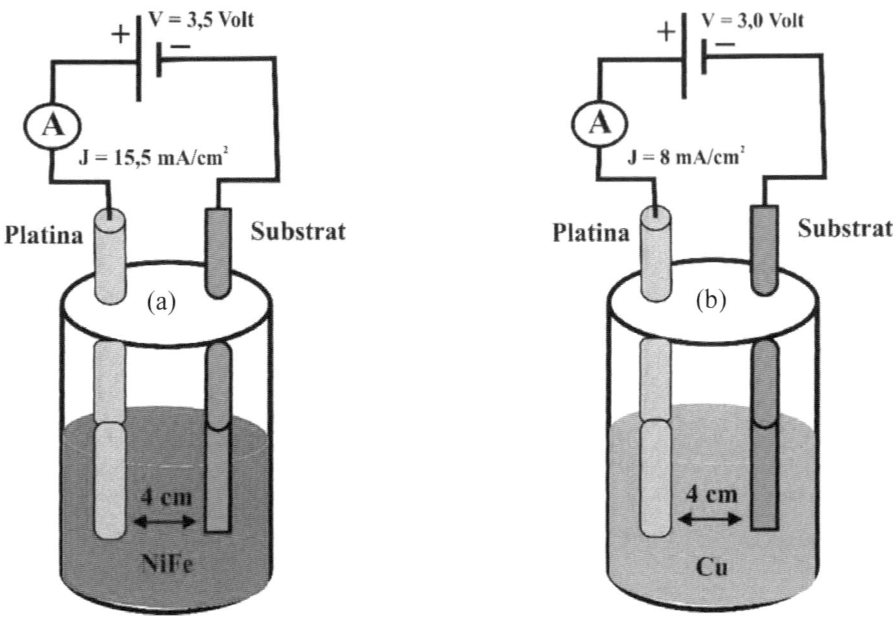

Figure 2. Schematic (a) NiFe and (b) Cu electrodeposition process

Magneto-impedance measurements in this study using LCR meter 819 GW Instek and performed at low frequency (see Fig. 3). Figure 3 shows the magneto-impedance measurement scheme. In this study, a large external magnetic field*(H)* is measured using Gaussmeter LZ-641 H made by Hunan Link Join Technology.

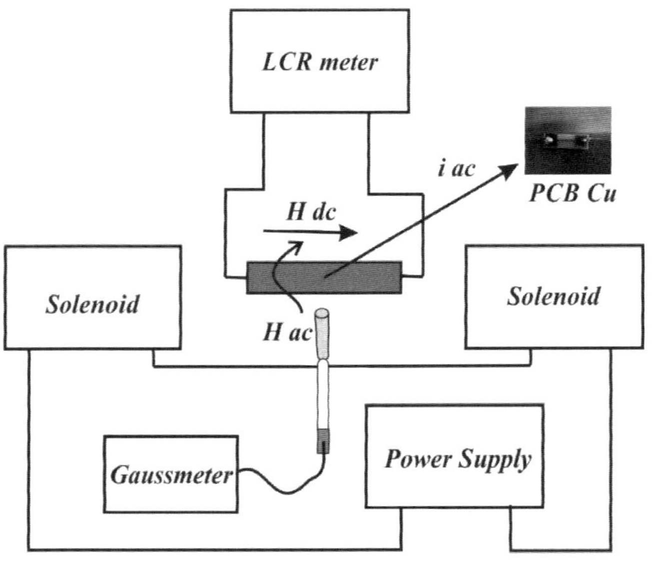

Figure 3. Schematic of the magneto-impedance measurement

3. Results and Discussion

Figure 4 (a) shows the effect of increasing the parallel-line number on Cu PCB substrate to magneto-impedance at a frequency of 100 kHz. The measurement results show the same typical magneto-impedance ratio reached a maximum at $H = 0$ mT. Figure 4(a) show that the magneto-impedance ratio will decrease significantly with increasing external magnetic field and achieve stable conditions in the external magnetic field $H = 20$ mT. The magneto-impedance effect reaches its peak as a consequence of the permeability contribution of both the domain wall movement and the magnetization rotation to the magneto-impedance [21]. Figure 4 (a) also shows that the more of the parallel-line number on the Cu PCB substrate the magneto-impedance ratio also increased.

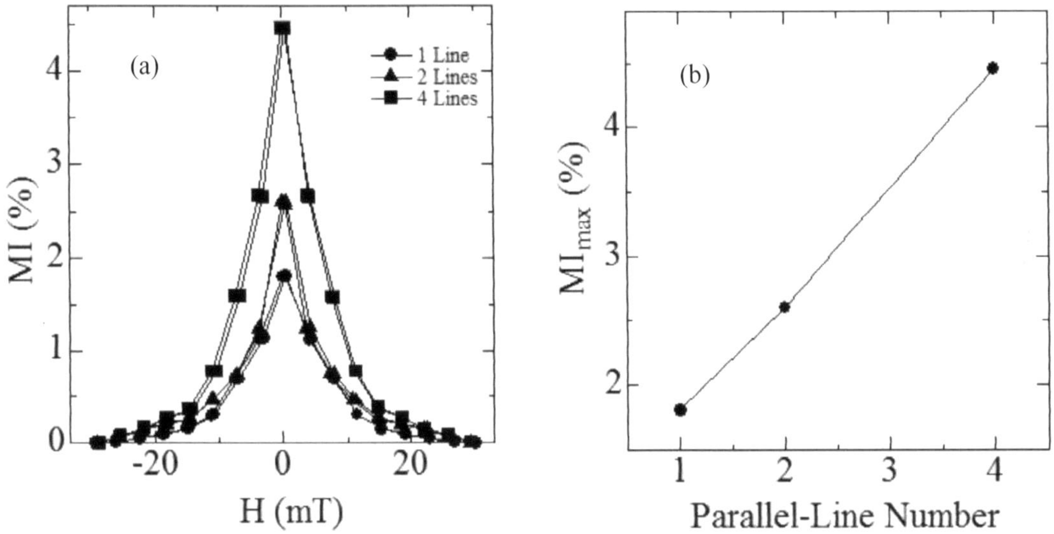

Figure 4. (a) A typical MI curve as a function of the magnetic field for variation in the parallel-line numbers on Cu PCB (b) Graph of MI variation of the parallel-line numbers on Cu PCB

Fig. 4(b) is a summary of Fig. 4(a). Figure 4 (b) show that the improved magneto-impedance ratio by increasing the number of parallel-line in more details Cu PCB substrate. In this study, the magneto-impedance ratio increased from 1.8% to 4.5% with an increase in the number of parallel-lines 1 to 4 lines. This study shows the more parallel-line on Cu PCB substrate, the magneto-impedance ratio also increases linearly. This can be explained as follows: when the conductor is flowed by current it will produce magnetic induction. Because the adjacent lines will mutually induce each other. This allegedly increases the magneto-impedance ratio.

In thin films, the maximum impedance decreases affected by the longitudinal field applied to the material that follows the dependence transverse permeability (μ_T) to the field [6]. Therefore, an increase in the parallel-line number will improve permeability transverse magnetic material in thin films causing magneto-impedance ratio will increase. As well as with the frequency variation, the magneto-impedance ratio increases with the increase in frequency.In general, low-frequency magneto-impedance influenced by magneto-inductive [6]. Impedance material that responds to the current distribution depends not only on the form of conductor and frequency but also on the permeability of the material concerning the current flowing in material.

Figure 5 (a) shows the influence of the frequency of AC current to magneto-impedance ratio multilayer $[Ni_{80}Fe_{20}/Cu]_5$ on the Cu PCB substrate with four parallel-line number. Because in this study used a low frequency, the frequency range used was between 20 kHz - 100 kHz. It can show that at each frequency variation shows a typical curve of the same magneto-impedance. Fig. 5(b) is a summary of Fig. 5(a). Improved magneto-impedance ratio with an increase in the frequency of AC current at Cu PCB substrate more details shown in Figure 5 (b). The result of measurement shows that the magneto-impedance ratio equal to 0.4% was obtained at frequency of 20 kHz and ratio magneto-impedance increase to 4.6% at frequency of 100 kHz.

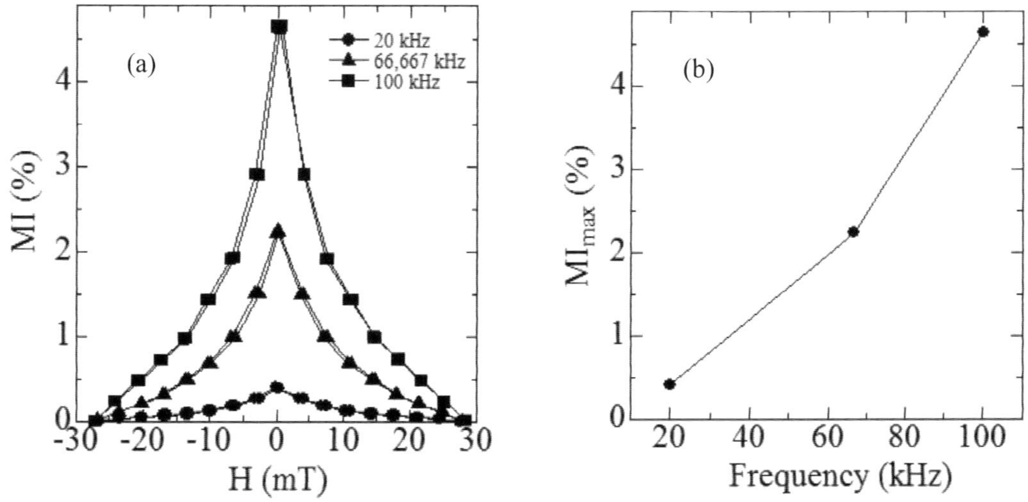

Figure 5. (a) Typical MI curve as a function of magnetic field for frequency variety b) Graph of MI variation of frequency on PCB Cu having four parallel lines

Increasing the frequency of AC current can cause the reactance value of the material also increased [22], thus improving magneto-impedance ratio. In addition to affecting the value of substance reactance, the frequency of AC current also affects the skin depth. In Eq. 2 show that the skin depth is inversely proportional to the root frequency. An increase in the frequency of material permeability can decrease the value of skin depth. So the smallest skin depth values are obtained at higher frequencies. Therefore, the highest magneto-impedance ratio is obtained at higher frequencies.

4. Conclusion

The phenomenon of magneto-impedance multilayer $[NiFe/Cu]_5$ was modified by varying of the parallel-line number on Cu PCB substrate. Coating multilayer $[NiFe/Cu]_5$ on the Cu PCB substrate formed from the electrodeposition with Pt as an electrode. The results showed that the increase in the parallel-line number on the Cu PCB $[NiFe/Cu]_5$ causes an increase in the magneto-impedance ratio. The experimental results show that the maximum magneto-impedance ratio on the Cu PCB substrate having four parallel-lines of 4.5%. Likewise, frequency variation, the magneto-impedance ratio increases with increasing frequency.

References

[1] Beach R S and Berkowitz 1994 *Appl. Phys.Lett.* **64** 3652
[2] Chaturvedi A, Ruiz A, Le A T, Tung M T, Mukherjee, Srikanth H and Phan M H 2014 *ScienceJet.* **3: 48** 1-4
[3] Cortes M, Peng T, Woytasik M and Moulin J 2015 *Journal. Of. The. Electrochemical. Society.* **7** pp.129-132
[4] Mishra A C, Sahoo T, Srinivas V and Thakur A K 2011 *Physica. B.* **406** 645-651
[5] Panina L V and Mohri K 1994 *Appl. Phys. Lett.* **65** 1189
[6] Phan M H and Peng H X 2008 *Progress. In. Materials. Science.* **53** 323-420
[7] Wang H, Qin F X, Xing D W, Cao F Y, Wang X D, Peng H X and Sun J F 2012 *Acta. Materialia.* **15** 5425-5436
[8] Zhang Y, Dong J, Sun X, Liu Q and Wang J 2013 *Nanoscale. Research. Letters.* **8** 471

[9] Tung M T, Hang Le T T, Tuan L A, Nghi N H and Phan M H 2014 *Physica. B. Elsevier.* **442** 16-20
[10] Atalay F E, Kaya H and Atalay S 2006 *J. Phys. D: Appl. Phys.* **39** 431-436
[11] Aydogmus E, Kaya H, Atalay F E, Atalay S and Avcu D 2014 *Proceedings. Of. The. 3rd. International. Congress. AP. MAS2013* vol 125 (Antalaya, Turkey) 227-229
[12] Bang W, Ko Y D, Lee H, Hong K, Chung J S and Lee H 2008 *Journal. Of. The. Electrochemical. Society.* **155** D429-D435
[13] Hu J, Qin H, Zhang L and Chen J 2004 *Materials. Science. And. Engineering.* **106** 202-206
[14] Kurlyandskaya G V, Jantaratana P, Bebenin N G and Vas'kovskiy V O 2012 *Solid. State. Phenomena.* **190** pp581-584
[15] Kollu P, Kim D Y and Kim G C 2007 *J. Of. Magnetics.* **12** 35-39
[16] Li X P, Zhao Z J, Seet H L, Heng W M, Oh T B and Lee J Y 2003 *Journal. Of. Applied. Physics.* **94** 6655-6658
[17] Cheng J K, Zhao Z J, Xin H L, Wu Z M, Yang X L and Yuan W Z 2011 *International. Journal. of. Modern. Physics B.* **25** 111-117
[18] Agra K, Mori T J A, Dorneles L S, Escobar, V M, Silva U C, Chesman C, Bohn F and Corrêa M A 2014 *Journal. of. Magnetism. and. Magnetic. Materials.* **355** 136-141
[19] Mishra A C, Sahoo T, Srinivas V and Thakur A K 2011 *Solid. State. Communications.* **151** 1787-1790
[20] Phan M H, Peng H X, Tung M T, Dung N V and Nghi N H 2007 *Journal. Of. Magnetism. And. Magnetic. Materials.* **316** 244-247
[21] Chen L, Zhou Y, Lei C, Zhou Z M and Ding W 2009 *J. Phys. D: Appl. Phys.* **42** 1-6
[22] Ismail, Nuryani and Purnama B 2016 *AIP.Conference.Proceedings* vol 1710 p. 030002

Effect of doping Strontium ions in co-precipitated cobalt ferrite

N Rikamukti, Utari and B. Purnama[1]

Department of Physics, Faculty of Mathematic and Natural Sciences, Universitas Sebelas Maret, Ir. Sutami Street 36 A Kentingan, Surakarta 57126, Indonesia

[1]Email : bpurnama@mipa.uns.ac.id

Abstract. The effect of doping strontium in structural and magnetic properties of co-precipitated strontium cobalt ferrite ($Sr_xCo_{1-x}Fe_2O_4$) nanoparticles are discussed in this paper. The mole of strontium ions was used for doping are 10% of mole cobalt. The synthesis was conducted at a temperature of 95°C with following post annealing at 600°C for five hours. The result of X-ray diffraction showed that the strongest peak intensity of sample decrease when available strontium ions doping. The crystallite size decreased from 23.50 to 22.65 nm while the strain increased. The IR spectra of the sample with strontium doping have lower transmittance than the sample without doping. The FTIR result showed that the strontium ions were suppressing the ions in the octahedral site. The magnetic properties showed that coercive field increased and the saturation magnetization decreased with the addition of strontium ions in cobalt ferrite nanoparticles.

1. Introduction

Nano crystallite size of spinel ferrite is the center of attention for researchers in recent years due to their excellent and tuneable (according to the need) electrical and magnetic properties [1]. Among the family of spinel ferrite, cobalt ferrite $CoFe_2O_4$ has attracted considerable attention due to their high Currie point (793 K), large magneto crystalline anisotropy, high coercive field, large Kerr effect and Faraday rotation, moderate saturation magnetization, large magnetostrictive coefficient, chemical stability and mechanical hardness [2-7]. These properties, along with their great physical and chemical stability, make $CoFe_2O_4$ nanoparticles suitable for magnetic recording applications and new generation electronic devices [8-10].

Pure and bulk cobalt ferrite has inversed spinel structure, where Co^{2+} ions have the preference for the octahedral site (B) while Fe^{3+} ions distribute equally between tetrahedral (A) and octahedral site [11]. The magnetic behavior of this nanoparticle strongly depends on shape, size and the purity of nanoparticles [12]. Magnetic properties of cobalt ferrite strongly depend upon the morphology, and hence greatly influenced by its particle size [13].

Many techniques have been developed for the synthesis of cobalt ferrite nanoparticles during the last year including combustion method [14], hydrothermal [15], thermal decomposition method [16], ball milling [17], sol-gel [18], hot injection thermolysis method [19] and co-precipitation method [10,20,22]. Among the processes, the co-precipitation method has been further developed because of its simplicity, suitable cost, nontoxic precursors, non-time-consuming, and industrial scale yielding

[20]. It is known that the particle size and crystallinity can be controlled by adjusting the super saturation during the nucleation and crystal growth processes, which it turn, it can strongly be affected solution chemistry and precipitation conditions such as reaction temperature and pH of process environment [9-11,20,21]

In this study, we used co-precipitation method to synthesize $Sr_xCo_{1-x}Fe_2O_4$ nanoparticles and focused on investigating the effect of doping strontium in cobalt ferrite. The addition of strontium in cobalt ferrite is expected only to suppress and not substitute ions of cobalt ferrite because of the greater strontium atomic size. In the group of alkaline earth metals, strontium (219 pm) is a better choice in terms of the atomic size because barium (253 pm) has a very large atomic size compared to cobalt (152 pm) and the atomic size of strontium is greater than calcium (194 pm). Strontium substitution will create more strain in lattice site compare to calcium and will create less problem for solid solution compare to barium, variations of mole and annealing temperature are related to stress induce magnetization [1]. The nanoparticles sample characterized by X-ray diffraction (XRD) to analyze the crystal structure and crystallite size, Fourier transform infrared (FTIR) used to analyze the force constant and then vibrating sample magnetometer (VSM) to analyze the magnetic properties.

2. Methods

2.1. Synthesis of $Sr_xCo_{1-x}Fe_2O_4$ nanoparticles

The nanoparticles of $Sr_xCo_{1-x}Fe_2O_4$ (x= 0.0 and 0.1) were synthesized by co-precipitation route as the same previouse procedure [23]. The high purity salts $Co(NO_3)_2.6H_2O$, $Sr(NO_3)_2.6H_2O$ and $Fe(NO_3)_3.9H_2O$ dissolved in 200 ml distilled water under stirring. The synthesis was conducted at a temperature of 95°C for three hours. The mixed salt solution was added dropwise to the 100 ml 4.8 M NaOH solution under constant stirring. The precipitate washed with ethanol and deionized water. The precipitates were dried overnight at 100°C to remove excess water. The following variation post annealing has been done at 600°C for five hours. The precipitates were ground using mortar to obtain fine powder samples. The powder samples were then characterized.

2.2. Characterization

The X-ray diffraction patterns of the samples were recorded with PANalytical X'pert Pro diffractometer using Cu Kα source (1.54 Å) radiation. The lattice constant a was calculated using the following relation:

$$a = \frac{\lambda}{2\sin\theta}\sqrt{h^2 + k^2 + l^2} \qquad (1)$$

Where λ is the wavelength of Cu ($K\alpha$), θ is Bragg's angle and (hkl) are the Miller indices. The X-ray densities (d_x) was determined by the following equation:

$$d_x = \frac{Z\,M}{N_A\,a^3} \qquad (2)$$

Where Z is number of molecule per unit cell, M is molecular weight, and N_A is Avogadro's number [20]. The average crystallite size of $Sr_xCo_{1-x}Fe_2O_4$ nanoparticles of the most prominent peak (311) is determined by using Scherrer's formula [24]:

$$D = \frac{0.9/\lambda}{\beta\cos\theta} \qquad (3)$$

Where β is the full width at half maximum of the peak. The strain in $Sr_xCo_{1-x}Fe_2O_4$ nanoparticles are determined by using Hall-Williamson plot using the following formula [24]:

$$\varepsilon = \frac{\beta_{hkl}}{4 \tan \theta} \quad (4)$$

Where ε is the strain and β_{hkl} is the full width at half maximum of the peak.

The IR spectra were recorded in the range of 350-4000 cm^{-1} by using Shimadzu IR Prestige 21 Fourier transform infrared spectrometer. The force constant for the tetrahedral site (k_t) and octahedral site (k_o) was calculated by the following relation [25]:

$$k_t = 7{,}62 \times M_1 \times k_1^2 \times 10^{-7} \frac{N}{m} \quad (5)$$

$$k_o = 10{,}62 \times \frac{M_2}{2} \times k_2^2 \times 10^{-7} \frac{N}{m} \quad (6)$$

Where M_1 and M_2 are the molecular weight of cations on the A- and B-sites, respectively. The magnetic properties were investigated by using vibrating sample magnetometer Dexing Magnet Ltd type VSM250.

3. Results and Discussion

3.1. Structural analysis

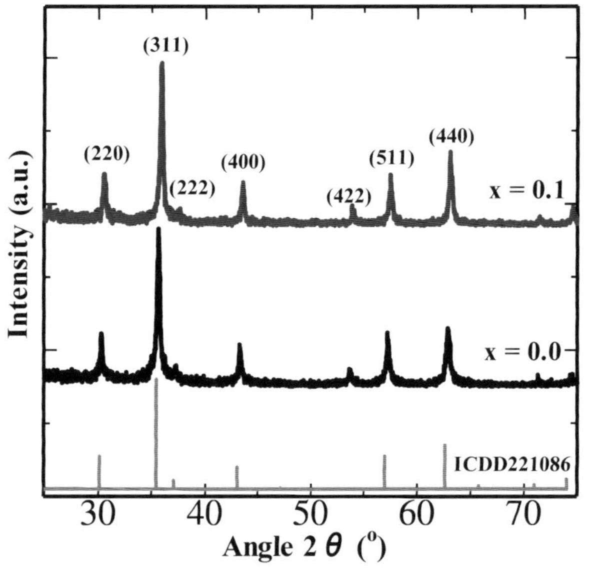

Figure 1. XRD patterns for $Sr_xCo_{1-x}Fe_2O_4$ synthesized at temperature of 95°C and annealed at 600°C

The X-ray diffraction patterns of cobalt ferrite without or with doping synthesized by co-precipitation method are shown in Fig. 1. This characterization shows that sample cobalt ferrite without and with doping strontium have a similarity to the data peaks of the International Centre for Diffraction Data (ICDD) number 221086. This similarity signifies that sample cobalt ferrite without or with doping strontium in inverse spinel face center cubic (fcc) structure. The similarity also attributes that there is no other impurities present and the high purity cobalt ferrite without and with doping strontium was successfully produced via co-precipitation method. The strongest peak (311) is located at an angle (2θ) 35.63° for sample cobalt ferrite without doping and 35.91° for sample cobalt ferrite with doping strontium. The strontium doping causes the strongest peak angle of cobalt ferrite slightly shift to the right of 0.28°. The intensity of cobalt ferrite sample decrease with the presence of strontium ions.

Table 1. XRD result of sample cobalt ferrite without and with strontium doping

	x = 0.0	x = 0.1
Crystallite size (nm)	23.50	22.65
Lattice parameter (Å)	8.34	8.28
Density (g/cm^3)	5.32	5.48
Strain	0.0048	0.0050

The crystallite sizes of cobalt ferrite are becoming smaller with the addition of strontium ions are shown in Table 1. The crystallite sizes decrease from 23.50 nm to 22.65 nm with addition strontium ions so the lattice parameter decrease. It causes the density of the nanoparticles cobalt ferrite increase with the addition strontium ions. The strains of cobalt ferrite increase from 0.0048 to 0.0050 with strontium ions doping. This may be because strontium has larger atomic radius than cobalt so that the strontium ions suppress in the cobalt ferrite sample. Kumar et al. report that the lattice strain increase with the addition of strontium ions in $Sr_xCo_{1-x}Fe_2O_4$ sample while the crystallite size decrease.

3.2. FTIR analysis

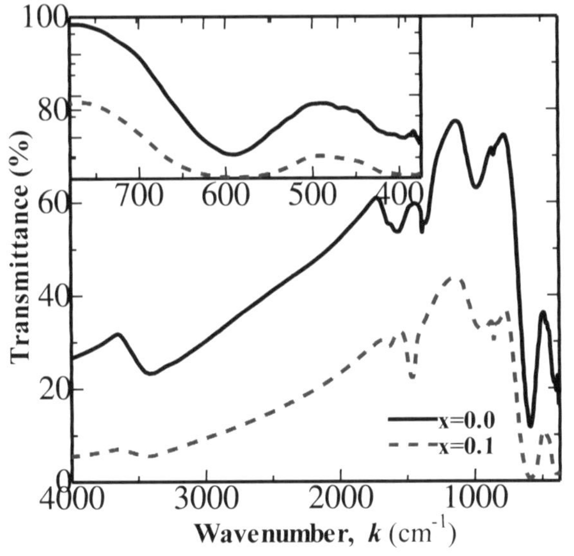

Figure 2. FTIR spectra for $Sr_xCo_{1-x}Fe_2O_4$synthesized at temperature of 95°C and annealed at 600°C

The formation of the spinel structure cobalt ferrite and its cations distribution is supported by FTIR analysis. Figure 2 shows the infrared transmittance spectra of cobalt ferrite sample without and with doping strontium ions. The transmittance of the nanoparticle cobalt ferrite decrease with the addition of strontium ions. The first characteristic absorption band is observed in the wavenumber range of 1500-1400 cm^{-1} that are thought to derive from the stretching and vibration modes of the H-O-H bending of free or absorbed water molecules [20]. The second characteristic of this absorption band is at wave numbers around 600-400 cm^{-1} (see inset Fig. 2) which is a common feature of spinel ferrite for both samples. The highest frequency band is located around 600 cm^{-1} which is regarded as a vibrating stretch of a tetrahedral metal complex consisting of bonds between oxygen ions and metal ions at the tetrahedral site [20]. Whereas the lower absorption bands located around wavenumber 400 cm^{-1} are regarded as bending vibrations of octahedral metal complexes consisting of bonds between oxygen ions and metal ions at the octahedral site [20]. This result confirms the appearance of $CoFe_2O_4$ in this sample with co-precipitation method.

Table 2. FTIR result of sample cobalt ferrite without and with strontium doping

x	k_1 (cm^{-1})	k_2 (cm^{-1})	k_t (N/m)	k_o (N/m)
0.0	590.24	393.50	14.87	10.11
0.1	589.28	380.96	14.82	9.48

Table 2 shows that the cobalt ferrite sample force constant at the tetrahedral site has increased from 14.87 to 14.82 N/m in the presence of strontium doping. But with the presence of strontium doping the force constants at the octahedral site decreased from 10.11 to 9.48 N/m. The change of the force constant at the tetrahedral site is 0.05 N/m while the octahedral site is 0.63 N/m. Based on these results, changes in the octahedral site were greater than at the tetrahedral site after doping of strontium ions. So it can be stated that the strontium ions is more pressing on the octahedral site. With the addition of doping ions can decrease and extend the absorption band length of the tetrahedral and octahedral site lattice. This indicates that the frequency difference between tetrahedral and octahedral sites is caused by changes in bond width between the two sites. So the strains that have been calculated in the previous chapter of the sample cobalt ferrite without and with doping strontium ions located at the octahedral site based on the FTIR result.

3.3. *Magnetic properties analysis*

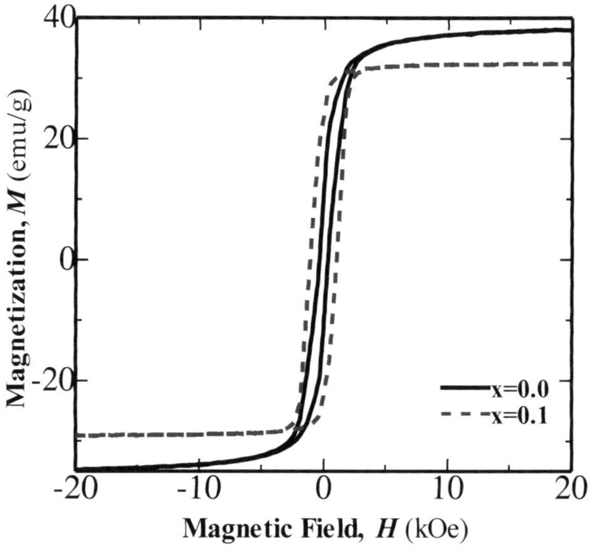

Figure 3. VSM result for $Sr_xCo_{1-x}Fe_2O_4$ synthesized at temperature of 95 °C and annealed at 600 °C

Figure 3 shows the hysteresis curve of cobalt ferrite without and with strontium doping synthesized at 95°C and annealed at 600°C for 5 hours. The picture showed that both samples have the same typical hysteresis curve but have different height and width of the curve. It is clear that the sample with strontium doping has shorter and wider hysteresis curves than the sample of cobalt ferrite without strontium doping. The presence of this strontium doping can affect the magnetic properties of the sample of cobalt ferrite as can be seen in Table 3. Table 3 shows changes in remnant magnetization, saturation magnetization, coercive field and the squareness ratio of the sample cobalt ferrite with addition strontium ions.

Table 3. VSM result of sample cobalt ferrite without and with strontium doping

x	M_r (emu/g)	M_s (emu/g)	H_c (kOe)	M_r/M_s
0.0	10.92	36.80	0.34	0.30
0.1	22.98	31.32	1.07	0.73

The remnant magnetization of cobalt ferrite samples was seen to increase from 10.92 to 22.98 emu/g with the presence of strontium ions. The presence of strontium ions causing the saturation magnetization of the ferrite cobalt sample decreased from 36.80 to 31.32 emu/g. This decrease in saturation magnetization may be attributed to the cation distribution of the sample. This decrease in saturation magnetization can be identified by the presence of strontium ions occupying site A (tetrahedral) or site B (octahedral). The redistribution of cations in tetrahedral and octahedral is related to the presence strains of force constants in metal bonds in tetrahedral or octahedral results of FTIR characterization. The decrease in saturation magnetization attributed to the strain induced magnetization by breaking the magnetic ion chains with the substitution of non-magnetic ions [1].

The coercive field of the cobalt ferrite sample increased from 0.34 to 1.07 kOe with the addition of strontium doping in the sample. An increased coercive field may be caused by the domain wall pinning on the surface of the nanoparticles [26]. So it can be ascertained that the strontium ions causes the magnetic moment of samples to be difficult to deliver because of the domain wall pinning which then causes the widening of the hysteresis curve from the sample. With a larger coercive field value this cobalt ferrite with strontium doping yields more a square hysteresis curve than sample cobalt ferrite without doping. It can be seen from the squareness ratio value (M_r/M_s) of sample cobalt ferrite with doping strontium larger than sample cobalt ferrite without doping. Improvement on coercive field may be influences to the improvement on squarness ratio (M_r/M_s). The coercive field indicates the magnetic field required for the demagnetization of the sample. So the cobalt ferrite sample with doping strontium is more difficult to remove the magnetic properties than sample without doping.

4. Conclusion

Cobalt ferrite nanoparticles without and with doping strontium were successfully synthesized using co-precipitation method. The addition of strontium ions in cobalt ferrite causes the crystallite sizes decrease and the strains increase. FTIR result showed that the strontium ions press cobalt ferrite in octahedral site. The magnetic properties showed that the addition strontium ions in cobalt ferrite successfully change the value of remnant magnetization, saturation magnetization, coercive field, and shape of the hysteresis curve.

References

[1] Kumar R and Kar M 2016 *JMMM* **416** 335
[2] Shokrollahi, H 2008 *JMMM* **320** 463
[3] Sangmanee M and Maensiri S (2009). *Appl. Phys. A: Mat. Science & Processing* **97** 167
[4] Pašukonienė V, Mlynska A, Steponkienė S, Poderys V, Matulionytė M, Karabanovas V, ... and Kurtinaitienė M 2014 *Medicina* **50** 237
[5] Kooti M and Afshari M 2012 *Scientia Iranica* **19** 1991
[6] Dey A, Singh R, and Purkait MK 2014 *J. Water Process Eng.* **3** 1
[7] Hannour A, Vincent D, Kahlouche F, Tchangoulian A, Neveu S, and Dupuis V 2014

J.Magn.Magn.Mater. **353** 29
[8] Sodaee T, Ghasemi A, Paimozd E, Paesano A, and Morisako A 2013 *J.Magn.Magn.Mater.* **330** 169
[9] Sharifi I, Shokrollahi H, Doroodmand MM, and Safi R 2012 *J.Magn.Magn.Mater.* **324** 1854
[10] Houshiar M, Zebhi F, Razi ZJ, Alidoust A, and Askari Z 2014 *J.Magn.Magn.Mater.* **371** 43
[11] Rana S, Philip J, and Raj B 2010 *Mat. Chem. and Phys.* **124** 264
[12] Song Q and Zhang, ZJ 2004 *J. American Chem. Soc.* **126** 6164
[13] Chiu WS, Radiman S, Abd-Shukor R, Abdullah MH, and Khiew PS 2008 *J. Alloys and Comp.* **459** 291
[14] Xiao SH, Jiang WF, Li LY, and Li XJ 2007 *Mat. Chem. and Phys.* **106** 82
[15] Liu Q, Sun J, Long H, Sun X, Zhong X, and Xu Z 2008 *Mat. Chem. and Phys.* **108** 269
[16] Carp O, Patron L, and Reller A 2007 *Mat. Chem. and phys.* **101** 142
[17] Chinnasamy CN, Narayanasamy A, Ponpandian N, Chattopadhyay K, Guerault H, and Greneche JM 2000 *J. Phys.: Condensed Matter* **12** 7795
[18] Sajjia M, Oubaha M, Hasanuzzaman M, and Olabi AG 2014 *Ceram. Int.* **40** 1147
[19] Abdulwahab K.O, Malik MA, O'Brien P, and Timco GA 2014 *Mat. Science in Semiconductor Processing* **27** 303
[20] Safi R, Ghasemi A, Shoja-Razavi R, and Tavousi M 2015 *J.Magn.Magn.Mater.* **396** 288
[21] Stoner EC and Wohlfarth EP 1948 *Philosophical Transactions of the Royal Society of London A: Math., Phys. and Eng. Sciences* **240** 599
[22] Chinnasamy CN, Jeyadevan B, Perales-Perez O, Shinoda K, Tohji K, and Kasuya A 2002 *IEEE Trans. on Magn.* **38** 2640
[23] Purnama B, Rahmawati R, Wijayanta AT, and Suharyana 2015 *J.Magn.* 20 207
[24] Rana K, Thakur P, Sharma P, Tomar M, Gupta V, and Thakur A 2015 *Ceram. Int.* **41** 4492
[25] Amiri S and Shokrollahi H 2013 *J.Magn.Magn.Mater.* **345** 18
[26] Maurya JC, Janrao PS, Datar AA, Kanhe NS, Bhoraskar SV, and Mathe VL 2016 *J.Magn.Magn.Mater.* **412** 164

IOP Conf. Series: Journal of Physics: Conf. Series **909** (2017) 012013 doi:10.1088/1742-6596/909/1/012013

The chemical bonds effect of anthocyanin and chlorophyll dyes on TiO_2 for dye-sensitized solar cell (DSSC)

A H Ahliha[1], F Nurosyid[1,a], A Supriyanto[1], and T Kusumaningsih[2]

[1]Department of Physics, Graduate Program, Universitas Sebelas Maret, St. Ir. Sutami 36A Kentingan Jebres Surakarta 57126, INDONESIA
[2]Department of Chemistry, Faculty of Mathematics and Natural Sciences, Universitas Sebelas Maret, St. Ir. Sutami 36A Kentingan Jebres Surakarta 57126, INDONESIA

E-mail: [a]nurosyid@yahoo.com

Abstract. Anthocyanin and chlorophyll dyes have been blended as the photosensitizer of *Dye-Sensitized Solar Cell* (DSSC). The results study showed the effect of chemical bond dyes on TiO_2 and the efficiency of DSSC. Ratio blend of the anthocyanin and chlorophyll dyes are 1:1. The absorbance of dyes and TiO_2 were characterized using UV-Vis Spectrophotometer. The chemical bonds contained in TiO_2- dyes were characterized using FT-IR spectrophotometer. The efficiency of DSSC was calculated using I-V meter. The absorption spectra of chlorophyll: anthocyanin blend dye solutions and TiO_2 films can increase after the dye adsorption. Absorbance characterization of anthocyanin and chlorophyll dye blend solutions showed three peaks at the wavelength of 412 nm; 535.5 nm; and 656.5 nm. Absorbance characterization of spinach before being blend with anthocyanin dyes solutions showed two peaks at the wavelength of 431 nm and 665.5 nm. The absorption spectra of TiO_2 films can increase after the dyes adsorption at the wavelength of 400 nm. FT-IR spectra of TiO_2 founded the functional groups C-Br, C=C, and O-H. The functional groups founded in anthocyanin: chlorophyll dye blended on the surface of TiO_2 are C-Br, C-O, O-H, C-H, C=C, C=O, and O-H. The result showed that the greatest efficiency of 0.0544% at dye red cabbage-spinach. Adsorption blends of anthocyanin and chlorophyll dyes on the surface of TiO_2 can be used as the photosensitizer for DSSC.

1. Introduction

Dye-sensitized solar cell (DSSC) is third generation solar cells that can convert photon energy into electrical energy[1 – 4]. In general of DSSC, these devices are based on a nanostructured metal oxide film, sensitized by an adsorbed dye molecule to harvest the visible light. These cells are composed of three basic components: a dye sensitizer, which absorb visible light to generate excitons and transfer electrons from the excited state into the metal oxide conduction band[5,6]; an electrons-transporting layer such as a metal oxide material[5,7], in which the injected electrons travel throughout the electrode. The dye is regenerated by an electron donor in the electrolyte solution[5,8].

In DSSC, the dye as a sensitizer plays a key role in absorbing sunlight and transforming solar energy into electrical energy. In DSSC, the sensitizers used such as the ruthenium complexes and the natural dyes [9]. The highest efficiency of DSSC sensitized by Ruthenium complexes absorbed on TiO_2 reached 11-12%. Dyes are not cheaper but have also been reported to reach an efficiency as high as 9.8%[10]. The natural dyes found in flowers, leaves, and fruits can be extracted by simple procedures. Several natural dyes have been employed as sensitizers in DSSC, such as

IOP Conf. Series: Journal of Physics: Conf. Series **909** (2017) 012013 doi:10.1088/1742-6596/909/1/012013

anthocyanin[11], chlorophyll[12], and carotene[13]. Many researches have been developed kinds of natural dyes. Pratiwi *et al.* employed anthocyanin dyes from dragon fruit and red cabbage as the sensitizer in DSSC and reported conversion efficiency of 0.024% and 0.054%[14]. Sengupta *et al.* reported that a conversion efficiency of 0.148% was obtained using spinach leaves dye as sensitizer[15].

In this study, natural dyes were extracted from pigments of anthocyanin and chlorophyll such as dragon fruit, black rice, red cabbage, and spinach. These extracted dyes were characterized by UV-Vis spectrophotometer. The chemical properties of the DSSC using these extracts as sensitizers were investigated. The anthocyanin and chlorophyll dyes blended as the photosensitizer of DSSC. The results study expected to find the effect of chemical bonds dyes on TiO_2 and the efficiency of DSSC.

2. Experiments

2.1. Preparation of TiO_2 Paste
TiO_2 pastes were prepared using TiO_2 powder in ethanol. The TiO_2 solution was stirred at 300 rpm for 30 minutes. TiO_2 pastes that have been formed are inserted into the bottle covered with aluminum foil and stored in an area free from direct sunlight.

2.2. Preparation of Working Electrode
TiO_2 pastes were deposited on the FTO conductive glass with an active area of 1 cm x 1 cm by spin coating technique. TiO_2 films were annealed in a furnace at 400°C for 10 minutes.

2.3. Preparation of Natural Dye Sensitizers
The materials used are natural dyes from dragon fruit, red cabbage, black rice, and spinach leaves which were cut into small pieces and mash in a mortar. The anthocyanin dyes were dissolved in acetic acid: quads: methanol solvents. The chlorophyll was dissolved in acetone solvent. It was stirred for 60 minutes at room temperature using vortex stirrer with a rotational speed of 300 rpm. It was kept for 24 hours. It was stored in closed containers and protected from the sun. Extract of anthocyanin and chlorophyll dyes solutions were filtered using filter paper quality no. 42, respectively. The dye solutions are evaporated using a rotary evaporator to get the more concentrated solutions. The anthocyanin dye solution blended with chlorophyll dye solution, respectively. Ratio blends of anthocyanin and chlorophyll dye are 1:1. The absorption spectra of dye solutions were characterized by using UV-Vis spectrophotometer 1601 PC.

2.4. Preparation of Counter Electrode
The counter electrode was prepared from platinum (Hexachloroplatinic (IV) acid 10%) catalyst on the FTO conductive glass. FTO glass is heated using a hot plate at a temperature 250°C for 10 minutes and then apply a platinum solution on the surface of the FTO glass. The glass that has been spilled by the platinum is directly cooled to room temperature.

2.5. Preparation of Assembly Sandwich DSSC
The working electrode and counter electrode were assembled into a sandwich type arrangement. The electrolyte solution has filled the space between the two electrodes. The DSSC were characterized using FT-IR spectrophotometer Shimadzu Prestige 21 and *I-V* meter Keithley 2602A.

3. Results and Discussions

3.1 Characterization of Dye
Characterization of absorbance and functional groups of dye were examined by using UV-Vis and FT-IR spectrophotometer. The results of absorbance spectra are shown in Figure 1 and 2.

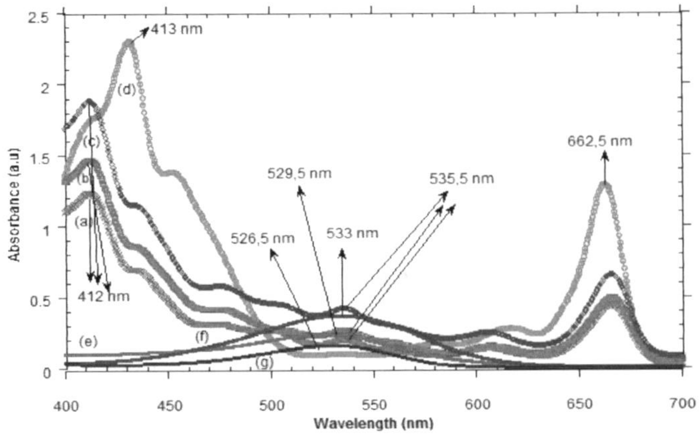

Figure 1. UV-Vis absorption spectrum of the dye solution from (a) dragon fruit-spinach blend, (b) red cabbage-spinach blend, (c) black rice-spinach blend, (d) spinach, (e) black rice, (f) dragon fruit, and (g) red cabbage

Figure 1 presents the absorption peaks of anthocyanin dye solutions from dragon fruit, black rice, and red cabbage at the wavelength of 533 nm; 526.5 nm; and 529.5 nm. Before blending, the dye of dragon fruit has higher absorption as compared to the black rice and red cabbage. The absorption peaks of spinach dye solution at the wavelength of 431 nm and 662.5 nm. The absorption peaks of anthocyanin and chlorophyll blended dye solution at the wavelength of 412 nm; 535.5 nm; and 665.5 nm. The wavelength of the blend indicates that the blend contains anthocyanin and chlorophyll pigments. The graph in Figure 1 showed that the dye solution of black rice and spinach blend resulted in widest and highest of absorbance. The dye of blend was resulted in the lowest of absorbance than the spinach dye before blended. The results of absorption wavelength showed that the dye solutions can be used as the photosensitizer for DSSC because they appeared the absorption of the visible region. The dye solutions contribute to the absorbance at the absorption peak each.

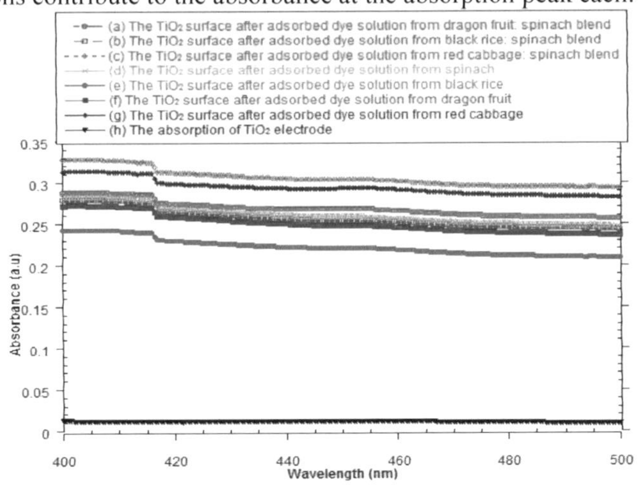

Figure 2. Absorption spectra of the TiO_2 after absorbed dye solution from (a) dragon fruit-spinach blend, (b) black rice-spinach blend, (c) red cabbage-spinach blend, (d) spinach, (e) black rice, (f) dragon fruit, (g) red cabbage, and (h) TiO_2 electrode

Figure 2 shows the absorption spectra of TiO_2 films can increase after the dye adsorption at the wavelength above 400 nm. The dye of red cabbage and spinach blended on the surface TiO_2 resulted in the highest of absorbance than the others of dyes was presented in Figure 2. The results of this study showed that the dye adsorption on the surface TiO_2 could be used as the photosensitizer for DSSC which absorbed the light from visible region. Danladi *et al.* reported that the chemical adsorption of these dyes is accepted to occur because of the formation of bond with the surface of nanostructured TiO_2[17].

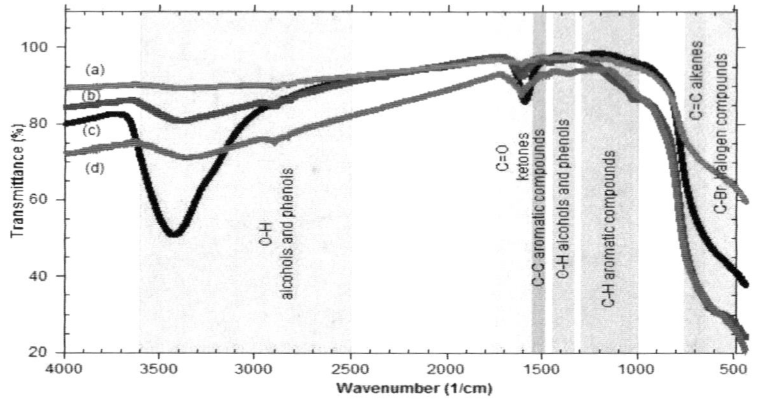

Figure 3. FT-IR spectra of (a) red cabbage, (b) dragon fruit, (c) TiO_2 nanoparticles, and (d) black rice adsorption on the surface TiO_2

The chemical structure of anthocyanin dyes from red cabbage, black rice, and dragon fruit before blended are shown in Figure 3. The dye of dragon fruit anthocyanin resulted in the lowest of intensity than the other of anthocyanin dyes and spinach chlorophyll dyes. The O-H bond is polar group appearing in the wave number at 3600-3200 cm^{-1} forming both intramolecular and intermolecular hydrogen bonded so that the absorption of widened and shifts in wave numbers.

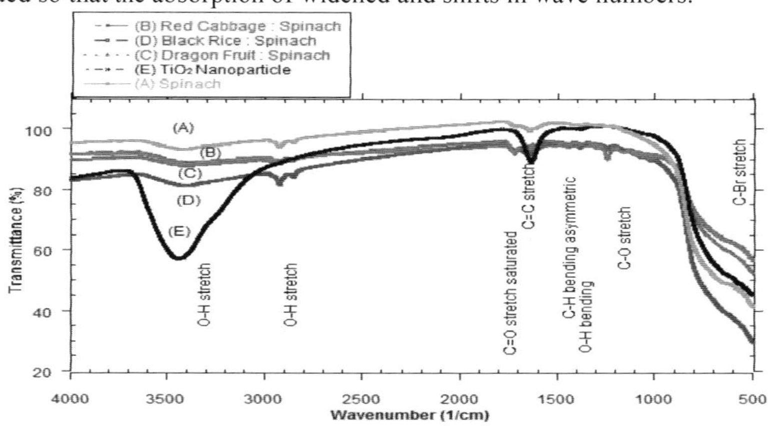

Figure 4. FTIR spetra of (a) TiO_2 nanoparticle (b) dragon fruit-spinach blend, (c) black rice-spinach blend, and (d) red cabbage-spinach blend adsorption on the surface TiO_2

The FT-IR spectra of extracted anthocyanin and chlorophyll dyes blended are shown in Figure 4. The broad absorption at wave number range 3600-3200 cm^{-1} arises due to the O-H stretching vibration of anthocyanin and chlorophyll dyes blended. The bond is represented by the appearance of weak bonds at wave numbers 1420-1330 cm^{-1} caused in the plane O-H bending. The dye of red cabbage and spinach blend resulted in the highest of intensity. The sharp peak at wave number range 2700-2900 cm^{-1} corresponds to the O-H stretching in aldehydes, where the O-H group of each molecule forms a hydrogen bond with the carbonyl group C=O molecules with dimer form. The hydrogen bond between the O atoms of the C = O aldehyde group will give rise to different OH bonds of strength, caused by the environment in which the hydrogen bonds are rather different for different molecules. The result will be a variety of OH stretching vibration at different frequencies. This peak at wave number range

3000-2900 cm^{-1} corresponds to the C-H stretching vibration of chlorophyll. The bending vibration of C-H asymmetric with medium intensity appeared on the region at 1485-1445 cm^{-1}.

Infrared absorption appears at wave number 1725-1705 cm^{-1} resulting in the C=O stretching vibration. The dye of dragon fruit and spinach resulted in the lowest of polarity. Stretching of carbonyl group showed changes of dipole moment large enough so that the high of the intensity. The dye of black rice and spinach blend at 94.714% of transmittance in the region of double bonds of the stretching vibration caused by located of C=C absorption at1680-1620 cm^{-1}. This area is used for identifying olefins. Olefins are unsaturated hydrocarbons with a double bond between the carbon atoms. This is because olefins have a much higher of acidity as compared with the alkanes.

The compound in the fingerprint region placed at 1260-1000 cm^{-1} wave number, which shows medium intensity into strong intensity. The dye of dragon fruit and spinach blended has the highest of transmittance of 93.219% than the other dye. This is due to the plane vibrations which include the C-O bonds the absorption of frequency has a variable value. The position of the tape in the region at 680-500 cm^{-1} indicates the absorption of alkyl halide caused by the stretching C-Br. The dye dragon fruit and spinach blended has the highest of transmittance of 64.555% when compared with other dye. In this area, absorption cannot be made a diagnosis because the thin layer and dye samples are mixed with KBr powder during FT-IR testing. Based on the result of a UV-Visible spectrum and FT-IR spectrum it is concluded that the extracted compound is a blended of anthocyanin and chlorophyll.

3.2 Performance of DSSC

Characterization of the current-voltage (*I-V*) is a method to determine how much ability DSSC can convert light into electrical energy. The samples were illuminated with Xe light source of *I-V* meter. The light intensity was kept constant at 1000 $W/_{m^2}$ and active area of the cell exposed to light was 1 cm^2 (1 cm x 1 cm). Photovoltaic test of DSSCs using these natural dyes as sensitizers are summarized Table 1.

Table 1. Parameters of DSSCs with anthocyanin dyes, anthocyanin and chlorophyll dyes blended.

Dye solution	V_{oc} (V)	I_{sc} (mA)	FF	η (%)
Spinach	0.639	0.33	0.337	0.0713
Black rice	0.580	0.126	0.272	0.0198
Dragon fruit	0.166	0.490	0.267	0.0217
Red cabbage	0.191	0.641	0.353	0.0432
Spinach leaves : Dragon fruits	0.460	0.19	0.0298	0.0258
Spinach leaves : Red cabbage	0.565	0.29	0.0328	0.0544
Spinach leaves : Black rice	0.565	0.19	0.0341	0.0358

The efficiency of the DSSC sensitized by the spinach was significantly higher than the DSSC sensitized by other natural dyes before blended. It gives a I_{sc} of 0.33 mA, V_{oc} of 0.639 V, *FF* of 0.337 and efficiency of 0.0713%. After blended, the DSSC sensitized by spinach leaves: red cabbage was higher than the DSSC sensitized by other natural dyes. It gives a I_{sc} of 0.29 mA, V_{oc} of 0.565 V, *FF* of 0.0328 and efficiency of 0.0544%. This is because there are available bonds between the dye and TiO_2 molecules through which electrons transport from the excited dye molecules to the TiO_2 film[18]. This result indicates that the interaction between the natural dye sensitizer and the TiO_2 film is significant in enhancing the energy conversion efficiency of DSSCs.

The results of the *I-V* test are supported by FT-IR test results (Figure 3 and 4) and UV-Vis spectra (Figure 2) showing that the absorption between the blended of dragon fruit and spinach dye, and black rice and spinach dye is almost the same. This is influenced by the few of the raised exciton and the low amount of charge that reaches the electrode when the cell is irradiated by low of intensity light. The few of charge reaches the electrode, caused by the length of the charge diffusion path so that many charges are recombined before it reaches the electrode. The measured parameters are influenced by each of the sample compiler's active layers. In DSSC made from organic, the stability of the voltage is still low and experiencing a rapid voltage drop when illuminated by light.

4. Conclusion

The absorption spectra of dye adsorption on the surface of TiO_2 which was resulted in the highest absorbance of red cabbage and spinach blended. Based on the results of a UV-Visible spectrum and FT-IR spectrum it is concluded that the extracted compound is a blended of anthocyanin and chlorophyll. The results were efficiency for DSSC that the solar cells using red cabbage and spinach blended as the sensitizer have highest values of efficiency. The few of the raised excitons and the low amount of charge that reaches the electrode when the cell is irradiated by low of intensity light. The few of charge reaches the electrode, caused by the length of the charge diffusion path so that many charges are recombined before it reaches the electrode. The efficiency was higher as the higher adsorption of the dye to the surface of TiO_2.

Acknowledgements

This work was supported by Institute for Research and Community services, Universitas Sebelas Maret by PUPT Grant Program (contract number. 1075/UN27.21/PP/2017).

References

[1] O'Regan B, and Gratzel M 1991 *Nature* **353** 737-740
[2] Dey A, Moyez Sk A, Mandal M K, and Roy S 2016 *Material Today: Proceedings* **3** 3498-3504
[3] Gomez-Ortiz N M, Vazquez-Maldonado I A, Perez-Espadas A R, and Mena-Rejon G J 2010 *Solar Energy Materials and Solar Cells* **94** 40-44
[4] Cakar S, Guy M, Ozacar M, and Findik F 2016 *Electrochimica Acta* **209** 407-422
[5] Gomez T, Jaramillo F, Schott E, Arratia-Perez R, and Zarate X 2017 *Solar Energy* 215-223
[6] Hardin B E, Snaith H J, and McGehee M D 2012 *Nat. Photonics* **6** 162-169
[7] Bai Y, Mora-Sero I, De Angelis F, Bisquert J, and Wang P 2014 *Chem. Rev.***114** 10095-10130
[8] Hagfeldt S, Boschloo G, Sun L, Kloo L, and Pettersson H 2010 *Chem. Rev.* **110** 6595-6663
[9] Nagata T and Murakami H 2009 *Ulvac Tech. J.***70E** 1-5
[10] Zhou H, Wu L, Gao Y, and Ma T 2011 Journal of Photochemistry and Photobiology A: Chemistry **219** 188-194
[11] Gokilamani N, muthukumarasamy N, Thambidurai M, Ranjitha A, and Velauthapillai D 2013 *J. Sol-Gel Sci. Technol.* **66** 212-219
[12] Krawczyk S and Zdyb A 2011 *J. Phys. Chem. C.* **115** 22328-22335
[13] Wang X F, Tamiaki H, Kitao O, Ikeuchi T, and Sasaki S 2013 *J. Power Sources* **242** 860-864
[14] Pratiwi D D, Nurosyid F, Supriyanto A, and Suryana R 2016 *J. Phys.: Conf. Ser.* **776** 012007
[15] Sengupta D, Mondal B, and Mukherjee K 2015 Spectrochimica Acta Part A: Molecular and Biomolecular Spectroscopy **148** 85-92
[16] Danladi E, Muhammad A, Idodo M, Danladi E, Aungwa F, and Sunday S 2016 *Journal of Energy and Natural Resources* **5** (1) 11-15
[17] Danladi E, Ezeoke J, Ahmad M S, Danladi E, Sarki S H, Ishaya I, and Gyuk P M 2016 *Physical Science International Journal* **10** (2) 1-7
[18] Adenike B, Okafor P, Ibrahim A, Oluwole S, Boyo H O, and Henry B 2013 *International Journal of Engineering Research and Development* **8** issue 12 38-42

IOP Conf. Series: Journal of Physics: Conf. Series **909** (2017) 012014 doi:10.1088/1742-6596/909/1/012014

Analysis of bend loss in loaded fiber coil for a circular and an elliptical shape

Ahmad Aftah Syukron[1], Ahmad Marzuki[1] and Ary Setyawan[2]

[1] Physics Department of Graduate Program Sebelas Maret University, Jl. Ir. Sutami 36A Kentingan Jebres Surakarta 57126, INDONESIA

[2] Civil Engineering of Engineering Faculty Sebelas Maret University, Jl. Ir. Sutami 36A Kentingan Jebres Surakarta 57126, INDONESIA

E-mail: dfata.aftah@gmail.com

Abstract
Bending in the optical fiber can result in a loss of optical fiber. Bending causes the output light intensity smaller than the intensity of the light source. Bending can be utilized to a sensor based on optical fiber. This paper presents to know the correlation between the optical fiber sensor utilizing bending effect with rubber coil to a circular and an elliptical shape.

The principle of this sensor is a detecting change in the output voltage through an optical fiber. The output voltage then converted to transmittance with the program of a computer. The optical fiber is divided into two arms, that is one reference arm and one modulate arm. Fiber bending sensor was made by coiling optical fiber in modulating arm. The displacement loaded to the optical fiber coil in a rubber thread, so that occur of bending. Rubber threaded used to have a 1.0 cm diameter for the circular shape, and 1.1 cm mayor axis and 0.9 cm minor axis for the elliptical shape. Tests are taken from variations in a number of turns, that are 1, 2, 3, and 4 turns and displacement load from 0.0–10.0 mm.

The results can be concluded that the loss is greater with the more number of turns. The larger of displacement load has also resulted in the greater loss. The ratio of loss generated to both of the rubber shapes is visible. The circular shape shows that the loss is a very small firstly. But, after the cross section coil turned into an elliptical shape with a minor axis, the loss increased significantly. While for the elliptical shape shows that the loss increased significantly from the first displacement loaded.

1. Introduction

The function of optical fiber is a light transmission line from one to another place. The optical fiber has been utilized since 1970 as a communication system. Many researchers have been used optical fiber to expanding research. One of this is an optical fiber sensor. The optical fiber has many advantages, such as a very sensitive, hold up with electromagnetic interference and radiation, and low noise [3].

Optical fiber sensor categorized by modulation principle, that are optical fiber sensor based on shift intensity, shift phase, and shift wavelength [2]. Optical fiber sensor based on shift intensity utilize occurrence attenuation signal output from the optical fiber. Attenuation carried out by bending on the optical fiber [1]. Bending is a mechanical disorder that causes redistribution of light power in an optical fiber. Bending generates transmission loss thus providing sensor sensitivity characteristics. Bending are classified into macro bending and micro bending. Macro bending is a bending of the optical fiber with a radius of curvature is greater than a radius of the optical fiber. Micro bending is a microscopic bending from the core of an optical fiber. This is due to the difference in thermal contraction between core and cladding [4].

2. Experiment

Figure 1 is the scheme of used in this experiment. Light source comes from light emitting Diode (LED, λ=±635 nm). Light has a split to 2 arms. One arm is coming in to optical fiber within modulation; it's called reference arm. And the other arm is coming in to optical fiber modulated; it's called modulation arm [1]. Modulation arm created by wrapping the optical fiber in the rubber cylinder coil. Fiber coil created from a rubber silicone RTV 588. The diameter of fiber coil is 1.00 cm for a circular shape, and 0.9 cm of the minor axis and 1.1 cm of the major axis for an elliptical shape. The optical fiber used in this experiment is a polymer optical fiber (POF). Light is coming out from the both of arms detected by a light detector. A light detector connected with the microcontroller arduino uno. Arduino uno is a microcontroller has function of conversion data analog to digital or otherwise. Arduino uno connected with a computer to display data output using software Labview 2012.

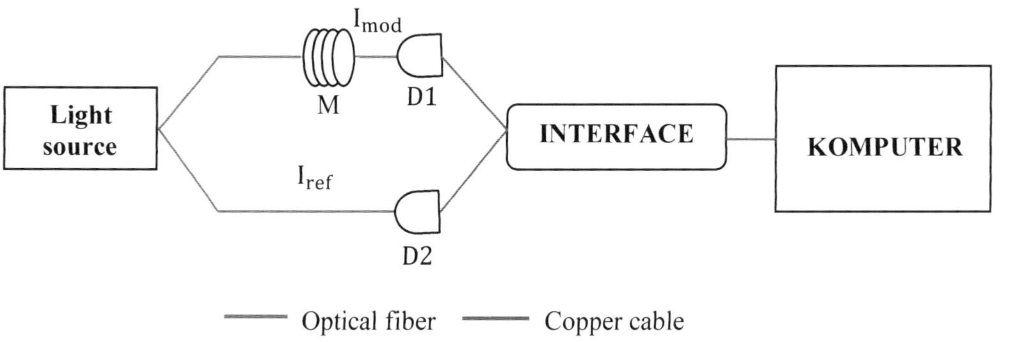

Figure 1. Experiment scheme of optical fiber sensor based on bend loss

First, taking data has carried out with a circular rubber coil shape. A light output intensity from the reference arm (I_{ref}) has set to as big as the modulation arm (I_{mod}), so obtained the transmittance value equal to 1 ($T=1$). As for the transmittance formula can be seen equation (1).

$$T = \frac{I_{\text{mod}}}{I_{ref}} \tag{1}$$

After the value $T=1$, then the modulation arm gave displacement load from 0.0–10.0 mm. Displacement load will change the cross section coil from circular shape to elliptical shape with minor axis ($x=y$–$axis$). Loss calculated from the transmittance data with the equation (2). As for the correlation of transmittance with loss is:

$$Loss(dB) = 10.^{10}\log\left(\frac{1}{T}\right)$$
$$Loss(dB) = 10.^{10}\log\left(\frac{I_{ref}}{I_{\text{mod}}}\right) \tag{2}$$

3. Results and Discussion

Data obtained from this experiment is a reference intensity (I_{ref}), modulation intensity (I_{mod}), and displacement load (d). Data used to calculation transmittance and loss with equation (1) and (2).

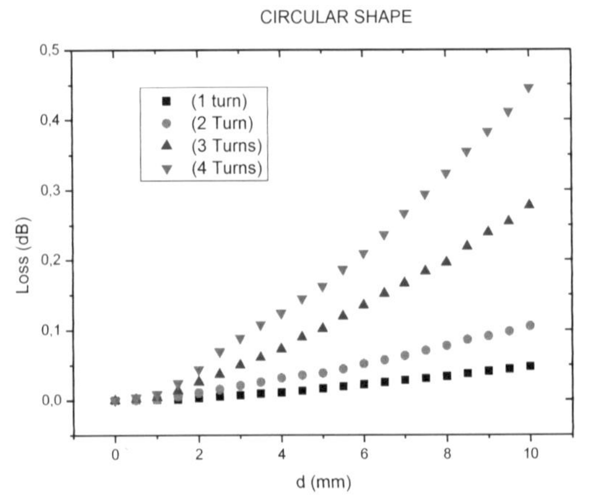

Figure 2. Bend loss from the circular shape

Figure 2 shows that loss is greater with the larger of displacement load. The more number of turns has also resulted in the greater loss. However, a loss is very small when the displacement load at 0.0–1.0 mm. This shows that in the circular shape coil less sensitivity on very small displacement load. Rubber coil requires displacement load along *d* prior to sensitivity.

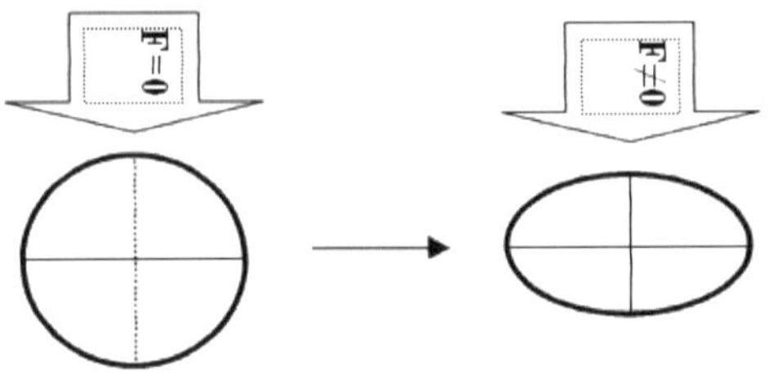

(a) (b)

Figure 3. Coil cross section change. (a) No external force. (b) External force

This experiment performed with the variation of turns from 1–4 turns, and variation of displacement load from 0.0–10.0 mm. Displacement load *M* obtained with emphasis. Emphasis carried out by using the micrometer screw. Figure 3 shows that the rubber coil shape. Figure 3a is no external force applied (*F=0*), rubber coil is a circular shape. Figure 3b is an external force applied (*F≠0*), rubber coil is changed to an elliptical shape. Both have the same value of perimeter *K* [5], Equation of an elliptical shape perimeter is

$$K_{el} = 2\pi \left(\frac{l^2+m^2}{2}\right)^{1/2} \quad (3)$$

where *l* is a radius of the major axis, *m* is a radius of the minor axis. The perimeter of an elliptical shape as big as the circular shape $K=2\pi r$, so can write

$$2\pi r = 2\pi \left(\frac{l^2+m^2}{2}\right)^{1/2} \quad (4)$$

So,

$$l = (2r^2 - m^2)^{1/2} \quad (5)$$

the radius of the minor axis is $m=r-d/2$ (d is a displacement), so equation (5) can write

$$l = \left(r^2 + rd - \frac{d^2}{4}\right)^{1/2} \quad (6)$$

From the equation 5, $m=r-d/2$ and $d = 1$ *mm,* accordingly obtained elliptical shape with $l = 0.9$ *cm,* and $m = 1.1$ *cm.*

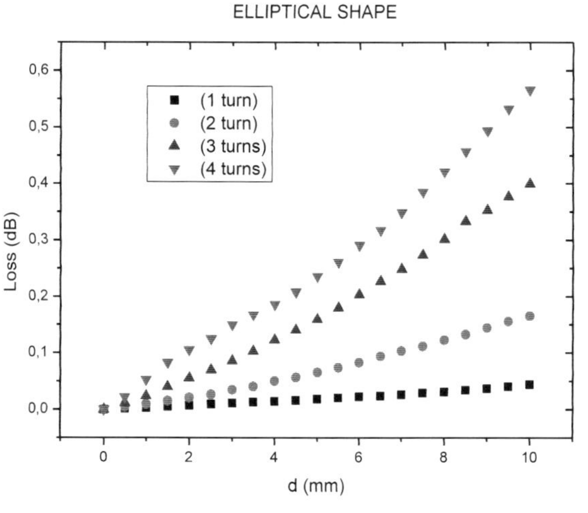

Figure 4. Bend loss from the elliptical shape

Figure 4 shows that loss is greater significantly from the displacement load 0.0 mm. This shows that rubber coil with the elliptical shape has the higher sensitivity from the circular shape.

4. Conclusions

The experiment to analysis bend loss created by wrapping the optical fiber in the rubber coil a circular and an elliptical shape. The diameter of rubber coil is 1.00 cm for a circular shape, and 0.9 cm of the minor axis and 1.1 cm of the major axis for an elliptical shape. The loss is greater with the more number of turns. The larger of displacement load has also resulted in the greater loss. The circular rubber coil shapes less sensitivity on very small displacement load. While the elliptical rubber coil shape sensitively. The rubber coil with the elliptical shape has the higher sensitivity from the circular shape.

References

[1] Marzuki A, Novianto H, Hidayat W and Setyawan A 2014 *SDIWC* ISBN: 978-0-9891305-4-7
[2] Kuang J-H, Chen P-C and Chen Y-C 2010 *Sensors* **10** 10198
[3] Bin M and Xinguo Z 2010 *ICICTA* **10** 1109
[4] Zendehnam A, Mirzaei M, Farashiani A and Farahadi H L 2010 *J. Phys.* **74** 591-603
[5] Marzuki A, Heriyanto M, Setiyadi I D and Koesuma S 2015 *Journal of Physics: Conference series* **622** 012059
[6] Marcuse D 1976 J. Opt. Soc. Am. 66 (3) 216-220.
[7] Lin F, Smith S T and Hussain G 2012 Precision Engineering 36 620–628

[8] Ren F, Li J, Li Z, He Y and Chen Z 2016 Optics Communications 360 15–19
[9] Yang D K W, Abakr Y A and Ghazali N M 2013 Procedia Engineering 56 (5) 842–848
[10] Chen Y, Meng F and Gong X 2016 Wear 368–369 470–484
[11] Hu H F, Sun S J, Lv R Q and Zhao Y 2016 Sensors and Actuators A 251 126–133
[12] Gao S F, Wang Y Y, Liu X L, Ding W and Wang P 2016 Optics Express 14801 24 (13)

IOP Conf. Series: Journal of Physics: Conf. Series **909** (2017) 012015 doi:10.1088/1742-6596/909/1/012015

Spectral of electrocardiographic RR intervals to indicate atrial fibrillation

Nuryani Nuryani*[1] and Anto Satrio Nugroho
*Physics Department, Sebelas Maret University,
Jl. Ir. Sutami 36A Kentingan Jebres Surakarta 57126, INDONESIA
**Information Department, Sebelas Maret University,
Jl. Ir. Sutami 36A Kentingan Jebres Surakarta 57126, INDONESIA

Email: nuryani@mipa.uns.ac.id

Abstract. Atrial fibrillation is a serious heart diseases, which is associated on the risk of death, and thus an early detection of atrial fibrillation is necessary. We have investigated spectral pattern of electrocardiogram in relation to atrial fibrillation. The utilized feature of electrocardiogram is RR interval. RR interval is the time interval between a two-consecutive R peaks. A series of RR intervals in a time segment is converted to a signal with a frequency domain. The frequency components are investigated to find the components which significantly associate to atrial fibrillation. A segment is defined as atrial fibrillation or normal segments by considering a defined number of atrial fibrillation RR in the segment. Using clinical data of 23 patients with atrial fibrillation, we find that the frequency components could be used to indicate atrial fibrillation.

1. Introduction

Atrial fibrillation is a serious heart diseases, which is associated on the risk of death. It is the problem in the heart atrial. During atrial fibrillation, the heart atrial beats chaotically and irregularly. A serious risk of AF is stroke [1]. This situation could increase mortality.

Early detection for atrial fibrillation is important to conduct appropriate actions which might reduce healthcare costs and improve patient care [2]. These actions are also necessary to prevent the serious complications of atrial fibrillation. Therefore, approaches have been investigated to detect atrial fibrillation early. Varied algorithms have been studied to find a sophisticated atrial fibrillation detection system accurately and quickly [3, 4]. Electrocardiographic features are often utilized for atrial fibrillation detection.

Neural network for an atrial fibrillation detection has been investigated by [5]. The approach detection employs probabilistic neural network and a wavelet feature of electrocardiogram. A neural network algorithm for atrial fibrillation is also investigated by (Mateo et al., 2013). This approach uses radial basis function (RBF) neural network and utilizes electrocardiographic features by cancelling the electrocardiographic QRST complex.

This article presents spectral features of electrocardiographic RR intervals for atrial fibrillation detection. RR interval is the time interval between two-consecutive R peaks of electrocardiogram [6]. RR interval represents heart rate. Heart rate of the heart in patient with atrial fibrillation fluctuates,

[1] Nuryani Nuryani

opposing the normal heart which beats stably. This happen as a result of that the atrial of the heart beats chaotically and irregularly.

2. Method

This article explores a study of the spectral of electrocardiographic RR interval for atrial fibrillation detection. Firstly, the data of electrocardiographic R peaks are obtained from Physionet [7]. The electrocardiograph is recorded from 23 patients. The recording is 10 hours in duration for each patient with sampling rate of 250 samples per second.

From the R peaks, RR interval is defined. It is the time interval between two consecutive R peaks. RR intervals is then segmented or windowed with a defined length. The length of segment is 60 RR intervals. A segment is defined as atrial fibrillation or normal by considering the number of atrial fibrillation happening in the segment. If the number of atrial fibrillation is more than a defined number, the segment is defined as atrial fibrillation. The other situation is defined as normal. The number of atrial fibrillation refers to the number of RR interval which is defined as atrial fibrillation.

The spectral of RR interval is defined from the defined segment. The spectral is obtained using fast Fourier transform (FFT). FFT computes Discrete Fourier Transform (DFT) defined as [8]:

$$X(k) = \sum_{n-0}^{N-1} x(n)e^{-j2\pi kn/N}$$

Using FFT, frequency components of the segment are presented. The amplitude of the components are investigated. The amplitude of the component is used to identify the related segment is associated to atrial fibrillation or non-atrial fibrillation. The identification (or the detection) is conducted using comparing the amplitude with a defined threshold. The performance of the detection is presented in term of accuracy. The accuracy is defined as the number of correct detection divided by the total number of data. In the other words, the accuracy is the correctness of both atrial fibrillation and normal events. Moreover, a segment is defined as atrial fibrillation or normal by considering the number of atrial fibrillation beats in the segment.

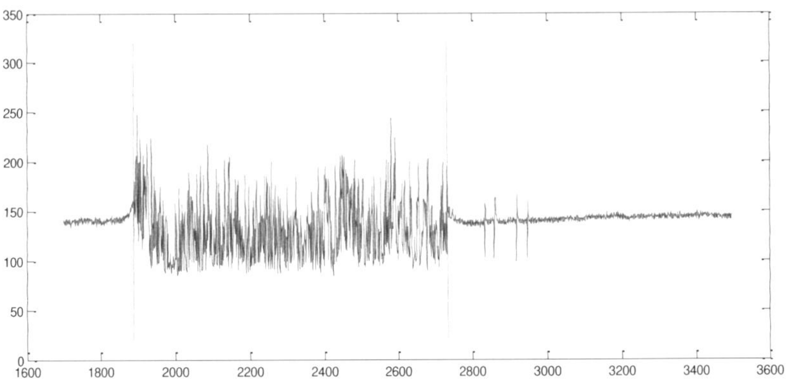

Figure 1. RR intervals pattern types of atrial fibrillation and normal. The high and low levels of square signal indicate the states of atrial fibrillation and normal, respectively.

3. Results

3.1. RR intervals during atrial fibrillation

IOP Conf. Series: Journal of Physics: Conf. Series **909** (2017) 012015 doi:10.1088/1742-6596/909/1/012015

We have obtained electrocardiographic RR intervals of all the patients of the MIT-BIH Atrial Fibrillation Database [9]. The RR intervals in the state of atrial fibrillation and normal of patients are presented in Figure 1. In the square signal, the high and low levels indicate the atrial fibrillation and normal states, respectively. In the atrial fibrillation state, the amplitudes of RR intervals are higher and fluctuate sharper than the normal one. The RR intervals of normal state appear as a signal with a relative small ripple. This indicates that the RR intervals are more stable in a certain value than that the atrial fibrillation one. However, in the normal state, sometimes it happens that RR intervals rise to a high magnitude and jump to a very low magnitude; it looks like a spark [10].

3.2. Frequency components of RR intervals during atrial fibrillation

The frequency components of the electrocardiographic RR intervals have ben obtained. The frequency components are obtained from each segment. Each segment of RR interval has a set of the associated frequency components, which has 10 components. The mean amplitude of the frequency components are summarized in Table 1. The first component is far higher than the other components, for both atrial fibrillation and normal. The amplitude of the first component in the normal state is higher than in the atrial fibrillation state. Conversely, the amplitude of the second and so forth components are higher in the atrial fibrillation than in the normal one, as presented in Figure 1.

Table 1: The frequency components of RR intervals during the atrial fibrillation and normal states (the average values are presented)

	Amplitudes for the atrial fibrillation	Amplitudes for the normal
1st Component	1567.70	1989.10
2nd Component	72.50	24.88
3rd Component	70.07	27.49
4th Component	68.06	28.23
5th Component	68.26	24.87
6th Component	55.30	21.34
7th Component	68.26	24.87
8th Component	68.06	28.23
9th Component	70.07	27.49
10th Component	72.50	24.88

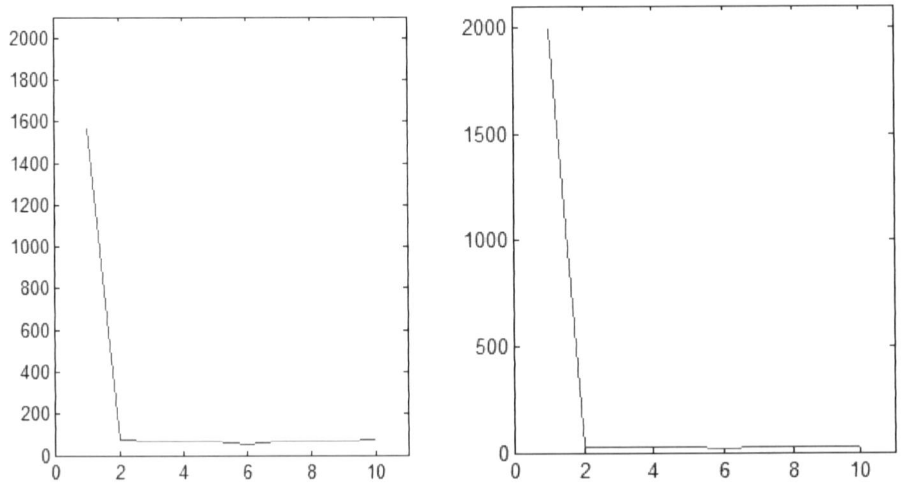

Figure 2. Frequency components of RR intervals of atrial fibrillation (the left) and normal (the right).

3.3. A simple atrial fibrillation detection using RR interval frequency components with a threshold

A simple detection or identification for atrial fibrillation using frequency components of RR intervals is explored. The detection method is based on the amplitude of the components. For the 1st components, the amplitudes which are higher than the threshold are defined as normal, and the lower one is defined as atrial fibrillation. Conversely, the other components, the amplitudes which are higher than the threshold are defined as atrial fibrillation, and the lower one is defined as normal. The threshold of the amplitude is varied. Figure 3 presents the accuracies of the detection using different components, which are the 1st, 2nd, 3rd and 4th components. The pattern for the other components are similar. The accuracies find the maximum value at certain thresholds and different components could have different threshold.

Among the single components, the detection using the 10th components find the highest accuracy, which is 78.99%. The lowest accuracy happen at which uses the first component; that is 72.72%. The detection which utilizes the mean of the 2nd to the 10th components find a significant higher than the other components, which is 82.23%. It implies that each component could contribute positively to improve the accuracy and when all the components are utilized the performance could find the higher. Note that the first component is not included to the mean value. The reason is that the amplitude of the 1st component is far higher than the other components and the characteristic is different. The difference is that the normal state associated the higher amplitudes, which is the opposite of the other components.

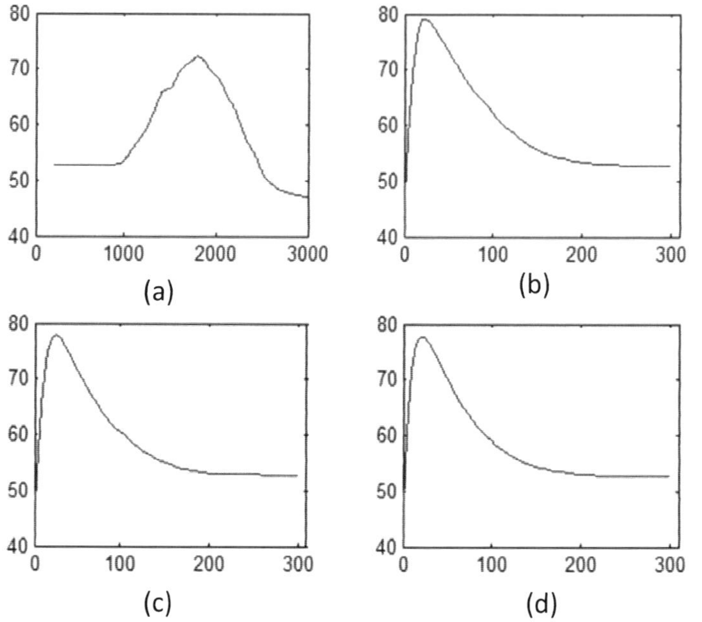

Figure 3. The accuracies of atrial fibrillation detection using the 1st, 2nd ,3rd and 4th frequency components with different thresholds, as shown in part a, b, c and d, respectively

Table 2: The maximum accuracies of the atrial fibrillation identification using the different frequency components of RR intervals during atrial fibrillation and normal states (values are in term of average)

The frequency componets	Accuracy (%)
1st Component	72.72
2nd Component	78.33
3rd Component	78.00
4th Component	77.72
5th Component	78.80
6th Component	75.44
7th Component	78.80
8th Component	77.72
9th Component	78.00
10th Component	78.99
Mean of the 2nd to the 10th components	82.23

4. Conclusion

A detection of atrial fibrillation utilizing the frequency components of RR intervals are presented. For the study, the detection approaches which utilize different components are compared. Each of the component has shown a positive contribution for the performance of the detection. The detection approach finds the highest performance when it utilizes mean the components. Using clinical data of the patients with atrial fibrillation, the highest accuracy found is 82.23%.

References

[1] J. S. Healey, S. J. Connolly, M. R. Gold, C. W. Israel, I. C. Van Gelder, A. Capucci, *et al.*, "Subclinical atrial fibrillation and the risk of stroke," *New England Journal of Medicine,* vol. 366, pp. 120-129, 2012.

[2] G. D. Clifford, I. Silva, B. Moody, Q. Li, D. Kella, A. Shahin, *et al.*, "The PhysioNet/Computing in Cardiology Challenge 2015: Reducing false arrhythmia alarms in the ICU," in *2015 Computing in Cardiology Conference (CinC)*, 2015, pp. 273-276.

[3] M. Sardana, J. Saczynski, N. Esa, K. Floyd, K. Chon, J. W. Chong, *et al.*, "Performance and usability of a novel smartphone application for atrial fibrillation detection in an ambulatory population referred for cardiac monitoring," *Journal of the American College of Cardiology,* vol. 67, pp. 844-844, 2016.

[4] T. Koivisto, T. Hurnanen, T. Vasankari, T. Kiviniemi, A. Saraste, and J. Airaksinen, "Automatic detection of atrial fibrillation using MEMS accelerometer," in *2015 Computing in Cardiology Conference (CinC)*, 2015, pp. 829-832.

[5] K. Daqrouq, A. Alkhateeb, M. Ajour, and A. Morfeq, "Neural network and wavelet average framing percentage energy for atrial fibrillation classification," *Computer methods and programs in biomedicine,* vol. 113, pp. 919-926, 2014.

[6] M. LIU, H.-h. WAN, S.-r. GONG, and X.-q. HAN, "detection of atrial fibrillation base on deep neural network of RR interval difference," *Laser Journal,* vol. 1, p. 025, 2015.

[7] A. L. Goldberger, L. A. N. Amaral, L. Glass, J. M. Hausdorff, P. C. Ivanov, R. G. Mark, *et al.*, "PhysioBank, PhysioToolkit, and PhysioNet," *Components of a New Research Resource for Complex Physiologic Signals,* vol. 101, pp. e215-e220, 2000.

[8] A. Das, "Signal Conditioning. An Introduction to Continuous Wave Communication and Signal Processing," in *Signals and Communication Technology*, ed: Springer Berlin Heidelberg, 2012, pp. 159-192.

[9] A. Goldberger, L. Amaral, L. Glass, J. Hausdorff, P. C. Ivanov, R. Mark, *et al.*, "PhysioBank, PhysioToolkit, and PhysioNet: Components of a new research resource for complex physiologic signals. Circulation [Online]. 101 (23), pp. e215–e220," ed, 2000.

[10] N. Nuryani, M. Solikhah, A. S. Nugoho, A. Afdala, and E. Anzihory, "RR-Interval variance of electrocardiogram for atrial fibrillation detection," in *Journal of Physics: Conference Series*, 2016, p. 012105.

IOP Conf. Series: Journal of Physics: Conf. Series **909** (2017) 012016 doi:10.1088/1742-6596/909/1/012016

Zero field reversal probability in thermally assisted magnetization reversal

E.B. Prasetya, Utari, B.Purnama[1]

Physics Department Sebelas Maret University,

Jl. Ir. Sutami 36A Kentingan Jebres Surakarta 57126, INDONESIA

E-mail:bpurnama@mipa.uns.ac.id

Abstract. This paper discussed about zero field reversal probability in thermally assisted magnetization reversal (TAMR). Appearance of reversal probability in zero field investigated through micromagnetic simulation by solving stochastic Landau-Lifshitz-Gibert (LLG). The perpendicularly anisotropy magnetic dot of 50×50×20 nm^3 is considered as single cell magnetic storage of magnetic random acces memory (MRAM). Thermally assisted magnetization reversal was performed by cooling writing process from near/almost Curie point to room temperature on 20 times runs for different randomly magnetized state. The results show that the probability reversal under zero magnetic field decreased with the increase of the energy barrier. The zero-field probability switching of 55% attained for energy barrier of 60 k_BT and the reversal probability become zero noted at energy barrier of 2348 k_BT. The higest zero-field switching probability of 55% attained for energy barrier of 60 k_BT which corespond to magnetif field of 150 Oe for switching.

1. Introduction

As one as a candidate of universal memory, magnetic random access memory (MRAM) is being studied over decades for reinforce performance and excellence. MRAM contain magnetic cell which magnetized as binary function. One element cell stored as one bit. Unlike sequential access magnetic memory, MRAM freely access data to reads in every single bit memory cell [1-2]. MRAM has multi advantages such as: non-volatile, low memory degradation, low energy consumption and less 10 nano second orde of access time made MRAM interesting to investigate furthermore [3].

In order to realize nano meter orde of MRAM, there are criteria to build reliable storing data, which barrier energy required exceed 60 k_BT to has thermal stability in temperature room [3]. High anisotropy magnetic cell necessary to ensure data not erased caused demagnetization. In addition anisotropy give advantage for patterning memory cell [4]. It was dilemma, because higher anisotropy have a higher energy barrier. Higher energy barrier mean higher magnetic field to reverse during writing scheme, later defined as reversal field [4]. Here, the limit of magnetic field to aligning a long to field direction (unstabilizing near magnetic saturation) defined as switching field, or can be defined as writing field in generally [1].

In this micromagnetic simulation, TAMR performed by heat up of the selected cell magnetic memory until almost Curie temperature. Then a constant field applied when cooling process from Curie Point to room temperature. Heat and magnetic field pulse given, named double pulse writing. Heat pulse surely decrease energy barrier with in result the drcrease of magnetic switching field.

Furthermore, the reduce switching field because the thermal fluctuation effect of TAMR cut coercive field to demagnetization [4]. Without TAMR, magnetic field required could be over 500%.

In this present paper, zero-magnetic field of the thermally assisted magnetization reversal is studied by the solve of the LLG equation. Randomly magnetized state and magnetic field fluctuation effect due to thermal calculate by using fluctuation dissipation energy. Finally, the modify of zero-magnetic field probability in TAMR scheme is evaluated by the change of the energy barrier

2. Simulation method

Attempted TAMR simulation begin with design magnetic nanodot appearance. Figure 2 show single magnetic element was designed in 50x50x20 nm^3 cubic shape with single domainly perpendicular anisotropy. Shape of simulation approached 15x15 grids with finite element approximation for 1 element cell. Reversal mode instancely represented binary as 1 when magnetization saturation go "up" and 0 when magnetization go "down". Required energy to opposite magnetization direction or other state closer equilibrium called energy barrier. [1].

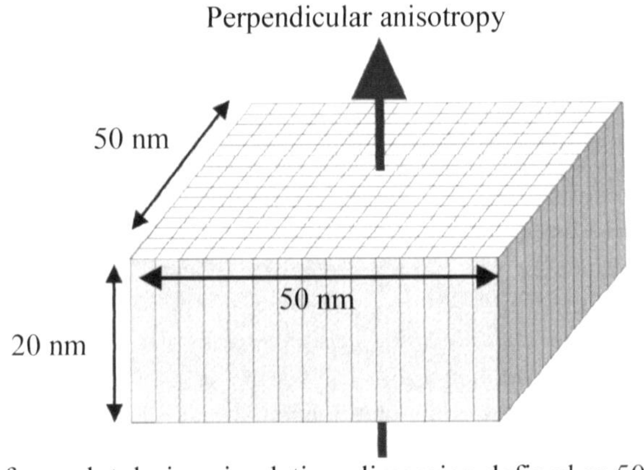

Figure 1. Design of nanodot during simulation, dimension defined as 50x50x20 nm^3 with single domain perpendicular anisotropy.

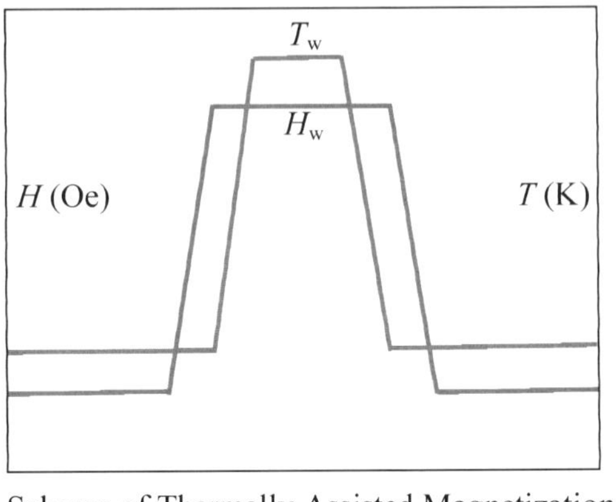

Figure 2. Scheme of Thermally Assisted Magnetization Reversal

Curie Temperature set on 372,9 Kelvin regarded as temperature of heat during double pulse scheme writing. Duration of heat pulse shorter than magnetic field pulse's duration. Pico second reversal magnetization occur in cooling proccess after Curie temperature reached and went 298 K room temperature state ilustrated in Figure 2. Randomness of magnetization cause demagnetization state before reversed. Energy barrier emerge during this state and reduced in TAMR scheme. Thermally assisted magnetization reversal was performed 20 times runs for different randomly magnetized state.

Micromagnetic simulation solving dynamic magnetization due Landau Lifshift-Gilbert stochastic equation [8,9]:

$$\left(\frac{dM^i}{dt}\right) = -|\Upsilon|M^i \times H_{eff} + \frac{\alpha}{M_s} M^i \times \frac{dM^i}{dt} \quad (1)$$

And fluctuation dissipation effect:

$$\sigma = \sqrt{\frac{2k_B T\alpha}{\Upsilon V M_s dt}} \quad (2)$$

To realize perpendicular anisotropy magnetic material there was material limit parameter Q > 1. Factor Q can be calculated due anisotropy constant (K_u) dan magnetization saturation (M_s).

$$factor\ Q = \frac{Ku}{2\pi M_s{}^{(0,5)}} \quad (3)$$

Anisotropy field (H_k) and energy barrier (ΔE) consist 2 main variable, M_s and K_u. Anisotropy field can be calculated using equation 4 and 5 where V is magnetic volume [6,9]:

$$H_k = \frac{2k_u}{\mu_0 M_s} \quad (4)$$

$$\Delta E = (K_u - 2\pi M_s{}^2)V \quad (5)$$

3. Result & Discussion

Figure 3a show a comparison of a time evolution of magnetization reversal for TAMR and non-TAMR schemes. When micro-magnetic simulation perform for non-TAMR scheme, step-like magnetization profile is observed. In contrary for TAMR scheme, the ripple of demagnetized state at arround zero magnetization realize due to thermal stochastic effect. A present of the demagnetized state is considered may decrease the magnetic field reversal for TAMS scheme. To ensure that the reversal magnetization process of the TAMR scheme takes place, the reversal magnetization micromagnetic simulation is performed for 20 different randomly magnetic states. Reversal magnetization occurs when the whole micromagnetic calculation of the 20 initial different randomly magnetic states gives the result of switching. Figure 3b show probability of reversal magnetization evaluated for three different magnetitude of energy barrier. For three cases, the zero field switching probability show different each and other.

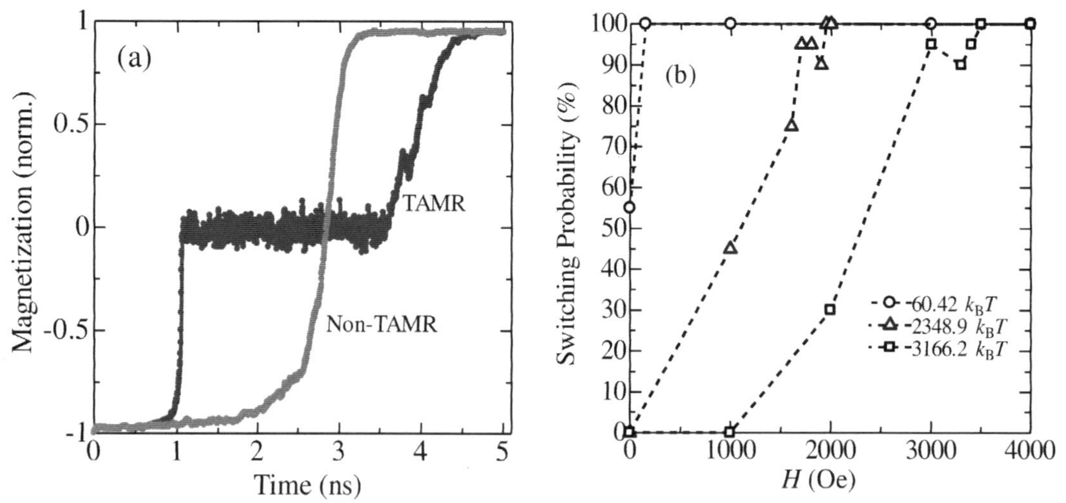

Figure 3. (a) Comparison magnetization reversal in TAMR (blue line) and non-TAMR (red line) schemes, (b) Probability switching evaluate for three different energy barrier as a function as energy barrier.

∧

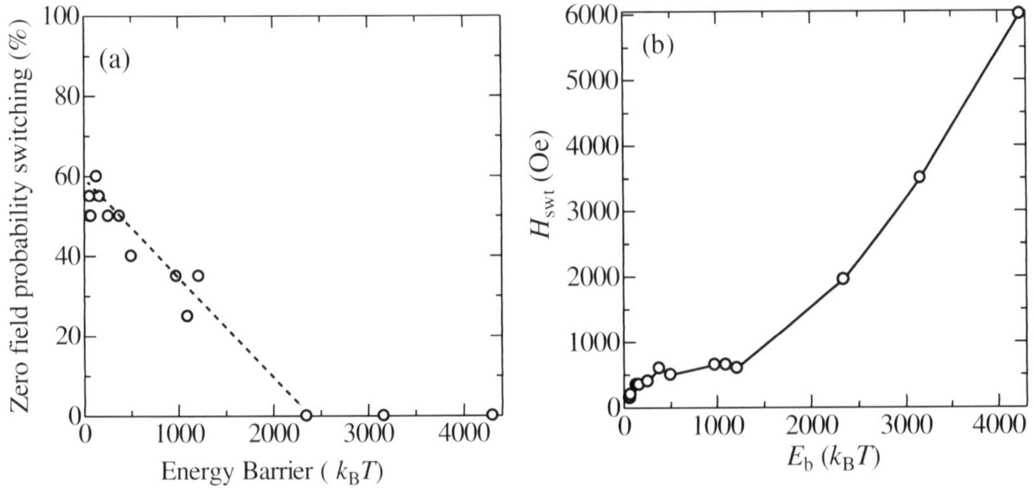

Figure 4. (a) Zero field probability switching and (b) switching field as a function as energy barrier.

The switching probability of 55% attain for energy barrier E_b of 60.42 k_BT. Then the switching probability become zero for both E_b of 2348.9 and 3166.2 k_BT. In order to find out furthermore the probability of switching in the zero field, the calculations are performed for the various barrier energy variable as shown in Figure 4. Figure 4a present the zero field probability switching evaluate for various energy barrier. It is clearly observed from the graph that probability decreases with increasing energy barrier and becomes zero when energy barrier E_b is much larger than 2349 k_BT. The result also indicate that there are some limit parameter for the zero field magnetization switching occurrence. In case for perpendicularly magnetic anisotropy material, it is mainly related to anisotropy magnetic field. However an other physical properties (i.e. damping constant) should attribute the available of the limit parameter for the zero field magnetization switching. Furthermore studies regarding this need to be done to clarify these predictions. Figure 4b show that the switching field as a function of energy barrier E_b. The magnetic field require for switching linearly increase with the increase of the E_b for lower energy barrier ($E_b < 500$ k_BT). Then, the switching field exponentially increase with the increase of the E_b ($E_b > 500$ k_BT). The minimum switching field of 150 Oe is obtained for material with $E_b = 60.58$ k_BT. From the results we can conclude that the switching field attributable to the existence of the zero field probability. In case a zero field probability is attained a high probability, a small magnetic field is required for switching, and vice versa.

4. Conclusion

Zero field reversal probability in TAMR-MRAM scheme have been presented by a micromagnetic simulation. The results show that the zero field reversal probability decreased with the increase of the energy barrier. The high probability in zero field reversal probability indicate the small magnetic field required for switching in TAMR scheme. The higest zero-field switching probability of 55% attained for energy barrier of 60 k_BT which corespond to magnetif field of 150 Oe for switching.

References

[1] Hubert, A and Schäfer R 2009 *Magnetic Domain : The Analysis of Magnetic Microstructures*. (New York: Springer Berlin Heidelberg) p. 533 & p. 189
[2] Ross C A, 2001 *Annu. Rev. Mater. Res.* 2001. 31:203–35
[3] Purnama B, Nozaki Y, and Matsuyama K. 2007 *JMMM* **310** 2683–2685
[4] Iskandarova I M., Ivanova A V, Knizhnik A A, Popkov A F, Potapkin B V, Skirdkov P N, Zvezdin K A, Stainer Q, Lombard L, and Mackay K. 2016 *Nanotechnologies in Russia* **11** 208 - 214
[5] Hanin F. Y. F, Utari and Budi P, 2016 *J. of Phys.: Conf. Series* **776**
[6] Craik D 1995 *Magnetism: Principles and Applications* (Nottingham,: John Wiley & Sons Ltd) p. 91 & p. 106
[7] Lakshmanan M, 2011 *An Overview. Phil. Trans. of The Royal Soc. A* **369**. 1280-1300.
[8] Cullity B D and Graham C D 2009 *Introduction to Magnetic Material 2nd edition*. (New Jersey: John Wiley & Sons) p. 419
[9] Purnama B. (2009). *Thermally assisted magnetization reversal in perpendicularly magnetized thin film* (Doctoral dissertation, Kyushu University).

Numerical study of proximity effect on critical field of rectangular type-II superconductor

M Naufal Indriatmoko, Fuad Anwar and Hery Purwanto[1]

[1] Physics Department of Sebelas Maret University,
Jl. Ir. Sutami 36A Kentingan Jebres Surakarta 57126,

Email: muhammadnaufal@student.uns.ac.id

Abstract. Study of proximity effect on critical field of a rectangular type-II superconductor has been done. It was a numerical study using with Time Dependent Ginzburg Landau (TDGL) equations as the basis. The superconductor conditioned in boundary with a non-superconducting material (i.e. metal). A uniform external field was applied to the superconductor in the direction of *z*-plane. Proximity effect or de Gennes boundary condition via extrapolation length (b) took account the properties caused by superconductor in contact with different material. As input to the simulation, we had three different dimensions each with a basis of 12×12 and 32×32 and five varying value of b. We found that the lower critical field H_{c1} will decrease as the increasing value of b for all the rectangular superconductor, except for the size of 16×64. Any increment in the proximity effect also triggered a raise in the third higher critical field H_{C3} for every grid size variations. We also concluded that the proximity effect is very effective to a superconductor with high number of circumference to area ratio.

1. Introduction

A lot of studies about magnetic properties of superconductor has done using Time Dependent Ginzburg-Landau equations as their foundation [1-8]. The Time Dependent Ginzburg-Landau equation, abbreviated and also widely known as TDGL equation, first developed by Gor'kov and Eliashberg now very successful in explaining properties and dynamics of superconductor [9]. Due to highly nonlinear nature of TDGL equations, it is common to solve the equation numerically [10]. One of several frequently used numerical solutions is ψU method. The method has been proven and worked great to solve the TDGL equation [1,3,4,5,6,7]. Previous study has succeeded acquiring the lower critical field and the third higher critical field using TDGL equations which were solved with ψU method [2].

In the practical usage of superconductor, it is very often to see that the superconductor is in contact with non-superconducting materials. This condition forces many studies about magnetic properties of superconductor to not abandoning the proximity effect [4,5,6].

The TDGL equations which govern the superconductivity order parameter ψ and the vector potential $\mathbf{A}$ in the zero-electric potential gauge are given by [1,6]:

$$\frac{\partial\psi(r,t)}{\partial t} = (\nabla - i\mathbf{A}(r,t))^2\psi(r,t) + \psi(r,t) - |\psi(r,t)|^2\psi(r,t) \qquad (1)$$

$$\sigma\frac{\partial\mathbf{A}(r,t)}{\partial t}=\frac{1}{2i}(\overline{\psi}(r,t)\nabla\psi(r,t)-\psi(r,t)\nabla\overline{\psi}(r,t)-2i|\psi(r,t)|^2\mathbf{A}(r,t))-\kappa^2(T)\nabla\times(\nabla\times\mathbf{A}(r,t)-\mathbf{H}_{\mathbf{ext}}(r,t)) \quad (2)$$

There are two equations that hold the boundary condition properties:

$$\hat{\mathbf{n}}\cdot[\nabla-i\mathbf{A}]\psi=\frac{i}{b}\psi \quad (3)$$

$$\hat{\mathbf{n}}\cdot[\nabla-i\mathbf{A}]\psi=0 \quad (4)$$

Equation (3) valid with the condition that superconductor is in contact with other materials. While equation (4) act if the superconductor is immersed in a vacuum. The proximity effect occurs when a superconductor is in contact with a normal metal. The proximity effect taken into account via de Gennes extrapolation length b in equation (3). All variables above are rescaled as: ψ in unit of $\psi_0=(|\alpha(T)|/\beta)^{1/2}$, $\mathbf{A}$ in unit of $A_0=\mu_0 H_{C2lin}(T)\xi(T)$, $\mathbf{H}_{\mathbf{ext}}$ in unit of $H_{C2lin}(T)$, σ in unit of $\sigma_0=1/(\mu_0\kappa(T)^2 D)$, r in unit of $\xi(T)$, t in unit of $\tau(T)=\xi(T)^2/D$. Quantity $\hat{\mathbf{n}}$ is the normal vector of the superconductor boundary, ψ is order parameter, b is real number determining the boundary condition, σ is electrical conductivity, $\alpha(T)$ and β are the Ginzburg-Landau free energy density coefficients, $\mathbf{A}$ is vector potential, $\mathbf{H}_{\mathbf{ext}}$ is the external magnetic field, $\xi(T)$ is coherence length, $\kappa(T)$ is Ginzburg-Landau parameter and D is phenomenological diffusion constant.

The main topic discussed in the paper was proximity effect on the critical fields of the rectangular type-II superconductor. The proximity effect was observed through de Gennes extrapolation length. The critical fields in context are the lower critical field H_{C1} and the third higher critical field H_{C3}.

2. Numerical Method

This study considered the superconductor as in figure 1. Rectangular shaped superconductor affected by spatially uniform, time-dependent external field pointing in z-plane direction. A normal metal situated in contact with the superconductor. In ψU method, it is assumed that the superconductor contained $N_x \times N_y$ cells of which have the size of $\Delta_x \times \Delta_y$. There are three complex link variables held inside each cell [1,3]. They are introduced to help obtain a better numerical convergence at high magnetic field [10]. These set of conditions were studied with TDGL equations which solved numerically by ψU method as have done in other previous studies [1,2,3,8]. A simulation program that has input and would yield desired output then generated based on this case.

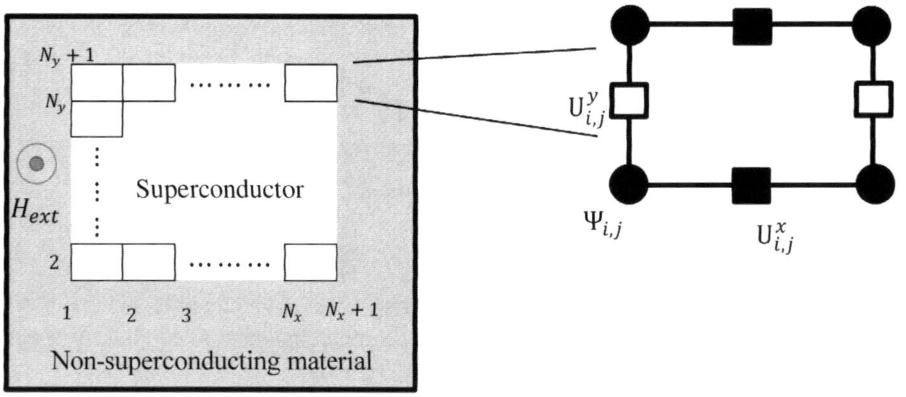

Figure 1. Rectangular shaped superconductor and the three important variables

We set the simulation input with time step of 0.010, mesh spacing 0.5×0.5, $\kappa=2.0$, initial applied external field 0 with increment of 1×10^{-6} and various grid sizes with a basis of 12×12 and

32×32. The two basis represent a small and large size superconductor. More details of the grid size input shown in table 1.

Table 1. Grid size variation

Input Name	$N_x \times N_y$	Area	Circumference	b
SA1	9×16	144	50	3
SA2	9×16	144	50	5
SA3	9×16	144	50	10
SA4	9×16	144	50	20
SA5	9×16	144	50	30
SB1	8×18	144	52	3
SB2	8×18	144	52	5
SB3	8×18	144	52	10
SB4	8×18	144	52	20
SB5	8×18	144	52	30
SR1	12×12	144	48	3
SR2	12×12	144	48	5
SR3	12×12	144	48	10
SR4	12×12	144	48	20
SR5	12×12	144	48	30
LA1	16×64	1024	160	3
LA2	16×64	1024	160	5
LA3	16×64	1024	160	10
LA4	16×64	1024	160	20
LA5	16×64	1024	160	30
LB1	8×128	1024	272	3
LB2	8×128	1024	272	5
LB3	8×128	1024	272	10
LB4	8×128	1024	272	20
LB5	8×128	1024	272	30
LR1	32×32	1024	128	3
LR2	32×32	1024	128	5
LR3	32×32	1024	128	10
LR4	32×32	1024	128	20
LR5	32×32	1024	128	30

3. Result and Discussion

The TDGL equations has been solved numerically and we obtained the magnetization, external field, order parameter and other contained quantities. Since our main discussion was the critical fields, we concerned only to $|\psi|^2$ and H_{ext}. From plotting those two variables we obtained the critical fields of the superconductor. The first local maximum determined the lower critical field H_{C1} and the zero-point determined the third upper critical field H_{C3} [2]. After we obtained the critical fields with the corresponding de Gennes extrapolation length b value, we plotted it to obtain two graphs shown in figure 2 and 3.

From the graphs obtained, we can see several interesting phenomena. In the group of small size superconductor, the lower critical field H_{C1} decreases when de Gennes extrapolation length b is raised. It is valid for either rectangular or square superconductor. However, in the large size group,

H_{C1} increases when b is raised only for the square (32×32) and 16×64. The proximity effect seems to give opposite response to 8×128 in term of H_{C1} value within the group. When it comes to the third higher critical field, all of the grid size variation gives the same response. The value of H_{C3} is increases when the value of proximity is raised. It is true for both square and rectangular superconductor. Regarding the geometry, one can discuss about the circumference to area ratio. The small group superconductor obviously has the bigger ratio over the large group. Here at figure 4 & 5 we shall see that varying value of b has less impact to the H_{C1} value of the large group. It means that the proximity effect is less effective to superconductor with an area much more bigger than its circumference. After all, the simulation is programmed to measure the order parameter ψ with the TDGL equations except for those near the interface which it used the boundary condition equation instead.

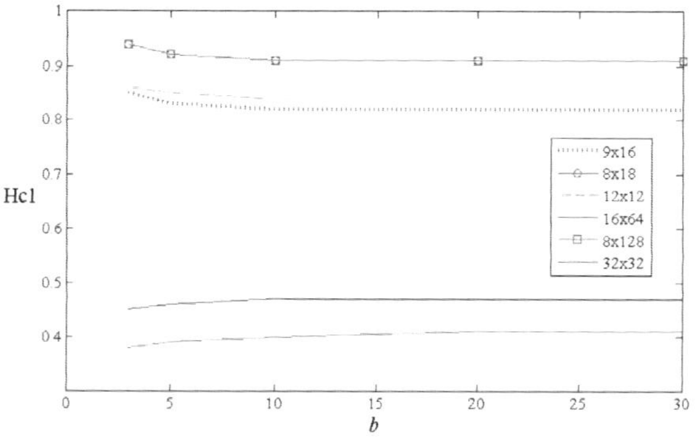

Figure 2. Graph of H_{C1} function b

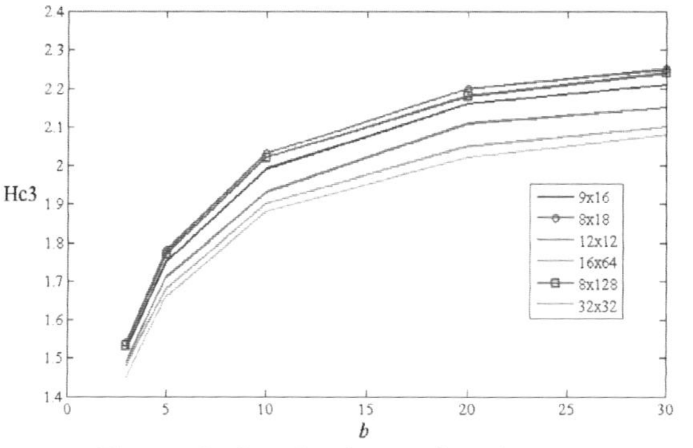

Figure 3. Graph of H_{C3} function b

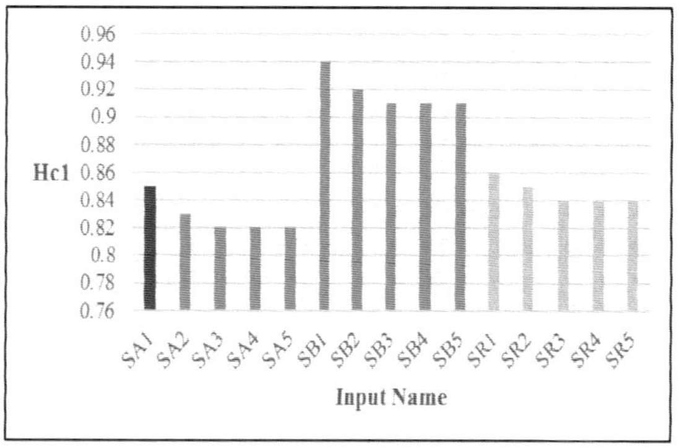

Figure 4. Value of H_{C1} for the small size group

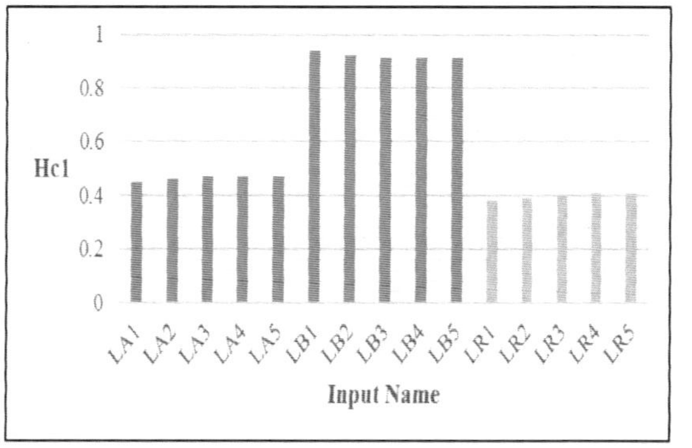

Figure 5. Value of H_{C1} for the large size group

4. Conclusion

The effect of de Gennes boundary condition (proximity effect) on the critical fields of rectangular shaped type-II superconductor has been examined by numerical simulation. We found that the lower critical field H_{C1} will decrease as the increasing value of b for all the rectangular superconductor, except for the size of 16×64. Any increment in the proximity effect also triggered a raise in the third higher critical field H_{C3} for every grid size variations. We also concluded that the proximity effect is very effective to a superconductor with high number of circumference to area ratio.

References

[1] Buscaglia G C, Bolech C and López A 2000 *On The Numerical Solution of The Time-Dependent Ginzburg-Landau Equations In Multiply Connected Domains* (Berlin: Springer)
[2] Anwar F, Nurwantoro P and Hermanto A 2013 *J. Nat. Sci. Res.* **3** 99-106
[3] Gropp W D, Kaper H G, Leaf G K, Levine D M, Palumbo M and Vinokur V M 1995 *J. Comp. Phys.* **123** 254-66
[4] Barba-Ortega J, de Souza Silva C C and Aguiar A J 2009 *Phys.* C **469** 852-56
[5] Barba Jose J, Silva C C de Souza, Carbal R E Leonardo and Aguiar A 2008 *Phys.* C **468** 718-721
[6] Barba-Ortega J, Sardella E, Aguiar J A and Brandt E H 2012 *Phys.* C **479** 49
[7] Barba-Ortega J and Joya M R 2012 *Dyna* **79** 46-50
[8] Winiecki T and Adams C S 2002 *J. Comp. Phys.* **179** 127
[9] Gor'kov L P and Eliashberg G M 1968 *Sov. Phys.* **54** 612
[10] Du Q 2005 *Journal of Mhatematical Physics* **46** 095109-1-22

A model of a nonlinear DNA-protein interaction system with general Hylleraas potential

Edy Syahroni[1], Suparmi[1,2], and Cari[1,2]
[1] Physics Department, Graduate Program, Sebelas Maret University,
Jl. Ir. Sutami 36A Kentingan Jebres Surakarta 57126, INDONESIA
[2] Physics Department, Faculty of Matematics and Nature Science, Sebelas Maret University,
Jl. Ir. Sutami 36A Kentingan Jebres Surakarta 57126, INDONESIA

E-mail: edy.syahroni1988@gmail.com

Abstract. All the functions of DNA depend on interactions with proteins. We proposed a model to describe the interaction between of them. For the simplest model, we took adenine and thymine molecule to represented each molecule. The interactions adapted from binding hydrogen atoms of the molecule. The General Hylleraas potential were used to describe both of the interactions. We proposed the Hamiltonian equation to describe the general model of interaction. The stability of the model from the DNA-protein interaction was investigated with Lyapunov function. We also have determined the equation of spectrum energy of the interaction model using WKB approximation method.

1. Introduction

DNA (deoxyribonucleic acid) is a repository of genetic information. In 1953, Frances Crick and James Watson discovered a molecular model of DNA as a double-stranded helical structure, or better known as the Watson-Crick double helix. DNA is a polynucleotide macromolecule polymer composed of nucleotides repeated, structured double, forming DNA double helix and twisting to the right [1, 2]. The specific structure was described in 1953 in his paper [2]. There have been many efforts to describe its complex dynamics. Some of them are both linear and nonlinear as described by Yakushevich [3]. The nonlinear effect might play an important role in DNA dynamic that was suggested in 1980 by Englander et al. [4].

The dynamic system of the DNA interactions can't be separated from how the DNA's molecule interacts with nearby molecules, either DNA molecule links directly or the others like protein molecule. Based on the chemical bond, there are three kinds of interaction that occurred among molecules, they are hydrogen, van der Waals, and electric bond. Hydrogen interaction among the three bonds interaction is the easiest one to observe.

In biological regulations, the interaction between DNA molecules and protein molecules provides a very important role in studying as making up the nucleosome, to control the gene expression, transcription factors, and cofactors [5]. Each process of transcription and replication will begin with the activity of a transcription factors (TFs) that triggers the signal [6]. Those proteins are often triggered in the cells by specific posttranslational modifications that allow TFs to bind to their specific sites of DNA. The activity status of these proteins also able to deducted by measuring such interaction of TFs. Such DNA-binding assay experiments sometimes able to combine with proteomics

experiments measuring specific phosphorylation events that can give a lot of information to the researchers about exact mechanisms of acting of this class of proteins [7]. While the process of DNA and protein binding, the regulatory proteins transmitted a high signal. This phenome become a triger to make an assumtion that the protein molecule induces conformational local distortion of the base pairs that cause respiratory soliton excitation. Some models put forward to show the symptoms [8-11]. The amplitude of respiratory soliton from model PB will be increase by protein regulator [12].
The spesific model of interactions between DNA and protein were described in 2016 by Dwiputra *et al* and Syahroni E *et al.* [12,13]. Their model described the model of nonlinear interactions with diferent potential. Their research includes several stages of the model from the model description, stability system analysis, and spectrum energy protein binding to the DNA molecule [14]. the assumtions while the protein binding to the DNA molecule, the protein molecule treated as an object of an atom in mecanicaly stable state and well isolated from its surroundings. In this condition the protein molecule will behave like an electron bound to an atom's nucleus [15]. At last part of this research, we also describe the energy of interaction by assumsing that the energy value of system was discrete. We used WKB (Wentzel, Kramer, and Brillouin) approximation method to determine the spectrum energy [16]. WKB approximation is often used to solve problems in quantum mechanics to determine either the wave function or energy [17]. Several studies using this approach have been proposed [18, 19]. In this research, Hylleraas potential [20] will be a potential to describe the interactions between protein molecule and DNA molecules. In molecular interaction between DNA and protein, while the DNA and protein make any interaction, the vibration of any molecular occur. The stabil proses while the vibration described by the harmonic osilator potential from the Hylleraas potential, the harmonic osilator potential represented with the expantions of eksponential form. By using Hyllerass potential, we propose a model of interaction DNA with protein that is inspired from the previous model and solutions of the Schrodinger equation.

2. Model Description

The model interaction between DNA and Protein can be observed with the focal point of Hydrogen's bond atom of the both elements. In this paper, we present the model of interaction between molecules of proteins bind to single of the DNA molecular. Our models only review the interaction of these two molecules on the x-axis and the *y-axis*. It means our model has two degrees of freedom such as on showed, describing the interaction of hydrogen atoms between molecules of proteins and DNA, and y_m, which describing the interaction of the hydrogen atom of the DNA chains. We use the glutamine molecule for example of the side chain of interaction.
Reviews are based on the Valence Shell Electron Pair Repulsion (VSEPR), molecular geometry of the bond between DNA molecules with protein molecules shows that the Nitrogen atom has a number of three electron pairs, which are bonded with two atoms of hydrogen and one atom of another of the constituent molecules of adenine. While on the bond, no free electrons obtained. Based on this, anailisi geometry on these molecules form a triangle planar or flat triangle [21], with the value of each angle formed from the three shared electrons are 120°.
In our research, the potential of the hydrogen bonds is estimated by the general Hylleraas potential. To distinguish the interaction between the chains of DNA and DNA-protein binding we propose a general Hamiltonian equation represented by the following equation.

$$H = H_{DNA} + H_{prot} + H_{Int} \tag{1}$$

where the first part is DNA chain interactions containing the kinetic energy, general Hylleraas potential and coupling interaction with the following equation.

$$\mathrm{H}_{\mathrm{DNA}} = \sum_{n} \frac{p_{yn}^{2}}{2m} + \left(V_1 \left[\frac{a + e^{\alpha y_n}}{b + e^{\alpha y_n}} \right] - V_2 \left[\frac{d + e^{\alpha y_n}}{b + e^{\alpha y_n}} \right] \right) + \frac{k}{2}(y_n - y_{n-1})^2 \tag{2}$$

We assume that the mass m of nucleotide bases and momentum p_y is homogenous. The second term of the Eq. (2) is the general equation of Hylleraas potential. The general form is given by equation [22-24]:

$$V = V_1\left[\frac{a+e^{\alpha y_n}}{b+e^{\alpha y_n}}\right] - V_2\left[\frac{d+e^{\alpha y_n}}{b+e^{\alpha y_n}}\right] \quad (3)$$

The parameters for Hylleraas potential consist of a, b, d and α, with $a \neq b$ and $d \neq b$. The third term is the harmonic coupling between two neighboring base pair and parameter k is coupling parameter. If we change any parameters as $A = \frac{aV_1}{b}, P_1 = \frac{V_1}{b}, D = \frac{dV_2}{b}, P_2 = \frac{V_2}{b}$, and $P = \frac{1}{b}$, then substituting into Eq. (2), we have

$$\mathrm{H}_{\mathrm{DNA}} = \sum_n \frac{p_{yn}^2}{2m} + \left[\frac{(A-D)+(P_1-P_2)e^{\alpha y_n}}{1+Pe^{\alpha y_n}}\right] + \frac{k}{2}(y_n - y_{n-1})^2 \quad (4)$$

Whereas for Hamiltonian equation of protein is represented by

$$H_{prot} = \frac{p_{xm}^2}{2M} + \left(V_3\left[\frac{f+e^{\beta x_m}}{g+e^{\beta x_m}}\right] - V_4\left[\frac{i+e^{\beta x_m}}{g+e^{\beta x_m}}\right]\right) \quad (5)$$

With the same case, if we change any parameters as $F = \frac{fV_3}{g}, Q_1 = \frac{V_3}{g}, I = \frac{iV_4}{g}, Q_2 = \frac{V_4}{g}$ and $Q = \frac{1}{g}$ and substituting into Eq. (4), we have

$$H_{prot} = \frac{p_{xm}^2}{2M} + \left[\frac{(F-I)+(Q_1-Q_2)e^{\beta x_m}}{1+Qe^{\beta x_m}}\right] \quad (6)$$

The indices m and n is the notation of the position of the chain molecules as DNA and protein. The coupling constant χ determines the sensitive and strong interaction. The value is not specified yet, but it could be set to obtain a good value for the interaction. The last part of Hamiltonian interaction about those two bonds is represented with general form

$$H_{\text{int}} = \sum_n \chi x_m^s y_n^t \quad (7)$$

We will set and retract the value of s and t in Eq. (7). It aims to determine the simplest model from the dynamic interactions system. The value of the parameter depends on biological requirement influenced by oscillator's initial conditions [12]. If we substituting Eq. (4), Eq. (6), and Eq. (7) into Eq. (1), we obtained the Hamiltonian equation as:

$$\mathrm{H} = \frac{p_{yn}^2}{2m} + \left[\frac{(A-D)+(P_1-P_2)e^{\alpha y_n}}{1+Pe^{\alpha y_n}}\right] + \frac{k}{2}(y_n - y_{n-1})^2 + \frac{p_{xm}^2}{2M} + \left[\frac{(F-I)+(Q_1-Q_2)e^{\beta x_m}}{1+Qe^{\beta x_m}}\right] + \chi x_m^s y_n^t \quad (8)$$

By the similar expectation with the previeous research, a pulses will be triggered by a protein molecule. That mean a shift in y_n requires an initial shift from x_m. It mean that this requires an initial shift in x_m to trigger a shift in y_n, or to alter the state. However, an initial shift in DNA, y_n, should not alter the state of a rest protein (i.e. $x_m = 0$), as if there is no protein. Otherwise, the protein would not steadily bind. These specific requirements trigger the result of special condition for integers: t=1 and s > 1. This can be understood by considering the equation of motion derived from Hamiltonian,

$$m\ddot{y}_m = -\left[\frac{(P_1 - P_2 - PA + PD)\alpha e^{\alpha y_n}}{(1+Pe^{\alpha y_n})^2} + k(2y_m - y_{m-1} - y_{m+1}) + t\chi x_m^s y_m^{t-1}\right]$$

$$m\ddot{x}_m = -\left[\frac{(Q_1 - Q_2 - QF + QI)\beta e^{\beta x_m}}{(1+Qe^{\beta x_m})^2} + s\chi x_m^{s-1} y_m^t\right] \tag{9}$$

We use the Hamiltonian interaction with the simplest form $H_{\text{int}} = \frac{1}{2}x_m^2 y_n$ to get good phase portraits.

In Eq. (9), part of the third was expected not to contain variable y. So the expected conditions were accomplished in which those conditions, the presence of the protein molecules would affect the dynamics of change in the motion of the DNA and was not influenced by the adjacent DNA. In this case, the value of each parameter s and t are fulfilling that b = 1 and a> 1. Thus, to form the simplest interaction equation researchers used the equation $H_{\text{int}} = \frac{1}{2}x_m^2 y_n$.

3. The Analysis of The Interaction Stability

This section discussed the stability of a system. Stability of a nonlinear system is required to define the system in accordance with the desired conditions. Lyapunov function is one of the frequently used functions in the process of analyzing the stability of the system [25, 26]. The stability analysis is calculated within the protein and the nearest base pair. Based on [12, 25], we only need to consider the Hylleraas potentials and the interaction term. Here the harmonic coupling is neglected because it is just longitudinally transferring an amplitude. Consider the potential function

$$U(x,y) = \frac{(A-D)+(P_1-P_2)e^{\alpha y}}{1+Pe^{\alpha y}} + \frac{(F-I)+(Q_1-Q_2)e^{\beta x}}{1+Qe^{\beta x}} + \frac{1}{2}\chi x^2 y \tag{10}$$

The local minima can be determined with $\nabla U(x_0, y_0) = 0$, we obtain

$$\frac{(P_1 - P_2 - PA + PD)\alpha e^{\alpha y}}{(1+Pe^{\alpha y})^2} + \frac{1}{2}\chi x^2 = 0, \tag{11}$$

The equilibrium and stability can be determined after we set the value of each parameter. Now we will determine the system in stable or unstable. It can be done by determining the determination of Hamiltonian (det H) using hessian matrix.

$$\textbf{Det H} = \left[\frac{(P_1 - P_2 - PA + PD)(1-Pe^{\alpha y})\alpha^2 e^{\alpha y}}{(1+Pe^{\alpha y})^3}\right]\left[\frac{(Q_1 - Q_2 - QF + QI)(1-Qe^{\beta x})\beta^2 e^{\beta x}}{(1+Qe^{\beta x})^3}\right] - [\chi x]^2 \tag{12}$$

$\det(H)_{(x0,y0)} > 0$ indicates the system is stable and the $\det(H)_{(x0,y0)} < 0$ indicates the system is unstable. In this stage, we can determine the system stability in general form, but we cannot claim that the system is already fully stable. To determine the stability, we pick a, b, d, f, g, h, V1, V2, V3, V4, α, β, and χ as an example when the hydrogen bonds between the base pair are stronger than those between the DNA-protein.

4. WKB approach method

In this research, Hylleraas potential V(x) is potential that represented the interaction between DNA and protein, while the energy of protein represented by E. We used the WKB approximation to determine the spectrum energy from the interaction as equation [16]:

$$\int_a^b k(x)dx = (n + \frac{1}{2})\pi \tag{13}$$

with n=0,1,2,3,..., $k(x) = \frac{1}{\hbar}\sqrt{2m(E - V(x))}$, a and b are the classical turning point, the value of a and b can be obtained with condition $E - V(x)$=0 (a and b are the root's value of x). If we substitute Hylleraas potential to eq. (1) we have the equation to determine the spectrum energy as:

$$\int_a^b \sqrt{\left(E-\left[\frac{(F-I)+(Q_1-Q_2)e^{\beta x}}{1+Qe^{\beta x}}\right]\right)}dx=\frac{\hbar}{\sqrt{2m}}(n+\frac{1}{2})\pi \tag{14}$$

The classical turning point $a=0$ and $b=\frac{1}{\beta}\ln(\frac{E-F+I}{Q_1-Q_2-EQ})$. By substituting a and b into Eq. (14), we obtained the general equation of Energy as:

$$E_n=\frac{\left(\frac{3\pi\hbar\beta}{2}\frac{(IQ-FQ+Q_1+Q_2)}{\sqrt{2m(1+Q)}}(n+\frac{1}{2})\right)^{\frac{2}{3}}-I+F-Q_2+Q_1}{(1+Q)} \tag{15}$$

Eq. (15) was the solution to obtained the energy interaction between DNA and protein. WKB approach can only be used for one-dimensional systems or a system that has been reduced to one dimension, and the calculation is simply using algebra.

5. Result and discussion

In this section, we discuss several findings in the previous section. By setting the value of each parameter. To obtain the phase portraits of the system we used Eq (8), and the stability analysis with Eq. (10), Eq. (11), and Eq. (12). Whereas and the spectrum energy with Eq. (15).

5.1. *Phase portraits of the system and stability analysis*

We set the value of each parameters and substituting its into Eq. (8), we obtain the phase portraits of the system as illustrated in Figure 1. The values of parameter on the hydrogen bond of the DNA chain and DNA with protein are $a=0.4$, $b=0.8$, $d=0.2$, $f=0.3$, $g=0.6$, $i=0.1$, $V1=4eV$, $V2=2eV$, $V3=2eV$, $V4=1eV$, $\alpha=2$, $\beta=2$ and $\chi=1eV.m^{-3}$. After we insert the parameter, we obtain the phase portraits of the system as illustrated in Figure 1 and 3,

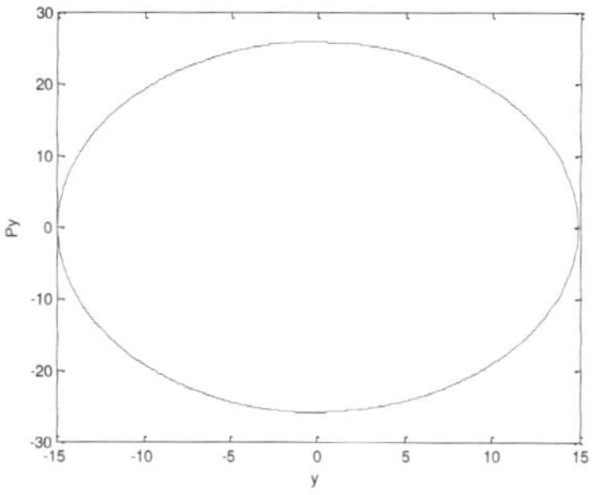

Figure 1a. Phase portraits of the systems with no initial amplitude $y_m=0$

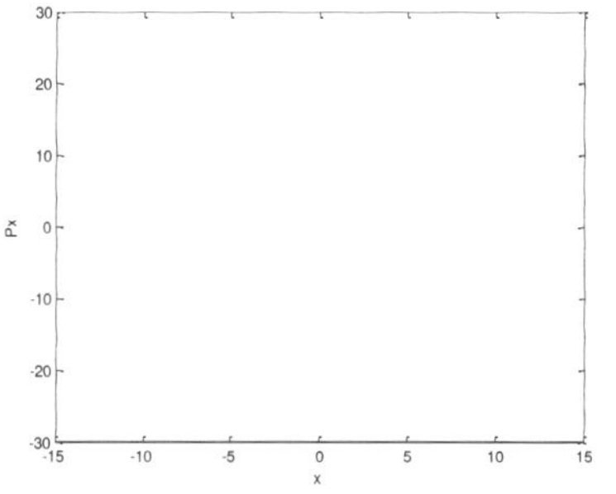

Figure 1b. Phase portraits of the systems with no initial amplitude $x_m=0$

By inspecting Fig 1a, we can be cocluded that the form of the trajectory of p_{yn} in wich y_n has no initial amplitude in x_m , this can be said that is a simple trajectory for DNA dynamic. The Fig 1b showns the trajectory ofamplitode of x_m, where the interaction of DNA with protein was not occurred yet. The momentum p_{xn} of protein can be approach as a stable state.

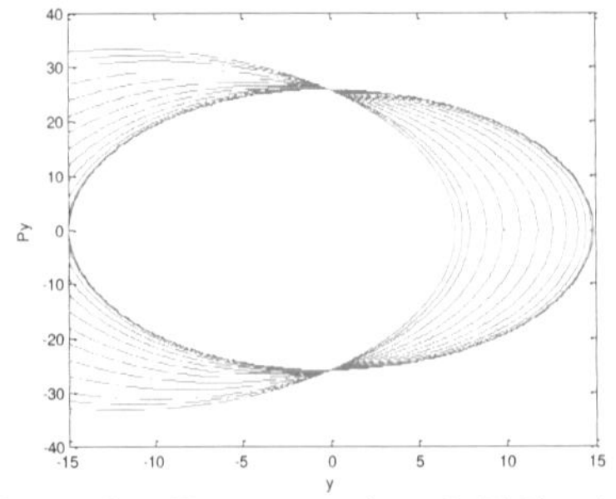

Figure 2a. Phase portraits of DNA chain system chang due to the molecule existences of a protein molecule.

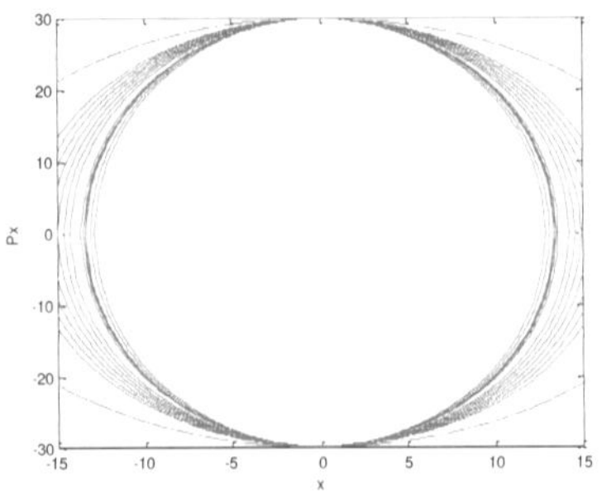

Figure 2b. Phase portraits of DNA binding with protein system changes due to the molecule existences of a protein molecule.

Figure 2a shows the change of trajectory as result of momentums p_y which is conection of y that was caused by the shifting of protein on the x direction. The change of y was affected by the shift of amplitude range in the x and also caused to the shift of center-line as well as affected the value of momentum. It's shown that the vibration on DNA chains will shift toward the negative *y-axis*.

Figure 2b shows trajectory changes as a result of the relationship between momentums p_x which is conection of x that was caused by the existing of DNA on the *y-axis*. However, the change of y is only affected to the maximum and minimum of amplitude, but it is not affected by the centre-line shifting. In accordance as the expected condition that initial shift in x_m will trigger a shift in yn, or to alter the state. However, an initial shift in DNA, y_n, should not alter the state of a resisting protein (i.e. x_m=0) as if there is no protein.

From Eq. (10) we obtain the equilibrium points: $\approx (0,0)$ with required condition and $\approx (-0.3427, 0.0033)$. After we insert the value of the equilibrium point into Eq. (11), we obtain the stability with $\approx (0,0)$ equilibrium has the determinans value (0.0012) and indicated the system at stable condition and $\approx (-0.3427, 0.0033)$ has the determinans value (-8.8286) and indicated the system at unstable condition. An unstable condition can be associated with the process of the separation of DNA molecules in strands of his bases pairs, and the separation prosses of the protein molecules from DNA-protein peptide grups [27].

5.2. Spectrum energy

In this section, we discuss several findings in the previous section of spectrum energy. We obtained some energy value with a variety of any parameters shown in the table below.

Table 1. The values of energy corresponding to several state n of molecule with the variation of V3 and V4. V3 and V4 are in eV.

n	Energy (eV)		
	V3=1 eV, V4= 5 eV	*V3=2 eV, V4= 5 eV*	*V3=2 eV, V4= 6 eV*
0	-1.2088	-0.3168	-0.8628
1	0.6583	1.6361	1.3104
2	2.1171	3.1621	3.0086
3	3.3882	4.4916	4.4881
4	4.5420	5.6984	5.8311
5	5.6126	6.8183	7.0773
6	6.6199	7.8719	8.2497
	7.5766	8.8726	9.3633

Table 2. The values of energy corresponding to several state *n* of molecule DNA under the influence of Hylleraas potential with variation of β.

N	Energy (eV)		
	$\beta=1$	$\beta=2$	$\beta=3$
0	-1.2088	-0.1934	0.6583
1	0.6583	2.7704	4.5420
2	2.1171	5.0862	4.5420
3	3.3882	7.1039	10.2205
4	4.5420	8.9354	12.6205
5	5.6126	10.6350	14.8475
6	6.6199	12.2339	16.9427
7	7.5766	13.7526	18.9327

Table 3. Energy values corresponding to several state *n* of molecule DNA under the influence of Hylleraas potential with the variation of mass. Mass is in atomic mass unit (AMU)

N	Energy (eV)		
	$m=6$	$m=12$	$m=24$
0	-1.2471	-1.7506	-2.1502
1	1.3887	0.3415	-0.4897
2	3.4482	1.9761	0.8077
3	5.2425	3.4003	1.9381
4	6.8713	4.6931	2.9642
5	8.3828	5.8927	3.9163
6	9.8047	7.0213	4.8121
7	11.1553	8.0932	5.6629

Based on Table 1, it is shown the value of energy corresponding to the condition of a particle-bound, as electrons bound to the nucleus. The energy turned toward positive corresponding with quantum state *n*. the change of potential V3 comparable with the change of energy. If we increase the value of V3, it will raise the value of energy. Meanwhile, by increasing the value of potential V4 will decrease the value of energy. The similar condition was shown in Table 2, by increasing the value of the parameter β will increase energy. We take three different value of β to prove that condition. In table 3 shown the value of energy correlation with several variations of mass. The value of bounding energy decreased. It means the bounding energy by DNA to protein was increased.

By inspecting the result of Energy bellow, all of the results have determined the condition of interaction corresponding to the condition of a particle-bound, as electrons level energy bound to the nucleus.

Based on the results of obtained the speckturm energy, some energy has a negative value. In quantum teory, these values indicate the condition of the protein molecule is still bound to the DNA molecule. The more negative the value of the energy obtained, the greater the energy needed protein for coud be separated from the DNA molecule, while the energy is positive showing protein molecule is not bound to the DNA molecule. Condition dependent DNA molecules can be associated with the condition at the time of reducing protein DNA molecule so that it can carry out its activities both replication and transcription, even doing both simultaneously [28].

6. Conclusion

By using Hyleras potential we have obtained a model of nonlinear interactions systems of DNA and protein. Based on Eq. (9), this potential can be use to show that DNA chains shift due to the presence of protein molecule, however, for the case of the DNA interation shows that the side chain of the DNA molecule will not give a significan effect to the interactios of protein. From the result, the stability of system can be obtained after we get the equations of Determinant of Hamiltonian. The Hamiltonian was treated as Lyapunov function. The stability was shown by substituting the value of each parameter into Eq. (13), phase space on Figure 1a., Figure 1b., Figure 2a., and Figure 2b. From the figure 1 and 2, we find the different change of form both trajectory before and after the interactions. Both trajectories are shifting, but a noticeable shift is apparent in changes in the trajectory of DNA occurring as a result of protein molecules.
We also determined the spectrum energy from the DNA-protein interaction. Its can be done by using the WKB approximation. The results obtained are adapted to the conditions of a molecule bound by the nucleus or condition of a particle-bound, as electrons bound to the nucleus. From the calculation of mathematically obtained the value of energy turned toward positive and its corresponding with quantum state condition.

Acknowledgments
This Research was partly supported by Higher Education Project Grant with contract no. 632/UN27.21/LT/2016

References

[1] Watson J. D., 1976 (California: W. A. Benjamin, Menlo Park).
[2] Watson J. D. and Crick F. H. C., 1953 *Nature* **171,** 737–738.
[3] Yakushevich L. V., 1998 (John Wiley, Chichester).
[4] Englander S. W., Kallenbach N. R., Heeger A. J., Krumhansl J. A. and Litwin S., 1980, Proc. Natl. Acad. Sci. (USA) **777,** 7222–7226.
[5] Bulyk M. L., Gentalen E., Lockhart D. J. and Church G. M., 1999 *Nat Biotechnol* **17** 573–577.
[6] Brivanlou A. H. and Darnell Jr J. E., 2002 *Science's Compass Review* **295,** 813-818.
[7] Kondrakhin Y., Valeev T., Sharipov R., Yevshin I., Kolpakov F. and Kel A. 2016, *EuPa Open Proteomics* **13** 14-23
[8] Ptashne M., 1985 Nature 322, 697–701.
[9] Wang J. C. and Giaever G. N., 1988 Science 240, 300–304
[10] Hogan M., Dattagupta N. and Crothers D., 1979 Nature 278, 521–524.
[11] Satari M. and Tuszy´nski J., 2002 Physical Review E 65; p. 051901.
[12] Dwiputra D. et al., 2016 J. Phys Conf. Series 694 012076
[13] Syahroni E., Suparmi A., Cari C. and Fuad A.,2016 J. Phys Conf. Series 776 012086
[14] Greiner W., 2000 (Springer-Verlag Berlin Heidelberg, New York).
[15] Peebles P. J. E.,1992 Quantum Mechanic (United Kingdom: Princeton University Press) p 17
[16] Suparmi, 2011 Mekanika Kuantum II (Surakarta : Sebelas Maret University) p 67-86
[17] Cari, Suparmi and Marini H., 2012 Indonesian Journal of Applied Physics 2 112
[18] Morsbøl J. O., Sorokin S. V. and Peake N., 2016 J. Sound and Vibration 375 162-186
[19] Cheng K. M., Leung P. T. and Pang C. S., 2003 J. Phys. A 36. 5044-5060
[20] Akaninyene D., Antia, and Eno E. Ituen, 2015 J. of physical science and aplication 5 400-406
[21] Gillespie R.J., 2008. Coordination Chemistry Reviews 252 1315–1327

[22] Akpan N.I., Maghsoodi E., Oladunjoye A. A., Zarrinkamar S., and Hassanabadi H., 2014 Quant. Phys. Lett. 3 7-14
[23] Akpan N.I., 2013 Communt. Theor. Phys. 59 268-272
[24] Hassanabadi H., Maghsoodi E., Ikpan N. I. and Zarrinkamar S., 2013 Applied Mathematics and Computation 219 9388-9398.
[25] Rink B. W. and Tuwankotta J. M., 2000 (Universiteit Utrecht. Department of Mathematics).
[26] Verhulst F., 1996 (Springer Science & Business Media).
[27] Bruce A., Alexander. J., Julian, L. Martin, R. Keith, R., and Peter W., 2008. (California: Jackie Harbor).
[28] Brivanlou, A. H. dan Darnell, Jr J. E. 2002. Science's Compass Review vol 295, hal 813-818.

IOP Conf. Series: Journal of Physics: Conf. Series **909** (2017) 012019 doi:10.1088/1742-6596/909/1/012019

Thermodynamic Functions of a Parabosonic System of Second Order

R Y A Sari[1] and W S B Dwandaru[2]

Department of Physics Education, Faculty of Mathematics and Natural Sciences,UniversitasNegeri Yogyakarta, Karangmalang Complex, 55281, Yogyakarta, Indonesia

E-mail: ryosia@uny.ac.id[1], wipsarian@uny.ac.id[2]

Abstract. This study aims to determine the thermodynamic properties of a paraboson system of order two. The thermodynamic properties to be searched include the grand canonical partition function (GCPF), Z, and the average number of particles, N. These parastatistics systems is a more general form compared to a quantum statistical distribution that has been known previously, i.e. Bose-Einstein statistics (BE). Starting from the recursion relation of the GCPF for paraboson system of order two, recursion linkages for some simple thermodynamic functions for paraboson system of order two may be derived. The recursion linkages are then used to calculate the thermodynamic function of the model system of identical particles with limited energy levels which is similar to the harmonic oscillator. The results from Z show that the thermodynamics properties of the paraboson system of order two can be derived and have a similar shape with paraboson system of order one (boson). The similarity of the graph shows similar thermodynamic properties.

1. Introduction

Given the principles of quantum mechanics, there is no necessity that the statistics of the particles must satisfy Bose-Einstein (BE) or Fermi-Dirac (FD) statistics. But both of the latter statistics have been proven through various experiments. Particles that satisfy Bose-Einstein statistics are called boson which has symmetrical wave functions under the exchange of any two particles. Examples include photons, alpha particles, and Helium atoms. On the other hand, fermions are particles that satisfy Fermi-Dirac statistics and Pauli's exclusion principle that is particles whose wave functions are anti-symmetric under the exchange of any two particles. Examples include protons, neutrons, and electrons.

Many physicists attempt to formulate a more general statistics than the above existing statistics, either by creating new types of statistics or by generalizing Bose and Fermi statistics. Some type of statistics other than Bose and Fermi statistics are already introduced, such as null statistics, ortho-fermi statistics, Hubbard statistics, and others [1]. Moreover, statistics resulted from a generalization of existing statistics include intermediate statistics, parastatistics, infinite statistics, paronic statistics, anyon statistics, and others [2].

Parastatistics which is first introduced by Green [3], is the first consistent generalization of BE and FD quantum statistics called paraboson and parafermion, respectively. Parastatistics satisfies the tri-linear commutation relation for the operator's creation and annihilation of particles and meets the principles of cluster decomposition [4]. Because of this, although there is no indication that the fundamental particles present today satisfy the parastatistics principles, this theory remains interesting for further investigation. Many physicists have attempted to obtain the corresponding physical quantities for the parastatistics system, in particular, the grand canonical partition function (GCPF). Obtaining the GCPF will reveal the thermodynamic properties of the parastatistics system, so it can serve as a basis for clarifying whether a physical system complies with these statistics or not.

Parastatistics is the first consistent generalization of the Bose and Fermi quantum statistics. In its formulation, the standard bi-linier commutation (Bose) and anti-commutation (Fermi) relationships for the particle creation and annihilation operators are

$$\left[\hat{a}_i, \hat{a}_j^\dagger\right]_q \equiv \hat{a}_i\hat{a}_j^\dagger - q\hat{a}_j^\dagger\hat{a}_i = \delta_{ij}, \tag{1}$$

with $q = +1$ for Bose statistics and $q = -1$ for Fermi statistics, which is substituted with the following tri-linear relation

$$\left[\hat{a}_i^\dagger\hat{a}_j \pm \hat{a}_j\hat{a}_i^\dagger, \hat{a}_k^\dagger\right] = \frac{2}{p}\delta_{jk}\hat{a}_i^\dagger, \tag{2}$$

with p is a positive integer which is the parastatistics order.

Hartle et al. (1970) have shown that parastatistics is a quantum statistical theory that satisfies the cluster decomposition principle. This principle states that the measurement of the physical quantities of isolated particles does not depend on the presence of other particles elsewhere far enough. Although it fulfills the cluster decomposition principle, there is no indication yet those fundamental particles present today satisfy parastatistics principles. But this theory remains interesting to investigate further. In particular, many physicists have attempted to obtain related physical quantities for the parastatistics system, in particular, the GCPF [5-9].

Consider a system of particles that do not interact with each other, with the Hamiltonian given by

$$\hat{H} = \textstyle\sum_\alpha^m E_\alpha \hat{N}_\alpha, \tag{3}$$

with E_α is the energy of a single quantum particle state $|i_\alpha\rangle$, $\hat{N}_\alpha$ is the operator that counts the number of particles in state $|i_\alpha\rangle$, and m is the number of different energy levels (may be degenerate states) denoted by $i_\alpha = 1, \cdots, m$, where m may be infinite.

The GCPF may be written as [10-12]:

$$Z(x_1, \cdots, x_m) = \mathrm{Tr}e^{\beta(\mu\hat{N} - \hat{H})}, \tag{4}$$

where $x_i = e^{\beta(\mu - E_i)}$, $\beta = 1/kT$, $T$is the absolute temperature, $k = 1{,}38 \times 10^{-23}$ J/K is the Boltzmann constant, and μ is the chemical potential. The trace covers all states in the system under consideration. GCPF for identical particle system which is invariant to particle permutation has been formulated in [7,13,14], in the form of Schur polynomials sums:

$$Z(x_1, \cdots, x_m) = \textstyle\sum_{\lambda\in\Lambda} s_\lambda(x_1, \cdots, x_m), \tag{5}$$

where Λ depends upon the type of particle statistics under consideration. Schur's polynomials, $s_\lambda(x_1, \cdots, x_m)$, is a symmetric polynomial which is defined as [15]:

$$s_\lambda(x_1, \cdots, x_m) = \begin{cases} \dfrac{\left|x_j^{m+\lambda_i - i}\right|}{\Delta(x_1, \cdots, x_m)}; \text{ for } \lambda_{m+1} = 0, \\ 0 \quad ; \quad \text{else,} \end{cases} \tag{6}$$

with $\Delta(x_1, \cdots, x_m) \equiv \prod(x_i - x_j)$is the Vandermonde polynomials. The form of Schur'spolynomials depend upon the λ parameter, which is a partition from an integer number n, that is $\lambda = (\lambda_1 \cdots \lambda_n)$ with $\lambda_i \geq \lambda_{i+1}$, and $\sum_i \lambda_i = n$. Every λ is related to Young's table which consists of boxes in left-justified rows with λ_i box on the i-th row. For example, for $n = 11$, and $\lambda = (5,3,1,1,1)$, Young's table is

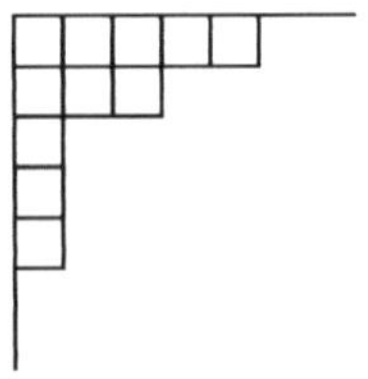

Figure 1. Young's table for $\lambda = (5,3,1,1,1)$.

For paraboson of order p, Λ in Eq. (5) consists of sets of all λ having Young's table of a maximum of p rows. For parafermi of order p, Λ consists of sets of all λ having Young's table of a maximum of p column.

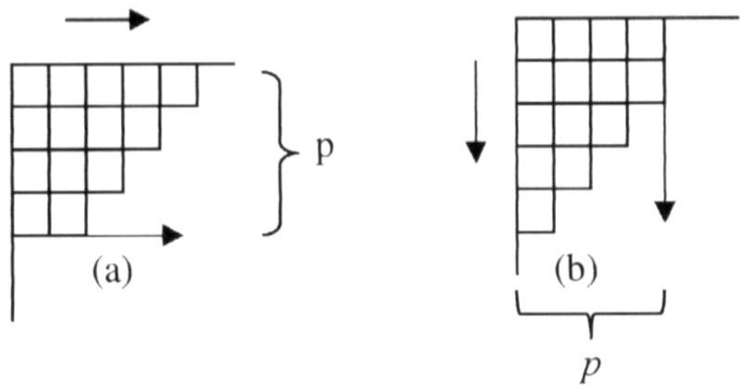

Figure 2. Young's table for $p = 4$ of paraboson (a) and parafermion (b).

GCPF for parafermion of order$p$in Eq. (5), may be simplify into a determinant of two ratios, viz. [5,14]:

$$Z^{pF}_{(p)}(x_1, \cdots, x_m) = \frac{\left|x_j^{m-i} - x_j^{m+p+i-1}\right|}{\left|x_j^{m-i} - x_j^{m+i-1}\right|}. \tag{7}$$

For paraboson of order$p$, the GCPF may also be stated as a determinant of two ratios as [8]:

$$Z^{pB}_{(p)}(x_1, \cdots, x_m) = \frac{|P_{(p)}(x_1, \cdots, x_m)|}{|P_{(0)}(x_1, \cdots, x_m)|}. \tag{8}$$

where $|P_{(p)}(x_1, \cdots, x_m)|$ is the determinant of a matrix with the elements given by

$$\left(P_{(p)}\right)_{ij} = \begin{cases} x_j^{m-i} & ;\text{for } 1 \leq i \leq p, \\ x_j^{m-i} + (-1)^{(p+1)} x_j^{m-p+i-1} & ;\text{untuk}(p+1) \leq i \leq m. \end{cases} \tag{9}$$

$Z^{pB}_{(p)}(x_1,\cdots,x_m)$ in Eq. (8) is obtained via solving a recursion relation which is derived using Schur's polynomial symmetrization formula of [16]:

$$s_\lambda(x_1,\cdots,x_m) = \sum_{i=1}^{m} s_\lambda(x_1,\cdots\hat{x}_i,\cdots,x_m)\prod_{\substack{j=1\\ j\neq i}}^{m}\frac{x_j}{(x_j-x_i)}. \quad (10)$$

Therefore, a recursion relation of parastatistic form is obtained as,

$$\begin{aligned} Z^{\text{ind}}_{(p,q)}(x_1,\cdots,x_m) &= \sum_{i=1}^{m} Z^{\text{ind}}_{(p,q)}(x_1,\cdots\hat{x}_i,\cdots,x_m)\prod_{\substack{j=1\\ j\neq i}}^{m}\frac{x_j}{(x_j-x_i)}, \\ &+x_1\cdots x_m Z^{\text{ind}}_{(p,q-1)}(x_1,\cdots,x_m) \end{aligned} \quad (11)$$

(with $\hat{x}_i$ means x_i is removed) for any λ partition with $\lambda_m = 0$. This equation is obtained from the expansion of minor determinant in Eq. (6). A simple case is exhibited when $p = m$, which previously may be simplified as [17]

$$\boldsymbol{Z^{pB}_{(p)}(x_1,\cdots,x_m)} = \frac{1}{\prod_{i=1}^{p}(1-x_i)}\frac{1}{\prod_{\substack{j=1\\ i<j}}^{p}(1-x_ix_j)}. \quad (12)$$

The above relation may be used as an initial input in the recursion relation of Eq. (11).

An energy level model which is similar to the harmonic oscillator is chosen because this system has energy levels with equal distance, making it easier to calculate. In addition, many physical systems have similar energy levels to harmonic oscillators, i.e.: $E_i = \left(i+\frac{1}{2}\right)\hbar\omega,\ i = 0,1,2,\cdots$ such as molecular vibrations, atoms in crystal lattices, and paramagnetic systems. The generalization to different energy levels can be easily made using the input values E_i, provided that the system is a non-interacting particle system.The thermodynamic quantities can be obtained from the complete canonical partition function, Z, which is obtained through its complete canonical potential,

$$\Omega(T,V,\mu) = -kTlnZ(T,V,\mu). \quad (13)$$

2. Research Method

This study is conducted in two stages, namely (i) computational method formulation and (ii) numerical casting in a computational programming language. In the first stage of computational method formulation, the study begins with the determination of boundary conditions. Then we analyze the thermodynamic properties of the parastatistics function. The next stage is numerical method casting into the computer program. Before the calculation, the numerical method is tested for its stability against the entered boundary conditions. This is important to be performed so that the result of computer calculation is not a set of data without physical meaning. Additionally, this procedure may also clarify the boundary conditions in the validity of the physical parameters that have been put into the numerics. The average number of particles can be derived as follows:

$$\boldsymbol{N(T,V,\mu)} = -\left.\frac{\partial\Omega}{\partial\mu}\right|_{T,V} = \frac{T}{Z}\left.\frac{\partial Z(x_1,\cdots,x_m)}{\partial\mu}\right|_{T}. \quad (14)$$

The form of $\boldsymbol{\partial Z/\partial\mu}$ is obtained by differentiating the recursion relation of Eq. (11) against $\boldsymbol{\mu}$ at constant temperature T and $\boldsymbol{\partial Z/\partial\mu}$ as an initial value in the recursion relation of Eq. (14) obtained from Eq. (11), that is

$$\left.\frac{\partial Z(x_1,x_2)}{\partial\mu}\right|_{T} = \boldsymbol{\beta Z(x_1,x_2)}\left[\frac{x_1}{(1-x_1)}+\frac{x_2}{(1-x_2)}+\frac{2x_1x_2}{(1-x_1x_2)}\right], \quad (15)$$

where $Z(x_1, x_2)$ is given as in Eq. (12).

3. Results and Discussion

From the computational calculations, we can obtain some graphs that can describe the properties of the thermodynamic functions of the second order paraboson model system. First of all, the area where the allowed values of μ are investigated. This is done by looking at the behavior of N which in general must be positive. Therefore, the N values arecalculated as a function of μ, which is given in the following graph.

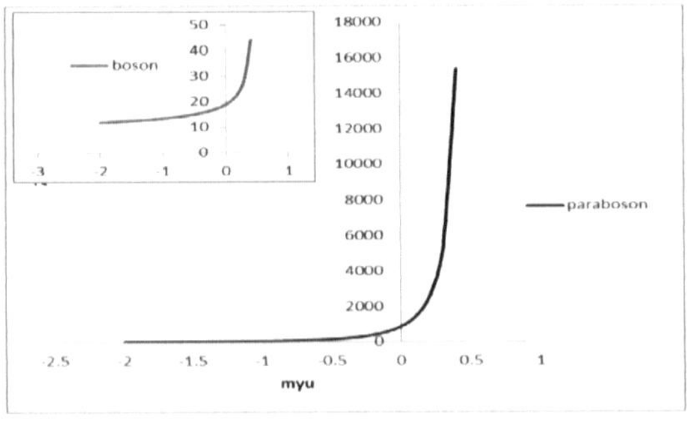

Figure 3. A complete canonical partition function, $Z(\mu)$for paraboson system. Inset: the same quantity for boson.

In Fig. 3 it appears that the complete canonical partition function, Z, for the paraboson system has the same pattern and the value of μ must also be smaller than the lowest energy level of the single particle, $\epsilon = \frac{1}{2}$, that is $\mu < \frac{1}{2}$. Moreover, the value of Z may be investigated, which in general should not be negative as well.

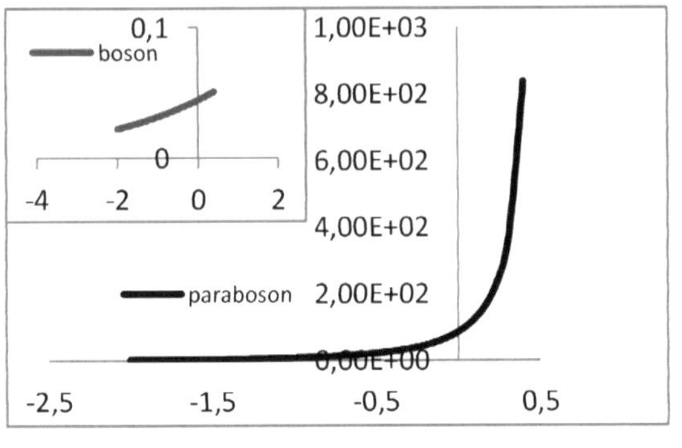

Figure 4. The average number of particles, $N(\mu)$, for paraboson system. Inset: the same quantity for boson.

Fig. 4 above shows that the value of μ is lower than the lowest ϵ. This means that the second-order paraboson system tends to absorb particles like the usual boson system, which tends to increase the number of particles into the system without losing energy (see e.g. Greiner et al. (1995)). From this property alone it can be expected that the second-order paraboson can experience condensation like a boson.

From Fig. 3 for the average number of particles, N, for paraboson and parafermion systems, the values of μ must be smaller than the lowest energy level of the single particle, $= \frac{1}{2}$, i.e. $\mu < \frac{1}{2}$. For $\mu \geq \frac{1}{2}$, the value of N can become infinitely negative (not shown in the graph).

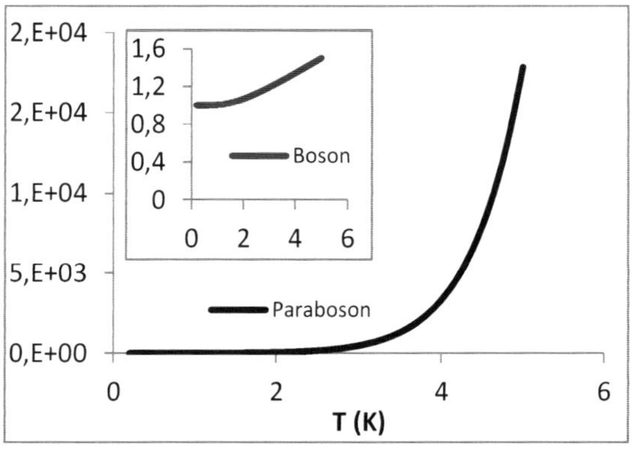

Figure 5. The complete canonical partition function, $Z(T)$, for paraboson system. Inset: the same quantity for boson.

Fig. 5 above shows that the $Z(T)$ pattern for paraboson and boson systems are similar. This means that in both of these systems the complete canonical partition function will increase along with the increase in temperature.

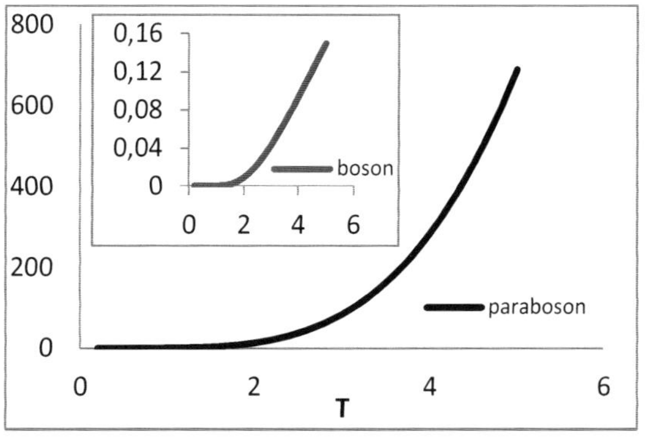

Figure 6. The average number of particles, $N(T)$, for parabosn system. Inset: the same quantity for boson.

Fig. 6 shows that as the temperature, T, increases, the value of N increases dramatically. This means that with an increase in T, the average total number of particles increases as well. The relation between N and T can be seen in Eq. (14), i.e.: $N \propto T$.

4. Conclusion

The complete canonical partition function, Z, for a paraboson system has a pattern similar to boson system. This means that the second-order paraboson system tends to absorb particles like the usual boson system, which tends to increase the number of particles into the system without losing energy. From this property alone it can be expected that the second-order para boson may experience condensation like a boson.

Acknowlegements

This research was supported by Hibah Bersaing Research provided by Ministry of Research, Technology, and Higher Education, contract number: 20/Hibah Bersaing/UN.34.21/2016, April 1, 2016.

References

[1] Mishra A K and Rajasekaran G 1995 *Pramana* **45** 91-93.
[2] Greenberg O W 1993 arXiv:cond-mat/9301002v1.
[3] Green H S 1953 *Phys. Rev.* **90** 270.
[4] Hartle J B et al. 1970 *Phys. Rev. D* **2** 1759-61.
[5] Chaturvedi S and Srinivasan V 1997 *Phys. Lett. A* **224** 249-52.
[6] Hama M et al. 1991 *Prog. Theor. Phys.* **86** 1141-48.
[7] Polychronakos A P 1996 *Nucl. Phys. B* **474** 529-539.
[8] Satriawan M 2002 *Generalized Parastatistics Systems* Dissertation University of Illinois at Chicago.
[9] Suranyi P 1990 *Phys. Rev. Lett.* **65** 2329.
[10] Huang K 1965 *Statistical Mechanics* (New York: John Wiley & Sons).
[11] Sears F W and Sallinger G L 1975 *Thermodynamics, Kinetic Theory, and Statistical Thermodynamics* (Massachusetts: Addison-Wesley).
[12] Greiner W et al. 1997 *Thermodynamics and Statistical Mechanics* (Heidelberg: Springer-Verlag).
[13] Meljanac S et al. 1996 *Partition Functions for General Multi-level Systems* (RBI-TH-06).
[14] Chaturvedi S et al. 1997 *Mod. Phys. Lett. A* **12** 1095-99.
[15] Fulton W 1991 *Young Tableaux* (New York: Cambridge University Press).
[16] Satriawan M 2004 *Physics Journal IPS Proceeding* **C8** 0515.
[17] Hama M et al. 1992 *Prog. Theor. Phys.* **88** 149-153.

Identification of atrial fibrillation using electrocardiographic RR-interval difference

M Eliana[1] and N Nuryani[1]

Department of Physics, Graduate Program, Universitas Sebelas Maret, Jl. Ir. Sutami 36A Kentingan Jebres Surakarta 57126, INDONESIA

Email : nuryani@mipa.uns.ac.id

Abstract. Automated detection of atrial fibrillation (AF) is an interesting topic. It is an account of very dangerous, not only as a trigger of embolic stroke, but it's also related to some else chronical disease. In this study, we analyse the presence of AF by determining irregularities of RR-interval.
We utilize the interval comparison to measure the degree of irregularities of RR-interval in a defined segment. The series of RR-interval is segmented with the length of 10 of them. In this study, we use interval comparison for the method. We were comparing all of the intervals there each other. Then we put the threshold to define the low difference and high difference (δ). A segment is defined as AF or Normal Sinus by the number of high δ, so we put the tolerance (β) of high δ there. We have used this method to test the 23 patients data from MIT-BIH. Using the approach and the clinical data we find accuracy, sensitivity, and specificity of 84.98%, 91.99%, and 77.85% respectively.

1. Introduction

Atrial fibrillation (AF) is an abnormality of atrial mechanical activity caused by very rapid and irregular electrical impulses that occur during atrial depolarization process [1][2]. Atrial depolarization is illustrated by P wave [2]. In the event of AF, P wave is not be visible because it replaced by the high frequency of fibrillation. [3]

AF could affect the increase in mortality and morbidity from stroke, heart failure, injury of neurology, and the greater disability. Patients with AF had five times higher stroke risk and three times higher risk of heart failure than patients without AF [4].

The most frightening complications of AF is the incidence of embolic stroke. The average annual risk is about 4,5 % in individuals with AF [1]. Atrial arrhythmias have varied proposed pathophysiology mechanism, they are one or more rapidly discharging foci, ordered reentry of short cycle length and random re-entry [1][10]. These processes lead to mechanical dysfunction with resultant thrombus formation in the complex [10].

The estimates prevalence of AF in the USA is 5.2 million, in the European is 8.8 million, in the Middle East is 4.9 million, and in the Asia is 4.2 million of the adult population [1]. In the Asian population, the described factors were: age above 80 years, history of heart disease, diminished glomerular filtration rate, and hypercholesterolemia. [11]. AF is also affects to 0,4% of the general population [8][9].

From the observational study of MONICA (Multinational Monitoring of Trend and Determinant In Cardiovascular Disease), it was found that in the urban population in Jakarta, the incidence of FA was

0.2% with the male and female ratio of 3 : 2. It also due to the significant increase of the percentage of elderly population in Indonesia by 7.74% (from 2000-2005) to 28.68% (WHO estimate of 2045-2050), the incidence of FA would also increase significantly [7].

International guidelines have classified AF as: (1) the detected AF for the first time (symptomatic or not, self-limited, or of unknown duration, or when the presence of previous episodes is unknown, being paroxysmal or persistent); (2) paroxysmal is characterized by recurrent episodes and spontaneous reversion; (3) persistent or lasting more than seven days and needing chemical or electrical cardioversion to re-establish the sinus rhythm; And (4) permanent or lasting more than one year, and refractory to cardioversion. [12]

Based on this classification, it needs continuously ECG recording to make a classification. However, based on the accuracy of some studies, at least AF monitoring performed over 72 hours [13]. This process could make it difficult for patients to be in bed for more than three days. Moreover, the physician has a big trouble to analyzing that much data. To help the physician in this job, a system of artificial intelligence who can read and diagnose whether a normal heartbeat or atrial fibrillation is required.

If we are referring to the initial definition of atrial fibrillation, irregularity of excitation, and atrial recovery, the basic of AF detection is read as to the loss of P waves as they are enclosed by Atrial Fibrillation. But in the ECG lead, there are at least two types of fibrillation that describe the characteristics of atrial fibrillation. First, coarse fibrillation. It is appearance of high-amplitude fibrillation waves, more than 1 mm. It is often seen in heart disease with large atria, the heart valve disease. Then fine fibrillation, it occurs in coronary heart disease [2][6]. Fine fibrillation is illustrated by the loss of P waves and the appearance of low-amplitude fibrillation waves, less than 1 mm. Sometimes Fibrillation waves do not appear significantly, so diagnostic enforcement is based on irregular QRS responses or arrhythmias [2][5]. In line with this statement, many researchers from biomedical use RR interval irregularity as a key feature.

There have been many investigations of AF automation detection using the degree of RR-irregularity interval. The used methods are CoSEN [15][16], RMSS, Coefficient Variance, Median absolute deviation [7], Detrended Fluctuation Analysis [16], Local Dynamic Score[16] and so on. In this study, we introduce a way to detect the irregularity of the RR interval by the interval comparison method. We were comparing all of RR-interval in the segment each other. The difference of the comparison yields a difference coefficient which can be used to classify the normality of a segment.

2. Material and Method

2.1. Data
In this study, we use data from MIT-BIH published by physionet [17]. The data are R peaks detected from 24 hours recording of 23 patients. From the presence of the R peaks, we detect the length of the overall RR intervals of the data. From the RR intervals, we get this research started.

2.2. Pre-Process
From the data, we get about millions of RR intervals. The series of RR-intervals is segmented with the length of 10 of them. This length is proportional to the 8-second recording, assuming a general heart rate period. This segmentation is almost identical to the research by Lake and Moorman in 2011[15]. They operate the method of the coefficient of sampe entropy for short segments, along 12 intervals.
Each beat has a basic description of normal or AF. In this encrypted research we attach it to each segment. Each segment has one identity description, its normal or AF. This identity is determined by considering the AF frequency of the base description on all data in that segment.

2.3. Process
In this study, we use the method we call the interval comparison operation. In the case of AF, the length of an interval is not the same as the length of the other interval.

In this study we have ten intervals in each segment. If in that segment there is a length of an interval is different from the length of nine other intervals, then we assume that the segment is AF.

The value of the difference between the first interval (I_1) and second interval (I_2) is not equal to the value of the difference between the first interval (I_1) and the 3rd interval (I_3). As described by equations 3.1 and figure 1. Furthermore, the value of the difference between the intervals we call the difference (δ).

$$\delta_{12} = |I_1 - I_2|$$
$$\delta_{12} \neq \delta_{13} \neq \cdots \neq \delta_{19} \qquad (2.1)$$

The next is also true:

$$|I_1 - I_2| \neq |I_2 - I_3| \qquad (2.2)$$

The first interval (I_1) have nine δ when operated with 9 other intervals. The second interval is also, when compared to 9 other intervals it would have 9 δ. Since the second interval (I_2) has already been operated with interval 1-on operation I_1-, then I_2 only needs to be operated with 8 intervals thereafter. Based on this description we can conclude that I_1 has 9 δ, I_2 has 8 δ, I_3 has 7 δ , etc. So, from all operations, overall we get as much as 45 δ in each segment.

I_1 = first interval
I_2 = second interval
I_n = n[th] interval
$\sigma = Threshold$
$\beta = Tolerance$

Input $I_1, I_2, I_3, \ldots, I_{10}$.

$\delta_{n,n\neq n} = |I_n - I_{n\neq n}|$

$high\ \delta = find\ (diff_{n,n\neq n} > \sigma)$

If number of $\delta > \sigma$

segment = AF

else segment = normal

Figure 1. Pseudocode of algorithm intervals comparison.

Back to the problem that one of the AF indications is the presence of arrhythmias or irregularities of RR-Interval on ECG recordings. Easily we can say that if a segment is a normal segment, then each of the 45 δ has a value of 0. That is, there is no difference in length at all between an interval with nine other intervals. But in practice, the difference of the intervals in the normal segment is not absolutely zero. Each interval still has differences with each other with small values. We assume that if a segment is called an arrhythmia when the number of difference is many or large. The number of difference can be said to be large or small when we have placed a threshold between the two.

In this study we tried to put the threshold values several times, i.e., 11, 13, 15, and so on up to 31. We find that value is from manual observation of some segment of AF and some segment of normal. Furthermore, we call the threshold (σ) for them. And remember that we call δ for the value of different.

Furthermore, the meaning of threshold selection is if the difference between intervals (δ) is smaller than σ then the δ is ignored, and the δ of interval is considered as a low δ. Conversely if the value of δ more than σ, then we assume that the δ is a high δ. To determine a normal or fibrillated segment then we calculate the number of high δ. If there is more higher δ than the tolerance β then we consider the segment is AF.

In this research, we put tolerance four times. It is 0 of high δ, 3 of high δ, 6 of high δ, and 10 of high δ.

3. Result

We tested this method on 23 patients. The smallest interval difference we get is 0, the meaning is that the interval with the other has the same length. We not only compare an interval with the closest interval, but also compare it with some intervals thereafter. Each interval in a segment relates to each other, to form a δ value. The method to calculate this δ is results more fluctuation of data. So, it gives us more opportunity to try the advanced method.

In this study, AF is defined as a state when the number of high δ in the segment is more than the number of tolerances β. In the process of threshold σ modification and tolerance β modification we get a variety of data as described in table 1.

Each modification of the maximum tolerance of the threshold value is also subject to change, as described in graphs in figure 2.

Based on these results, in this study, we determined the best threshold σ is 23, and the best tolerance β is 3. The results we get in the process are accuracy, sensitivity, and specificity of 84.98%, 91.99%, and 77.85%, respectively.

Table 1. Table of accuracy with the variance of threshold (σ) and tolerance (β).

σ	*Accuracy (%)*			
	β=0	**β =3**	**β =6**	**β =10**
11	78.43	79.44	80.50	83.13
13	80.70	81.47	82.26	84.18
15	82.09	82.70	84.32	84.77
17	83.05	83.54	84.45	84.95
19	83.67	84.07	84.44	84.94
21	84.03	84.33	84.36	84.72
23	84.25	84.98	84.16	84.41
25	84.38	84.50	83.97	84.00
27	84.41	84.42	83.32	83.53
29	84.38	84.30	83.92	83.00
31	84.34	84.17	83.59	82.37

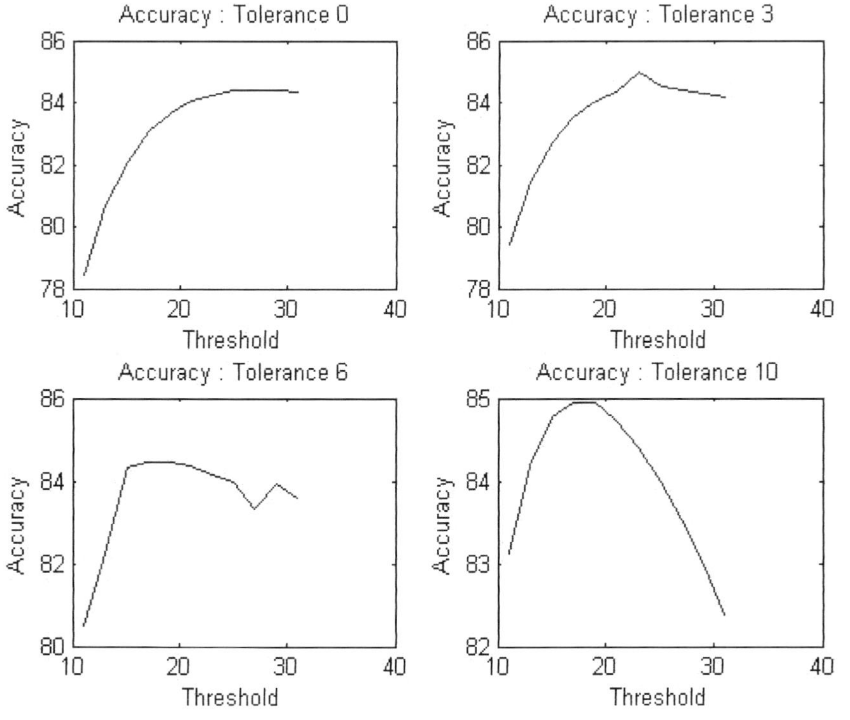

Figure 2. Accuracy with Variation of Tolerance

4. Discussion

After all of the data is segmented, then we operate the interval comparison method to each segment. Since each segment produces 45 δ, then, overall we have about 47.456.640 δ. The smallest δ of them is 0, and the biggest one is about 1638.

To determine a δ including low or high δ we place a threshold. In this experiment, we modified several times of thresholds ranging from 15 to 31. Then to determine a segment to be AF or normal, in each trial threshold, we test it on some variations of high δ tolerance β. In this experiment, at least we tried four times the number of tolerance, 0, 3, 6, and 10.

For the tolerance number 0, when the threshold is rise, the accuracy rise as well until the threshold reaches 27. After that, if the value of threshold is increased, then the accuracy is down. Such trends is also apply to tolerance numbers 3, 6, and 10. The accuracy is gets higher as the threshold rises, then stops at a certain point, then accuracy falls back. The stop point, or the point when we get the best accuracy we call the maximum threshold point. It is described by Figure 2.

For the tolerance number 0, the best threshold is 27. For the tolerance number 3, the best threshold is 23. For the tolerance number 6, the best threshold is 17. For the tolerance number 10, the best threshold is 17.

Based on these data, after the 3 (6 or 10) tolerance number, the trend of accuracy is getting down. This trend is due to the high tolerance, precisely some segments that actually AF cannot be recognized by the program. This means the program considers the segment as normal. So, the answer of the program is wrong. Hence, the accuracy is low.

The data presented by MIT-BIH is not only consisting of Normal Sinus and AF, but there is also Fluter atrium (AFL) and AV Junction (J) on the recording of some patients. Because AF and normal use binary classification, we have to make a decision whether AFL and J are included in the normal or not.

Knight (1999) [14] revealed that medically the Fluter atrium has regular RR intervals. Based on the statement, the identity of AFL segment should be zero, of course, it's because the value of the δ must be low. So, with this method, the AFL segment should be detected as a normal segment. However, our results show that the AFL segment belongs to the fibrillated segment.

It is the same as what Lake's done with his crew (2011) [15]. That they were evaluated this measure with a clinician's eye. Especially regarding the atrial flutter (AFL) ECG recordings. This rhythm of RR intervals is very regular when in a stable conduction, but is clinically treated much like AF. Thus, they were categorized AFL as AF, even though its regular. This regularity is make an entropy of RR interval time series are low. It's very different from irregular RR interval. It has high-entropy AF. Throughout, they were analyzed AFL and AF together. Although the AFL algorithm would improve, if they excluded the AFL.

5. Conclusion

We have conducted a study of atrial fibrillation detection using the difference of electrocardiographic RR-intervals. The difference is obtained by comparing all intervals in a segment each other. We varied the threshold of the difference and tolerance. Using the approach and the clinical data we find accuracy, sensitivity, and specificity of 84.98%, 91.99%, and 77.85%, respectively. Thus, it implies that the RR interval difference is quite effective to detect the atrial fibrillation.

References

[1] Rahman, F. and Benjamin, E.J. 2015. *Classification and Epidemiology of Atrial Fibrillation Classification.* Page 9. In Estes, N.A.M. (ed), *Atrial Fibrillation : A Multidisciplinary Approach to Improving Patient Outcomes*. Minneapolis : Cardiotext Publishing, LLC.

[2] Das, M.K., and Zipes, D.P. 2012. *Electrocardiography of Arrhythmias : A Comprehensive Review.* Philadelphia : Elsevier Saunders.

[3] Yap, Y.G. and Camm, A.J. 2014. *Essential of Atrial Fibrillation*. London : Springer Healthcare.

[4] Shenasa, M. and Camm, A.J. 2015. *Management of Atrial Fibrillation : A Practical Approach.* New York : Oxford University Press.

[5] Munawar, M., and Sutandar, H. 2003. *Elektrokardiografi*. Page : 41-64. in Tjokronegoro, A. (edt), *Buku Ajar Kardiologi.* Jakarta : Gaya Baru.

[6] Trisnohadi, H.B. 2009. *Gangguan Jantung yang Spesifik*. Page : 1605-1612. In Sudoyo, A.W. (edt). *Buku Ajar Ilmu Penyakit Dalam*. Pusat Penerbitan Departemen Ilmu Penyakit Dalam Fakultas Kedokteran Universitas Indonesia, Jakarta.

[7] Yuniadi, Y. *et al.* 2014. Pedoman Tata Laksana Fibrilasi Atrium. Perhimpunan Dokter Spesialis Kardiovaskular Indonesia. Jakarta : Centra Communication.

[8] Haralabos P., Ramesh B.C., and Al-Alao B. 2014. Surgery for Atrial Fibrillation: An Overview Chapter from the Book Cardiac Arrhythmias-Mechanism, Pathophysiology, and Treatment. Cardiothoracic Department, Royal Victoria Hospital, Belfast, UK.

[9] Kannel W.B., Abbott RD, Savage DD, Mc Namara PM. *Coronary hearth disease and atrial fibrilation: the Framingham Study.* Am Hearth J 1983; Vol 106: 389-96.

[10] Khan, M. and Daniel J.Miller. 2013. Detection of Paroxysmal Atrial Fibrillation in Stroke/Tia Patients. Department of Neurology, University of Massachusetts, USA. Hindawi Publishing Corporation. Stroke Research and Treatment.Volume 2013, Article ID 840265

[11] Iguchi Y., Kimura K, Shibazaki K, *et al.* Annual incidence of atrial fibrillation and related factors in adults. *Am J Cardiol*. 2010; 106: 1129–1133.

[12] Fuster V, Rydén LE, Cannon DS, *et al.* ACC/AHA/ESC 2006 Guidelines for the Management of Patients with Atrial Fibrillation.A Report of the American College of Cardiology/American Heart Association Task Force on Practice Guidelines and the European Society of Cardiology Committee for Practice Guidelines (Writing Committee to Revise the 2001 Guidelines for the Management of Patients With Atrial Fibrillation). *J Am Coll Cardiol* 2006; 48: 854–906.

[13] Hariri, E., Hachem, A., Sarkis, G., dan Nasr, S. 2016. Optimal Duration of Monitoring for Atrial Fibrillation in Cryptogenic Stroke : A Nonsystematic Review. *BioMed Research International* of *Lebanese American University*. Volume 2016, Article ID 5704963.

[14] Knight BP, Michaud GF, et al. 1999. Electrocardiographic Differentiation of Atrial Flutter from Atrial Fibrillation by Physicians. J Electrocardiol 1999;32:315-9.

[15] Lake D.E. and Moorman J.R. 2011. Accurate Estimation of Entropy in very Short Physioloical Time Series: the problem of atrial fibrillation detectionin implanted ventricular devices. Am. J. Physiol. Hearth Circ. Physiol. 300 H319-25.

[16] M. Carrara, L. Carozzi, T. J. Moss, M. de Pasquale, S. Cerutti, M. Ferrario, D. E. Lake, and J. R. Moorman. 2015. "Heart rate dynamics distinguish among atrial fibrillation, normal sinus rhythm, and sinus rhythm with frequent ectopy, "PhySiol. Meas., vol. 36, no. 9, p. 2873, 2015.

[17] Goldberger AL, Amaral LAN, Glass L, Hausdorff JM, Ivanov PCh, Mark RG, Mietus JE, Moody GB, Peng C-K, Stanley HE. PhysioBank, PhysioToolkit, and PhysioNet: Components of a New Research Resource for Complex Physiologic Signals. Circulation 101(23):e215-e220 [Circulation Electronic Pages; http://circ.ahajournals.org /content/101/23/ e215.full]; 2000 (June 13).

IOP Conf. Series: Journal of Physics: Conf. Series **909** (2017) 012021 doi:10.1088/1742-6596/909/1/012021

Study of the Variation of Material layer Compotition and Thickness Related Neutron Flux and Gamma Radiation

Yuliana Dian Nirmalasari[1], Suparmi[1], Y. Sardjono[2]

[1]Physics Department of Graduate Program Sebelas Maret University, Jl. Ir. Sutami 36A Kentingan Jebres Surakarta 57126, INDONESIA

[2]PSTA-Batan Yogyakarta

E-mail: yulians.dn@gmail.com

Abstract. Optimation of simulation design of collimator is corresponding to 30 MeV cyclotron generator. The simulation has used the variation of the thickness materials layers that was applied at treatment room's door. The purpose of the variation and thickness of the material in this simulation to obtain optimum results for the shielding design in the irradiation chamber. The layers that we used are Pb-Fe and Pb-SS312. Simulation on cancer treatment is used with monte carlo simaulation MCNPX. The spesifications that we used for cyclotron is the spesification of the HM-30 Proton Cyclotron from Sumitomo Heavy Industries Ltd. The variation of the thickness materials layers that was applied at treatment room's door are Pb remains 4cm while Fe and SS312 varies between 2 cm, 4 cm, 6 cm respectively. This simulation of Fe layer on Pb was give good result in measurement simulation at 4 cm thickness.

1. Introduction

Radiation in its development has been widely applied in various fields, one of which is in the field of health as in the pilot plant Boron Neutron Capture Therapy (BNCT). BNCT is a method of cancer therapy which is the development of long-established NCT. BNCT is a combination method of chemotherapy and radiotherapy that was first tested in 1951 [1]. Radiation is the emission energy generated by an unstable transformation of atoms or nuclei. Radiation of energy emitted by radiation sources, can cause changes in physical, chemical and biological properties in its path [2]. The spread of radiative energy can occur through materials and spaces in the form of electromagnetic waves or particles. Radiation of charged particles can ionize the material either directly or indirectly because the radiated energy of charged particles exceeds the ionise potential energy material [3 - 5].

It is well known that nuclear reactions are divided into several types depending on the type of bombarded particles. The following reactions may occur: by the action of neutrons n, proton p, deuteron d, heavier particles and α particles, and γ-rays [6]. Photons (energy above some MeV) that can stimulate the nucleus and (γ, n) interactions can cause radiation protection problems. Photonuclear interactions do not depend on dosimetry. The power of photon radiation increases as the plasma is rich in baryon [7]. Photon interacts with the material (as an x-ray beam or gamma radiation that passes through a material) and will be completely absorbed by depositing its energy, it will penetrate the material part without any interaction, or will interact and spread or bend from the original. In this research used lead (Pb) because Pb can hold radiation well and used variation of Ferit (Fe) and 312 of Stainless Stell type (SS312) that coating material because this material is considered easy to have economic value.

2. Material and Methode

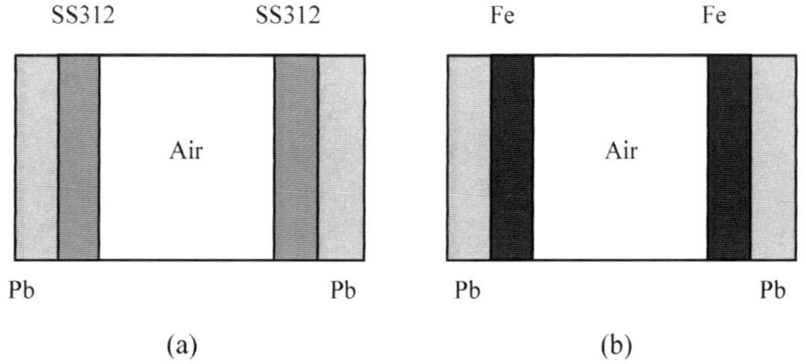

Figure 1. Treatmen room's door materials (a). With SS312 layers, (b). With Fe layers

The design optimization of the collimator simulation corresponds to the 30 MeV cyclotron generator and 1 mA proton current. Simulations have used a variety of layers of thickness materials applied to the treatment room door. The layers we use are Pb-Fe and Pb-SS312. Simulations on cancer treatment are used with the monte carlo version X (MCNPX) simaulation code. Monte Carlo simulation is a very helpful tool for investigating complex radiation fields [8]. The specifications we use for cyclotron are the specifications of HM-30 Proton Cyclotron from Sumitomo Heavy Industries Ltd. The thickness of the thickness material applied to the treatment room door is fixed Pb 4 cm for each segment, this thickness was given better absorb for neutron flux whereas Fe and SS312 vary between 2 cm with the air segmen is 242 cm, 4 cm with the air segmen is 238 cm and 6 cm with the air segmen is 236 cm for each segment layer. The neutron flux and gamma radiation were tallied using F4:n and F4:p, while the dose rate was tallied with the DE and DF tallies.

3. Result and Discusion

Figure 2. discribe the spectrum energy that producd by the simulation with Alumunium (Al) moderator and its can be seen, the neutrons flux that produced by HM-30 almost in epithermal energy range with the maximum energy of 3,8 MeV.

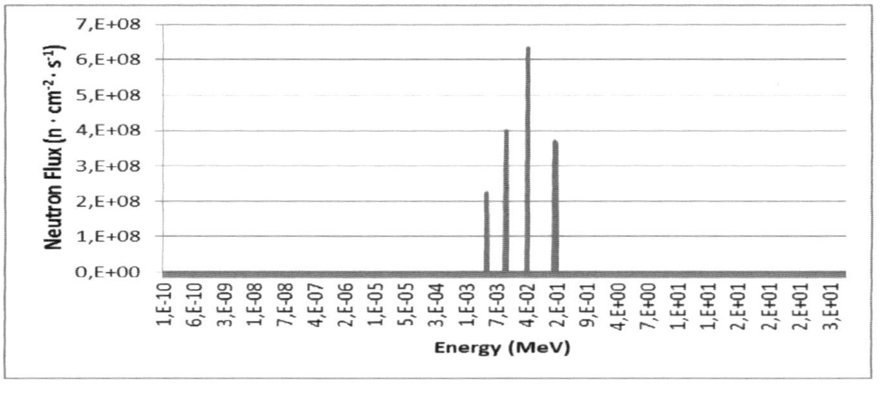

Figure 2. Neutron energy spectrum produced by 1 mA 30 MeV

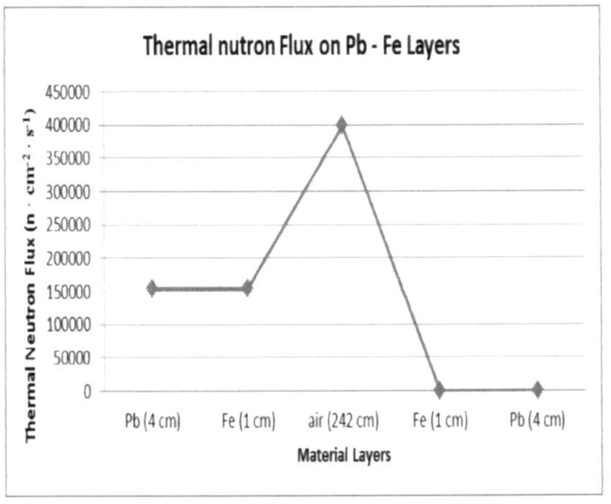

(a)

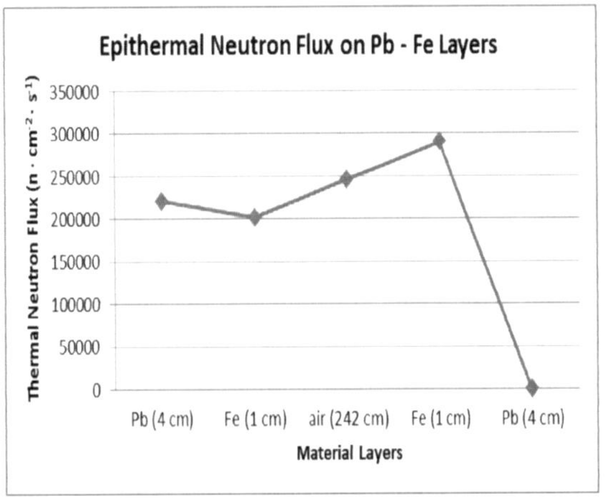

(b)

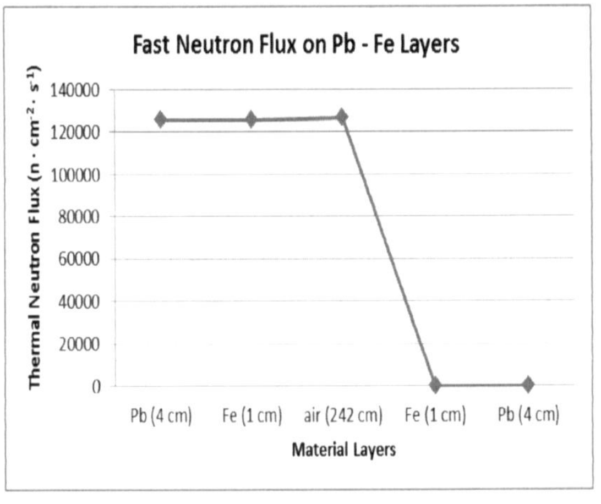

(c)

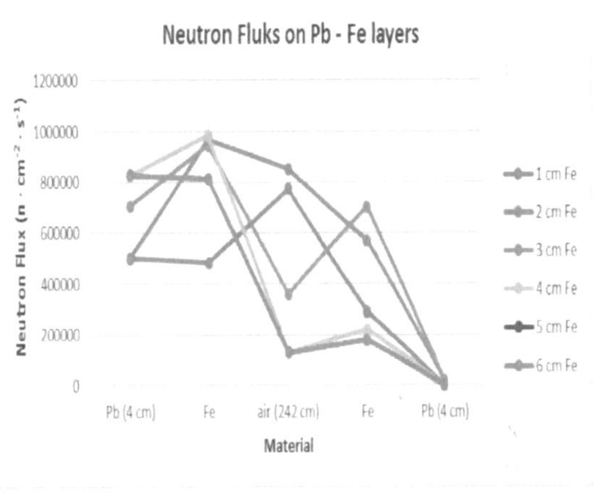

(d)

Figure 3. (a), (b), and (c) Neutron flux on energy range in Pb – Fe material layers, (d) total neutron fluxes on Pb – Fe layers

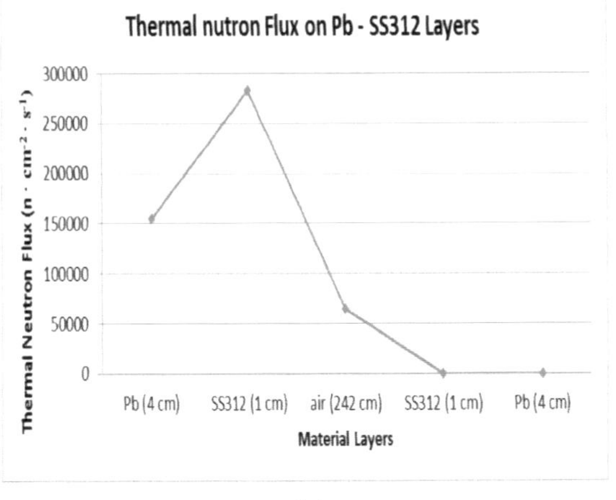

(a)

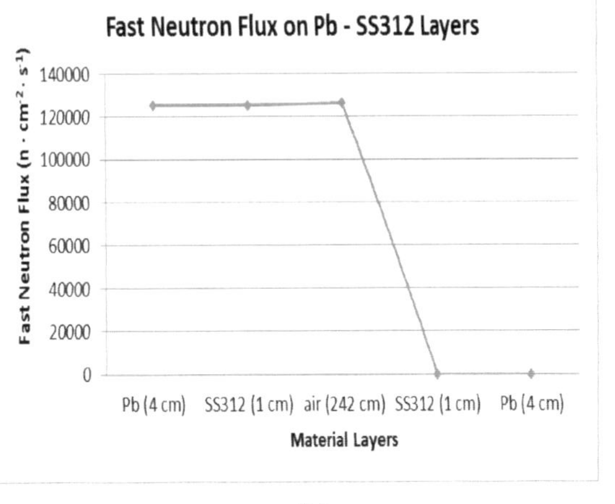

(b)

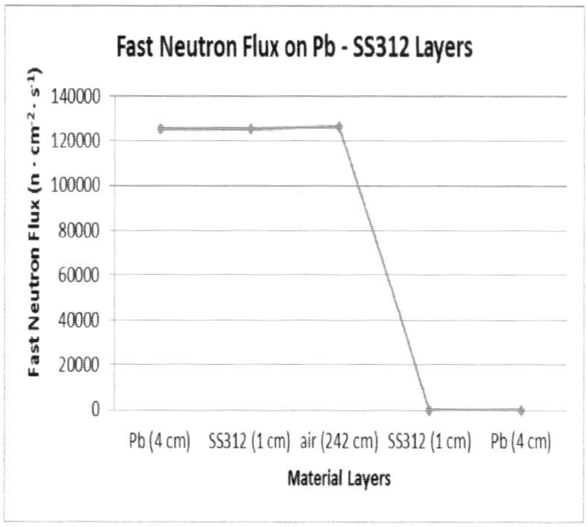

(c)

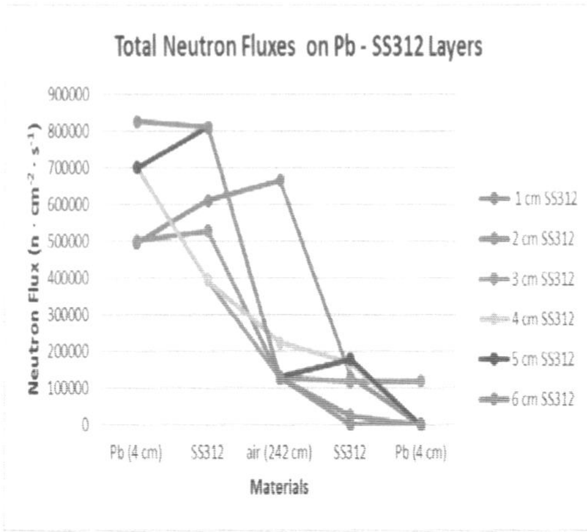

(d)

Figure 4. (a), (b), and (c) Neutron flux on energy range in Pb – SS312 material layers, (d) total neutron fluxes on Pb – SS312 layers

The Neutron flux passing through the Pb-Fe door averages a significant decrease as it reaches the Pb door in the second segment (Fe-Pb). On the other hand the value of neutron flux that passes from this material is zero.

From the comparison the figure 2 and figure 3, it can be concluded that the SS312 layers can absorb neutrons better than denan using Fe layers. With 4 cm SS321 neutron layer that reaches the second Pb layer has a value of zero while using a 4 cm Fe layer there is still a neutron escaped that is 2.03484 x 10^4 n. cm^{-2} s^{-1}. For neutron flux, which is detected through the majority Pb layer is a slow neutron.

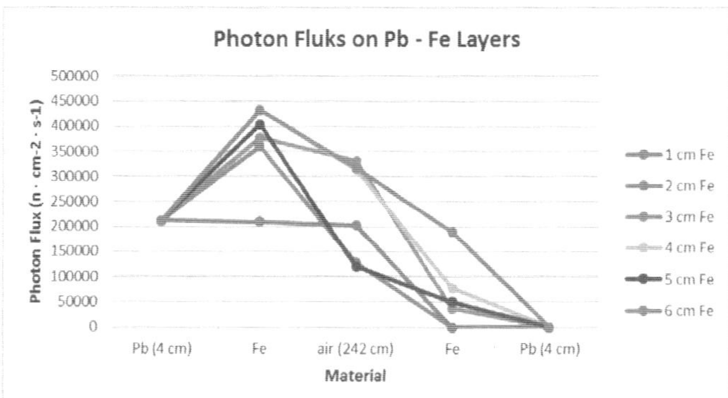

Figure 5. Total photon Flux for Pb – Fe layers with Fe variation of thickness

IOP Conf. Series: Journal of Physics: Conf. Series **909** (2017) 012021 doi:10.1088/1742-6596/909/1/012021

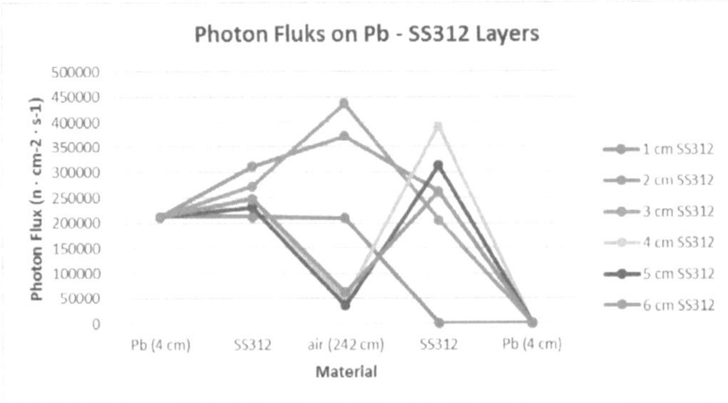

Figure 6. Total photon Flux for Pb – SS312 layers with Fe variation of thickness

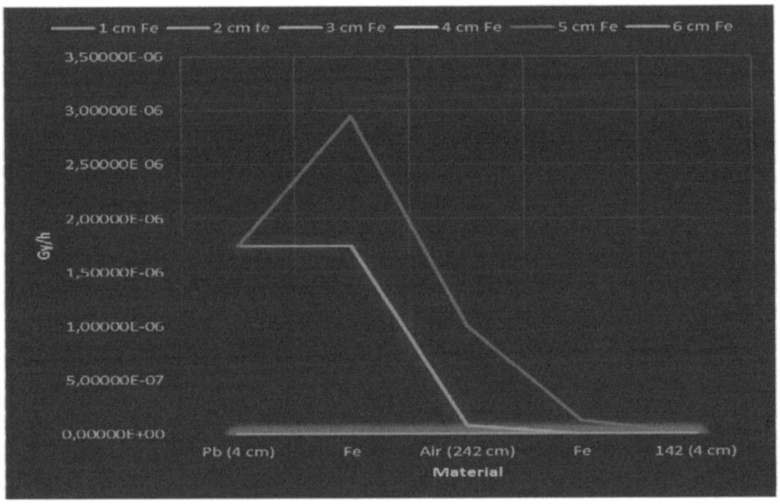

Figure 7. Gamma dose of Pb – Fe layers, for 2,3 and 6 cm Thickness of Fe, the line are looks stacked

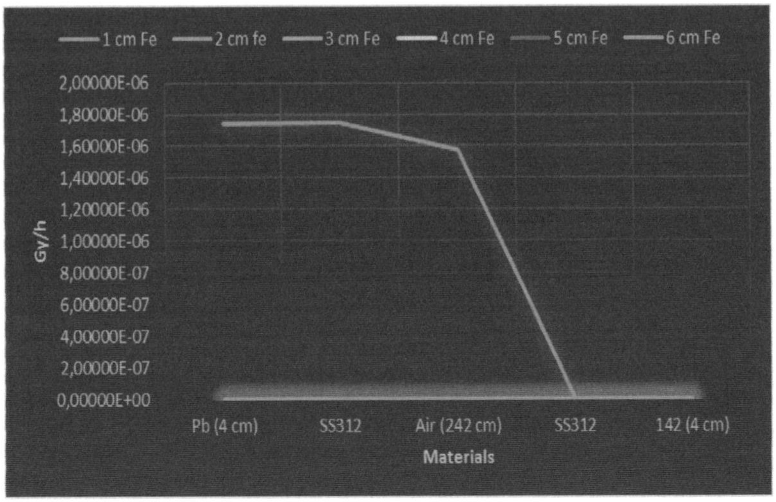

Figure 8. Gamma dose of Pb – Fe layers, for 2 - 6 cm Thickness of Fe, the line are looks stacked

For Photon flux, it can be absorbed on coatings both for SS312 and for Fe, but from the simulation results, the comparison of these two coating materials is best shown by the Fe coating. A review is required for this flux photon. Associated with the gamma dose rate emitted at the time of irradiation, with this coated Pb material obtained, the best value of this simulation is with Fe layer is on the first layer of Pb (6 cm of Fe) is 5,12862 x 10^{-22} Gy/h and on the second layer of Fe is 3,40739 x 10^{-22} Gy/h, while the gamma dose rate on the second of Pb layers is zero. However, it is necessary to re-examine gamma absorbing doses that escape from other shielding parts such as concrete walls and gamma filters in the moderator section to reduce the risk for radiation workers outside the room.

4. Conclusion

From this simulation that has been done can be concluded that the layer material Pb door treatment room obtained that Fe coated Pb has been able to reduce neutron flux and photon flux. However, for gamma-rate doses emitted in the therapy room it is necessary to review again even if the gamma dose can also be absorbed so that no gamma escapes. Then the use of Fe to coat Pb can be applied.

Acknowledgments
Thanks to The Head of PSTA, which has given permition for condacting reaserch and also to Prof. Dra. A. Suparmi, M.A, Ph.D and Prof. Ir. Yohannes Sardjono for supervising in the reasech.

References

[1]. W. Sauerwein, A. Witting, R. Moss, and Y. Nakagawa, 2012, *Neutron Chapture Therapy*, Spreanger, New York
[2]. Filano R., Hidayanto E., Arifin Z., 2014, *Youngster Physics Journal*, Vol. 3. Pp. 317-328
[3]. E.B. Prodgorsak, 2005, *Radiation Oncology Physics: A Handbook for Teacher and Student*, International Atomic Energy Agency, Vienna
[4]. International Atomic Energy Agency, 2001, IAEA-TECDOC-1223.Vienna
[5]. Pusat Pendidikan dan Pelatihan Badan Tenaga Nuklir Nasional, 2004, *Dokumen Proteksi Radiasi*, BadanTenaga Nuklir Nasional.
[6]. Boris I.K., Oxana V.K, and Osbaldo Ortiz Méndez, 2013, *Radiation Synthesis of Materials and Compounds*, CRC Press, New York.
[7]. Y. Kumar, A Goel and P Jain, 2015, Journal of Physics: Conference Series 636, 012018
[8]. M.G. Pia, T. Basaglia, Z. W. Bell, P. V. Dressendorfer, 2010, *Joint International Confrence on Supercomputing in Nuclear Applications and Monte Carlo* 2010 (SNA+ MC2010) Hitotsubashi Memorial Hall, Tokyo, Japan, October17-21

Diffusion processes for planar waveguides fabrication in soda-lime glasses

Muchlas Yulianto[1], **Ahmad Marzuki**[1], **and Venty Suryanti**[2]

[1]Physics Department of Graduate Program, Sebelas Maret University
Jl. Ir. Sutami 36A Kentingan, Surakarta 57126, Indonesia.
[2]Department of Chemistry, Faculty of Mathematics and Natural Science Sebelas Maret University Jl. Ir. Sutami 36A Kentingan, Surakarta 57126, Indonesia.

E-mail: muchlasyulianto@gmail.com

Abstract. This paper reports about the fabrication of planar waveguides made of the ion-exchanged technique on a soda-lime glass. The process of the ion-exchanged from molten salt $AgNO_3$+KNO_3. Fabrication is conducted at 70 mol% of KNO_3 and 30 mol% of an $AgNO_3$ melt takes place for the realization of planar waveguides. The duration of ion-exchanged has been varied in the range of 25-900 minutes at the temperature of 315°C and 350°C to diffusion processes. The purpose of our experiments is especially to study the effect temperature and duration of the ion-exchange process to the main parameters used to characterized optical waveguides. The refractive index change, the number of modes waveguides, and diffusion depth had been investigated. The *m*-lines spectroscopy with standard prism-coupling technique at 632.8 nm was used to measure the number of modes and depth of waveguides. Refractive index was measured with refractometer ABBE. The Ocean Optics USB4000 Spectrometer is used to obtain absorption spectra. The values of penetration depth from glass surface increased with increasing ion-exchange temperature and time.

1. Introduction

Planar waveguide is a device that has a geometric shape structure which light are guided only on one dimension. Fabrication of planar waveguides using thin transparent glass with refractive index on the substrate, or that is between two layers of substrate. There are several planar waveguide fabrication techniques that researchers often use such as sputtering, chemical vapor deposition (CVD), sol gel coating, ion-exchange, and ion implantation [1]. Among these techniques, an ion-exchange method is the most popular technique for fabricating planar waveguides. The ion-exchange method is based ion displacement on a glass substrate with different ions in a molten salt bath. Some materials are used for integrated optical fabrication such as glass, dielectrics, organic materials, semiconductors, silica based materials [2]. For obtaining optical waveguides, glass is an excellent material as a planar substrate in which to fabricate waveguides. Due to its compatibility with optical fibers in material properties, the fabrication process are very simple and low cost [3]. In this paper, the planar waveguides of soda-lime glass are prepared by silver ions from $AgNO_3$ and potassium ions from KNO_3 has been used to produce localized variations in refractive index. The diffusion processes of waveguides fabricated at different temperatures and varying time of diffusion. Planar waveguides are characterized by their optical properties. There are absorption spectra, a number of modes, refractive index change, and penetration depth.

2. Experimental

The fabrication of planar waveguides involved the sample preparations and diffusion processes. The soda-lime glass was chosen due to its high optical homogeneity, specifically the nitrate salts. Ions from the glass diffused out of the sample, while ions from the source diffused into the sample. Moreover, the transition temperature of the glass was also higher than the temperature required for the ion-exchange processes [4].

2.1. Sample preparation

The substrates used a commercially sail brand microscope slide which has dimensions 76.2 mm × 25.4 mm × 1 mm. The chemical composition of the glass was specified (wt.%) SiO_2 73.1%, Na_2O 13.8%, CaO 8.6%, MgO 4.0%, SO_3 0.3%, Al_2O_3 0.1%, and Fe_2O_3 0.1%. The glass surface was cut into dimensions of 20 mm x 20 mm. After that, cleaning was realized using ethanol and aquades in an ultrasonic cleaner for 15 minutes to remove any contaminants.

2.2. Diffusion processes

The first crucible containing 30% mol $AgNO_3$ and 70% mol KNO_3 were used as starting materials. The temperature of the treatment 315°C and 350°C was set using an electric furnace. Diffusion processes were carried out for the time intervals duration ranging from 25 to 900 minutes. In the crucible Ag^+/K^+ ions were driven into the glass substrate from the mixed salt bath. After diffusion processes, the glasses were pulled out of the salt mixture in a crucible and put at room temperature. Furthermore glass substrate washed using ethanol liquid and aquades by cleaning in an ultrasonic cleaner for 15 minutes to remove any residues from the surface and dried in the air. Finally the glass changed to transparent and orange caused by this process continues for different varying time of diffusion into the soda-lime glass.

2.3. Characterization methods

The characterization obtained to determine the optical properties of the waveguide and the practical use of this component in integrated optical circuits (IOC). The ion-exchanged planar waveguides were characterized by the absorption spectra, refractive index profile and a number of wave modes. The absorption by Ocean Optics USB4000 Spectrometer. The refractive index is one of the most important properties of an optical waveguide. Refractive index profile obtained using a refractometer ABBE. The *m*-line with standard prism-coupling technique at 632.8 nm has been used for the measurement of the penetration depth of waveguide glass [5].

2.3.1. Number of modes measurements

Measurements number of wave modes were made by the *m*-line spectroscopy technique. The incident on a prism whose refractive index (n_f) as the core material higher than substrate index (n_s) and cover film index (n_c) as cladding. The polarization of the laser controlled by a polarizer.

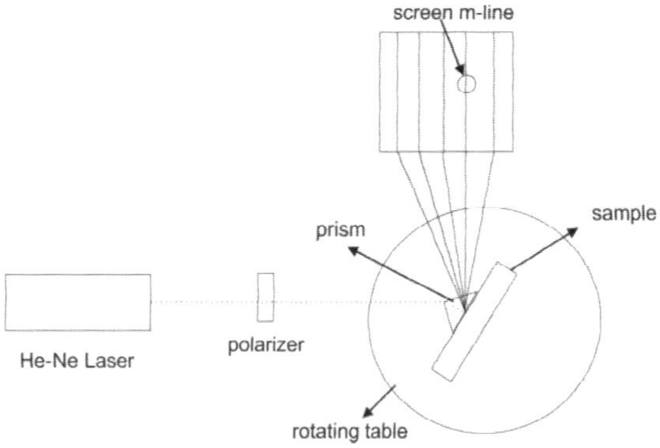

Figure 1. Schematic of the prism coupling method with in *m*-line.

Figure 1 shows schematic of the experimental prism set-up used to modes planar waveguides obtained. Experimental set-up He-Ne laser, a polarizer, a prism, planar waveguide, and substrate [6]. The light source of He-Ne laser with wavelength 632.8 nm is passed through a polarizer with prism is placed on a spectrometer rotation stage [7]. Then light enters into the prism and will be refracted first into a substrate and reflected. A prism which has the refractive index (n_p) higher than the refractive index substrate materials (n_s), or cover material (n_c) [8]. Finally the excitation of guided modes is realized throughout the evanescent waves created in the air gap between the prism and guiding layer on the black screen.

3. Results and Discussion

The dependence of the planar waveguide parameters of different diffusion process temperatures and varying time of diffusion were illustrated in table 1.

Table 1. Characterization of planar waveguides fabricated in molten salt with varying time and temperature diffusion.

Waveguides	Temperature of the process (°C)	Time of diffusion (s)	Number of modes	Refractive index change	Penetration depth (μm)
Sample Ag/K 1	315	25	6	2×10^{-3}	24.30
Sample Ag/K 2		100	8	2×10^{-3}	32.40
Sample Ag/K 3		225	11	3×10^{-3}	381.
Sample Ag/K 4		400	13	2×10^{-3}	52.66
Sample Ag/K 5		625	16	5×10^{-3}	332.50
Sample Ag/K 6		900	17	5×10^{-3}	353.28
Sample Ag/K 1	350	25	6	2×10^{-3}	24.30
Sample Ag/K 2		100	8	3×10^{-3}	26.53
Sample Ag/K 3		225	10	3×10^{-3}	33.16
Sample Ag/K 4		400	12	3×10^{-3}	311.82
Sample Ag/K 5		625	15	4×10^{-3}	389.78
Sample Ag/K 6		900	20	4×10^{-3}	519.71

Table 1 reports that the values of penetration depth from glass surface increased with increasing ion-exchange temperature and time at wavelength 632.8 nm. The number of modes obtained was higher than refractive index of the glass on the same wavelength.

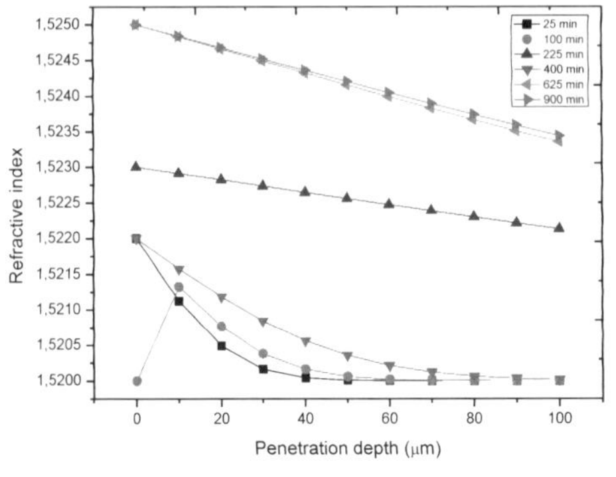

Figure 2. The refractive index in a different time and temperature 315°C diffusion of $Ag^+/K^+–Na^+$ ions.

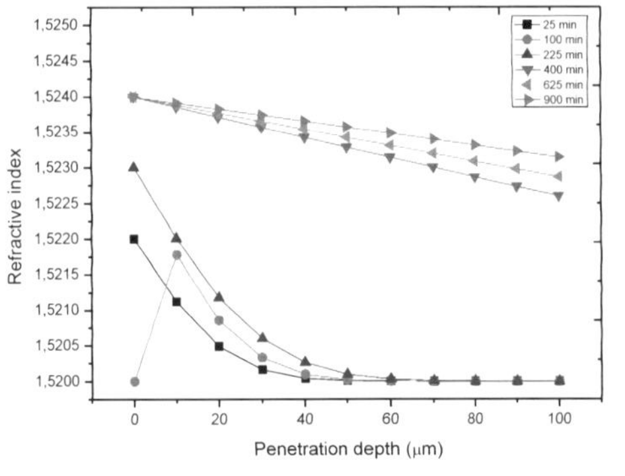

Figure 3. The refractive index in a different time and temperature 350 °C diffusion of $Ag^+/K^+–Na^+$ ions.

Figure 2 shows the refractive index for six planar waveguides where the points represent refractive index profile of the planar waveguide calculated made the number of wave modes. The value of the refractive index glass waveguide by ion-exchange at 315°C for 25–900 minutes. The waveguide depths from glass surface increased with increasing ion-exchange time. The n_{surf} of the waveguides were 1.522–1.525. Refractive index change from glass substrate index was 0.002–0.005. Figure 3 shows the refractive index profiles planar waveguide using ion-exchanged at 350°C for 25–900 minutes. The n_{surf} of the waveguides was 1.522–1.524. Refractive index changed from glass substrate index were 0.002–0.005. Planar waveguide at temperature 315°C had higher refractive index than planar waveguide for 350°C at the same wavelength. Where indicate the number of wave modes obtained for each waveguide as well as the refractive index changed Δn.

3.1. UV-Vis Spectrophotometer

The Ocean Optics USB4000 Spectrometer was used to obtain absorption spectra between 200 and 1100 nm. Experiments were performed in the wavelength range 350–600 nm, with a resolution 5.5 μm.

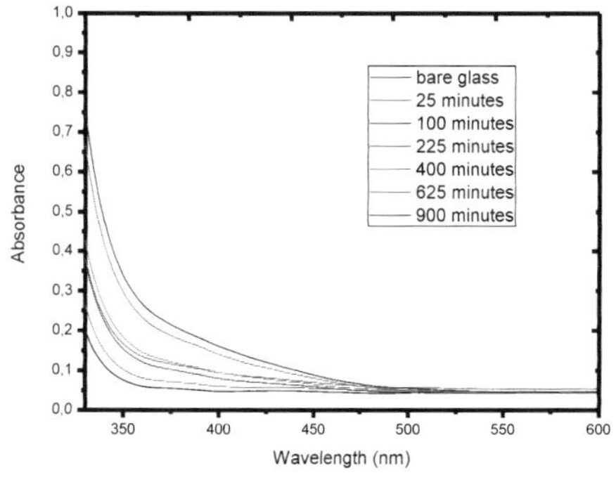

Figure 4. The absorption spectra of planar waveguide produced in soda-lime glass in different time and temperature 315°C diffusion of $Ag^+/K^+–Na^+$ ions.

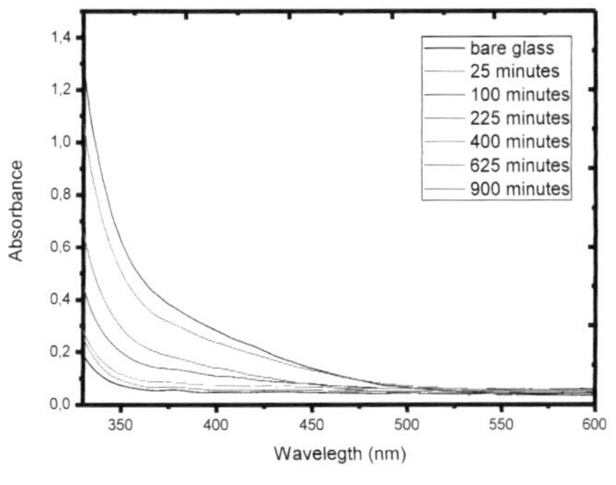

Figure 5. The absorption spectra of planar waveguide produced in soda-lime glass in different time and temperature 350°C diffusion of $Ag^+/K^+–Na^+$ ions.

Figure 4 shows the absorption spectra in the UV and visible spectral regions of all exchanged soda-lime glass planar waveguide $Ag^+/K^+–Na^+$ ion exchanged glasses processed within the time of diffusion range of 25 to 900 minutes at temperature 315°C. Figure 5 shows the absorption spectra in the UV and visible spectral regions of all exchanged soda-lime glass planar waveguide $Ag^+/K^+–Na^+$ ion exchanged glasses processed within the time of diffusion range of 25 to 900 minutes at temperature 350°C.

3.2. Diffusion parameters

The approximate the planar waveguide index profile by the function $n(x)$, where erfc is the complementary error function [9].

$$\begin{aligned} n(x) &= n_{sub} + (n_{surf} - n_{sub})\,\mathrm{erfc}(x/d) \\ &= n_{sub} + \Delta n\,\mathrm{erfc}(x/d) \end{aligned} \quad (1)$$

Where x is the penetration depth from the surface of glass substrate, n_{sub} is the refractive index of glass substrate, n_{surf} is the refractive index of glass surface, Δn is the refractive index change at the surface of the substrate, and d is the penetration depth of the planar waveguide [10]. d is used as a representative penetration depth to calculate the diffusion parameters of $Ag^+/K^+–Na^+$ ions in the planar waveguide, since the gradual refractive index change of the planar waveguide with the penetration depth makes the determination of waveguide difficult. d could be given by the following equation that described the diffusion process:

$$d = M\pi/k\,(n_{surf} - n_{sub})^{-1/2} \quad (2)$$

where M is number of modes, and k is $2\pi/\lambda$.

4. Conclusion

The planar waveguide has been successfully fabricated on soda-lime glass substrate by $Ag^+/K^+–Na^+$ ion-exchange under various conditions. Glass samples are colored sligthly, so a decrease was recorded in the visible region absorption for all of the varying time diffusion. The optical properties waveguide is already characterized. Based on the experiment obtained the planar waveguides of the soda-lime glass have been able to fabricated by $Ag^+/K^+–Na^+$ ion-exchange at temperatures 315°C and 350°C for 25–900 minutes. *M*-line spectroscopy also uses determination of the refractive index and penetration depth. The refractive index change give the value of different diffusion process temperatures and varying time of diffusion. This shows that n_{surf} and refractive index change is the independent of the diffusion process temperatures and time of diffusion.

5. Acknowledgments

The authors would like to thank the assistance at Optics and Photonics Research Groups, Sebelas Maret University. We also like to acknowledge for Edi Prasetyo.

References

[1] Najafi S I 1992 *Introduction to Glass Integrated Optics (Artech House)*
[2] Azzedine B 2009 *Photonic Waveguides (ISTE and John Wiley & Sons)*
[3] Shinichi S, Tokuro N, and Yoshinari M 2007. *Opt. Mater.* **30** 586–593
[4] Roman R 2012 *Ion Exchange Technologies* (InTech) chapter 7 155–190
[5] Melo de R P, P da Silva B J, Falcao, E F da Silva, D V Petrov and Cid B de A 1995. *American Institute of Physics.*
[6] Tong X C 2014. *Advanced Materials for Integrated Optical Waveguides.* **46** 552
[7] Ulrich R and Torge R 1973. *Applied Optics.* **12** 2901–2908
[8] Jiri C, Jiri J and Miroslav S. 1994 *Radio Engineering.* **3** 5–8
[9] Lotfi R, Chartier GH and Samra AS 1992. *Engineering Bulletin.* **27** 229–240
[10] Kimura K, Sakida S, Benino.Y and Nanba. T 2011. *Mater. Sci. Eng.* **18** 1–4

IOP Conf. Series: Journal of Physics: Conf. Series **909** (2017) 012023 doi:10.1088/1742-6596/909/1/012023

Solution of Klein Gordon equation for hyperbolic cotangent potential in the presence of a minimal length using Hypergeometric method

Isnaini Lilis Elviyanti, A Suparmi, C Cari, Dewanta Arya Nugraha, and Beta Nur Pratiwi

Physics Department, Graduate Program, Sebelas Maret University

Email: isna.elviyanti@gmail.com

Abstract. This research was to investigate the solution of Klein Gordon equation within minimal length formalism. The Hypergeometric method was used to solve Klein Gordon equation to obtain the relativistic energy and wave functions within minimal length formalism. The relativistic energy was calculated numerically using Matlab software and un-normalized wave function was expressed in hypergeometric terms. The results showed that the relativistic energy within minimal length formalism was greater than the original relativistic energy and the relativistic energy increase by the increasing of the minimal length parameter. The minimal length parameter caused the shape change in wave functions.

1. Introduction

The study of behavior microscopic spin-zero particles in relativistic quantum mechanics was described using Klein Gordon equation [1,2]. An exact solution of Klein Gordon equation with a potential vector (V) and scalar potential (S) was obtainable in the case of exact symmetric spin condition when S=V and $S = -V$ for exact symmetric pseudospin condition. In these two cases, Klein Gordon equation reduces to Schrodinger like equation thus a state solution was obtained using methods in nonrelativistic quantum mechanics [3]. Klein Gordon equation can be solved by Supersymmetric Quantum Mechanics (SUSY) [4], Algebraic solution [5], Nikiforov Uvarov (NU) method [6], and Asymptotic Iteration Method (AIM) [7-9,20].

The minimal length idea in quantum mechanics was to modify Generalized Uncertainty Principle (GUP) [10]. This modified version of quantum mechanics was based on the following deformed commutation relations between position and momentum operators [10]. The minimal length may be viewed as an intrinsic scale characterizing the system [11]. The minimal length to Klein Gordon equation had done by Bouaziz [12], Janna and Roy [11], and Merad et al. [13]. The solution of minimal length had been done by Alimmohamadi and Hassanabadi in 2017 for Bohr Mottelson equation [14].

In this paper, we solved symmetric spin condition of Klein Gordon equation with equal scalar and vector potentials in the presence of a minimal length. The radial part solution of Klein-Gordon equation in the presence of a minimal length for hyperbolic cotangent function potential has been studied. The hyperbolic cotangent potential was a part hyperbolic Manning Rosen potential. The hyperbolic Manning Rosen potential was used to study quark-gluon dynamic [15-17].The hypergeometric method has been applied to obtain the relativistic energy and radial wave functions in the presence of a minimal length. The work was organized as follows. In Section 2, Klein Gordon equation with minimal length formalism was introduced. In Section 3, we described briefly Hypergeometric method which was used to solve Klein Gordon equation. In Section 4, the relativistic energy and wave function of Klein Gordon equation was presented. The conclusion of Klein Gordon equation with minimal length was presented in Section 5.

2. Klein Gordon Equation with Minimal Length Formalism

The uncertainty of particle position was explained using commutation relations between position and momentum operators as expressed in Heisenberg's uncertainty principle. The general equationis given as [18,19],

$$[X,P] \geq i\hbar \tag{1}$$

For the case of minimal length occurring in position uncertainty and momentum uncertainty was influenced by a field then equation (1) can be written as [18,19]

$$[X,P] \geq i\hbar\left(1+\alpha_{ML}(\Delta P)^2\right) \tag{2}$$

where X was position and P was the corresponding momentum. By equation (2), the position and momentum operators [18-20] were defined as,

$$\hat{X}_i = \hat{x}_i \tag{3}$$

$$\hat{P}_i = \left(1+\alpha_{ML}\hat{p}^2\right)\hat{p}_i \tag{4}$$

where α_{ML} was a minimal length parameter that has value $0 \leq \alpha_{ML} \leq 1$. Klein Gordon equation was given by [9,11]

$$\left(E-V(r,\theta,\varphi)\right)^2\psi(r,\theta,\varphi) = \left[P^2c^2+\left(M_oc^2+S(r,\theta,\varphi)\right)^2\right]\psi(r,\theta,\varphi) \tag{5}$$

where $V(r,\theta,\varphi)$ and $S(r,\theta,\varphi)$ were vector and scalar potentials, respectively. E was relativistic energy and M_o was rest mass. By setting $S(r,\theta,\varphi)=V(r,\theta,\varphi)$ in equation (5) and substituting equation (4) into equation (5) with $\hat{p}=-i\hbar\nabla$ and $c=\hbar=1$ (natural unit),we had

$$\left(E^2-M_o^2-(E+M_o)V(r,\theta,\varphi)\right)\psi(r,\theta,\varphi) = -\left(\Delta-2\alpha_{ML}\Delta_0^2\right)\psi(r,\theta,\varphi) \tag{6}$$

where we must set $2V(r,\theta,\varphi)$ from equation (5) becomes $V(r,\theta,\varphi)$, so we obtained equation (6).

In equation (6), α_{ML}^2 was ignored because α_{ML}^2 had very small value. In the case for Klein Gordon without minimal length formalism with $\alpha_{ML}=0$ [13], then equation (6) was reduced into

$$\left(E_0^2-M_o^2-(E_0+M_o)V(r,\theta,\varphi)\right)\psi(r,\theta,\varphi) = -(\Delta)\psi(r,\theta,\varphi) = -(\Delta_0)\psi(r,\theta,\varphi) \tag{7}$$

E_0 was relativistic energy at any level without the presence of a minimal length, by rewriting equation (7) as

$$\Delta_0^2 = \left(E_0^2-M_o^2-(E_0+M_o)V(r,\theta,\varphi)\right)^2 \tag{8}$$

In equation (8) the expression of Δ_0^2 depends only on the potential $V(r,\theta,\varphi)$ and it was not considered to be spherical Laplacian, thus by inserting equation (8) into equation (6) we got

$$\left(E^2-M_o^2-(E+M_o)V(r,\theta,\varphi)\right)\psi(r,\theta,\varphi) = -\left(\Delta-2\alpha_{ML}\left(E_0^2-M_o^2-(E_0+M_o)V(r,\theta,\varphi)\right)^2\right)\psi(r,\theta,\varphi) \tag{9}$$

By applying spherical Laplacian operator was given as,

$$\Delta = \frac{1}{r^2}\frac{\partial}{\partial r}r^2\frac{\partial}{\partial r}+\frac{1}{r^2}\left[\frac{1}{\sin\theta}\frac{\partial}{\partial\theta}\sin\theta\frac{\partial}{\partial\theta}+\frac{1}{\sin^2\theta}\frac{\partial^2}{\partial\varphi^2}\right] \tag{10}$$

And using variable separation method by setting $\psi(r,\theta,\varphi)=R(r)\Theta(\theta)\Phi(\varphi)$, so we had radial part,

$$\frac{1}{r^2}\frac{\partial}{\partial r}r^2\frac{\partial R(r)}{\partial r}+\left[\begin{array}{l}\left(E^2-M_o^{\ 2}-(E+M_o)V(r)\right)\\ -2\alpha_{ML}\left(\begin{array}{l}\left(E_0^{\ 2}-M_o^{\ 2}\right)^2\\ -2\left(E_0^{\ 2}-M_o^{\ 2}\right)(E_0+M_o)V(r)\\ +(E_0+M_o)^2V^2(r)\end{array}\right)\end{array}\right]R(r)=\frac{\lambda}{r^2}R(r) \tag{11}$$

By setting $R(r)=\chi(r)/r$, so equation (11) becomes,

$$\frac{d^2\chi(r)}{dr^2}-\frac{L(L+1)}{r^2}\chi(r)+\left[\begin{array}{l}\left(E^2-M_o^{\ 2}-(E+M_o)V(r)\right)\\ -2\alpha_{ML}\left(\begin{array}{l}\left(E_0^{\ 2}-M_o^{\ 2}\right)^2\\ -2\left(E_0^{\ 2}-M_o^{\ 2}\right)(E_0+M_o)V(r)\\ +(E_0+M_o)^2V^2(r)\end{array}\right)\end{array}\right]\chi(r)=0 \tag{12}$$

Equation (12) was Klein Gordon equation in the presence of a minimal length for the case symmetric spin condition.

3. Hypergeometric Method

The principles of Hypergeometric method is substituted with new variable and parameter to obtain a second-order differential equation of hypergeometric function which is expressed [22,23],

$$z(1-z)\frac{d^2\phi}{dz^2}+\left(c-(a+b+1)z\right)\frac{d\phi}{dz}-ab\phi=0 \tag{13}$$

with the solution of a wave function is given as [22]

$$\phi_1(z)={}_2F_1(a,b,c;z)=\sum\nolimits_{n=0}\frac{(a)_n(b)_n}{(c)_n n!}z^n \tag{14}$$

The energy eigenvalue is obtained from the condition in equation (13), [20]

$$a=-n \text{ or } b=-n \tag{15}$$

where n=0,1,2,3.... Equation(14) can be finite series of polynomials of rank n by equation(15).

4. Klein Gordon Equation with Minimal Length Formalism

By applying the hyperbolic cotangent potential [24] was given as,

$$V(r)=V_o\coth(\omega r)+V_1 \tag{16}$$

where V_o and V_1 were constant of potential ω was a range of potential into equation (12), then equation (12) becomes

$$\left[\begin{array}{l}\frac{d^2\chi(r)}{dr^2}-\frac{L(L+1)}{r^2}\chi(r)-2\alpha_{ML}\gamma^2V_o^2\,\mathrm{csch}^2(\omega r)\chi(r)\\ +\left(\left(4\alpha_{ML}\mu-(E+M_o)\right)V_o-2\alpha_{ML}\gamma^2V_2\right)\coth(\omega r)\chi(r)+\left(\begin{array}{l}\left(4\alpha_{ML}\mu-(E+M_o)\right)V_1\\ +\left(E^2-M_o^{\ 2}-2\alpha_{ML}\delta^2\right)\\ -2\alpha_{ML}\gamma^2V_1^2-2\alpha_{ML}\gamma^2V_o^2\end{array}\right)\chi(r)\end{array}\right]=0 \tag{17}$$

where,

$$\gamma = \left(E_0 + M_o\right), \quad \mu = \left(E_0^{\ 2} - M_o^{\ 2}\right)\left(E_0 + M_o\right), \quad \delta = \left(E_0^{\ 2} - M_o^{\ 2}\right) \tag{18}$$

By applying the centrifugal approximation [25] in equation(19) was given as,

$$\frac{1}{r^2} = \frac{\omega}{\sinh^2(\omega r)} \tag{19}$$

Then equation (17) change into

$$\left[\begin{aligned}&\frac{d^2\chi(r)}{dr^2} - \left(\frac{\left(\omega L(L+1) + 2\alpha_{ML}\gamma^2 V_o^2\right)}{\sinh^2(\omega r)}\right)\chi(r) + \left(\left(4\alpha_{ML}\mu - (E+M_o)\right)V_o - 2\alpha_{ML}\gamma^2 V_2\right)\coth(\omega r)\chi(r)\\ &+\left(\left(4\alpha_{ML}\mu - (E+M_o)\right)V_1 + \left(E^2 - M_o^2 - 2\alpha_{ML}\lambda^2\right) - 2\alpha_{ML}\gamma^2 V_1^2 - 2\alpha_{ML}\gamma^2 V_o^2\right)\chi(r)\end{aligned}\right] = 0 \tag{20}$$

By setting,

$$v(v-1) = \left(\omega L(L+1) + 2\alpha_{ML}\gamma^2 V_o^{\ 2}\right) \tag{21}$$

$$2q = \left(\left(4\alpha_{ML}\mu - (E+M_o)\right)V_o - 2\alpha_{ML}\gamma^2 V_2\right) \tag{22}$$

$$-k^2 = \left(\left(4\alpha_{ML}\mu - (E+M_o)\right)V_1 + \left(E^2 - M_o^{\ 2} - 2\alpha_{ML}\lambda^2\right) - 2\alpha_{ML}\gamma^2 V_1^2 - 2\alpha_{ML}\gamma^2 V_o^{\ 2}\right) \tag{23}$$

Then equation (20) becomes,

$$\frac{d^2\chi(r)}{dr^2} - \left[\frac{v(v-1)}{\sinh^2(\omega r)} - 2q\coth(\omega r) + k^2\right]\chi(r) = 0 \tag{24}$$

Equation (24) was a differential equation which had to be reduced to hypergeometric differential equation type. By using the suitable variable change $\coth(\omega r) = 1-2z$ so we got,

$$z(1-z)\frac{d^2\chi(r)}{dz^2} + (1-2z)\frac{d\chi(r)}{dz} + \left[v'(v'-1) - \frac{4\alpha^2}{4z} - \frac{4\beta^2}{4(1-z)}\right]\chi(r) = 0 \tag{25}$$

with

$$\frac{-2q+k^2}{\omega^2} = 4\alpha^2 \tag{26}$$

$$\frac{2q+k^2}{\omega^2} = 4\beta^2 \tag{27}$$

$$v'(v'-1) = \frac{v(v-1)}{\omega^2} \tag{28}$$

Equation (25) was an intermediate of the hypergeometric differential equation. We reduced equation (25) by using new wave function,

$$\chi(r) = z^{\alpha}(1-z)^{\beta} f(z) \tag{29}$$

and by inserting equations (26-29) into equation (25), we had

$$z(1-z)\frac{f^2(z)}{dz^2} + \left[(2\alpha+1) - (2\alpha+2\beta+2)z\right]\frac{f(z)}{dz} + \left[v'(v'-1) - (\alpha+\beta)(\alpha+\beta+1)\right]f(z) = 0 \tag{30}$$

Equation (30) shows the hypergeometric differential equation therefore with the solution in the form of $f(z) = {}_2F_1(a,b,c,z)$, with

$$a = \alpha + \beta - (v'-1) \tag{31}$$

$$b = \alpha + \beta + v' \tag{32}$$

$$c = 2\alpha + 1 \tag{33}$$

Using equations (18), (26-28), and (31-33), we obtained the relativistic energy equation, was given as,

$$\left(E^2 - M_o^2\right) = -\omega^2\left[\left(\delta_L - \frac{1}{2} - n\right)^2 + \frac{\left(\frac{\left(4\alpha_{ML}\mu - \left(E+M_o\right)V_o\right) - 2\alpha_{ML}\gamma^2 V_2}{2\omega^2}\right)^2}{\left(\delta_L - \frac{1}{2} - n\right)^2}\right] - \left(4\alpha_{ML}\mu - \left(E+M_o\right)\right)V_1 + \eta_{\alpha_{ML}} \tag{34}$$

with

$$\delta_L = \sqrt{\frac{\omega L\left(L+1\right) + 2\alpha_{ML}\gamma^2 V_o^2}{\omega^2} + \frac{1}{4}} \tag{35}$$

$$\eta_{\alpha_{ML}} = 2\alpha_{ML}\lambda^2 + 2\alpha_{ML}\gamma^2 V_1^2 + 2\alpha_{ML}\gamma^2 V_o^2 \tag{36}$$

where L was angular momentum and n was a quantum number. Equation (34) was the relativistic energy of Klein Gordon equation in the presence of a minimal length for hyperbolic cotangent potential. From equation (34), we calculated either the relativistic energy E_0 for $\alpha_{ML} = 0$ and E for $\alpha_{ML} \neq 0$. The relativistic energy value was calculated numerically by using Matlab software which was shown in Table 1.

Table 1. The relativistic energy for $M_o=1$, $V_o=0.01$, $V_1=0.1$ $V_2=0.001$, and $L=0$.

n	E_n (eV)					
	$\omega = 0.1$			$\omega = 0.2$		
	$\alpha_{ML} = 0$	$\alpha_{ML} = 0.03$	$\alpha_{ML} = 0.09$	$\alpha_{ML} = 0$	$\alpha_{ML} = 0.03$	$\alpha_{ML} = 0.09$
1	1.0900	1.0901	1.0904	1.0795	1.0796	1.0798
2	1.0795	1.0796	1.0799	1.0205	1.0214	1.0231
3	1.0556	1.0560	1.0567	0.9115	0.9161	0.9252
4	1.0205	1.0214	1.0232	0.7300	0.7494	0.7825
5	0.9731	0.9753	0.9796	0.3701	0.4532	0.5818

Table 1 shows that the relativistic energy was influenced by quantum numbers, the range of potentials and minimal length parameters. The value of relativistic energy decreased by the increasing of quantum number n and range of potential ω. The increasing of relativistic energy value causes the increasing of theminimal length parameter.

Then, for the wave function was obtained from equations (26-29), was given as

$$\chi\left(r\right) = \frac{\left(1 - \coth\left(\omega r\right)\right)^{\alpha}}{2}\frac{\left(1 + \coth\left(\omega r\right)\right)^{\beta}}{2}\ {}_2F_1\left(a,b,c,z\right) \tag{37}$$

with

$${}_2F_1\left(a,b,c,z\right) = 1 + \frac{ab}{c}z + \frac{a\left(a+1\right)b\left(b+1\right)}{c\left(c+1\right)}\frac{z^2}{2!} + \frac{a\left(a+1\right)\left(a+2\right)b\left(b+1\right)\left(b+2\right)}{c\left(c+1\right)\left(c+2\right)}\frac{z^3}{3!} + \ldots \tag{38}$$

The hypergeometric parameters from equations (31-33) were substituted into equation (13) and equation (14), so we got,

$$a = -n \tag{39}$$

$$b = 2\vartheta - n \tag{40}$$

$$c = \tau - \sigma \tag{41}$$

with,

$$\vartheta=\sqrt{\frac{\omega L(L+1)+2\alpha_{ML}\gamma^2V_o^2}{\omega^2}+\frac{1}{4}} \tag{42}$$

$$\tau=\sqrt{\frac{\omega L(L+1)+2\alpha_{ML}\gamma^2V_o^2}{\omega^2}+\frac{1}{4}}-\frac{1}{2}-n \tag{43}$$

$$\sigma=\frac{\left(4\alpha_{ML}\mu-(E+M_o)V_o\right)-2\alpha_{ML}\gamma^2V_2}{2\omega^2\sqrt{\frac{\omega L(L+1)+2\alpha_{ML}\gamma^2V_o^2}{\omega^2}+\frac{1}{4}}-\frac{1}{2}-n} \tag{44}$$

By inserting equations (39-41) into equation (37), we had the wave functions of quantum number *n=0,1,2* that were shown in Table 2.

Table 2. The wave function of Klein Gordon equation in the presence of minimal length

n	Wave Function
0	$\chi_0(r)=(-1)^{\left(\frac{\tau}{2}+\sigma\right)}\dfrac{\left(\frac{\operatorname{csch}(\omega r)}{2}\right)^{(\tau)}}{\left(\frac{\sinh(\omega r)-\cosh(\omega r)}{2}\right)^{(\sigma)}}$
1	$\chi(r)=(-1)^{\left(\frac{\tau}{2}+\sigma\right)}\dfrac{\left(\frac{\operatorname{csch}(\omega r)}{2}\right)^{(\tau)}}{\left(\frac{\sinh(\omega r)-\cosh(\omega r)}{2}\right)^{(\sigma)}}\left(1+\dfrac{(-n)(2\vartheta-n)}{(\tau-\sigma)}\left(\dfrac{(1-\coth(\omega r))}{2}\right)\right)$
2	$\chi_2(r)=(-1)^{\left(\frac{\tau}{2}+\sigma\right)}\dfrac{\left(\frac{\operatorname{csch}\omega r}{2}\right)^{(\tau)}}{\left(\frac{\sinh\omega r-\cosh\omega r}{2}\right)^{(\sigma)}}\left(1+\dfrac{(-n)(2\vartheta-n)}{(\tau-\sigma)}\left(\dfrac{(1-\coth\omega r)}{2}\right)+\dfrac{(-n)(-n+1)(2\vartheta-n)(2\vartheta-n+1)}{(\tau-\sigma)(\tau-\sigma+1)}\dfrac{\left(\frac{(1-\coth\omega r)}{2}\right)^2}{2!}\right)$

5. Conclusion

We investigated Klein Gordon equation in the presence of a minimal length for hyperbolic cotangent potential. The hypergeometric method was used to obtain the relativistic energy and radial wave functions for Klein-Gordon equation in the presence of minimal length. The minimal length parameter gives the influence in the relativistic energy value which by increasing the minimal length parameter causes the greater value of energy. The value of relativistic energy decreases respect to various quantum number and range of potential.

Acknowledgement

This research was partly supported by Sebelas Maret University Higher Education Project Grant 623/UN27.21/PP/2017.

References

[1] Ikhdair S M and Sever R 2008 *arXiv:0808.1002v1 [quant-ph]*14.

IOP Conf. Series: Journal of Physics: Conf. Series **909** (2017) 012023 doi:10.1088/1742-6596/909/1/012023

[2] Momtazi E, Rajabi A A, and Yazarloo B H 2014 *Turkish J. Phys* **38** 1 81–85.
[3] Greiner W 2000 Springer-Verlag Berlin Heidelberg, New York.
[4] Setare M R and Nazari Z 2009 *ActaPhys. Pol. B***40** 10 2809–24.
[5] Poszwa A 2014 *Acta Physica PolonicaA* **126** 6.
[6] Ikhdair S M 2011 *arXiv:1110.0943v1 [quant-ph]* 1–25.
[7] Barakat T 2006 *Int. J. Mod. Phys. A* **21** 19 4127–35.
[8] Das T 2014 *arXiv:1409.1457v1 [quant-ph]* 1–11.
[9] Nugraha D A, Suparmi A, Cari C and Pratiwi BN 2017 *J. Phys. Conf. Ser***795**.
[10] Hassanabadi H, Maghsoodi E, Ikot N A, and Zarrinkamar S 2013 *Hindawi Publishing Corporation.*
[11] Jana K T and Roy P 2009 *Physics Letters A***373** 1239–41.
[12] Bouaziz D 2013 *arXiv:1311.7405v1 [quant-ph].*
[13] Merad M, Zeroual F, and Benzair H 2010 *Electronic Journal of Theoretical Physics* **7** 23 41-56.
[14] Alimohammadi M and Hassanabadi H 2017 *Nuclear Physics A* **957** 439-449.
[15] A N Ikot, L E Akpabio, and E J Uwah, *Electron. J. Theor. Phys* **25** 25 225-232.
[16] Agboola D 2011 *J.Phys* **76** 6 875-885.
[17] D E Alvarez , C B Compean and M Kirchbach 2011 *arXiv:1105.1354v1[quant-ph].*
[18] Sprenger M, Nicolini P, and Bleicher M 2012 *European Journal Of Physics* **33** 853-862.
[19] Garay J L 1994 *International Journal of Modern Physics A***10** 2 145-165.
[20] Chabab M, ElBatoul A, Lahbas A, and Oulne M 2016 *Physics Letters B* 212-216.
[21] Nugraha D A, Suparmi A, Cari C, and Pratiwi B N 2017 *J. Phys. Conf. Ser* **820**.
[22] Suparmi, Department of Physics Sebelas Maret University, Surakarta. 2011.
[23] Morales J, Martínez G J, Ravelo G J and Peña J 2015 *Journal of Applied Mathematics and Physics.*
[24] Cari, Suparmi, Deta U A, and Werdiningsih I S 2013 *Makara J. Sci* **17**.
[25] Naderia L and Hassanabadi H 2016 *The European Physical Journal Plus* 131: 133.

IOP Conf. Series: Journal of Physics: Conf. Series **909** (2017) 012024 doi:10.1088/1742-6596/909/1/012024

Nonzero θ_{13} and CP violation from cobimaximal neutrino mixing matrix

Asan Damanik

Department of Physics Education, Sanata Dharma University
Kampus III USD Paingan, Maguwoharjo, Sleman, Yogyakarta, Indonesia

Email: asandamanik11@gmail.com

Abstract. The nonzero mixing angle θ_{13} have been reported and confirmed by many collaborations. If the mixing angle θ_{13} is nonzero, then the possiblity of the CP violation existence on neutrino sector become an attractive research subject both from experimental and theoretical sides. We evaluate the power predictions of cobimaximal neutrino mixing matrix on mixing angle θ_{13} and CP violation for neutrino sector by determining the Jarlskog invariant as a measure of CP violation.

1. Introduction

Now, we have confidence that neutrino have a tiny mass based on the experimental facts of neutrino oscillations. Neutrino oscillation is a term addressed to a changes of neutrino flavor from one flavor of neutrino to another flavor during its propogation in vacuum. From neutrino oscillation experiments, we have only got two squared-mass difference (not an absolute mass of neutrino) and three mixing angles. The problems of neutrino mass existence together with the process or mechanism to generate them and neutrino mass absolute values become an interesting research subject both from theoretical and experimental sides. The problem of absolute value of neutrino mass and the mechanism to generate them are still unsolved problems till today, another interesting challenge arise from the experimental fact i.e. the nonzero and relatively large mixing angle θ_{13} were reported by many collaborations [1-4].

The nonzero and relatively large mixing angle θ_{13} become an attractive subject to theorists because it give us a hint that CP violation could be exist in neutrino sector as well as in quark sector. There have been several attempts to explain the nonzero and relative large mixing angle θ_{13} and the possibility existence of CP violation in neutrino sector, see for example Ref. [5,6] and references there in. Most of the attempts are to modify the previously three well-known neutrino mixing matrix i.e. tribimaximal, bimaximal, and democratic. An alternative and interesting mixing angle, which is known as cobimaximal mixing, was introduced by Ma [7] that he claimed that cobimaximal mixing is achieve rigorously in a renormalizable model of radiative charged-lepton and neutrino masses. Authors in Ref. [8] used a scotogenic model i.e one loop neutrino mass model with dark right-handed neutrino gauge singlets and one inert dark scalar gauge doublet η which has symmetries that lead to cobimaximal mixing matrix.

In this paper we evaluate the power prediction of cobimaximal mixing matrix on mixing angle θ_{13} and CP violation in neutrino sector by determining the Jarlskog invariant as a measure of CP violation. The paper is organized as follow: in section 2, we discuss the cobimaximal neutrino mixing matrix and its brief formulation. In section 3, we evaluate the power prediction of cobimaximal mixing on mixing angle θ_{13} and Jarlskog invariant as a measure of CP violation. Finally, the section 4 is devoted for conclusions.

2. Cobimaximal neutrino mixing matrix

We have already known that neutrino come in three flavor eigenstates (ν_e, ν_μ, ν_τ) correspond to its charged partner (e, μ, τ) in the frame of weak interactions. When only using the flavor concept, we cannot explain the neutrino oscillation phenomena, then theorists formulated neutrino mass eigenstates (ν_1, ν_2, ν_3) concept which are different from neutrino flavor eigenstates. The flavor eigenstates are linked to neutrino mass eigenstates via a neutrino mixing matrix as follow

$$\begin{pmatrix} \nu_e \\ \nu_\mu \\ \nu_\tau \end{pmatrix} = V \begin{pmatrix} \nu_1 \\ \nu_2 \\ \nu_3 \end{pmatrix}, \tag{1}$$

where V is neutrino mixing matrix. The standard parameterization of mixing matrix without Majorana phase read [9]

$$V = \begin{pmatrix} c_{12}c_{13} & s_{12}c_{13} & s_{13}e^{-i\delta} \\ -s_{12}c_{23} - c_{12}s_{23}e^{i\delta} & c_{12}c_{23} - s_{12}s_{23}e^{i\delta} & s_{23}c_{13} \\ s_{12}s_{23} - c_{12}c_{23}e^{i\delta} & -c_{12}s_{23} - s_{12}c_{23}e^{i\delta} & c_{23}c_{13} \end{pmatrix} \tag{2}$$

where $c_{ij} = \cos\theta_{ij}$, $s_{ij} = \sin\theta_{ij}$, and δ is the Dirac phase.

As stated in section 1, there are three already well-known of neutrino mixing matrix, i.e. bimaximal mixing (BM), tribimaximal mixing (TBM), and democratic mixing (DM). But, all of these mixing matrices predict the mixing angle $\theta_{13} = 0$ which are contrary to the recent experimental facts that mixing angle $\theta_{13} \neq 0$ and relatively large as reported by T2K Collaboration [2]

$$5^\circ \leq \theta_{13} \leq 16^\circ, \tag{3}$$

for neutrino masses in normal hierarchy (NH) and

$$5.8^\circ \leq \theta_{13} \leq 17.8^\circ, \tag{4}$$

for inverted hierarchy (IH) with Dirac phase $\delta = 0^\circ$. The reported results of T2K Collaboration on nonzero mixing angle θ_{13} was also confirmed by Daya Bay Collaboration [4]

$$\sin^2 2\theta_{13} = 0.092 \pm 0.016\,(\text{stat.}) \pm 0.005\,(\text{syst.}). \tag{5}$$

Now, for a new mixing matrix which is known as cobimaximal mixing, one put the mixing angle $\theta_{23} = \pi/4$ and the Dirac phase $\delta = \pm\pi/2$ as suggested by authors of Ref. [7,8] and within this scenario the neutrino mixing matrix in Eq. (2) reads

$$V_{CBM} = \begin{pmatrix} c_{12}c_{13} & s_{12}c_{13} & \mp i s_{13} \\ -\frac{\sqrt{2}}{2}(s_{12} \mp i c_{12}) & \frac{\sqrt{2}}{2}(c_{12} \mp i s_{12}) & \frac{\sqrt{2}}{2}c_{13} \\ \frac{\sqrt{2}}{2}(s_{12} \mp c_{12}) & -\frac{\sqrt{2}}{2}(c_{12} \pm i s_{12}) & \frac{\sqrt{2}}{2}c_{13} \end{pmatrix} \tag{6}$$

The cobimaximal neutrino mixing matrix may be achieved in the context of non-Abelian discrete symmetry by soft breaking A_4 to Z_3 and soft breaking of $S_3 \times Z_2$ to $Z_2 \times Z_2$ [10].

It is clear from Eq. (6) that cobimaximal mixing matrix depend on the mixing angles θ_{12} and θ_{13} values. Thus, when concerning with the neutrino oscillation experiment, we need to make a precise measurement of mixing angles θ_{12} and θ_{13} in the experiment. It is also apparent from Eq. (6) that cobimaximal mixing is not an orthogonal matrix and the value of mixing angle θ_{13} can be arbitrary. Meanwhile, the recent value of mixing angle θ_{12} based on global analysis of neutrino oscillation experiments was reported in Ref. [11] as follow

$$\theta_{12} = 34.5 \pm 1.0\left(^{+3.2}_{-2.8}\right)^{\circ} \tag{7}$$

at $1\sigma\,(3\sigma)$ level. We will use the mixing angle value in Eq. (7) as input for determining the cobimaximal mixing matrix predictions on mixing angle θ_{13} and CP violation.

3. Predictions of cobimaximal mixing on θ_{13} and J_{CP}

After introducing cobimaximal neutrino mixing as an alternative mixing angle and quote the recent experimental value of mixing angle θ_{12}, now we are in position to evaluate the power predictions of cobimaximal mixing on mixing angle θ_{13} and Jarlskog invariant (J_{CP}) as a measure of CP violation in neutrino sector.

From a cobimaximal neutrino mixing matrix in Eq. (6) we can see that the value of mixing angle θ_{13} can be arbitrary if the mixing matrix is a non-orthogonal matrix. But, if we put the mixing matrix to be an orthogonal matrix, then we must put mixing angle θ_{13} to be zero. Based on the cobimaximal mixing matrix, that the value of mixing angle θ_{13} can be arbitrary, we suggest the experimentalist to measure the precise value of mixing angle θ_{13} in the future experiments.

The Jarlskog invariant (J_{CP}) as a parameter for determining the CP violation is given by [12]

$$J_{CP} = \mathrm{Im}\left[(V_{CBM})_{11}(V_{CBM})_{22}(V_{CBM})^{*}_{12}(V_{CBM})^{*}_{21}\right]. \tag{8}$$

From Eqs. (6) and (8), for Dirac phase $\delta = \pi/2$, we have the Jarlskog invariant

$$J_{CP} = \tfrac{1}{2}\left(\left(1-s_{12}^2\right)^{3/2}\left(1-s_{13}^2\right)s_{12}s_{13}\right) - \tfrac{1}{2}\left(\left(1-s_{12}^2\right)^{1/2}\left(1-s_{13}^2\right)s_{12}^3 s_{13}\right). \tag{9}$$

It is apparent from Eq. (9) that the value of Jarlskog invariant in the scheme of cobimaximal neutrino mixing matrix depend on the mixing angles θ_{12} and θ_{13}.

To get the qualitative prediction of the cobimaximal mixing on CP violation we can plot J_{CP} as function of s_{13} and s_{12}. If we plot J_{CP} as function of s_{13} and s_{12}, then we have a figure as shown in Figure 1.

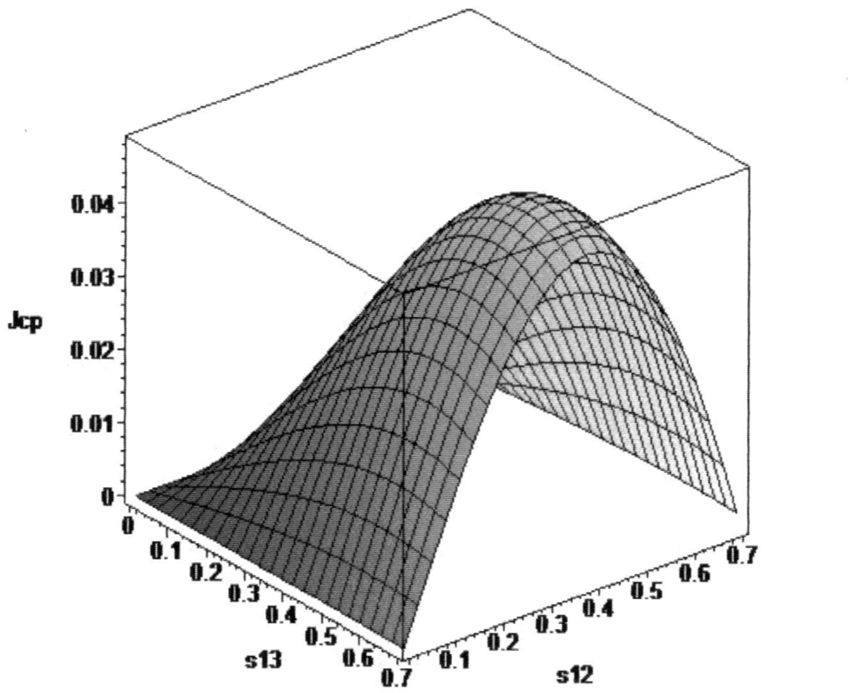

Figure 1. Plot of Jarlskog invariant (J_{CP}) as function of $\sin\theta_{13}$ and $\sin\theta_{12}$

From Fig. 1 we can see that the parameter J_{CP} has a maximun value when $\sin\theta_{13} = 0.55$ and $\sin\theta_{12} = 0.4$ that give the value of $J_{CP} \approx 0.05$. But, if we insert the center experimental value of mixing angles $\theta_{12} = 34.5^\circ$ and $\theta_{13} = 10^\circ$, then we have $J_{CP} \approx 0.013$.

4. Conclusions

We have evaluate the power predictions of cobimaximal neutrino mixing matrix on mixing angle θ_{13} and the Jarlskog invariant as a parameter of CP violation by using the central value of the experimental data of neutrino mixing angle $\theta_{12} = 34.5^\circ$ as input. It is also apparent that the value of mixing angle θ_{13} in the cobimaximal neutrino mixing matrix can be arbitrary. The Jarlskog invariant can be put as function of mixing angle θ_{13} and θ_{12} and its has a maximum value around $J_{CP} \approx 0.013$. When the central value of mixing angles $\theta_{12} = 34.5^\circ$ is inserted into the Jarlskog invariant we have $J_{CP} \approx 0.05$.

Acknowledgement
Author would like to thank the ICSAS 2017 Organizer for a nice hospitality during the conference. This researh is partly supported by Hibah Penelitian Internal LPPM USD with contract No. 070/Penel./LPPM-USD/IV/2017 and Department of Physics Education Faculty of Teacher Training and Education Sanata Dharma University Yogyakarta.

References

[1] Adamsom P et al. (MINOS Collab.) 2011 Measurement of neutrino mass splitting and flavor mixing by MINOS *Phys. Rev. Lett.* **107** 181801 (*Preprint* arXiv: 1103.0340 [hep-ex])

[2] Abe K et al. (T2K collab.) 2011 Indication of electron neutrino appearance from an accelerator-produced off-axis muon neutrino beam *Phys. Rev. Lett.* **107** 041801 (*Preprint* arXiv: 1106.2822 [hep-ex])

[3] Ahn J K et al. (RENO Collab.) 2012 Observation of reactor electron antineutrino disappearance in the RENO experiment *Phys. Rev. Lett.* **108** 191802 (*Preprint* arXiv: 1204.0626 [hep-ex])

[4] An F P et al. (Daya Bay Collab.) 2012 Observation of electron-antineutrino disappearance at Daya Bay *Phys. Rev. Lett.* **108** 171803 (*Preprint* arXiv: 1203.1669 [hep-ex])

[5] Damanik A 2012 Nonzero θ_{13} and neutrino masses from modified neutrino mixing matrix *Int. J. Mod. Phys.* A **27** 1250091 (*Preprint* arXiv: 1201.2747 [hep-ph])

[6] Damanik A 2014 *Chinese J. Phys.* **52** 1439

[7] Ma E 2016 Soft $A_4 \rightarrow Z_3$ symmetry breaking and cobimaximal neutrino mixing *Phys. Lett.* B **755** 348 (*Preprint* arXiv: 1601.00138 [hep-ph])

[8] Ferreira P M, Grimus W, Jurciukonis D and Lavoura L 2016 Scotogenic model for co-bimaximal mixing *J. High Energy Phys.* JHEP07(2016) 010

[9] Harrison P F, Perkins D H, and Scott W G 2002 Tri-bimaximal mixing and neutrino oscillation data *Phys. Lett.* B **530** 167 (*Preprint* hep-ph/0202074)

[10] Ma E 2017 Cobimaximal neutrino mixing from $S_3 \times Z_2$ *Preprint* arXiv: 1707.03352v1 [hep-ph]

[11] Gonzales-Garcia M C, Maltoni M and Salvado J 2010 Updated global fit to three neutrino mixing: status of the hints of $\theta_{13} > 0$ *J. High Energy Phys.* JHEP04(2010)056

[12] Branco C G, Gonzales Felipe R, Joaquim F R, Masina I, Rebelo M N and Savoy C A 2002 Minimal scenarios for leptogenesis and CP violation *Phys. Rev.*D **67** 073025 (*Preprint* hep-ph/0211001v2)

Performance improvement of dye-sensitized solar cells (DSSC) by using dyes mixture from chlorophyll and anthocyanin

D D Pratiwi, F Nurosyid, Kusumandari, A Supriyanto and R Suryana

Department of Physics, Graduate Program, Universitas Sebelas Maret, Jl. Ir. Sutami 36A Kentingan Jebres Surakarta 57126, Indonesia

E-mail: nurosyid@yahoo.com

Abstract. This article showed the effect of single and mixture natural dyes on the DSSC performance. The single dyes extracted from moss chlorophyll and mangosteen peels anthocyanin. The dyes mixture was prepared by mixing from both chlorophyll and anthocyanin. The absorbance of dyes solution and the adsorption of the dye onto the working electrode were analyzed using UV-Vis spectroscopy. The photocurrent-photovoltage of DSSCs were measured using *I-V* meter. The dyes mixture has an increased absorption at visible spectrum range as compared to single dye. The adsorption of the dyes mixture onto the TiO_2 electrode has higher absorbance than single dye. The DSSC with single dye from moss chlorophyll and mangosteen peels anthocyanin resulted the conversion efficiency of 0.049% and 0.042% respectively. The dyes mixture of chlorophyll and anthocyanin improved the conversion efficiency of 0.154%.

1. Introduction

A dye-sensitized solar cell is the type of solar cells based on the dye of semiconductor that converts the photon energy into electricity[1,2]. Advantages of DSSC are simple and low-cost in fabrication, can be coated on the flexible substrate, and also achieve the high conversion efficiency for the future[3 – 5]. The standard structure of DSSC consists of conducting glass substrate made of Transparent Conducting Oxide (TCO). TCO usually used Fluorine doped tin oxide (FTO) or Indium doped tin oxide (ITO)[6]. Working electrodes are obtained from the semiconductor material (e.g., TiO_2, SnO_2 or ZnO)[7]. The material such as platinum (Pt) or carbon (C) are usually chosen as counter electrodes. Sensitizers can be made from synthetic and/or natural dyes, while solution of iodide/triiodide (I^-/I^{-3}) redox couple can be used as electrolyte[8]. Many researchs developed the modification of DSSC components to obtain the high efficiency and low-cost DSSC. The photosensitizer is one of the main components which are widely developed by researchers because it affects in working mechanism of DSSC. The role of the sensitizer is to absorb the light, inject the electrons into the semiconductor, and convert the energy of DSSC[9,10].

Many synthetic dyes are employed as a sensitizer for DSSC. The most common of synthetic dye used ruthenium based metal complexes like N719 and black dye[11]. They are susceptible to absorb the broad spectrum of light and can achieve the high conversion efficiency. It were reported the conversion efficiency of 11-12%[12 – 13]. However, the synthetic dyes have the demerits like toxicity, high cost, and poor variations of color. The natural dyes are alternatively a photosensitizer with environmentally friendly.

The advantages of natural dyes are easy to obtain, having many variations of color, easily available from the environment, low-cost, and environmental friendliness. The natural dyes can be explored from the part of plants including leaves, flowers, fruits, trunk, roots, seeds, and rind[14]. Source of plants are used as natural dyes from pigment molecule consists of chlorophyll, anthocyanin, carotene, tannin, and flavonoid[14 - 15].

The research has been successfully used pigments molecule as DSSC photosensitizer. Arifin *et al.* (2017) reported the efficiencies from 0.040% to 0.094% for papaya leaves as a photosensitizer in DSSC[16]. Taya *et al.* (2013) explored the green algae and found 0.01% efficiency[17]. Pratiwi *et al.* (2016) extracted the anthocyanin from dragon fruits, mangosteen peels, and red cabbage with efficiency of 0.024%, 0.042%, and 0.054% respectively[18]. Utilization of flower extract also used like Oleander, hibiscus, and Ixora with 0.59%, 1.02%, and 0.30% efficiencies, respectively[10]. Flavonoid source was extracted from *Jatropha curcas* and achieved 0.12% efficiency[19]. El-Agez *et al.* (2012) obtained the carotene source from canarium nut extract. It resulted the conversion efficiency of 0.01%[20].

The research of dye modification has been done by mixing two dyes with different region of light absorption. The mixing dyes aim to widen the wavelengths range of light absorption. The wider light absorption can enhance the conversion efficiency of DSSC. The research by Chang *et al.* (2013) extracted the chlorophyll dye from wormwood and anthocyanin dye from purple cabbage with efficiency of 0.54% and 0.75% respectively, while the dyes mixture from both dyes can enhance the efficiency of 1.47%[15]. Pratiwi *et al.* (2017) tried by mixing the natural dye from red cabbage anthocyanin with synthetic dye from N719. The mixing dyes aim to improve the efficiency of natural dye from red cabbage anthocyanin. DSSC based on red cabbage anthocyanin resulted in the efficiency of 0.024%, while by addition 20% synthetic dye improve the conversion efficiency of natural dye from red cabbage anthocyanin of 0.054%[21].

Based on previous research, the anthocyanin dye showed absorption in the visible wavelength range from 480-580 nm, while the chlorophyll dye in the range 500-600 nm and 600-700 nm. Hence, the dyes mixture from two different dyes needs to be studied to improve the DSSC performance. This article will investigate the dyes mixture on the DSSC performance.

2. Experiments

Natural dyes were extracted from moss chlorophyll and mangosteen peels anthocyanin. The dyes mixture was obtained from both chlorophyll and anthocyanin (1:1 volume ratio). The substrates used FTO glass from Dyesol with a size of 2 cm x 2,5 cm and active cell area of 2 cm^2. Working electrodes were coated using TiO_2 paste from Aldrich 21 nm by the spin coating technique,counter electrodes were coated using carbon, and electrolyte solution used EL-HPE Dyesol.

Moss powder was dissolved in the aceton solvent and stirred to get homogeneous solution. Its solution was filtered using filter paper. The solution of filter result was held chromatography process for moss chlorophyll extract was obtained. Mangosteen peels powder was dissolved in the ethanol solvent and stirred for 12 hour. Extract of mangosteen peels anthocyanin was filtered using whatman filter paper.

Characterization includes optical properties (absorbance) and electrical properties. Optical characterization using UV-Visible Spectrophotometer Lambda 25 to analyze the absorbance of dye solutions and dye after adsorbed onto the TiO_2 electrode. Electrical properties of the DSSCs were measured using Keithley *I-V* Meter 2602A.

3. Results and Discussion

3.1. Optical Characterization of Different Natural Dye Solutions and the Dye After Adsorbed onto the TiO_2 Electrode

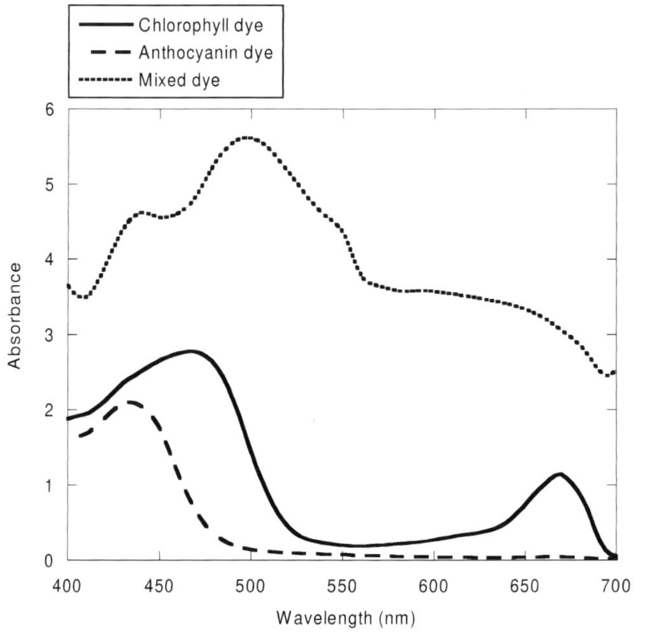

Figure 1. Absorption spectra of chlorophyll extracted from moss, anthocyanin extracted from mangosteen peels, and their mixtures

Fig. 1 revealed the UV-Vis absorption spectra of single moss; mangosteen peels extract as well as their mixtures. The chlorophyll extract exhibited the first absorption spectra at wavelength 400-490 nm and the second peak at wavelength 630-700 nm. The dye solution of anthocyanin extract can absorb the light at wavelength 400-480 nm. These results represented the dyes solution of moss contain the chlorophyll pigment and mangosteen peels contain the anthocyanin pigment, so that they can be used as DSSC photosensitizers because they have the light absorption in the visible spectrum. The chlorophyll and anthocyanin dyes have the different light absorption, therefore to improve and widen the region of light absorption by mixing of both chlorophyll and anthocyanin dyes at a volume ratio of 1:1.

The result of dyes mixture from chlorophyll and anthocyanin can improve the absorbance and widen the region of light absorption. The curve in Fig.1 revealed the dyes mixture has higher absorbance than single dye. The dyes mixture performed the increase in absorbance from the single dye. It indicated the increase in the intensity of light absorption (hyperchromic effect). The result of dyes mixture has two absorption peaks at wavelength 430 nm (the contribution of anthocyanin dye) and 500 nm (the contribution of the first peak from chlorophyll dye). The second absorption peak from chlorophyll dye was not contribution the absorption peak. This situation can be caused by both chlorophyll and anthocyanin dyes used the different solvent. Each of the solvents was affected by the level of polarity. It involved the shift in the absorption peak and level of absorbance. The shape of absorption peak can be shifted to longer wavelengths (a red shift/bathochromic) and shorter wavelengths (a blue shift/hypsochromic).

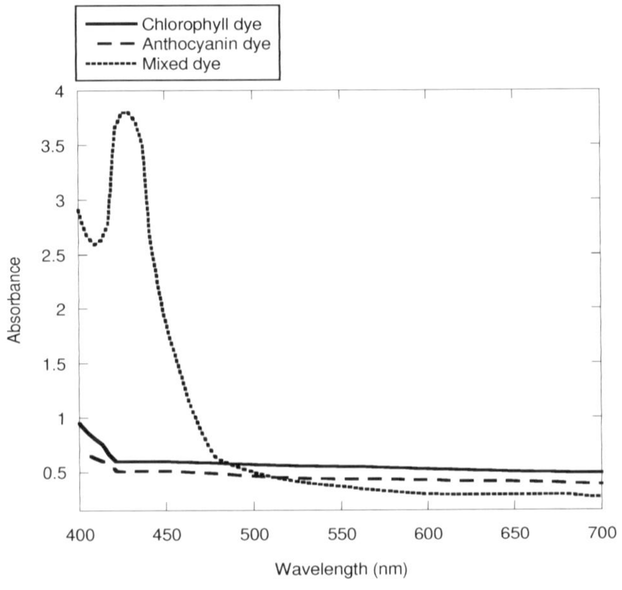

Figure 2. Absorption spectra of dye after adsorbed onto the TiO_2 electrode

The UV-Vis characteristic for absorption spectra of dye after adsorbed onto the TiO_2 electrode is shown in Fig.2. The curve can be observed the absorbance peak of dye after adsorbed onto the TiO_2 electrode increased and approached into the absorption of UV region. This situation indicated that the absorption spectra of dye after adsorbed onto the TiO_2 electrode was not good to absorb in the visible region.

The absorbance of TiO_2 electrode using dyes mixture from chlorophyll and anthocyanin performed the significant peak in absorption spectra compared with single dye. It was comparable to the UV-Vis characteristic of dyes solution that dye solution using the mixture from chlorophyll and anthocyanin has higher absorbance than single dye. The increased peak of absorbance and widened region of light absorption can improve the intensity of absorption onto the TiO_2 semiconductor.

3.2. I-V Characterization

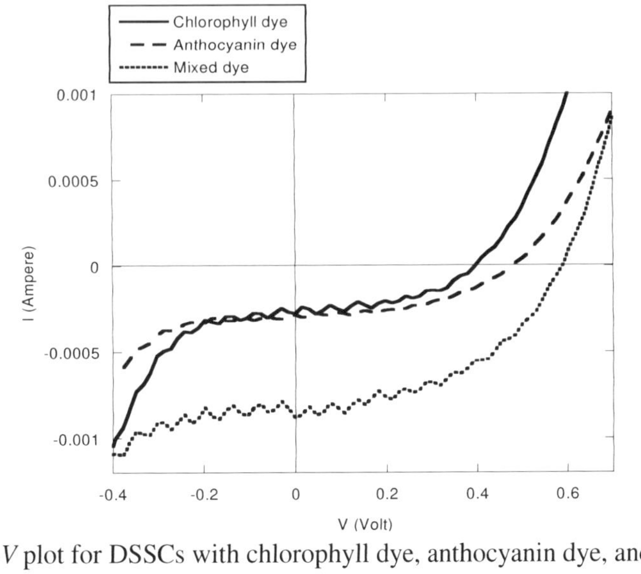

Figure 3. *I-V* plot for DSSCs with chlorophyll dye, anthocyanin dye, and dyes mixture

Fig.3 shows *I-V* characteristic curve for DSSC with single dye and dyes mixture from chlorophyll and anthocyanin. From *I-V* curve in Fig.3 the parameters of DSSC were determined, and its values were observed in Table 1.

The conversion efficiency of DSSC significantly improved by using dyes mixture. It was due to the absorption of visible light spectrum over a wider range as compared to single dye. This result was also supported by UV-Vis characteristic of dye after adsorbed onto the TiO_2 electrode which dyes mixture from moss chlorophyll and mangosteen peels anthocyanin extract has higher absorbance than single dye.

Table 1. Parameters of DSSCs sensitized by chlorophyll, anthocyanin, and mixed dyes.

Dye Source	Photos	I_{max} (mA)	V_{max} (V)	I_{sc} (mA)	V_{oc} (V)	FF	η (%)
Moss	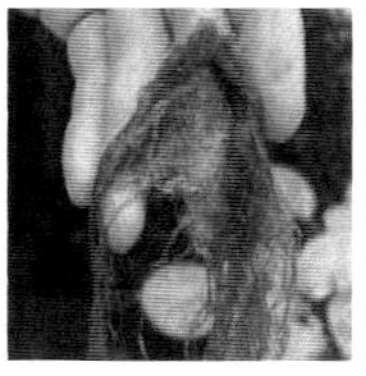	0.30	0.32	0.46	0.39	0.56	0.049
Mangosteen peels		0.25	0.34	0.38	0.46	0.48	0.042
Moss + mangosteen peels	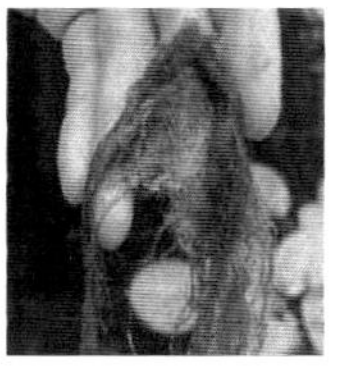 + 	0.83	0.37	1.13	0.60	0.46	0.154

Table 1 represented that DSSC by the dyes mixture can achieve better performance for open-circuit voltage (V_{oc}), short-circuit current (I_{sc}), and fill factor (*FF*). It exhibited great adsorption of dyes mixture from chlorophyll and anthocyanin onto the TiO_2 electrode, the intensity of light absorption was high with over a wider range. Theoretically, more higher light absorption will increase the number of excited electron from the HOMO to the LUMO state of dye. Thus, the measured currents will also increase. Therefore, the efficiency DSSC based on dyes mixture from chlorophyll and anthocyanin also improved.

4. Conclusion

The performance of dye-sensitized solar cells based on the single and mixture narutal dyes has been reported in this article. The dyes mixture from chlorophyll and anthocyanin extract has higher light absorption at visible spectrum range as compared to single dye. Moreover, the adsorption of the dyes mixture onto the TiO_2 electrode has higher absorbance than single dye. The DSSC based on dyes mixture performed an open-circuit voltage (V_{oc}), short-circuit current (I_{sc}), and conversion efficiency of 600 mV, 1.13 mA, and 0.154%, respectively. It suggest, the dyes mixture can improved the DSSC performance.

Acknowledgments

This work was supported by Higher Education and Institute for Research and Community services, Universitas Sebelas Maret by PU UNS Grant Program (Contract Number: 1075/UN27.21/PP/2017)

References

[1] Grätzel M 2005 *J. Inorg. Chem.* **44** 20 6841-6851
[2] Maurya I C, Srivastava P, and Bahadur L 2016 *Optical. Materials.* **52** 150-156
[3] Bhogaita M, Shukla A D, and Nalini R P 2016 *Solar. Energy.* **137** 212-224
[4] Chen J-Z, Yan Y-C, and Lin K-J 2010 *J. Chin. Chem. Soc.* **57** 5B 1180-1184.
[5] Nogueira AF, Longo C, and De Paoli M-A 2004 *J. Coord. Chems. Rev.* **248** 13 1455–1468.
[6] Bauer C, Boschloo G, Mukhtar E, and Hagfeldt A 2002 *J. Phys. Chem. B* **106** 12693–12704
[7] O'Regan B C and Grätzel M 1991 *Nature.* **353** 737-740
[8] Torchani A, Saadaoui S, Gharbi R, and Fathallah M 2015 *Curren. Applied. Phy.* **15** 307-312
[9] Calogero G, Citro I, Crupi C, and Marco G D 2014 Spectrochimica. Acta. Part A: Molecular. and Biomolecular. Spectroscopy. **132** 477-484
[10] Jeon B C, Kim M S, Cho M J, Choi D H, Ahn K-S, and Kim J H 2014 *J. Synthetic. Metals.* **188** 130-135.
[11] Ellis H 2013 *Characterization of Dye-Sensitized Solar Cells* (Sweden: Uppsala University)
[12] Chiba Y, Islam A, Watanabe Y, Komiya R, Koide N, and Han L Y 2006 *Japanese. J. Appl. Phy.* 24-28 **45** 1638-1640
[13] Buscaino R, Baiocchi C, and Barolo C 2008 *Inorganic. Chimica. Acta.* 3 **361** 798-805
[14] Shalini S, Prabhu R B, Prasanna S, Mallick T K, and Senthilarasu S 2015 *Renewable. and. Sustainable. Energy. Reviews.* **51** 1306-1325
[15] Chang H, Kao M, Chen T, Chen C, Cho K, and Lai X 2013 *Int. J. Photoenergy.* **2013** 159502 8 pages
[16] Arifin Z, Soeparman S, Widhiyanuriyawan D, and Suyitno 2017 *Int. J. Photoenergy.* **2017** 2704864 5 pages
[17] Taya S, El-Agez T M, El-Ghamri HS, and Abdel-Latif M S 2013 *Int. J. Mater. Sci. Appl.* **2** 37-42
[18] Pratiwi D D, Nurosyid F, Supriyanto A, and Suryana R 2016 *J. Phy. Conf. Series.* **776** 012007
[19] Boyo A O, Abdulsalami I O, Oluwa T, Oluwole S O, and Umar A 2013 *Sci. J. Phys.* **182** 4 Pages
[20] El-Agez T M, El Tayyan A A, Al-Kahlout A, Taya S A, and Abdel-Latif M S 2012 *Int. J. Mater. Chem.* **2** 105–110
[21] Pratiwi D D, Nurosyid F, Supriyanto A, and Suryana R 2017 *J. Phy. Conf. Series. Mater. Sci. Eng.* **176** 012012

Influence of electrode spacing on the efficiency of dye-sensitized solar cell

S R Alfidharisti, F Nurosyid*, A Supriyanto, R Suryana and Y Iriani

Department of Physics, Faculty of Mathematics and Natural Sciences, Universitas Sebelas Maret, Ir. Sutami Street No. 36 A Kentingan, Surakarta, Indonesia 57126

*Email : nurosyid@yahoo.com

Abstract. This study reported the fabrication of Dye-Sensitized Solar Cell (DSSC) with optimization of electrode spacing. Optimization of electrodes in DSSC will affect the amount of electrolyte solution provided. The DSSC fabrication in this study consisted of coating the TiO_2 semiconductor as the working electrode, Platinum as the catalyst on the counter electrode, the Dye Ruthenium Complex as a photosensitizer, and the electrolyte solution as the electron transport media. The spacer between the working electrode and the counter electrode was varied by five variations with the thickness of 38 μm, 76 μm, 114 μm, 152 μm, and 190 μm. The DSSC was characterized using Keithley *I-V* Meter to know the efficiency of DSSC. The characterization showed that the best DSSC efficiency was at 76 μm electrode spacing.

1. Introduction

Dye-Sensitized Solar Cell (DSSC) is the third generation of solar cells which synthetic or organic dye are sensitized inside [1-2]. The advantages of DSSC are the low-cost production but the efficiency resulted is still relatively small so many research is done to improve it [3-6]. DSSC consists of several components that include Transparent Conductive Oxide (TCO) substrate as a working electrode and a counter electrode. A semiconductor layer is coated on the conductive side of the TCO glass as working electrode and the catalyst is coated on the TCO glass as counter electrode, dye as a photosensitizer and electrolyte solution as electron transport media (via redox reactions) [7-10]. Since 1991, the efficiency of DSSC obtained has increased from 7.1% to 13% [11].

Several studies have been reported to improve the efficiency and lifetime of DSSC, the optimization of electrolyte solution for example. Grätzel research group reported that one of the difficulties to keep the lifetime of a solar cell is the decomposition of sealing and electrolytes [12]. Electrolyte optimization can be investigated by varying the type or composition of the electrolyte. El-Shater *et al.* (2015) reported the effect of spacer thickness on the efficiency of DSSC. They used 10 μm, 20 μm, 30 μm, 40 μm, and 50 μm spacer thickness variations. The efficiency resulted was 2.47 % at 10 μm spacer and it increased to 7.21 % at 40 μm and then it decreased to 5.92 % at 50 μm [13]. Taleb *et al.* (2015) reported a study to improve the DSSC performance by using 25 μm, the polymer spacer between the working electrode and counter electrode and it was filled by iodide/tri iodide (I^- / I_3^-) liquid electrolytes. The efficiency resulted was 4.3% [14].

Optimization of electrode spacing in DSSC affected the amount of electrolyte solution provided. Electrolyte filled the space between the working electrode, and the counter electrode should be optimal if it is excessive, it caused the dye degradation and counter electrode corrosion. On the contrary, less electrolyte was easier to dry and resulted in the lower performance of DSSC.

2. Methods

Substrates used for working and counter electrodes were fluoride-doped tin oxide (FTO) glass. TiO_2 transparent paste was used for semiconductor layer on working electrode. Platinum was used for a catalyst layer on the counter electrode. Ruthenium Complex Dye was synthetic dye which used for photosensitizer. Electrolyte solution was used for electron transport media and 38 μm spacer.

Working electrode preparation using TiO_2 paste was mixed with ethanol in ratio 1 gram: 1 ml and it was coated on conductive side of FTO glass in 0.5 cm^2 active area by screen printing method. TiO_2 films were dried on the hotplate at 130 ^{0}C for 10 minutes. It was annealed in a furnace at 500 ^{0}C for 10 minutes. Working electrode was soaked in Ruthenium Complex Dye for 24 hours.

Counter electrode preparation using platinum liquid was coated by brush painting method. Counter electrode coated was dried on the hotplate at 250 ^{0}C for 10 minutes. The electrolyte solution was made by using potassium iodide (KI), poly ethylene glycol (PEG) 400, and iodine (I_2). The first solution was made by the mixture of 10 ml PEG and 0,8 gram KI. It was mixed with the second solution made by the mixture from 0.127 gram I_2 and 10 ml PEG. The electrolyte solution was stirred for 30 minutes to get the homogenous solution [16].

DSSC consisted of the working and counter electrode was assembled by 38 μm, 76 μm, 114 μm, 152 μm, and 190 μm spacers into sandwich type and it was filled by the electrolyte solution. The DSSC was characterized using Keithley *I-V* Meter measurement with *Xenon* 1000 W/m^2 illumination.

3. Result and Discussion

Figure 1 shows *I-V* curves for DSSC with spacer variations.

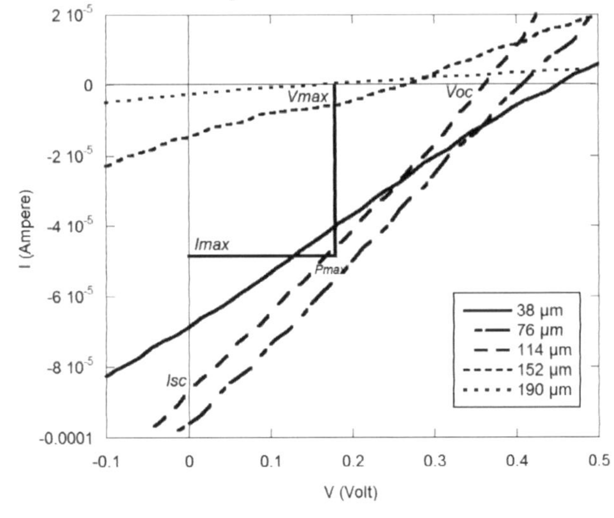

Figure 1. *I-V* curves for DSSC with spacer variations

Table 1. The parameters of DSSC with spacer variations

Electrode spacing (μm)	V_{oc} (V)	I_{sc} (A) (10^{-5})	V_{max} (V)	I_{max} (A) (10^{-5})	P_{max} (Watt) (10^{-5})	*FF*	η (%) (10^{-2})
38	0.44	6.37	0.20	3.93	0.80	0.28	1.61
76	0.40	10.00	0.23	4.91	0.11	0.29	2.31
114	0.35	7.94	0.19	4.87	0.92	0.32	1.85
152	0.29	1.60	0.17	1.00	0.17	0.37	0.34
190	0.14	2.61	0.13	0.09	0.01	0.33	0.02

Based on Table 1, the highest efficiency of DSSC was at 76 μm spacer, and it decreased at 114 μm until 190 μm. The thicker the spacer, the more the electrolyte was added, and it influenced the efficiency of DSSC. Electrolyte solution consisted of a redox pair served as a substitute for excited electrons of the dye. The redox pair on the electrolyte consisted of I^- and I_3^-. Tri iodide was formed on the TiO_2 surface while the redox reaction diffused to the catalyst. The redox reaction was accelerated by a catalyst deposited on the counter electrode [17]. Tri iodide reduced back to iodide by electrons arrived from the external circuit [18-19].

The less electrolyte in DSSC would lead to dry and its amount was probably not equal to number of excited electron so it was not able to fill the electron hole in dye as 38 μm spacer which had the lower efficiency. The excessive electrolyte caused the dye degradation and counter electrode corrosion with decreasing DSSC performance [2, 20]. Dye degradation caused its function to be not optimal to inject electrons into the TiO_2 semiconductor layer [21]. The optimal electrode spacing was at 76 μm with the efficiency of 0.0231 %. It can be seen from the value of V_{max} and I_{max} produced are higher than the others that are 0.23 Volt and (4.91 x 10^{-5}) Ampere.

4. Conclusion

It reported the influence of electrode spacing on the efficiency of the dye-sensitized solar cell by varied the spacers. The highest efficiency of DSSC was at 76 μm spacer, and it decreased at 114 μm until 190 μm. It explained that the optimal spacer for this DSSC was 76 μm.

5. Acknowledgment

This work is supported by Higher Education and Institute for Research and Community services, Universitas Sebelas Maret by PUPT Grant Program (Contract Number : 1075/UN27.21/PP/2017).

References

[1] Wang YC and Cho CP 2017 Journal. of. Photochemistry. and. Photobiology. A. Chemistry. **332** 1-9
[2] Li K, Wang Y, Sun Y, and Yuan C 2010 *Materials. Science. and. Engineering. B.* **175** 44-47
[3] Senthil RA, Theerthagiri J, Madhavan J, Ganesan S, and Arof AK 2017 *Journal. of. Physics. and. Chemistry. of. Solids.* **101** 18-24
[4] Wu J, Lan Z, Hao S, Li P, Lin J, Huang M, Fang L, and Huang Y 2008 *Pure. and. Applied. Chemistry.* **80** 2241-2258
[5] Ela SE, Ocakoglu K, Tarnowska A, Vakuliuk O, and Gryko DT 2015 *Dyes. and. Pigments.* **114** 129-137
[6] Jalali M, Moakhar RS, Kushwaha A, Goh GKL, Noori NR, and Sadrnezhaad SK 2015 Electrochimica. Acta. **169** 395-401
[7] Pablo CV, Enrique RR, Jose SG, Enrique MP, Juan LH, and Eddie AM 2016 *Materials. Today. Proceedings.* **3** 194-200
[8] Prasetyo YH, Wahyuningsih S, and Suryana R 2014 *Jurnal. Fisika. Indonesia.* **18** 1410-2994
[9] Bahramian A and Vashaee D 2015 Solar. Energy. Materials. and. Solar. Cells. **143** 284-295
[10] Chen L, Tan W, Zhang J, Zhou X, Zhang X, and Lin Y 2010 *Electrochimica. Acta.* **55** 3721-3726

[11] Wu C, Chen B, Zheng X, and Priya S 2016 *Solar. Energy. Materials. and. Solar. Cells.* **157** 438-446

[12] Alwani MA, Mohamad AB, Ludin NA, Kadhum AA, and Sopian K 2016 *Renewable. and. Sustainable. Energy. Reviews.* **65** 183-215

[13] El-Shater RE, Galeil MMA, Kawamura G, and Matsuda A 2015 *Journal. of. Nanomaterials.* **2015**

[14] Taleb A, Mesguich F, Herissan A, Justin CC, Yanpeng X, and Dubot P 2015 *Solar. Energy. Materials. Solar. Cells.* **148** 52-59

[15] Su'ait MS, Rahman MY, and Ahmad A 2015 *Solar. Energy.* **115** 452-470

[16] Pratiwi D, Nurosyid F, Supriyanto A, and Suryana R 2016 *Journal. of. Physics. Conference.* Series. **776**

[17] Deb SK 2004 Solar. Energy. Materials. and. Solar. Cells. **88** 1-10

[18] Macaira J, Andrade L, and Mendes A 2013 *Renewable. and. Sustainable. Energy. Reviews.* **27** 334-349

[19] Mahalingam S and Abdullah H 2016 Renewable. and. Sustainable. Energy. Reviews. **63** 245-255

[20] Kim SJ, Kwon J, Nam JK, Kim W, and Park JH 2016 *Electrochimica. Acta.* **227** 95-100

[21] Susanthy D, Gusnedi, and Kamus Z 2014 *Pillar. of. Physics.* **1** 33-40

IOP Conf. Series: Journal of Physics: Conf. Series **909** (2017) 012027 doi:10.1088/1742-6596/909/1/012027

Optimization of process parameters in the RF-DC plasma N_2-H_2 for AISI420 *molds* and *dies*

Hengky Herdianto[1], D.J. Djoko H. Santjojo[2,*)] and Masruroh[3]

Department of Physics, Faculty of Mathematics and Natural Science,
University of Brawijaya, Veteran Street, Malang 65145, INDONESIA

E-mail: [1]hengkyherdianto@gmail.com; [2]dsantjojo@ub.ac.id; [3]ruroh@ub.ac.id
*) Corresponding author

Abstract. The RF-DC plasma N_2-H_2 was used to make precise AISI420 *molds* and *dies* have complex textured geometry. The quality of the *molds* and *dies* directly affect the quality of the produced parts. The excellent examples of *molds* were used for injection molding lenses and *dies* used for the precision forging of automotive drive train components. In this study, a temperature, DC bias, and duration as process parameters of the RF-DC plasma N_2-H_2 have been optimized for *molds* and *dies* fabrication. The mask-less micro-patterned method was utilized to draw the initial 2D micro patterns directly onto the AISI420 substrate surface. The unprinted substrate surfaces were selectively nitrided by the RF-DC plasma N_2-H_2 at 673 K for 5400 s by 70 Pa with hollow cathode device. Energy Dispersive X-ray was utilized to describe the nitrogen content distribution at the vicinity of the border between the unprinted surfaces. This exclusive nitrogen mapping proves that only the unprinted parts of the substrate have high content nitrogen solutes. XRD analysis was performed to investigate whether the iron nitrides were precipitated by RF-DC plasma N_2-H_2 in the AISI420.

1. Introduction

The quality of *molds* and *dies* directly affect the quality of the produced parts. The excellent examples of *molds* were used for injection molding lenses and *dies* used for the precision forging of automotive drive train components [1]. The case-hardened AISI420, graded by STAVAX in the commercial is used as a mold for injection molding [2]. Micro- and nano-textures on the metal and polymer surfaces of parts and components have functions to reduce the friction and wear in [3,4] and to improve the joining strength in [5,6].

EDM (Electric Discharge Machining), laser machines [7], LIGA and photolithography process is the way to create micro and nano textures onto a substrate [8]. These four techniques have advantages in their applications but are expensive on an industrial. In the present study, the original 2D micro patterns were directly drawn onto the *molds* and *dies* surfaces. This printed *molds* and *dies* were subjected to the plasma N_2-H_2 by using the printed area as a mask. Since the masked area is free from the plasma N_2-H_2, the unmasked or unprinted area is selectively plasma nitrided and hardened.

After [2,7,9,10], the non-traditional processing was proposed to make micro-texturing onto *molds* and *dies* with aid of the RF-DC Plasma N_2-H_2. The process parameters of plasma N_2-H_2 are important factors for obtaining desired material properties. The species of plasma are a key point in the functionalization process of the metal surface for *molds* and *dies* micro pattern.

Properly selected parameters can improve species of plasma and material properties e.g. surface hardness, wear resistance, corrosion resistance [11,12], and suitability [13]. The process parameters of plasma N_2-H_2 consist of temperature, gas flow rate, stress, pressure, and duration [14]. In those previous works, was found optimal parameters [7,9,10] that can be explained in Table 1.

Table 1. Summarizes the experimental process parameters for RF-DC plasma N_2-H_2 in those previous works [7,9,10]

Process	Parameters
Pre-sputtering	DC (-500 V), RF (0 V), pressure (75 Pa), temperature (693 K), duration (1800 s), and carrier gas (N_2 only)
Nitriding	DC (-300 V), RF (250 V), pressure (75 Pa), temperature (693 K), duration (14400 s), carrier gas (N_2 + H_2), and with partial pressure (N_2 = 100 ml/min; H_2=20 ml/min)

The used DC bias at -500 voltages and 1800 s in the pre-sputtering process have a negative impact on the quality of masks that have been printed using Ink-Jet printing technology with carbon ink. Masks undergo layer vanished so that it looks faded both at the edge and on the micro pattern. The temperature was used at 693 K could potentially increase the precipitation of CrN and Cr_2N. According to [15] and [16], the precedence of CrN and Cr_2N were hard and wear-resistant characteristics but reduces corrosion resistance. So the low-temperature nitriding process can be avoided and minimize chromium migration.

The used gas pressure 75 Pa and 14400 s also increased the rate of vanished on the mask layer. In this study, a temperature, DC bias, and duration as process parameters of the RF-DC plasma N_2-H_2 have been optimized for *molds* and *dies* fabrication. The possibility of fine micro-texturing by the plasma N_2-H_2 must be demonstrated by using the well-aligned micro-patterns.

2. Experiment

2.1. Maskless patterning with use of nano-carbon ink

The 2D micro patterns were drawn onto the AISI420 specimen and *molds*. Figure 1 depicts a typical micro-pattern drawn on the AISI420 sheet with use of the nano-carbon ink. Even when using the same CAD data for micro patterning, there were two manners in drawing: the positive-image drawn as depicted in Figure 1a and the negative-image drawn as shown in Figure 1b.

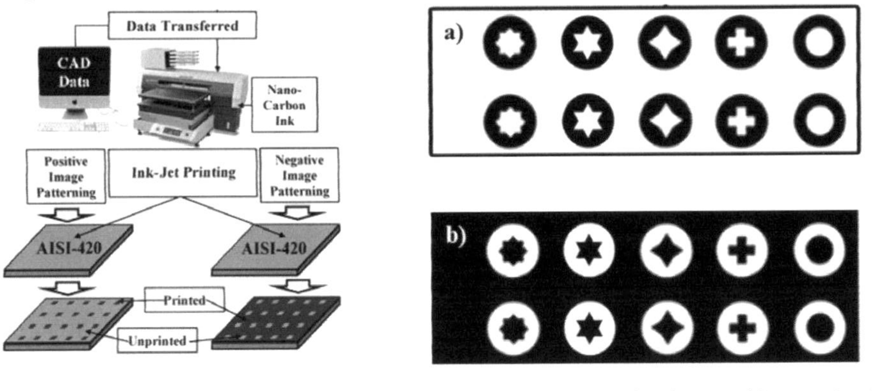

Figure 1. Micro patterned onto the AISI420 sheet surface: a) positive image, b) negative image

At present, the drawing area was limited by 25 x 25 mm^2 with the spatial resolution down to 1 μm. The various kinds of micro-units with geometric complexity were straight forwardly designed by CAD and drawn on the AISI420 substrates. This mask-less micro-patterning was useful to draw any designed micro-head pattern with the complex geometries and within the high spatial resolution.

2.2. *RF-DC plasma N_2-H_2 system*

The RF-DC plasma N_2-H_2 was employed for selective nitrogen concentration into their unprinted part. RF-DC plasma N_2-H_2 system was set-up for selective nitriding and hardening of AISI420. Different from the PECVD (Plasma Enhanced Chemical Vapor Deposition) or DC-pulse nitriding, where the plasmas were ignited and generated at the frequency of 13.56 MHz or its multiples, the present RF-DC plasma N_2-H_2 system has no mechanical matching box with a slow response time of 1 s to 10 s to adjust the applied power. Since both the input and output powers were automatically matched by frequency adjustment around 2 MHz, the matched response time was only limited to 1 ms at most. This prompt power control provides to make full use of mesoscopic plasma pressure range over 50 Pa.

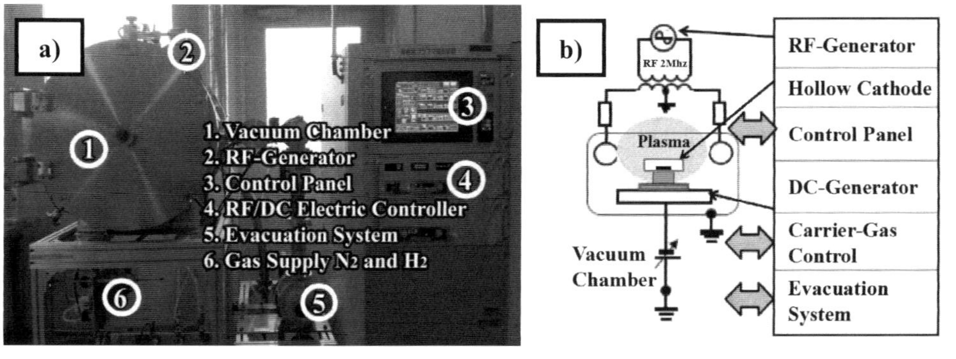

Figure 2. RF-DC plasma N_2-H_2 system: a) outlook of system, and, b) schematic illustration of system

The differences from the conventional processes, the vacuum chamber was electrically neutral so that RF-power and DC-bias should be controlled independently from each other. A dipole electrode was utilized to generate RF-plasma; DC bias was directly applied to the specimens. Heating unit was located under this DC-biased cathode plate. In the following nitriding experiments, the specimens were located on the cathode table before evacuation down to the base pressure of 0.1 Pa.

Then, nitrogen gas was first introduced as a carrier gas for heating. After heating to the specified holding temperature, the nitrogen pre-sputtering was started at the constant pressure. After pre-sputtering, the hydrogen gas was added to nitrogen gas with the specified partial pressure ratio. In particular, the partial pressure ratio of nitrogen to hydrogen gases is constant by 5 to 1 as a standard condition by controlling the gas flow rate to be 100 ml/min for N_2 gas and 20 ml/min for H_2 gas. Both of pressure (P) and temperature (T) controls were automatically performed with the tolerance of $\Delta P <$ 1 Pa in deviation of partial pressure and $\Delta T <$ 1 K in temperature fluctuation.

2.3. *Observation and measurement*

The plasma condition was observed and identified by using the optical emissive-light spectroscopy (EOS; PMA-11, Hamamatsu, Co. Ltd). SEM (Scanning electron microscope; Shimadzu, Corp) was used to observe printed mask pattern of the specimen before nitriding. EDX (Energy Dispersive X-ray; SSX-550, Shimadzu, Co. Ltd.) was utilized to analyze the nitrogen content distribution at the vicinity of the border between the printed and unprinted surface. The hardness of nitride specimen was measured by Vickers micro hardness testing (Mitsutoyo, Co. Ltd) while XRD (Smartlab, Rigaku, Co., Ltd.) was used for analyzing the microstructure of specimen.

3. Results and discussion

3.1. Species diagnostic of RF-DC plasma N_2-H_2

Through the quantitative plasma diagnosis, the generated species in the plasmas were detected and monitored by EOS. Figure 3 and Table 2 shows an emissive light spectrum measured in the plasmas for the pre-sputtering process (DC with N_2 only) and nitriding process (RF-DC with N_2 and H_2). NH, as well as the activated nitrogen atom and molecules, were detected in both spectra. With the use of the mixed nitrogen and hydrogen gases, every peak of intensity was enhanced even with a little addition of hydrogen gas. The species of N_2^+, N_2, N_2^*, N-1, and NH might be responsible for high infiltration of nitrogen atoms into the AISI420.

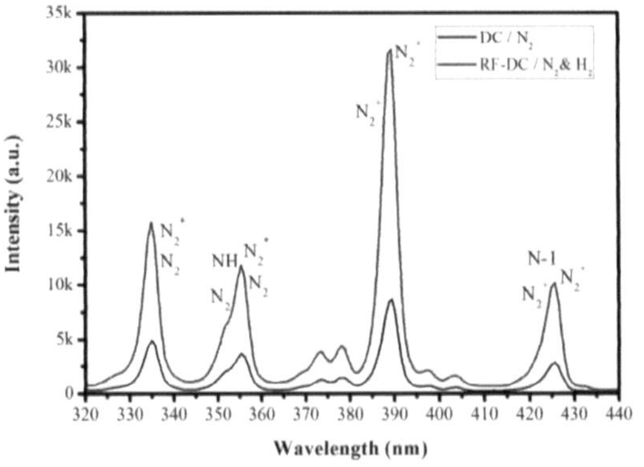

Figure 3. The EOS of plasmas N_2-H_2

Table 2. Species of plasma N_2-H_2 [17-21]

Species	Transition	λ (nm)	Reference
N_2^+	$B^2\Sigma_u^+ \rightarrow X^2\Pi_g^+$ (1→1)	388.43	[17]
	$B^2\Sigma_u^+ \rightarrow X^2\Pi_g^+$ (0→0)	391.4	[18]
	$B^2\Sigma_u^+ \rightarrow X^2\Pi_g^+$ (1→2)	423.65	[19]
	$B^2\Sigma_u^+ \rightarrow X^2\Pi_g^+$ (0→1)	427.81	[19]
N_2	$C^3\Pi_u \rightarrow B^3\Pi_g$ (0→0)	337.1	[18]
	$C^3\Pi_u \rightarrow B^3\Pi_g$ (1→2)	353.67	[17]
	$C^3\Pi_u \rightarrow B^3\Pi_g$ (0→1)	357.69	[19]
N_2^*		337.1	[20]
		357.6	[21]
N–1	$2s^22p^2(^3P)3p \rightarrow 2s^22p^2(^1D)5d$	425.01	[17]
NH	$A^3\Pi \rightarrow X^3\Sigma^-$ (0→0)	336.0	[18]

3.2. Optimization of the DC bias and duration in the pre-sputtering process

The combination of RF and DC sources produce high state plasma because RF field creates extended plasmas and used together with the enhanced sputtering and biasing effect of the DC source. This effect is expected to increase the effectiveness of nitriding process and save energy consumption [22]. DC bias at -500 voltage and 1800 s in the pre-sputtering process [7,9,10] have a negative impact on the quality of masks that have been printed using Ink-Jet printer with carbon ink. In the present study, shows that pre-sputtering parameters with DC bias at -450 volts and duration at 400 s can minimize the carbon layer degradation. The content of carbon layer lasts in the range of 10-15 wt%.

3.3. Optimization of the temperature and duration in the nitriding process

The variations in the pre-sputtering and nitriding process temperatures were used 573 K, 593 K, 613 K, 633 K, 653 K, 673 K, and 693 K. The results were shown in Figure 4, same as previous research resulted [22] with increasing temperature, the deposited nitrogen atom at substrate surface has enough energy and longer chance to form homogeny of nitride layer, and therefore the maximum surface hardness increases as depicted in Figure 5. As nitriding temperature is increased the vibrations of atoms in the substrate will increase thereby increasing the diffusivity of nitrogen into the substrate so that the thickness of the nitride layer increases.

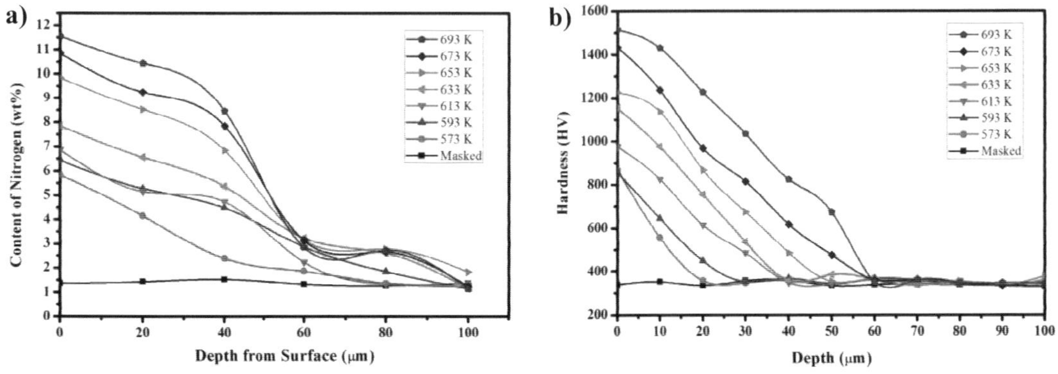

Figure 4. a) Distributions of nitrogen solute in the depth from the surface of the masked and unmasked region on the cross-section, b) micro-hardness depth profiles of RF-DC plasma N_2-H_2 onto AISI420 specimens for different temperature

Figure 4a shows concentration depth profiles of nitrogen elements mapping measured by Energy Dispersive X-ray (EDX) on samples the cross-section. The concentration of the nitrogen near the surface reached up to 11.54 wt% and decreased drastically at a depth of 60 μm. Micro-Vickers hardness testing was performed to describe the surface hardness distribution as well as the hardness depth profile. The applied load was constant by 0.50 N (or 50 g). Decreasing of nitrogen concentration has an impact on hardness reduction as depicted in Figure 4b.

Figure 5b depicts the measured micro-Vickers hardness profile along the scanning yellow line in Figure 5a. The measured hardness was nearly equal to the original matrix hardness of 348 Hv in the masked regions of the specimen surface. The average hardness reaches 1468 Hv in the unmasked or the nitrided regions. A sharp change in hardness from 348 Hv up to 1468 Hv was detected in Figure 5b across the border between the masked and unmasked regions.

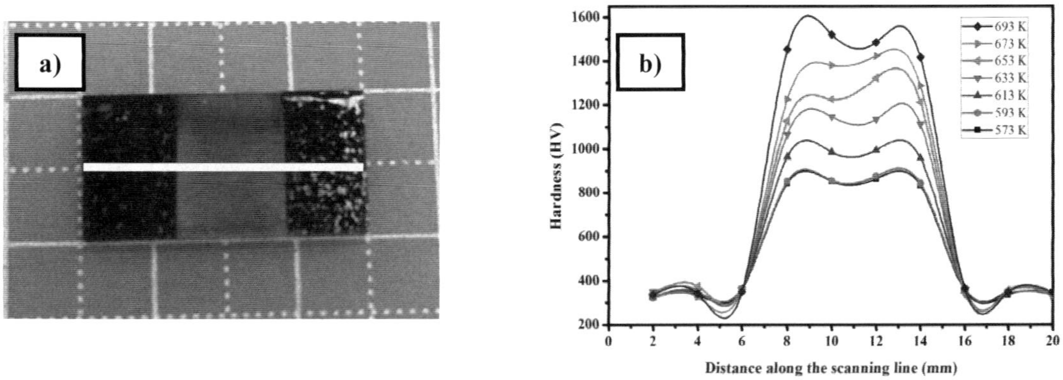

Figure 5. a) AISI420 specimen after nitriding and polishing, b) measured hardness profile along the yellow scanning line in Figure 5a

The diffusion of nitrogen significantly increase and nitriding process more effective if temperature approaching 500°C. This result [22] was due to the fact that Cr precipitates and reacts with N to form CrN at temperatures above 400°C and it increases when the temperature approaches 500°C. At higher temperature Cr precipitates stronger so that nitrogen has longer chance to react with Cr to produce CrN and Cr_2N. Although at a temperature of 693 K is more hardness than 673 K, this optimization selects 673 K due to minimizing CrN and Cr_2N. The precipitation of CrN and Cr_2N not only have a hard and wear-resistant characteristic but also low corrosion resistance [23].

The duration of RF-DC plasma N_2-H_2 is the most effective at 5400 s and condition of the micro pattern was maintained with good quality on this parameter. In the present study, shows that nitriding parameters with duration at 5400 s can minimize the carbon layer degradation. The content of carbon layer lasts in the range of 7.63 wt%. Figure 6b depicts the measured micro-hardness depth profiles the original hardness that near surface reached up to 1214 Hv and while it remained to be 335 Hv at a depth of 60 μm from the surface.

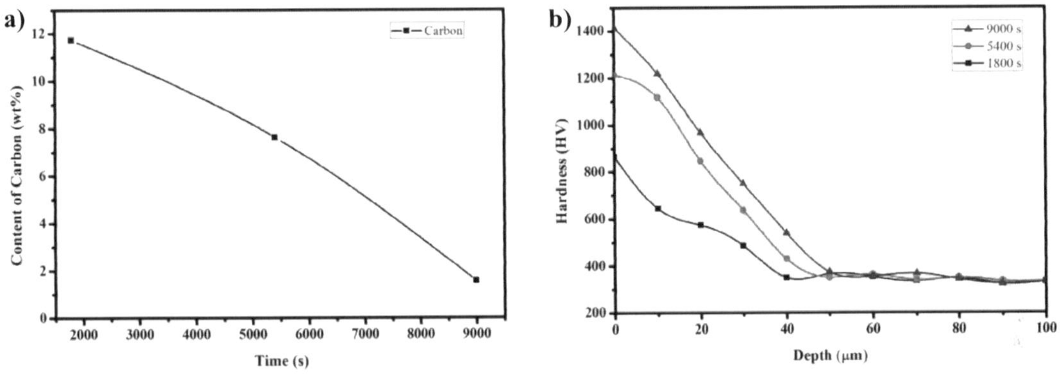

Figure 6. a) Content of carbon layer for different nitriding duration, b) micro-hardness depth profiles of RF-DC plasma N_2-H_2 onto AISI420 specimens for different nitriding duration

Figure 7 shows that N_2^+ species (388.43 nm) and N_2^+ (391.40 nm) have consistently increased when nitriding process temperature was increased. In addition, NH species tend to be stable at a high intensity if compared with N_2, N_2^*, and N-1 species. So the most plausible plasma species in the nitriding process along with temperature optimization are N_2^+ and NH.

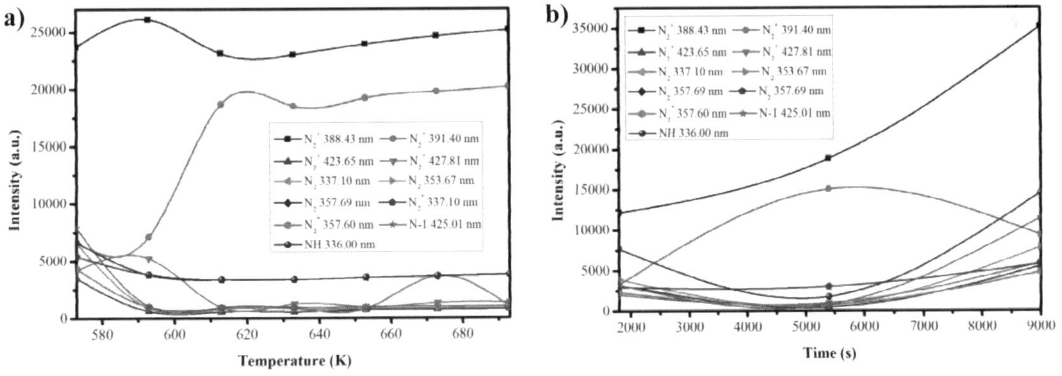

Figure 7. a) Effect of nitriding temperature on intensity of plasma species, b) effect of nitriding duration on intensity of plasma species

The formation of N_2^+ species occurs by the ionization reaction chain $N_2 + e^- \rightarrow N_2^+ + 2e^-$. The N_2^+ species act directly on the Fe substrates surfaces at AISI420 through a diffusion process. The NH species are the result of the dissociation process by reaction $N + H_2 \rightarrow NH + H$. In addition, NH is formed by ionization by reaction $N^+ + H_2 \rightarrow NH + H^+$. The role of NH as a damper from the oxide surface will increase the diffusion of nitrogen atoms. Additionally, saturates trapped nitrogen to increase the diffused species. Species formed between nitrogen and hydrogen neutralizes the barrier effect of surface oxidation during diffusion.

Geometry and dimensions were observed and measured by SEM after ink jet printing and RF-DC plasma N_2-H_2 in the present parameters that were optimized. Energy Dispersive X-ray analysis was utilized to describe the nitrogen content distribution at the vicinity of the border between the unprinted surface. As depicted in the first group of Figure 8a, 8b, 8c and second group of Figure 8g, 8h, 8i, the nitrogen maps selectively on the unprinted surface area both in the positive and negative images. No nitrogen was presented in the printed areas. On the other hand, the carbon from the ink maps only on the printed surface area in both images. This exclusive carbon-nitrogen mapping proves that only the unprinted parts of the substrate should have high content nitrogen solutes.

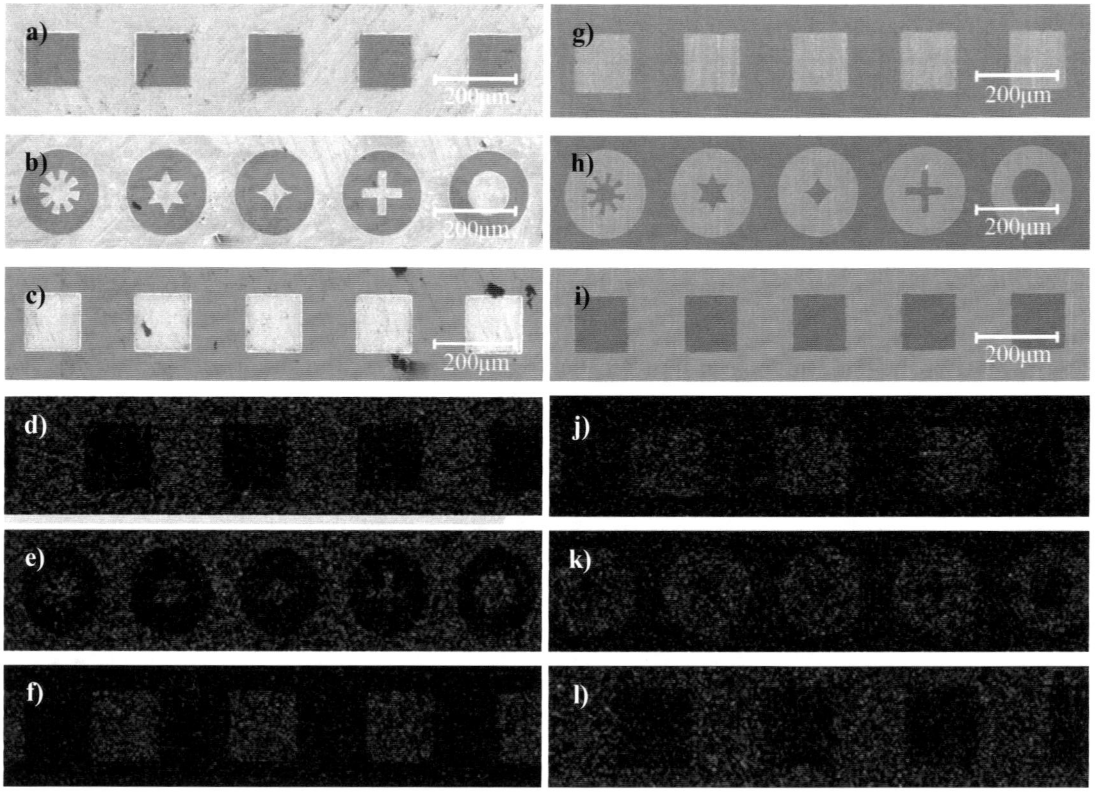

Figure 8. Printed micro positive model onto the AISI420 (a, b, c), printed micro negative model onto the AISI420 (g, h, i), nitrogen distribution of micro-patterns positive model onto the AISI420 (d, e, f), and nitrogen distribution of micro-patterns negative model onto the AISI420 (j, k, l)

The hardness profile was measured along a single yellow scanning horizontal line across the micro-pattern between the printed and unprinted regions. In order to demonstrate that the original printed pattern should be homogeneously transformed into the hardness profile pattern, a single square region was only left unprinted at the center of AISI420 specimen as shown in Figure 9a and 9b.

Figure 9c compares the hardness profiles measured both in the lateral and longitudinal directions across the mask. The less significant difference was seen in both hardness profiles; i.e. the hardness in the unprinted regions was 1220 Hv, and, it remains to be 335 Hv in the printed region. The steep change of hardness across the edge of masks reveals that the unprinted regions were nitrided to have much higher hardness than matrix hardness of AISI420. The printed regions were free from infiltration of nitrogen atoms into the matrix.

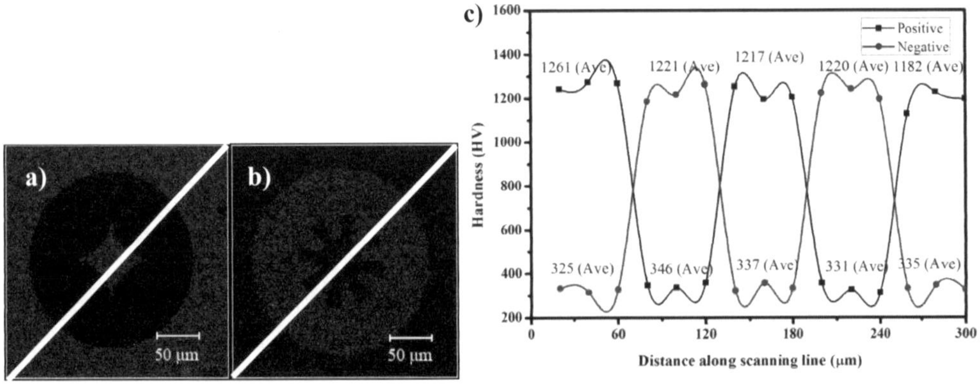

Figure 9. Transformation of 2D micro patterns to the hardness profile by the present RF-DC plasma N_2-H_2 for 5400 s at 673 K: a) micro-patterns positive model onto the AISI420 surface after nitriding, b) micro-patterns negative model onto the AISI420 surface after nitriding, c) measured hardness profile along the yellow scanning horizontal line in Figure 9a and 9b

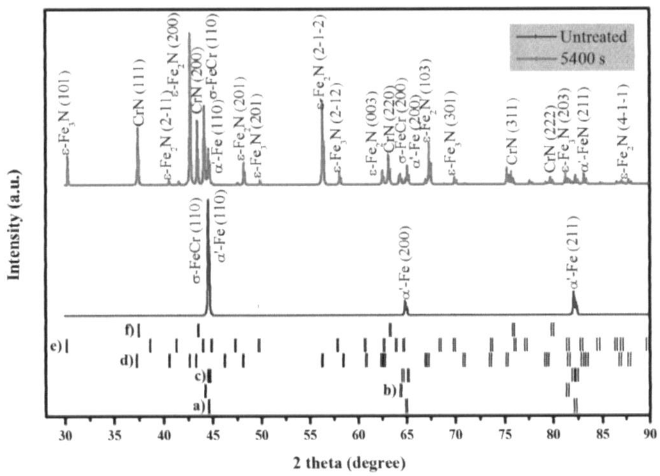

Figure 10. Diffractogram of RF-DC plasma N_2-H_2 during 5400 s (optimize of duration), untreated condition, and Bragg position of a) α'-Fe, b) σ-FeCr, c) α'-FeN, d) ε-Fe_2N, e) ε-Fe_3N, and f) CrN

The powdered diffractogram was analyzed result according to *FullProf Suite* program [24] of *Rietveld* [25,26] method. The results of lattice parameters of the crystal structure detailed were presented in Table 3. After [2], no nitrides were synthesized in the lower temperature plasma nitriding; the nitrogen atoms were present as a solute in the Fe/Cr lattices by occupying their vacancy sites.

XRD analysis was performed to investigate whether the iron nitrides were precipitated by RF-DC plasma N_2-H_2 at 673 K for 5400 s for AISI420 sheet-sample. Hardening mechanism of nitrided AISI420 generally caused by precipitation formation or solid solution atom in matrix specimen. Firstly, precipitation hardening which is usually generated by CrN (37,62°, 43,72°, 63,37°, 76,25°, and 80,31°) precipitation formation in the matrix specimen. Nitrogen atoms diffusing into matrix reacts with chromium in AISI420 [27]. Besides for the original α'-Fe (44,76°, 65,16°, and 82,52°) and σ-FeCr (44,24°) a new peak were detected at α'-FeN (82,51°), ε-Fe_2N (42,90°, 43,57°, 48,42°, 56,78°, 63,26°, and 67,62°), and ε-Fe_3N (30,12°, 49,67°, 57,81°, 73,49°, and 81,72°). Peak was related to the extended martensitic lattice by high nitrogen extraordinary solid solution just as stated in [2] and [28].

Tabel 3. Detailed lattice parameters due to *FullProf Suite* program [24] of *Rietveld* [25,26] analysis

Phase	Crystal system	Space group	a (Å)	b (Å)	c (Å)	Volume (Å^3)	α (°)	β (°)	γ (°)	Ref.
α'-Fe	cubic	Im-3m (229)	2.86802	2.86802	2.86802	23.590964	90	90	90	[29]
σ-FeCr	cubic	Im-3m (229)	2.90000	2.90000	2.90000	24.389002	90	90	90	[30]
α'-FeN	tetragonal	I4/mmm (139)	2.86350	2.86350	2.89330	23.723998	90	90	90	[31]
ε-Fe_2N	trigonal	P 12 (149)	4.79300	4.79300	4.41700	87.876532	90	90	120	[29]
ε-Fe_3N	trigonal	P312 (149)	4.66800	4.66800	4.36200	82.314819	90	90	120	[29]
CrN	cubic	Fm-3m (225)	4.14800	4.14800	4.14800	71.370078	90	90	90	[32]

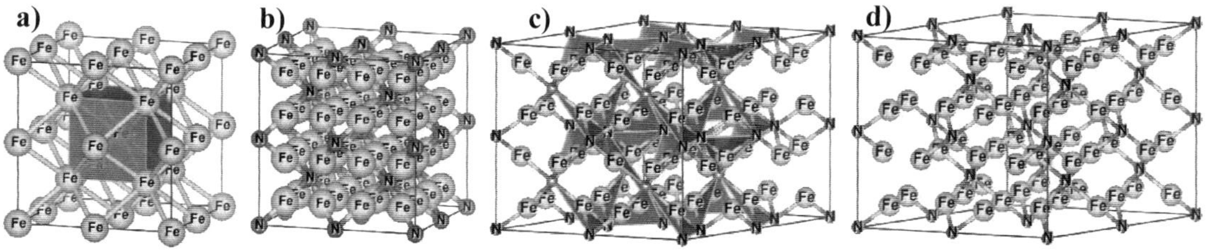

Figure 11. Schematic picture of the structure obtained using *Visualization for Electronic and STructural Analysis (VESTA) v.3.3.2* 64 bit [33]: a) α'-Fe (Im-3m), b) α'-FeN (I4/mmm), c) ε-Fe_2N (P312), and d) ε-Fe_3N (P312)

4. Conclusions

Through optimization of the nano-carbon ink contents, the printed patterns have sufficient heat resistivity even during RF-DC plasma in this present works, was found optimal parameters that can be explained in Table 4. Besides for the original α'-Fe and σ-FeCr a new peak were detected at α'-FeN, ε-Fe_2N, and ε-Fe_3N. The hardness on the unprinted surface was 1220 Hv while it remained to be 335 Hv on the printed surface.

Table 4. Summarizes the experimental process parameters for RF-DC plasma N_2-H_2 in this present works

Process	Parameters
Pre-sputtering	DC (-450 V), RF (0 V), pressure (70 Pa), temperature (room standard), duration (400 s), and carrier gas (N_2 only)
Nitriding	DC (-300 V), RF (250 V), pressure (70 Pa), temperature (673 K), duration (5400 s), carrier gas (N_2 + H_2) and with partial pressure (N_2 = 100 ml/min; H_2=20 ml/min)

Acknowledgment

The authors would like to thank Prof. Tatsuhiko Aizawa from Shibaura Institute of Technology (Japan) for the research facilities provided as well as guidance for the implementation of the research in LLC Nanofilm & Coat Laboratory, Ota-ku, Tokyo (Japan). The authors are also thankful to ASMAT Research Collaboration, the University of Brawijaya, which is funded through *Program Hibah Kerjasama Luar Negeri JSPS Dikti* with the contract number 063/SP2H/LT/DRPM/IV/2017 so that this research can be accomplished.

The authors would like to express gratitude to Dr. E.E. Yunata, Mr. Y. Seki, Mr. A. Farghari, and Mr. S. Kurozumi from Shibaura Institute of Technology (Japan), Dr. K. Wasa from TECDIA, Co. Ltd. (Japan), and Mr. H. Morita from LLC Nanofilm & Coat Laboratory, Ota-ku, Tokyo (Japan) for their help in experiments.

References

[1] Taylan A, Blaine L and Y.C Yen 2001 *J. of Manu. Tech.* **50**(2) 404
[2] Tatsuhiko A, Tatsuya F and Hiroshi M 2016 *J. of Manu. Rev.* **3**(5) 1
[3] I Etsion 2004 *J. of Tribo. Lett.* **17** 733
[4] T Czerwiec, G Marcos, T Thiriet, Y Guo and T Belmonte 2009 *Proc. of IOP Conf. Series: Mater. Sci. & Eng.* **5** 012012-1
[5] Tatsuhiko A, Saya S and Tetsuya Y 2014 *Proc. 9th ICOMM* **15** 26D4-15-p.1
[6] A.V Filippov, S-Y Tarasov, O.A Podgornyh, N.N Shamarin and E.O Filippova 2016 *AIP Conf. Proc.* **1783** 020057-1
[7] Takahisa K, Tatsuhiko A and Tetsuya Y 2015 *J. of Manu. Rev.* **2**(2) 1
[8] Tatsuhiko A, Masahiro T and Tatsuya F 2014 *J. of Procedia Eng.* **81** 1427
[9] Tatsuhiko A and Tetsuya Y 2014 *Proc. 9th IWMF* **12** 55
[10] Tatsuhiko A, Hiroaki S and Tetsuya Y 2015 *Proc. MATEC* **21** 09002-p.1
[11] I Alphonsa, A Chainani, P.M Raole, B Ganguli and P.I John 2002 *J. of Surf. & Coat. Tech.* **150**(2-3) 263
[12] V Valasamudram, S.S.M Nazirudeen, P Chandramohan and K.P Thenmozhi 2008 *J. Anti-Cor. Meth. & Mater.* **55**(2) 73
[13] C.A Figueroa, F Alvarez, D.R.G Mitchell, G.A Collins and K.T Short 2006 *J. Vac. Sci. Technol. A* **24**(5) 1795
[14] Pilch O and Hruby V 2016 *J. of Sol. Sta. Phen.* **258** 395
[15] M.J Baldwin, G.A Collinss, M.P Fewell, S.C Haydon, S Kumar, K.T Short and J Tendys 1997 *J. Appl. Phys.* **36** 4941
[16] Yun-tao X, Dao-xin L and Dong H 2008 *J. of Surf. & Coat. Tech.* **202**(12) 2577
[17] A Saeed, A.W Khan, F Jan, H.U Shah, M Abrar, M Zaka-Ul-Islam, M Khalid and M Zakaullah 2014 *J. of Plas. Sci. & Tech.* **16**(5) 462
[18] L Petitjean and A Ricard 1984 *J. of Phys. D: Appl. Phys.* **17** 923
[19] M Hannemann, S Hamann, I Burlacov, K Börner, H-J Spies and J Röpcke 2013 *J. of Surf. & Coat. Tech.* **235** 564
[20] Yoon-Ho C, Ji-Hun K, Kwang-Hyun P, Won-Tae J and Y.S Hwang 2005 *J. of Surf. & Coat. Tech.* **193** 321
[21] K.S Suraj, P Bharathi, V Prahlad and S Mukherjee 2007 *J. of Surf. & Coat. Tech.* **202** 303
[22] Istiroyah, I.N.G Wardana and Dionysius J.D.H.S 2014 *J. of App. Mech. & Mater.* **493** 755
[23] M.J Baldwin, G.A Collinss, M.P Fewell, S.C Haydon, S Kumar, K.T Short and J Tendys *Jpn. J. of Appl. Phys.* **36** 4941
[24] Juan R-C 1993 *J. of Phys. B* **192** 58
[25] Vikash K, Swati K, Pawan K, Manoranjan K and Lawrence K 2015 *J. of Adv. Mater. Lett.* **6** 141
[26] A.L Ortiz, F.L Cumbrera, F Sanchez-Bajo, F Guiberteau and R Caruso 2000 *J. of the Euro. Cera. Soci.* **20** 1846
[27] Istiroyah, I.N.G Wardana and Dionysius J.D.H.S 2016 *J. of App. Mech. & Mater.* **836** 214
[28] Abdelrahman F and Tatsuhiko A 2017 *J. of Mater. Trans.* **58**(4) 699
[29] S.B Hendricks and P.R Kosting 1930 *Int. J. of Crys. Geo. Crys. Phys. Chem. Crys.* **74** 520
[30] T.I Badjuk, G.P Kushma and O.I Rybajlo 1974 *Proc. of Chern. Metal.* **17** 126
[31] Jette O, Thomas L.C, Kenny S and Marcel A.J.S 2008 *J. of Mater. Sci.* **43** 5358
[32] M.N Eddine, E.F Bertaut, M Roubin and J Paris 1977 *J. of Acta Cryst. B* **33** 3010
[33] Koichi M and Fujio I 2011 *J. of App. Crys.* **44** 1272

IOP Conf. Series: Journal of Physics: Conf. Series **909** (2017) 012028 doi:10.1088/1742-6596/909/1/012028

The construction of partner potential from the general potential Rosen-Morse and Manning Rosen in 4 dimensional Schrodinger system

Kristiana Nathalia Wea, A. Suparmi, C. Cari and Wahyulianti

Graduate School Program of Sebelas Maret University
Jln. Ir. Sutami 36A, Kentingan, Surakarta 57126, Indonesia

Email: nataliawea@gmail.com

Abstract. The solution of the Schrodinger equation with physical potential is the important part in quantum physics. Many methods have been developed to resolve the Schrodinger equation. The Nikiforov-Uvarov method and supersymmetric method are the most methods that interesting to be explored. The supersymmetric method not only used to solve the Schrodinger equation but also used to construct the partner potential from a general potential. In this study, the Nikiforov-Uvarov method was used to solve the Schrodinger equation while the supersymmetric method was used to construction partner potential. The study about the construction of the partner potential from general potential Rosen-Morse and Manning Rosen in *D*-dimensional Schrodinger system has been done. The partner potential was obtained are solvable. By using the Nikiforov-Uvarov method the eigenfunction of the Schrodinger equation in *D*-dimensional system with general potential Rosen-Morse and Manning Rosen and the Schrodinger equation in *D*-dimensional system with partner potential Rosen-Morse and Manning Rosen are determined. The eigenfunctions are different between the Schrodinger equation with general potential and the Schrodinger potential with the partner potential.

1. Introduction

The Schrodinger equation is the important part in quantum physics. The Schrodinger equation can describe the characteristics of particles in the quantum potential field. Many methods have been developed to solve the Schrodinger equation. For example the Nikiforov-Uvarov method[1-3], the supersymmetric method [4-6], the Ansatz wave function method [7-9], the asymptotic iteration method (AIM)[10-12], and other [13-14]. In this study, we used the Nikiforov-Uvarov method and the supersymmetric method. The Nikiforov-Uvarov method was used to solve the Schrodinger equation while the supersymmetric method was used to construction partner potential. The Nikiforov-Uvarov differential equation was the equation that formulated from the general hypergeometric differential equation.

Some study about construction partner potential has been done. For example the construction partner potential from the Hulthen potential [15] and the construction partner potential from the Hylleraas potential [16]. The main idea of the construction partner potential is using the supersymmetric operators, the ground state eigenfunction and the original potential. In this study, we choose potential Rosen-Morse and potential Manning Rosen as the original potential. The general form of Rosen-Morse and Manning Rosen potential are [5]

$$V_1(\theta_1)=\frac{\hbar^2}{2m}\left(\frac{\upsilon(\upsilon+1)}{\cos^2\theta_1}-2q\tan\theta_1\right)$$
$$V_i(\theta_i)=\frac{\hbar^2}{2m}\left(\frac{\upsilon(\upsilon-1)}{\sin^2\theta i}-2q\cot\theta_i\right) \tag{1}$$

where υ and q are potential parameters, m is mass and $\hbar$ is reduced Planck constant.

The first section of this paper we will review about Nikiforov-Uvarov method and supersymmetric method. The next section we will explain about the Schrodinger equation in D-dimensional system. The third section we will present about the solution of *D*-dimensional Schrodinger equation with potential Rosen-Morse and Manning Rosen using Nikoforov-Uvarov method. The next section we will present about construction partner potential from origin potentials Rosen-Morse and Manning Rosen in *D*-dimensional Schrodinger system. The next section we will present about the solution of *D*-dimensional Schrodinger Equation with partner potential Rosen-Morse and Manning Rosen. In this study, we obtained the eigenfunction of Schrodinger equation with potential original and Schrodinger equation with the partner potential.

2. Method

2.1. Nikiforov-Uvarov method

The Schrodinger equation with physical potential field reduced become hypergeometric differential equation type Nikiforov-Uvarov by the variable substitution method. The general equation of Nikiforov-Uvarov differential equation as follows [17]

$$\frac{\partial^2\Psi(s)}{\partial s^2}+\frac{\bar{\tau}(s)}{\sigma(s)}\frac{\partial\Psi(s)}{\partial s}+\frac{\bar{\sigma}(s)}{\sigma^2}\Psi(s)=0 \tag{2}$$

with $\sigma(s)$ and $\bar{\sigma}(s)$ are second order polynomials, $\bar{\tau}(s)$ is first orde polynomials. By set $\Psi=\varphi(s)y(s)$, Eq. (2) become

$$\sigma\frac{\partial^2 y}{\partial s^2}+\tau(s)\frac{\partial y}{\partial s}+\lambda y(s)=0 \tag{3}$$

The first part eigenfunction as follows

$$\frac{\phi'}{\phi}=\frac{\pi}{\sigma} \tag{4}$$

and we obtained the equation that will use to determined energy spectra and the second part eigenfunction y_n as follow

$$\pi=\left(\frac{\sigma'-\tau}{2}\right)\pm\sqrt{\left(\frac{\sigma'-\tau}{2}\right)^2-\sigma+k\sigma} \tag{5}$$

and

$$\lambda=k+\pi',\ \tau=\tau+2\pi,\ \lambda=\lambda_n=-n\tau'-\frac{n(n-1)}{2}\sigma'',n=0,1,2,\cdots \tag{6}$$

The k in Eq. (5) obtained from the condition that the value under square-root of Eq. (5) was first orde polynomials and perfect quadratic so the discriminate of the quadratic equation is zero. The energy in Eq. (4) obtained from the condition $\tau'<0$. The second eigenfunction $y_n(s)$ satisfy the Rodrigues relation, as follow

$$y_n(s)=\frac{C_n}{\rho(s)}\frac{d^n}{ds^n}\left[\sigma^n(s)\rho(s)\right] \tag{7}$$

where C_n is normalization constant, and $\rho(s)$ satisfy the condition

$$\frac{\partial(\sigma\rho)}{\partial s}=\tau(s)\rho(s) \tag{8}$$

If the general hypergeometric equation that we get as [18]

$$\frac{\partial^2\psi(s)}{\partial s^2}+\frac{(c_1-c_2 s)}{s(1-c_3 s)}\frac{\partial\psi(s)}{\partial s}+\frac{(-\varepsilon_1 s^2+\varepsilon_2 s-\varepsilon_3)}{s^2(1-c_3 s)^2}\psi(s)=0 \tag{9}$$

by compared Eq. (9) with Eq. (2), we obtained the eigenvalue as

$$c_2 n-(2n+1)c_5+(2n+1)\left(\sqrt{c_9}+c_3\sqrt{c_8}\right)+n(n-1)c_3+c_7+2c_3c_8+2\sqrt{c_8c_9}=0 \tag{10}$$

and the eigenfunction as follows

$$\psi(s)=N_{nl}s^{c_{12}}(1-c_3 s)^{-c_{14}}P_n^{\left(c_{10}-1,\left(c_{11}/c_3\right)-c_{10}-1\right)}(1-2c_3 s) \tag{11}$$

with N_{nl} is normalization constant P is Jacobi polynomials and

$$\begin{gathered}c_4=\frac{1}{2}(1-c_1),c_5=\frac{1}{2}(c_2-2c_3),c_6=c_5^2+\varepsilon_1,c_7=2c_4c_5-\varepsilon_2,c_8=c_4^2+\varepsilon_3,\\ c_9=c_3c_7+c_3^2c_8+c_6,c_{10}=c_1+2c_4+2\sqrt{c_8},c_{12}=c_4+\sqrt{c_8},\\ c_{11}=c_2-2c_5+2\left(\sqrt{c_9}+c_3\sqrt{c_8}\right),c_{13}=c_5-\left(\sqrt{c_9}+c_3\sqrt{c_8}\right),c_{14}=c_{12}+\frac{c_{13}}{c_3}\end{gathered} \tag{12}$$

2.2. *Supersymmetric method*

Definition of supersymmetric Hamiltonian H_{ss} partner for change operator that commute with H_{ss} by Witten as follow[19]

$$H_{ss}=\begin{pmatrix}-\frac{d^2}{dx^2}+\frac{d\varphi(x)}{dx}+\varphi^2(x) & 0\\ 0 & -\frac{d^2}{dx^2}-\frac{d\varphi(x)}{dx}+\varphi^2(x)\end{pmatrix}=\begin{pmatrix}H_+ & 0\\ 0 & H_-\end{pmatrix} \tag{13}$$

The Hamiltonian partner $H_-=H_1$ and $H_+=H_2$ and partner potential $V_-=V_1$ and $V_+=V_2$ are

$$V_-(x)=V_1=\varphi^2(x)-\varphi'(x)\text{ and }V_+(x)=V_2=\varphi^2(x)+\varphi'(x) \tag{14}$$

The relation between effective potential V_{ef} and first potential V_1 as follow

$$V_{ef}(x)=V_-(x;a_0)+E_0=V_1(x;a_0) \tag{15}$$

with E_0 is ground state energy for effective potential. By setting the new supersymmetric operator are the lowering operator A and the rising operator $A^\dagger$ as

$$\begin{gathered}A^\dagger=-\frac{d}{dx}+\varphi(x),\\ A=\frac{d}{dx}+\varphi(x)\end{gathered} \tag{16}$$

so the Hamiltonian partner in Eq. (13) can be rewritten as

$$\begin{gathered}H_-(x)=H_1=A^\dagger A,\\ H_+(x)=H_2=AA^\dagger\end{gathered} \tag{17}$$

and also

$$A\psi_0^{(-)} = A\psi_0 = 0 \tag{18}$$

The new partner potential V_2 from Eq. (15) and Eq. (18) we get the reation [16]

$$V_2(x) = V_1 - 2\frac{d^2}{dx^2}\ln\psi_0 = V_{ef} - E_0 - 2\frac{d^2}{dx^2}\ln\psi_0 \tag{19}$$

This equation will be used to construct the partner potential from the original potential in Eq. (1).

3. Results and discussion

3.1. *The Schrodinger equation in D-dimensional system*

The general Schrodinger equation in D-dimensional system [20]

$$-\frac{\hbar^2}{2m}\nabla_D^2\Psi(r,\Omega) + V(r,\Omega)\Psi(r,\Omega) = E\Psi(r,\Omega) \tag{20}$$

with D is dimensional and the Laplacian operator as [21]

$$\nabla_D^2 = \frac{1}{r^{D-1}}\frac{\partial}{\partial r}(r^{D-1}\frac{\partial}{\partial r}) + \frac{1}{r^2}\sum_{j=1}^{D-2}\frac{1}{\sin^2\theta_{j+1}\sin^2\theta_{j+2}\cdots\sin^2\theta_{D-1}}\times$$
$$\left\{\frac{1}{\sin^{j-1}\theta_j}\left(\frac{\partial}{\partial\theta_j}\sin^{j-1}\theta_j\frac{\partial}{\partial\theta_j}\right)\right\} + \frac{1}{r^2}\left\{\frac{1}{\sin^{D-1}\theta_{D-1}}\left(\frac{\partial}{\partial\theta_{D-1}}\sin^{D-2}\theta_{D-1}\frac{\partial}{\partial\theta_{D-1}}\right)\right\} \tag{21}$$

By using variable separation method from Eq. (20) we get the D-dimensional Schrodinger equation radial part and polar part. The polar part D-dimensional Schrodinger equation for $D = 4$ as follow

$$\left[\begin{array}{l}\frac{1}{\sin^2\theta_2\sin^2\theta_3}\left(\frac{\partial^2}{\partial\theta_1^2}\right) + \\ \frac{1}{\sin^2\theta_3}\frac{\partial}{\partial\theta_2}\left\{\frac{1}{\sin\theta_2}\left(\sin\theta_2\frac{\partial}{\partial\theta_2}\right)\right\} + \\ \frac{1}{\sin^3\theta_3}\frac{\partial}{\partial\theta_3}\left(\sin^3\theta_3\frac{\partial}{\partial\theta_3}\right)\end{array}\right] Y_l^m(\Omega_D) - \frac{2m}{\hbar^2}\left\{\begin{array}{l}\frac{V(\theta_1)}{\sin^2\theta_2\sin^2\theta_3} + \\ \frac{V(\theta_2)}{\sin^2\theta_3} + V(\theta_3)\end{array}\right\} Y_l^m(\Omega_D) = \lambda_1 Y_l^m(\Omega_D) \tag{22}$$

To simplify calculation we set $\hbar = 1, 2m = 1$. By setting $Y_l^m(\Omega_D) = P_1(\theta_1), P_2(\theta_2), P_3(\theta_3)$, and $V_1(\theta_1), V_2(\theta_2), V_3(\theta_3)$ from Eq. (1) and By using algebra method, from Eq. (22) we have the Schrodinger equation for θ_1 as follow

$$\frac{1}{P_1(\theta_1)}\left(\frac{\partial^2 P_1(\theta_1)}{\partial\theta_1^2}\right) - V(\theta_1) + \lambda_1 = 0, \tag{23}$$

the Schrodinger equation for θ_2 as follow

$$\frac{1}{P_2(\theta_2)}\left\{\frac{1}{\sin\theta_2}\frac{\partial}{\partial\theta_2}\left(\sin\theta_2\frac{\partial P_2(\theta_2)}{\partial\theta_2}\right)\right\} - V(\theta_2) - \frac{\lambda_1}{\sin^2\theta_2} = -\lambda_2, \tag{24}$$

and the Schrodinger equation for θ_3 as follow

$$\frac{1}{P_3\theta_3}\left\{\frac{1}{\sin^2\theta_3}\frac{\partial}{\partial\theta_3}\left(\sin^2\theta_3\frac{\partial P_3\theta_3}{\partial\theta_3}\right)\right\}-V\left(\theta_3\right)-\frac{\lambda_2}{\sin^2\theta_3}+\lambda_3=0 \tag{25}$$

where $V(\theta)$ in Eq. (1). The eq. (23), (24) and (25) are the formula that we will solve at next section.

3.2. The Solution of 4 dimensional Schrodinger equation with potential Rosen-Morse and Manning Rosen

First calculation, the solution Schrodinger equation for θ_1 with Rosen-Morse potential by using the Nikiforov-Uvarov method. By substituting Eq. (1) into Eq. (23) and multiplied with $P_1\left(\theta_1\right)$ we have

$$\left(\frac{\partial^2 P_1(\theta_1)}{\partial\theta_1^2}\right)-\left(\frac{\upsilon\left(\upsilon+1\right)}{\cos^2\theta_1}-2q\tan\theta_1\right)P_1(\theta_1)+\lambda_1 P_1(\theta_1)=0 \tag{26}$$

By set

$$i\left(1-2s_1\right)=\tan\theta_1 \tag{27}$$

from Eq. (26) we get

$$\begin{aligned}&\frac{\partial^2 P_1\left(s_1\right)}{\partial s_1{}^2}+\frac{\left(1-2s_1\right)}{s_1\left(1-s_1\right)}\frac{\partial P_1\left(s_1\right)}{\partial s_1}P_1\left(s_1\right)-\frac{\left(-\upsilon\left(\upsilon+1\right)\right)s_1{}^2}{s_1{}^2\left(1-s_1\right)^2}P_1\left(s_1\right)\\&+\frac{\left(-\left[\upsilon\left(\upsilon+1\right)+qi\right]\right)s_1}{s_1{}^2\left(1-s_1\right)^2}P_1\left(s_1\right)-\frac{\left(-\frac{1}{2}qi-\frac{1}{4}\lambda_1\right)}{s_1{}^2\left(1-s_1\right)^2}P_1\left(s_1\right)=0\end{aligned} \tag{28}$$

By compared Eq. (28) with Eq. (9) we get

$$\begin{aligned}&c_1=1, c_1=2, c_3=1, \varepsilon_1=\left(-\upsilon\left(\upsilon+1\right)\right),\\&\varepsilon_2=\left(-\left[\upsilon\left(\upsilon+1\right)+qi\right]\right), \varepsilon_3=\left(-\frac{1}{2}qi-\frac{1}{4}\lambda_1\right)\end{aligned} \tag{29}$$

By using the relation in Eq. (12) and Eq. (11) we get the eigenfunction as follow

$$\psi\left(\theta_1\right)=\left[\begin{array}{l}N_{nl}\left(\frac{1+i\tan\theta_1}{2}\right)^{\sqrt{\left(-\frac{1}{2}qi-\frac{1}{4}\lambda_1\right)}}\left(\frac{1-i\tan\theta_1}{2}\right)^{\sqrt{\left(\frac{1}{2}qi-\frac{1}{4}\lambda_1\right)}}\\P_n^{\left(2\sqrt{\left(-\frac{1}{2}qi-\frac{1}{4}\lambda_1\right)},2\sqrt{\left(\frac{1}{2}qi-\frac{1}{4}\lambda_1\right)}\right)}\left(-i\tan\theta_1\right)\end{array}\right], \; n=0,1,2,3,\ldots \tag{30}$$

with N_{nl} is normalization constant and P is Jacobi polynomials. By using Eq. (10) we get

$$\lambda_1=\frac{4q^2-\frac{1}{4}\left\{-\left(2n+1\right)\pm\sqrt{1-4\upsilon\left(\upsilon+1\right)}\right\}^4}{\left\{-\left(2n+1\right)\pm\sqrt{1-4\upsilon\left(\upsilon+1\right)}\right\}^2}, \; n=0,1,2,3,\ldots \tag{31}$$

Second calculation, the solution of Schrodinger equation for θ_2 with Manning Rosen potential using Nikiforov-Uvarov method. Substituting q. (1) into Eq. (24) and then multiplied by $Y_2\left(\theta_2\right)$ and by setting $Y_2(\theta_2)=P_2\left(\theta_2\right)\sin^{-1/2}\theta_2$ we have

$$P_2''(\theta_2)-\left(\frac{\upsilon(\upsilon-1)+\lambda_1-\frac{1}{4}}{\sin^2\theta_2}-2q\cot\theta_2\right)P_2(\theta_2)+\left(\frac{1}{4}+\lambda_2\right)P_2(\theta_2)=0 \quad (32)$$

by setting $i(1-2s_2)=\cot\theta_2$, and by using the steps used to solve the first part of angular Schrodinger equation in Eq (26), we obtain the the solution of Eq. (34) as follow

$$\psi(\theta_2)=\left[N_{nl}\left(\frac{1+i\cot\theta_2}{2}\right)^{\sqrt{\left(\frac{1}{8}+\frac{\lambda_2}{4}+\frac{1}{2}qi\right)}}\left(\frac{1-i\cot\theta_2}{2}\right)^{\sqrt{\left(\frac{1}{8}+\frac{\lambda_2}{4}-\frac{1}{2}qi\right)}} P_n^{\left(2\sqrt{\left(\frac{1}{8}+\frac{\lambda_2}{4}+\frac{1}{2}qi\right)},2\sqrt{\left(\frac{1}{8}+\frac{\lambda_2}{4}-\frac{1}{2}qi\right)}\right)}(-i\cot\theta_2)\right],\ n=0,1,2,3,\ldots \quad (33)$$

and also

$$\lambda_2=\left[\frac{-4q^2+\frac{1}{4}\left\{-(2n+1)\pm2\sqrt{\upsilon(\upsilon-1)+\lambda_1}\right\}^4}{\left\{-(2n+1)\pm2\sqrt{\upsilon(\upsilon-1)+\lambda_1}\right\}^2}-\frac{1}{2}\right]\ n=0,1,2,3,\ldots \quad (34)$$

with N_{nl} is normalization constant, P is Jacobi polynomials and λ_1 as shown in Eq. (31)

Third calculation, the solution of Schrodinger equation for θ_3 with Manning Rosen potential using Nikiforov-Uvarov method. Substituting Eq. (1) into Eq. (25) and by setting $Y_3(\theta_3)=P_3(\theta_3)\sin^{-1}\theta_3$ lead us to

$$\left(\frac{\partial^2}{\partial\theta_3^2}P_3(\theta_3)\right)-\left(\frac{\upsilon(\upsilon-1)+\lambda_2}{\sin^2\theta_3}-2q\cot\theta_3\right)P_3(\theta_3)+(1+\lambda_3)P_3(\theta_3)=0 \quad (35)$$

In the similar way, we obtain the solution of third part of angular Schrodinger we have the solution of Eq. (35) as follow

$$\psi(\theta_3)=\left[N_{nl}\left(\frac{1+i\cot\theta_3}{2}\right)^{\sqrt{\left(\frac{1}{2}qi+\frac{1}{4}(1+\lambda_3)\right)}}\left(\frac{1-i\cot\theta_3}{2}\right)^{\sqrt{\left(-\frac{1}{2}qi+\frac{1}{4}(1+\lambda_3)\right)}} P_n^{\left(2\sqrt{\left(\frac{1}{2}qi+\frac{1}{4}(1+\lambda_3)\right)},2\sqrt{\left(-\frac{1}{2}qi+\frac{1}{4}(1+\lambda_3)\right)}\right)}(-i\cot\theta_3)\right],\ n=0,1,2,3,\ldots \quad (36)$$

and we get

$$\lambda_3=\left[\frac{-4q^2+\frac{1}{4}\left\{-(2n+1)\pm\sqrt{1+4(\upsilon(\upsilon-1)+\lambda_2)}\right\}^4}{\left\{-(2n+1)\pm\sqrt{1+4(\upsilon(\upsilon-1)+\lambda_2)}\right\}^2}-1\right],\ n=0,1,2,3,\ldots \quad (37)$$

with N_{nl} is normalization constant, P is Jacobi polynomials. The eigenfunction that we obtained will be used to consruct the new partner potential of the potential Rosen-Morse and Manning Rosen.

3.3. The construction of partner potential from origin potential Rosen Morse and Manning Rosen in Schrodinger 4 dimensional system

To construction the partner potential we use Eq. (19). So we obtained the new partner potential as follow

$$V_2(\theta_1)=\left\{\begin{matrix}\left(\dfrac{\upsilon(\upsilon+1)}{\cos^2\theta_1}-2q\tan\theta_1\right)-E_{01}\\ -2\sqrt{\left(-\dfrac{1}{2}qi-\dfrac{1}{4}\lambda_1\right)}\left[i\sec^2\theta_1\tan\theta_1\left(\dfrac{1+i\tan\theta_1}{2}\right)^{-1}+\left(\dfrac{\sec^2\theta_1}{2}\right)^2\left(\dfrac{1+i\tan\theta_1}{2}\right)^{-2}\right]\\ +2\sqrt{\left(\dfrac{1}{2}qi-\dfrac{1}{4}\lambda_1\right)}\left[i\sec^2\theta_1\tan\theta_1\left(\dfrac{1-i\tan\theta_1}{2}\right)^{-1}-\left(\dfrac{\sec^2\theta_1}{2}\right)^2\left(\dfrac{1-i\tan\theta_1}{2}\right)^{-2}\right]\end{matrix}\right\}, \quad (38a)$$

$$V_2(\theta_2)=\left\{\begin{matrix}\left(\dfrac{\upsilon(\upsilon-1)+\left(\lambda_1-1/4\right)}{\sin^2\theta_2}-2q\cot\theta_2\right)-E_{02}\\ -2\sqrt{\left(\dfrac{1}{8}+\dfrac{\lambda_2}{4}+\dfrac{1}{2}qi\right)}\left[i\csc^2\theta_2\cot\theta_2\left(\dfrac{1+i\cot\theta_2}{2}\right)^{-1}-\left(\dfrac{i\csc^2\theta_2}{2}\right)^2\left(\dfrac{1+i\cot\theta_2}{2}\right)^{-2}\right]\\ +2\sqrt{\left(\dfrac{1}{8}+\dfrac{\lambda_2}{4}-\dfrac{1}{2}qi\right)}\left[i\csc^2\theta_2\cot\theta_2\left(\dfrac{1-i\cot\theta_2}{2}\right)^{-1}+\left(\dfrac{i\csc^2\theta_2}{2}\right)^2\left(\dfrac{1-i\cot\theta_2}{2}\right)^{-2}\right]\end{matrix}\right\}, \quad (38b)$$

$$V_2(\theta_3)=\left\{\begin{matrix}\left(\dfrac{\upsilon(\upsilon-1)-\lambda_2}{\sin^2\theta_3}-2q\cot\theta_3\right)-E_{03}\\ -2\sqrt{\left(\dfrac{1}{2}qi+\dfrac{1}{4}(1+\lambda_3)\right)}\left[i\csc^2\theta_3\cot\theta_3\left(\dfrac{1+i\cot\theta_3}{2}\right)^{-1}-\left(\dfrac{i\csc^2\theta_3}{2}\right)^2\left(\dfrac{1+i\cot\theta_3}{2}\right)^{-2}\right]\\ +2\sqrt{\left(-\dfrac{1}{2}qi+\dfrac{1}{4}(1+\lambda_3)\right)}\left[i\csc^2\theta_3\cot\theta_3\left(\dfrac{1-i\cot\theta_3}{2}\right)^{-1}-\left(\dfrac{i\csc^2\theta_3}{2}\right)^2\left(\dfrac{1-i\cot\theta_3}{2}\right)^{-2}\right]\end{matrix}\right\} \quad (38c)$$

where $E_{01}=\left[(\upsilon+1)^2-\dfrac{q^2}{(\upsilon+1)^2}\right]$ dan $E_{02}=E_{03}=\left[(\upsilon-1)^2-\dfrac{q^2}{(\upsilon-1)^2}\right]$.

We use the new partner potential in Eq. (38) as a new potential. Then, the new partner potential solved in Schrodinger equation sistem by using the same steps that we used to solved the angular part Schrodinger equation with the original potential. The solution of the angular part Schrodinger equation with partner potential are

$$\psi'(\theta_1)=\left[\begin{matrix}N_{nl}\left(\dfrac{1+i\tan\theta_1}{2}\right)^{\sqrt{\left(-\frac{1}{2}qi-\frac{1}{4}E_{01}-\frac{1}{4}\lambda_1'\right)}}\left(\dfrac{1-i\tan\theta_1}{2}\right)^{\sqrt{\left(\frac{1}{2}qi-\frac{1}{4}E_{01}-\frac{1}{4}\lambda_1'\right)}}\\ P_n^{\left(2\sqrt{\left(-\frac{1}{2}qi-\frac{1}{4}E_0-\frac{1}{4}\lambda_1'\right)},2\sqrt{\left(-\frac{1}{2}qi-\frac{1}{4}E_{01}-\frac{1}{4}\lambda_1'\right)}\right)}(-i\tan\theta_1)\end{matrix}\right] \quad (39a)$$

$$\psi'(\theta_2)=\left[N_{nl}\left(\frac{1+i\cot\theta_2}{2}\right)^{\sqrt{\left(\frac{1}{2}qi-4c_{12}+\frac{1}{16}+\frac{\lambda_2'}{4}+\frac{1}{4}E_{02}\right)}}\left(\frac{1-i\cot\theta_2}{2}\right)^{\sqrt{\left(-\frac{1}{2}qi-4c_{12}+\frac{1}{16}+\frac{\lambda_2'}{4}+\frac{1}{4}E_{02}\right)}} P_n^{\left(2\sqrt{\left(\frac{1}{2}qi-4c_{12}+\frac{1}{16}+\frac{\lambda_2'}{4}+\frac{1}{4}E_{02}\right)},2\sqrt{\left(-\frac{1}{2}qi-4c_{12}+\frac{1}{16}+\frac{\lambda_2'}{4}+\frac{1}{4}E_{02}\right)}\right)}\left(-i\cot\theta_2\right)\right] \tag{39b}$$

$$\psi'(\theta_3)=\left[N_{nl}\left(\frac{1+i\cot\theta_3}{2}\right)^{\sqrt{\left(\frac{1}{2}qi+\frac{1}{4}E_{03}+\frac{1}{4}(1+\lambda_3')\right)}}\left(\frac{1-i\cot\theta_3}{2}\right)^{\sqrt{\left(-\frac{1}{2}qi+\frac{1}{4}E_{03}+\frac{1}{4}(1+\lambda_3')\right)}} P_n^{\left(2\sqrt{\left(\frac{1}{2}qi+\frac{1}{4}E_{03}+\frac{1}{4}(1+\lambda_3')\right)},2\sqrt{\left(-\frac{1}{2}qi+\frac{1}{4}E_{03}+\frac{1}{4}(1+\lambda_3')\right)}\right)}\left(-i\cot\theta_3\right)\right] \tag{39c}$$

with the separated variable contant are

$$\lambda_1'=\left[\frac{4(q)^2-\frac{1}{4}\left\{-(2n+1)\pm\sqrt{1-4\upsilon(\upsilon+1)+8(c_{121}+c_{141})}\right\}^4}{\left\{-(2n+1)\pm\sqrt{1-4\upsilon(\upsilon+1)+8(c_{121}+c_{141})}\right\}^2}-2E_0\right], \tag{40a}$$

$$\lambda_2'=\left[-\frac{4q^2}{\left\{-(2n+1)\pm\sqrt{4\left(\upsilon(\upsilon-1)+\lambda_1'+\lambda_1-2c_{122}+2c_{142}-\frac{1}{4}\right)}\right\}^2}+\frac{1}{4}\left\{-(2n+1)\pm\sqrt{4\left(\upsilon(\upsilon-1)+\lambda_1'+\lambda_1-2c_{122}+2c_{142}-\frac{1}{4}\right)}\right\}^2-\left(\frac{1}{4}+E_0-16c_{122}\right)\right], \tag{40b}$$

$$\lambda_3'=\left[-\frac{4q^2}{\left\{-(2n+1)\pm\sqrt{1-4\left(-\left(\upsilon(\upsilon-1)-\lambda_2+\lambda_2'\right)+2c_{123}-2c_{143}\right)}\right\}^2}+\frac{1}{4}\left\{-(2n+1)\pm\sqrt{1-4\left(-\left(\upsilon(\upsilon-1)-\lambda_2+\lambda_2'\right)+2c_{123}-2c_{143}\right)}\right\}^2-\left(E_0+1\right)\right] \tag{40c}$$

where λ_1 in Eq. (31), λ_2 in Eq. (34), and

$$c_{121}=\sqrt{\left(-\frac{1}{2}qi-\frac{1}{4}\lambda_1\right)},\ c_{141}=-\sqrt{\left(\frac{1}{2}qi-\frac{1}{4}\lambda_1\right)},\ c_{122}=\sqrt{\left(\frac{1}{8}+\frac{\lambda_2}{4}+\frac{1}{2}qi\right)},$$

$$c_{142}=-\sqrt{\left(\frac{1}{8}+\frac{\lambda_2}{4}-\frac{1}{2}qi\right)},\ c_{123}=\sqrt{\left(\frac{1}{2}qi+\frac{1}{4}(1+\lambda_3)\right)},\ c_{143}=-\sqrt{\left(-\frac{1}{2}qi+\frac{1}{4}(1+\lambda_3)\right)}$$

And then, to see the form of eigenfuction angular part of Schrodinger equation with original potential and Schrodiger equation with new partner potential we make visualization by using matlab program, for example we present the eigenfunction of the angular part θ_3 in Table 1.

Table 1. The eigenfunction of angular part θ_3 Schrodinger equation for potential Manning Rosen and partner potential Manning Rosen

n_l	**Original Potential Manning Rosen**	**Partner Potential Manning Rosen**
0	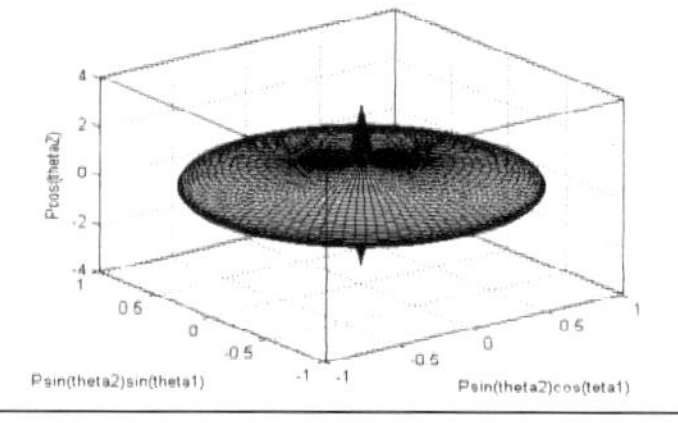	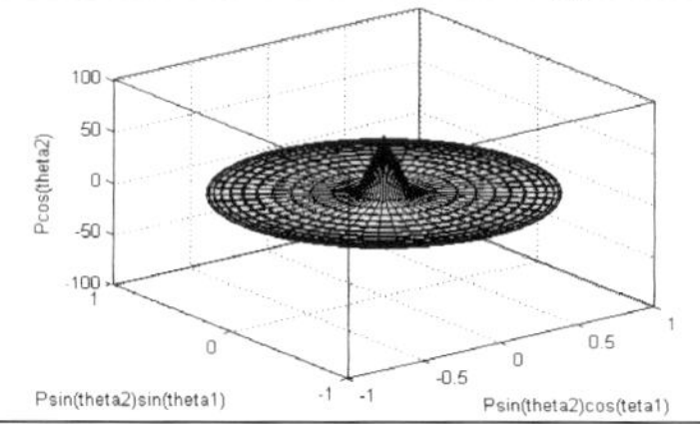
1	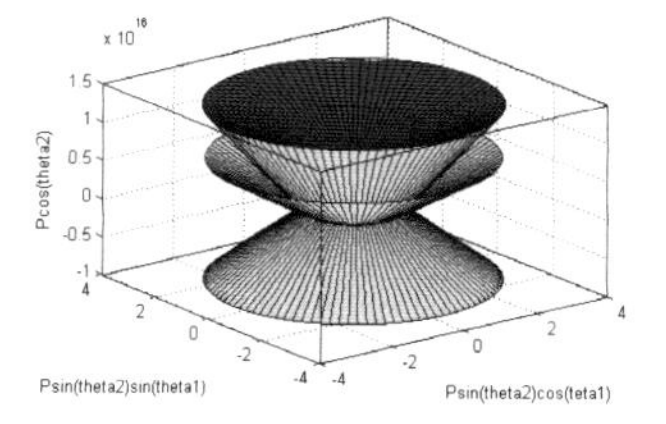	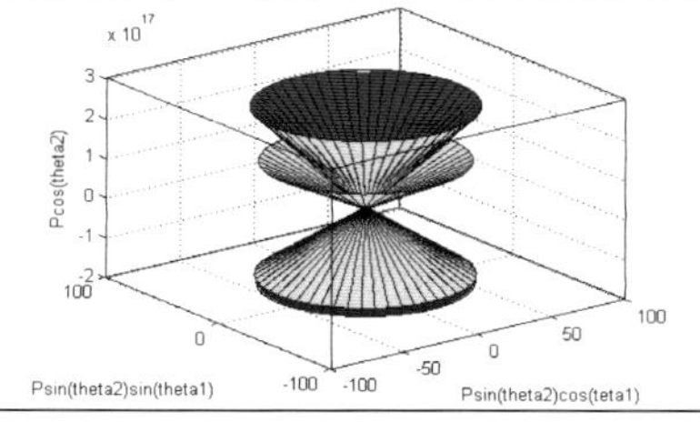
2	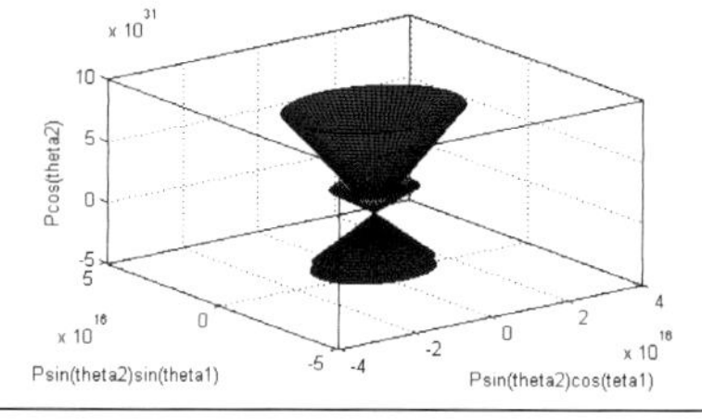	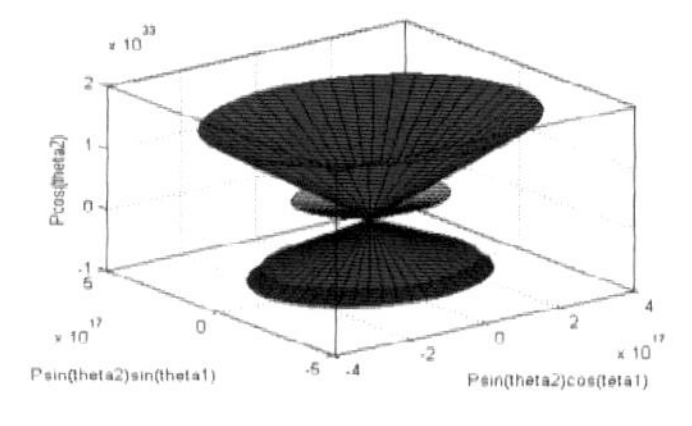

In Table 1 we can see that the eigenfunction of the original potential Manning Rosen and the partner potentials Manning Rosen have the same shape. The higher the number of angular quantum the more waves are formed. The significant difference in their amlitude. The amplitude of the eigenfunction of partner potentential bigger than original potential.

From all of the calculations that we have done above, we can see that from the solution of quantum potential we can construct the new potentials partner. The new partner potentials are solvable so can also be solved in quantum mechanics system. The eigenfunction of original potential is different compared with the new partner potential.

4. Conclusion

We have constructed the partner potential from original potential Rosen-Morse and Manning Rosen in Schrodinger 4 dimensional system using the supersymmetric method. The partner potentials that we obtained are solvable. By using the Nikiforov-Uvarov method we obtained the eigenfunction of angular part of Schrodinger equation with partner potential Rosen-Morse and Manning Rosen. The eigenfunctions of Schrodinger equation with the potential original and the partner potential have same shape, but have the difference amplitude. . The higher the number of angular quantum the more waves are formed and the amplitude of the eigenfunction of partner potentential bigger than original potential.

Acknowledgments

This work partly supported by PUT-MRG of LPPM Sebelas Maret University.

References

[1] Nikiforov A. F. and Uvarov V. B. 1998. *Special Fnctions of Mathematical Physics*

(Birkhauser, Basel).
[2] Cari C. and Suparmi A. 2012. *IOSR Journal of Applied Physics*. **2** 13-23.
[3] Cari C. and Suparmi A. 2013. Journal of Physics: Conference Series. **423** 012031.
[4] Mitftahul M., Suparmi A. adnd Cari C. 2016. *Journal of Physics: Conference Series* **776** 012079.
[5] Suparmi. 2011. *Mekanika kuantum II*.(Surakarta: Sebelas Maret University). 57-65.
[6] Siregar A., Suparmi A., Cari C. and Yuliani H. 2013. *Internationan Journal of Engineering and Science*. **2** 14-26
[7] Child M. S., Dong S. H. And Wang X. G. 2000. *Journal of Physics A: Math Gen*. **33**.
[8] Wea K. N., Suparmi A., Cari C. And Anwar F. 2017. *Journal of Physics: Conference Series*. **795** 012023
[9] Wahyulianti, Suparmi A., Cari C. And Anwar F. 2017. *Journal of Physics: Conference Series*. **795** 012022.
[10] Bayrak O., Bostoosum I. And Cifti H. 2006. *International Journal of Quantum Chemistry*. **107**.
[11] Pratiwi B. N., Suparmi A., Cari C. and Anwar F. 2016. *Journal of Physics : Conference Series* **776** 012090.
[12] Barakat T. 2005. *Physics Letter A*. **344** 411-417.
[13] Ikot A. N. 2012. *Chinese Physics Letter*. **29** 090302.
[14] Ikhdair S. M. and Sever R. 2008. *Journal of Molecular Sructure*. **855** 12-17
[15] Bhoi J. And Laha U. 2016. *Brazilian Journal of Physics*. **49** 129-132.
[16] Suparmi A., Cari C., Pratiwi B. N. And Nugraha D. A. 2017. *Journal of Physics: Conference Series*. **820** 012023.
[17] Nikiforov A. F and Uvarov V.B. 1988. *Mathematics and Computers in Simulation* . **30** 4.
[18] Tezcan C. And Sever R. 2008. *International Journal of Theorytical Physics*. **48** 337-350.
[19] Ikot A. N., Awoga O. A. and Antia D. A. 2013. *Chinese Physics B*. **22** 020304.
[20] Witten E. 1981. *Nuclear Physics B*. **188** 513.
[21] Dong. S. H. 2011. *Wave Equation in Higer Dimensions*. (New York: Spinger). 42.

IOP Conf. Series: Journal of Physics: Conf. Series **909** (2017) 012029 doi:10.1088/1742-6596/909/1/012029

Optimization Properties of Environmentally Friendly Paper Coating Based Starch-Polyethylene glycol (PEG) Mixture

Diani Galih Saputri[1], Khairuddin [2*], Nanik Dwi Nurhayati [3], [4]Trinh Pham

[1,2*]Physics Department of Sebelas Maret University, Indonesia 57126
[3]Chemistry (FKIP) Department of Graduate Program Sebelas Maret University, Indonesia 57126
[4]Danang University of Science and Technology, Vietnam 550000

*corresponding author (Khairuddin): khairuddin_fmipa@staff.uns.ac.id

Abstract. The use of starch as biodegradable base material for packaging application was of great interest as an environmentally friendly alternative to the present use of polyethylene and polyvinyl chloride. However, starch tended to be brittle and had a lack of stability due to exposure to water. Several aproaches have been done to improve shellac properties including through chemical modification, mixing with polymers, clays, and plasticizers. The present study related to optimization of starch properties when mixing with polyethylene glycol (PEG) coated on the paper. The aim was to obtain the temperature and mixing time between starch and PEG so produced composites with optimal barrier properties. The composites of PEG/starch 10 % w/w were prepared using solvent casting and coated on paper surface, and dried in the oven for 12 hours at 40°C. Water Vapour Transmitter Rate (WVTR) (Payne cup method) showed that 70 °C was the optimum temperature when mixing time was 30 minutes. Moreover, it showed that the optimum mixing time was 30 minutes when mixing temperature was 80 and 70 °C. Fourier Transform Infra Red (FTIR) showed a strong interaction between PEG400 and starch.

1. Introduction

Packaging was of great concern as it played important role in food and pharmaceutical industry. The issue was driven by the need to develop new material as an alternative to present material synthesized from a chemical compound [1]. One type of such packaging was conventional paper made from polyvinyl chloride (PVC), polyethylene (PE) [2]. This material was difficult to degradation naturally when it was disposed in nature leading to environmental problem [2]. This synthetic polymers have been widely used for food packaging because they were affordable and good for their mechanical and chemical properties.

Biodegradable polymers are polymers having ability to degrade naturally by breaking the polymer chains or inter-monomer bonds on the main chain of polymers into short chains by living organism such as fungi, bacteria, and enzymes [3]. This type of polymer exists in nature and derived from renewable natural such as tubers, corn, and others [4]. One of the most potential natural polymeric is starch. It is a mixture of amylose and amylopectin which is a complex carbohydrate in the form of white powder, tasteless and odorless. It is a potensial material to use because of it's relative cheap, abundent, renewable, and biodegradable [4], so it can be applied in food packaging. However, it has poor mechanical and water barrier properties [5].

Many effort have been done to improve the properties of starch. One of the effort was to mix starch with Poly vinyl alcohol (PVA) [1,6]. Another aproach is to mix starch with clays [7]. However, it produced brittle composite. Recently, plasticizers also have been applied to starch. Those included sorbitol [8,9], glycerol [10], glycerol and urea [11].

In this work plasticizers polyethylene glycol (PEG) was applied to improve the properties of starch. It water soluble, non-toxic, and safe for pacgkaging. The addition of PEG may affect the permeability of water vapor in the coating [12]. Starch has gelatinating temperature where the properties of starch change drastically. The gelatinization temperature also depends on the types of starch [13]. Therefore, the purpose of this study is to obtain time and temperature of starch and PEG mixture giving optimum properties. The interaction between starch and plasticizer PEG was explored using Fourier Transform Infrared (FTIR).

2. Materials and methods

2.1. Sample Preparation

The starch was cassava (amylum manihot) from local resources and plasticizer was PEG having molecular weight 400 supllied by Sigma Aldrich. Starch of 0.9 gram was dissolved in 10 ml of distilled water and stirred for 30 minutes from 55°C to 80°C and then held for a heating rate of 1 hour.

PEG 400 of 0.1gram was dissolved in 5 ml of distilled water and stirred for 2 hours. Starch and PEG 400 solution were mixed and heated for 30 minutes at different temperature i.e. 60°C, 70°C, 80°C. The PEG 400 starch suspension was then coated on the surface of paper and dried for 12 hours at 40°C.

2.2. Sample characterization

2.2.1. Water Vapor Transmitter Rate (Payne cup method)

The nature of barrier properties of the paper coating was determined by Water Vapor Transmission Rate (Payne cup Analysis method) at 85% humidity. The humidity was achieved by saturaded KCl solution placed in humidity chamber.

The value of the water vapor transmission rate (WVTR) was calculated by the following equation (Standart british, 1996):

$$WVTR = \frac{240xm}{sxt} \quad (1)$$

Where:
m = The weight of gram in a certain time
s = Surface area of the test film (cm)
t = The time between the last weighing

2.3. Fourier Transform Infrared

A spectrometer along with a mercury cadmium telluride detector cooled by liquid nitrogen was used for all the FTIR analysis. The spectrometer was coupled with a single reflection diamond ATR cell. The film was placed on top of the crystal diamond and the spectra of the films taken. The wavelenght of interest ranges from 700 to 4000 cm^{-1} and was collected with resolution of 4 cm^{-1} and 64 scans.

3. Results and discussion

Figure 1. shows that the accumulation of water transmitted through the samples as a function of time. The samples are paper (with no composites, starch heated 70°C for 30 minutes (coated on paper), K+PEG400/70C/30' (starch and PEG400 mixture heated at 80°C for 30 minutes), K+PEG 400/70C/30' (starch and PEG400 mixture heated at 70C for 30 minutes), and K+PEG 400/60C/30' (starch and PEG400 mixture heated at 60C for 30 minutes) as a function of time. It shows that the water accumulation increases linearly with time within the period of experiment under studies. The WVTR value is calculated using Equation 1 and given in Table 1. It showed that the WVTR value of the samples compared to the paper on its own reduced ~ 26 % when coated with starch heated at 70C. However, adding PEG400 to starch reduced the WVTR value only ,~ 1.9 % when heated at 60C, ~20.5 % when heated at 70C, and ~16.9 % when heated at 80C. These result suggested that the gelatinization of starch did not occur when heating temperatue was 60°C. This was supported by previous study that gelatinization only occured above temperature 60°C [7] The the WVTR value compared to starch on its own increased ~ 6.7 % when PEG400 incorporated into starch. Plasticizer tended to increase the WVTR values of polymers as shown by previous studies (Ghasemlou et al. 2011; Hernández-Munoz et al. 2003; Kristo and Biliaderis 2006; Zhang and Han 2006) In overal, heating temperature at 70 C gave the most optimal barres properties when PEG400 incorporated with starch.

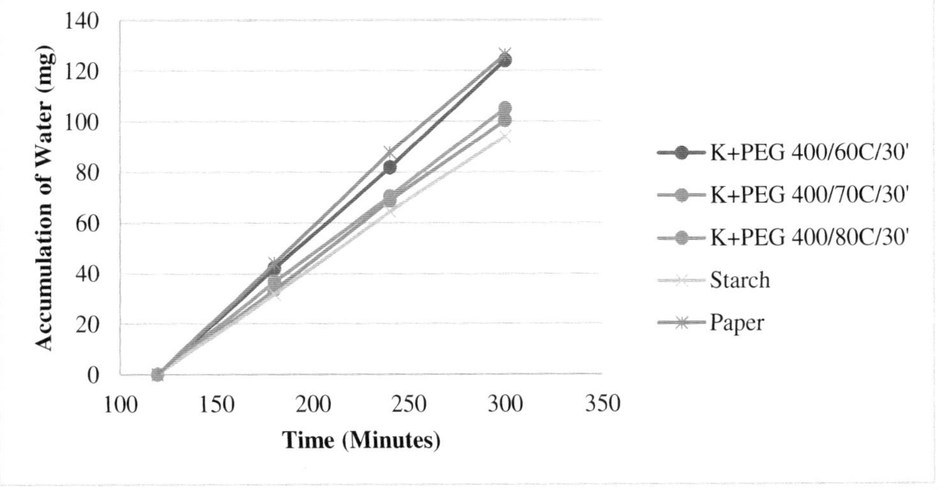

Figure 1. Water accumulation as a function of time of paper, starch, and PEG 400/starch mixture.

Altough our results showed that the addition of plasticizers into the starch reduced the water barier properties, however, the paper coated with starch only produced brittle composite and easy to crack and break, so it is not applicable in food packaging. In the other side, the incorporation of PEG400 into starch produced flexible composite and readily to use for packaging application. Another studies also showed the same trend [1,6]. The starch was sensitive to water leading to damage of the structure [1].

Table 1. WVTR value of the samples

Cooking Temperature (°C)	Value of WVTR (g/m^2day)
Starch	752,00
Paper	1010,67
K+PEG 400/60C/30'	992,00
K+PEG 400/70C/30'	802,67
K+PEG 400/80C/30'	840,00

Figure 2 shows the FTIR spectra of PEG400, starch and PEG400/starch mixture. It shows that four main peaks exists in those samples. The intensity peak in starch sample at 3300 cm-[1] atributed to hydroxyl groups [16], 2926 cm-[1] attributed to methyl groups [15], and 1149 cm-[1] attributed to ether groups. There three peaks in PEG-starch mixture is observed at 3300 cm-[1] (hydroxyl), 2881 cm-[1] (methyl), and 1145 cm-[1] (ether) which are only slightly different from starch. Similar FTIR trends also were shown by Ortega-Toro et al [14] in –glycerol starch coated with PEG.

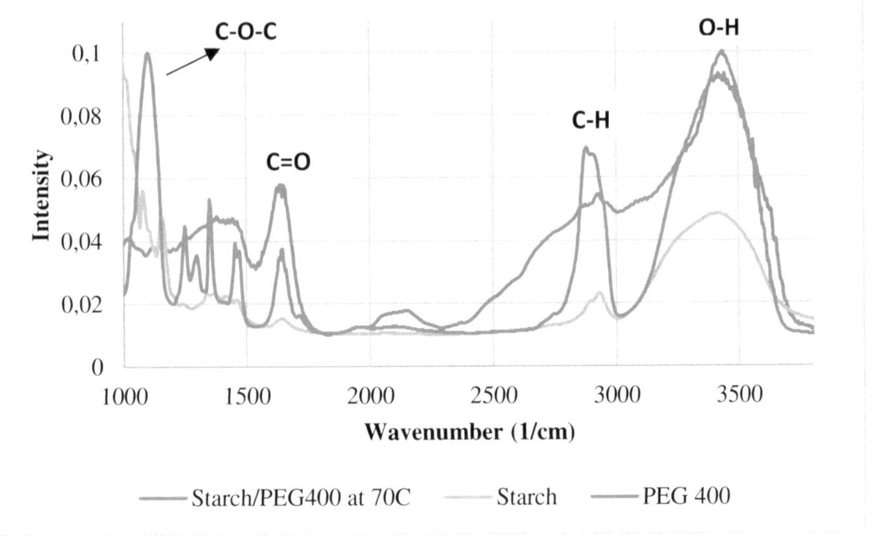

Figure 2. FTIR spectra of starch, PEG 400, and starch/PEG 400.

FTIR spectra of starch/PEG 400 heated at 60°C, starch/PEG 400 heated at 70°C, and starch/PEG 400 heated at 80°C is presented in Figure 3. It shows that the four group (C-O-C, C=O, C-H, O-H also present in PEG/starch mixture heated at 60,70, and 80. The band peak intensities are presented in Table 2.

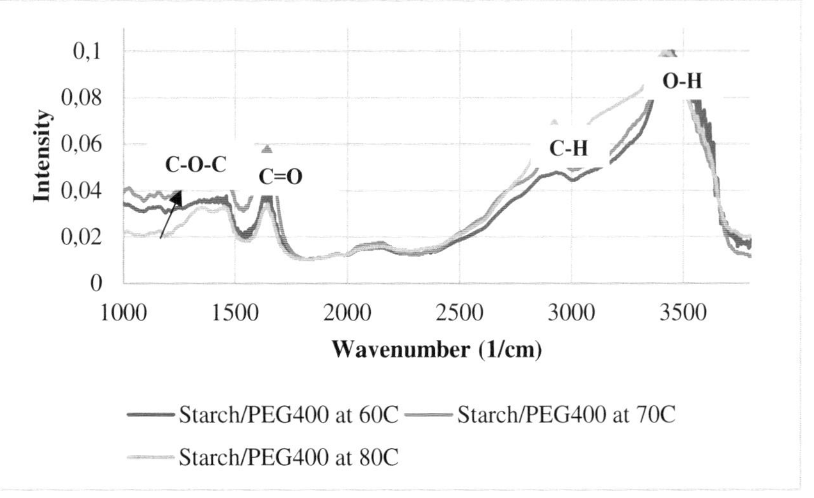

Figure 3. FTIR spectra of starch/PEG 400 heated at 60°C, starch/PEG 400 heated at 70°C, and starch/PEG 400 heated at 80°C.

The result in table 2 indicated that the band peak position in starch and starch/PEG400 mixture show no dissilimarity with those obtained by Jagadish et al [15] and Yu et al [16] with only slightly peak shift, in the functional group. The incorporation of PEG400 into starch shifted the band position of O-H from 3418.97 cm-1 into 3456.59 cm-1 as shown in PEG/starch mixture heated at 70C. indicating there is interaction between starch and PEG. The PEG molecule may disturb the interaction between starch molecules. in starch. The highest shift of peak position of O-H was shown by PEG/starch mixture heated at 70 C. This result is coinsident with the trend of WVTR value in table 1. The graph of the data is shown in Figure 3.

Table 2. Peak absorbance in starch, PEG 400, starch-PEG400 (60°C), starch-PEG400 (70°C), starch-PEG400 (80°C).

Groups	Starch	PEG 400	S+PEG400 (60°C)	S+PEG400 (70°C)	S+PEG400 (80°C)
O-H Peak Intensity Abs	3418,97 0,048421	3469,12 0,088319	3449,84 0,1	3456,59 0,096221	3417,04 0,1
C-H Peak Intensity Abs	2930,00 0,023248	2902,03 0,067364	2931,93 0,047665	2925,17 0,054608	2921,32 0,069982
C=O Peak Intensity Abs	1642,46 0,015229	1635,71 0,034638	1653,07 0,041122	1637,64 0,058114	1646,32 0,034113
C-O-C Peak Intensity Abs	1157,34 0,047919	1102,37 0,1	1166,02 0,032949	1155,41 0,039149	1166,02 0,022622

The change of peak position when PEG was present in starch was also observed C-H group where it was shifted from 2930.00 cm^{-1} to 2925.17 cm^{-1}, the C=O group changed from 1642,46 cm-1 to 1637,64 cm^{-1}, and the C-O-C group change from 1157.34 cm-1 to 1155.41 cm -1. These results again clearly showed that there was a strong interaction between starch and PEG400.

The intensity values of the peaks also increased when PEG400 was added to starch indicating that adding PEG 400 to starch increasing the number of bonds in the group, and similar result also shown by other study [16]. However, the incoporation of PEG into starch did not create a new peak indicating there was no new chemical bonding during the process of coating. Ortega [14] also reported similar result.

4. Conclusions

The processing of starch-plasticizer (PEG 400) for 30 minutes and temperature variation (60, 70, 80) ^{o}C is obtained optimum temperature at 70°C, the starch processing and the addition of PEG 400 strongly influenced the increase of moisture barrier properties and made the paper have more flexible on the processing of starch.

In the mixing of plasticizer molecules with starch there was a strong interaction with the peak shift in the four groups (O-H, C-H, C=O, and C-O-C) and increased number of bonds in the OH, CH, and C=O groups for mixing starch-PEG1000 shown by increased intensity. Both the starch-PEG1000 and starch-glycerol samples do not form new bonds.

References

[1] Sin, L.T., Rahman, W.A., Rahmat, A.R., Khan, M.I. 2010. Detection of synergistic interaction of polyvinyl alcohol-cassava starch blends through DSC. *Carbohydr. Polym* **79** 224-226

[2] Elizondo, N.J., Sobral, P.J.A., & Menegalli, F.C. 2009. Development of film based on blends of am aranthus cruentus flour and poly vinyl alcohol. *Carbohydrat Polymer* **75** 592-598

[3] Noezar. I, V.S. Praptowidodo, R. Nugraheni and M.H. Nasution. 2008. Biodegradable Polymer from Lactic Acid. *Indonesian Chemical Engineering Journal* **6** (2): 627-628

[4] Gonera, A., & Cornillon, P. 2002. Gelatinization of starch/gum/sugar system studied by using DSC, NMR, and CSLM. *Starch/starke* **54** 508-516

[5] Tang. X.Z., Alavi. S., Herald. T.J. 2008. Barrier and mechanical properties of starch-clay nanocomposite films. *Cereal Chem* **85** 433-439

[6] Zhou. Y.X., Cui. F.Y., Jia. D.M., Xie. D. 2009. Effect of complex plasticizer on the structure and properties of the thermoplastic PVA/ starch blends. *Polym-Plast. Technol. Eng* **48** 489-495

[7] Shodik, I. S. J. 2017. Fabrication and Characterization of Paper Coatings with Starch and Clay Montmorillonite as Environmentally Friendly Packaging Materials. Essay. FMIPA, Physics. Sebelas Maret University: Surakarta

[8] Isotton, F. S., Bernardo, G. L., Baldasso, C., Rosa, L. M., Zeni, M. 2015. *The Plasticizer Effect on Preparation and Properties of Etherified Corn Starchs Films*. Industrial Crops and Products. **76** 717-724

[9] Gaudin, S., Lourdin, D., Forssell, P. M., Colonnna, P. 2000. Antiplasticisation and Oxygen Permeability of Strach Sorbitol Films. *Carbohydr. Polym.* **43** 33-37

[10] Mali, S., Grossman, M. V. E., Garcia, M. A., Martino, M. N., Zaritzky, N. E. 2004 Barrier, Mechanical and Octical Properties of Plasticized Yam Strach Films *Carbohyd. Polym.* **56** 129-135

[11] Zullo, R., Lannace, S. 2009. The Effects of Different Starch Sources and Plasticizers on Film Blowing of Thermoplastic Starch: Correlation Among Process, Elongational Properties and Macromolecular Structure. *Carbohyd. Polym.* **77** 376-383

[12] Anan, M., Limmatvapirat, S., Nunthanid, J., & Wanawongthai, C. 2007 Effect of Salts and Plasticizers on Stability of Shellac Lapisan. *Journal of Agricultural and Food Chemistry* **55** 687-692

[13] Alvarez, M.D., Fuentes, R., & Canet, W. 2015 Effects of pressure, temperature, treatment time, and storage on rheological, texture, and structural properties of heat-induced chickpea gels. *Foods* **4** 80-114

[14] Ortega-Toro. R., Munoz, A., Talens, P., Chiralt, A. 2016. Improvement of Properties of Glycerol Plasticized Starch Films by Blending with a Low Ratio of Polycaprolactone and/or Polyethylene Glycol. *Food Hydrocolloid* **56** 9-19

[15] Jagadish, R. S., Raj, B., Parameswara, P., & Somashekar, R. 2012. Structure property relations in polyethylene oxide/starch blended films using WAXS. *Techniques Journal of Applied Polymer Science* **127(2)** 1191-1197

[16] Yu, F., Prashanthaa, K., Soulestin, J., Lacrampea, M. F., & Krawczak, P. 2013. Plasticized-starch/poly(ethylene oxide) blends prepared by extrusion. *Carbohydrate Polymers* **91**(1) 253-261

Magneto-impedance in Multilayered $[Ni_{80}Fe_{20}/Cu]_4$ with modification of the line-length pattern on Cu printed circuit board

Wahyu Eko Prastyo[1], **F Maulana**[1], **N. Nuryani**[1] **and B. Purnama**[1,a)]

Graduate Program, Physics Department, Sebelas Maret University,
Jl. Ir. Sutami 36A Kentingan Jebres Surakarta 57126, INDONESIA

E-mail: prastyowahyoe@gmail.com, [1,a)]bpurnama@mipa.uns.ac.id

Abstract. Magneto-impedance of multilayer $[Ni_{80}Fe_{20}/Cu]_4$on Cu PCB substrate has successfully studied. To enhance the magneto-impedance we modify the geometry the Cu PCB substrate. The multilayer is made by electro-deposition with Pt (Platinum) as elctrodes. Electro-deposition is made at room temperature. Magneto-impedance total is evaluated under the external magnetic field. The results shows that magneto-impedance curve is symmetry. The difference of line-length result in the difference of the magneto-impedance. The samples is with the longest line-length has the largest magneto-impedance ratio. The homogeneity of the samples is on account estimated the increase magneto-impedance ratio.

Keyword: magneto-impedance ratio, electro-deposition, $[Ni_{80}Fe_{20}/Cu]_4$

1. Introduction

The change of the impedance magnitude in a magnetic material (magnetic films, magnetic multilayer films or magnetic ribbon) due to the external magnetic field is called magneto-impedance. This phenomenon is a complement to the magneto-resistance which open spin electronic (spintronic) era. The magneto-impedance effect is given as different impedance when sample under and without magnetic field (ΔZ) and is formulated as follows :

$$\frac{\Delta Z}{Z}(100\%) = (Z(H) - Z(H_{max}))/(Z(H_{max})) \times 100\% \quad (1)$$

In general, the total impedance consists of resistance and reactance following formula $Z^2 = R^2 + X^2$. Where the magnitude of reactance include capacitance and impedance which depend on frequency. In the case of magnetic material, the magnetic permeability also influences the impedance magnitude.

The magneto-impedance effect realizes on amorphous metal wire (amorphous metallic wire), amorphous metal ribbon (amorphous metal ribbon), magnetic thin film, and nano crystalline magnetic alloy [1]. Furthermore, the impedance magnitude is a geometrically physics-property, so that the phenomenon of magneto-impedance possible to create in two dimensions of micro-conductor [2,3] such kind of Cu printed circuit board. The ease of this design is one of the advantages of the magneto-impedance sensor that meets the flexibility aspects of sensor fabrication on the substrate.

Therefore, this magneto-impedance effect can be apply on a wide of magnetic sensors [4]. Moreover, magneto-impedance has a high sensitivity. So that a current magneto-impedance much developed in the manufacture of biomedical magnetic sensors [5]. Magneto-impedance sensors [6] are relatively easy, cost-efficient, the ability of lightweight, reliable, rapid response, and can be cread as portable sensors [7].

In this paper, an attempt to increase the value of the magneto-impedance ratio on Cu PCB substrate are evaluationwith a modification of the geometry of the track line-length on Cu PCB substrate. Multilayer $[Ni_{80}Fe_{20}/Cu]_4$ Permalloy deposited with electro-deposition method [8,9] use the sample a magneto-impedance sensor.

2. Methods

A thin film of multilayer $[Ni_{80}Fe_{20}/Cu]_N$ produces by using an electro-deposition method using Pt (Platinum) electrodes [10]. The substrate used is a Cu Printed Circuit Board (PCB) with modification of geometry. Before use, the substratesare clean with an ultrasonic cleaner.

The electrolyte solutions in electro-depositedmultilayer procedure are listin Table 1. Electrolyte solution adjustsa range 2,5-3 of pH with H_2SO_4 0,1 M.

Table 1. Electrolyte solutions used to depositionthe sample of multilayer$[Ni_{80}Fe_{20}/Cu]_4$

Electrolyte	Materials	Total of Number
For deposited NiFe	$NiSO_4.6H_2O$	0,099 M
	$FeSO_4.7H_2O$	0,012 M
	H_3BO_4	0,149 M
	$C_6H_8O_3$	0,002 M
For deposited Cu	$CuSO_4.5H_2O$	0,065 M
	$C_6H_8O_3$	0,002 M

The current density of 15.5 mA/cm^2 is used to deposited $Ni_{80}Fe_{20}$ and 8 mA/cm^2 for Cu spacer layer. The procedure is repeated to obtain a multilayer of $[Ni_{80}Fe_{20}(800\ nm)/Cu(300\ nm)]_4$.Multilayer samples are do at room temperature.The magneto-impedance characteristic is evaluated with the measure of the total impedance for the various magnetic field as depicted in Figure 1.

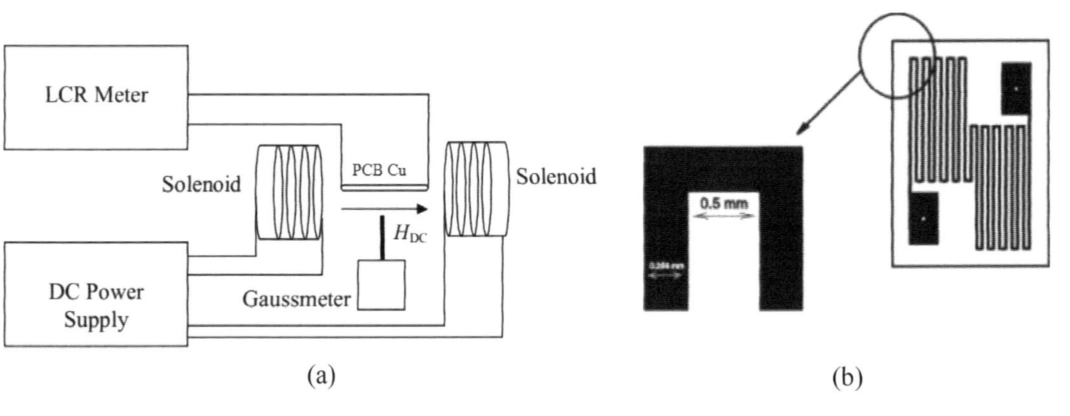

Figure 1. The scheme of the characterization of magneto-impedance (MI)

The value of the impedance total is $Z^2=R^2+X^2$ with resistance as areal component and donation of the imaginary component of reactance. The schematic of the samples is shown in Fig1(b). The distance between of the line track is 500 μm, and the width of the line track is 250 μm. Here, the MI ratio modify by the change of the line-length total,*l*should be 14.2 (later call A), 21.2 (B), 25.7 (C) and 28.7 cm (D).

3. Results and Discussion

Figure 2 shows the multilayer samples of $[Ni_{80}Fe_{20}/Cu]_4$ on a patterned PCB. Observed from the Figure that the geometrical design of the samples is same but different in the length-line. Without magnetic field, the initial impedance of sample A is 0.4598 Ω. Then the total impedance change become 0.7789 Ω and 0.7957 Ω for sample B and C. Finally, the total impedance is 1.056 Ω for sample of D. From this result confirm that the change of the line-length modify the total impedance of the multilayer $[Ni_{80}Fe_{20}/Cu]_4$. It is prediction that the reactance component of the impedance total will response the magnetic field in the electrical measurement under magnetic field. So that, it also modifies the ratio magneto-impedance of the samples.

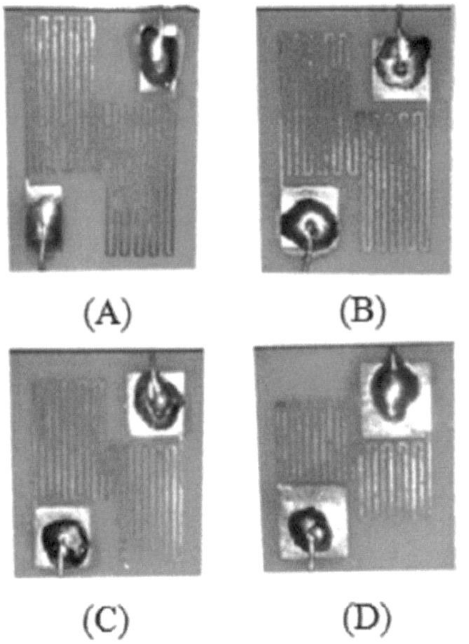

Figure 2. The modified of the multilayers $[Ni_{80}Fe_{20}/Cu]_4$ on a patterned Cu PCB substrates

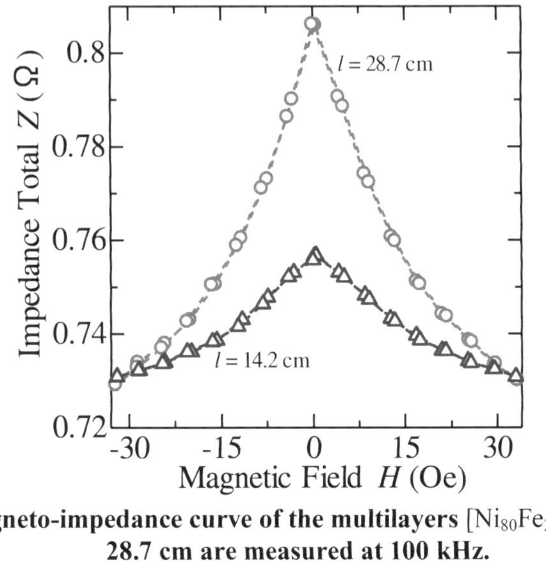

Figure 3. The magneto-impedance curve of the multilayers $[Ni_{80}Fe_{20}/Cu]_4$ for *l* = 14.2 cm and 28.7 cm are measured at 100 kHz.

The typical magneto-impedance curve in electrodeposited $[Ni_{80}Fe_{20}/Cu]_4$for l = 14.2 cm and 28.7 cm are performed at a frequency of 100 kHz. The magnetic field strength variations from 0 mT to ±40 mT generate by electromagnetic field source of a Helmholtz coil. The symmetry characteristic of the magneto-impedance curve is obtained. The Z maximum when zero magnetic fields and the Z become saturate at some magnitude magnetic field. In case the longest sample, magneto-impedance effect realize much a larger than the shorter one. It is indication that reactance component significantly attributes to MI effect. The ratio of the MI calculate by using formula, MI ratio (%)= $[Z(H)-Z(H_{max})]/Z(H_{max})\times 100\%$ and is depicted as Figure 4. The figure is other expression of magneto impedance ratio as a function of the length. It clearly that the MI ratio increase with the increase of the length samples.This result is in accordance with the results that have been published by Vasques [11,12]. In Fig. 4, the MI ratio of 5.98% is obtained for sample D and the largest one of 7.84% for sample A.

Also, the magneto-impedance ratio reflectsthe sensitivity of the samples . Using the equations $\eta = 2\frac{[\Delta Z/_{Z}(\%)]max}{\Delta H}$ so that the value of the greatest sensitivity for the multilayer samples (Sample A) is 932.7 %/T. Thus the sample with the longest line-length has the largest of the magneto-impedance ratio as well as a sensitivity value.

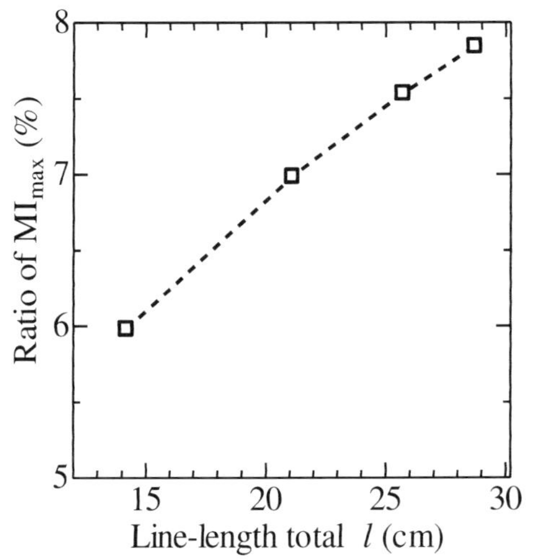

Figure 4. The ratio of MI max (%) as function of the line-length samples

4. Conclusion

Multilayer $[Ni_{80}Fe_{20}/Cu]_N$ is successfully produced with an electro-deposition method on a patterned Cu PCB substrates. The results show that the MI ratio increase with the increase of the line-length samples. The largest MI ratio of 7.84% is obtained for the sample with l = 28.7 cm.

Reference

[1] Phan M.H, Peng H.X 2008 *Progress in Materials Science* **53** 323-420
[2] Panina L.V 1993 *Journal of Applied Physics* **76** 6198-6203
[3] Hanafi E.P 2013 *Jurnal Fisika dan Aplikasinya* **9** 63-65
[4] Alfredo Garcia-Arribas, Jon Gutierrez, Galina V. Kurlyandskaya, Jose M. Barandiaran 2014
[5] Devkota J, Ruiz A, Mukherjee P, Srikanth H, Phan M.H 2013 *IEEE Transactions On Magnetics* **49** 295-302
[6] Valentina Zhukova, Mihail Ipatov, Arcady Zhukov 2009 *Sensors* **9** 9216-9240
[7] Devkota J, Mai T.T.T, Stojak K, Pham H.N, Nguyen X.P, Mukherjee P, Srikanth H, Phan M.H 2014 *Sensors and Actuators*. B **190** 715-722

[8] Robertson N 2012 *Electrodeposition of Magnetic Materials* a Western Digital Company
[9] Mishra A.C 2011 *Physica B* **406** 645-651
[10] Ismail, Nuryani and Purnama B 2016 *AIP.Conference.Proceedings* vol 1710 p. 030002
[11] Vasques M, Li Y.F, Chen D.X 2002 *Journal of Applied Physics* **91** 6539-6544
[12] Qin F.X, Peng H.X, Phan M.H 2010 *Materials Science and Engineering B* **167** 129-132

IOP Conf. Series: Journal of Physics: Conf. Series **909** (2017) 012031 doi:10.1088/1742-6596/909/1/012031

Seismic microzonation by using microtremor method in Pandan volcano, Bojonegoro, East Java

Samsul Hidayat[1], Dwa Desa Warnana[2], Sorja Koesuma[1] and Cari[1]

[1] Physics Department of Graduate Program, Sebelas Maret University, Jl. Ir. Sutami 36A Kentingan Jebres Surakarta 57126, INDONESIA

[2] Civil Engineering Department, Institut Teknologi Sepuluh Nopember, Jl. Arief Rahman Hakim Sukolilo Surabaya 60111, INDONESIA

E-mail: cak.syam.hidayat@gmail.com

Abstract. Pandan volcano geographically is located in the north of Madiun and Nganjuk district, and south Bojonegoro district. The status of this volcano is longer not active, but it still emerges hot spring in some areas. This research aims to understand the distribution of the natural frequency value, the amplification factor value, and the vulnerability index value by using mictrotremor horizontal to vertical spectrum ratio (HVSR) method, that is believed to have an important contribution to earthquake disaster mitigation. There are 19 measurement sites where geographically ranges between 7°26'36.60"–7°27'16.49"S and 111°46'31.76"–111°47'11.98"E. The survey used portable seismograph 3-components SL07 SARA single station. For each site of measurement needs 30 minutes of ambient noise recording with a sampling rate of 100 Hz. The data processing was done by HVSR analysis method on Easy HVSR software. The natural frequency (f_o) value ranges between 1,1 and 9,95 Hz, and amplification factor (A_g) value varies from 2,1 to 7,2. The vulnerability index (K_g) value ranging from 0,48 to 25,34. The highest risk index is located in the north part of the research area where have high potential to damage when the earthquake happens.

1. Introduction

Microtremor is one of the passive seismic method which utilizes ambient noise from nature to understand the structure character of the subsurface layer without causing a disturbance on its structure. Nakamura [1] introduce mictrotremor horizontal to vertical spectral ratio (HVSR) method or called Nakamura technique to determine local site effects on seismic ground motion, expressed by natural frequency (f_o) and amplification factor (A_g) of sediment. Local site effects play an important role in the earthquake damage distribution and refer to the effect of the local geological surface and subsurface structure characteristics on the seismic ground motion [2]. In terms of seismic risk, the unconsolidated sediment materials could amplify ground motion during an earthquake and this causes building on these grounds to become more vulnerable to earthquake damage than those build above the hard rocks [3]. Nakamura (e.g. [4], [5], [6]) proposed the vulnerability index (K_g) to identified weak zone, a potential to damage when the earthquake occurs. The magnitude of vulnerability index value is related to ground amplification factor and natural frequency. The vulnerability index is indicated the soil damage level due to ground motions [7].

Now, the microtremor HVSR method has been widely used for seismic microzonation purposes (e.g. [8], [9], [10]). Seismic microzonation is the process to dividing a zone into smaller zone based on the response local geology to earthquakes [11]. Seismic microzonation is usefull for earthquake disaster mitigation (e.g. [12], [13]).

Earthquake events around Pandan volcano during 2016 have increased significantly [14]. Recorded earthquake data from Indonesian agency for Meteorological, Climatological, and Geophysical (Figure 1) shows that there are 73 earthquake events with a magnitude range below 4 Richter scale and the depth ranges is mostly below 30 km (shallow earthquake). Related to the occurrence of the earthquake, Santoso et al [15] explain there are two possibilities, first it was related to thrust fault activation (in the north of Pandan volcano a thrust fault is found), and secondly is related to sub magmatic activation. There is also possibility that due to active fault movement may trigger upwelling magma.

Figure 1. earthquake events during 2016 in Pandan volcano (blue circle) and surrounding (modified [16])

To reduce the impact of disaster risk, disaster mitigation efforts should be undertaken. One of the earthquake disaster mitigation efforts can be done by assessing local site effect. Microtremor measurement was conducted in the research area in order to know site characteristics based on natural frequency and amplification factor, and vulnerability index values to predict area has high potential to damage due to ground motions.

2. Location and geological setting

Pandan volcano geographically is located in north of Madiun and Nganjuk district, and south Bojonegoro district. It is approximately 40 km northeast of Madiun town. The status of this volcano is longer not active, but it still emerges hot spring in some areas. Location of site survey is on the western flank of Pandan volcano, Klino village, Bojonegoro district or geographically ranges between 7°26'36.60" – 7°27'16.49" S and 111°46'31.76" – 111°47'11.98"E (Figure 2). The elevation ranges between 474 m and 578 m.

Based on the geologic map of the Bojonegoro Quadrangle, East Java **[17]**, Pandan volcano is a Quaternary volcano formed on Early Pleistocene. The rock unit of the research area is dominated by the volcanic breccia (Pandan breccia) (Figure 3).

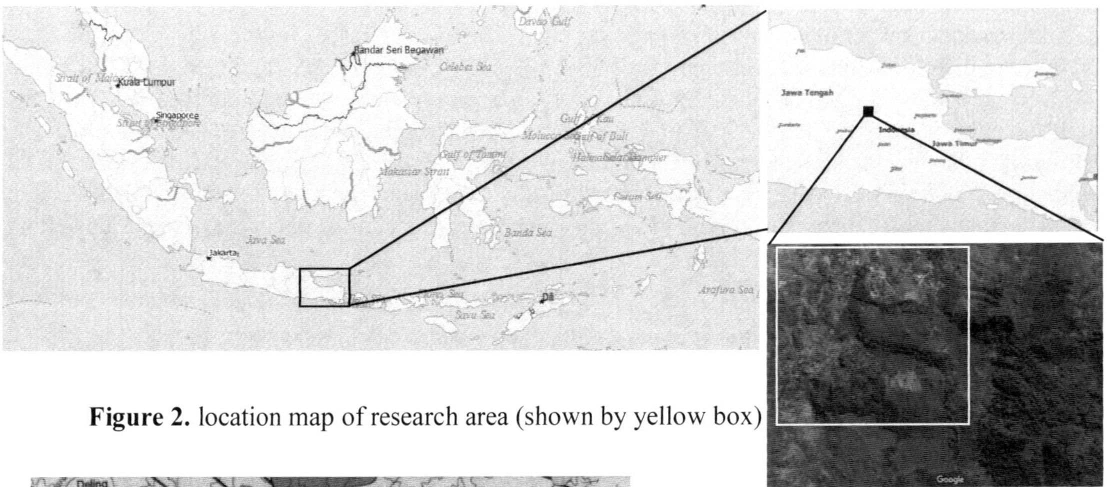

Figure 2. location map of research area (shown by yellow box)

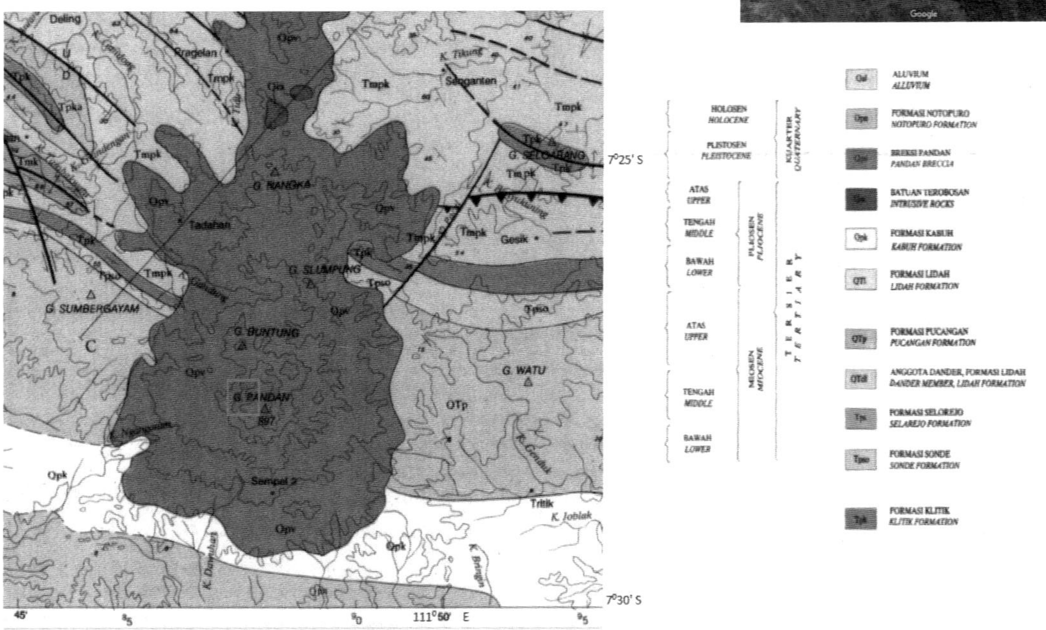

Figure 3. geological map Pandan volcano and surrounding (site area shown by red box)

3. Microtremor measurement and processing

The Nakamura technique is very useful tool for subsurface soil structure and site response studies, especially recommended in areas of low and moderate seismicity, owing to the lack of significant earthquake recording **[18]**. Measurements microtremor data was carried out at 19 sites on the western flank of Pandan volcano. The survey to record the ambient noise was used portable seismograph 3-components SL07 SARA single station, that recorded signal in three perpendicular directions: vertical (Z) and two horizontal (NS and EW). For each site of measurement needs 30 minutes of ambient noise recording with a sampling rate of 100 Hz (dt=0.01 s). The guidelines of the SESAME European research project **[19]** was used for the implementation of the Nakamura technique for microtremors. These H/V user guidelines recommend the procedure for observation point selection, field experiment design, data processing, and interpretation of result data **[20]**. The data processing was done by HVSR analysis method on Easy HVSR software.

In order to obtain the HVSR curve at each site, the collected data was performed processing data in the following steps: each component of the recorded signal was windowed in a time series of 40 sec length non overlapping and for each selected time window the Fast Fourier Transform (FFT) was calculated and smoothed using Konno & Ohmachi logarithmic window function. Recorded data of FFT and spectrums of each component were obtained using frequency-domain analysis. For each selected windows the spectral ratio between the mean square average spectrum of the horizontal component over the spectrum of the vertical component was calculated. Finally, the HVSR curve was generated. The resulting HVSR curve satisfies the criteria defined by SESAME European research project **[18]**. The criteria for a reliable HVSR curve and clear HVSR peak for reliable measurements are shown in Tabel 1.

Table 1. Criteria for a reliable HVSR curve and clear HVSR peak defined by SESAME European research project.

Criteria for a reliable HVSR curve
(all criteria should be fulfilled)
1) $f_o > 10/I_w$
2) $n_c(f_o) > 200$
3) $\sigma_A(f) < 2$ for $0.5f_o < f < 2f_o$ if $f_o > 0.5$Hz or
$\sigma_A(f) < 2$ for $0.5f_o < f < 2f_o$ if $f_o < 0.5$Hz

Criteria for clear HVSR peak
(at least 5 out of 6 criteria fulfilled)
1) $\exists f^- \in [f_o/4, f_o] \mid A_{H/V}(f^-) < A_o/2$
2) $\exists f^+ \in [f_o, 4f_o] \mid A_{H/V}(f^+) < A_o/2$
3) $A_o > 2$
4) $f_{peak}[A_{H/V}(f) \pm \sigma_A(f)] = f_o \pm 5\%$
5) $\sigma_f < \varepsilon(f_o)$
6) $\sigma_A(f_o) < \theta(f_o)$

- I_w = window length
- n_c = number of significant cycles
- $n_c = I_w.n_w.f_o$
- n_w = number of window selected for the average HVSR curve
- f = current frequency
- f_o = HVSR peak frequency
- $\sigma_A(f)$ = standard deviation of $A_{H/V}(f)$
- $A_{H/V}(f)$ = HVSR curve amplitude at frequency
- A_o = HVSR peak amplitude at frequency f_o
- σ_f = standard deviation of HVSR peak frequency ($f_o \pm \sigma_f$)
- $\varepsilon(f_o)$ = threshold value for the stability condition $\sigma_f < \varepsilon(f_o)$
- f^- = frequency between $f_o/4$ and fo for which $A_{H/V}(f^-) < A_o/2$
- f^+ = frequency between f_o and 4/fo for which $A_{H/V}(f^+) < A_o/2$
- $\theta(f_o)$ = threshold value for the stability condition $\sigma_A(f) < \theta(f_o)$
- $\sigma_{logH/V}(f)$ = standard deviation of log $A_{H/V}(f)$ curve

Threshold value for σ_f and $\sigma_A(f_o)$

Frequency range (Hz)	<0.2	0.2 – 0.5	0.5 – 1.0	1.0 – 2.0	> 2.0
$\varepsilon(f_o)$ (Hz)	0.25 f_o	0.2 f_o	0.15 f_o	0.10 f_o	0.05 f_o
$\theta(f_o)$ for $\sigma_A(f_o)$	3.0	2.5	2.0	1.78	1.58
log $\theta(f_o)$ for $\sigma_{logH/V}(f_o)$	0.48	0.40	0.30	0.25	0.20

Reference **[21]** explained that three criteria for reliable HVSR curve are based on the relation of peak frequency to the window length, a number of significant cycles and standard deviation of the peak amplitude. Six criteria for a clear peak are based on the relation of the peak amplitude to the level of the HVSR curve elsewhere, and standard deviations of the peak frequency and of its amplitude (the amplitude should decrease rapidly on each side). An example of microtremor HVSR data has been processed on Easy HVSR software and fulfill nine criteria defined by SESAME project shown in Figure 4.

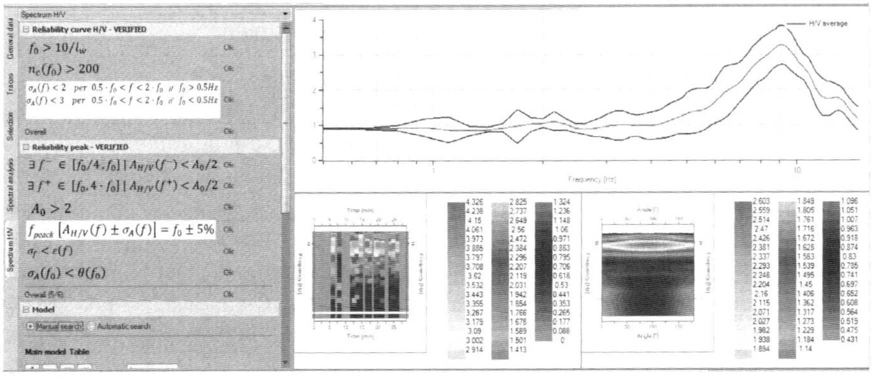

Figure 4. an example HVSR curve fulfill nine criteria defined by SESAME project

4. Result and discussion

The HVSR curve contains information of natural frequency and an amplification factor of a site. Figure 5 shows the distribution of natural frequency, ranging from 1,1 to 9,95 Hz. The formula $f_o=C_s/4h$, where C_s is shear wave velocity of a surface layer and h is depth of the basement (or thickness of sediment) **[5]**, shows that the natural frequency associated with the thickness of sediment. The greater of natural frequency value indicates smaller of thickness of sediment. Previous studies has proved the validity of this formula (e.g. **[21]**, **[22]**, **[23]**, **[24]**).

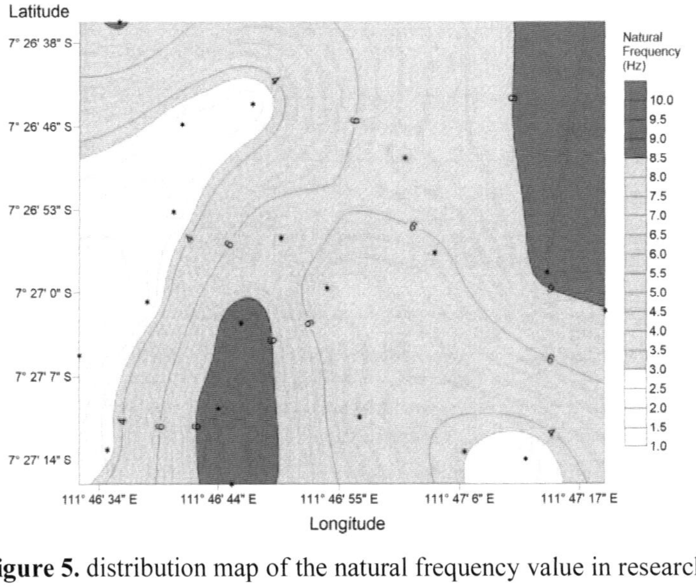

Figure 5. distribution map of the natural frequency value in research area

The amplification factor (or peak HVSR curve) value in research area shown in Figure 6. It ranges between 2,1 and 7,2, though most areas having 2,01 to 3,5 A_g value. The large amplification factor is related to high impedance contras between the sedimentary cover and the bedrock **[22]**. Reference **[7]** explained that no correlation between natural frequency and amplification factor. The variation of amplification factor value is not strongly effected by the soil depth, but the geological factors are more dominant to the variation of amplification factor value.

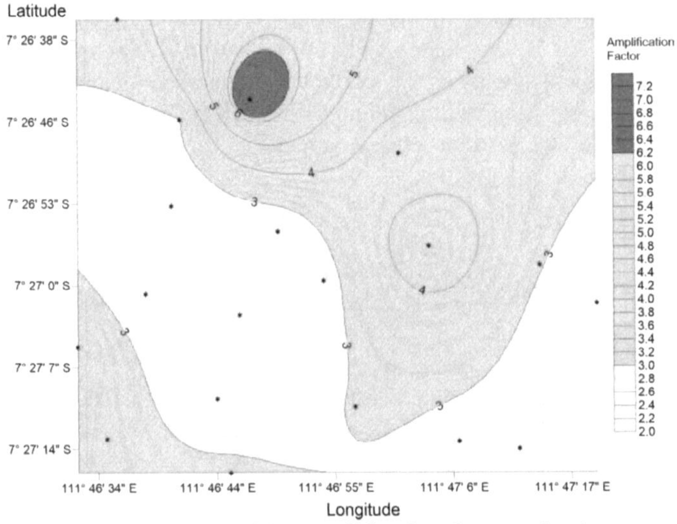

Figure 6. distribution map of the amplification factor value in research area

The natural frequency (f_0) value and the amplification factor (A_g) value were used as the input of parameters to generate seismic vulnerability index (K_g) to the formula: $K_g=A_g^2/f_o$ **[4]**. Reference **[4]** also explained that vulnerability index useful as an indicator weak points of the ground due to ground motion. The distribution of vulnerability index in research area shown in Figure 7. Its shows K_g value ranges between 0,48 and 25,34. The highest risk index is located in the north part of the research area where have high potential to damage when the earthquake occurring.

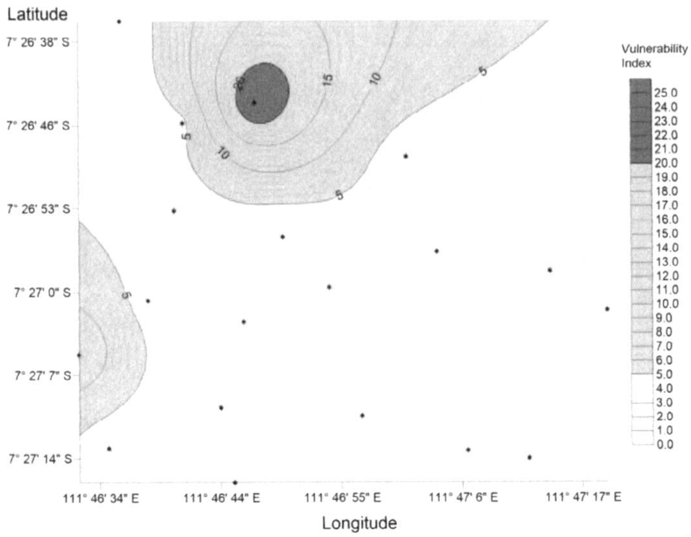

Figure 7. distribution map of the vulnerability index value in research area

5. Conclusion

The microtremor HVSR method has been widely used in seismic micro-zonation studies and site effect studies. Past earthquake information and the geological map of the research area is very valuable information for seismic microzonation at the research area. Microtremor measurements performed on the western flank of Pandan volcano, Bojonegoro district. The data analysis used Nakamura technique (horizontal to vertical spectral ratio method) on Easy HVSR software. The result data showed that the natural frequency value varies from 1,1 to 9,95 Hz and the amplification factor value ranges between 2,1 and 7,2. The vulnerability index determined based on the natural frequency value and the amplification factor value. In this study, the vulnerability index value ranging from 0,48 to 25,34. The highest K_g value is located in the north part of the research area, as indicate as weak point and potential to damage when the earthquake happens.

References

[1] Nakamura Y 1989 *Quarterly Report of Railway Technical Research Institute (RTRI)* **30** 25
[2] Stanko D, Markusic S, Strelec S, Gazdek M 2017 *Soil Dynamics and Earthquake Engineering* **92** 666
[3] Mohamed A A, Helal A M A, Mohamed A M E, Shokry M M F, Ezzelarab M 2016 *NRIAG Journal of Astronomy and Geophysics* **5** 55
[4] Nakamura Y 1997 *World Congress on Railway Research* (Italy: Florence) p 1
[5] Nakamura Y 2000 *World Congress on Earthquakes Engineering* (New Zealand) 2656
[6] Nakamura Y 2008 *14th World Congress on Earthquakes Engineering* (China: Beijing)
[7] Warnana D D, Soemitro R A A, Utama W 2011 *International Journal of Basic & Applied Sciences IJBAS-IJENS* **11** 73
[8] Leyton F, Ruiz S, Sepulveda S A, Contreras J P, Rebolledo S, Astroza M 2013 *Engineering Geology* **161** 26
[9] Herak M 2011 *Geofizika* **28** 21
[10] Akbari M, Ghafoori M, Moghaddas N H, Lashkaripour G R 2011 *Annals of Geophysics* **227** 354
[11] P Nur Ayu Diana Citra Dewi S, Lestari R T, Soemitro R A A, Warnana D D 2016 *Procedia Social an Behavioral Sciences* **227** 354
[12] Mokhberi M 2015 *International Journal of Disaster Risk Reduction* **13** 369
[13] Suita J, P Maria L, Bautista, Nakamura Y 2004 *13th World Congress on Earthquakes Engineering* (Canada: Vancouver) 905
[14] Hidayat S, Cari, Warnana D D, Koesuma S 2017 *Pertemuan Ilmiah XXXI HFI Jateng & DIY* (Yogyakarta: STTN Batan)
[15] Santoso D, Wahyudi E J, Alawiyah S, Nugraha A D, Widiyantoro S, Kadir W G A, Supendi P, Wiyono S, Zulkafriza 2017 *IOP Conf. Series: Earth and Environmental Sciences* **62** 1
[16] http://repogempa.bmkg.go.id/index_peta.php?id=101&session_id=J93zN0LZ
[17] Pringgoprawiro H and Sukido 1992 *Geologic Map of the Bojonegoro Quadrangle, East Java* (Bandung: Pusat penelitian dan pengembangan geologi)
[18] Kyaw Z L, Pramumijoyo S, Husein S, Fathani T F, Kiyono J 2015 *Procedia Earth and Planetary Science* **12** 31
[19] SESAME 2004 Guidelines for the Implementation of the H/V Spectral Ratio Thecnique on Ambient Vibrations: Meassurement Processing and Interpretation *SESAME European research project* WP12
[20] Akbari M, Ghafoori M, Moghaddas N H, Lashkaripour G R 2011 *Annals of Geophysics* **54** 424
[21] Gosar A 2009 *Acta Geotechnica Slovenica* **2** 31
[22] Leyton F, Ruiz S, Sepulveda S A, Contreras J P, Rebolledo S, Astroza M 2015 *Engineering Geology* **161** 26
[23] Walling M Y, Mohanty W K, Nath S K, Mitra S, John A 2009 *Engineering Geology* **106** 123
[24] Herak M 2011 *Geofizika* **28** 21

IOP Conf. Series: Journal of Physics: Conf. Series **909** (2017) 012032 doi:10.1088/1742-6596/909/1/012032

Annealing temperature dependence of magnetoimpedance effect in electrodeposited $[Ni_{80}fe_{20}/Cu]_3$ multilayers

Frendi Maulana[1], W Eko Prastyo[1], Nuryani[1] , B Purnama[1,a)]

Physics Department, Post Graduate program Universitas Sebelas Maret, Jl. Ir. Sutami 36A Kentingan Surakarta INDONESIA 57126

E-mail: frendimaulana270@gmail.com, [1,a)]bpurnama@mipa.uns.ac.id

Abstract. We have conducted an experiment of magnetoimpedance with a variation of annealing temperature of $[Ni_{80}Fe_{20}/Cu)]_3$ multilayers. The multilayer is electrodeposited on Cu-PCB substrate. Magnetoimpedance effect is impedance measure on account of external magnetic field. The found MI (magnetoimpedance) ratio is 7,63 % (without annealing) and 4,75 % (using annealing) of 100 °C. We find that MI ratio depends on to annealing temperature and current frequence. MI ratio decreases due to rising temperature and identified increase due to the frequency. The highest MI ratio is on a sample without annealing temperature and measurement at 100 kHz frequence.

1. Introduction

Since the effect magnetoimpedance (MI) was first discovered by Mohri et al. have a lot of growing techniques and methods were applied to the sample MI, one of which with annealing. During the annealing process, in general a sample is given the heat (temperature) which is maintained until a several time, and after that the sample can be measured [1]. Magnetoimpedance (MI) is a state of impedance change of magnetic material due to an external magnetic field when AC (alternative current) passes [1-5]. The phenomenon of impedance changes associated with the modification of the value of skin depth which is dominated by circumferential permeability and to the wire transverse permeability in the ribbon-thin layer [1-3]. In the fact, the phenomenon of MI effect observed on the wire with the high-frequency range (several MHz) with sufficiently high annealing temperature (300°C). Better MI is observed with the increase in permeability and saturation magnetization of the sample. Annealing process affects the magnetic anisotropy and magnetic softness in the form of permeability and saturation magnetization have been reported [1,6]. This new step with annealing is quite important considering the better results will be oriented to more effective sensor applications.

In recent years, fluxgate sensors, giant magnetoresistive sensors, hall sensors and sensor-based MI is an example of a magnetic sensor that continues to grow along with the development of research results. The MI-based sensors are very attractive to develop because these sensor-based solutions offer several advantages over other sensors such as better sensitivity, stability and linearity. A good sensor is characterized by high sensitivity. The sensitivity of the sensor can be expressed by the equation $\xi = \{ 2(\Delta Z/Z)_{max})/ \Delta H) \}$, where $H\Delta$ a full width at half maximum ratio curve MI [7].

In the purpose of this study, the effect of MI on samples multilayers $[Ni_{80}Fe_{20}/Cu)]_3$ was evaluated by the addition of annealing temperature variation and measurement in the low-frequency range (f<100 kHz). The sample $[Ni_{80}Fe_{20}/Cu)]_3$ is produced through deposition NiFe and Cu using electroplating or electrodeposition method. The magnitude of the impedance is represented as the sum of the values of resistance and reactance $Z = \sqrt{R^2 + X^2}$. While the change in a value of MI ratio is expressed using the equation:

$$\frac{\Delta Z}{Z}(\%) = 100\% \frac{Z(H) - Z(H)\max}{Z(H)\max}$$

with Z (H) is the impedance measured at the time given the external magnetic field, and Z (H) max is the measured impedance value at the time of maximum H [1].

2. Methods

Multilayers is grown by adopting a multi-layered configuration of NiFe/Cu/NiFe through electrodeposition method. A non-magnetic layer is inserted between the magnetic layer to keep the deposition rate for 400 seconds NiFe layer and a layer of Cu 50 seconds. The potential difference is used to permalloy $Ni_{80}Fe_{20}$ was 3.5 V at the current density of 15.5 mA / s while Cu as a spacer coating with 3 V at the current density of 8 mA/s. During this study, we keep the set-up device in accordance procedural previous experiments [5, 8,9]. The solution of ethanol and acetone used to cleanerization process Cu-PCB substrate using Ultrasonic cleaner.

The composition of the electrolyte solution forming successive magnetic layers is $NiSO_4.6H_2O$ (0,009 M), $FeSO_4.7H_2O$ (0,012 M), H_3BO_3 (0,149 M), $C_6H_8O_3$ (0,002 M), and 120 ml of distilled water. While the composition of the solution forming a conductive spacer layer is Cu as $CuSO_4.5H_2O$ (0,065 M), $C_6H_{12}O_6$ (0,002 M), and 120 ml of distilled water [10]. In this study, the additive liquid H_2SO_4 concentration of 0.1 M (± one drop) each added to two of the electrolyte solution to maintain the acidity of pH = 2.6 [10,11]. In his book, Kanani (2004) explains that the function of the additive in electrodeposition process is as stabilizers, reducing agents, complexing agents and to maintain the pH of the electrolyte solution. Both electrolyte solution to be used pursued in condition fresh (new) and the room temperature is kept constant during the fabrication process progresses. The mechanism of the electrodeposition process is shown in Figure 1.

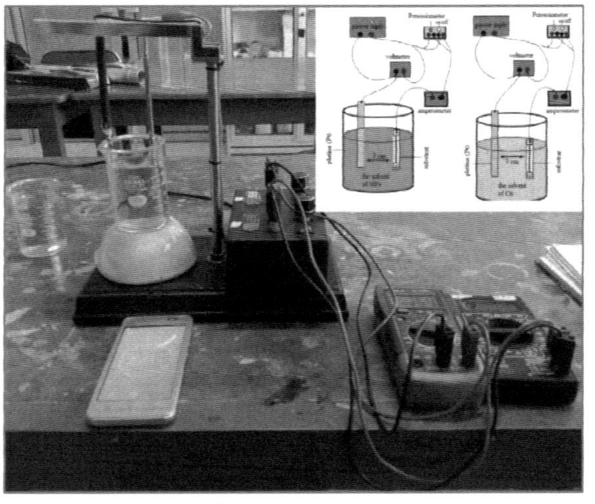

Figure 1. Experimental set up for electrodeposition of the multilayers $[Ni_{80}Fe_{20}/Cu)]_3$.

The multilayered sample of electrodeposition results in annealing using an oven with a 10-minute stand-up time setup as well as the selection of temperature variations at 28 °C , 50°C, 75°C and 100°C. The characterization structure of $[Ni_{80}Fe_{20}(800\ nm)/Cu(200\ nm)]_3$ is entirely determined by XRF and XRD analysis (not shown here). The set-up of MI measurement is presented by Figure 2.

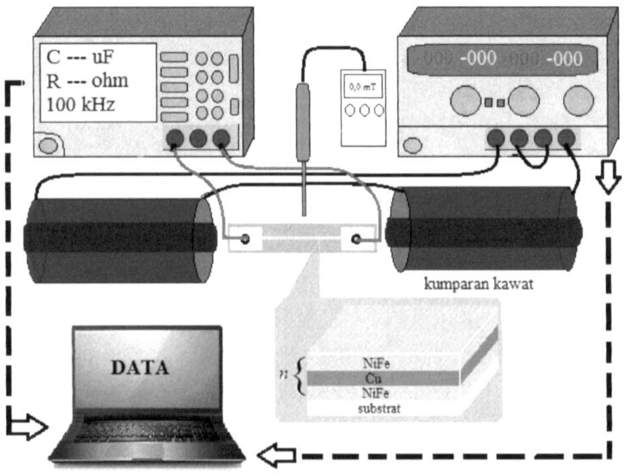

Figure 2. The setup of magneto-impedance measurement for multilayer samples of $[Ni_{80}Fe_{20}/Cu)]_3$.

3. Results and Discussion

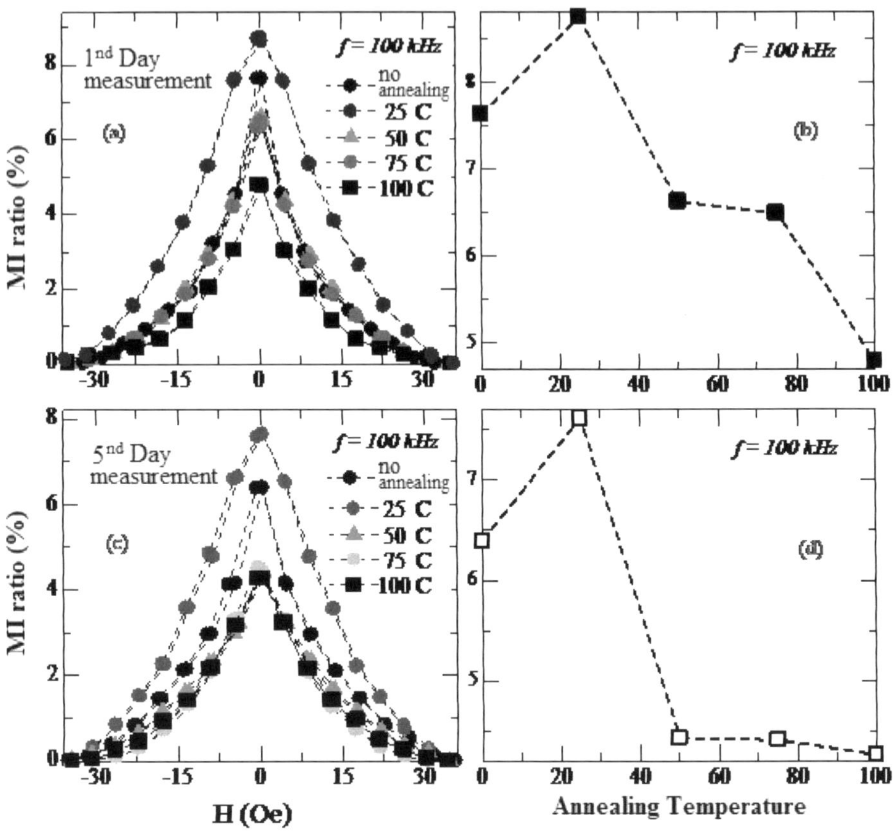

Figure 3: (a),(c) the MI ratio curve at 100 kHz and (b),(d) MI ratio as fungtions of annealing temperature (C^0) .

Figure 3 (a) (c) expressing the phenomenon of MI ratio as a function of the external magnetic field (*H*). In this study, the phenomenon of MI is observed in samples annealed at various temperature by looking at the change impedance (*Z*) caused by a magnetic field. The results are presented in the form of MI ratio curves. In the samples, the MI maximum is achieved when no external magnetic field is given (when *H* is equal to zero). And then the increasing external magnetic field, MI ratio is obtained gradually decreases to approach the relative saturation value after ± 34.36 Oe (measurements of the first day) and 34.62 ± Oe (fifth day).

Overall MI ratio decreased significantly from samples without treatment of annealing with samples after annealing treatment up to 100 °C. The effect of this annealing treatment resulted in a decreased MI ratio of the first day measured from ± 7.6 % to ± 4.7 % and measured the fifth day from ± 6.39 % to ± 4.25 %. The identified MI ratios decreased by 0.61 and 0.66-fold during the measurement (samples with 100 °C annealing or without annealing treatment). The decrease MI ratio is believed to be due to transverse anisotropy fluctuations in the transverse direction of the sample [6]. In this discussion, the presence of the magnetic anisotropy of the sample may be related to the change in MI ratio values during the measurement activity. Samples without annealing proved to have greater MI ratios due to their small magnetic anisotropy. While the annealing process will reduce the magnetic transverse anisotropy in the sample and further increase the permeability according to Somer and Chein (1995).

In general, samples with high permeability is required to achieve a greater MI effect [3]. If the terms of this assumption given the annealing process will tend to increase the ratio of MI. However, this condition is not fully applicable since the fact that the ratio of MI was identified in this study is identical to decline and only on the sample at a temperature of 28 °C which MI ratio values greater than the given sample without annealing. The results were found is believed to be due to the consequences of the low frequencies (f < 100 kHz) measurement. The result is likely to be different if the sample is measured in the high-frequency range (f > 1MHz) [1].

MI ratio analysis shows the effect of different sample measurements on MI behavior. The MI ratio decreased significantly in the samples measured on the first day compared with measurements on the fifth day. The trend of MI ratio decline is believed to be due to degraded or contaminated samples by the environment causing microstructural changes in the sample. The measurement of MI ratio and the complete difference is shown in Figure 4

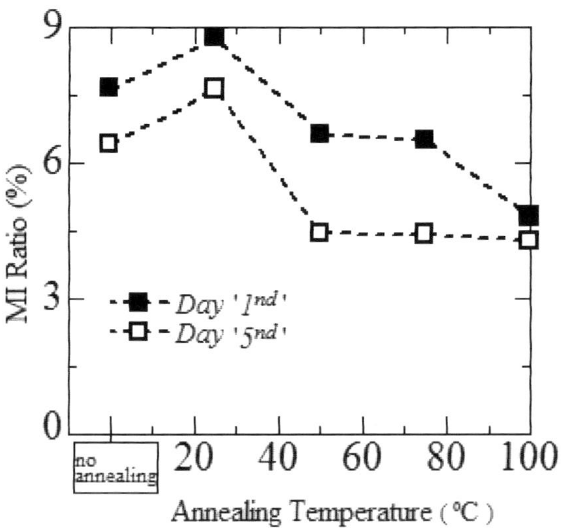

Figure 4. The difference in the measurement of the ratio of MI as a function of annealing temperature variations.

Another expression is shown in Figure 5 as the ratio of MI against frequency curves. Observed very clearly shows an MI curve typically the same as in the previous annealing temperature variation curve. The measured results obtained at several frequencies ranging from the MI ratio increased during the addition of frequence AC current. The maximum MI ratio was found 0.03 % during measurement at a frequency of 20 kHz. the MI ratio increased dramatically to 7.63 % when the frequency of 100 kHz. This means that the increase in the ratio of MI responded very well during the addition of the frequency ratio. Overall the ratio of MI increased ± 254.3 times over four times the measurement range (20 kHz to 100 kHz). This is indicated by the dominance of skin depth [1,2]. The smaller of the skin depth will be increased value of the sample resistance marked by increasing frequency. The resistance contributes predominantly in increasing the Z oriented impedance value of the MI ratio [8].

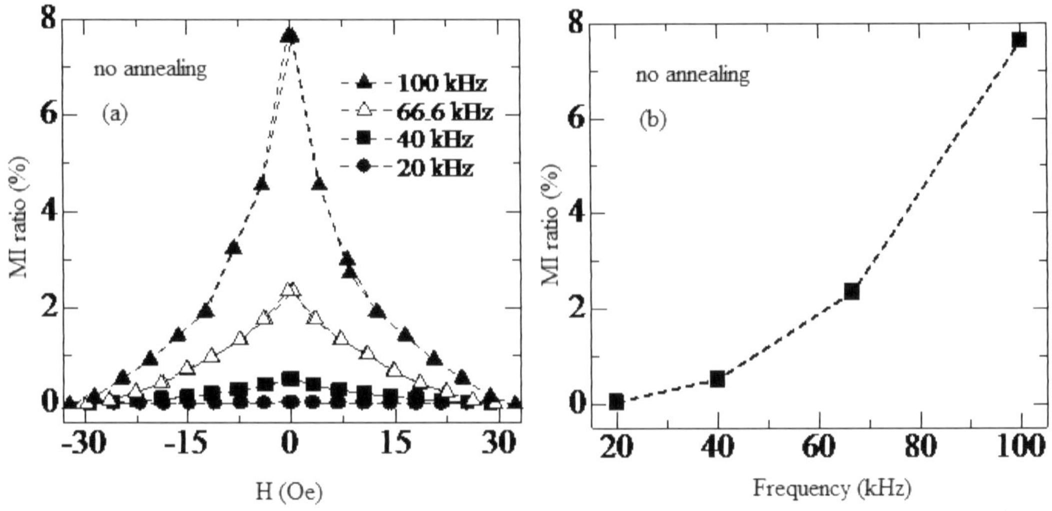

Figure 5: (a) MI curve $[Ni_{80}Fe_{20}/Cu)]_3$ of the 20 kHz to 100 kHz on a frequency measurement, (b) MI ratio as function of frequency.

4. Conclussion

The phenomenon of magnetoimpedance has been observed in thin film multilayers $[Ni_{80}Fe_{20}/Cu)]_3$ in low frequency range (f < 100 kHz). The MI ratio can be improved by increasing frequency. analysis of the measurement showed that the MI ratio decreased by increasing the annealing temperature (°C) resulted the transverse anisotropy fluctuation in the sample.

References

[1] Phan M H and Peng H X 2008 *Progress in Materials Science* 53 323-420
[2] Borge, A. (2005). Giant magneto-impedance effect in thin film layered structures. Thesis, Department of Electrical and Computer Engineering University of Central Florida.
[3] Knobel M, Vazquez M, Kraus L. 2003. Giant magnetoimpedance. In: Buschow KH, editor. Handbook of magnetic materials, vol. 15. Amsterdam: Elsevier Science B.V; p. 1–69
[4] T. Morikawa, Y. Nishibe, H. Yamadera, Y. Nanomura, M. Takeuchi, Y. Taga, (1997). Giant Magneto-Impedance in Layered thin films. IEEE Transactions on Magnetics vol. 33, No. 5.
[5] Ismail, Nuryani, and Purnama B 2016 *AIP conference Proceedings* 1710 .030002
[6] Sommer RL, dan Chien CL. 1995. Role of magnetic anisotropy in the magnetoimpedance effect in amorphous alloys. Appl Phys Lett; 67:857–9.
[7] Tuan L H, Huy N T, and Huy P T 2009 *journal of physics*, 187(1), 1-5
[8] Wicakson B A, Nuryani, and Purnama B 2015 *Jurnal fisika dan aplikasinya* 11, no.3

[9] Maulana F, Nuryani, and Purnama B, 2017 IOP Conf. Series: Journal of Physics: Conf. Series 795, 012018
[10] Aydogmus E, Kaya H, Atalay F E, Atalay s, and Avcu D, 2014 *Acta physica Polanica A* 125, 227-229
[11] Hanafi E P, Utari, and Purnama B, 2013 *Jurnal Fisika & Aplikasinya*, 9 No.2 63-65
[12] Kanani, N. (2004). Electroplating Basic Principles, Processes and Practice. Berlin: Elsevier.
[13] Marques M S, Mori T J A, Schelp L F, Chesman C, Bohn F, and Corrêa M A 2012 *Thin solid films* 520 2173-2177

IOP Conf. Series: Journal of Physics: Conf. Series **909** (2017) 012033 doi:10.1088/1742-6596/909/1/012033

Fabrication and Analysis Signal Optical Fiber Sensor Based On Bend Loss for Weight in Motion Applications

Ahmad Aftah Syukron[1], Ahmad Marzuki[1] and Ary Setyawan[2]

[1] Physics Department of Graduate Program Sebelas Maret University, Jl. Ir. Sutami 36A Kentingan Jebres Surakarta 57126, INDONESIA

[2] Civil Engineering of Engineering Faculty Sebelas Maret University, Jl. Ir. Sutami 36A Kentingan Jebres Surakarta 57126, INDONESIA

E-mail: dfata.aftah@gmail.com

Abstract

Road network plays very important role in economic development. Overweight is one of the main factors contributing to road damage. To minimize this factor, road authority has to make sure that all vehicles operate in according to maximum vehicle regulation set by the government. The one solution can use from this problem is Weight in motion (WIM) technology. WIM technology allows measuring vehicle weight quickly. The sensor is one of the important components in the WIM system.

This paper presents a model of WIM fiber sensor work based on bend loss. Fiber sensor has made by coiling optical fiber. Coiling optical fiber has managed in the elliptical shape rubber coil. Rubber coil then is planted in the pad of sensor. The principle of this sensor is a detecting of the shift light intensity output of optical fiber when the vehicles a passing through on fiber sensor. Loading was carried out using loaded truck model. Data was carried out with variations of load and load positions in the truck.

The results can be concluded that the shift light intensity is greater with the more shift loads. The loader of the truck has also resulted in the greater loss. Loads in the truck distributed on the axles due to the position of loads.

1. Introduction

Road network plays very important role in economic development. However, many roads quickly damaged. That has resulted in the distribution of goods are obstructed. Overweight is one of the main factors contributing to road damage. To minimize this factor, road authority has to make sure that all vehicles operate in according to maximum vehicle regulation set by the government. The one solution can use from this problem is Weight in motion (WIM) technology. WIM technology allows measuring vehicle weight quickly [5].

The principle of WIM technology is a detecting of the shift light intensity output of optical fiber when the vehicles a passing through on fiber sensor [1]. Different with the weighbridge method, vehicles must be stopped to measure vehicle weight. So, this is causes of the traffic congestion. Some advantages of WIM technology are more efficient and saving time, especially when the traffic is busy. Some WIM technologies are there current now use piezoelectric cable, capacitive mats, hydraulic, and load cell system [6]. However, these sensors have some weakness are corrosion, can be affected electromagnetic interference and has a low accuracy.

IOP Conf. Series: Journal of Physics: Conf. Series **909** (2017) 012033 doi:10.1088/1742-6596/909/1/012033

Sensor technology based on optical fiber seems to be a good solution. The optical fiber is the transmission media or cylindrical light waveguide. Optical fiber sensor has some advantages, such as hold up with electromagnetic interference and radiation, high sensitively, and hold up to corrosion [3].

2. Experiment

This experiment aim is to designing and fabricating optical fiber sensor for weight in motion application. Fiber sensor created by coiling optical fiber in the elliptical rubber shape with the specific diameter. Rubber coil then is planted in the pad sensor. Then, the analysis optic signal shift in the fiber sensor when the weight in motion effect occurs. This experiment consists of some main components: a light source, optical fiber sensor, a light detector, analog digital converter (ADC), and the software of taking and data processing.

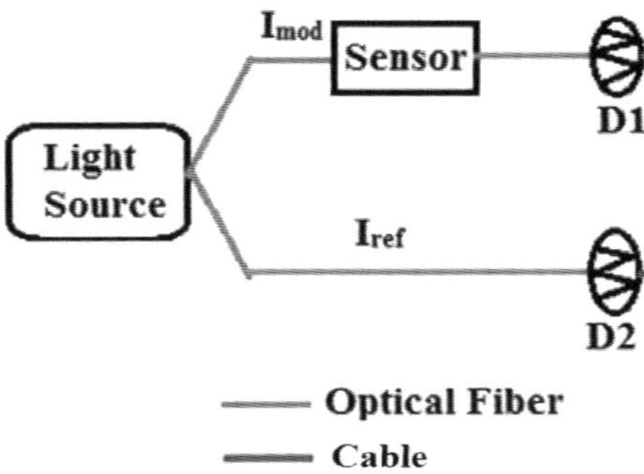

Figure 1. Schematic of the experiment model

Figure 1 is the schematic model of used in this experiment. Light from the light source has split into 2 arms. One arm is coming in to optical fiber as the modulation intensity. One arm again is coming in to optical fiber as the reference intensity. Fiber sensor will be through on loss light or disappearance of light transmittance. The transmitted light signal then detected by a light detector. This light signal still in the analog signal. Analog signal compared to a digital signal with the analog digital converter (ADC). Digital data from the ADC process then connected to a personal computer to display with the LABVIEW 2012 program. Figure 2 is the Weight in motion data acquisition using LABVIEW 2012 program.

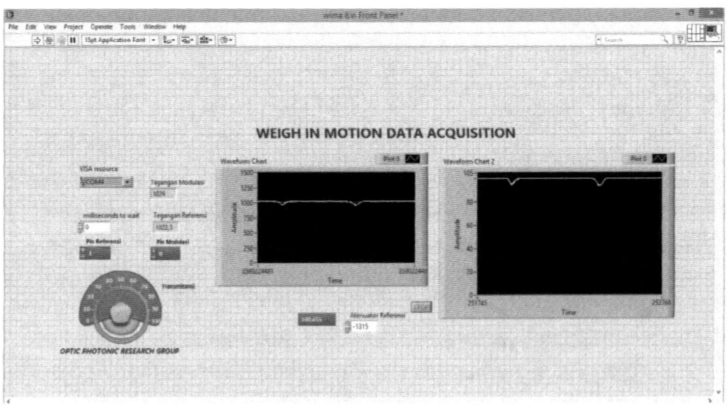

Figure 2. Weight in motion data acquisition program

3. Results and Discussion

The taking data carried out by variations of load positions in the truck. The load positions are in the front, in the middle, and in the back of a truck. Measurement data carried out with the toy truck. Figure 3 is the measurement data result shown on the weight in motion data acquisition.

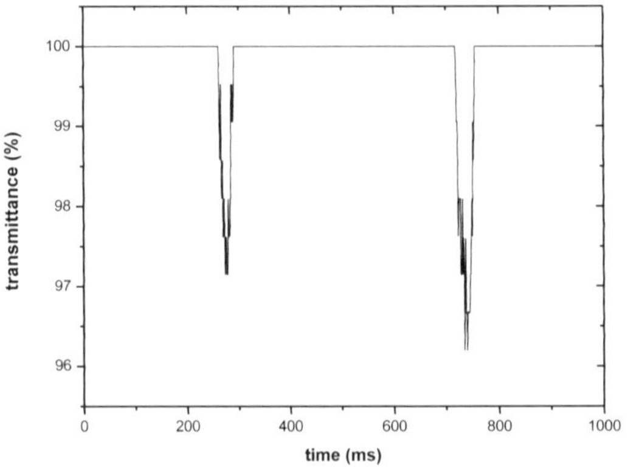

Figure 3. Chart of transmittance fiber sensor when passing through on the truck

The chart on figure 3 shows light transmittance value at the time. The chart there are 2 valleys depth. Valleys depth are shows light transmittance value in the fiber sensor when a truck is passing through on fiber sensor. First valley depth is in the front tyre of the truck. Second valley depth is in the rear tyre of the truck.

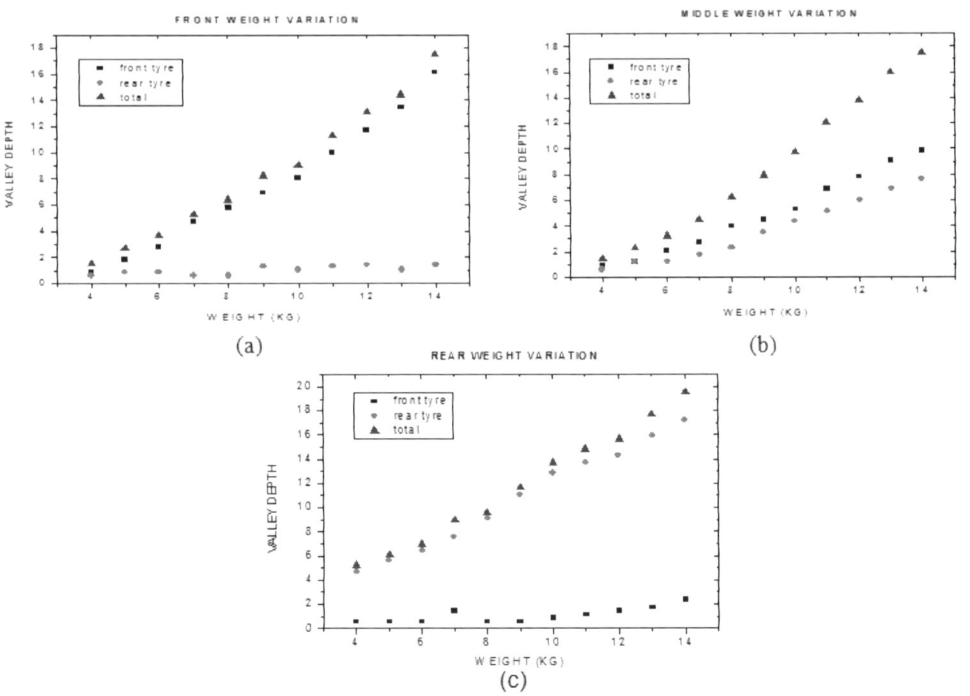

Figure 4. Data results in this experiment (a) Front tyre variation, (b) middle tyre variation, (c) rear tyre variation

Figure 4 shows that the valley depth is greater with the greater of loads is passing through on fiber sensor. The great of the valley depth shows the optical fiber loss in the fiber sensor. The loss is greater with the greater of loads. Figure 4 also shows that the loads distribution is following the position of the load in the truck. On the front load variation, the loads distribution on the front tyre is greater than on the rear tyre. The greater of the loads has resulted in the greater valley depth. In the otherwise, The valley depth on the rear tyre constant at the front variation of the loads. On the rear load variation, the loads distribution on the rear tyre is greater than on the front tyre. While on the middle load variation, the loads distribution is evenly distributed on the both tyre. It can be explained with the physical illustration below.

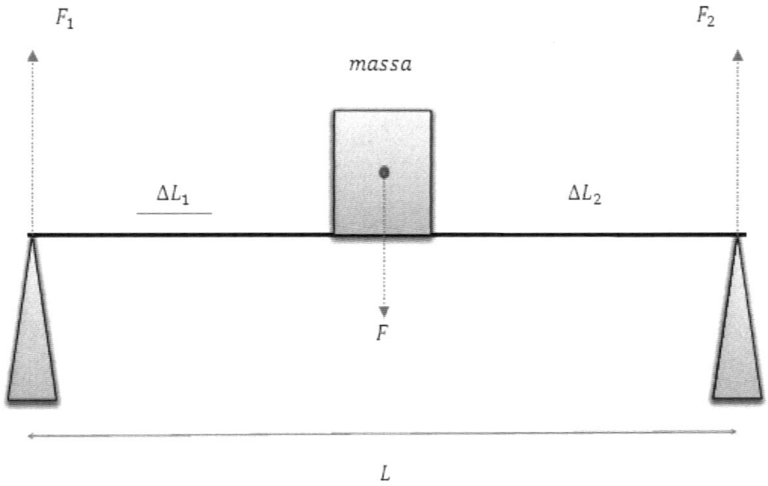

Figure 5. The illustration of distribution loads in the weight in motion

Figure 5 shows that the loads distribution has set in the buffer. The buffer is analogous as the front tyre and rear tyre of the truck. The mass has set in the middle of the buffer. It is giving a force F downward direction. That is due to Newton's Third Law where the action force same the reaction force and opposite direction, shown in the equation (1).

$$F_{action} = -F_{reaction} \tag{1}$$

F_{action} is the weight of the mass acting downward direction. $F_{reaction}$ is the force exerted by the ground on the mass in the upward direction. The reaction force is the total force from the both buffers, shown in the equation (2).

$$F_{action} = -(F_1 + F_2) \tag{2}$$

F_1 is the reaction force in the front tyre, F_2 is the reaction force in the rear tyre. Therefore, each of the buffers will give the same of the reaction force. It is shown that the reaction force distributed in the both buffers. The magnitude of reaction mass on the both buffers depends on the position of the loads. The magnitude of each force from the both buffers can write:

$$F_1 = F\frac{\Delta L_2}{L}$$

$$F_2 = F\frac{\Delta L_1}{L} \tag{3}$$

4. Conclusions

The fabrication of fiber sensor to weight in motion application has made. Fiber sensor has made by coiling optical fiber sensor in the elliptical shape rubber coil. Rubber coil then is planted in the pad of sensor. This experiment consists of some main components: a light source, optical fiber sensor, a light detector, analog digital converter (ADC), and the software of taking and data processing. The results can be concluded that the shift light intensity is greater with the more shift loads. The loader of the truck has also resulted in the greater loss. Loads in the truck distributed on the axles due to the position of loads.

References

[1] Batenko A, Grakovski A, Kabashkin I, Petersons E, Sikerzhicki Y 2011 Transport and Telecommunication **12** (4) 27–33
[2] Kuang J-H, Chen P-C, Chen Y-C 2010 Sensors **10** 10198
[3] Bin M, Xinguo Z 2010 ICICTA **10** 1109
[4] Zendehnam A, Mirzaei M, Farashiani A and Farahadi H L 2010 *J. Phys.* **74** 591–603
[5] Malla R B, Sen A, Garrick N W 2008 Sensors **8** 2551–2568
[6] Cheng L, Zhang H, Li Q 2007 sensors **7** 1530–1544
[7] Wang L, Hu X, Huang Y and Xu H 2013 ICCICN **5** 156–159
[8] Yuan S, Ansari F, Liu X and Zhao Y 2005 Sensors and Actuators A **120** 53–58
[9] Wang J and Wu M 2004 Proc. World Congress on Int. Con. Aut. **5** 5241
[10] Pinheiro M J 2009 New Journal of Phys.
[11] Hellingman C 1992 Physics Education **27** (2) 011
[12] Itza-Ortiz S F, Rebello S and Zollman D 2004 Eur. J. Phys. **25** 81–89

IOP Conf. Series: Journal of Physics: Conf. Series **909** (2017) 012034 doi:10.1088/1742-6596/909/1/012034

The solution of 4-dimensional Schrodinger equation for Scarf potential and its partner potential constructed By SUSY QM

Wahyulianti, A. Suparmi, C. Cari, and Kristiana N. Wea

Physics Department of Post Graduate Program Sebelas Maret University, Jl. Ir. Sutami 36A Kentingan Jebres Surakarta 57126, INDONESIA

E-mail: wahyuliantiyuli@yahoo.com

Abstract. The angular part of 4-dimensional Schrodinger equation for Scarf potential was solved by using the Nikiforov-Uvarov method and Supersymmetric Quantum Mechanic method. The determination of the ground state wave function has been used Nikiforov-Uvarov method and by applying the parametric generalization of the hypergeometric type equation. By using manipulation of the properties and operators of the Supersymmetric Quantum Mechanic method the partner potential was constructed. The ground state wave functions of original Scarf potential is different than the ground state wave functions of the construction result potential.

1. Introduction

The D-dimensional Schrodinger equation is one of the wave equations for higher dimensions in quantum mechanics [1-2]. The D-dimensional Schrodinger equation for some potentials are exactly solved by some methods such as Nikiforov-Uvarov (NU) method [3-6], Supersymmetric Quantum Mechanic (SUSY QM) [4,8], Romanovski polynomial method [9], Asymptotic Iteration Method (AIM) [10] and Factorization methods [11]. The NU method is one of the methods with common application, and NU differential equations have been formulated with general parameters of the general hypergeometric type equation [4]. By applying and manipulating the operators of SUSY QM, the properties of SUSY QM and the ground state wave functions of the original potential, the partner potential was constructed.[4]

The purpose of this research is to construct partner potential which is the new potential of the generalized Scarf potential by manipulating the partner potential equations with the ground state wave function of Scarf potential. In this paper, the solution of angular D-dimensional Schrodinger equation for Scarf potential was studied using Nikiforov-Uvarov method with hypergeometric type equations and SUSY QM. The paper is organized as follows. The method is presented in Section 2. The results and discussion are presented in Section 3 and a conclusion in Section 4.

2. Method

In this section, we review the solution of D-dimensional Schrodinger equation with Nikiforov-Uvarov method and Supersymmetry Quantum Mechanics method. The Generalized Schrodinger equation in D-dimensional shown as followed equation [1,12] given as

$$-\frac{\hbar^2}{2\mu}\nabla_D^2\psi(r,\Omega)+V(r,\Omega)\psi(r,\Omega)=E\psi(r,\Omega) \quad (1)$$

with

$$\psi(r,\Omega)=\frac{1}{r^{\frac{D-1}{2}}}U(r)Y^{l}_{l_{D-1}\cdots l_1}(\hat{x}) \qquad ;\hat{x}=\theta_1,\theta_2,\cdots\theta_{D-1} \tag{2}$$

The 4-dimensional Schrodinger equation obtained by inserted Eq. (2) into Eq. (1) and solve of using variable separation method with $Y_{\ell_1,\ldots\ell_3}(\theta_1,\theta_2,\theta_3)=P(\theta_1)P(\theta_2)P(\theta_3)$ is given as [13]:

$$\frac{\partial^2 U}{\partial r^2}-\left(\lambda_3+\frac{3}{4}\right)\frac{1}{r^2}U+\frac{2\mu}{\hbar^2}\left[E-V(r)\right]U=0 \tag{3}$$

$$\frac{\partial^2 P(\theta_1)}{\partial\theta_1}-V(\theta_1)P(\theta_1)+\lambda_1 P(\theta_1)=0 \tag{4}$$

$$\frac{1}{P(\theta_2)}\left\{\frac{1}{\sin\theta_2}\left(\frac{\partial}{\partial\theta_2}\sin\theta_2\frac{\partial P(\theta_2)}{\partial\theta_2}\right)\right\}-V(\theta_2)-\frac{\lambda_1}{\sin^2\theta_2}+\lambda_2=0 \tag{5}$$

$$\frac{1}{P(\theta_3)}\left\{\frac{1}{\sin^2\theta_3}\left(\frac{\partial}{\partial\theta_3}\sin^2\theta_3\frac{\partial P(\theta_3)}{\partial\theta_3}\right)\right\}-V(\theta_3)+\lambda_3-\frac{\lambda_2}{\sin^2\theta_3}=0 \tag{6}$$

where λ_1,λ_2 and λ_3 is variable separation constant.

In this paper the potential energi of Scarf potential generally given as [14]:

$$V_i(\theta_i)=\left(\frac{A_i}{\sin^2\theta_i}+\frac{B_i\cos\theta_i}{\sin^2\theta_i}\right),\ \text{for } i=1,2 \text{ and } 3\ . \tag{7}$$

with $A_i=b_i^2+a_i(a_i-1)$ and $B_i=-2b_i\left(a_i-\frac{1}{2}\right)$.

2.1 Review of NU method

The solution of Schrodinger equation is reduced to the hypergeometric type differential equation by using variable transformation. By using the NU method, the solution of the hypergeometric type differential equation is expressed as

$$\frac{\partial^2\psi(s)}{\partial s^2}+\frac{\tilde{\tau}(s)}{\sigma(s)}\frac{\partial\psi(s)}{\partial s}+\frac{\tilde{\sigma}(s)}{\sigma^2(s)}\psi(s)=0 \tag{8}$$

where $\tilde{\tau}(s)$ is first order polynomial, $\sigma(s)$ and $\tilde{\sigma}(s)$ are mostly second order polynomials. By using variable separation method, substitution equation (4) with

$$\psi(s)=\phi(s)\,y(s) \tag{9}$$

we obtain hypergeometric type as

$$\sigma\frac{\partial^2 y(s)}{\partial s^2}+\tau(s)\frac{\partial y(s)}{\partial s}+\lambda y(s)=0 \tag{10}$$

where $\phi(s)$ is logarithmic derivative $\frac{\phi'}{\phi}=\frac{\pi}{\sigma}$ with $\pi(s)$, λ, and τ are

$$\pi=\left(\frac{\sigma'-\tilde{\tau}}{2}\right)\pm\sqrt{\left(\frac{\sigma'-\tilde{\tau}}{2}\right)^2-\tilde{\sigma}+k\sigma} \tag{11}$$

$$\tau=\tilde{\tau}+2\pi,\quad \lambda=k+\pi'\text{ and}\qquad \lambda=\lambda_n=-n\tau'-\frac{n(n-1)}{2}\sigma'',\quad n=0,1,2,\ldots \tag{12}$$

The solution of the part $y_n(s)$ by using Rodrigues relation, is given as

$$y_n(s) = \frac{C_n}{w(s)} \frac{d^n}{ds^n}\left(\sigma^n(s)w(s)\right) \tag{13}$$

with C_n is normalization constant, and the weight function $w(s)$ must satisfy the condition

$$\frac{\partial(\sigma w)}{\partial s} = \tau(s)w(s) \tag{14}$$

The parametric generalization of the hypergeometric type equation in equation (8) is written as

$$\frac{\partial^2\psi(s)}{\partial s^2} + \frac{(c_1 - c_2 s)}{s(1-c_3 s)}\frac{\partial\psi(s)}{\partial s} + \frac{(-\varepsilon_1 s^2 + \varepsilon_2 s - \varepsilon_3)}{s^2(1-c_3 s)^2}\psi(s) = 0 \tag{15}$$

By comparing equation (8) and equation (15), we obtained the energy value equation and eigen function are respectively given as

$$c_2 n - (2n+1)c_5 + (2n+1)\left(\sqrt{c_9} + c_3\sqrt{c_8}\right) + n(n-1)c_3 + c_7 + 2c_3c_8 + 2\sqrt{c_8c_9} = 0 \tag{16}$$

and

$$\psi(s) = N_{n1} s^{c_{12}} \left(1-c_3 s\right)^{-c_{12}-(c_{13}/c_3)} P_n^{(c_{10}-1,\,(c_{11}/c_3)-c_{10}-1)}\left(1-2c_3 s\right) \tag{17}$$

where

$$\begin{aligned} &c_4 = \frac{1}{2}(1-c_1),\; c_5 = \frac{1}{2}(c_2 - 2c_3),\; c_6 = c_5^2 + \varepsilon_1,\; c_7 = 2c_4 c_5 - \varepsilon_2,\; c_8 = c_4^2 + \varepsilon_3 \\ &c_9 = c_3 c_7 + c_3^2 c_8 + c_6,\; c_{10} = c_1 + 2c_4 + 2\sqrt{c_8},\; c_{12} = c_4 + \sqrt{c_8} \\ &c_{11} = c_2 - 2c_5 + 2\left(\sqrt{c_9} + c_3\sqrt{c_8}\right), c_{13} = c_5 - \left(\sqrt{c_9} + c_3\sqrt{c_8}\right) \end{aligned} \tag{18}$$

By using equations (16) and (17) for $n = 0$ the ground state energy and wave function are determined [3-6].

2.2 Review of SUSY QM

Witten defined two charge operators are commute with Hamiltonian (H_{ss}) of the supersymmetry quantum system [4,8] is geven as

$$H_{ss} = \begin{pmatrix} -\frac{d^2}{dx^2} + \frac{d\varphi(x)}{dx} + \varphi^2(x) & 0 \\ 0 & -\frac{d^2}{dx^2} - \frac{d\varphi(x)}{dx} + \varphi^2(x) \end{pmatrix} = \begin{pmatrix} H_+ & 0 \\ 0 & H_- \end{pmatrix} \tag{19}$$

with partner Hamiltonian $H_- = H_1$ and $H_+ = H_2$. By setting the new SUSY operators for raising operator $A^+ = -\frac{d}{dx} + \varphi(x)$ and lowering operator $A = \frac{d}{dx} + \varphi(x)$.

The SUSY Hamiltonians of equation (19) are

$$H_-(x) = H_1 = A^+A\text{, and } H_+(x) = H_2 = AA^+ \tag{20}$$

and

$$A\psi_0^{(-)} = A\psi_0 = 0 \tag{21}$$

The partners of potential in the SUSY QM method is $V_- = V_1$ and $V_+ = V_2$ are

$$V_-(x) = V_1 = \varphi^2(x) - \varphi'(x) \text{ and } V_+(x) = V_2 = \varphi^2(x) + \varphi'(x) \tag{22}$$

where φ is superpotential. The partner potential V_1 is determined as

$$V_-(x;a_0) = V_1(x;a_0) = V_{ef}(x) - E_0 \tag{23}$$

with V_{ef} is effective potential and E_0 is the ground state energy.

By manipulating equation (22) and by applying equation (17) and equation (23), we construct the new partner potential V_2 as

$$V_2(x) = V_{ef}(x) - E_0 - 2\frac{d}{dx}\left(\frac{d\psi_0/dx}{\psi_0}\right) \tag{24}$$

where ψ_0 is the ground state wave function [4,8]. Then the new eigen function of new partner potential is also determined by using Nikiforov-Uvarov method.

3. Results and Discussion

In this section, the solution of the Schrodinger equation in 4-dimensional with Nikiforov-Uvarov method and Supersymmetry Quantum Mechanics method .

The solution of the Schrodinger equation in 4-dimensional for θ_1 obtained by inserting Eq. (7) to Eq. (4) is given as

$$\frac{\partial^2 P(\theta_1)}{\partial\theta_1} - \left(\frac{A_1}{\sin^2\theta_1} + \frac{B_1\cos\theta_1}{\sin^2\theta_1}\right)P(\theta_1) + \lambda_1 P(\theta_1) = 0 \tag{25}$$

By substituting Eq. (25) with $\cos\theta_i = (1-2s)$ we obtain

$$\frac{d^2P_1(s)}{ds^2} + \frac{\left(\frac{1}{2}-s\right)}{s(1-s)}\frac{dP_1(s)}{ds} + \frac{1}{s^2(1-s)^2}\left[-(\lambda_1)s^2 + \left(\lambda_1 + \frac{B_1}{2}\right)s - \left(\frac{A_1+B_1}{4}\right)\right]P_1(\mathrm{s}) = 0 \tag{26}$$

By comparing Eq. (26) and Eq. (15) and then by applying Eq. (16), Eq. (17) and Eq. (18) we obtained the λ_1 and eigen function are respectively given as

$$\lambda_1 = \frac{1}{2}\left(\frac{3}{4}+A_1\right) + n(n+1) + (2n+1)\left(\frac{1}{2}\sqrt{\frac{1}{4}+A_1-B_1} + \frac{1}{2}\sqrt{\frac{1}{4}+A_1+B_1}\right) + \frac{1}{2}\sqrt{\left(\frac{1}{4}+A_1+B_1\right)\left(\frac{1}{4}+A_1-B_1\right)} \tag{27}$$

and

$$P(\theta_1) = N_0\left(\frac{1-\cos\theta_1}{2}\right)^{\frac{1}{4}+\frac{1}{2}\sqrt{\frac{1}{4}+A_1+B_1}}\left(\frac{1+\cos\theta_1}{2}\right)^{\frac{1}{4}+\frac{1}{2}\sqrt{\frac{1}{4}+A_1-B_1}} P_{n1}^{\left(\sqrt{\frac{1}{4}+A_1+B_1},\,\sqrt{\frac{1}{4}+A_1-B_1}\right)}(\cos\theta_1) \tag{28}$$

with

$$c_1 = \frac{1}{2},\ c_2 = c_3 = 1,\ \varepsilon_1 = \lambda_1,\quad \varepsilon_2 = \lambda_1 + \frac{B_1}{2},\quad \varepsilon_3 = \frac{1}{4}(A_1 + B_1) \tag{29}$$

The solution of Eq. (27) and Eq. (28) for the ground state $n = 0$, we obtained

$$\lambda_{01} = \frac{1}{2}\left(\frac{3}{4}+A_1\right) + \frac{1}{2}\sqrt{\frac{1}{4}+A_1-B_1} + \frac{1}{2}\sqrt{\frac{1}{4}+A_1+B_1} + \frac{1}{2}\sqrt{\left(\frac{1}{4}+A_1+B_1\right)\left(\frac{1}{4}+A_1-B_1\right)} \tag{30}$$

and

$$P_0(\theta_1) = N_0\left(\frac{1-\cos\theta_1}{2}\right)^{\frac{1}{4}+\frac{1}{2}\sqrt{\frac{1}{4}+A_1+B_1}}\left(\frac{1+\cos\theta_1}{2}\right)^{\frac{1}{4}+\frac{1}{2}\sqrt{\frac{1}{4}+A_1-B_1}} \tag{31}$$

The partner potential V_2 as new potential constructed from Eq. (31) and Eq. (24) is

$$V_2(\theta_1) = \frac{A_1}{\sin^2\theta_1} + \frac{B_1\cos\theta_1}{\sin^2\theta_1} - E_{01} + \frac{c_{12}}{\left(\frac{1-\cos\theta_1}{2}\right)} - \frac{c_{14}}{\left(\frac{1+\cos\theta_1}{2}\right)} \tag{32}$$

with

$$E_{01} = \lambda_{01},\quad c_{12} = \frac{1}{4} + \frac{1}{2}\sqrt{\frac{1}{4} + A_1 + B_1} \text{ and } c_{14} = \frac{1}{4} + \frac{1}{2}\sqrt{\frac{1}{4} + A_1 - B_1} \tag{33}$$

The solution of the Schrodinger equation in 4-dimensional for θ_1 partner potensial obtained by inserting Eq. (32) to Eq. (4) is given as

$$\frac{\partial^2 P(\theta_1)}{\partial\theta_1} - \left(\frac{A_1}{\sin^2\theta_1} + \frac{B_1\cos\theta_1}{\sin^2\theta_1} - E_{01} + \frac{c_{12}}{\left(\frac{1-\cos\theta_1}{2}\right)} - \frac{c_{14}}{\left(\frac{1+\cos\theta_1}{2}\right)}\right) P(\theta_1) + \lambda_{01}^{'} P(\theta_1) = 0 \tag{34}$$

By substituting Eq. (34) with $\cos\theta_i = (1-2s)$ we get

$$\frac{d^2P_1(s)}{ds^2} + \frac{\left(\frac{1}{2} - s\right)}{s(1-s)}\frac{dP_1(s)}{ds} + \frac{1}{s^2(1-s)^2}\left[\begin{array}{l} -\left(E_{01} + \lambda_{01}^{'}\right)s^2 + \left(E_{01} + \lambda_{01}^{'} + c_{14} + c_{12} + \frac{B_1}{2}\right)s \\ -\left(\frac{A_1 + B_1}{4} + c_{12}\right) \end{array}\right] P_1(s) = 0 \tag{35}$$

By comparing Eq. (35) and Eq. (15) and then by applying Eq. (16), Eq. (17) and Eq. (18) we obtained $\lambda_{01}^{'}$ and eigen function are respectively given as

$$\begin{aligned} \lambda_{01}^{'} = {} & \frac{3}{8} - E_{10} - c_{14} + \frac{1}{2}\left(A_1 + 2c_{12}\right) + \frac{1}{2}\sqrt{\frac{1}{4} + A_1 - B_1 - 4c_{14}} + \frac{1}{2}\sqrt{\frac{1}{4} + A_1 + B_1 + 4c_{12}} \\ & + \frac{1}{2}\sqrt{\left(\frac{1}{4} + A_1 + B_1 + 4c_{12}\right)\left(\frac{1}{4} + A_1 - B_1 - 4c_{14}\right)} \end{aligned} \tag{36}$$

and

$$P_0^{'}(\theta_1) = N_{01}^{'}\left(\frac{1-\cos\theta_1}{2}\right)^{\frac{1}{4}+\frac{1}{2}\sqrt{\frac{1}{4}+A_1+B_1+4c_{12}}}\left(\frac{1+\cos\theta_1}{2}\right)^{\frac{1}{4}+\frac{1}{2}\sqrt{\frac{1}{4}+A_1-B_1-4c_{14}}} \tag{37}$$

Secondly, the similarly with the steps of θ_1 for solution of θ_2 by applying Eq. (5) as the Schrodinger equation. The solution of θ_2 for ground state *n* = *0*, we obtained λ_{02} and eigen function are respectively given as:

$$\lambda_{02} = \frac{1}{2}\left(A_2 + \lambda_{01}\right) + \frac{1}{2}\sqrt{A_2 - B_2 + \lambda_{01}} + \frac{1}{2}\sqrt{A_2 + B_2 + \lambda_{01}} + \frac{1}{2}\sqrt{\left(A_2 + B_2 + \lambda_{01}\right)\left(A_2 - B_2 + \lambda_{01}\right)} \tag{38}$$

and

$$P_0(\theta_2) = N_0 \frac{1}{\sqrt{\sin\theta_2}} \left(\frac{1-\cos\theta_2}{2}\right)^{\frac{1}{4}+\frac{1}{2}\sqrt{A_2+B_2+\lambda_{01}}} \left(\frac{1+\cos\theta_2}{2}\right)^{\frac{1}{4}+\frac{1}{2}\sqrt{A_2-B_2+\lambda_{01}}} \tag{39}$$

By inserting Eq. (39) to Eq. (24) We get the partner potential θ_2 as

$$V_2(\theta_2) = \frac{A_2}{\sin^2\theta_2} + \frac{B_2\cos\theta_2}{\sin^2\theta_2} + \frac{\lambda_{01} - \frac{1}{4}}{\sin^2\theta_2} - E_{02} + \frac{c_{12}}{\frac{1-\cos\theta_2}{2}} - \frac{c_{14}}{\frac{1+\cos\theta_2}{2}} \tag{40}$$

with

$$E_{02} = \lambda_{02} + \frac{1}{4},\quad c_{12} = \frac{1}{4} + \frac{1}{2}\sqrt{A_2 + B_2 + \lambda_{01}} \text{ and } c_{14} = -\frac{1}{4} - \frac{1}{2}\sqrt{A_2 - B_2 + \lambda_{01}} \tag{41}$$

By substituting Eq. (40) to Eq. (5) and then by applying Eq. (15) - Eq. (18) we obtain

$$\begin{aligned}\lambda'_{02} &= c_{12} - c_{14} - E_{02} + \frac{1}{2}\left(A_2 + \lambda_{01} + \lambda'_{01} - \frac{1}{4}\right) + \frac{1}{2}\sqrt{A_2 - B_2 + \lambda_{01} + \lambda'_{01} - 4c_{14} - \frac{1}{4}} + \\ &\frac{1}{2}\sqrt{A_2 + B_2 + 4c_{12} + \lambda_{01} + \lambda'_{01} - \frac{1}{4}} + \frac{1}{2}\sqrt{\left(A_2 + B_2 + 4c_{12} + \lambda_{01} + \lambda'_{01} - \frac{1}{4}\right)\left(A_2 - B_2 + \lambda_{01} + \lambda'_{01} - 4c_{14} - \frac{1}{4}\right)}\end{aligned} \tag{42}$$

and

$$P'_0(\theta_2) = N'_{02} \frac{1}{\sqrt{\sin\theta_2}} \left(\frac{1-\cos\theta_2}{2}\right)^{\frac{1}{4}+\frac{1}{2}\sqrt{A_2+B_2+4c_{12}+\lambda_{01}+\lambda'_{01}-\frac{1}{4}}} \left(\frac{1+\cos\theta_2}{2}\right)^{\frac{1}{4}+\frac{1}{2}\sqrt{A_2-B_2+\lambda_{01}+\lambda'_{01}-4c_{14}-\frac{1}{4}}} \tag{43}$$

Finally, the similarly with the steps of θ_1 for solution of θ_3 by applying Eq. (6) as the Schrodinger equation. Then λ_{03} and eigen function for ground state $n = 0$ are respectively given as:

$$\begin{aligned}\lambda_{03} &= -\frac{5}{8} + \frac{1}{2}(\lambda_{02} + A_3) + \frac{1}{2}\sqrt{\frac{1}{4} + \lambda_{02} + A_3 - B_3} + \frac{1}{2}\sqrt{\frac{1}{4} + \lambda_{02} + A_3 + B_3} \\ &+ \frac{1}{2}\sqrt{\left(\frac{1}{4} + \lambda_{02} + A_3 + B_3\right)\left(\frac{1}{4} + \lambda_{02} + A_3 - B_3\right)}\end{aligned} \tag{44}$$

and

$$P_0(\theta_3) = N_0 \frac{1}{\sin\theta_3} \left(\frac{1-\cos\theta_3}{2}\right)^{\frac{1}{4}+\frac{1}{2}\sqrt{\frac{1}{4}+\lambda_{02}+A_3+B_3}} \left(\frac{1+\cos\theta_3}{2}\right)^{\frac{1}{4}+\frac{1}{2}\sqrt{\frac{1}{4}+\lambda_{02}+A_3-B_3}} \tag{45}$$

By Substituting Eq. (45) to Eq. (24) we obtain the partner potential θ_3 as

$$V_2(\theta_3) = \frac{A_3}{\sin^2\theta_3} + \frac{B_3\cos\theta_3}{\sin^2\theta_3} + \frac{\lambda_{02}}{\sin^2\theta_3} - E_{03} + \frac{c_{12}}{\frac{1-\cos\theta_3}{2}} - \frac{c_{14}}{\frac{1+\cos\theta_3}{2}} \tag{46}$$

with

$$E_{03} = \lambda_{03} + 1,\quad c_{12} = \frac{1}{4} + \frac{1}{2}\sqrt{\frac{1}{4} + \lambda_{02} + A_3 + B_3} \text{ and } c_{14} = -\frac{1}{4} - \frac{1}{2}\sqrt{\frac{1}{4} + \lambda_{02} + A_3 - B_3} \tag{47}$$

IOP Conf. Series: Journal of Physics: Conf. Series **909** (2017) 012034 doi:10.1088/1742-6596/909/1/012034

By substituting Eq. (46) to Eq. (6) and then by applying Eq. (15) - Eq. (18) we get:

$$\lambda_{03}' = -\frac{3}{4} - E_{03} - \frac{B_3}{2} - c_{12} - c_{14} + \frac{1}{2}\sqrt{\frac{1}{4} + A_3 - B_3 + \lambda_{02} + \lambda_{02}' - 4c_{14}} + \frac{1}{2}\sqrt{\frac{1}{4} + A_3 + B_3 + \lambda_{02} + \lambda_{02}' + 4c_{12}}$$
$$+ \frac{1}{2}\sqrt{\left(\frac{1}{4} + A_3 + B_3 + \lambda_{02} + \lambda_{02}' + 4c_{12}\right)\left(\frac{1}{4} + A_3 - B_3 + \lambda_{02} + \lambda_{02}' - 4c_{14}\right)} \quad (48)$$

and

$$P_0'(\theta_3) = N_{03}' \frac{1}{\sin\theta_3}\left(\frac{1-\cos\theta_3}{2}\right)^{\frac{1}{4}+\frac{1}{2}\sqrt{\frac{1}{4}+A_3+B_3+\lambda_{02}+\lambda_{02}'+4c_{12}}}\left(\frac{1+\cos\theta_3}{2}\right)^{\frac{1}{4}+\frac{1}{2}\sqrt{\frac{1}{4}+A_3-B_3+\lambda_{02}+\lambda_{02}'-4c_{14}}} \quad (49)$$

The generalized wave functions are

$$P(\theta_1) = N_{n1}\left(\frac{1-\cos\theta_1}{2}\right)^{\frac{1}{4}+\frac{1}{2}\sqrt{\frac{1}{4}+A_1+B_1}}\left(\frac{1+\cos\theta_1}{2}\right)^{\frac{1}{4}+\frac{1}{2}\sqrt{\frac{1}{4}+A_1-B_1}} P_{n1}^{\left(\sqrt{\frac{1}{4}+A_1+B_1},\sqrt{\frac{1}{4}+A_1-B_1}\right)}(\cos\theta_1)$$

$$P(\theta_2) = N_{n2}\frac{1}{\sqrt{\sin\theta_2}}\left(\frac{1-\cos\theta_2}{2}\right)^{\frac{1}{4}+\frac{1}{2}\sqrt{A_2+B_2+\lambda_{01}}}\left(\frac{1+\cos\theta_2}{2}\right)^{\frac{1}{4}+\frac{1}{2}\sqrt{A_2-B_2+\lambda_{01}}} P_{n2}^{\left(\sqrt{A_2+B_2+\lambda_{01}},\sqrt{A_2-B_2+\lambda_{01}}\right)}(\cos\theta_2) \quad (50)$$

$$P(\theta_3) = N_{n3}\frac{1}{\sin\theta_3}\left(\frac{1-\cos\theta_3}{2}\right)^{\frac{1}{4}+\frac{1}{2}\sqrt{\frac{1}{4}+\lambda_{02}+A_3+B_3}}\left(\frac{1+\cos\theta_3}{2}\right)^{\frac{1}{4}+\frac{1}{2}\sqrt{\frac{1}{4}+\lambda_{02}+A_3-B_3}} P_{n3}^{\left(\sqrt{\frac{1}{4}+\lambda_{02}+A_3+B_3},\sqrt{\frac{1}{4}+\lambda_{02}+A_3-B_3}\right)}(\cos\theta_3)$$

and

$$P'(\theta_1) = N_{n1}'\left(\frac{1-\cos\theta_1}{2}\right)^{\frac{1}{4}+\frac{1}{2}\sqrt{\frac{1}{4}+A_1+B_1+4c_{12}}}\left(\frac{1+\cos\theta_1}{2}\right)^{\frac{1}{4}+\frac{1}{2}\sqrt{\frac{1}{4}+A_1-B_1-4c_{14}}} P_{n1}^{\left(\sqrt{\frac{1}{4}+A_1+B_1+4c_{12}},\sqrt{\frac{1}{4}+A_1-B_1-4c_{14}}\right)}(\cos\theta_1)$$

$$P'(\theta_2) = N_{n2}'\frac{1}{\sqrt{\sin\theta_2}}\left(\frac{1-\cos\theta_2}{2}\right)^{\frac{1}{4}+\frac{1}{2}\sqrt{A_2+B_2+4c_{12}+\lambda_{01}+\lambda_{01}'-\frac{1}{4}}}\left(\frac{1+\cos\theta_2}{2}\right)^{\frac{1}{4}+\frac{1}{2}\sqrt{A_2-B_2+\lambda_{01}+\lambda_{01}'-4c_{14}-\frac{1}{4}}} P_{n2}^{\left(\sqrt{A_2+B_2+4c_{12}+\lambda_{01}+\lambda_{01}'-\frac{1}{4}},\sqrt{A_2-B_2+\lambda_{01}+\lambda_{01}'-4c_{14}-\frac{1}{4}}\right)}(\cos\theta_2) \quad (51)$$

$$P'(\theta_3) = N_{n3}'\frac{1}{\sin\theta_3}\left(\frac{1-\cos\theta_3}{2}\right)^{\frac{1}{4}+\frac{1}{2}\sqrt{\frac{1}{4}+A_3+B_3+\lambda_{02}+\lambda_{02}'+4c_{12}}}\left(\frac{1+\cos\theta_3}{2}\right)^{\frac{1}{4}+\frac{1}{2}\sqrt{\frac{1}{4}+A_3-B_3+\lambda_{02}+\lambda_{02}'-4c_{14}}} P_{n3}^{\left(\sqrt{\frac{1}{4}+A_3+B_3+\lambda_{02}+\lambda_{02}'+4c_{12}},\sqrt{\frac{1}{4}+A_3-B_3+\lambda_{02}+\lambda_{02}'-4c_{14}}\right)}(\cos\theta_3)$$

By using the same value of potential parameters, the wave functions for the original Scarf potential and the partner potential as potential construction result are different. For details of the wave function of Equation (50) and Equation (51) can be shown Table 1.

Table 1 Three-dimensional representations of the angular wave function for variation of n_l, a and b.

a	b	n_l	$P(\theta_2)$	$P'(\theta_2)$
4	4	0	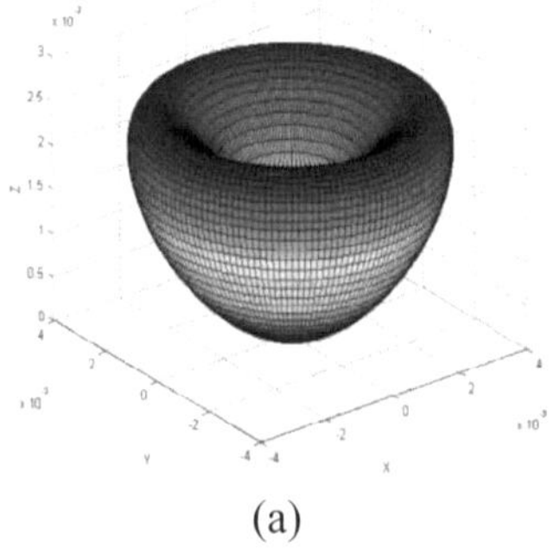 (a)	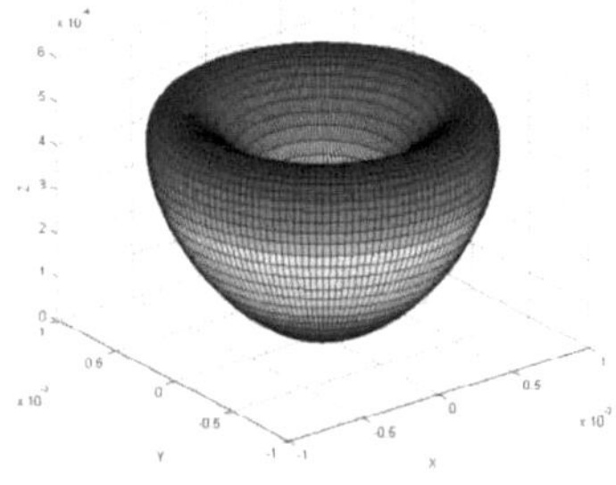 (b)
4	4	1	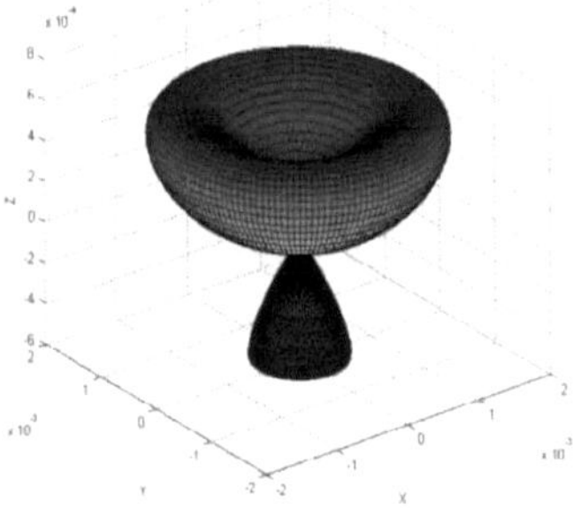 (c)	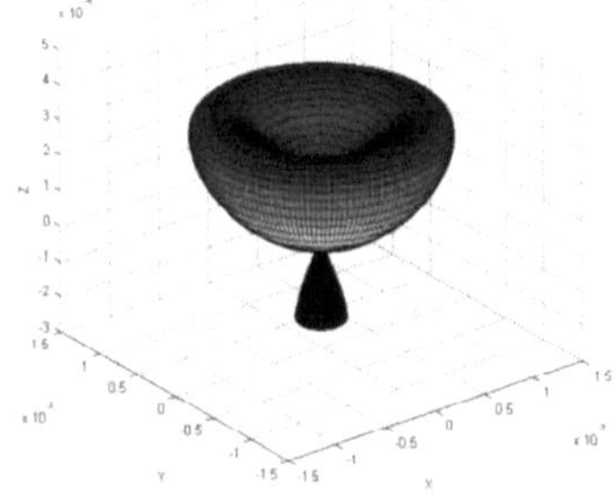 (d)
6	6	1	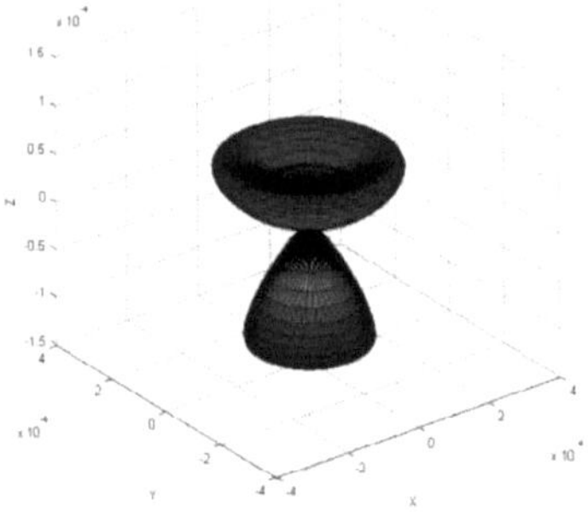 (e)	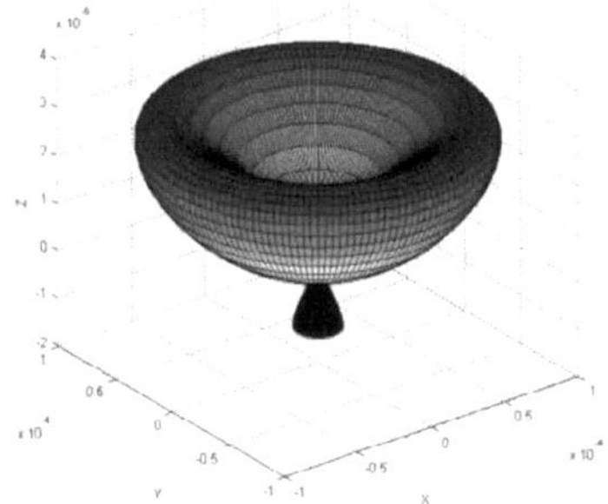 (f)
8	8	1	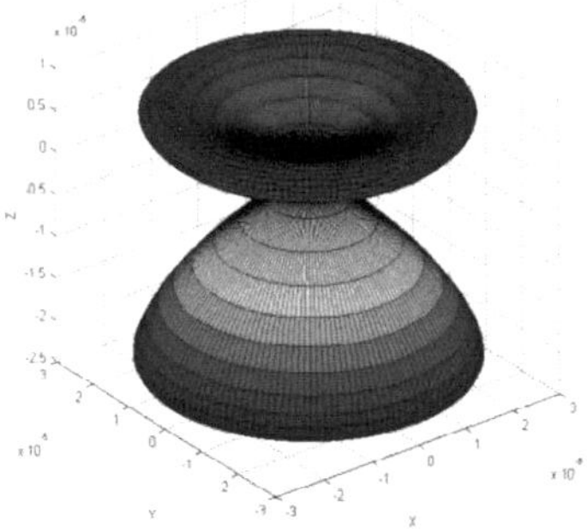 (g)	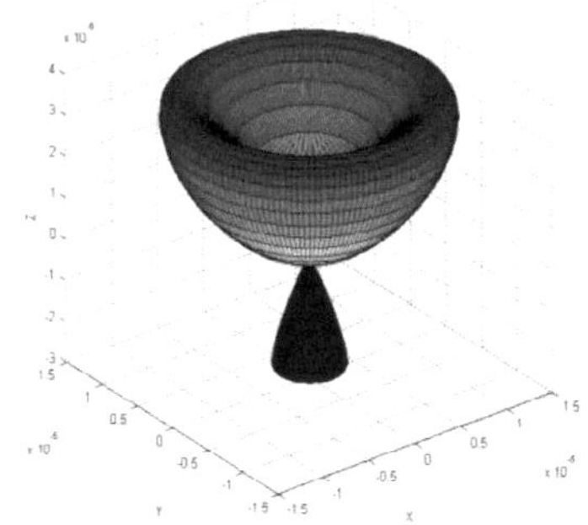 (h)

From Table 1, we have wave function for variations of a, b, and n_l. The increases of n_l value is given the effect in form of wave functions and the motion range of particle. The wave functions for the original Scarf potential and its the partner potential are different. There is a decrease in the amplitude value at $P'(\theta_2)$ beside $P(\theta_2)$. An increase in the value of n_l causes a decrease in the amplitude value. The increase of a and b values causes a decrease in the amplitude value.

4. Conclusion

In this paper, we have presented the solution of 4-dimensional Schrodinger equation of angular part for Scarf potential and partner potential using Nikiforov-Uvarov method and Supersymmetric quantum mechanics method. We obtained the wave functions from angular part solution, which the wave functions depend on the parameters of all components of the composed potential. The partner potential have different wave functions compared to the ground state wave functions of the original potential. This partner potential was considered to be new potential.

Acknowledgement

This research was partly supported by PUT-MRG of Sebelas Maret University

References

[1] Shi-Hai Dong 2011 Wave Equations in Higher Dimensions Springer New York P97
[2] Wahyulianti, Suparmi, Cari, and Fuad Anwar 2017 *IOP Conf. Series: Journal of Physics: Conf. Series* **795** 012022
[3] B H Yazarloo, L Lu, G Liu, S Zarrinkamar, and H Hassanabadi 2013 *Advances in High Energy Physics* 2013 **5**
[4] A. Suparmi, C. Cari, Beta Nur Pratiwi, and Dewanta Arya Nugraha 2017 *IOP Conf. Series: Journal of Physics: Conf. Series* **820** 012023
[5] A. N. Ikot 2013 *Communications in Theoretical Physics* **59** 268-272
[6] Benedict Iserom Ita, Alexander Immaanyikwa Ikeuba 2013 *Applied Mathematics* **4** 1-6
[7] D. Agboola 2009 Phys. Scr. **80** 065304
[8] Suparmi dan Cari 2014 *J. Math. Fund. Sci.* 46 205
[9] A Suparmi, C Cari, and U A Deta 2014 *Chin. Phys. B* **23** 090304
[10] R A Sari, A Suparmi, and C Cari 2016 *Chin. Phys. B* **25** 010301
[11] Ituen B Okon, Eno E Ituen, Oyebola Popoola, and Akaninyene D Antia 2013 *International Journal of Recent Advances in Physics (IJRAP)* **2** 1
[12] Akpan N Ikot, O A Awoga, and A D Antia 2013 *Chin. Phys. B* **22** 020304
[13] A Suparmi, C Cari, Beta Nur Pratiwi, and Dewanta Arya Nugraha 2017 *IOP Conf. Series: Journal of Physics: Conf. Series* **795** 012001
[14] Luqman H, Suparmi, Cari, and U.A. Deta 2014 *IOP Conf. Series: Journal of Physics: Conference Series* **539** 012019

IOP Conf. Series: Journal of Physics: Conf. Series **909** (2017) 012035 doi:10.1088/1742-6596/909/1/012035

Anealling dependence of structural and magnetic properties in aluminum substituted cobalt-ferrite nano particle

Budi Purnama[1], Suharyana

Physics Department Sebelas Maret University,
Jl. Ir. Sutami 36A Kentingan Jebres Surakarta 57126, INDONESIA

E-mail:bpurnama@mipa.uns.ac.id

Abstract. Anealling dependence of structural and magnetic properties in aluminum substituted cobalt ferrite nano particle are presented here. The composition is defined by the mole-stociometry calculation. The co-precipitation method is performed to produce of the aluminum substituted cobalt ferrite. The anealling temperature are selected to be 600°C, 800°C and 1000°C for 5 hours at atmosphere condition. The XRD results show that the XRD-peak increase with the increase of the anelling temperature. It mean that the fine crystalline structure realize at the higher temperature. The The VSM results indicate that the saturated magnetisation increase with increasing anealling temperature. This is can be explained that the refine of crystalline structure is also promote re-distribution cation at crystalline cites which the final of the magnetic moment become large at a higher temperature anealling.

1. Introduction

Within the last decade, magnetic nanoparticles have become one of the interesting research topic considering superior physical characteristics compared to the bulk phase [1,2]. Of the many nanoparticles of magnetic, cobalt ferrite becomes the main material. One important property of the magnetic nanoparticles are not lost the magnetic property when nano particles are less than 100 nm. The presence of a single magnetic domain on a nanoparticle is a target for researchers considering being able to open up opportunities in the medical application. Especially as contrast agents on MRI as well as drug carriers in cancer treatment systems [3].

Cobalt ferrite is classified as inverse spinel oxide. With Co^{2+} cations occupying at the octahedral site and Fe^{3+} cations occupy the tetrahedral site and octahedral sites [4]. Conventionally, the nano magnetic properties in cobalt ferrite modify by the redistribution of cations such as post annealing treatment, and also by the substitution of transition metal ions (Ni, Al) and rare earth metals (Sr, Gd). Currently, the various cobalt ferrite syntheses that have been developed include coprecipitation [5], sol-gel, micro-emultion, autocombution [6] and reverse coprecipitation [7]. Among the various methods, the coprecipitation method is one of the most widely used methods of synthesis of ferric cobalt nanoparticles [8].

In this paper, a study of the synthesis in aluminum substituted cobalt ferrite nanoparticles is presented. The samples were synthesized using coprecipitation method followed by post annealing treatment. The crystalline structure of the obtained nano particle samples were characterized by XRD. Where as the magnetic property are characterized by a vibrating sample magnetometer (VSM) at room temperature.

2. Experimental

The synthesis of cobalt ferrite nanoparticles is carried out by co-precipitated method as in the previous study [10]. In order to modify the magnetic characteristic of cobalt ferrite nanoparticles, 0.001 mole $Al(NO_3)_3.9H_2O$ add to precursor solution. The synthesis is performed at temperatures of 95°C. After cool to room temperature, the product wash with ethanol and deionised water. The washing procedure is repeated for several time so that the clean product is obtained. Then the obtained product are dried overnight at 100°C. The following post annealing has been done at 600°C, 800°C and 1000°C for 5 hours. The X-ray diffraction (XRD) patterns of the samples are evaluated by Bruker D8 Advanced system X-ray powder diffractometer using Cu Kα source (1.54 Å) radiation. The scans of the selected diffraction peaks are selected in the step mode 0.02° from 15° to 70°. Finally the magnetic properties are characterized by vibrating samples of magneto meters at room temperature using magnetic field scan range from -20 kOe to 20 kOe.

3. Results and Discussion

Figure 1 presents the FTIR result of alumunium substituted cobalt ferrite which vary in the anealling temperature i.e 600°C, 800°C dan 1000°C for 5 hours. The absorption band around 3400 cm^{-1} and 1600 cm^{-1} are ascribed to the stretching mode of O-H and H-O-H bending vibration, respectively [11]. It is clearly observed from the graph that the characteristic absorption peaks disappear by increasing the anealling temperature. This means that the heat energy from the anealing treatment is used to break the O-H bonds in both stretching and bending modes. Whereas the characteristic of absorption peak at wave number k = 596.99 cm^{-1}, 590.24 cm^{-1}, 586.39 cm^{-1}, which correspond to temperature anealling 600°C, 800°C and 1000°C is the stretching vibration of aluminum substituted cobalt ferrite [11].

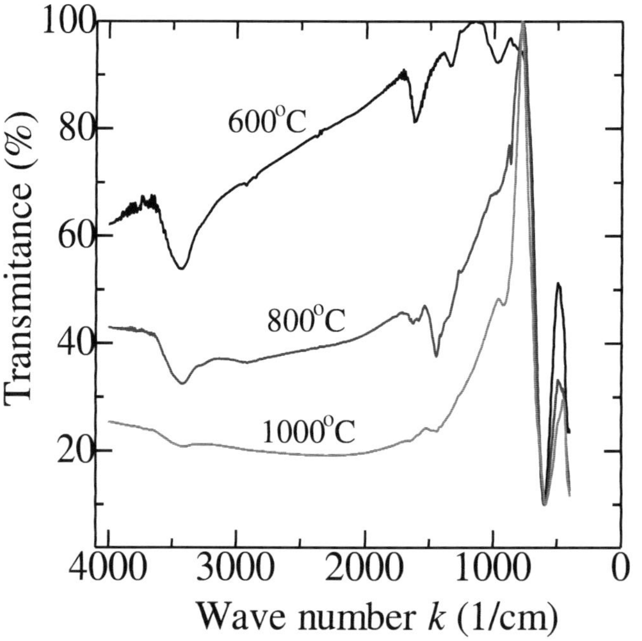

Figure 1. FTIR spectra of the cobalt ferrite nanoparticles synthesized at 95°C and annealed at temperature of 600°C, 800°C and 1000°C

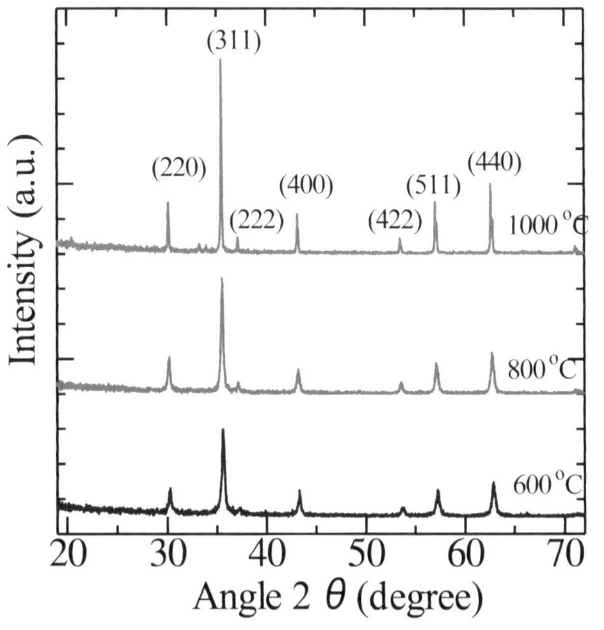

Figure 2. XRD patterns for aluminum substituted cobalt ferrite synthesized at temperature of 95°C and annealed at 600°C, 800°C and 1000°C for 5 h.

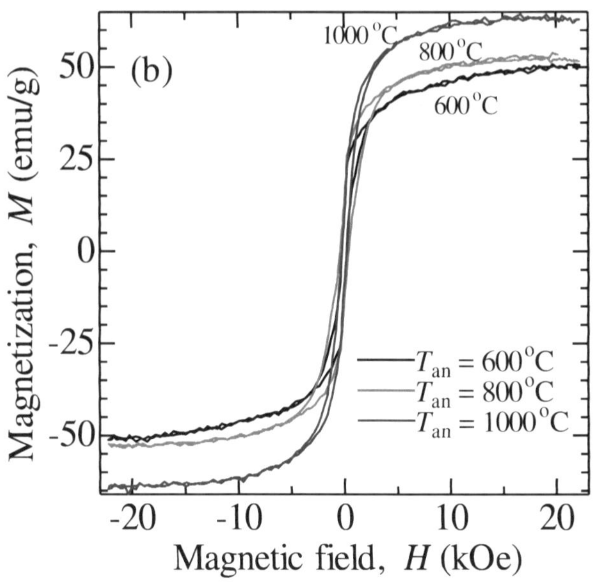

Figure 3. Hysteresis curve of aluminum substituted cobalt ferrite synthesized at temperature of 95°C measure for samples with anealling temperature of 600°C, 800°C and 1000°C for 5 h.

Figure 2 shows the increase in the intensity of x-ray diffraction with the increase of the anealling temperature. It mean that the fine crystalline structure realize at the higher temperature. Furthermore, the increase of the peak XRD pattern also indicate the change of crystallite size. Calculation of the crystallite size at the strongest peak in XRD pattern using by Scherer formula is obtained 32.6 nm for anealling temperature of 600°C. The crystallite size change to become 34.9 nm and 37.5 nm when annealed at 800°C and 1000°C respectively. It seem to be the higher anealling temperature owing is enough two or more nano particles merge realize a larger particles size. These result consistent to study published by Huixia et al [12] and Swatsitang et al [13].

Magnetic properties of aluminum cobalt ferrite nano particle is depicted as Fig 3. The application of magnetic field from -20 kOe to 20 kOe appears to be enough to cause the magnetization of the nano particle sample to be saturated. It is clearly observed that saturated magnetization increases with the increase of the anealling temperature. The saturated magnetization M_S slighly change from 32.37 emu/g to 38.71 emu/g when anealling temperature increase from 600°C to 800°C. The increase of anealling temperature 1000°C, the M_S significantly increase of 48.88 emu/g. This is can be explained that the refine of crystalline structure is also promote re-distribution cation at crystalline cites which the final of the magnetic moment become large at high temperature anealling.

4. Conclusion

The anealling dependence of structural and magnetic properties in aluminum substituted cobalt ferrite nano particle are presented here. The composition is defined by the mole-stociometry calculation. The co-precipitation method is performed to produce of the aluminum substituted cobalt ferrite. The anealling temperature are selected to be 600°C, 800°C and 1000°C for 5 hours at atmosphere condition. The XRD results show that the aluminum is not significant change characteristic peak of the cobalt ferrite pattern. It should be expected that the aluminum ion may substituted in original crystalline structure of the cobalt ferrite. The whole of the XRD-peak increase with the increase of the anelling temperature. It mean that the fine crystalline structure realize at the higher temperature. The The VSM results indicate that the saturated magnetisation increase with increasing anealling temperature. This is can be explained that the refine of crystalline structure is also promote re-distribution cation at crystalline cites which the final of the magnetic moment become large at high temperature anealling.

Acknowledgments

This research was funded by DIPA Universitas Sebelas Maret of Republic of Indonesia (Penelitian Mandatory Contract No. 1075/UN27.21/PP/2017).

References

[1] Buzea, C., Blandino, I.I.P., & Robbie, K. 2007. *Biointerphases* Vol. **2**, issue 4, MR17- MR172.
[2] Velasquez, E.A., Altbir, D., Mazo-Zuluaga, J., Duque, L.F., & Mejia-Lopez, J. 2013. *J. Magn. Magn. Mater* **10**.
[3] Wirmanda, Adi., Dahlan T, Nurlaela R. 2015. Sintesis dan penentuan sifat struktur cobalt ferrite (CoFe2O4) menggunakan metode kopresipitasi dengan memvariasikan dengan memvariasikan suhu sintesis. Program sarjana Universitas Hasanuddin. Makassar
[4] R. Safi, A. Ghasemi, S. R. Razavi and M. 2015. *J. Magn. Magn. Mater* vol. **396**, pp. 288-294.
[5] F. Huixia, C. Baiyi, Z. Jiangiang and L. J. Tan, 2014. *J. Magn. Magn. Mater*, vol. **356**, pp. 68-72.
[6] R. &. S. J. Kotnala, 2015. Ferrite Materials: Nano to spintronic regime, vol. 23, New Delhi, India.: Handbook of Magnetic Materials.
[7] K. Maaz, S. Karim, A. Mumtaz, S. Hasanain, J. Liu and J. Duan, 2009. *J. Magn. Magn. Mater*, vol. **321**, pp. 1838-1842.
[8] Setiadi, E.A., Nanda Shabrina, Hesti R.B.U., N.F., Nur Fadhilah F., Takeshi K., Satoshi I., Edi S. 2013. *Indonesian Journal of Applied Physics* Vol.**3** No.1

[9] Zaki, H.M., Saleh. H. Al-Heneti, A. Hashhash. 2016. *J. Magn. Magn. Mater.* Vol **401**, pp. 1027-1032.
[10] Purnama B, Rahmawati R, Wijayanta AT, and Suharyana 2015 *J.Magn.* 20 207
[11] Safi R, Ghasemi A, Shoja–Razavi R, and Tavousi M 2015 *J. Magn. Magn. Mater* 396 288
[12] F. Huixia, C. Baiyi, Z. Deyi, Z. Jianqiang, and T. Lin, *J. Magn. Magn. Mater.* vol. 356 (2014) pp. 68–72.doi:10.1016/j.jmmm.2013.12.033.
[13] E. Swatsitang, S. Phokha, S. Hunpratub, B. Usher, A. Bootchanont, S. Maensiri, and P. Chindaprasirt, 2016. *J. Alloys. Compd*, vol. **664**, pp. 792–797. doi:10.1016/j.jallcom.2015.12.230

Schrödinger equation solution for q-deformed Scarf II potential plus Pöschl-Teller potential and trigonometric Scarf potential

Fery Widiyanto, A Suparmi, C Cari, Fuad Anwar and Mohtar Yunianto

Physics Department, Sebelas Maret University, INDONESIA

Email: widiyantofery@student.uns.ac.id

Abstract. The Schrödinger equation for q-deformed Scarf II potential plus Pöschl-Teller potential and trigonometric Scarf potential was solved using Asymptotic Iteration Method (AIM). The combination of three potentials was substituted into the Schrödinger equation, then we separated the three-dimensional Schrodinger equation into three one-dimensional differential equation, consist of the radial part, polar part and azimuth part. The radial, polar and azimuth part differential equations were solved by reducing them into the hypergeometric intermediaries equation, to further we resolved it using AIM. By using AIM, the energy equation and the variable separation constant equations, λ_1 for the polar part and λ_2 for the azimuth part were obtained, where both are interrelated between the quantum numbers. The value of energies were calculated numerically using Matlab software, where the increasing in the radial quantum number n_r causes increase and decrease in the energy. Radial, polar and azimuth wave functions were defined as hypergeometric functions and were visualized using Matlab software. The results show that the disturbance of Poschl-Teller potential and trigonometric Scarf potential give effect in probability

1. Introduction

Schrodinger equation is a wave equation as a representation of electrons or microscopic particles in the non-relativistic case. The solution of the Schrodinger equation was obtained wave function equation and energy equation which will be used to find out how the behavior of electrons. The various potentials have been investigated using Schrodinger equation to explain the behavior of the particle. Several types of potentials, especially the exact settlement type of the Schrodinger equation are possible only for angular momentum = 0. But when the angular momentum is ≠ 0 the Schrodinger equation can only be solved by an appropriate method approach such as Asymptotic Iteration method, 1/N expansion method, the Nikiforov-Uvarov method [1-2], and the Supersymmetry method [3-4] in quantum mechanic. In quantum mechanics, different approaches are used in determining the quantities associated with particle motion. Completion of the wave function of moving particles can be obtained using the Schrodinger equation. [5] Approximate solution of Schrodinger equation for modified Poschl-Teller plus trigonometric Rosen-Morse non-central potentials in terms of finite Romanovski polynomial has been investigated by Suparmi, et al. in [6]. Deta, et al. in 2014 solved the solution of Schrodinger equation in D-dimensions for Scarf Hyperbolic plus non-central Poschl-Teller potential using Nikiforov-Uvarov method [7]. The solution of three-dimensional Schrodinger equation for Eckart and Manning-Rosen non-central potential using asymptotic iteration method has been investigated by Farizky, et al. [8]. From the previous research, it gave information that the certainty method can be used to solve the Schrodinger equation which refers to the potential.

The method to solve the Schrodinger equation for a charged particle at a central and noncentral potential has been developed using Asymptotic Iteration Method [8]. Asymptotic iteration method is one of a method to solve the second orde differential equation by using the asymptotic approach [9]. AIM has been used in various solutions, not only for Schrodinger equation but also for any solution in the Dirac equation [10-12] and Klein-Gordon equation[13-14].

In this study, we used the q-deformed Scarf II potential ($V_{SFII}(r)$), Pöschl-Teller potential ($V_{PT(\theta)}$) and trigonometric Scarf potential ($V_{SF}(\varphi)$) in Schrodinger equation to find the solution using AIM. The presence of trigonometric Scarf potential in azimuth part gives influence in the value of energy, which it is still rare in solution the three-dimensional Schrodinger equation. The three-dimensional Schrodinger equation was written in section 2, which was followed by the brief review of AIM method. The results and discussion can be found in section 3, and for the conclusion in section 4.

2. Basic Theory

2.1. Schrodinger equations in 3 dimensions

The Schrodinger equation independent of time is given as[5]

$$-\frac{\hbar^2}{2m}\nabla^2\psi(r,\theta,\varphi)+V\psi(r,\theta,\varphi)=E\,\psi(r,\theta,\varphi) \tag{1}$$

where the non-central potential is written as:

$$V(r,\theta,\varphi)=V_{SFII}(r)+\frac{V_{PT(\theta)}}{r^2}+\frac{V_{SF}(\varphi)}{r^2\sin^2\theta} \tag{2}$$

In this study, the potential of the radial part is the q-deformed Scarf II potential, is given as

$$V_{SFII(r)}=\frac{\hbar^2}{2m}\alpha^2\left(\frac{b^2+a(a+1)}{\sinh_q^2\alpha r}-\frac{2b(a+\frac{1}{2})\cosh_q\alpha r}{\sinh_q^2\alpha r}\right) \tag{3}$$

where b, a are potential constants, q is deformation constant and α is range of q-deformed Scarf II potential in system. while, for non-central potential consist of Poschl-Teller potential and Trigonometric Scarf potential, are written as, respectively

$$V_{PT(\theta)}=\frac{\hbar^2}{2m}\left(\frac{k(k-1)}{\sin^2\theta}+\frac{\lambda(\lambda-1)}{\cos^2\theta}\right) \tag{4}$$

$$V_{SF}(\varphi)=\frac{\hbar^2}{2m}\left(\frac{B^2+A(A-1)}{\cos^2\varphi}+\frac{2B\left(A-\frac{1}{2}\right)\sin\varphi}{\cos^2\varphi}\right) \tag{5}$$

where λ, k, B, A are potential constants. The Pöschl–Teller potential is used to explain spectrum vibration and interaction of atomic systems [15], while the Scarf potential is applied to explain the atomic or molecular force [16].

By substituting equations (3-5) into equation (1), and by setting the wave function $\psi(r,\theta,\varphi)=R(r)\Theta(\theta)\phi(\varphi)$ for the spherical coordinates, and using the separation variable, we obtained the three one-dimensional differential equation, consist of radial part, polar part and azimuth part, respectively are given as,

$$\frac{1}{R(r)}\frac{\partial}{\partial r}\left(r^2\frac{\partial}{\partial r}R(r)\right)-\frac{2m}{\hbar^2}r^2\left(\frac{\hbar^2}{2m}\alpha^2\left(\frac{b^2+a(a+1)}{\sinh_q^2\alpha r}-\frac{2b(a+\frac{1}{2})\cosh_q\alpha r}{\sinh_q^2\alpha r}\right)-E\right)-\lambda_1=0 \tag{6}$$

$$\sin\theta\frac{1}{\Theta(\theta)}\frac{\partial}{\partial\theta}\left(\sin\theta\frac{\partial}{\partial\theta}\Theta(\theta)\right)-\sin^2\theta\left(\frac{k(k-1)}{\sin^2\theta}+\frac{\lambda(\lambda-1)}{\cos^2\theta}\right)+\lambda_1\sin^2\theta-\lambda_2=0 \tag{7}$$

$$\frac{1}{\phi(\varphi)}\frac{\partial^2}{\partial\varphi^2}\phi(\varphi)-\left(\frac{B^2+A(A-1)}{\cos^2\varphi}+\frac{2B\left(A-\frac{1}{2}\right)\sin\varphi}{\cos^2\varphi}\right)+\lambda_2=0 \tag{8}$$

where λ_1, λ_2 are variable separation constants. Equation (6-8) are the basis equation which will be solved using Asymptotic Iteration Method (AIM) to obtain the energy equation and wavefunction equation.

2.2. *Asymptotic Iteration Method (AIM)*

The Asymptotic Iteration Method (AIM) has the second order differential equation in term,

$$y_n'' = \lambda_0(x)y_n' + s_0(x)y_n \tag{9}$$

If equation (9) is decreased to *x* by *n* times, then equation (9) become,

$$y_n^i(x) - \lambda_{i-2}(x)y_n'(x) - s_{i-2}(x)y_n(x) = 0 \tag{10}$$

for *i=1,2,3,..,*where

$$\lambda_i(x) = \lambda_{i-1}' + \lambda_{i-1}\lambda_o + s_{i-1} \tag{11}$$

$$s_i(x) = s_{i-1}' + s_o\,\lambda_{i-1} \tag{12}$$

By using the asymptotic aspect of the iteration method for a sufficiently large value of *i,* we had

$$\frac{s_i}{\lambda_i} = \frac{s_{i-1}}{\lambda_{i-1}} = \alpha \tag{13}$$

The eigen value of equation (9) can be obtained using the termination conditions, is given as

$$\lambda_i(x)s_{i-1}(x) - \lambda_{i-1}(x)s_i(x) = 0 = \Delta_i, i = 1,2,3 \tag{14}$$

To obtain the wave function of the Schrodinger equation, we can use the second order differential equation as follows,

$$y_n''(x) = 2\left(\frac{tx^{N+1}}{1-bx^{N-2}} - \frac{c+1}{x}\right)y'(x) - \frac{Wx^N}{1-bx^{N+2}} \tag{15}$$

Equation (15) is the other term of AIM-type differential equation in equation (9) to obtain the wave function simpler. Refer to equation (15), the wave function solution can be obtained using,

$$y_n(x) = (-1)^n C'^{(N+2)^n(\sigma)_n}\cdot{}_2F_1(-n, p+n, \sigma, bx^{N+2}) \tag{16}$$

where

$$(\sigma)_n = \frac{\Gamma(\sigma+n)}{\Gamma(\sigma)}\ (13)\ , \sigma = \frac{2c+N+3}{N+2}\ (14)\ , p = \frac{(2c+1)b+2t}{(N+2)b} \tag{17}$$

So by reducing the Schrodinger equation to AIM-type differential equation similar to equations (9,15), we can obtain the solution by using AIM.

3. Solution and Discussion

3.1. *The solution of radial part*

By using equation (6), and by setting $R(r) = \frac{U(r)}{r}$, we obtained

$$U''(r) - \frac{2m}{\hbar^2}U(r)\left(\frac{\hbar^2}{2m}\alpha^2\left(\frac{b^2+a(a+1)}{\sinh_q^2\alpha r} - \frac{2b(a+\frac{1}{2})\cosh_q\alpha r}{\sinh_q^2\alpha r}\right) - E\right) - \frac{\lambda_1}{r^2}U(r) = 0 \tag{18}$$

Equation (18) can be solved by using the approximation $\frac{1}{r^2} \approx \frac{\alpha^2}{\sinh_q^2\alpha r}$, and by setting the suitable variable $\cosh_q\alpha r = (1-2z)\sqrt{q}$, yields

$$z(1-z)\frac{\partial^2}{\partial z^2}U(z) + \frac{1}{2}(1-2z)\frac{\partial}{\partial z}U(z) + \left[\frac{\frac{(b+2ab)\sqrt{q}-(b^2+a(a+1)+\lambda_1)}{q}}{4z} + \frac{\frac{(-b-2ab)\sqrt{q}-(b^2+a(a+1)+\lambda_1)}{q}}{4(1-z)} - \frac{2mE}{\hbar^2\alpha^2}\right]U(z) = 0 \tag{19}$$

Equation (19) must be transformed to AIM-type equation by setting the new wave function,

$$U(z) = z^\rho(1-z)^\beta f(z) \tag{20}$$

so, equation (19) become

$$f''(z) + \left[\frac{\left(2\rho+\frac{1}{2}\right)-(2\rho+2\beta+1)z}{z(1-z)}\right]f'(z) + \left[\frac{-\frac{2mE}{\hbar^2\alpha^2}-(\rho+\beta)^2}{z(1-z)}\right]f(z) = 0 \tag{21}$$

where,

$$\rho = \frac{1}{4} \pm \frac{1}{2}\sqrt{\frac{-(b+2ab)\sqrt{q}+(b^2+a(a+1)+\lambda_1)}{q} + \frac{1}{4}} \tag{22}$$

$$\beta = \frac{1}{4} \pm \frac{1}{2}\sqrt{\frac{(b+2ab)\sqrt{q}+ (b^2+a(a+1)+\lambda_1)}{q} + \frac{1}{4}} \quad (23)$$

Equation (21) is similar to the AIM-type equation in equation (9), where

$$\lambda_0(z) = \frac{(2\rho+2\beta+1)z-\left(2\rho+\frac{1}{2}\right)}{z(1-z)} \quad (24)$$

$$s_0(z) = \frac{(\rho+\beta)^2+\frac{2mE}{\hbar^2\alpha^2}}{z(1-z)} \quad (25)$$

By using equations (11,14) corresponding to equations (21,24-25), we obtained the eigenvalue of equation (21), is given as in energy equation,

$$E = -\alpha^2\left(\frac{1}{2} - \frac{1}{2}\sqrt{\frac{-(b+2ab)\sqrt{q}+(b^2+a(a+1)+\lambda_1)}{q} + \frac{1}{4}} - \frac{1}{2}\sqrt{\frac{(b+2ab)\sqrt{q}+(b^2+a(a+1)+\lambda_1)}{q} + \frac{1}{4}} + n_r\right)^2 \quad (26)$$

where λ_1, λ_2 are variable separation constants, which were obtained from the polar part solution in equation (36) and azimuth part solution in equation (45) Equation (26) is energy equation, which was influenced by quantum number n_r, n_θ, n_φ. From the numerically results, we visualized the energy value in radial quantum number n_r function, was shown in Figure 1.

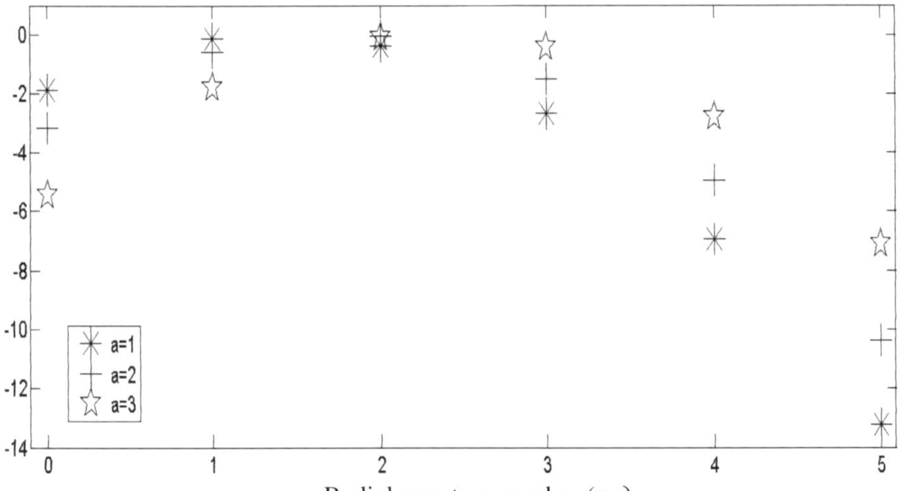

Figure 1. Energy value in radial quantum number n_r function

Figure 1 gives information about the energy value for variation radial quantum number. It shows that in the increasing radial quantum number cause the increasing and decreasing value of energy. The variation of q-deformed Scarf II potential constant a give influence in the value of energy.

For the radial part wave function, we used equation (16-17) and by using equation (20), we obtained the radial wave function, is given as

$$U(r) = \left(\frac{1}{2}\right)^{\rho+\beta}\left(1 - \frac{\cosh_q \alpha r}{\sqrt{q}}\right)^{\rho}\left(1 + \frac{\cosh_q \alpha r}{\sqrt{q}}\right)^{\beta}(-1)^{n_r}C'(1)^{n_r}\left(2\rho + \frac{1}{2}\right)_{n_r}\, {}_2F_1\left(-n_r, 2\rho + 2\beta + n_r, 2\rho + \frac{1}{2}, \frac{1}{2} - \frac{1}{2}\frac{\cosh_q \alpha r}{\sqrt{q}}\right) \quad (27)$$

Where C' is the radial normalization constant and ${}_2F_1$ is a hypergeometric function. Equation (27) is the radial wave function equation depending on the distance of the particles as far as r on the potential existence. The visualization of radial wave function from equation (27) was shown in Figure 2. Figure 2 shows that for the increasing value of potential constant a causes the increasing in amplitude of wave function. It shows that for the increasing value of potential constant causes the particle moves away from the nucleus, so can be said that the increasing potential constant give effect in lowering the binding energy of particle.

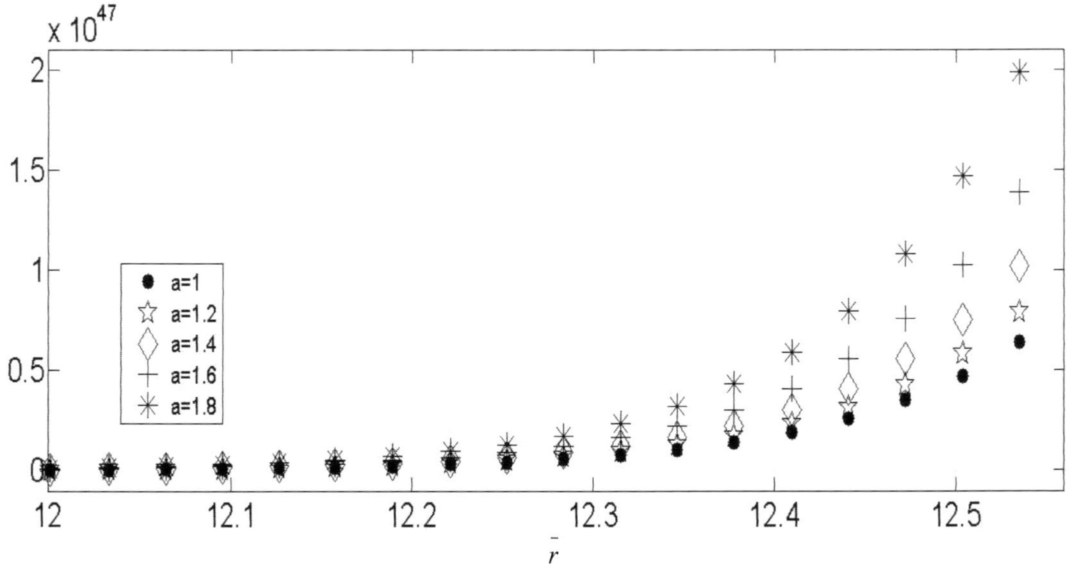

Figure 2. Radial wave function in influence of the q-deformed Scarf II potential constant *a*

3.2. Solution of polar part

For the angular part, we used equation (7) by setting $\Theta(\theta) = \frac{T(\theta)}{sin^{1/2}\theta}$, so equation (7) becomes

$$T''(\theta) + \left(\frac{\frac{1}{4}-\lambda_2-k(k-1)}{sin^2\theta} - \frac{\lambda(\lambda-1)}{cos^2\theta} + \lambda_1 + \frac{1}{4}\right)T(\theta) = 0 \quad (28)$$

By setting $sin^2\theta = z$ in equation (28), yield

$$z(1-z)\frac{\partial^2 T(z)}{\partial z^2} + \frac{1}{2}(1-2z)\frac{\partial T(z)}{\partial z} + \left(\frac{\frac{1}{4}-\lambda_2-k(k-1)}{4\,z} - \frac{\lambda(\lambda-1)}{4(1-z)} + \frac{\lambda_1+\frac{1}{4}}{4}\right)T(z) = 0 \quad (29)$$

Equation (29) must be reduced to AIM-type differential equation by substituting the new wave function,

$$T(z) = z^{\gamma}(1-z)^{\omega}f(z) \quad (30)$$

in equation (29), so we got

$$f''(z) + \left[\frac{\left(2\gamma+\frac{1}{2}\right)-(2\gamma+2\,\omega+1)z}{z(1-z)}\right]f'(z) + \left[\frac{\frac{\lambda_1+\frac{1}{4}}{4}-(\gamma+\omega)^2}{z(1-z)}\right]f(z) = 0 \quad (31)$$

where,

$$\gamma = \frac{1}{4} \pm \frac{1}{2}\sqrt{\lambda_2 + k(k-1)} \quad (32)$$

$$\omega = \frac{1}{4} \pm \left(\frac{1}{2}\lambda - \frac{1}{4}\right) \quad (33)$$

Equation (31) is AIM-type differential equation, with

$$\lambda_0(z) = \frac{(2\gamma+2\,\omega+1)z-\left(2\gamma+\frac{1}{2}\right)}{z(1-z)} \quad (34)$$

$$s_o(z) = \frac{(\gamma+\omega)^2-\frac{\lambda_1+\frac{1}{4}}{4}}{z(1-z)} \quad (35)$$

By using equations (11,14) corresponding to equations (31,34-35), we obtained the eigenvalue of equation (31), is given as the polar part variable separation constant,

$$\lambda_1 = -\frac{1}{4} + 4\left(\frac{1}{2} \pm \frac{1}{2}\sqrt{\lambda_2 + k(k-1)} \pm \left(\frac{1}{2}\lambda - \frac{1}{4}\right) + n_\theta\right)^2 \quad (36)$$

IOP Conf. Series: Journal of Physics: Conf. Series **909** (2017) 012036 doi:10.1088/1742-6596/909/1/012036

And then for polar wave function, by using equations (15-17), equations (30-31), we obtained the polar wave function is given as,

$$T(\theta) = (sin^2\theta)^{\gamma}(cos^2\theta)^{\omega}(-1)^{n_\theta}C''(1)^{n_\theta}(2\gamma + \tfrac{1}{2})_{n_\theta}\, {}_2F_1\left(-n_\theta, 2\gamma + 2\omega + n_\theta, 2\gamma + \tfrac{1}{2}, sin^2\theta\right) \quad (37)$$

where C " is the normalization constant of polar part and ${}_2F_1$ is hypergeometric function. By using equation (37) to obtain the polar wave function, where $\Theta(\theta) = \frac{T(\theta)}{sin^{1/2}\theta}$, we visualized in Figure 3 for polar wave function in $n_\theta = 2$ (left side) and for $n_\theta = 3$ (right side).

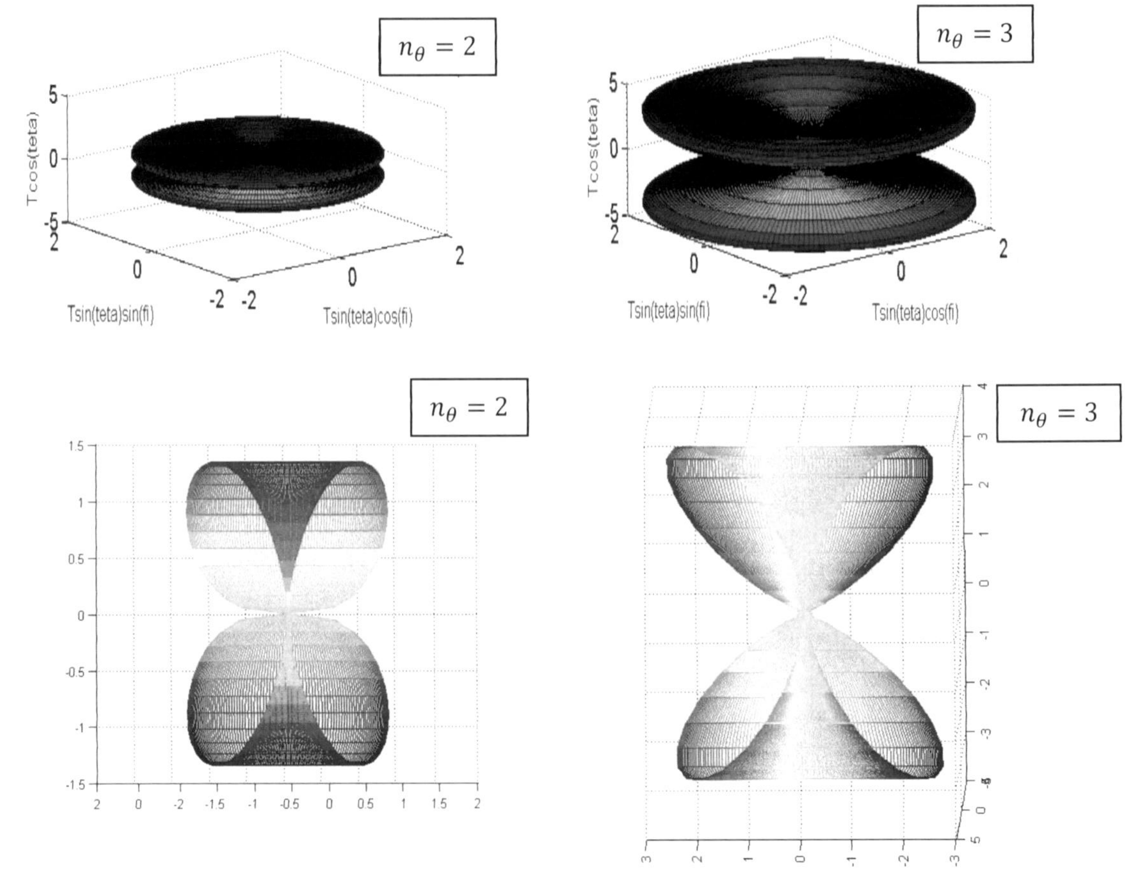

Figure 3. The visual of polar wave function in 3-dimensions (up side) and 2-dimension (bottom side) for $n_\theta = 2$ (left side) and for $n_\theta = 3$ (right side)

Figure 3 (up side) are three-dimensional wave function in spherical coordinates, it shows that for the increasing polar quantum number n_θ causes the increasing amplitude of the wave function. It means that the probability to find particle in $n_\theta = 3$ is wider than in $n_\theta = 2$. The amplitude of polar wave function can be seen more clear from Figure 3 (bottom side) in visual 2-dimension.

3.3. *Solution of azimuth part*

In this part, we used equation (8) to obtain the azimuth quantum number and wave function. By setting $sin\,\varphi = 1 - 2z$ in equation (8), yield

$$z(1-z)\frac{\partial^2\phi(z)}{\partial z^2}+\left(\frac{1}{2}-z\right)\frac{\partial\phi(z)}{\partial z}+\left(\lambda_2-\frac{2B\left(A-\frac{1}{2}\right)+B^2+A(A-1)}{4z}-\frac{-2B\left(A-\frac{1}{2}\right)+B^2+A(A-1)}{4(1-z)}\right)\phi(z)=0 \quad (38)$$

Equation (38) must be reduced to AIM type differential equation by substituting the new wave function, is given as

$$\phi(z)=z^{\varepsilon}(1-z)^{\mu}f(z) \quad (39)$$

so equation (38) becomes

$$f''(z)+\left[\frac{\left(2\,\varepsilon+\frac{1}{2}\right)-(2\varepsilon+2\mu+1)z}{z(1-z)}\right]f'(z)+\left[\frac{\lambda_2-(\varepsilon+\mu)^2}{z(1-z)}\right]f(z)=0 \quad (40)$$

where

$$\varepsilon=\frac{1}{4}\pm\frac{1}{2}\sqrt{(A+B)^2-(A+B)+\frac{1}{4}} \quad (41)$$

$$\mu=\frac{1}{4}\pm\frac{1}{2}\sqrt{(A-B)^2-(A-B)+\frac{1}{4}} \quad (42)$$

Equation (40) is AIM-type differential equation, where

$$\lambda_0(z)=\frac{(2\varepsilon+2\mu+1)z-\left(2\,\varepsilon+\frac{1}{2}\right)}{z(1-z)} \quad (43)$$

$$s_o(z)=\frac{-\lambda_2+(\varepsilon+\mu)^2}{z(1-z)} \quad (44)$$

By using AIM method to solve equation (40), we obtained the eigenvalue by using equations (11-12,14), which was expressed,

$$\lambda_2=\left(\frac{1}{2}\pm\frac{1}{2}\sqrt{(A+B)^2-(A+B)+\frac{1}{4}}\pm\frac{1}{2}\sqrt{(A-B)^2-(A-B)+\frac{1}{4}}+n_{\varphi}\right)^2 \quad (45)$$

Equation (45) is the variable separation constant of azimuth part which influences the value of energy. And then for the azimuth wave function, we used equations (15-17) and equations (39-42) to obtain the general equation of azimuth wave function, is given as

$$\phi(\varphi)=\left(\frac{1-\sin\varphi}{2}\right)^{\varepsilon}\left(\frac{1+\sin\varphi}{2}\right)^{\mu}(-1)^{n_\varphi}C'''(2\varepsilon+\tfrac{1}{2})_{n_\varphi}\,{}_2F_1\left(-n_\varphi,2\varepsilon+2\mu+n_\varphi,2\varepsilon+\tfrac{1}{2},\frac{1-\sin\varphi}{2}\right) \quad (46)$$

Here C '''is the normalization constant of azimuth part and ${}_2F_1$ is hypergeometric function. By using equation (46), the visual of azimuth wave function can be obtained which was shown in Figure 4.

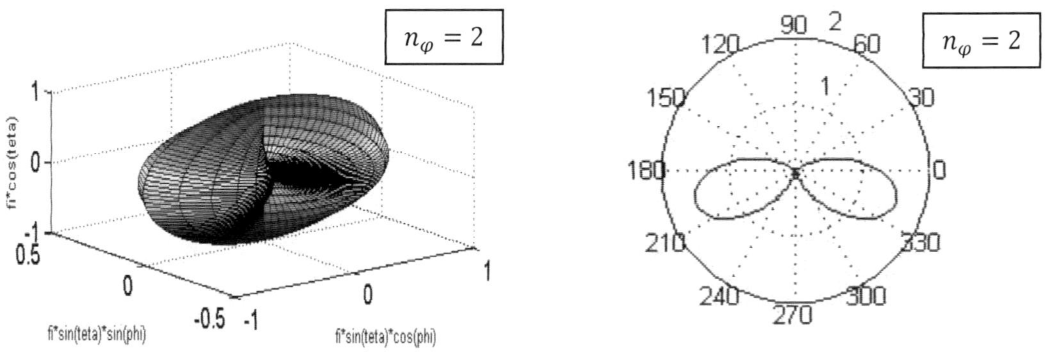

Figure 4. The visual of azimuth wave function in 3-dimensions (left side) and 2-dimension (right side) for $n_\varphi = 2$

Figure 4 (left side) for 3-dimension visual wave function give information that the particle was bounded in certainty area refer to that figure. For accurately, we can see in Figure 4 (right side) in visualization 2-dimensions, it shows that for azimuth wave function for $n_\varphi = 2$, the most probability to find the particle is approximately on the range value of $\varphi = 180 - 210$ and of $\varphi = 300 - 360$, while on the others area weren't found the existence particle.

4. Conclusion

In this research, we solved the Schrodinger equation for potential combination, consist of q-deformed Scarf II potential, Poschl-Teller potential and Trigonometric Scarf potential using asymptotic iteration method. The energies value were calculated numerically using Matlab software, and it was found that the q-deformed Scarf II potential constant *a* influenced the value of energies. The wave functions were obtained in terms of hypergeometric function and were visualized using Matlab software. The wave function gives information about the probability to find particle in the system.

Acknowledgement

This work is partly funded by Hibah PUT-MRG, Sebelas Maret University, No.623/UN27.21/PP/2017

References

[1] Cari C and Suparmi A 2012 *IOSR Journal of Applied Physics (IOSR-JAP)* **2** (3) 13-23
[2] Deta U A, Suparmi and Cari 2013 *AIP Conference Proceedings* **1554** (1) 190-193
[3] Cari C, Suparmi A, Yunianto M and Pratiwi B N 2016 *Journal of Physics: Conference Series* **739** (1) 012023
[4] Suparmi A, Cari C and Pratiwi B N 2016 *Journal of Physics: Conference Series* **710** (1) 012026
[5] Suparmi 2011 *Mekanika Kuantum I* Jurusan Fisika Fakultas Pascasarjana: Universitas Sebelas Maret Surakarta.
[6] Suparmi A, Cari C, Handhika J, Yanuarief C and Marini H 2012 *IOSR Journal of Applied Physics (IOSR-JAP)* **2** (2) 43-51
[7] Deta U A, Suparmi A and Cari C 2014 *Journal of Physics: Conference Series* **539** (1) 012018
[8] Farizky M N, Suparmi A, Cari C and Yunianto M 2016 *Journal of Physics: Conference Series* **776** (1) 012085
[9] Ciftci H, Hall R I and Saad N 2005 *arXiv:0505069v1 [math ph]*
[10] Sari R A, Suparmi A and Cari C 2015 *Chinese Physics B* **25** (1) 010301
[11] Pramono S, Suparmi A and Cari C 2016 *Advances in High Energy Physics* **2016**
[12] Pratiwi B N, Suparmi A, Cari C and Husein A S 2017 *Pramana* **88** (2) 25
[13] Nugraha D A, Suparmi A and Cari C 2017 *Journal of Physics: Theories and Applications* **1** (1) 42-48
[14] Nugraha D A, Suparmi A, Cari C and Pratiwi B N 2017 *Journal of Physics: Conference Series* **820** (1) 012014
[15] Suparmi A, Cari C, Yuliani H and Yuniati D 2013 *J. Fisika Indonesia* **XVII** 51 41
[16] Nurhayati, Suparmi, Variani V I, Cari C and Wahyudi 2012 Prosiding Pertemuan Ilmiah **XXVI** HFI Jateng & DIY, April 14, 2012 Purworejo.p.229

IOP Conf. Series: Journal of Physics: Conf. Series **909** (2017) 012037 doi:10.1088/1742-6596/909/1/012037

Synthesis and characterization of high-purity gold nanoparticles by laser ablation method using low-energy Nd:YAG laser 1064 nm

Ali Khumaeni[1*], Wahyu Setia Budi[1], Heri Sutanto[1]

[1]Department of Physics, Faculty of Science and Mathematics, Diponegoro University Jl. Prof. Soedharto, SH, Tembalang, Semarang 50275, Indonesia

[*]E-mail: khumaeni@fisika.undip.ac.id

Abstract. High-purity gold nanoparticles (GNPs) has been successfully synthesized by using laser ablation method utilizing low-power neodymium yttrium aluminum garnet (Nd:YAG)laser at the fundamental wavelength. Experimentally, pulse laser beam (Nd:YAG laser, 1064 nm, 7 ns, 30 mJ) was directed and focused onto a high-purity gold sheet (99.95%), which was placed into a pure liquid of deionized water, to produce GNPs colloid. Dark-red color colloid of high-purity GNPs was successfully synthesized. The GNPs had a spherical shape with an average diameter of 23.5 nm and standard deviation of 6.4 nm. The surface plasma resonance was centered at wavelength maximum at 520 nm.

1. Introduction

Nanoparticles (NPs) have become scientific products, which are intensively being explored for various uses fields. NPs are microscopic particles with a dimension between 1 to 100 nm. One of the NPs intensively studied was gold nanoparticles (GNPs) [1-3]. GNPs have become a great interest for many scientists to be applied in various fields as catalysis [4], surface-enhanced Raman scattering [5], photonics device [6], electrochemical sensors [7], and medical applications [8]. In the medical field, GNPs were applied as drug carriers, contrast agents, thermal therapy, and radiosensitisers due to its peculiar characteristics compared to the bulk material including electronic, optic, and chemical properties [9,10].

Various methods have been employed to synthesize GNPs such as Turkevich, Brust, sol gel and chemical methods [11,12]. However, those methods required very delicate sample pretreatment, time-consuming, and labor intensive. Furthermore, ultrahigh-purity nanoparticles could not be produced due to the impurity coming from additional agents and stabilizers during their synthesis. The other method was physical technique such as pulse laser ablation (PLA), which involves high-power pulse laser as an energy source.

PLA method is very potential for synthesizing GNPs with high-purity, which is required for medical applications such as drug delivery, contrast agents and radiosensitizer [8]. Compared to other synthesis methods, PLA only uses very simple procedure and free from chemical agents and stabilizers, and therefore high-purity nanoparticles can be simply and rapidly produced. In this technique, the pulse laser was directed and focused onto a high-purity metal, which is placed into a pure liquid such as deionized water, to produce GNPs colloid. The colloid contained high-purity GNPs ablated from the pure gold metal due to laser bombardment. In PLA, the femtosecond laser is

IOP Conf. Series: Journal of Physics: Conf. Series **909** (2017) 012037 doi:10.1088/1742-6596/909/1/012037

commonly used as an energy source for material ablation, which results in gold particle at various sizes. The other laser used in PLA is pulse nanosecond laser. However, the laser with high-power (more than 50 mJ) is usually employed to ablate the high-purity gold metal immersed in the pure water [10].

In this study, pulse Nd:YAG laser with low energy (around 30 mJ) was employed as an ablation source for GNPs production. The GNPs was then characterized morphologically by scanning electron microscopy (SEM). Size distributions and surface plasma resonances were also studied. Nanoparticle size of gold with low standard deviation was successfully produced. The present technique using low-energy Nd:YAG laser potentially suppress the production cost and effectively produces the GNPs with the low size distribution of nanoparticles.

2. Experimental procedure

Basic experimental setup used in this research is shown in Fig. 1(a). A high-purity gold metal sheet (99.95%) was placed at the bottom of a glass petri dish, which was filled with 7 ml pure aquades. A Nd:YAG laser beam (New Polaris II at the fundamental wavelength of 1064 nm, 50 mJ) was focused on the surface of the sheet by a convex lens to ablate the sheet to produce GNPs. The repetition rate and pulse width of the laser beam were 15 Hz and 7 ns, respectively. The laser beam was bombarded on the sample surface for the duration of 10 minutes consecutively. It should be pointed that during laser bombardment, the gold sheet and solution were periodically moved to achieve a homogeneous GNPs colloid and new-sheet position.

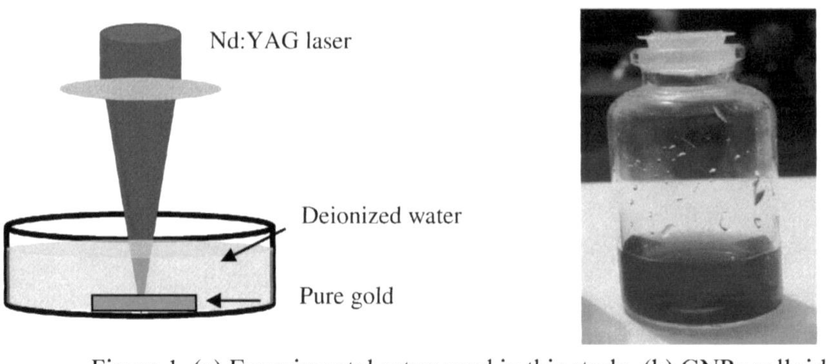

Figure 1. (a) Experimental setup used in this study, (b) GNPs colloid

The GNPs produced in this research was then characterized by using various techniques. Morphology study was made by Scanning Electron Microscopy – Energy Diffraction X-ray (SEM-EDX, JEOL JED-2300). Size distribution of nanoparticles was characterized by particle size analyzer (DelsaNano) and Surface plasma resonance was analyzed by using Ultraviolet-Visible (UV-Vis) light spectroscopy.

3. Results and discussion

Gold nanoparticles (GNPs) colloid produced in this study is shown in Fig. 1(b). Dark-red color GNPs was successfully synthesized by using the present technique of PLA utilizing low-energy Nd:YAG laser. The color of GNPs colloid is similar to the results obtained by other researchers published here [13]. It should be emphasized that the GNPs produced in this study have very high-purity because no additional agents and stabilizers used in the synthesis process.

Figure 2(a) shows the morphological characterization of GNPs obtained by using SEM-EDX. The magnification of obtained figure was set at 20000, resulting in smallest particle measurement of 1 μm. It can clearly be seen that the spherical shape of GNPs was produced with various size distribution. To

IOP Conf. Series: Journal of Physics: Conf. Series **909** (2017) 012037 doi:10.1088/1742-6596/909/1/012037

obtain this photograph, 1 mL GNPs colloid was poured on Si sheet and placed at a cleaned room with a temperature of around 30^{0}C for 24 hours. To measure the particle size of GNPs, particle size analyzer was used. The average diameter of GNPs is 23.5 nm with a standard deviation of 6.4 nm as shown in Fig. 2(b). It should be mentioned when the laser energy was increased from 30 mJ (Fig. 2b) to 40 mJ, the diameter of GNPs also increased to 29.5 nm. This result has the same tendency with the result published by Piriyawonget. al. in nanoparticle alumina [14], namely the diameter of nanoparticle increased with increasing the laser energy.

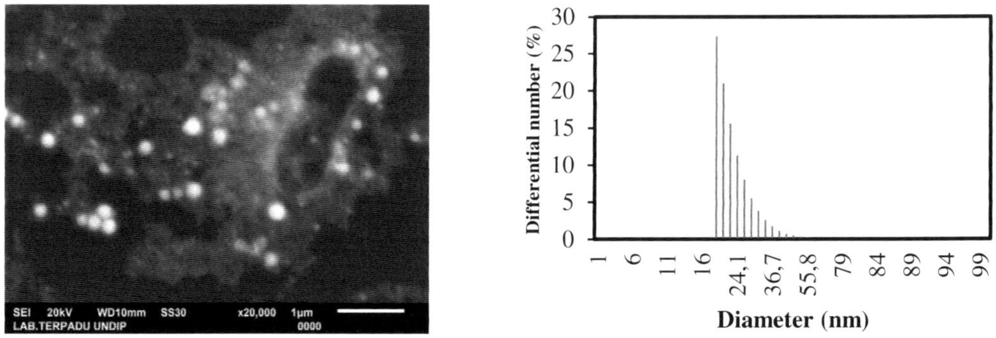

(a) **(b)**

Figure 2.(a) Morphology of GNPs by SEM-EDX, (b) average diameter of nanoparticles measured by PSA method.

Figure 3 shows the absorption spectrum of GNPs obtained by using UV-Vis. The surface plasmon resonance (SPR) can clearly be seen at center wavelength of 520 nm. Single SPR certified that the gold nanoparticles were produced in the form of spherical shape.

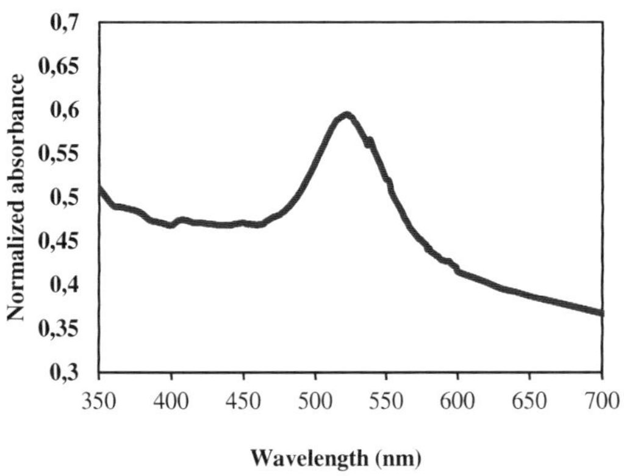

Figure 3.Absorption spectrum of gold nanoparticles produced by using the present method

4. Conclusion

This study has demonstrated that pulse laser ablation method using low-power Nd:YAG laser can be employed to synthesize high-purity gold nanoparticles (GNPs) in aquades. The GNPs were characterized by using SEM-EDX, particle size analyzer (PSA), and UV-Vis spectroscopy to identify the morphology, size distribution, and surface plasmon resonance of GNPs, respectively. GNPs with the form of spherical shape and average diameter of 23.5 nm were produced. The results certified that low-power Nd:YAG laser can be successfully used to produce gold nanoparticles with high purity and quiet homogenous in size distribution. Therefore, the use of low-power laser can potentially suppress the cost of GNPs production.

Acknowledgment
The present study includes the results of project financially supported by Ministry of Research and Technology and Higher Education, Indonesia under the project of PUPT scheme (Contract No. 344-33/UN7.5.1/PP/2017).

References
[1] Amendola V and Meneghetti M 2009 *Phys. Chem. Chem. Phys.* **11** 3805
[2] Paciotti GF, Kingston DGI, Tamarkin L 2006 *Drug Dev. Res.* **67** 47
[3] Paciotti GF, Myer L, Weinreich D, Goia D, Pavel N, McLaughlin RE, and Tamarkin L 2004 *Drug Deliv.* **11** 169
[4] Cataliotti RS, Compagnini G, Crisafulli C, Minico S, Pignataro B, Sassi P, and Scire S 2001 *Surf. Sci.* **494** 75
[5] Hu Y, Liao J, Wang D, and Li G 2014 *Analyt. Chem.* **86** 3955
[6] Zhang X 2006 *Nano Lett.* **6** 651
[7] Guo S and Wang E 2007 *Anal. Chim. Acta* **598** 181
[8] Jain S, Hirst DG, O`Sullivan JM 2012 *The British J. Radiol.* **85** 101
[9] Herrera GM, Padilla AC, and Hernandez-Rivera P 2013 *Nanomaterials* **3** 158
[10] Shukri WNW, Bidin N, Affandi S, Bohari SP 2015 *Jurnal Teknologi* **78** (3) 267
[11] Pillai ZS and Kamat PV 2004 *J. Phys. Chem.* **108**(3) 945
[12] Pastoriza-Santos I and Liz-Marzan LM 1999*Langmuir* **15**(4) 948
[13] Ghosh D and Chattopadhyay N 2015 *J Luminescence* **160** 223
[14] Piriyawong V, Thongpool, Asanithi P, and Limsuwan P 2012 *Procedia Engineer.* **32** 1107

IOP Conf. Series: Journal of Physics: Conf. Series **909** (2017) 012038 doi:10.1088/1742-6596/909/1/012038

A simple but accurate explicit finite difference method for the advection-diffusion equation

Febi Sanjaya[1] and Sudi Mungkasi[2]

[1] Department of Mathematics Education, Faculty of Teacher Training and Education, Sanata Dharma University, Mrican, Tromol Pos 29, Yogyakarta 55002, INDONESIA
[2] Department of Mathematics, Faculty of Science and Technology, Sanata Dharma University, Mrican, Tromol Pos 29, Yogyakarta 55002, INDONESIA

E-mail: febi@usd.ac.id, sudi@usd.ac.id

Abstract. We consider the advection-diffusion equation in one dimension. The equation is solved using an explicit finite difference method. As the method is explicit, it is simple to implement. We investigate the performance of the method. We obtain that the method is indeed accurate.

1. Introduction

The advection-diffusion equation is a model that can be used for simulation of the spreading of pollutant. It governs the process of advection and diffusion concurrently. A large number of authors (to mention some of them, see [1-15]) have attempted to solve and use this equation in their simulations.

In this paper, we investigate the performance of a finite difference method due to Karahan [4] for solving the advection-diffusion equation of pollutant transports. The finite difference approach is appropriate because the equation is a partial differential equation having the parabolic type. We use the MATLAB software in the implementation of the finite difference method. The iterations are started from a given initial condition using an arbitrary space width and time step.

This paper is organised simply as follows. Sections 2-5 contains the mathematical model, numerical method, numerical results, and conclusion, respectively.

2. Advection-diffusion equation

The mathematical model describing the pollutant transport is the one-dimensional advection-diffusion equation [4]

$$\frac{\partial c}{\partial t}+u\frac{\partial c}{\partial x}=D\frac{\partial^2 c}{\partial x^2},\ 0<x<L,\ 0<t\leq T \tag{1}$$

with the initial condition

$$c(x,0)=f(x),\quad 0\leq x<L \tag{2}$$

and boundary conditions

$$c(0,t)=g(t),\quad 0<t\leq T \tag{3}$$

$$c(L,t)=h(t),\quad 0<t\leq T \tag{4}$$

in a finite space domain of x and a finite time domain of t. The variable $c = c(x,t)$ denotes the concentration of a pollutant. In addition, f, g and h are known functions; u, x, D respectively represent the velocity, the direction, and the dispersion coefficient. Here u and D are assumed to be positive constants. As advection-diffusion equation contains advection and diffusion phenomena, the solution should reveal those two phenomena.

3. Finite difference method

We use the finite difference method due to Karahan [4] to solve the problem. Note that the equation is a partial differential equation of the parabolic type, so finite difference methods should be able to solve the problem. This is because finite difference methods are based on the direct-discretisation of differential equations. Solutions of parabolic partial differential equations are always continuous even when the initial condition is discontinuous.

We discretise the problem by choosing a stepsize Δx as the spatial step in x and stepsize Δt as the temporal step in t. We also approximate the concentration c on a grid of point in the xt-plane. Moreover,

$$x_i = i\Delta x,\ i = 1, 2, 3, ..., M,\ \Delta x = \frac{L}{M} \tag{5}$$

$$t_n = n\Delta t,\ t = 1,2,3,\ldots,N,\ \Delta t = \frac{T}{N} \tag{6}$$

Furthermore, we take the following discretisations

$$\frac{\partial c}{\partial t} = \frac{c(i,n+1) - c(i,n)}{\Delta t} \tag{7}$$

$$u\frac{\partial c}{\partial x} = u\left(\frac{1}{2\Delta x}\right)\left(c(i,n+1) - c(i-1,n+1) + c(i+1,n) - c(i,n)\right) \tag{8}$$

$$D\frac{\partial^2 c}{\partial x^2} = D\left[\frac{\theta}{(\Delta x)^2}\right][c(i-1,n+1) - c(i,n+1) + c(i+1,n) - c(i,n)] + D\left[\frac{1-\theta}{(\Delta x)^2}\right][c(i-1,n) - 2c(i,n) + c(i+1,n)] \tag{9}$$

where θ is the weighting factor. Here the time integration (7) is forward difference, so it has the first order of accuracy.

Substituting the equations above into equation (1), we have

$$\frac{c(i,n+1) - c(i,n)}{\Delta t} + u\left(\frac{1}{2\Delta x}\right)[c(i,n+1) - c(i-1,n+1) + c(i+1,n) - c(i,n)] = D\left[\frac{\theta}{(\Delta x)^2}\right] [c(i-1,n+1) - c(i,n+1) + c(i+1,n) - c(i,n)] + D\left[\frac{1-\theta}{(\Delta x)^2}\right][c(i-1,n) - 2c(i,n) + c(i+1,n)] \tag{10}$$

for $1 \le i \le M$ and $1 \le n \le N$. Equation (10) is equivalent to

$$\left[1 + \frac{Cr}{2} + \theta\left(\frac{Cr}{Pe}\right)\right]c(i,n+1) - \left[\frac{Cr}{2} + \theta\left(\frac{Cr}{Pe}\right)\right]c(i-1,n+1) = \left[\left(\frac{Cr}{Pe}\right) - \theta\left(\frac{Cr}{Pe}\right)\right]c(i-1,n) + \left[1 + \left(\frac{Cr}{2}\right) - 2\left(\frac{Cr}{Pe}\right) + \theta\left(\frac{Cr}{Pe}\right)\right]c(i,n) + \left[\left(\frac{Cr}{Pe}\right) - \left(\frac{Cr}{2}\right)\right]c(i+1,n) \tag{11}$$

where the Courant number $Cr = u\Delta t / \Delta x$, and the Peclet number $Pe = u\Delta x / D$. The boundary condition, equation (3), makes the equation is explicit, so we obtain

$$c(i,n+1)=\left\{\left[\left(\frac{Cr}{Pe}\right)-\theta\left(\frac{Cr}{Pe}\right)\right]c(i-1,n)+\left[1+\left(\frac{Cr}{2}\right)-2\left(\frac{Cr}{Pe}\right)+\theta\left(\frac{Cr}{Pe}\right)\right]c(i,n)+\left[\left(\frac{Cr}{Pe}\right)-\left(\frac{Cr}{2}\right)\right]\right.$$
$$\left.c(i+1,n)+\left[\frac{Cr}{2}+\theta\left(\frac{Cr}{Pe}\right)\right]c(i-1,n+1)\right\}\left[1+\frac{Cr}{2}+\theta\left(\frac{Cr}{Pe}\right)\right]^{-1} \quad (12)$$

Equation (12) is the explicit finite difference method due to Karahan [4] for solving the advection-diffusion equation (1).

4. Results and discussion

In this section, we present our numerical results and discuss them. We assume that all quantities are in SI units with the MKS system.

As the numerical test, we consider the following settings. The initial condition is

$$c(x,0)=\exp\left(-\frac{(x-1)^2}{D}\right). \quad (13)$$

This initial condition is illustrated in Figure 1. Parameters being used are $D = 0.005$ and $u = 0.8$. The spatial step is $\Delta x = 0.025$, and the temporal step is $\Delta t = 0.0125$. We consider the spatial domain $0 \le x \le 9$. We seek the solution at time $t = 2.5$ and $t = 5$.

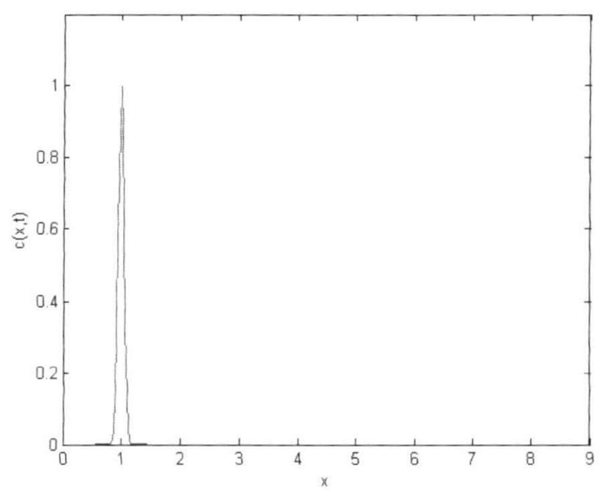

Figure 1. Initial condition ($t = 0$) of an advection-diffusion problem.

The analytical solution for this problem is [4]

$$c(x,t)=\frac{1}{\sqrt{4t+1}}\exp\left(-\frac{(x-1-ut)^2}{D\,(4t+1)}\right). \quad (14)$$

Note that this problem is chosen, because the analytical solution has been found (see Sankaranarayanan et al. [11]). In general, the analytical solution to the advection-diffusion equation is not available. Therefore, we do need numerical methods to solve the advection-diffusion equation.

Numerical results show that the method is simple to implement, yet gives accurate solutions. This is shown in Figure 2 as representatives of our numerical results. This figure shows the solution at time $t = 2.5$ and $t = 5$, respectively. Notice that our solutions reveal advection and diffusion phenomena at the same time. That is, as time evolves the solution translates to the right (advection phenomenon) and it diffuses from the centre of the concentration position (diffusion phenomenon).

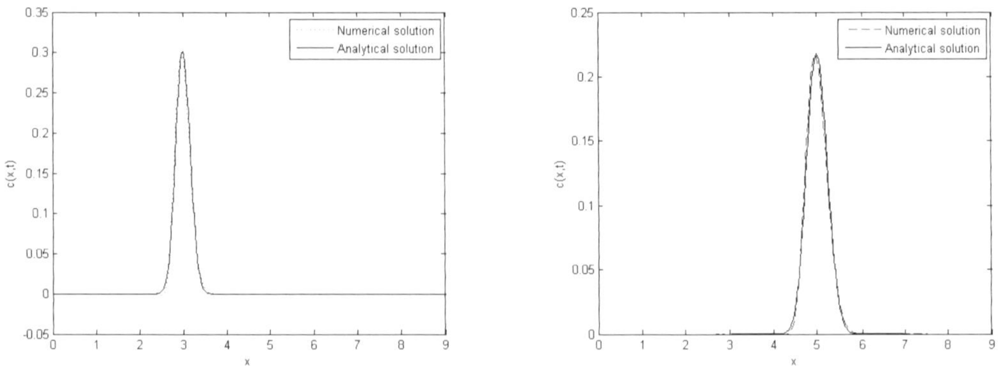

Figure 2. Numerical and analytical solutions using $\Delta x = 0.025$ and $\Delta t = 0.0125$ at time $t = 2.5$ (left figure) and $t = 5.0$ (right figure) respectively. We observe that the numerical solution is very accurate, as the error is quite small with respect to the analytical solution.

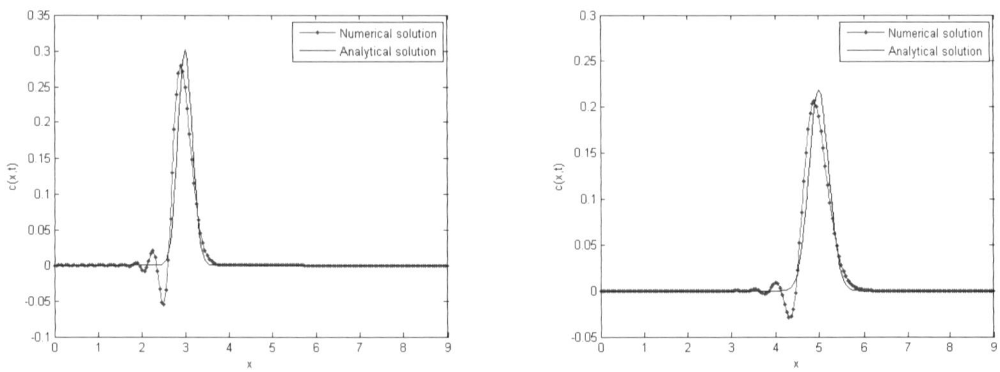

Figure 3. Numerical and analytical solutions using $\Delta x = 0.05$ and $\Delta t = 0.0125$ at time $t = 2.5$ (left figure) and $t = 5.0$ (right figure) respectively. We observe that artificial oscillations occur if the spatial step is taken too large.

We should not take the spatial step Δx too large, as it may not be able to cover the accuracy of our numerical discretisation. Taking the spatial step Δx too large may lead to unphysical solutions. That is, artificial oscillations may occur in the numerical solutions if the spatial step Δx is too large. This is shown in Figure 3. This figure shows the solution at time $t = 2.5$ and $t = 5$, respectively using the same numerical settings as before, except the spatial step here is $\Delta x = 0.05$.

5. Conclusion

We have investigated the performance of an explicit finite difference method for solving the advection-diffusion equation. As we use the forward difference for the time derivative, numerical solutions should have the first order of accuracy at best. The general analytical solution to the advection-diffusion equation is not available, so we need the numerical method to solve the advection-diffusion equation for the general case. The proposed method can be extended for solving the advection-diffusion equation with a source term.

Acknowledgment

This research was financially supported by Universitas Sanata Dharma (Sanata Dharma University) and Direktorat Riset dan Pengabdian kepada Masyarakat of Direktorat Jenderal Penguatan Riset dan Pengembangan of Kementerian Riset, Teknologi, dan Pendidikan Tinggi of Republik Indonesia in the financial year 2017 with Research Contract Number 051/HB-LIT/IV/2017 dated 14 April 2017 (under DIPA-042.06-0.1.401516/2017 funds). Both authors are very grateful for the financial support.

References

[1] Benito M, Charpentier C and Kojouharov H V 2013 An unconditionally positivity preserving scheme for advection-diffusion reaction equations *Mathematical and Computer Modelling* **57** 2177
[2] Company R, Ponsoda E, Romero J V and Roselló M D 2009 A second order numerical method for solving advection-diffusion models *Mathematical and Computer Modelling* **50** 806
[3] Karahan H 2007 A third-order upwind scheme for the advection-diffusion equation using spreadsheets *Advances in Engineering Software* **38** 688
[4] Karahan H 2007 Unconditional stable explicit finite difference technique for the advection-diffusion equation using spreadsheets *Advances in Engineering Software* **38** 80
[5] Mohebbi A and Dehghan M 2010 High-order compact solution of the one-dimensional heat and advection-diffusion equations *Applied Mathematical Modelling* **34** 3071
[6] Mojtabi A and Deville M O 2015 One-dimensional linear advection-diffusion equation: Analytical and finite element solutions *Computers & Fluids* **107** 189
[7] Nazir T, Abbas M, Ismail A I M, Majid A A and Rashid A 2016 The numerical solution of advection-diffusion problems using new cubic trigonometric B-splines approach *Applied Mathematical Modelling* **40** 4586
[8] Nishikawa H 2014 First, second, and third order finite-volume schemes for advection-diffusion *Journal of Computational Physics* **273** 287
[9] Prieto F U, Muñoz J J B, Corvinos L G 2011 Application of the generalized finite difference method to solve the advection-diffusion equation *Journal of Computational and Applied Mathematics* **235** 1849
[10] Ricchiuto M, Villedieu N, Abgrall R, and Deconinck H 2008 On uniformly high-order accurate residual distribution schemes for advection-diffusion *Journal of Computational and Applied Mathematics* **215** 547
[11] Sankaranarayanan S, Shankar N J and Cheong H F 1998 Three-dimensional finite difference model for transport of conservative pollutants *Ocean Engineering* **25** 425
[12] Saul'Yev V K 1964 *Integration of Equations of Parabolic Type by the Method of Nets* (Oxford: Pergamon Press)
[13] Savovic S and Djordjevich A 2012 Finite difference solution of the one-dimensional advection-diffusion equation with variable coefficients in semi-infinite media *International Journal of Heat and Mass Transfer* **55** 4291
[14] Sousa E 2009 Finite difference approximations for a fractional advection diffusion problem *Journal of Computational Physics* **228** 4038
[15] Witek M L, Teixeira J and Flatau P J 2008 On stable and explicit numerical methods for the advection-diffusion equation *Mathematics and Computers in Simulation* **79** 561

IOP Conf. Series: Journal of Physics: Conf. Series **909** (2017) 012039 doi:10.1088/1742-6596/909/1/012039

On the local metric dimension of a lollipop graph, a web graph, and a friendship graph

A N Cahyabudi and T A Kusmayadi

Mathematics Department, Faculty of Mathematics and Natural Sciences, Sebelas Maret University,
Jl. Ir. Sutami 36A Kentingan Jebres Surakarta 57126, INDONESIA

ancahyabudi@student.uns.ac.id; tri.atmojo.kusmayadi@gmail.com

Abstract. Let G be a simple connected graph with the vertex set $V(G)$ and the edge set $E(G)$. The distance between two vertices u and v, denoted by $d(u,v)$, is the length of a shortest $u-v$ path. If $W=\{w_1,w_2,w_3,\ldots,w_n\}$ is a finite set of vertices of G and $v\in V(G)$, then the representation of v with respect to W is an ordered n-tuple $r(v|W)=(d(v,w_1),d(v,w_2),d(v,w_3),\ldots,d(v,w_n))$. The set W is called a local metric generator for G if every two adjacent vertices of G has distinct representations. A minimum local metric generator is called a *local metric basis* for G and its cardinality is called the *local metric dimension* of G. A lollipop graph $L_{m,n}$ for $m\geq 3$ and $n\geq 2$ is the graph obtained by joining a complete graph K_m to a path graph P_n with a bridge. A web graph W_n for $n\geq 3$ is a generalized prism graph $Y_{n+1,3}$ with the edge of the outer cycle removed. A friendship graph f_n for $n\geq 2$ is a graph constructed by joining n copies of the cycle graph C_3 with a common vertex. In this paper, we determine the local metric dimension of a lollipop graph, a web graph, and a friendship graph.

1. Introduction

All graphs considered in this paper are finite, simple, and non-null graphs. If G is a graph, we denote by $V(G)$ and $E(G)$ its vertex and edge sets respectively. If u and v are the end vertices of an edge, we write uv for the edge itself. If $uv\in E(G)$ we say that u is adjacent to v (in G).

In 2010, Okamoto et al. [6] introduced the local metric dimension of a graph. For $u,v\in V(G)$ we define the distance between u and v, denoted by $d(u,v)$, as the length of the shortest $u-v$ path. If $W=\{w_1,w_2,w_3,\ldots,w_n\}$ is a finite set of vertices of G and $v\in V(G)$, then the representation of v with respect to W is an ordered n-tuple $r(v|W)=(d(v,w_1),d(v,w_2),d(v,w_3),\ldots,d(v,w_n))$. The set W is called a local metric generator for G, if every two adjacent vertices of G has distinct representations. A minimum local metric generator is called a *local metric basis* for G, and its cardinality is called the *local metric dimension* of G and denoted by $dim_l(G)$.

Many researchers have investigated the local metric dimension of graph. In 2013 Rodriguez-Velazquez et al. [8] investigated the local metric dimension of corona product graphs. In 2015 Rodriguez-Velazquez et al. [7] investigated computing the local metric dimension of a graph from the local metric dimension of primary subgraphs. In 2014 Cynthia et al. [2] investigated the local metric dimension of cyclic split graph. In 2014 Salman et al. [10] investigated minimum fault-tolerant, local and strong metric dimension of graphs. In 2013 Rodriguez-Velazquez et al. [9] investigated on the (adjacency) metric dimension of corona and strong product graphs and their local variants :

commbinatorial and computational results. Okamoto et al. [6] found that $\dim_l(G) = 1$ if only if G is a connected bipartite graph, and Chartrand and Lesniak [1] proved that a nontrivial connected graph is bipartite if only if it contains no odd cycles. In this paper, we determine the local metric dimension of a lollipop graph, a web graph, and a friendship graph.

2. Main Results

2.1. Local Metric Dimension of A Lollipop Graph

Hoory et al. [4] defined a lollipop graph is a graph obtained by joining a complete graph K_m to a path graph P_n with a bridge.

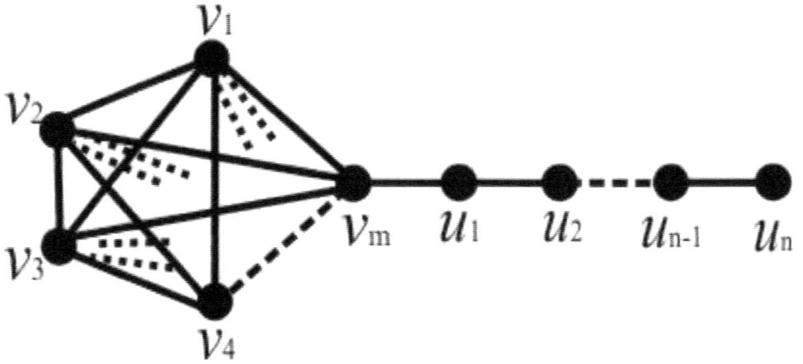

Figure 1. Lollipop Graph $L_{m,n}$.

Theorem 1. *Let $L_{m,n}$ be a lollipop graph with $m \geq 3$ and $n \geq 2$. Then $dim_l(L_{m,n}) = m - 1$.*

Proof Let $L_{m,n}$ be a lollipop graph with $m \geq 3$ and $n \geq 2$. We will show $dim_l(L_{m,n}) \leq m - 1$. Assume $W = \{v_1, v_2, v_3, \dots, v_{m-2}, v_{m-1}\}$. Cardinality of W is $m - 1$. Then the representations of all vertices $L_{m,n}$ with respect to W are

$$r(v_1|W) = (0,1,1,\dots,1,1); \quad r(u_1|W) = (2,2,2,\dots,2,2);$$
$$r(v_2|W) = (1,0,1,\dots,1,1); \quad r(u_2|W) = (3,3,3,\dots,3,3);$$
$$\dots \quad \dots$$
$$\dots \quad \dots$$
$$r(v_{m-1}|W) = (1,1,1,\dots,1,0); \quad r(u_{n-1}|W) = (n,n,n,\dots,n,n);$$
$$r(v_m|W) = (1,1,1,\dots,1,1); \quad r(u_n|W) = (n+1,\dots,n+1).$$

All vertices in $V(L_{m,n})$ have distinct representations with respect to W so that W is a local metric generator with $m - 1$ elements. Then $dim_l(L_{m,n}) \leq m - 1$.

Next we show $dim_l(L_{m,n}) \geq m - 1$. Assume W is a local metric generator of a lollipop graph $L_{m,n}$ with $|W| < m - 1$. There are three possibilities to choose the vertex of W.

(1) If all vertices of W in $V(K_m) = \{v_i | 1 \leq i \leq m\} \subset V(L_{m,n})$, then at least two vertices $x, y \in V(K_m)$ are adjacent such that
$$r(x|W) = r(y|W) = (1, 1, \dots, 1).$$

(2) If some vertices of W in $V(K_m) = \{v_i | 1 \leq i \leq m\} \subset V(L_{m,n})$ and other vertices in $V(P_n) = \{u_j | 1 \leq j \leq n\} \subset V(L_{m,n})$, then at least two vertices $x, y \in V(K_m)$ are adjacent such that
$$d(x, v_i) = d(y, v_i) = 1, \text{ for } v_i \in W;$$
$$d(x, u_j) = d(y, u_j) = j + 1, \text{ for } u_j \in W.$$

(3) If all vertices of W in $V(P_n) = \{u_j | 1 \leq j \leq n\} \subset V(L_{m,n})$, then at least two vertices $x, y \in V(K_m)$ are adjacent such that

$$d(x, u_j) = d(y, u_j) = j + 1, \text{ for } u_j \in W.$$

From all possibilities to choose the vertex of W there are at least two adjacent vertices with the same representations, so W is not a local metric generator. It contradicts with the fact that W is a local metric generator of $L_{m,n}$. Hence $dim_l(L_{m,n}) \geq m - 1$.
It completes the proof of the theorem. □

2.2. Local Metric Dimension of A Web Graph
Koh et al. [5] defined a web graph W_n for $n \geq 3$ as a generalized prism graph $Y_{n+1,3}$ with the edge of the outer cycle removed.

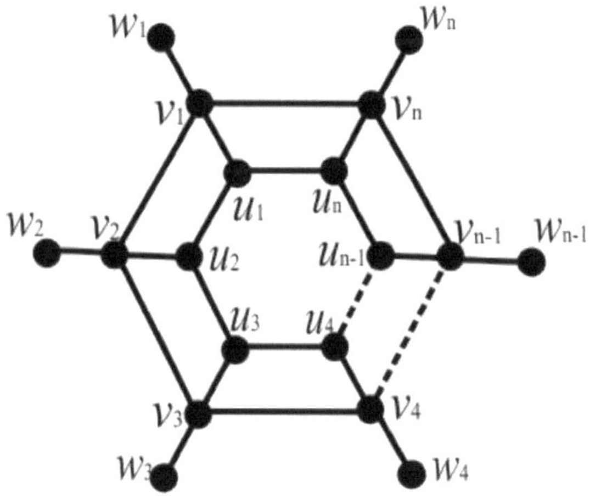

Figure 2. Web Graph W_n.

Theorem 2. *Let W_n be a web graph. Then for $n \geq 3$*

$$dim_l(W_n) = \begin{cases} 1, n \text{ } even; \\ 2, n \text{ } odd. \end{cases}$$

Proof Given a web graph W_n with the set of vertices $V(W_n) = \{u_1, u_2, \ldots, u_n, v_1, v_2, \ldots, v_n, w_1, w_2, \ldots, w_n\}$. We consider two cases based on the values of n.

Case 1. n is even, $n \geq 3$.
Since web graph W_n is a bipartite graph, then by using the results of Okamoto et al. [6] and Chartrand and Lesniak [1], we have $dim_l(W_n) = 1$.

Case 2. n is odd, $n \geq 3$.
Let $W = \{u_1, u_{\frac{n+1}{2}}\} \subset V(W_n)$, then representations of all vertices $V(W_n)$ with respect to W are

$$r(u_i|W) = \begin{cases} \left(i - 1, \frac{n+1}{2} - i\right), & i = 1,2, \ldots, \frac{n+1}{2}; \\ \left(n - i + 1, i - \frac{n+1}{2}\right), & i = \frac{n+3}{2}, \frac{n+5}{2}, \ldots, n; \end{cases}$$

$$r(v_i|W) = \begin{cases} \left(i, \frac{n+3}{2} - i\right), & i = 1,2,\dots,\frac{n+1}{2}; \\ \left(n-i+2, i-\frac{n-1}{2}\right), & i = \frac{n+3}{2}, \frac{n+5}{2}, \dots, n; \end{cases}$$
$$r(w_i|W) = \begin{cases} \left(i+1, \frac{n+5}{2} - i\right), & i = 1,2,\dots,\frac{n+1}{2}; \\ \left(n-i+3, i-\frac{n-3}{2}\right), & i = \frac{n+3}{2}, \frac{n+5}{2}, \dots, n. \end{cases}$$

All vertices in $V(W_n)$ have distinct representations with respect to W so that W is a local metric generator with two elements. In the case of a web graph W_n, we do not have a local metric generator with one element. Hence $dim_l(W_n) = 2$. □

2.3. Local Metric Dimension of A Friendship Graph

Gallian [3] defined a friendship graph f_n for $n \geq 2$ is a graph constructed by joining n copies of the cycle graph C_3 with the set of vertices $V(f_n) = \{v_1, v_2, \dots, v_{2n}\}$ with a common vertex u.

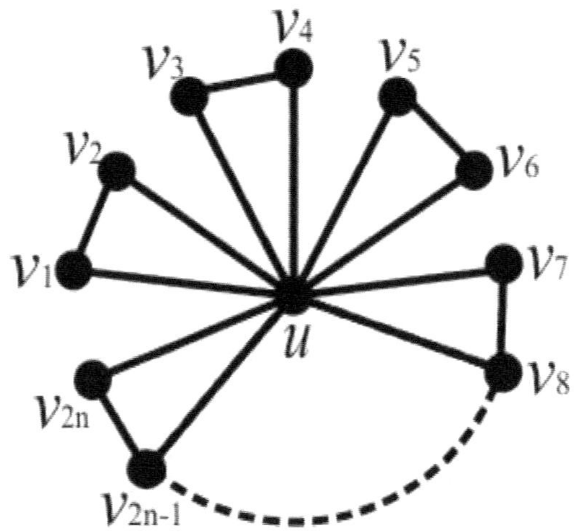

Figure 3. Friendship Graph F_n.

Lemma 1. *If W is a local metric generator for a friendship graph f_n, then $|W| \geq n$.*

Proof By a contradiction, we will show that $|W| \geq n$. Assume that W is a local metric generator with $|W| < n$. The f_n graph has a vertex u where u is adjacent with all vertices in graph $V(f_n) - \{u\}$, so $d(v_i, u) = d(v_j, u)$ for $i \neq j$. Then there are two vertices $v_{2n-1}, v_{2n} \notin W \subset V(f_n) - \{u, v_{2n-1}, v_{2n}\}$ such that $r(v_{2n-1}|W) = r(v_{2n}|W) = (2,2,\dots,2)$ where vertex v_{2n-1} and vertex v_{2n} are adjacent. It contradicts with the fact that W is a local metric generator of f_n and so $|W| \geq n$. □

Lemma 2. *If $W = \{v_1, v_3, v_5, \dots, v_{2n-3}, v_{2n-1}\} \subset V(f_n)$, then W is a local metric generator for a friendship graph f_n.*

Proof All vertices of W are not adjacent each other, then the representations of all vertices of $V(f_n)$ with respect to W are

$$r(u|W) = (1,1,1,\dots,1)$$

IOP Conf. Series: Journal of Physics: Conf. Series **909** (2017) 012039 doi:10.1088/1742-6596/909/1/012039

$$r(v_i|W) = \begin{cases} (i-1,2,2,2,\dots,2) & i = 1,2; \\ (2,i-3,2,2,\dots,2) & i = 3,4; \\ (2,2,i-5,2,\dots,2) & i = 5,6; \\ \vdots & \\ \vdots & \\ (2,2,2,\dots,2,i-(2n-1)) & i = 2n-1,2n. \end{cases}$$

Clearly, all vertices of $V(f_n)$ have distinct representations with respect to W so that W is a local metric generator with n elements. □

Theorem 3. *Let f_n be a friendship graph. Then for $n \geq 2$*

$$dim_l(f_n) = n.$$

Proof Given a f_n graph with $n \geq 2$. By Lemma 1 and Lemma 2 we have $dim_l(f_n) = n$. □

References

[1] Chartrand G and L Lesniak, *Graphs and digraphs*, 3rd ed., Chapman and Hall, United States of America, 2000.

[2] Cynthia J A and Ramya, *The Local Metric Dimension of Cyclic Split Graph*, Annals of Pure and Applied Mathematics **8** (2014), 201-205.

[3] Gallian J A, *A Dynamic Survey of Graphs Labeling*, The Electronic Journal of Combinatorics #DS6, 2016.

[4] Hoory S, N Linial and A Wigderson, *Expander Graphs and Their Application*, Bulletin of the American Mathematical Society **43** (2006), 439-561.

[5] Koh K M, D G Rogers, H K Teo and K Y Yap, *Graceful Graph: Some Further Results and Problems*, Congress. Numer., **29** (1980), 559-571.

[6] Okamoto F, B Phinezy and P Zhang, *The local metric dimension of a graph*, Mathematica Bohemia **135** (2010), 239-255.

[7] Rodriguez-Velazquez J A, C G Gomez and G A Barragan-Ramirez, *Computing the Local Metric Dimension of a Graph from the Local Metric Dimension of Primary Subgraphs*. Int. J. Comput. Math. **92** (2015), 686-693.

[8] Rodriguez-Velazquez J A, G A Barragan-Ramirez and C G Gomez, *On the Local Metric Dimension of Corona Product Graphs*, arXiv: 1308.6689 (2013).

[9] Rodriguez-Velazquez J A and H Fernau, *On the (adjacency) Metric dimension of Corona and Strong Product Graphs and Their Local Variants : Commbinatorial and Computational Results*, arXiv: 1309.2275 (2013)

[10] Salman M, I Javaid and M A Chaudry, *Minimum Fault-Tolerant, Local and Strong Metric Dimension of Graphs,* arXiv: 1409.2695 (2014).

Estimating the Value-at-Risk for some stocks at the capital market in Indonesia based on ARMA-FIGARCH models

Sukono[1], E. Lesmana[1], D. Susanti[1], H. Napitupulu[1*], Y. Hidayat[2*]
[1]Department of Mathematics, FMIPA, Universitas Padjadjaran
[2]Department of Mathematics, FMIPA, Universitas Padjadjaran
Jl. Raya Bandung-Sumedang Km 21, Jatinangor 45363, Jawa Barat, INDONESIA

E-mail: sukono@unpad.ac.id: man.msie@gmail.com; dwi.susanti@unpad.ac.id; napitupuluherlina@gmail.com; yuyun.hidayat@unpad.ac.id

[*]corresponding author: napitupuluherlina@gmail.com; yuyun.hidayat@unpad.ac.id

Abstract. Value-at-Risk has already become a standard measurement that must be carried out by the financial institution for both internal interest and regulatory. In this paper, the estimation of Value-at-Risk of some stocks with econometric models approach is analyzed. In this research, we assume that the stock return follows the time series model. To do the estimation of mean value we are using ARMA models, while to estimate the variance value we are using FIGARCH models. Furthermore, the mean value estimator and the variance are used to estimate the Value-at-Risk. The result of the analysis shows that from five stock PRUF, BBRI, MPPA, BMRI, and INDF, the Value-at-Risk obtained are 0.01791, 0.06037, 0.02550, 0.06030, and 0.02585 respectively. Since Value-at-Risk represents the maximum risk size of each stock at a 95% level of significance, then it can be taken into consideration in determining the investment policy on stocks.

1. Introduction

There are various financial assets that offer different levels of profit and risk in the capital market. Risks are often associated with variability or dispersion. The asset is said to have no risk if its return does not have a variability. The greater the return variability of an asset, the more the probability of return is different from the expected result [1]. Because investors are facing a risky investment opportunity, investment decisions cannot just consider the expected level of returns, but also the extent of losses that may occur [4]. The commonly used measurement of asset loss risk is the standard variance or deviation. Both measure how far the actual return is different from the average return. Variance measures the average of the quadratic difference between the actual returns and the mean returns. The greater the value of variance, the farther the actual returns is different from the average return. Since variance is a measure of average risk, so it cannot accommodate all risk events. There is an idea of measuring risk by using quintile, which is further known as Value-at-Risk (VaR) [7].

Value-at-Risk measures typically consist of two parameters, namely mean and volatility as well as a certain confidence level percentage [6]. The mean of asset return data is often nonconstant over time, so it is following the time series model, and the mean can be estimated using *Autoregressive Moving Average* (ARMA) model. Similarly, the volatility of asset return data is often non constant and has a long-term correlation structure. Characteristics of such volatility can be estimated using *Fractionally Integrated Generalized Autoregressive Conditional Heteroscedasticity* (FIGARCH) models. We have done the research on Value-at-Risk estimation using time series model approach. Makiel [11], estimates market risk using Value-at-Risk based on ARIMA-GARCH models. Kasman [10], estimates Value-at-Risk for the stock price index in Turkey which concern to the long-term correlation structure in volatility. Balibey and Turkyilmaz [5], conducting Value-at-Risk analysis that concern to the asymmetry and long memory structure. Aloni and Hamida [2], estimates and measures Value-at-Risk performance based on long memory models from GARCH-Class.

According to the previous description, this paper analyzes the estimation of Value-at-Risk of some stocks with econometric models approach. The aim is to estimate Value-at-Risk of some stock return traded on the capital market in Indonesia of which has characteristics that follow ARMA-FIGARCH models. As a case study, stock return data of PRUF, BBRI, MPPA, BMRI, and INDF are analyzed. So that can be used as a consideration for investors who make investments on the five shares.

2. Risk Measurement Model

The intention of this section is to describe the models used in the Value-at-Risk estimation. The models used include stock return, mean model, volatility model, and Value-at-Risk model with its performance measurement method. The discussions begin by following explanation of stock return.

2.1. Stock Return

In this section, the aims are to determine the stock returns. Suppose that the stock price on a day-t is X_t. To analyze financial data within *horizon* daily time, *return* r_t often given in the form of *continuously compound return* or log *return* in following formula:

$$r_t = \ln\left(\frac{X_t}{X_{t-1}}\right), \tag{1}$$

with $t = 1,2,...,T$ where T is the number of observation data, and it is assumed that $X_0 = 1$ [7]. Thus, we used this return for modeling in the following sections.

2.2. Mean Model

The aim of this section is to discuss the mean model in time series data. Suppose that r_t is a log *return* stock at time t, generally, the *Autoregressive Moving Average* model, ARMA(p,q), can be expressed by the following equation [13], [14]:

$$r_t = \phi_0 + \sum_{i=1}^{p}\phi_i r_{t-i} + a_t - \sum_{j=1}^{q}\theta_j a_{t-j}\,. \tag{2}$$

Where $\{a_t\}$ is a residual sequence assumed to be normal white noise distribution with mean 0 and variance σ_a^2. A non-negative integer p and q are the order of ARMA. AR and MA model are the special cases of ARMA(p,q) model. Using *back-shift* operator, model (2) can be written as:

$$(1-\phi_1 B - ... - \phi_p B^p)r_t = \phi_0 + (1-\theta_1 B - ... - \theta_q B^q)a_t\,. \tag{3}$$

Polynomial $1-\phi_1 B - ... - \phi_p B^p$ is from AR model and polynomial $1-\theta_1 B - ... - \theta_q B^q$ is from MA model. If the solution of characteristic equations is absolutely less than 1, then the ARMA model is stationary weak. In this case, the non-conditional mean of the model is $E(r_t) = \phi_0 / (1-\phi_1 - ... - \phi_p)$ [13].

Mean Modeling Steps. Broadly speaking, according to Tsay [14], the steps of modeling the mean models are as follows: (i) Identification of the model, determining order value of p and q by autocorrelation function (ACF) and partial autocorrelation function (PACF) plot. (ii) Parameter

Estimation can be done by least squares method or *maximum likelihood*. (iii) Diagnostic test by *white noise* test and non-correlation serially against residual a_t. (iv) Forecasting, if the model is suitable it can be used for prediction done recursively [7], [15].

2.3 Volatility Model

The aims of this section are to discuss the volatility model of time series data. Engle in 1982 proposed the *autoregressive conditional heteroscedasticity* (ARCH) model. In ARCH model, the following conditional variance is an *autoregressive* process. ARCH(p) model assumes that log *return* $r_1, r_2, ...$ described by the process [8], [14]:

$$r_t = \mu_t + a_t,\ a_t = \sigma_t \varepsilon_t,\ \text{and}\ \sigma_t^2 = \alpha_0 + \sum_{i=1}^{p} \alpha_i a_{t-i}^2 + \varepsilon_t\ , \tag{4}$$

where μ_t is the expectation of r_t, σ_t^2 is the variance of r_t, ε_t is the random *residual shock* with mean 0 and variance 1. The general assumption are that $\{\varepsilon_t\} \sim iid\ N(0,1)$, α_0 and α_i are constant, $\alpha_0 > 0$, and $\alpha_i \geq 0$, $i = 1,..., p$.

Testing of ARCH Effect. The test that widely used for the detection of ARCH effect is the Lagrange Multiplier (LM) test, which was introduced by Engle in 1982. Refer to the equation (4), test for the ARCH(p) effect is based on the null hypothesis $H_0 : \alpha_0 = \alpha_1 = ... = \alpha_p = 0$ against alternative hypothesis $H_1 : \exists \alpha_0 \neq \alpha_1 \neq ... \neq \alpha_p \neq 0$ [7], [14].

Statistical LM test. This hypothesis shows that it is assymtotically equivalent to the statistical test $T \times R^2$, where T is the sample size and R^2 is calculated from regression [13]:

$$\hat{\varepsilon}_t^2 = a_0 + a_1 \hat{\varepsilon}_{t-1}^2 + ... + a_p \hat{\varepsilon}_{t-p}^2 + v_t\ .$$

Under the null hypothesis, there is no ARCH effect, LM and statistical test $T \times R^2$ asymptotically $\chi^2(p)$ distributed. As an alternative to the LM test form, by asymptotically equivalent can used the portmanteau test, as the Ljung and Box statistical test for ε_t^2 [14].

GARCH Model. GARCH model was introduced by Bollerslev in 1986, is a common form or generalizations of ARCH models. In general, GARCH (p, q) models can be written as follows [2]:

$$a_t = \sigma_t \varepsilon_t,\ \ \sigma_t^2 = \alpha_0 + \sum_{i=1}^{p} \alpha_i a_{t-i}^2 + \sum_{j=1}^{q} \beta_j \sigma_{t-j}^2 + \varepsilon_t\ . \tag{5}$$

Based on equations (4) and (5), the conditional expectation and variance of ε_t is:

$$E(\varepsilon_t | F_{t-1}) = 0 \tag{6}$$

$$Var(\varepsilon_t | F_{t-1}) = E(\varepsilon_t^2 | F_{t-1}) = \sigma_t^2 \tag{7}$$

Compared to ARCH model, GARCH model is considered to provide simpler results because it uses less parameter [2],[14].

FIGARCH Model. FIGARCH model empirically is a volatility *long memory* in the financial market [14]. To demonstrate the properties of long memory of volatility in the financial markets, Baillie, Bollerslev & Mikkelsen in 1996, expand the IGARCH model by replacing the first difference operator $(1-B)$ to fractional difference operator $(1-B)^d$ with $0 < d < 1$. The FIGARCH(p,d,q) model is expanded as follows [12], [15]:

$$\phi(B)(1-B)^d\, \varepsilon_t^2 = \omega + (1-\beta(B))v_t\ ;\ v_t = \varepsilon_t^2 - \sigma_t^2\ . \tag{8}$$

It is obvious that FIGARCH model is GARCH and IGARCH special models in the case $d = 0$ or $d = 1$. Generally, the formula of FIGARCH(*p,d,q*) model variance is:

$$\sigma_t^2 = \omega(1-\beta(B))^{-1} + (1-\beta(B))^{-1}(1-\phi(B)(1-B)^d)\varepsilon_t^2. \tag{9}$$

where $(1-B)^d$ can be shown by the expansion of Maclaurin series:

$$(1-B)^d = \sum_{k=0}^{\infty} \frac{\Gamma(k-d)}{\Gamma(k+1)\Gamma(-d)} B^k . \tag{10}$$

$\Gamma(k-d)/\Gamma(k+1) \approx k^{-d-1}$ if k is large, then the above coefficients in the infinite polynomials are hyperbolic decreased [3], [9].

Volatility Modeling Steps. In general, according to Tsay [14], the steps of volatility modeling are as follows: (i) Estimate the mean model by time series model (for example ARMA model). (ii) Use the residual of the mean model to test the ARCH effect. (iii) If there is an ARCH effect, estimate the volatility model, and construct the combined estimation form the mean model and volatility model. (iv) Conduct diagnostic tests to observe the suitability of the model. (v) If the model is fit, use to predict by recursive prediction [13], [15].

2.4 Value-at-Risk Model

In this section, the aim is to discuss the Value-at-Risk as a measure of risk. Suppose that the portfolio depends only on one risk factor. The standard method assumes that the distribution of assets return is univariate normally distributed, has two parameters: mean μ, and standard deviation σ. The Value-at-Risk estimation issue is how to determine the $(1-\alpha)\%$ percentile of the standard normal distribution $z_{1-\alpha}$ [7], [14]:

$$1-\alpha = \int_{-\infty}^{q} f(r)dr = \int_{-\infty}^{z_{1-\alpha}} \Phi(z)dz = N(z_{1-\alpha}), \text{ with the quantile } q = z_{1-\alpha}\sigma + \mu , \tag{11}$$

where $\Phi(z)$ is the density function of standard normal distribution, $N(z)$ is a normal cumulative distribution function, R is a random variable of stock *returns* with value r, $f(r)$ is a density function of normal distribution for log *returns* with a mean μ and standard deviation σ, and ξ is the smallest log *return* if given the level of confidence is α. The estimate Value-at-Risk *(*VaR*)* is performed by the formula:

$$VaR = -S_0 \times \xi = -S_0(z_{1-\alpha}\sigma_t + \mu_t), \tag{12}$$

where S_0 is the initial investment, and the minus sign (-) state the losses [7].

Back Test Model. To measure the performance of VaR which has been estimated, can be perform by using the back test method. Suppose that r_t is a profit or loss incurred over a period of time t, and VaR_t is a prediction of *VaR* in time t. Lopez in 1998 introduced the *size-adjusted frequency* model approach as:

$$C_t = \begin{cases} 1+(r_t - VaR_t)^2; & r_t > VaR_t \\ 0; & r_t \le VaR_t \end{cases}. \tag{13}$$

A performance of *VaR* is said to be good if the following *Quadratic Probability Score* (*QPS*) function:

$$QPS = \frac{2}{n}\sum_{i=1}^{n}(C_t - p)^2 , \tag{14}$$

has a small value near to zero. Where p is a probability or a convidence level [7].

3. Results and Discussion

This section is intended to perform the data processing and the discussion, which includes: analyzed data, estimating the mean model, estimating the volatility model, predicting one period ahead, estimating Value-at-Risk, and discussion. It starts with a discussion of the data being analyzed, as follows.

3.1. Data Analyzed

This section purposed in giving the explaination about the analyzed data. The data analyzed here include stock data of PRUF, BBRI, MPPA, BMRI, and INDF, from the period of January 2013 to April 2016. The data is accessed through the website http://www.finance.go.id//. The rate of return (return) of each stock data are specified using equation (1), and then used for modeling the mean and variance below.

3.2. Estimating the Average Model

In this section the aims is to estimate the mean model of the return data following the time series pattern. Return data that has been determined based on the equation (1), then used to estimate the mean model using statistical software Eviews-8. We did the estimation by the following steps. First, the stock return data is tested stationarity by using Augmented Dickey Fuller (ADF) statistic. The test results, indicating that the data return of the five stocks analyzed all have been stationary. Second, we performed the identification and estimation of the mean model. We do the if identification through the sample autocorrelation function (ACF) and partial autocorrelation function (PACF). Based on the pattern of ACF and PACF each return of five stocks, determined tentative model. Further, an estimate is made according to the tentative model, concerning the equation (2). Based on the estimation result, the return data from five stocks follow the following models: ARMA (0,1), ARMA (1,1), ARMA (2,0), ARMA (1,1) and ARMA (2,0) is best. Third, the results of estimators of these models are tested for verification and diagnosis. Verification and diagnostic test performed using correlogram residuals data and Ljung-Box hypotheses test. The test results showed that residuals of these models are white noise. Normality test results for each residual a_t shows normal distribution. So there is no need find other alternative models. The residual data of the five stock returns were analyzed, then it is used for the following volatility modeling.

3.3. Estimating the Volatility Model

In this section, the aim is to estimate the volatility models of return data that follows the time series pattern. The steps description are as follows.

Identification of long memory effects; in the volatility models, the residual data of the mean model on the return of five stocks analyzed are used. Before the estimation of the volatility models, it is necessary to detect the effect of long memory by referring to equations (8), (9), and (10).

For this purpose, we do the fractional differentiation of $\hat{d}$ towards residual data of mean model. The parameter $\hat{d}$ is estimated using GPH method (Geweke Potter and Huddak). This process is performed using the R software, and in Table 1 are the results. To assure that the pattern of *long memory* exists, the hypothesis $H_0 : \hat{d}_i = 0$ is done againts $H_1 : \hat{d}_i \neq 0$, $i = 1,...,5$. Based on the calculation, statistics are obtained z_i, while for the level of significance $\alpha = 0.05$. From the table of standard normal distribution , we obtained the values of $z_{0.05/2} = -1.96$ and $z_{1-0.05/2} = 1.96$. Since z_1, z_2, z_4 and z_5 are greater than $z_{1-0.05/2}$, can be concluded that the test results are significant, means that data of stock return of PRUF, BBRI, BMRI, and INDF have a *long memory* effect. The confident interval is within $-0.5 < d < 0.5$. In the other hand, MPPA stock return rate has no long memory effect because that interval exceeded $-0.5 < d < 0.5$. Next step, the fractionally diferentiated data $\hat{d}_i$ is used to estimate the mean and variance models.

Table 1. The results of identification of long memory effects

Stocks	$\hat{d}_i$	$SE(\hat{d}_i)$	Interval of Convidence	z_i	*Long Memory* Effects
PRFU	-0.096	0.1345	$-0.317<\hat{d}_1<0.125$	5.87	Significant
BBRI	-0.064	0.1331	$-0.283<\hat{d}_2<0.155$	2.48	Significant
MPPA	0.043	0.3453	$-0.525<\hat{d}_3<0.611$	3.12	Not Significant
BMRI	0.132	0.1152	$-0.058<\hat{d}_4<0.322$	3.62	Significant
INDF	0.238	0.1343	$-0.017<\hat{d}_5<0.459$	3.54	Significant

Estimation of Volatility Models; firstly, the detection of *autoregressive conditional heteroscedasticity* (ARCH) element against residual a_t for each stock returns, using ARCH-LM method by using Eviews-8 *software*. The obtained results of χ^2 (obs**R-Square*) of each stock return of PRUF, BBRI, MPPA, BMRI, and INDF are respectively 112.8209; 52.8883; 9.1234; 17.7026 and 9.12340 with probability of each is 0.0000 or less than 5%, means that those have ARCH element.

Secondly, the identification and estimation of the volatility models. Here, we use the models of generalized autoregressive conditional heterscedasticity (GARCH) refer to the equation (5). Based on the squared residuals correlogram a_t^2, that is the ACF and PACF, we select each possible tentative volatility model. We estimate volatility model of each stock return simultaneously with its mean model. Resulted that the best model obtained respectively as follows:

- PRUF follows ARMA(0, 1)-FIGARCH(1, $\hat{d}_1$,1) model which is formulated as:

  $r_t = a_t - 0.070448\, a_{t-1}$;

  $\sigma_t^2 = 0.00000674 + 0.072812 a_{t-1}^2 + 0.897956 \sigma_{t-1}^2 + \varepsilon_t$;

- BBRI follows ARMA(1, 1)-FIGARCH(1, $\hat{d}_2$,1)-M model which is formulated as:

  $r_t = -0.359453 r_{t-1} + 0.465529 a_{t-1} + 0.676944 \sigma_t^2 + a_t$;

  $\sigma_t^2 = 0.0000196 + 0.054776 a_{t-1}^2 + 0.915829 \sigma_{t-1}^2 + \varepsilon_t$;

- MPPA follows ARMA(2, 0)-GARCH(1,1) model which is formulated as:

  $r_t = -0.15440\, r_{t-1} - 0.126754\, r_{t-2} + a_t$;

  $\sigma_t^2 = 0.0000774 + 0.177777 a_{t-1}^2 + 0.488259 \sigma_{t-1}^2 + \varepsilon_t$;

- BMRI follows ARMA(1,1)-FIGARCH(1, $\hat{d}_4$, 1)-M model which is formulated as:

  $r_t = -0.646429 r_{t-1} + 0.700476 a_{t-1} + 3.105569 \sigma_t^2 + a_t$;

  $\sigma_t^2 = 0.0000212 + 0.056086 a_{t-1}^2 + 0.912034 \sigma_{t-1}^2 + \varepsilon_t$;

- INDF follows ARMA(2, 0)-FITGARCH(1, $\hat{d}_5$, 1) model which is formulated as:

  $r_t = -0.013454\, r_{t-1} - 0.004895\, r_{t-2} + a_t$;

  $\sigma_t^2 = 0.0000753 - 0.25404 a_{t-1}^2 + 0.033608 I_{t-1} + 0.545633 \sigma_{t-1}^2 + \varepsilon_t$;

Based on ARCH-LM test, residual ε_t of volatility models of stocks PRUF, BBRI, MPPA, BMRI, and INDF, there is no element of ARCH, and also have white noise. This mean model and volatility, then used to predict the value of $\hat{\mu}_t = \hat{r}_t(1)$ and $\hat{\sigma}_t^2 = \sigma_t^2(1)$ one step ahead.

3.4. The Prediction of Mean and Volatility.

In this section, the aims is to predict the mean values and volatility of stock returns for one-period ahead. By using the mean models and volatility of the stock returns of PRUF, BBRI, MPPA, BMRI, and INDF mentioned previously, we calculate the values of $\hat{\mu}_t = \hat{r}_t(1)$ and $\hat{\sigma}_t^2 = \sigma_t^2(1)$ recursively. The results are given in columns $\hat{\mu}_i$ and $\hat{\sigma}_i^2$ in Table 2.

Table 2. The mean, variance, *VaR*, *QPS* , and ratio

Stock	$\hat{\mu}_i$	$\hat{\sigma}_i^2$	VaR_i	QPS	$\hat{\mu}_i / VaR_i$
PRUF	0.000112	0.00012	0.01791	0.016368	0.006254
BBRI	0.001840	0.00143	0.06037	0.005721	0.030481
MPPA	0.000511	0.00025	0.02550	0.042813	0.020040
BMRI	0.002980	0.00148	0.06030	0.008252	0.049416
INDF	0.000158	0.00025	0.02585	0.013623	0.006112

3.5. The Calculation of Value-at-Risk

This section aims to calculate Value-at-Risk of the stock returns analyzed. The estimator of mean value $\hat{\mu}_i$ $(i = 1,...,5)$ and volatility (variance) $\hat{\sigma}_i^2 (i = 1,...,5)$ from Table 2 are used to calculate the Value-at-Risk. If we determine a 95% confidence level, it is obtained in a standard normal distribution table that $z_{0.05} = -1.645$, and by assuming initial investment is S_0=1 unit, then we do the Value-at-Risk calculation by using equation (12). Value-at-Risk calculation results are given in Table 2, column VaR_i .

Furthermore, the estimator of Value-at-Risk performance needs to be evaluated. If the probability is determined to be $p = 0.05$, then using equations (13) and (14), the values of performance indicators of Value-at-Risk are obtained as given in columns *QPS* in Table 2. By notice to the *QPS* values in Table 2, it is appears that the values are relatively small and tend to be close to zero. This shows that the performance of Value-at-Risk is good. This means that the measurement of risk using Value-at-Risk based on the ARMA-FIGARCH model is good to be used on the five stock returns analyzed.

3.6. Discussion

In this section aims to hold a discussion of the results of analysis of stock return data are analyzed. By taking into account the estimator of the ratio values between the mean and Value-at-Risk in Table 2, the column $\hat{\mu}_i / VaR_i$ the values vary from the largest to the smallest. The largest ratio value is for BMRI stock return that is 0.049416, it illustrates that by investing stocks to BMRI will provide a relatively large expected return compared to the risk value faced. The second largest ratio value is for stock return of BBRI of 0.030481, and the third largest value ratio is for stock return of MPPA of 0.020040. While the smallest ratio value is for INDF stock return of 0.006112, it illustrates that by investing in INDF stock will give relatively small expected return compared to risk value encountered. The second smallest ratio value is for PRUF share return of 0.006254.

These values of ratio are very useful for investors, in deciding of which stock the investors will invest. If investors choose three out of five shares to invest, then the priority order of choice is on stocks of BMRI, BBRI, and MPPA. However, if in the stock selection the investor merely sees large profit, then the priority of choice to invest funds is the stocks of BMRI, BBRI, and MPPA. Meanwhile, if the investors in the stock selection simply see a small risk, then the priority of choice is the stocks of PRUF, INDF, and MPPA. So this analysis of risk investment using Value-at-Risk is useful for investors to assist in the selection of stocks as a means of investment.

4. Conclusions

In this paper, we have analyzed of Value-at-Risk estimation using economics time series approach. The data analyzed is the return of five shares, namely: PRUF, BBRI, MPPA, BMRI, and INDF. Based on the previous analysis, we draw the following conclusions. Return of the five stocks analyzed, respectively follows the models: ARMA(0, 1)-FIGARCH(1, $\hat{d}_1$,1), ARMA(1, 1)-FIGARCH(1, $\hat{d}_2$,1)-M, ARMA(2, 0)-GARCH(1,1), ARMA(1,1)-FIGARCH(1, $\hat{d}_4$, 1)-M, and ARMA(2, 0)-FITGARCH(1, $\hat{d}_5$, 1). Value-at-Risk of the return of five stocks that have been analyzed, respectively are as follows: 0.01791, 0.06037, 0.02550, 0.06030, and 0.02585. Based on the values of QPS that relatively small and close to zero, indicating that the measurement of risk using Value-at-Risk at five stocks analyzed was have a good performance. This research certainly very useful in making decisions for investors in selection of stocks as a means of investment.

Acknowledgment
Further thanks to the Rector, Director of DRPMI, and Dean of FMIPA, Universitas Padjadjaran, which has a grant program of Competence Research of Lecturer of Unpad (*Riset Kompetensi Dosen Unpad*/RKDU) under the coordination of Dr. Sukono, which is a means to increase research activities and publications to researchers at the Universitas Padjadjaran.

References

[1] Al Janabi, M.A.M. 2007. A Value at Risk Approach to Measuring Equity Trading Risk Exposure in Emerging Stock Markets. *AAMJAF*, Vol. 3, No. 1, 1–19, 2007.

[2] Aloui, C. and Hamida, H.B. 2015. Estimation and Performance Assessment of Value-at-Risk and Expected Shortfall Based on Long-Memory GARCH-Class Models. Finance a úver-Czech *Journal of Economics and Finance*, 65, 2015, no. 1

[3] Baillie, R.T. and Morana, C. 2009. Modelling long memory and structural breaks in conditional variances: An adaptive FIGARCH approach. *Journal of Economic Dynamics & Control* 33 (2009) 1577–1592.

[4] Bogdan, S., Baresa, S., and Ivanovic, Z. 2015. Estimating Risk on the Capital Market with VaR Method. *UTMS Journal of Economics* 6 (1): 165–175.

[5] Balibey, M. and Turkyilmaz, S. 2014. Value-at-Risk Analysis in the Presence of Asymmetry and Long Memory: The Case of Turkish Stock Market. *International Journal of Economics and Financial Issues,* Vol. 4, No. 4, 2014, pp.836-848.

[6] Cheung, Y.H. and Powell, R.J. 2009. Anybody can do Value at Risk: A Teaching Study using Parametric Computation and Monte Carlo Simulation, *Australasian Accounting, Business, and Finance Journal*, 6(5), 2012, 101-118. http://ro.uow.edu.au/aabfj

[7] Dowd, K. 2002. *An Introduction to Market Risk Measurement.* United State American: John Wiley & Sons Inc.

[8] Gökbulut, R.I. and Pekkaya, M. 2014. Estimating and Forecasting Volatility of Financial Markets Using Asymmetric GARCH Models: An Application on Turkish Financial Markets. *International Journal of Economics and Finance*; Vol. 6, No. 4; 2014. ISSN 1916-971X E-ISSN 1916-9728 . Published by Canadian Center of Science and Education.

[9] Goudarzi, H. 2010. Modeling Long Memory in The Indian Stock Market using Fractionally Integrated Egarch Model. *International Journal of Trade, Economics and Finance*, Vol.1, No.3, October, 2010. 2010-023X.

[10] Kasman, A. 2009. Estimating Value-at-Risk for the Turkish Stock Index Futures in the Presence of Long Memory Volatility. Central Bank Review ISSN 1303-0701 print / 1305-8800 online. © 2009 Central Bank of the Republic of Turkey. http://www.tcmb.gov.tr/research/review/

[11] Makiel, M. 2012. ARIMA-GARCH Models in Estimating Market Risk Using Value at Risk for the WIG20 Index. *Financial Internet Quarterly, "e-Finanse"* 2012, vol. 8, no. 2. www.e-finanse.com

[12] Naeem, M., Hao Ji, and Liseo, B. 2014. Negative Return-Volume Relationship in Asian Stock Markets: FIGARCH-COPULA Approach. *Eurasian Journal of Economics and Finance, 2(2), 2014, 1-20.* DOI: 10.15604/ejef.2014.02.02.001.

[13] Sukono, Hidayat, Y., Suhartono, Sutijo, B., Bon, A.T., and Supian, S. 2016. Indonesian Financial Data Modeling and Forecasting By Using Econometrics Time Series and Neural Network. *Global Journal of Pure and Applied Mathematics*. ISSN 0973-1768 Volume 12, Number 4 (2016), pp. 3745-3757. © Research India Publications. http://www.ripublication.com/ gjpam.htm

[14] Tsay, R. S., 2005. *Analysis of Financial Time Series. Second Edition*. Hoboken, New Jersey: John Wiley & Sons, Inc.

[15] Turkyilmaz, S. and Balibey, M. 2014. Long Memory Behavior in the Returns of Pakistan Stock Market: ARFIMA-FIGARCH Models. *International Journal of Economics and Financial Issues,* Vol. 4, No. 2, 2014, pp.400-410.

Analysis of Chi-square Automatic Interaction Detection (CHAID) and Classification and Regression Tree (CRT) for Classification of Corn Production

Yuliana Susanti [1, a)], Etik Zukhronah[2, b)], Hasih Pratiwi[3, c)], Respatiwulan[4, d)], and Sri Sulistijowati H.[5, e)]

[1,2,3,4,5]Study Program of Statistics, Universitas Sebelas Maret, Surakarta 57126, Indonesia

[a)]Corresponding author: yulianasusanti@staff.uns.ac.id
[b)]etikzukhronah@staff.uns.ac.id
[c)]hasihpratiwi@staff.uns.ac.id
[d)]respatiwulan@staff.uns.ac.id
[e)]rr.ssh@staff.uns.ac.id

Abstract. To achieve food resilience in Indonesia, food diversification by exploring potentials of local food is required. Corn is one of alternating staple food of Javanese society. For that reason, corn production needs to be improved by considering the influencing factors. CHAID and CRT are methods of data mining which can be used to classify the influencing variables. The present study seeks to dig up information on the potentials of local food availability of corn in regencies and cities in Java Island. CHAID analysis yields four classifications with accuracy of 78.8%, while CRT analysis yields seven classifications with accuracy of 79.6%.

1. Introduction

Corn is primary staple food widely produced in Indonesia. In reference to the 2015 data of Central Agency of Statistics (*Badan Pusat Statistik—BPS*) [3], national corn production of 19.61 million tons increased 3.18% compared to the 2014 data. The increase can be seen from the rise of corn production in regencies/ cities in Java of 0.46 million tons. The regencies or cities producing corn in Java are spread over such provinces as Banten, West Java, Central Java, Yogyakarta, and East Java. If corn production in Java cannot meet the needs of corn, it will depend on the corn production in regions outside Java.

Food availability is one of important instruments to create national food safety. Realizing the importance of attempts to optimize the potential of local food, we require an appropriate method to classify, predict, create production map, and formulate distribution patterns of food in Indonesia (Susanti et al., [11]). With regard to the importance of attempts to optimize corn production in regencies/ cities in Java, methods of exploring data to classify data by building such classification trees as chi Square Automatic Interaction Detection (CHAID) and Classification and regression trees (CRT).

The present study applies CHAID and CRT methods to classify corn production in regencies/ cities in Java in the year of 2015. The classification will result in tree diagrams and variables influencing corn production in regencies/ cities in Java. The classification of corn production in Java will provide positive inputs for policy strategies related to the optimization of corn production in regencies/ cities in Java.

2. Research Method

2.1. CHAID Method

There are three important things in CHAID methods (Sharp, et al.[10], (Bagozzi [2]):

1. The chi-Square test of independence is performed to identify significant independent variables in the data.
2. The Bonferroni correction is carried out when several statistical tests for independence are being performed simultaneously.
3. The CHAID algorithms are used to combine the influencing variable categories.

According to Gallagher [6], CHAID is a method which uses the chi-Square test of independence to examine categorical independent variables in classification individually towards the categorical dependent variables. The statistic for the chi-Square test of independence is

$$\chi^2 = \sum_{i,j} \frac{(O_{ij} - E_{ij})^2}{E_{ij}}$$

where$i$ = 1,2, ...,r, j = 1,2, ...,c, O_{ij} represents the observed frequency of the cell andE_{ij} represents the expected frequency of the cell. Bonferroni correction is a method used when several statistical tests for independence are being performed simultaneously (Sharp, et al.[10]). Galanger [6] defined it as the number of the ways c categories can be merged into r categories if independent variables having c categories exist and can be alleviated into r categories in procedure of merging. Therefore, the obtained p-value of the chi-Square test of independence is multiplication result of p-value and Bonferroni correction in accordance with the variable type.

The procedures in CHAID method is divided into three, i.e. merging, splitting, and stopping (Bagozzi [2], Alkhasawneh [1]).

1. Merging

   This stage involves a significance test of each categorical independent variable towards the dependent variable. The procedures include:

   a. Draw a contingency table for each categorical independent variable and dependent variable.

   b. Compute the chi-Square test of independence for each pair of categorical independent variables and dependent variables by creating 2xJ table contingency, where J is the number of categorical dependent variables.
2. Splitting

   Splitting is carried out by selecting independent variables having the smallest or the most significant p-value which will be used as a node separator. The node separator is selected by comparing p-values obtained in previous stage in each independent variable. In case that there is no independent variable having the most significant p-value, splitting is not carried out and node is determined as the last node due to the absence of branch.
3. Stopping

   It is performed by repeating the stage of merging to analyze the following subgroups. If all of the subgroups have been analyzed, the process is stopped.

The results of segment formation in CHAID method are displayed in a classification tree diagram (Lehmann and Eherler [8]). In the diagram, a dependent variable is represented by a root. The independent variable directly associated with the root is that having significant smallest significant corrected p-value. The classification tree diagram follows top-down rule; the tree diagram is constructed from parents to subgroups split from the parents based on criterion of the number of independent variable categories after merging procedure (Myers [9]). Bagozzi [2] mentioned that a classification tree diagram in CHAID can illustrate:

1. Categorical independent variables which exert influence on merged categorical dependent variables.
2. Data size in a group which is represented by N

2.2. CRT Method

CRT is one of methods of decision tree applied to yield data set as attribute of a classification and to illustrate the relationship between response variables and one or more predictor variables. According to Breiman, et al. [4], if response variables are continuous, regression tree method is employed. Meanwhile, if response variables are categorical, classification tree method is used. The present study involves continuous response variables and therefore regression tree method was applied.

The regression tree method is technically known as binary recursive partitioning. It involves binary splitting since data set called node is always split into multiple partitions called child nodes. Breiman, et al. [4] pointed out that the main elements of the regression tree cover rules for splitting of each node, determining of each node, and determining of value with response estimation for each last node. The CRT method procedures include:

1. Rules of splitting
   According to Breiman et al. [4], a regression tree is constructed through splitting of data groups and a series of binary splitting until the last node is yielded. To split a node into two child notes, the following rules should be followed:
   a. Each of the splitting depends on the value of one predictor variable.
   b. For continuous variables, median of two consecutive observed values of variables is used in the splitting.
   c. For categorical predictor variables, the splitting is derived from splitting probability based on the formation of two child nodes and of two disjoint sets.
2. Process of splitting
   According to Breiman et al., [4] each node's process of splitting is conducted by:
   a. Determining all splitting probabilities in each predictor variable.
   b. Selecting the best splitting in each predictor variable and in the best splitting sets.
   The best splitting is determined with regard to the sum of squared deviation difference of each child node and parent node. The biggest difference will be selected as the best splitting. The error sum of squares on the *t*-th node is determined as a criterion of homogeneity in each node. Mean response in the node is predicted by the mean response in the node and is formulated by
   $$\bar{Y}(t) = \frac{1}{n(t)}\sum_{X_n \in t} Y_n$$
   The sum squares of error on node t is expressed as
   $$SSE(t) = \sum_{X_n \in t} [Y_n - \bar{Y}(t)]^2$$
   For instance, the splitting s splits t into left child tL and right child tR. The function of the splitter is
   $$\emptyset(s,t) = SSE(t) - [SSE(t_L) + SSE(t_R)]$$
   where $\Phi(s,t)$ is the function of splitter in regression tree, SSE(t) is the sum squares of error of parent nodes, SSE(tL) the sum squares of error of left child, and SSE(tR) is the sum squares of error of right child. The best splitting s* is
   $$\emptyset(s^*,t) = max_{s \in S}\emptyset(s,t)$$
   where S denotes a set of splitting probability.
3. Rules of stopping
   The splitting is determined by splitting data into two parts. The parts have both left child and right child. The process of splitting is repeated towards both children until it is impossible to do and finally it is stopped. Breiman, et al. [4] stated that the recursive process will discontinue regarding the minimum number of observations. In addition, the regression tree is discontinued by maximizing the homogeneity of the variance of each node, and it is influenced by variables affecting responses. Unsplittable nodes are termed the last nodes.
4. Rules of determining the estimate value of response

In a regression tree method, if a node is determined as the last node, the response estimation of the observation of the last node is the mean response. Therefore, the mean response is used as the response estimator.

3. Results and Discussion

The data used in the present study involve corn production in regencies and cities in Java Island in the year of 2015 which were obtained from the BPS [3] and Department of Agriculture [5]. The dependent variables include corn production in regencies and cities in Java, while the independent variables include the harvested area, temperature, rain fall, and area altitude of each regency and city.

The areas with corn production in Java in 2015 included 113 regencies and cities. Dependent variable (corn production) is categorized based on median of corn production in all regencies/ cities in Java. The categories are: low (1) (< 6,067 tons) and high (2) (≥ 60,670 tons). For independent variables in CHAID method, the width of harvested area is classified as narrow (1) (< 35,000 ha), moderate (2) (35 ha – 100,000 ha) and wide (3) (> 100,000 ha), rainfall is classified as low (1) (< 85 mm/month), moderate (2) (85 = 200 mm/month and high (3) (> 200 mm/month), while temperature as low (1) (<230C), moderate (2) (230C - 230C) and high (3) (>230C) and area altitude as low (1) (<100 m), moderate (2) (100-600 m), and high (3) (>600 m). In CRT method, independent variables are continuous.

A tree diagram of CHAID method is illustrated in Figure 1and the classification, along with the characteristics (Table 1), as well as the percentage of each classification are demonstrated (Table 2).Table 2 indicated that the biggest percentage of total production of corn in regencies/cities in Java of less than 6,067 tons (low corn production) is on the 1st classification (the total production of corn with low and high rainfall is 84.8%). Meanwhile, the total production of corn in regencies/ cities in Java which is greater than or equal to 6,067 tons (high corn production) is on the 4th classification (corn production with moderate rainfall, moderate and wide harvested area is 79.1%).

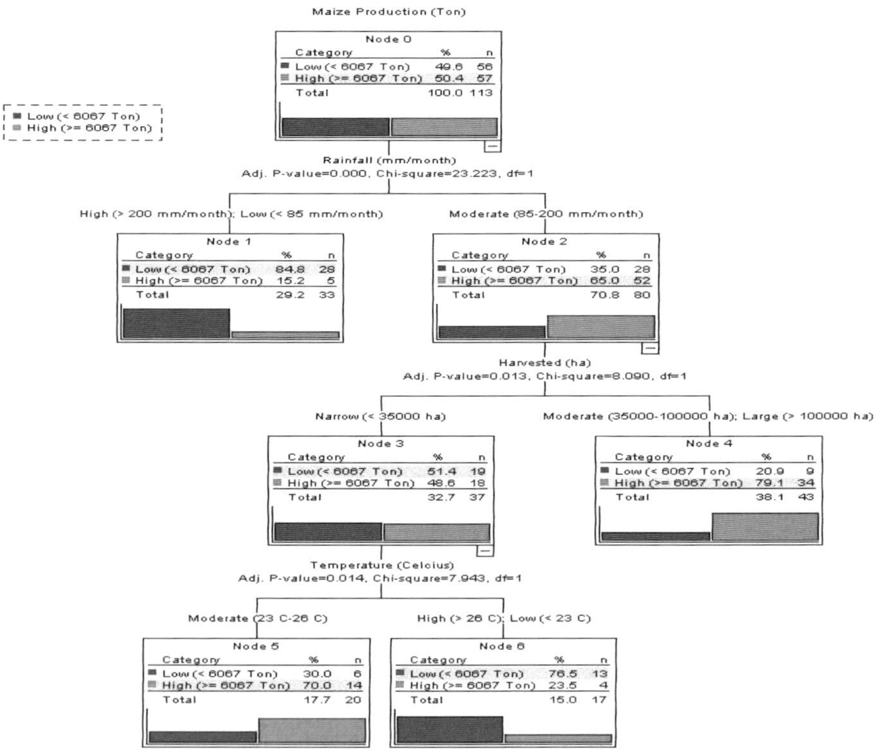

Figure 1. The Classification of Corn Production in Java Using CHAID method

IOP Conf. Series: Journal of Physics: Conf. Series **909** (2017) 012041 doi:10.1088/1742-6596/909/1/012041

Table 1. Classification of Corn Production Based on Tree Diagram of CHAID

Classification	Node	Characteristics
1st	1	areas with low or high rainfall will yield low corn production
2nd	2,3,6	areas with moderate rain fall, narrow harvested area, and high or low temperature will yield low corn production
3rd	2,3,5	areas with moderate rain fall, narrow harvested area , and moderate temperature will yield high corn production
4th	2,4	areas with moderate rain fall, wide harvested area will also yield high corn production

Table 2. Percentage of Classification of Corn Production Using CHAID Method

Classification	Corn Production			
	Low		High	
	Number of regencies/cities	Percentage	Number of regencies/cities	Percentage
1	28	84.8%	5	15.2%
2	13	30,0%	14	70,0%
3	6	76,5%	4	23,5%
4	9	20.9%	34	79.1%

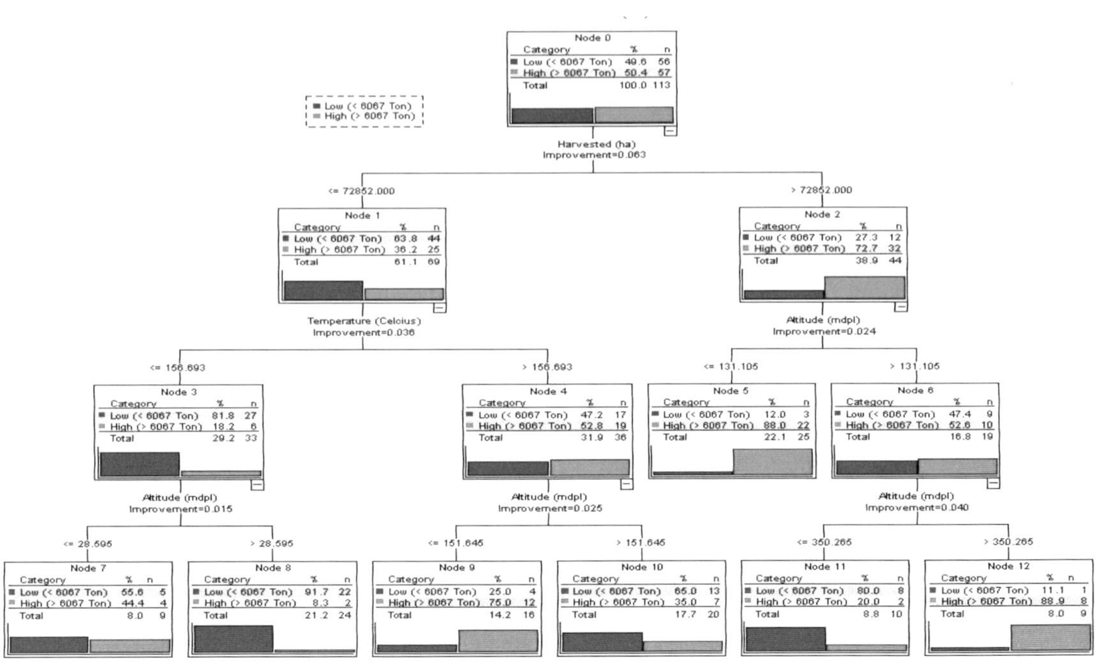

Figure 2. The Classification of Corn Production in Java Using CRT method

Table 3. The Classification based on CRT Tree Diagram

Classification	Node	Characteristics
1st	1,3,7	harvested area (< 72,852 ha), rainfall (< 156,693 mm/month) and area altitude of < 28,595 m will yield low corn production.
2nd	1,3,8	low harvested area, low rainfall and area altitude > 28,595 m will result in lower maize production
3rd	1,4,9	narrow harvested area, high rainfall, and area altitude < 151,645 m will yield high corn production
4th	1,4,10	narrow harvested area and area altitude > 151,645 m will yield low corn production
5th	2,5	wide harvested area and area altitude < 131,105 m will yield high corn production
6th	2,6,11	wide harvested area, area altitude of between 131,105 m and 350,265 m will yield low corn production
7th	2,6,12	wide harvested area and area altitude of > 350,265 m will yield high corn production

Table 4. The Percentage of Classification of Corn Production Using CRT Method

Classification	Corn Production			
	Low (< 6,067 tons)		High (≥ 6,067 tons)	
	Number of regencies/cities	Percentage	Number of regencies/citie	Percentage
1	5	55.6%	4	44.4%
2	22	91,7%	2	8,3%
3	4	25%	12	75%
4	13	65%	7	35%
5	3	12%	22	88%
6	8	80%	2	20%
7	1	11,1%	8	88,9%

The classification of corn production in Java using CRT method is shown by Figure 2. Figure 2 explains the classification and their characteristics while Table 3 and Table 4 demonstrates the percentage. It is clear from Table 4 that the biggest percentage of total production of corn in regencies/ cities in Java of less than 6,067 tons is on the 2st classification. The classification involves corn production in regencies/ cities in Java having width of harvested area less than or equal to 72,852 hectares, rainfall less than 156,693 mm and area altitude more than 28,595 m with percentage of 81.8%. The total production of corn in regencies/ cities in Java with greater than or equal to 6,067 tons is on the 7th classification (corn production in regencies/ cities in Java with width of harvested area greater than 72,852 hectares and area altitude greater than 350.265 m) with percentage of 88,9%. By using CRT method, the classification of data of corn production in regencies/ cities in Java is found out.

CHAID analysis results in four classifications: 1) the first classification indicates that areas with low or high rainfall will yield low corn production, 2) the second classification shows that areas with moderate rainfall, narrow harvested area, and high or low temperature will yield low corn production, 3) the third classification demonstrates that areas with moderate rainfall, narrow harvested area, and moderate temperature will also yield high corn production, and 4) the fourth classification shows that areas with moderate rainfall and wide harvested area will yield high corn production. Variables influencing corn production are rainfall, harvested area, and temperature.

Meanwhile, CRT analysis identifies seven classifications: 1) the first classification shows that areas with narrow harvested area less than 72,852 ha, low rainfall less than 156,693 mm/month and area altitude less than 28,595 m will yield low corn production, 2) the second classification demonstrates that areas with narrow harvested area, low rainfall and area altitude greater than 28,595 m will yield low corn production, 3) the third classification indicates that areas with narrow harvested area, high rainfall and area altitude less than 151,645 m will yield high corn production, 4) the forth classification shows that areas with narrow harvested area, high rainfall and area altitude greater than 151,645 m will yield low corn production, 5) the fifth classification points out that areas with wide harvested area and area altitude less than 131,105 m will yield high corn production , 6) the sixth classification shows that areas with wide harvested area, area altitude between 131,105 m and 350,265 m will yield low corn production, and 7) the seventh classification indicates that areas with wide harvested area and area altitude greater than 350,265 m will yield high corn production.

Table 5. The risk of classifying corn production by CHAID and CRT methods

Method	Accuracy	Standard error
CHAID	0.788	0.038
CRT	0.796	0.038

Table 6. Prediction of classification of corn production by CHAID method

Observed	Predicted		
	Low (< 6067 Ton)	High (>= 6067 Ton)	Percent Correct
Low (< 6,067 tons)	41	15	73.2%
High(>= 6,067 tons)	9	48	84.2%
Overall Percentage	44.2%	55.8%	78.8%

Table 7. Prediction of classification of corn production by CRT method

Observed	Predicted		
	Low (< 6067 Ton)	High (>= 6067 Ton)	Percent Correct
Low (< 6,067 tons)	48	8	85.7%
High(>= 6,067 tons)	15	42	73.7%
Overall Percentage	55.8%	44.2%	79.6%

Based on Table 5, the CHAID and CRT methods provide the same standard error of 0.038. The CHAID method can predict the classification of corn production in Java with a precise maximum of 78.8%, while the CRT method can predict the classification of corn production in Java with a precise maximum of 79.6%.Table 6 shows the percentage for predicting a lower corn production or less than 6067 tons correctly equal to 73.2%, meanwhile for predicting higher corn production or more than 6067 tons correctly equal to 84.2%. This shows that the CHAID model is the best to predict higher corn production or more than 6067 tons with a percentage of 84.2%. Table 7 shows the percentage for predicting lower corn production or less than 6067 tons in Java correctly equal to 85.7%, meanwhile for predicting higher corn production or more than 6067 tons in Java correctly equal to 73.7%. This suggests that the CRT model is the best to predict lower corn production or less than 6067 tons with a percentage of 85.7%.

4. Conclusion
CHAID analysis yields four classifications and variables that affect corn production are rainfall, harvested area, and temperature. CRT analysis yields seven classifications and influencing variables are harvested area, temperature, and area altitude. The CHAID and CRT methods can predict the classification of corn production in Java with a precise maximum of 78.8% and 79.6% respectively.

5. Acknowledgments
The author would like to thank the Indonesian Ministry of Research, Technology and Higher Education as well as Universitas Sebelas Maret for financial support through Grant of Leading Research of University 2017.

6. References
[1] M.Sh. Alkhasawneh, Umi, K.N., Lea, T.T., Nor Ashidi, M.I., and Mohammad, S.A., *Modeling and Testing Landslide Hazard Using Decision Tree*, Journal of Applied Mathematics, Hindawi Pub. Corp., 2014

[2] R.P. Bagozzi, *Advanced Methods of Marketing Research*, Blackwell Publishers Ltd, Oxford, 1994

[3] Badan Pusat Statistik, *Production of Paddy Maize and Soybeans*, www.bps.go.id/release/Production of Paddy Maize and Soybeans, 2015

[4] L. Breiman, Friedman, J.H, Olshen, R.A and Stone, C.J. *Classification And Regression Tree*, Chapman And Hall, New York, 1993

[5] Dinas Pertanian, *Suhu dan Curah Hujan dalam Wilayah Regional Pulau Jawa*, www.distan.go.id/download/ outlooksuhudancurahhujan.pdf, 2015

[6] C.A. Gallagher, *An Iterative Approach to Classification Analysis*, Journal of Applied Statistics,**29**(2000), pp 256-266

[7] G.V. Kass, *An Exploratory Technique for Investigating Large Quantities of Categorical Data*, Journal of Applied Statistics **29** (1980), pp 119-127

[8] T. Lehmann, and Eherler, D., *Responder Profiling with CHAID and Dependency Analysis*, Jena University, Jena, 2001

[9] J.H. Myers, *Segmentation and Positioning for Strategic Marketing Decisions*, American Marketing Association, Chicago, 1996

[10] A. Sharp and J. Romaniuk, *The Performance of Segmentation Variables: A Comparative Study*, www.anzmac1998/Sharp222.pdf, 2002

[11] Y. Susanti, Pratiwi, H., and Handayani, S.S., *Paddy Availability Modeling in Indonesia Using Spatial Regression*, International Journal of Applied Mathematics, **45** (2015) pp 349-360

[12] A. Widiastuti, *Penelusuran Karakteristik Kepemilikan Kartu Kredit dengan Metode CHAID dan Exhaustive CHAID*, Skripsi, Institut Pertanian Bogor, 2001

[13] W. Yuniar, *Penentuan Penciri Keragaman Konsumsi Pangan Rumah Tangga Rawan Pangan dengan Metode CHAID dan Exhaustive CHAID*, Skripsi, Institut Pertanian Bogor, 2014

[14] J. Zhang and Chikaraisih, M., *Interdepenendces Household Residential and Car Ownership Behavior*, Journal of Transport Geography **34** (2014), pp 165-174

IOP Conf. Series: Journal of Physics: Conf. Series **909** (2017) 012042 doi:10.1088/1742-6596/909/1/012042

A staggered grid Jin–Xin relaxation method used to solve the Burgers equation

Maria Anjelina Irawati Ule[1] and Sudi Mungkasi[2]

[1] Department of Mathematics Education, Faculty of Teacher Training and Education, Sanata Dharma University, Mrican, Tromol Pos 29, Yogyakarta 55002, INDONESIA
[2] Department of Mathematics, Faculty of Science and Technology, Sanata Dharma University, Mrican, Tromol Pos 29, Yogyakarta 55002, INDONESIA

E-mail: eleven.ira@gmail.com, sudi@usd.ac.id

Abstract. Burgers equation is considered. Burgers equation has applications in fluid dynamics, traffic flows, etc. It mimics the Navier–Stokes equations. Therefore, in this paper, an accurate method to solve the Burgers equations is sought. Jin–Xin relaxation method has been an active research area for collocated grids. However, the literature suggests that it has never been tested to solve models on staggered grids. In this paper, we test the performance of the Jin–Xin relaxation method in solving a hyperbolic problem on staggered grids. Our research results show that the Jin–Xin method is successful in solving the Burgers equations on staggered grids.

1. Introduction

Burgers equation is a partial differential equation that possesses the behaviour of conservation laws. Conservation laws have been used to model a number of physics problems, which involve the dynamics of mass, momentum, and energy of a physical system [1-2]. The Burgers equation is nonlinear and hyperbolic. Its solution can be either smooth or nonsmooth. It mimics the characteristics of the Navier–Stokes equations [3].

Conservation laws are governed by

$$u_t + f(u)_x = 0 \tag{1}$$

where the time variable $t \in [0, \infty)$ and the space variable $x \in (-\infty, \infty)$. The initial condition is denoted by

$$u(x, 0) = u_0(x) \tag{2}$$

Here, $u = u(x, t)$ is the variable of the conserved quantity which is dependent on the space and time variables. The function $f(u(x, t))$ represents the flux at the space point x and time t. The flux function for the Burgers equation is $f(u) = 1/2u^2$. The function $u_0(x)$ gives the function satisfies the Burgers equation at time $t = 0$. Some previous studies on the Burgers equation have been conducted, for example, see [4-7].

Staggered grid numerical methods have been active research areas for solving conservation law problems. On the other hand, the Jin–Xin relaxation method [8] has been successful to solve conservation laws on collocated grids [9-10]. The Burgers equation has also been attempted to be solved numerically [11-12]. However, the Jin–Xin relaxation method has never been tested to solve the Burgers equation via relaxation system, such as the Jin–Xin relaxation system. The goal of this paper is to implement a staggered grid Jin–Xin relaxation method for solving the Burgers Equation.

The rest of this paper is structured as follows. We recall the Jin–Xin relaxation method on collocated grids in Section 2. We propose the Jin–Xin relaxation method on staggered grids for solving the Burgers equation in Section 3. Numerical results are presented in Section 4. We conclude the paper with some remarks inferred in Section 5.

2. Problem formulation

Conservation laws (1) can be written in a relaxation form. Jin and Xin proposed that its relaxation system is

$$u_t + v_x = 0 \tag{3}$$

$$v_t + au_x = -\frac{1}{\varepsilon}(v - f(u)) \tag{4}$$

where the initial condition is now given by

$$u(x,0) = u_0(x) \tag{5}$$

$$v(x,0) = f\big(u_0(x)\big). \tag{6}$$

Here, $v = v(x,t)$ is a dummy variable and a is a positive constant satisfying

$$-\sqrt{a} \le f'(u) \le \sqrt{a} \tag{7}$$

for all u. The parameter ε is a small positive constant, which is called the relaxation rate. For smooth solutions, equations (3) and (4) is equivalent to

$$u_t + f(u)_x = \varepsilon(au_{xx} - u_n). \tag{8}$$

For $\varepsilon \to 0^+$, equation (8) approaches equation (1). This means that equations (3) and (4) is about equivalent to equation (1). We note that here for $\varepsilon \to 0^+$, we have approximated equation (1) which is nonlinear using the system of equations (3) and (4) which are linear.

3. Numerical method

In this section, we propose the staggered grid Jin–Xin relaxation method used to solve the Burgers equation. To do so, let us consider $u_i^n \approx u(x_i, t^n)$,

$$u_t|_i \approx \left.\frac{\Delta u}{\Delta t}\right|_i = \frac{u_i^{n+1} - u_i^n}{\Delta t}, \tag{9}$$

$$v_x|_n \approx \left.\frac{\Delta v}{\Delta x}\right|_n = \frac{v_{i+1/2}^n - v_{i-1/2}^n}{\Delta x}. \tag{10}$$

Equation (3) can then be written in a fully discrete form as [8, 11-12]

$$\frac{u_i^{n+1}-u_i^n}{\Delta t}+\frac{v_{i+1/2}^n-v_{i-1/2}^n}{\Delta x}=0\,, \tag{11}$$

or

$$u_i^{n+1}=u_i^n-\frac{\Delta t}{\Delta x}\left(v_{i+\frac{1}{2}}^n-v_{i-\frac{1}{2}}^n\right). \tag{12}$$

Equation (4) can be written is a fully discrete form as

$$\frac{v_{i+1/2}^{n+1}-v_{i+1/2}^n}{\Delta t}+a\frac{u_{i+1}^n-u_i^n}{\Delta x}=-\frac{1}{\varepsilon}(v_{i+1/2}^n-f(u_{i+1/2}^n))\,, \tag{13}$$

or

$$v_{i+1/2}^{n+1}=v_{i+1/2}^n-a\frac{\Delta t}{\Delta x}\left[(u_{i+1}^n-u_i^n)-\frac{1}{\varepsilon}(v_{i+1/2}^n-f\big(u_{i+1/2}^n\big)\right]. \tag{14}$$

Here x_i denotes the i –th spatial point with $i=1,2,3,\ldots$ and t^n is the n –th temporal point with $0,1,2,\ldots$ In addition, Δt is the time step and Δx is the cell width. Note that the subscript i is written for the index of the space variable and the superscript n is intended for the index of the time variable.

We obtain that the staggered grid numerical scheme to solve the Burgers equation via its relaxation system is given by equations (12) and (14) simultaneously. All values at the right-hand sides of equations (12) and (14) are known, except the value of $f\big(u_{i+1/2}^n\big)$. The value of $f\big(u_{i+1/2}^n\big)$ is approximated using the average of $f(u_{i+1}^n)$ and $f(u_i^n)$.

4. Numerical results

In this section, we present our numerical results. We implement the proposed staggered grid Jin–Xin relaxation method to solve the (inviscid) Burgers equation

$$u_t+\left(\frac{1}{2}u^2\right)_x=0\,. \tag{15}$$

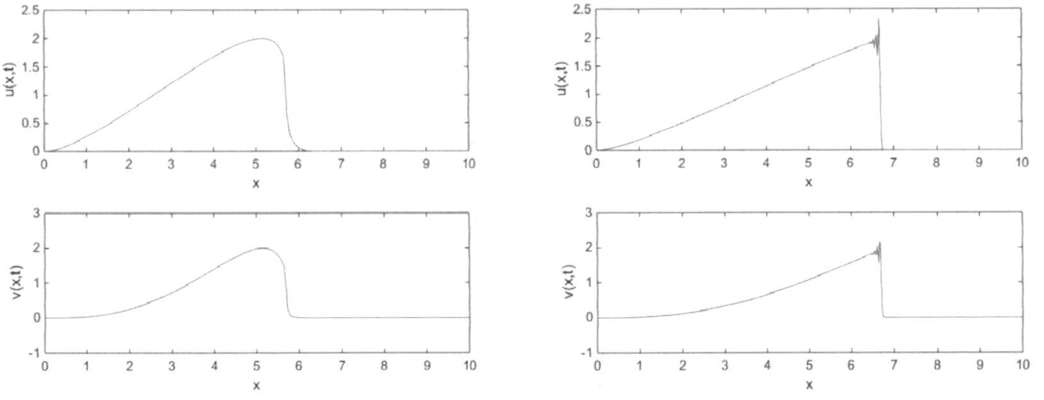

(a). Jin–Xin staggered grid solution at $t=1$. (b). Jin–Xin staggered grid solution at $t=2$.

Figure 1. Jin–Xin solution for the Burgers problem on staggered grids.

Let us consider that the Burgers equation (15) is defined for $0 \leq x \leq 10$, with the initial condition

$$u_0(x) = \begin{cases} 1 - \cos(x) \text{ if } 0 \leq x \leq 2\pi \\ 0 \qquad\qquad \text{if } 2\pi < x \leq 10 \end{cases} \tag{16}$$

and boundary condition

$$u(0,t) = u(10,t) = 0 \tag{17}$$

for all $t > 0$. The cell width is calculated as $\Delta x = L/(N-1)$ where N is the number of spatial points x, and L is the length of the space domain. In our simulation, we take $N = 1000$ and $L = 10$. We choose to take $\Delta t = 0.1\,\Delta x$ and $\varepsilon = 10^{-3}$ for stability.

Figure 1 shows the numerical results of the Burgers problem. For an initially given hyperbolic bump, as time evolves we obtain that the solution moves to the right based on its characteristics until a discontinuity is reached. Then this solution keeps moving to the right with its amplitude gets smaller as time evolves. These results are consistent with those in [11-12].

5. Conclusion

We have proposed a staggered grid Jin–Xin relaxation method for solving the Burgers equation. The Burgers equation is indeed solved numerically in a stable behaviour. As the Burgers equation is a nonlinear hyperbolic conservation law, so our proposed method should be able to be extended for other types of hyperbolic conservation laws. This could be a future research direction regarding this numerical method.

Acknowledgment

Expenses of this research were covered by Sanata Dharma University. Both authors are very grateful to Sanata Dharma University for the financial support.

References

[1] LeVeque R J 1992 *Numerical Methods for Conservation Laws* (Basel: Springer)

[2] Banda M K and Herty M 2012 Adjoint IMEX-based schemes for control problems governed by hyperbolic conservation laws *Computation Optimization and Applications* **51** 909

[3] Landajuela M 2011 Burgers equation *BCAM Internship-Summer 2011* (Bilbao: Basque Center for Applied Mathematics)

[4] Basto M, Semiano V and Calheiros F L 2007 Numerical study of modified Adomian's method applied to Burgers equation *Journal of Computational and Applied Mathematics* **206** 927

[5] Saka B 2008 A numerical study of the Burgers' equation *Journal of the Franklin Institute* **345** 328

[6] Joseph K T and Sachdev P L 2010 Initial boundary value problems for scalar and vector Burgers equations *Studies in Applied Mathematics* **106** 481

[7] Mungkasi S 2015 Some advantages of implementing an adaptive moving mesh for the solution to the Burgers equation *IOP Conference Series: Materials Science and Engineering* **78** 012031

[8] Jin S and Xin Z 1995 The relaxation schemes for systems of conservation laws in arbitrary space dimensions *Communication on Pure and Applied Mathematics* **48** 235

[9] Murthy A S V 2007 An alternate relaxation approximation to conservation laws *Journal of Computational and Applied Mathematics* **203** 437

[10] Zou Q, Zhao H and Wang T 2012 The Jin–Xin relaxation approximation of scalar conservation

laws in several dimensions with large initial perturbation *Journal of Differential Equations* **253** 563

[11] Mungkasi S and Laras F 2017 Jin–Xin relaxation solution to the spatially-varying Burgers equation in one dimension *Journal of Physics: Conference Series* **795** 012043

[12] Mungkasi S, Harini B H and Wiryanto L H 2017 Jin–Xin relaxation method used to solve the one-dimensional inviscid Burgers equation *Journal of Physics: Conference Series* **856** 012007

IOP Conf. Series: Journal of Physics: Conf. Series **909** (2017) 012043 doi:10.1088/1742-6596/909/1/012043

Estimation of Mean Population in Small Area with Spatial Best Linear Unbiased Prediction Method

Syahril Ramadhan[1], Titin Siswantining, and Saskya Mary Soemartojo

Department of Mathematics, Faculty of Mathematics and Natural Sciences, Universitas Indonesia (UI)
Kampus Baru UI Depok West Java –16424, INDONESIA

E-mail: syahril.ramadhan@sci.ui.ac.id, titin@sci.ui.ac.id, saskya@sci.ui.ac.id

Abstract. Survey is aimed at estimating population parameters such as the total as well as the mean of an area with a large sample size. One approach in estimating population parameters are obtained through direct estimation methods. However, there is a problem when the direct estimation is used for a small area, which will cause a large standard error. This problem was addressed by developing a method of parameter estimation known as the Small Area Estimation (SAE) method. In this paper, we will describe the procedure to find the mean population estimate in a small area using Spatial Best Linear Unbiased Prediction (Spatial BLUP) method that follows Simultaneously Autoregressive (SAR) model. In general, this procedure begins with the definition of an area-level model. Then, the area-level model is expanded by the addition of spatial effect into the random effects of the area. In the end, the spatial model of the area level is used as the basis for estimating the mean population in small areas.

1. Introduction

Surveys are an important part of the data. Survey implementation is aimed at estimating population parameters. The results of this survey can be used to obtain a reliable estimate of the total as well as the mean population value of an object with a large sample size. The object used as a population is called a domain (area).

One approach to estimating population value parameters is based on the application of design-based models such as simple random sampling, stratified random sampling, etc. where the estimators generated through this approach are called direct estimation. Direct estimation is a method of estimation in an area based on information of sample data from that area. However, there are problems where such direct estimates are used for an area with a small sample size, which will lead to a large standard error (Ghosh & Rao, 1994). An area is regarded as "small" if the area-specific sample is not large enough to support direct estimates of adequate precision (Rao, 2003). Or in other words, a small area can be expressed as an area with a small sample size. We use the term *small area* to denote any area for which direct estimates of adequate precision cannot be produced.

In addition to causing a large standard error, Rao (2003) also adds that direct estimation methods cannot be performed on non-selected areas as samples, in the absence of data that can be used to make the estimation. These problems were addressed by developing a method of parameter estimation known as the Small Area Estimation (SAE) method. According to Rao (2003), the SAE method can be defined as a method for estimating population parameters in a relatively small area of the survey sampling by utilizing information from outside the area in the form of other information from a similar area, from within the area itself in the form of a previous survey in the same area, and from outside the survey in the form of supporting variables related to the variable to be expected. Chand & Alexander (1995) also mentioned that the SAE procedure essentially harnesses the power of the surrounding area information and data sources outside of the area whose statistical value is to be obtained through the establishment of appropriate models to improve the effectiveness of the sample size. This statement means that using a small sample size can get a good estimator. The estimation of the parameters and their inferences by using such additional information is called indirect estimation. Indirect estimation is a method of estimating population parameters in an area by involving additional information from within the area of concern, other areas, or other surveys (Rao, 2003).

There are two basic assumptions in the development of a model for the estimation of small areas, namely that the variability within the observed small area of the variable can be explained by the corresponding variability relationships in the supplemental information called fixed effects. Another assumption is that the small area-specific variability cannot be explained by additional information and is a small area random effect. The combined two assumptions form a linear mixed model.

Fay & Herriot (1979) is the first researcher to develop SAE based on a mixed linear model. This model came to be known as the Fay-Herriot model and became the basis in SAE which assumed that the random effects of the areas are independent. But in some cases, these assumptions are often violated. General, the variance of an area is affected by the surrounding area so that spatial effect can be incorporated into random effects. Spatial effect is common between one area and another, or it can be said that one area affects the other. Moreover, there is no reason to exclude the assumption that the random effect between the neighboring areas are correlated, and that correlation decays to zero as distance increases. Based on this, spatial information can be used in the SAE model. By incorporating the effect of spatial dependence into the random effects of the area, SAE will be obtained by considering the effect of spatial dependence across areas (Asfar, 2016).

One method in SAE is the best linear unbiased prediction method (BLUP). The BLUP estimator taking into account the random effects of spatially dependent areas is known as the spatial best linear unbiased prediction (Spatial BLUP). Researchers have previously mentioned that the use of spatial information can minimize variance.

The paper is organized as it follows. The spatial model of the area level is described in Section 2. In Section 2 discussed: small area model, area-level model, Fay-Herriot model, SAR model, and the spatial model of the area level. Section 3 discusses the BLUP method, and Section 4 shows the Spatial BLUP estimator. The conclusion can be found in Section 5.

2. The spatial model of the area level

In this section, we will first discuss the understanding of the area-level model that is the basis of the models to be discussed in this paper. Furthermore, the area-level model is expanded with the addition of spatial effects to the random effects of the area. Finally, the mean population parameter in a small area is presumably using the Spatial Best Linear Unbiased Prediction (Spatial BLUP) method.

2.1. Small area model

Small-area estimation models are classified into two types: the area-level model and the unit-level model (Rao, 2003). Area-level model is used if the auxiliary data corresponding to observed variable data is not available at the unit level of observation, whereas unit-level model is used if the auxiliary data corresponding to observed variable data is available up to the unit level of observation.

2.1.1. Area-level model. Area-level model is models based on the availability of the auxiliary data that exist only for a given area level. The auxiliary variable is a collection of variables that contain information from each unit of population and not a variable of concern. If the auxiliary variables are included in the model, it can increase the information in predicting the parameters in the SAE (Small Area Estimation) method, so that the resulting estimates will be better. A critical assumption for area-level models, we assume the absence of informative sampling of the areas in situations where only some of the areas are selected to the sample, that is, the sample area value (the direct estimates) obey the assumed population model (Rao, 2003).

Suppose a population of U is divided into the M area, where there is only a $m \leq M$ area that allows surveying and obtains an auxiliary variables $\mathbf{x}_i = (x_{i1}, x_{i2}, \ldots, x_{ip})^T$ which is the auxiliary variables of the i-th area. If the parameter to be expected θ_i (θ_i represents the parameter of the i-th area) to know its value, then the area-level model assumes that θ_i will be interconnected with the auxiliary variables $\mathbf{x}_i$ through the following model:

$$\theta_i = \mathbf{x}_i^T \boldsymbol{\beta} + b_i v_i, \qquad i = 1,2, \ldots, m \qquad (1)$$

Where $\boldsymbol{\beta}$ is the $p \times 1$ vector of regression coefficients, b_i is the weight of the random effects for the i-th area, m is the number of small areas, and v_i is the random effects of the area in the i-th area with assumption $v_i \sim_{iid} N(0, \sigma_v^2)$.

From now on, the model of equation (1) is referred to as the *linking model* between parameter θ_i and the auxiliary variables $\mathbf{x}_i$. Availability of good auxiliary data and determination of suitable linking models are crucial to the formation of indirect estimators.

The assumption in making the conclusions about the population based on the model in equation (1) is the availability of direct estimator $\hat{\theta}_i$ and can be written as follows:

$$\hat{\theta}_i = \theta_i + e_i, \qquad i = 1,2, \ldots, m \qquad (2)$$

Where e_i is the independent sampling error in the i-th area with assumption $e_i | \theta_i \sim_{ind} N(0, \psi_i)$. From now on, the model of equation (2) is referred to as the *sampling model* showing the relationship between direct observation of $\hat{\theta}_i$ and parameter θ_i.

Equations (1) and (2) can be combined so that the model is obtained:

$$\hat{\theta}_i = \mathbf{x}_i^T \boldsymbol{\beta} + b_i v_i + e_i, \qquad i = 1,2, \ldots, m \qquad (3)$$

With the additional assumption that v_i and e_i are independent. From now on, the model of equation (3) is referred to as the *area-level model*.

2.2 Fay-Herriot Model

Research on SAE was first performed by Fay and Herriot (1979) i.e., by estimating the per capita income of a small area based on US Census Bureau survey data. This model is then known as the Fay-Herriot model, where in this model the weight of random effect between areas b_i is assumed to have the same effect, that is value $b_i = 1$ in equation (3).

Similar to the area-level model in SAE, the Fay-Herriot model consists of two components, the linking model (1) and the sampling model (2) where the sampling error variance ψ_i in the sampling model is assumed to be known. Through this definition, the Fay-Herriot model is defined as follows:

$$\hat{\theta}_i = \mathbf{x}_i^T \boldsymbol{\beta} + v_i + e_i, \qquad i = 1,2, \ldots, m \qquad (4)$$

Where assumptions on the linking model and sampling model are maintained in the Fay-Herriot model. In this paper, the model in equation (4) is used as the basis of the area-level model.

The effect of random area v_i on equation (4) is still assumed to be independent between areas. In fact, the spatial relationship between small areas becomes an interesting thing to observe. Therefore, in this paper will be discussed further about the expansion of v_i as a random effect area that is no longer independent but has a spatial dependence.

2.3. SAR Model

In fact, it is reasonable to say that there is a dependency between the adjacent area and the dependence will decrease as the distance between the areas increases. Inter-area spatial dependence needs to be explained through a model known as spatial modeling. In this paper, the spatial model that focuses on this paper is the *simultaneous autoregressive model* (SAR model).

The SAR model was introduced by Anselin in 1992 where the vector of random effects of area **v** is defined as follows:

$$\mathbf{v} = (\mathbf{I} - \rho\mathbf{W})^{-1}\mathbf{u} \tag{5}$$

Where **v** is the $m \times 1$ vector of dependent random variables according to an SAR error process with mean **0** and variance-covariance matrix $\mathbf{G} = \sigma_u^2\left[(\mathbf{I} - \rho\mathbf{W})(\mathbf{I} - \rho\mathbf{W}^{\mathrm{T}})\right]^{-1}$. In this matrix, ρ is the spatial autoregressive coefficient, **W** is the $m \times m$ proximity matrix, **I** is a $m \times m$ identity matrix, and **u** is the $m \times 1$ vector of independent error term with zero mean and constant variance σ_u^2.

A variance-covariance matrix **G** can be derived directly with some matrix manipulation:

$$\begin{aligned}\mathbf{G} = E[\mathbf{v}\mathbf{v}^T] &= E[(\mathbf{I} - \rho\mathbf{W})^{-1}\mathbf{u}\mathbf{u}^T((\mathbf{I} - \rho\mathbf{W})^{-1})^T] \\ &= (\mathbf{I} - \rho\mathbf{W})^{-1}E[\mathbf{u}\mathbf{u}^T]((\mathbf{I} - \rho\mathbf{W})^{-1})^T \\ &= (\mathbf{I} - \rho\mathbf{W})^{-1}\sigma_u^2\mathbf{I}((\mathbf{I} - \rho\mathbf{W})^{-1})^T \\ &= \sigma_u^2(\mathbf{I} - \rho\mathbf{W})^{-1}((\mathbf{I} - \rho\mathbf{W})^T)^{-1} \\ &= \sigma_u^2\left[(\mathbf{I} - \rho\mathbf{W})(\mathbf{I} - \rho\mathbf{W}^{\mathrm{T}})\right]^{-1}\end{aligned} \tag{6}$$

The spatial autoregressive coefficient ρ in equation (6) defines the strength of the spatial relationship among the random effects associated with neighboring areas. The value of ρ ranges from -1 to 1. The value of $\rho > 0$ indicates that an area with a high parameter value tends to be surrounded by another area with a high parameter value and vice versa. On the other hand, the value of $\rho < 0$ indicates that an area with a high parameter value is surrounded by another area with a low parameter value or vice versa.

The proximity matrix **W** is an significant component in spatial data modeling in which the data contain spatial dependence. This matrix declares neighboring structures of small areas (such as correlation, distance, threshold, etc.). A general proximity matrix spatial weight matrix can be defined by a symmetric binary contiguity matrix, which can be generated from topological information provided by the geographical information system (GIS) based on adjacency criteria: the element of the proximity matrix $\{w_{ij}\}$ is one if area i is adjacent to area j, and zero otherwise.

2.4. The spatial model of the area level

This section will explain the extension of the area-level model with the addition of spatial random effects.

The SAR model (5) is associated with the parameter you want to know θ_i through the linking model, in the form of a matrix can be written as follows:

$$\boldsymbol{\theta} = \mathbf{X}\boldsymbol{\beta} + \mathbf{Z}(\mathbf{I} - \rho\mathbf{W})^{-1}\mathbf{u} \tag{7}$$

Where **X** is the $m \times p$ matrix of the area specific auxiliary covariates $\mathbf{x}_i = \left(x_{i1}, x_{i2}, \dots, x_{ip}\right)^T$, $\boldsymbol{\beta}$ is the $p \times 1$ vector of regression coefficients, **Z** is a $m \times m$ matrix of known positive constant, and $\mathbf{v} = (\mathbf{I} - \rho\mathbf{W})^{-1}\mathbf{u}$ is the $m \times 1$ vector of dependent random variables according to an SAR error process with mean **0** and variance-covariance matrix $\mathbf{G} = \sigma_u^2\left[(\mathbf{I} - \rho\mathbf{W})(\mathbf{I} - \rho\mathbf{W}^{\mathrm{T}})\right]^{-1}$. Moreover, it assumes that direct estimators $\hat{\theta}_i$ are available and follow the following sampling model:

$$\hat{\boldsymbol{\theta}} = \boldsymbol{\theta} + \mathbf{e} \tag{8}$$

Where **e** represents the $m \times 1$ matrix of independent sampling errors with mean **0** and known $m \times m$ variance matrix $\mathrm{diag}(\psi_i)$. Combining (7) and (8), with **e** independent of **v**, the resulting model is:

$$\hat{\boldsymbol{\theta}} = \mathbf{X}\boldsymbol{\beta} + \mathbf{Z}(\mathbf{I} - \rho\mathbf{W})^{-1}\mathbf{u} + \mathbf{e} \tag{9}$$

Where $m \times m$ variance-covariance matrix for **v** and **e**, respectively, as follows:

$$\mathbf{G} = \sigma_u^2\left[(\mathbf{I} - \rho\mathbf{W})(\mathbf{I} - \rho\mathbf{W}^{\mathrm{T}})\right]^{-1} \tag{10}$$

and

$$\mathbf{R} = \mathrm{diag}(\psi_i). \tag{11}$$

Then, the covariance matrix of studied variable is:

$$\mathbf{V} = \mathbf{R} + \mathbf{ZGZ}^T = \mathrm{diag}(\psi_i) + \mathbf{Z}\sigma_u^2\left[(\mathbf{I} - \rho\mathbf{W})\left(\mathbf{I} - \rho\mathbf{W}^{\mathrm{T}}\right)\right]^{-1}\mathbf{Z}^T \tag{12}$$

The model in equation (9) is then later used in this paper as the spatial model of the area level, from now on called the spatial Fay-Herriot model.

3. Best linear unbiased prediction (BLUP) method

In the previous section, we have introduced a small area model that can be viewed as a special case of a general mixed linear model involving both fixed and random effects. In fact, small area parameters can be written as linear combinations of fixed effects and random effects. One of the interesting properties of a mixed linear model is its ability to estimate linear combinations of fixed effects and random effects. One method for estimating parameters based on a linear mixed model has been developed by Henderson (1975), the Best Linear Unbiased Prediction (BLUP) method.

The BLUP estimator minimizes the Mean Square Error (MSE) among the class of linear unbiased estimators and do not depend on the normality of the random effects. But, they depend on the variances (and covariances) of random effects which can be estimated. Maximum likelihood method can be used to estimate the variance and covariance components, assuming normality. Using these estimated components in the BLUP estimator, we obtain a two-stage estimator which is referred to as empirical BLUP or EBLUP estimator (Harville, 1991).

The BLUP context is the following mixed linear model:

$$\widehat{\boldsymbol{\theta}} = \mathbf{X}\boldsymbol{\beta} + \mathbf{Zv} + \mathbf{e} \tag{13}$$

where $\widehat{\boldsymbol{\theta}}$ is a vector of n observable random variables, $\boldsymbol{\beta}$ is the vector of p unknown parameters having fixed values (fixed effects), $\mathbf{X}$ and $\mathbf{Z}$ are known matrices, and $\mathbf{v}$ and $\mathbf{e}$ are vectors of q and n, respectively, unobservable random variables (random effects) such that $E(\mathbf{v}) = \mathbf{0}$, $E(\mathbf{e}) = \mathbf{0}$, and

$$\mathrm{Var}\begin{bmatrix}\mathbf{v}\\ \mathbf{e}\end{bmatrix} = \begin{bmatrix}\mathbf{G} & \mathbf{0}\\ \mathbf{0} & \mathbf{R}\end{bmatrix}\sigma^2 \tag{14}$$

where $\mathbf{G}$ and $\mathbf{R}$ are known positive definite matrices, and σ^2 is a positive constant.

In the BLUP estimator, *best* states that the resulting estimates have a minimum MSE value in other unbiased estimators, *linear* states that the resulting estimator is a linear function of the data $\widehat{\boldsymbol{\theta}}$, and *unbiased* states that the expected value of the estimate is equal to the value of the parameter.

Mathematically, the BLUP estimates $\widehat{\boldsymbol{\beta}}$ of $\boldsymbol{\beta}$ and $\hat{\mathbf{v}}$ of $\mathbf{v}$ are defined as solutions to the following simultaneous equations which were given by Henderson (1950) and is known as the Henderson linear mixed model. The Henderson linear mixed model is used to find $\widehat{\boldsymbol{\beta}}$ and $\hat{\mathbf{v}}$. Here is the Henderson linear mixed model:

$$\mathbf{X}^{\mathrm{T}}\mathbf{R}^{-1}\mathbf{X}\boldsymbol{\beta} + \mathbf{X}^{\mathrm{T}}\mathbf{R}^{-1}\mathbf{Z}\mathbf{v} = \mathbf{X}^{\mathrm{T}}\mathbf{R}^{-1}\widehat{\boldsymbol{\theta}} \tag{15}$$

$$\mathbf{Z}^{\mathrm{T}}\mathbf{R}^{-1}\mathbf{X}\boldsymbol{\beta} + \left(\mathbf{Z}^{\mathrm{T}}\mathbf{R}^{-1}\mathbf{Z} + \mathbf{G}^{-1}\right)\mathbf{v} = \mathbf{Z}^{\mathrm{T}}\mathbf{R}^{-1}\widehat{\boldsymbol{\theta}} \tag{16}$$

the details of the calculation are reported in Appendix A.

3.1. Estimator for $\boldsymbol{\beta}$

First, there will be an elimination $\mathbf{v}$ between equation (15) and equation (16). To do this, multiply equation (16) with $\mathbf{X}^{\mathrm{T}}\mathbf{R}^{-1}\mathbf{Z}\left(\mathbf{Z}^{\mathrm{T}}\mathbf{R}^{-1}\mathbf{Z} + \mathbf{G}^{-1}\right)^{-1}$ and perform a subtraction operation. The following results are obtained:

$$\begin{aligned}\mathbf{X}^{\mathrm{T}}\mathbf{R}^{-1}\mathbf{X}\boldsymbol{\beta} - \mathbf{X}^{\mathrm{T}}\mathbf{R}^{-1}\mathbf{Z}\left(\mathbf{Z}^{\mathrm{T}}\mathbf{R}^{-1}\mathbf{Z} + \mathbf{G}^{-1}\right)^{-1}\mathbf{Z}^{\mathrm{T}}\mathbf{R}^{-1}\mathbf{X}\boldsymbol{\beta} = \\ \mathbf{X}^{\mathrm{T}}\mathbf{R}^{-1}\widehat{\boldsymbol{\theta}} - \mathbf{X}^{\mathrm{T}}\mathbf{R}^{-1}\mathbf{Z}\left(\mathbf{Z}^{\mathrm{T}}\mathbf{R}^{-1}\mathbf{Z} + \mathbf{G}^{-1}\right)^{-1}\mathbf{Z}^{\mathrm{T}}\mathbf{R}^{-1}\widehat{\boldsymbol{\theta}}.\end{aligned} \tag{17}$$

Now using a matrix identity which is commonly used in this subject area:

$$\mathbf{V}^{-1} := (\mathbf{ZGZ}^T + \mathbf{R})^{-1} = \mathbf{R}^{-1} - \mathbf{R}^{-1}\mathbf{Z}(\mathbf{Z}^T\mathbf{R}^{-1}\mathbf{Z} + \mathbf{G}^{-1})^{-1}\mathbf{Z}^T\mathbf{R}^{-1} \tag{18}$$

gives

$$\mathbf{X^TR^{-1}X}\boldsymbol{\beta} - \mathbf{X^T\{R^{-1} - V^{-1}\}X}\boldsymbol{\beta} = \mathbf{X^TR^{-1}}\widehat{\boldsymbol{\theta}} - \mathbf{X^T\{R^{-1} - V^{-1}\}}\widehat{\boldsymbol{\theta}}$$

or can be written as:

$$\mathbf{X^TV^{-1}X}\boldsymbol{\beta} = \mathbf{X^TV^{-1}}\widehat{\boldsymbol{\theta}},$$

So that the BLUP estimator for $\boldsymbol{\beta}$ is:

$$\widehat{\boldsymbol{\beta}} = \left(\mathbf{X^TV^{-1}X}\right)^{-1}\mathbf{X^TV^{-1}}\widehat{\boldsymbol{\theta}}. \tag{19}$$

3.2. *Estimator for* v

Based on equation (16), where there is no fixed effect, the estimator for random effects is given as follows:

$$\mathbf{Z^TR^{-1}X}\boldsymbol{\beta} + \left(\mathbf{Z^TR^{-1}Z} + \mathbf{G^{-1}}\right)\mathbf{v} = \mathbf{Z^TR^{-1}}\widehat{\boldsymbol{\theta}}$$

or can be written as:

$$\mathbf{v} = \left(\mathbf{Z^TR^{-1}Z} + \mathbf{G^{-1}}\right)^{-1}\mathbf{Z^TR^{-1}}\left(\widehat{\boldsymbol{\theta}} - \mathbf{X}\boldsymbol{\beta}\right) \tag{20}$$

Now using a matrix identity which is commonly used in this subject area:

$$\mathbf{GZ}^T\mathbf{V}^{-1} = (\mathbf{Z}^T\mathbf{R}^{-1}\mathbf{Z} + \mathbf{G}^{-1})^{-1}\mathbf{Z}^T\mathbf{R}^{-1} \tag{21}$$

gives

$$\hat{\mathbf{v}} = \mathbf{GZ^TV^{-1}}\left(\widehat{\boldsymbol{\theta}} - \mathbf{X}\boldsymbol{\beta}\right). \tag{22}$$

If $\boldsymbol{\beta}$ in equation (22) is substituted by $\widehat{\boldsymbol{\beta}}$, then the estimator for **v** is the BLUP estimator. So that the BLUP estimator for **v** is:

$$\hat{\mathbf{v}} = \mathbf{GZ^TV^{-1}}\left(\widehat{\boldsymbol{\theta}} - \mathbf{X}\widehat{\boldsymbol{\beta}}\right). \tag{23}$$

Then, the results of these two BLUP estimators will be used on the Fay-Herriot spatial model that has been defined earlier in this paper.

4. Implementation of BLUP method in the spatial model of the area level

Based on the model of equation (9) and the predicted result in Section 3, it is substituted by the covariance matrix (12) on the resultant estimator. The Spatial Best Linear Unbiased Predictor (Spatial BLUP) estimator of θ_i is:

$$\begin{aligned}\tilde{\theta}_i^S(\sigma_u^2,\rho) &= \mathbf{x}_i\widehat{\boldsymbol{\beta}} \\ &+ \mathbf{b}_i^T\left\{\sigma_u^2\left[(\mathbf{I}-\rho\mathbf{W})\left(\mathbf{I}-\rho\mathbf{W}^{\mathrm{T}}\right)\right]^{-1}\right\}\mathbf{Z}^T\left\{\mathrm{diag}(\psi_i)\right. \\ &\left. + \mathbf{Z}\sigma_u^2\left[(\mathbf{I}-\rho\mathbf{W})\left(\mathbf{I}-\rho\mathbf{W}^{\mathrm{T}}\right)\right]^{-1}\mathbf{Z}^T\right\}^{-1}\left(\widehat{\boldsymbol{\theta}} - \mathbf{X}\widehat{\boldsymbol{\beta}}\right)\end{aligned} \tag{24}$$

where $\widehat{\boldsymbol{\beta}} = \left(\mathbf{X^TV^{-1}X}\right)^{-1}\mathbf{X^TV^{-1}}\widehat{\boldsymbol{\theta}}$ and $\mathbf{b}_i^T$ is $1 \times m$ vector $(0,0,\dots,0,1,0,\dots,0)$ with 1 referred to i-th area. Spatial BLUP will be the same value as BLUP under the model of random effect area when $\rho = 0$ (Pratesi & Salvati, 2005).

In this paper, the estimator in equation (24) is the mean population parameter estimator in a small area of the spatial Fay-Herriot model based on the BLUP method.

5. Final Remark

Our result shows that the Spatial BLUP estimator obtained concerning spatial dependence across areas, where the use of the available spatial auxiliary information can reduce both the bias and the sampling error in small area estimation. Moreover, give additional help in borrowing strength from the related area in the estimation of small area parameters.

However, the SBLUP estimator is still dependent on the component of the variance of the random area error σ_u^2 and spatial autoregressive coefficient ρ whose value is predictable.

Next step is to explore what happens when random area effects follow a *conditional autoregressive* (CAR) process or when the Fay-Herriot model with correlated sampling errors.

Acknowledgments
The authors are grateful to our campus for support. We also thank the referees for the many suggestions which improved the article.

Appendix A
It is assumed that $\mathbf{v}$ and $\mathbf{e}$ are normally distributed, and will maximize the joint probability density function (pdf) of $\widehat{\boldsymbol{\theta}}$ (which depends on $\boldsymbol{\beta}$ and $\mathbf{v}$) and $\mathbf{v}$.

Noted that

$$E(\mathbf{v}) = \mathbf{0}$$
$$\mathrm{Var}(\mathbf{v}) = \mathbf{G}\sigma^2$$

and

$$\begin{aligned} E(\widehat{\boldsymbol{\theta}}|\boldsymbol{\beta},\mathbf{v}) &= E(\mathbf{X}\boldsymbol{\beta} + \mathbf{Zv} + \mathbf{e}|\boldsymbol{\beta},\mathbf{v}) \\ &= \mathbf{X}\boldsymbol{\beta} + \mathbf{Zv} \\ \mathrm{Var}(\widehat{\boldsymbol{\theta}}|\boldsymbol{\beta},\mathbf{v}) &= \mathrm{Var}(\mathbf{X}\boldsymbol{\beta} + \mathbf{Zv} + \mathbf{e}|\boldsymbol{\beta},\mathbf{v}) \\ &= \mathbf{R}\sigma^2. \end{aligned}$$

Based on the multivariate normal distribution, pdf for $\mathbf{v}$ and $\widehat{\boldsymbol{\theta}}|(\boldsymbol{\beta},\mathbf{v})$, respectively is:

$$\begin{aligned} f(\mathbf{v}) &= \frac{1}{\sqrt{(2\pi)^q}|\mathbf{G}\sigma^2|^{\frac{1}{2}}} \exp\left[-\frac{1}{2}(\mathbf{v})^T(\mathbf{G}\sigma^2)^{-1}(\mathbf{v})\right] \\ &= \frac{1}{(2\pi)^{\frac{q}{2}}(\sigma^2)^{\frac{q}{2}}|\mathbf{G}|^{\frac{1}{2}}} \exp\left[-\frac{1}{2\sigma^2}(\mathbf{v})^T(\mathbf{G})^{-1}(\mathbf{v})\right] \\ &= \frac{1}{(2\pi\sigma^2)^{\frac{q}{2}}|\mathbf{G}|^{\frac{1}{2}}} \exp\left[-\frac{1}{2\sigma^2}(\mathbf{v})^T(\mathbf{G})^{-1}(\mathbf{v})\right] \end{aligned} \tag{A.1}$$

and

$$f(\widehat{\boldsymbol{\theta}}|\boldsymbol{\beta},\mathbf{v}) = \frac{1}{(2\pi\sigma^2)^{\frac{n}{2}}|\mathbf{R}|^{\frac{1}{2}}} \exp\left[-\frac{1}{2\sigma^2}(\widehat{\boldsymbol{\theta}} - \mathbf{X}\boldsymbol{\beta} - \mathbf{Zv})^T(\mathbf{R})^{-1}(\widehat{\boldsymbol{\theta}} - \mathbf{X}\boldsymbol{\beta} - \mathbf{Zv})\right] \tag{A.2}$$

Furthermore, with the additional assumption that $\mathbf{v}$ and $\widehat{\boldsymbol{\theta}}|(\boldsymbol{\beta},\mathbf{v})$ are independent, then joint pdf for $\mathbf{v}$ dan $\widehat{\boldsymbol{\theta}}|(\boldsymbol{\beta},\mathbf{v})$ is

$$(2\pi\sigma^2)^{-\frac{1}{2}q-\frac{1}{2}n}\left(\det\begin{pmatrix}\mathbf{G} & \mathbf{0}\\ \mathbf{0} & \mathbf{R}\end{pmatrix}\right)^{-\frac{1}{2}} . \exp\left\{-\frac{1}{2\sigma^2}\begin{pmatrix}\mathbf{v}\\ \widehat{\boldsymbol{\theta}} - \mathbf{X}\boldsymbol{\beta} - \mathbf{Zv}\end{pmatrix}^T\begin{pmatrix}\mathbf{G} & \mathbf{0}\\ \mathbf{0} & \mathbf{R}\end{pmatrix}^{-1}\begin{pmatrix}\mathbf{v}\\ \widehat{\boldsymbol{\theta}} - \mathbf{X}\boldsymbol{\beta} - \mathbf{Zv}\end{pmatrix}\right\} \tag{A.3}$$

To maximize the joint pdf respect to $\boldsymbol{\beta}$ and $\mathbf{v}$, that is by minimizing

$$\min_{\boldsymbol{\beta},\mathbf{v}} \mathbf{v}^T\mathbf{G}^{-1}\mathbf{v} + (\widehat{\boldsymbol{\theta}} - \mathbf{X}\boldsymbol{\beta} - \mathbf{Zv})^T\mathbf{R}^{-1}(\widehat{\boldsymbol{\theta}} - \mathbf{X}\boldsymbol{\beta} - \mathbf{Zv}) \tag{A.4}$$

In the end, will be minimized $\mathbf{v}^T\mathbf{G}^{-1}\mathbf{v} + (\widehat{\boldsymbol{\theta}} - \mathbf{X}\boldsymbol{\beta} - \mathbf{Zv})^T\mathbf{R}^{-1}(\widehat{\boldsymbol{\theta}} - \mathbf{X}\boldsymbol{\beta} - \mathbf{Zv})$ by doing the derivative of $\boldsymbol{\beta}$ and $\mathbf{v}$ then equating the derived result with zero in such a way as to obtain:

$$-2\widehat{\boldsymbol{\theta}}^T\mathbf{R}^{-1}\mathbf{X} + 2\mathbf{v}^T\mathbf{Z}^T\mathbf{R}^{-1}\mathbf{X} + 2\boldsymbol{\beta}^T\mathbf{X}^T\mathbf{R}^{-1}\mathbf{X} = \mathbf{0} \tag{A.5}$$
$$-2\widehat{\boldsymbol{\theta}}^T\mathbf{R}^{-1}\mathbf{Z} + 2\boldsymbol{\beta}^T\mathbf{X}^T\mathbf{R}^{-1}\mathbf{Z} + 2\mathbf{v}^T(\mathbf{Z}^T\mathbf{R}^{-1}\mathbf{Z} + \mathbf{G}^{-1}) = \mathbf{0} \tag{A.6}$$

or can be written as follows:

$$\mathbf{X}^\mathbf{T}\mathbf{R}^{-1}\mathbf{X}\boldsymbol{\beta} + \mathbf{X}^\mathbf{T}\mathbf{R}^{-1}\mathbf{Zv} = \mathbf{X}^\mathbf{T}\mathbf{R}^{-1}\widehat{\boldsymbol{\theta}} \tag{A.7}$$
$$\mathbf{Z}^\mathbf{T}\mathbf{R}^{-1}\mathbf{X}\boldsymbol{\beta} + (\mathbf{Z}^\mathbf{T}\mathbf{R}^{-1}\mathbf{Z} + \mathbf{G}^{-1})\mathbf{v} = \mathbf{Z}^\mathbf{T}\mathbf{R}^{-1}\widehat{\boldsymbol{\theta}} \tag{A.8}$$

where the last two equations are referred to as Henderson's linear mixed model.

References

[1] Anselin L 1992 *Spatial Econometrics: Method and Models* (Boston: Kluwer Academic Publishers)
[2] Anton H 1993 *Elementary Linear Algebra* (New York: John Wiley & Sons)
[3] Asfar 2016 *Study of Determining Optimum Spatial Weighting Matrix in Small Area Estimation (thesis)* (Bogor: Bogor Agricultural University)
[4] Bingham N H and Fry J M 2010 *Regression: Linear Models In Statistics* (London: Springer-Verlag, Inc)
[5] Chand N and Alexander C H 1995 *Using administrative record for small area estimation in the American community survey* (US Bureau of the Census)
[6] Fay R E and Herriot R A 1979 Estimates of income for small places: an application of James Stein procedures to census data. *J Am Stat Assoc.* **74**:269-277
[7] Ghosh M and Rao J N K 1994 Small Area Estimation: An Appraisal. *Statistical Science.* **9**(1): 55-76
[8] Harville D A 1991 That That BLUP is a Good Thing: The Estimation of Random Effects. *Statistical Science.* **6**, 1, 35-39.
[9] Harville D A 1997 *Matrix Algebra From a Statistician's Perspective* (New York: Springer-Verlag, Inc)
[10] Henderson C R 1975 Best linear unbiased estimation and prediction under a selection model. *Biometrics.* **31**:423-447
[11] Hogg R V, McKean J W, and Craig A T 2005 *Introduction to Mathematical Statistics 6th ed* (Boston: Pearson Education, Inc)
[12] Jennrich R I and Sampson P F 1976 Newton-Rapson and related algorithms for maximum likelihood variance component estimation. *Technometrics.* **18**, 11-17
[13] Kackar R N and Harville D A 1984 Approximations for Standard Error of Estimator for Fixed and Random Effect in Mixed Models. *Journal of the American Statistical Association.* **79**, 853-862.
[14] Kurnia A 2009 *An Empirical Best Prediction Method for Logarithmic Transformation Model in Small Area Estimation with Particular Application to Susenas Data (dissertation)* (Bogor: Bogor Agricultural University)
[15] Matualage D 2012 *Spatial Empirical Best Linear Unbiased Prediction Methods for Small Areas to Estimate monthly Expenditure per capita (Case Study: The Province of East Java District Jember)* (Bogor: Bogor Agricultural University)
[16] Montgomery C, Peck A, and Vining G 2001 *Introduction to Linear Regression Analysis* (New York: John Wiley & Sons)
[17] Petrucci A and Salvati N 2006 Small Area Estimation for Spatial Correlation in Watershed Erosion Assessment. *Journal of Agricultural, Biological, and Environmental Statistics.* **11**(2): 169-182
[18] Pratesi M and Salvati N 2005 Small area estimation: the EBLUP estimator with autoregressive random area effects. *Report no 261 Departement of Statistics and Mathematics, University of Pisa*
[19] Pratesi M and Salvati N 2008 Small area estimation: the EBLUP estimator based on spatially correlated random area effects. *Stat. Meth. & Appl.* **17**:113-141
[20] Rao J N K 2003 *Small Area Estimation* (New York: John Wiley and Sons)
[21] Rencher A C 2002 *Method of Multivariate Analysis 2th ed* (New York: John Wiley & Sons, Inc)
[22] Robinson G K 1991 That BLUP is a Good Thing: The Estimation of Random Effects. *Statistical Science.* **6**, 1, 15-51
[23] Salvati N 2004 Small area estimation by spatial models: the spatial empirical best linear unbiased prediction (spatial EBLUP) *Working Paper 2004/03. Firenze: Universitá degli Studi di Firenze*
[24] Searle S R 1982 *Matrix algebra useful for statistics* (John Wiley & Sons, Inc)

[25] Susvitasari K 2014 *Searching for Point Estimation for Mean of Fay-Herriot Model and CAR Fay-Herriot Model Using Hierarchical Bayesian (HB) Approach on Small Area Estimation (SAE) (mini thesis)* (Depok: Universitas Indonesia)
[26] Yudistira 2014 *Point Estimation of Population Parameter of Small Area with Spatial Empirical Bayes based on Area Level Model (mini thesis)* (Depok: Universitas Indonesia)

IOP Conf. Series: Journal of Physics: Conf. Series **909** (2017) 012044 doi:10.1088/1742-6596/909/1/012044

Finite difference and Runge–Kutta methods for solving vibration problems

Scolastika Lintang Renganis Radityani[1] and Sudi Mungkasi[2]

[1] Department of Mathematics Education, Faculty of Teacher Training and Education, Sanata Dharma University, Mrican, Tromol Pos 29, Yogyakarta 55002, INDONESIA
[2] Department of Mathematics, Faculty of Science and Technology, Sanata Dharma University, Mrican, Tromol Pos 29, Yogyakarta 55002, INDONESIA

E-mail: justlintang@gmail.com, sudi@usd.ac.id

Abstract. The vibration of a storey building can be modelled into a system of second order ordinary differential equations. If the number of floors of a building is large, then the result is a large scale system of second order ordinary differential equations. The large scale system is difficult to solve, and if it can be solved, the solution may not be accurate. Therefore, in this paper, we seek for accurate methods for solving vibration problems. We compare the performance of numerical finite difference and Runge–Kutta methods for solving large scale systems of second order ordinary differential equations. The finite difference methods include the forward and central differences. The Runge–Kutta methods include the Euler and Heun methods. Our research results show that the central finite difference and the Heun methods produce more accurate solutions than the forward finite difference and the Euler methods do.

1. Introduction

Vibration problems have been an interesting research area in physics, engineering, as well as applied mathematics. Modelling vibrations on the structural system, such as storey buildings, have the same principles as modelling vibrations on the spring-mass system (for more details see [1-6]). This is done by assuming that each floor of the building is a mass, and pillars of the building are considered as a spring which has a stiffness. Therefore, a vibration in storey buildings can be modelled into a system of second order ordinary differential equations.

Furthermore, if the floor of the building is large, then we can model it to a large scale system of second order ordinary differential equations [7-8]. This makes the system of ordinary differential equations difficult to solve [9]. In this case, we need some solving methods that produces accurate solutions.

Therefore, in this paper, we compare the performance of numerical finite difference and Runge–Kutta methods to solve large scale systems of second order ordinary differential equations. From our research results, we shall know which numerical method has a higher degree of accuracy, so it produces accurate solutions. We note that in engineering applications, vibration studies on the structural system complement elasticity studies on the materials of the structures [10-11]. This means that our work in this paper shall be useful.

The remainder of this paper consists of three sections. Section 2 presents mathematical models and methods. Section 3 provides numerical results and discussion. We conclude the paper in Section 4.

2. Mathematical models and methods

The vibration of a one-storey building can be modelled like a vibration problem of a spring-mass system, that is, to be a second-order ordinary differential equation. In this paper, we assume that vibration problems do not involve friction. This can be done for problems with negligible damping.

The mathematical model for the spring-mass problem without friction is

$$mx'' + kx = 0, \quad x \in \mathbb{R}^1, \tag{1}$$

with the vibration frequency $f = \frac{1}{2\pi}\sqrt{\frac{k}{m}}$. Here m represents the mass of the object and k denotes the stiffness of the pillar of the building. Furthermore, t is the time variable and x is the space variable. If the building has n levels, then the model is a large scale system of second order ordinary differential equations. Afterwards, the large scale system can be written in the form of matrix equations

$$MX'' + KX = 0, \quad X \in \mathbb{R}^n, \tag{2}$$

where M is the mass matrix, K is the stiffness matrix, and X is the vector of space variables. Before solving equation (2), we first review that equation (1) can be solved using existing numerical methods. The numerical methods that we consider in this paper are the finite difference and the Runge–Kutta types. The finite difference methods are the forward and central differences, and the Runge–Kutta methods are the Euler and Heun methods. By comparing these methods, we shall know which method has better performance than the others.

2.1. Case 1: If equation (1) is solved using forward difference method

Based on forward difference method algorithm, equation (1) can be changed to

$$m\frac{(x')_{t=t^{n+1}}-(x')_{t=t^n}}{\Delta t} + kx|_{t=t^n} = 0, \tag{3}$$

$$m\frac{\frac{x|_{t=t^{n+2}} - x|_{t=t^{n+1}}}{\Delta t} - \frac{x|_{t=t^{n+1}} - x|_{t=t^n}}{\Delta t}}{\Delta t} + kx|_{t=t^n} = 0, \tag{4}$$

$$m\frac{\frac{x^{n+2}-x^{n+1}}{\Delta t} - \frac{x^{n+1}-x^n}{\Delta t}}{\Delta t} + kx^n = 0, \tag{5}$$

$$m\frac{x^{n+2}-2x^{n+1}+x^n}{\Delta t^2} + kx^n = 0, \tag{6}$$

or can be changed to

$$m\frac{x^{n+1}-2x^n+x^{n-1}}{\Delta t^2} + kx^{n-1} = 0, \tag{7}$$

so that

$$m(x^{n+1} - 2x^n + x^{n-1}) = -kx^{n-1}\Delta t^2, \tag{8}$$

$$x^{n+1} - 2x^n + x^{n-1} = \frac{-kx^{n-1}\Delta t^2}{m}, \tag{9}$$

$$x^{n+1} = \frac{-kx^{n-1}\Delta t^2}{m} + 2x^n - x^{n-1}. \tag{10}$$

Equation (10) is the forward difference scheme for equation (1). Without loss of generality, we assume that the value of $k = 1$ and $m = 1$, then we obtain

$$x^{n+1} = -x^{n-1}\Delta t^2 + 2x^n - x^{n-1}. \tag{11}$$

2.2. Case 2: If equation (1) is solved using central difference method
Equation (1) can be changed to

$$m\frac{(x')_{t=t^{n+1}}-(x')_{t=t^{n-1}}}{2\Delta t}+kx|_{t=t^n}=0, \tag{12}$$

$$m\frac{\frac{x|_{t=t^{n+2}}-x|_{t=t^n}}{2\Delta t}-\frac{x|_{t=t^n}-x|_{t=t^{n-2}}}{2\Delta t}}{2\Delta t}+kx|_{t=t^n}=0, \tag{13}$$

$$m\frac{\frac{x^{n+2}-x^n}{2\Delta t}-\frac{x^n-x^{n-2}}{2\Delta t}}{2\Delta t}+kx^n=0, \tag{14}$$

$$m\frac{x^{n+2}-2x^n+x^{n-2}}{4\Delta t^2}+kx^n=0, \tag{15}$$

or can be written as

$$m\frac{x^{n+1}-2x^n+x^{n-1}}{\Delta t^2}+kx^n=0, \tag{16}$$

so we get

$$m(x^{n+1}-2x^n+x^{n-1})=-kx^n\Delta t^2, \tag{17}$$

$$x^{n+1}-2x^n+x^{n-1}=\frac{-kx^n\Delta t^2}{m}, \tag{18}$$

$$x^{n+1}=\frac{-kx^n\Delta t^2}{m}+2x^n-x^{n-1}. \tag{19}$$

Equation (19) is the central difference scheme for equation (1). For $k=1$ and $m=1$, we obtain

$$x^{n+1}=-x^n\Delta t^2+2x^n-x^{n-1}. \tag{20}$$

Notice that the central difference scheme (20) is not the same as the forward difference scheme (11).

2.3. Case 3: If equation (1) is solved using the first order Runge–Kutta method (Euler method)
Given equation (1) where the value of $k=1$ and $m=1$, we obtain

$$x''+x=0. \tag{21}$$

Furthermore, we derived a system of first order differential equations from equation (21). Let $x_1=x$ and $x_2=x'$, so the system of equations (21) becomes

$$x_1'=x'=x_2, \tag{22}$$

$$x_2'=x''=-x=-x_1. \tag{23}$$

The system of equations (22) and (23) can be solved using the Euler method as follows

$$X^{n+1}=X^n+f(t_n,X^n)\Delta t, \tag{24}$$

or can be written in a matrix equation

$$X^{n+1}=X^n+\begin{pmatrix}0&1\\-1&0\end{pmatrix}\begin{pmatrix}{x_1}^n\\{x_2}^n\end{pmatrix}\Delta t. \tag{25}$$

2.4. *Case 4: If equation (1) is solved using a second order Runge–Kutta method (Heun method)*

The Heun method for the system of equations (22) and (23) is

$$X^{n+1} = X^n + \frac{f(t_n, X^n) + f(t_{n+1}, X^{n+1})}{2} \Delta t, \tag{26}$$

With $f(t_{n+1}, X^{n+1}) \approx f(t_{n+1}, \overline{x_{n+1}})$, where $\overline{x_{n+1}} = X^n + f(t_n, X^n)\Delta t$ or in other words $\overline{x_{n+1}} = X^{n+1}$ in the Euler method. Therefore, the algorithm for Heun method becomes

$$X^{n+1} = X^n + \frac{f(t_n, X^n) + f(t_{n+1}, \overline{x_{n+1}})}{2} \Delta t, \tag{27}$$

or it can be written in a matrix equation

$$X^{n+1} = X^n + \left[\begin{pmatrix} 0 & 1 \\ -1 & 0 \end{pmatrix}\begin{pmatrix} {x_1}^n \\ {x_2}^n \end{pmatrix} + \begin{pmatrix} 0 & 1 \\ -1 & 0 \end{pmatrix}\begin{pmatrix} {x_1}^{n+1*} \\ {x_2}^{n+1*} \end{pmatrix}\right]\frac{\Delta t}{2}, \tag{28}$$

where ${x_1}^{n+1*}$ and ${x_2}^{n+1*}$ are obtained from the one-step Euler method.

3. Results of finite difference methods and Runge–Kutta methods

This section compares results of forward and central difference methods, as well as those of the Euler and Heun methods in solving equation (1). Simulations of the four methods are done using MATLAB, with values $0 \leq t \leq 10$ and $\Delta t = 0.5; 0.25; 0.125; 0.0625$. Based on the algorithm that is made in MATLAB, we present a graphic image showing the comparison results of the four methods together with analytical results.

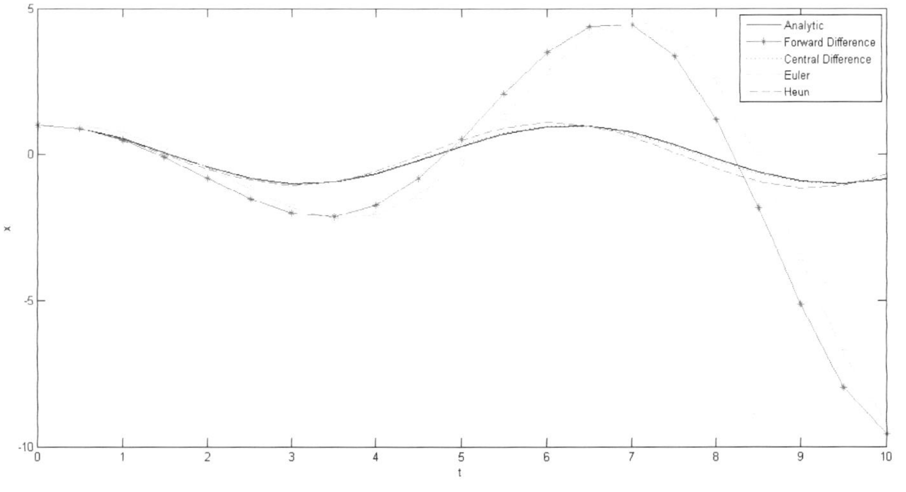

Figure 1. Analytical and numerical solutions using time step $\Delta t = 0.5$.

Representatives of our numerical results are shown in Figures 1-4. Respectively these figures show analytical solution together with numerical solutions using time steps $\Delta t = 0.5; 0.25; 0.125; 0.0625$. We observe that the central difference method and the Heun method approximate the analytical solution better than the forward difference method and the Euler method do.

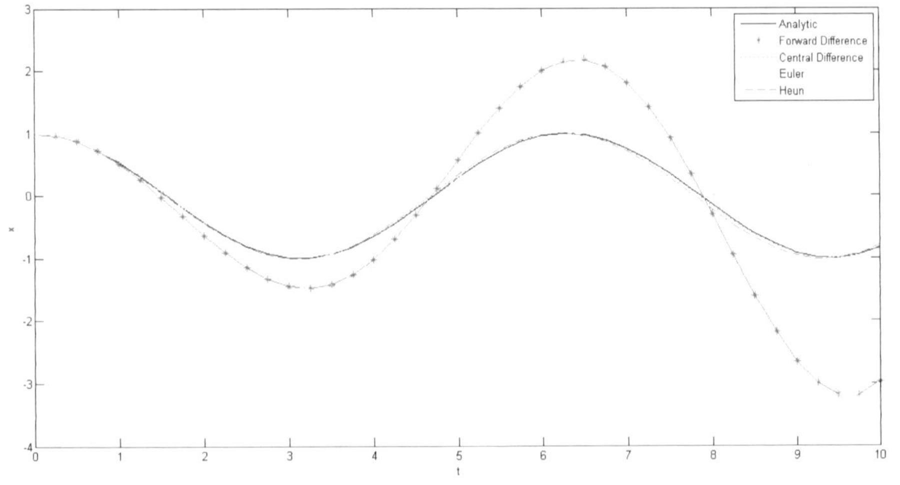

Figure 2. Analytical and numerical solutions using time step $\Delta t = 0.25$

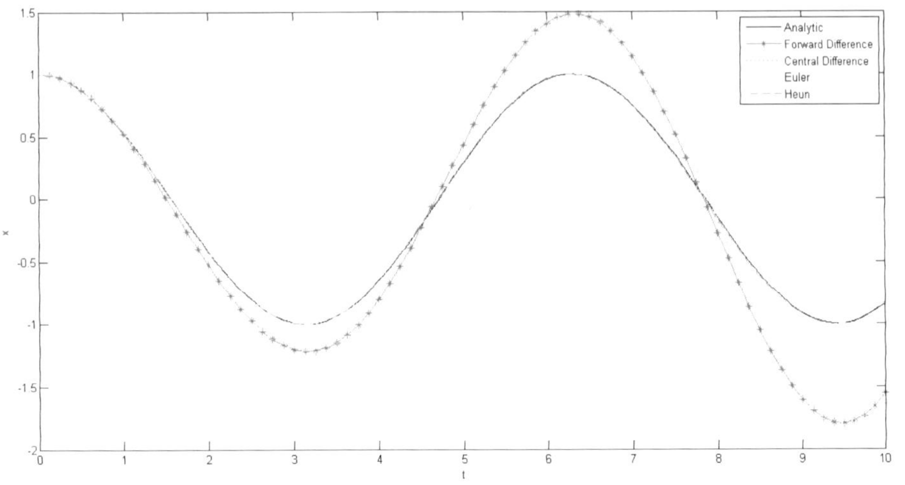

Figure 3. Analytical and numerical solutions using time step $\Delta t = 0.125$

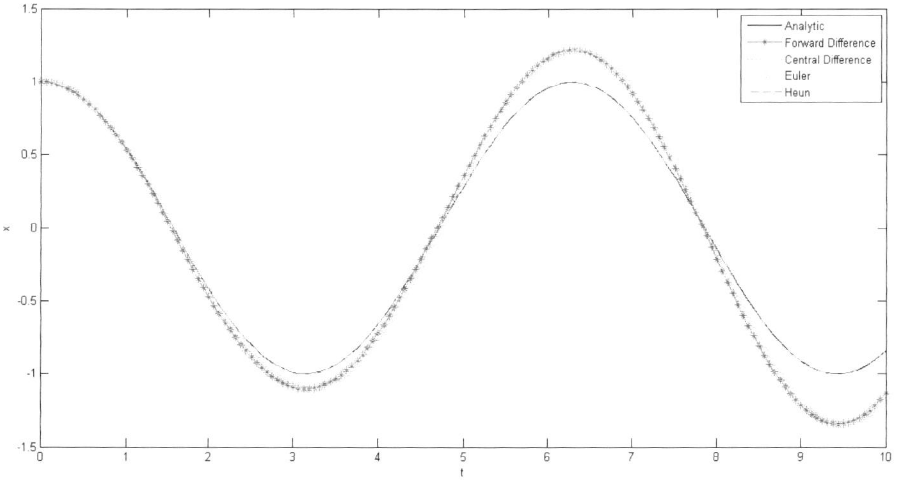

Figure 4. Analytical and numerical solutions using time step $\Delta t = 0.0625$

Table 1. Error comparison of the forward difference, central difference, Euler, and Heun methods.

Δt	Error			
	Forward difference	Central difference	Euler method	Heun method
0.5	1.9924	0.0291	1.8732	0.1284
0.25	0.6659	0.0074	0.6510	0.0313
0.125	0.2671	0.0019	0.2611	0.0078
0.0625	0.1199	0.0005	0.1169	0.0019

Furthermore, error comparison with varying time step is summarised in Table 1. We obtain that for central difference and the Heun methods, as the time step is halved, the error gets quartered. For forward difference and Euler methods, as the time step is halved, the error gets halved too. These mean that central difference and the Heun methods have the second order of accuracy. Forward difference and Euler methods have the first order of accuracy.

4. Conclusion

We have compared the performance of the forward finite difference, central finite difference, Euler, and Heun methods for solving vibration problems. We obtain that the central finite difference and the Heun methods have the second order of accuracy to solve vibration problems modelled in ordinary differential equations. These two methods produce more accurate solutions than the forward finite difference and the Euler methods having only the first order of accuracy.

Acknowledgment

This research was supported financially by Sanata Dharma University. Both authors are very thankful to Sanata Dharma University for the financial support.

References

[1] Sakr T A 2015 Vibration control of buildings by using partial floor loads as multiple tuned mass dampers *HBRC Journal* (in press) DOI: 10.1016/j.hbrcj.2015.04.004

[2] Connolly D P, Marecki G P, Kouroussis G, Thalassinakis I and Woodward P K 2016 The growth of railway ground vibration problems – A review *Science of The Total Environment* **586** 1276

[3] Floden O, Persson K and Sandberg G 2016 Numerical methods for predicting vibrations in multi-storey buildings *World Conference on Timber Engineering* **WCTE 2016** 124667

[4] Shimoda M and Liu Y 2016 Node-base free-form optimization method for vibration problems of shell structures *Computers & Structures* **177** 91

[5] Shirokov V S, Kholopov I S and Solovejv 2016 Determination of the frequency of natural vibration of a modular building *Procedia Engineering* **153** 655

[6] López-Mendoza D, Romero A, Connolly D P and Galvin P 2017 Scoping assessment of building vibration induced by railway traffic *Soil Dynamics and Earthquake Engineering* **93** 147

[7] Kalaycioglu T and Ozguven H N 2014 Nonlinear structural modification and nonlinear coupling *Mechanical Systems and Signal Processing* **46** 289

[8] Supriyadi B and Mungkasi S 2016 Structural dynamic modification using matrix perturbation for vibrations without friction *Journal Physics: Conference Series* **776** 012078

[9] Xie W C 2010 *Differential Equations for Engineers* (Cambridge: Cambridge University Press)

[10] LeVeque R J 2002 Finite-volume methods for non-linear elasticity in heterogeneous media *International Journal for Numerical Methods in Fluids* **40** 93

[11] Supriyadi B and Mungkasi S 2016 Finite volume numerical solvers for non-linear elasticity in heterogeneous media *International Journal for Multiscale Computational Engineering* **14** 479

IOP Conf. Series: Journal of Physics: Conf. Series **909** (2017) 012045 doi:10.1088/1742-6596/909/1/012045

Reconstruction of initial conditions of the Burgers equation in conservation law problems

Sudi Mungkasi and Fioretta Laras

Department of Mathematics, Faculty of Science and Technology, Sanata Dharma University, Mrican, Tromol Pos 29, Yogyakarta 55002, INDONESIA

E-mail: sudi@usd.ac.id

Abstract. In this paper, we solve the Burgers equation, which is a representation of nonlinear conservation laws. The research question is that after we solve the Burgers equation for the final time, how can we reconstruct the initial condition if we need the initial condition, whereas the initial condition has evolved to the final solution. To answer this research question, we propose the use of the Lax–Friedrichs finite volume method to reconstruct the initial condition of the Burgers equation in one dimension. To demonstrate our proposal, we solve the Burgers equation first using the Jin–Xin relaxation method until we accomplish the final solution. After the final solution is obtained, we reconstruct the initial condition backward in time and compare it with the exact initial condition which is used earlier. We find that the proposed numerical method is successful in reconstructing the initial condition accurately.

1. Introduction

Burgers equation contains some characteristics of the Navier–Stokes equation for fluid dynamics [1]. The Burgers equation is a nonlinear hyperbolic partial differential equation [2-3]. Literature has been available largely in correspondence with solving the Burgers equations, for example, see References [4-8]. Nevertheless, how to track the history of a solution to the Burgers equation is still questionable. In this paper, we concern on reconstructing the initial condition of the Burgers equation in one dimension.

First, the Jin–Xin relaxation method [9-11] is used to solve the Burgers equation until we discover the final numerical solution. After that, using the final numerical solution, we calculate backward in time the estimation of initial condition using the Lax–Friedrichs finite volume method. We choose the Jin–Xin relaxation method because of its linearity in the models, so we do not need too much efforts to calculate the solution, while the Lax–Friedrichs finite volume method presents the simplicity in the models.

To see the accuracy of our reconstruction, we compare the reconstructed initial condition with the exact initial condition. We examine the error between them and observe the effectiveness of the Lax–Friedrichs finite volume method to reconstruct the initial condition. This research can be used to estimate the history of wave if all we have is present data without any background from the past.

The paper is organised as follows. Section 2 recalls the defined model and the implemented methods. Results and discussion are presented in Section 3. Conclusion is expressed in Section 4.

2. Model and methods

We consider conservation laws in one dimension

$$u_t + f(x,u)_x = 0 \tag{1}$$

where $f(x,u)$ is a flux function. Here, the flux for the Burgers equation is

$$f(u) = \frac{u^2}{2}. \tag{2}$$

We assume that the initial condition for u when $1 < x < 1 + 2\pi$ is

$$u(x,0) = 1 + \sin(x - \frac{\pi}{2} - 1) + \epsilon\,, \tag{3}$$

otherwise the value is ϵ, which is a sufficiently small positive number. Here x is a spatial variable, t is the time variable and u is a conserved quantity depending on x and t.

There are two methods which are used in this paper. They are the Jin–Xin relaxation method and the Lax–Friedrichs finite volume method. The Jin–Xin relaxation method is implemented to obtain the final numerical solution of the Burgers equation. After we obtain the final numerical solution, the research question is:

If the initial condition has gone and unknown, can we reconstruct the initial condition?

In this paper, we propose the Lax–Friedrichs finite volume method backward in time to reconstruct the initial conditions. Note that the equation solved using the Lax–Friedrichs finite volume method is different from those solved using the Jin–Xin relaxation method.

2.1 Jin–Xin relaxation method

Equation (1) can be written in a linear system [9]

$$\frac{\partial u}{\partial t} + \frac{\partial v}{\partial x} = 0 \tag{4}$$

and

$$\frac{\partial v}{\partial t} + a\frac{\partial u}{\partial x} = -\frac{1}{\epsilon}(v - f(u)) \tag{5}$$

where v is defined by $v = f(u)$, while ϵ is a small positive constant and a is a constant satisfying $a = \max\,(f'(u))^2$.

We can write equation (4) and (5) in the discretised space of the conservative form

$$\frac{\partial}{\partial t}u_i + \frac{1}{\Delta x}\left(v_{i+\frac{1}{2}} - v_{i-\frac{1}{2}}\right) = 0 \tag{6}$$

and

$$\frac{\partial}{\partial t}v_i + \frac{1}{\Delta x}a\left(u_{i+\frac{1}{2}} - u_{i-\frac{1}{2}}\right) = -\frac{1}{\epsilon}(v_i - f(u_i)). \tag{7}$$

The fully explicit schemes of the Jin–Xin relaxation method in one dimension are

$$u_i^{n+1} = u_i^n - \frac{\Delta t}{2\Delta x}\left(v_{i+1}^n - v_i^n - a^{\frac{1}{2}}(u_{i+1}^n - 2u_i^n + u_{i-1}^n)\right), \tag{8}$$

and

$$v_i^{n+1} = v_i^n - \frac{\Delta t}{2\Delta x}\left(a(u_{i+1}^n - u_i^n) - a^{\frac{1}{2}}(v_{i+1}^n - 2v_i^n + v_{i-1}^n)\right) - \frac{1}{\epsilon}\left(v_i^n - f(u_i^n)\right). \tag{9}$$

2.2 Lax–Friedrichs finite volume method

The Lax–Friedrichs finite volume method is used for the reconstruction of initial condition. First, we are given the final solution $u(x,t^M)$, where t^M is assumed to be the final time. We use this final solution as the initial value of p, where p is the variable for the reconstruction of the history of the solution:

$$u(x,t^M) = p(x,t^M). \tag{10}$$

We can write equation (4) and (5) in adjoint equations [11]

$$-q_t - p_x = -\frac{q}{\epsilon}, \tag{11}$$

and

$$-p_t - aq_x = \frac{q}{\epsilon} f'(u)\,. \tag{12}$$

Based on equations (11) and (12), we obtain [12]

$$-p_t - f'(u)p_x = \epsilon a p_{xx} \tag{13}$$

and solve equation (13) using the Lax–Friedrichs finite volume method backward in time. The explicit numerical scheme of the finite volume method backward in time is [3]

$$P_i^n = P_i^{n+1} - \frac{\Delta t}{\Delta x}\left(F_{i+\frac{1}{2}}^{n+1} - F_{i-\frac{1}{2}}^{n+1}\right) \tag{14}$$

where

$$\begin{aligned} F_{i+\frac{1}{2}}^{n+1} &= \frac{1}{2}\left(F_{i+1}^{n+1} + F_i^{n+1}\right) - \frac{\Delta x}{2\Delta t}\left(P_{i+1}^{n+1} - P_i^{n+1}\right) \\ &= \frac{1}{2}\left(\frac{1}{2}\left(P_{i+1}^{n+1}\right)^2 + \frac{1}{2}\left(P_i^{n+1}\right)^2\right) - \frac{\Delta x}{2\Delta t}\left(P_{i+1}^{n} - P_i^{n}\right), \end{aligned} \tag{15}$$

and

$$\begin{aligned} F_{i-\frac{1}{2}}^{n+1} &= \frac{1}{2}\left(F_i^{n+1} + F_{i-1}^{n+1}\right) - \frac{\Delta x}{2\Delta t}\left(P_i^{n+1} - P_{i-1}^{n+1}\right) \\ &= \frac{1}{2}\left(\frac{1}{2}\left(P_i^{n+1}\right)^2 + \frac{1}{2}\left(P_{i-1}^{n+1}\right)^2\right) - \frac{\Delta x}{2\Delta t}\left(P_i^{n} - P_{i-1}^{n}\right). \end{aligned} \tag{16}$$

Here $F_{i+1/2}^{n+1} \approx f\big(p(x_{i+1/2}, t^{n+1})\big)$ is the flux, and $P_i^{n+1} \approx p(x_i, t^{n+1})$ is an approximation of the conserved quantity, Δx is the cell-width, Δt is the time step, i represents the index of space, and n denotes the index of time.

To see the accuracy of our reconstructed initial conditions, numerical error and convergence rate should be calculated. We define N as the number of cells based on how large the cell-width is and E_i as the error of the solution based on the cell-width Δx_i. Therefore, the formulas for the numerical error (E) and the convergence rate (CR) are as follows [13]:

$$E = \frac{1}{N}\sum_{i=1}^{N} |u(x,0) - p(x,0)| \tag{17}$$

and

$$\text{CR} = \frac{1}{N-1}\sum_{i}^{N-1} R_i \tag{18}$$

where

$$R_i = \frac{\log\left(\frac{E_i}{E_{i+1}}\right)}{\log\left(\frac{\Delta x_i}{\Delta x_{i+1}}\right)}. \tag{19}$$

3. Results and discussion

In this section, we present our results and give a discussion.

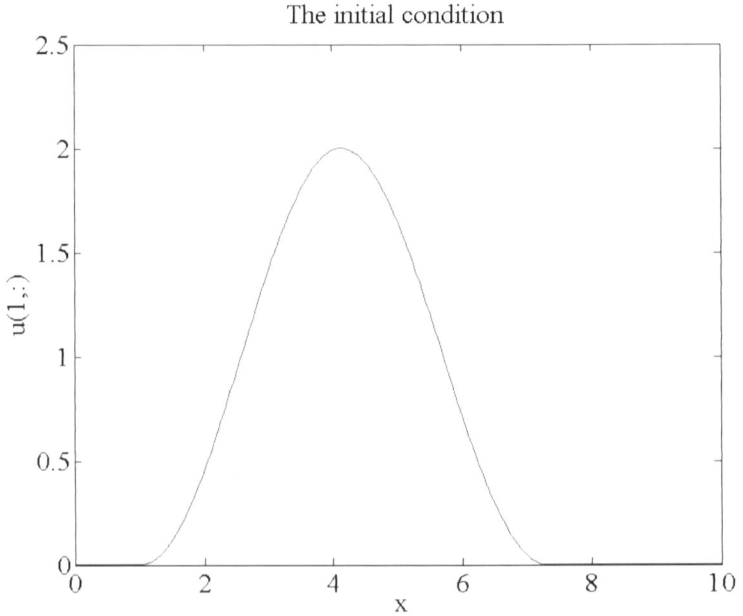

Figure 1. The initial condition that is used as a test case in the Burgers equation. It is evolved for the final solution of the Burgers equation.

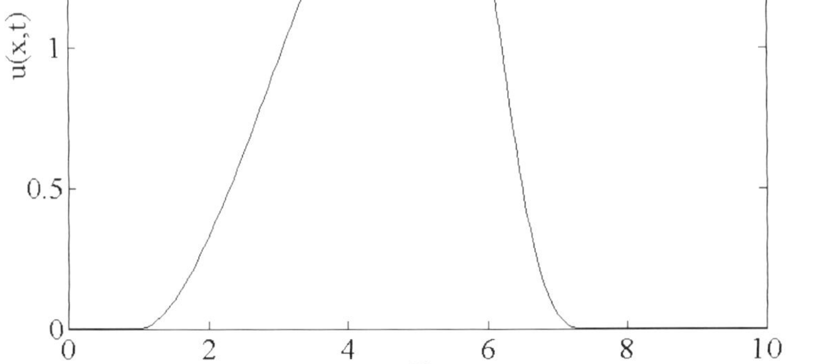

Figure 2. Numerical solution of Burgers equation using Jin–Xin relaxation method forward in time for case 1.

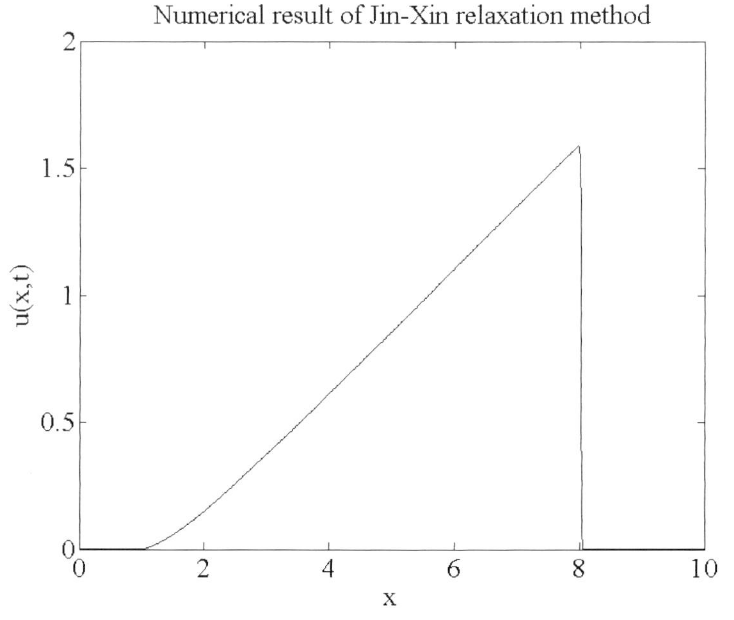

Figure 3. Numerical solution of Burgers equation using Jin–Xin relaxation method forward in time for case 2.

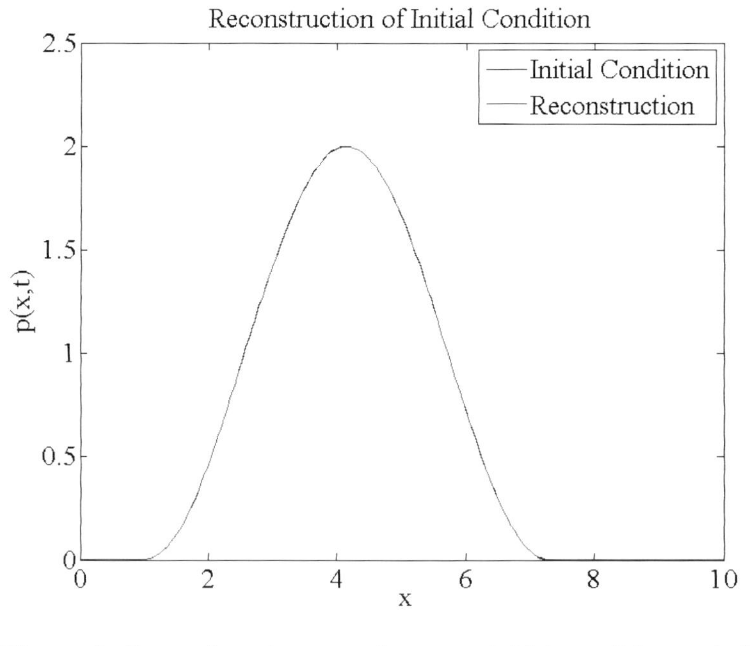

Figure 4. Comparison between the exact initial condition and the reconstruction for case 1.

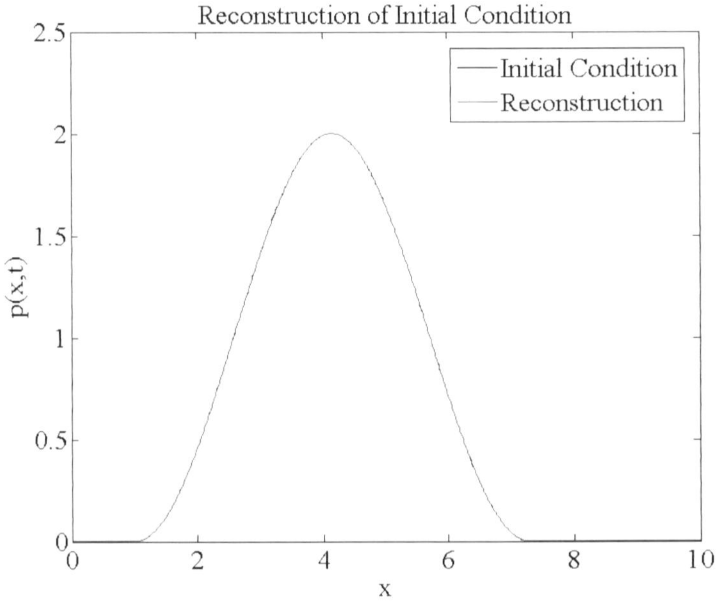

Figure 5. Comparison between the exact initial condition and the reconstruction for case 2.

3.1. Initial condition of Burgers equation

For computation in MATLAB, we take $x \in [0,10]$ with cell-width $\Delta x = 0.05$ and $t \in [0,0.5]$ with time step $\Delta t = 0.025\, \Delta x$. The initial condition has been mentioned in equation (3). Figure 1 illustrates the exact initial condition that we will reconstruct later.

3.2. Results using Jin–Xin relaxation method forward in time

The Jin–Xin relaxation method is used to solve the Burgers equation with the boundary condition $u(0,t)$ and $u(10,t)$ are set to be transmissive and $\epsilon = 10^{-3}$.

Case 1: We take spatial domain $0 \le x \le 10$ and time domain $0 \le t \le 0.5$ with $\Delta x = 0.05$ and time step $\Delta t = 0.025\Delta x$. Figure 2 shows the results at the final time.

Case 2: We take spatial domain $0 \le x \le 10$ and time domain $0 \le t \le 3$ with $\Delta x = 0.0125$ and time step $\Delta t = 0.025\Delta x$. Figure 3 draws the results at the final time.

We notice that as time evolves the final solution becomes steeper to form (almost) a discontinuity.

3.3. Reconstruction of initial conditions using Lax–Friedrichs finite volume method backward in time

To reconstruct the initial condition, we set the final solution which was obtained earlier as the initial condition in the Lax–Friedrichs finite volume method backward in time. The boundary conditions are set to be transmissive and $\epsilon = 10^{-3}$.

Case 1: We use the same Δx and Δt as in the previous calculation in case 1 on Subsection 3.2. Figure 4 shows the reconstruction results.

Case 2: We select numerical settings the same as the previous simulation in case 2 on Subsection 3.2. Figure 5 illustrates the reconstruction results.

As we observe in Figures 4 and 5, the error between the initial condition and its reconstruction is insignificant.

3.4. Results of convergence rate

We take variety of Δx which are 0.4, 0.2, 0.1, 0.05, 0.025 and 0.0125, and $x \in [0,10]$, $t \in [0,0.5]$, and Δt are the same as the previous simulation. Table 1 contains the errors and convergence rate for problems of either case 1 or case 2 in the reconstruction of the initial condition.

Table 1. The numerical error and convergence rate of either case 1 or case 2.

Δx	Error	CR
0.4	0.0319	-
0.2	0.0090	1.825560
0.1	0.0030	1.584963
0.05	0.0012	1.321928
0.025	5.3692 x 10^-4	1.160255
0.0125	2.5736 x 10^-4	1.060919
	Rate	1.738406

From Table 1, we notice that the smaller Δx gives the smaller error. The convergence rate of our proposed method is one as Δx approaches to zero.

4. Conclusion

We have proposed a finite volume approach to reconstruct the initial condition for problems of conservation laws. The conservation law under investigation is the Burgers equation. The numerical method being proposed is the Lax–Friedrichs finite volume method. Our convergence rate of the numerical method is one, as the cell width is taken smaller approaching to zero.

Acknowledgments

This research was financially supported by DRPM of Ditjen Penguatan Riset dan Pengembangan of Kementerian Riset, Teknologi, dan Pendidikan Tinggi of Republik Indonesia in 2017 with Research Contract Number 051/HB-LIT/IV/2017 dated on 14 April 2017 (under DIPA-042.06-0.1.401516/2017 funds).

References

[1] Landajuela M 2011 Burgers equation *BCAM Internship-Summer 2011* (Bilbao: Basque Center for Applied Mathematics)
[2] LeVeque R J 1992 *Numerical Methods for Conservation Laws* (Basel: Springer)
[3] LeVeque R J 2004 *Finite Volume Method for Hyperbolic Problems* (Cambridge: Cambridge University Press)
[4] Basto M, Semiano V and Calheiros F L 2007 Numerical study of modified Adomian's method applied to Burgers equation *Journal of Computational and Applied Mathematics* **206** 927
[5] Saka B 2008 A numerical study of the Burgers' equation *Journal of the Franklin Institute* **345** 328
[6] Mungkasi S 2015 Some advantages of implementing an adaptive moving mesh for the solution to the Burgers equation *IOP Conference Series: Materials Science and Engineering* **78** 012031
[7] Mungkasi S, Harini B H and Wiryanto, L H 2017 Jin–Xin relaxation method used to solve the one-dimensional inviscid Burgers equation *Journal of Physics: Conference Series* **856** 012007
[8] Mungkasi S and Laras F 2017 Jin–Xin relaxation solution to the spatially-varying Burgers equation in one dimension *Journal of Physics: Conference Series* **795** 012043
[9] Jin S and Xin Z 1995 The relaxation schemes for systems of conservation laws in arbitrary space dimensions *Communications on Pure and Applied Mathematics* **48** 235

[10] Banda M K and Seaid M 2005 Higher-order relaxation schemes for hyperbolic systems of conservation laws *Journal of Numerical Mathematics* **13** 171
[11] Yohana E 2012 Adjoint-based Optimization for Optimal Control Problems Governed by Nonlinear Hyperbolic Conservation Laws *MSc Thesis* (Johannesburg: University of the Witwatersrand)
[12] Banda M K and Herty M 2012 Adjoint IMEX-based schemes for control problems governed by hyperbolic conservation laws *Computational Optimization and Applications* **51** 909
[13] Hidayat N, Suhariningsih, Suryanto A and Mungkasi S 2014 The significance of spatial reconstruction in finite volume methods for the shallow water equations *Applied Mathematical Sciences* **8** 1411

A staggered grid finite difference method for solving the gravity wave-model equations

Catharina Mara Apriani[1] and Sudi Mungkasi[2]

[1] Department of Mathematics Education, Faculty of Teacher Training and Education, Sanata Dharma University, Mrican, Tromol Pos 29, Yogyakarta 55002, INDONESIA
[2] Department of Mathematics, Faculty of Science and Technology, Sanata Dharma University, Mrican, Tromol Pos 29, Yogyakarta 55002, INDONESIA

E-mail: catharinamara@gmail.com, sudi@usd.ac.id

Abstract. The gravity wave-model equations are considered. The equations govern shallow water flows, where the gravity effect is significant. The equations form a system of partial differential equations with the hyperbolic type. When the gravity wave-model equations are solved using a finite difference method with staggered grids, artificial oscillations may occur in the numerical solution. In this paper, we propose a numerical treatment of a staggered grid finite difference method for solving the gravity wave-model equations. Using our proposed treatment, artificial oscillations can be eliminated, so the resulting numerical solution mimics the physics of the problem.

1. Introduction

The gravity wave-model equations have been used to model water flows, such as dam break problems [1] and flood simulations [2]. These equations can also be used to model river flows at their steady states [3]. Gravity waves have also been researched by other authors [4-5] relating to internal water waves.

The gravity wave-model equations considered in this paper are simplifications of the Saint-Venant shallow water wave equations, where the convective term is assumed to be negligible. This assumption makes the model easier to be used for flow simulations, yet the conservation of mass is still enforced, and the conservation of momentum based on gravity term still exists in the system.

Solving the gravity wave-model equations numerically is our goal in this paper. We shall use a staggered grid finite difference method. This method is chosen, because fluxes do not need to be approximated. Using the staggered grid technique, fluxes are known. This method is applicable for solving smooth problems. We note that staggered grid numerical methods have had a great success in solving some problems relating to elasticity, water, seismic, and other wave models [6-10].

However, a staggered grid method may generate unphysical oscillations when we implement it to solve nonsmooth problems. In this paper, a numerical treatment is proposed, so that the unphysical oscillations can be eliminated for nonsmooth problems. Numerical tests confirm our claims.

The rest of this paper is written as follows. We recall the mathematical model and numerical methods in Section 2. We provide numerical results in Section 3. Conclusion is drawn in Section 4.

2. Model and method

The gravity wave-model equations are

$$\frac{\partial h}{\partial t}+\frac{\partial q}{\partial x}=0\,, \tag{1}$$

$$\frac{\partial q}{\partial t}+\frac{g}{2}\frac{\partial h^2}{\partial x}=0\,. \tag{2}$$

where $h(x,t)$ is water depth, $q(x,t)$ is unit-discharge, and g is the acceleration due to gravity. The gravity wave-model equations (1) and (2) are simplifications of shallow water equations (the Saint-Venant shallow water wave equations). Ancey et al. [11] and Mungkasi et al. [12] provide some forms of shallow water equations.

To solve equation (1), we use discrete forms as follows

$$\left.\frac{\partial h}{\partial t}\right|_{\substack{x=x_j\\ t=t^n}}\approx\frac{h_j^{n+1}-h_j^n}{\Delta t}, \tag{3}$$

$$\left.\frac{\partial q}{\partial x}\right|_{\substack{x=x_j\\ t=t^n}}\approx\frac{q_{j+1/2}^n-q_{j-1/2}^n}{\Delta x}. \tag{4}$$

Here x_j is the j-th spatial point, t^n is the n-th temporal point, $j=1,2,3,\ldots$ and $n=0,1,2,3,\ldots$, Δx is uniform spatial distance, Δt is the time step, and h_j^n is water depth at time t^n and point x_j. The notations h_j^{n+1}, $q_{j+1/2}^n$, $q_{j-1/2}^n$ are understood analogously. Substitution of equations (3) and (4) to equation (1) results in

$$h_j^{n+1}=h_j^n-\frac{\Delta t}{\Delta x}\left(q_{j+\frac{1}{2}}^n-q_{j-\frac{1}{2}}^n\right). \tag{5}$$

To solve equation (2), we use discrete approximations as follows

$$\left.\frac{\partial q}{\partial t}\right|_{\substack{x=x_{j+1/2}\\ t=t^n}}\approx\frac{q_{j+1/2}^{n+1}-q_{j+1/2}^n}{\Delta t}, \tag{6}$$

$$\left.\frac{\partial h^2}{\partial x}\right|_{\substack{x=x_{j+\frac{1}{2}}\\ t=t^n}}\approx\frac{\left(h_{j+1}^n\right)^2-\left(h_j^n\right)^2}{\Delta x}. \tag{7}$$

Substitution of equations (6) and (7) to equation (2) leads to

$$q_{j+1/2}^{n+1}=q_{j+1/2}^n-\frac{g}{2}\frac{\Delta t}{\Delta x}\left((h_{j+1}^n)^2-(h_j^n)^2\right). \tag{8}$$

The system of equations (5) and (8) is the staggered grid numerical finite difference scheme. It is an explicit scheme. It is stable as long as the time step is taken sufficiently small.

3. Numerical results

In this section, we provide three test cases. The first is a problem having a smooth solution. The second is a problem having a discontinuous solution. The first and the second problems are solved using the standard staggered finite difference scheme. The third is a problem having a discontinuous solution, but is solved using a modified staggered finite difference scheme. The modification is the numerical treatment that has been aforementioned and shall be discussed. All quantities are assumed to have SI units with the MKS system. The acceleration due to gravity is 9,81.

3.1. Problem having smooth solution

Assume that at time $t=0$, water depth is given by

$$h(x,0)=\begin{cases}11+\cos(x), & \text{if}-\pi\le x\le\pi,\\ 0, & \text{if } x<-\pi \text{ or } x>\pi,\end{cases}$$

and its velocity is zero everywhere. Water surface and velocity at the initial time are shown in Figure 1.

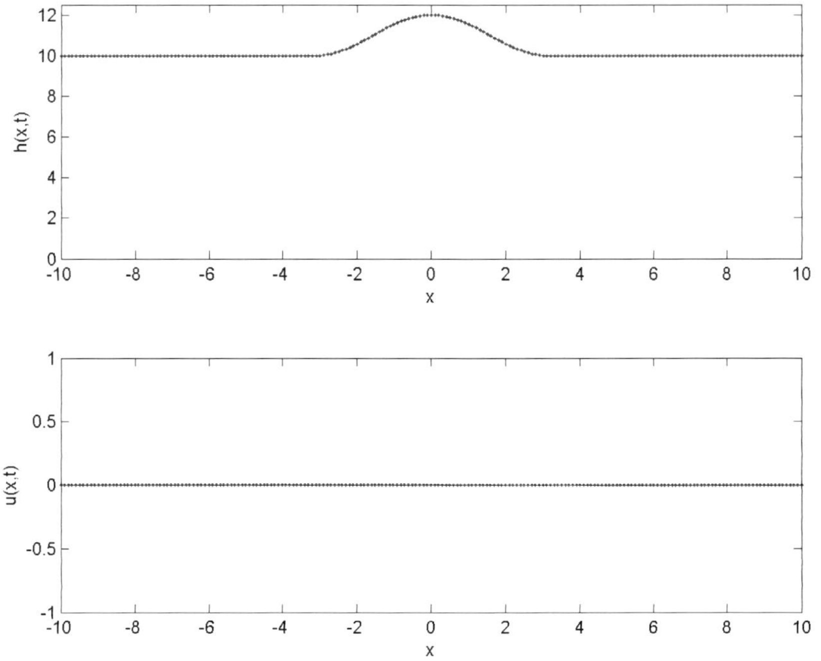

Figure 1. Water surface and velocity at time $t = 0$.

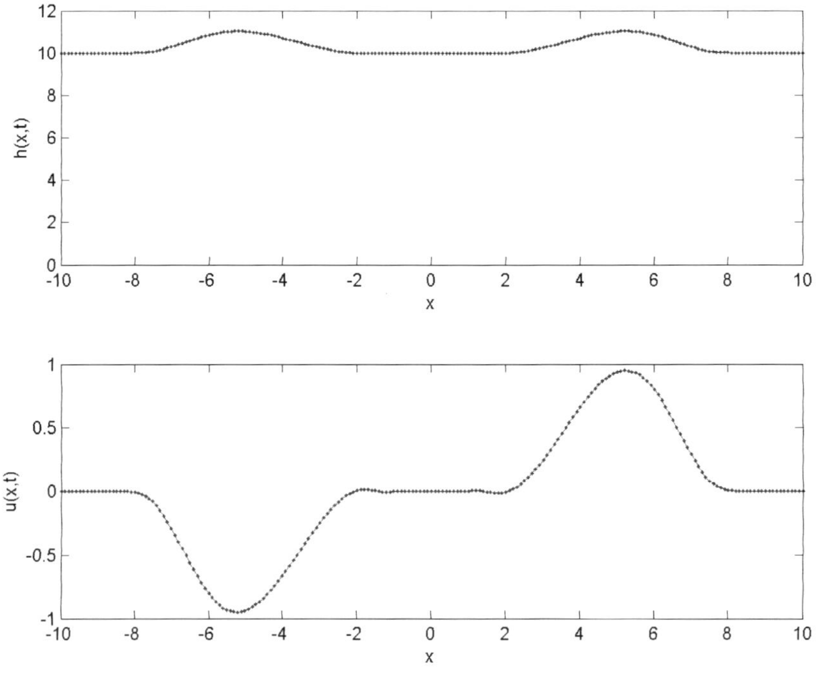

Figure 2. Water surface and velocity at time $t = 0.5$.

Figure 2 shows an illustration of the simulated fluid depth and fluid velocity for the considered problem. This simulation uses 201 grid points, $\Delta x = 0.1$, and $\Delta t = 0.005\Delta x$. The results of this simulation are correct with respect to its physics. That is, if the velocity is positive, then water wave moves to the right. If the velocity is negative, then water moves to the left.

3.2. Problem having nonsmooth solution without numerical treatment

Suppose that the water depth from the point $x = -10$ until $x = 0$ is 10 and the water depth from the point $x = 0$ until $x = 10$ is 5. Initially, water is still. These are illustrated in Figure 3.

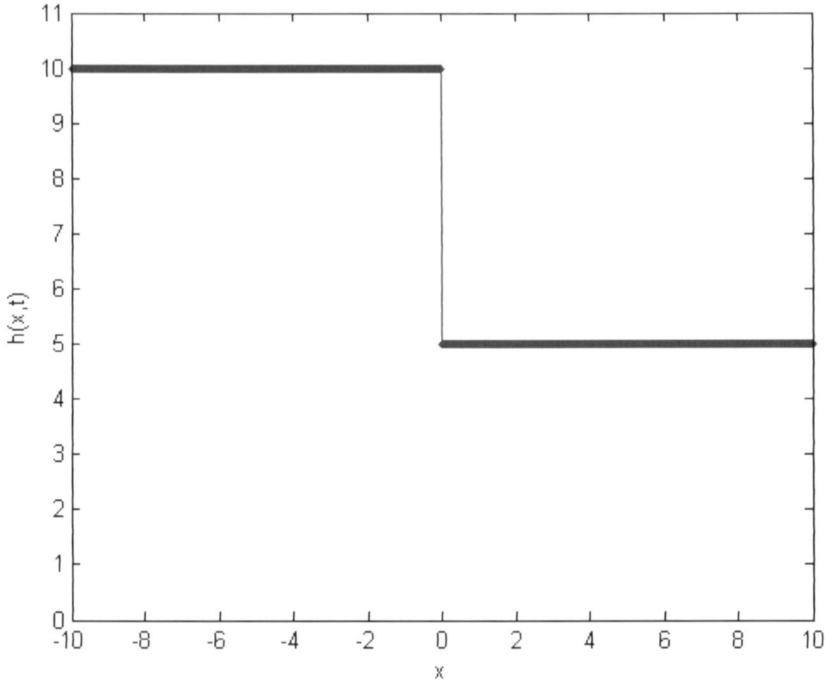

Figure 3. Initial water surface of the dam break problem.

Figure 4 shows an illustration of simulated water depth and velocity at time $t = 0.1$ using the scheme (5) and (8). This simulation uses 2001 spatial points, $\Delta x = 0.01$, and $\Delta t = 0.005\Delta x$. The simulation shows that artificial oscillations occur in the numerical results. To avoid oscillation, numerical treatment is used, as follows: h_j^n in equation (5) is changed to the average $(h_{j+1}^n + h_j^n + h_{j-1}^n)/3$, and q_j^n in equation (8) is replaced by the average $(q_{j+3/2}^n + q_{j+1/2}^n + q_{j-1/2}^n)/3$, so we obtain the following alternative scheme

$$h_j^{n+1} = \frac{h_j^{n-1} + h_j^n + h_j^{n+1}}{3} - \frac{\Delta t}{\Delta x}\left(q_{j+\frac{1}{2}}^n - q_{j-\frac{1}{2}}^n\right), \tag{9}$$

$$q_{j+1/2}^{n+1} = \frac{q_{j+3/2}^n + q_{j+1/2}^n + q_{j-1/2}^n}{3} - \frac{g}{2}\frac{\Delta t}{\Delta x}\left((h_{j+1}^n)^2 - (h_j^n)^2\right). \tag{10}$$

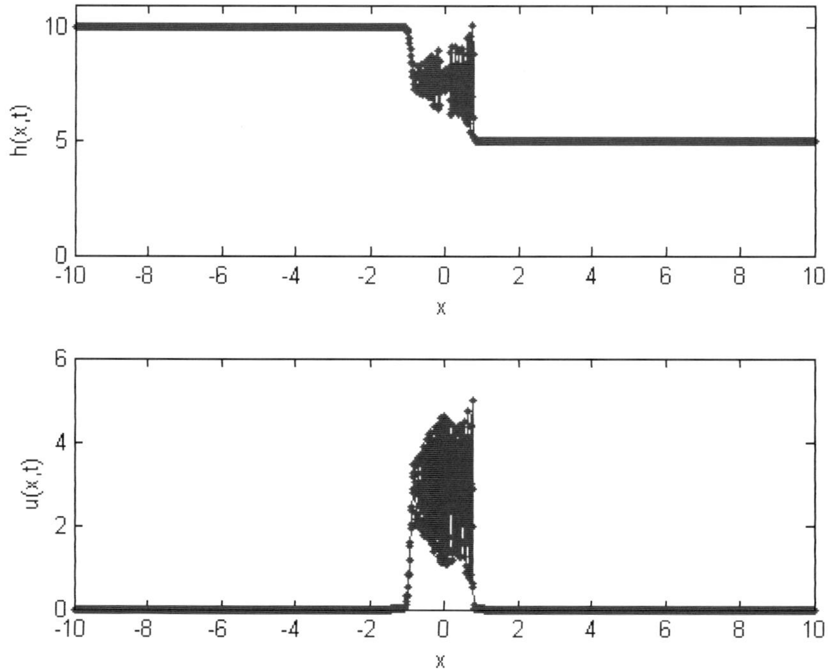

Figure 4. Staggered grid solution without scheme modification.

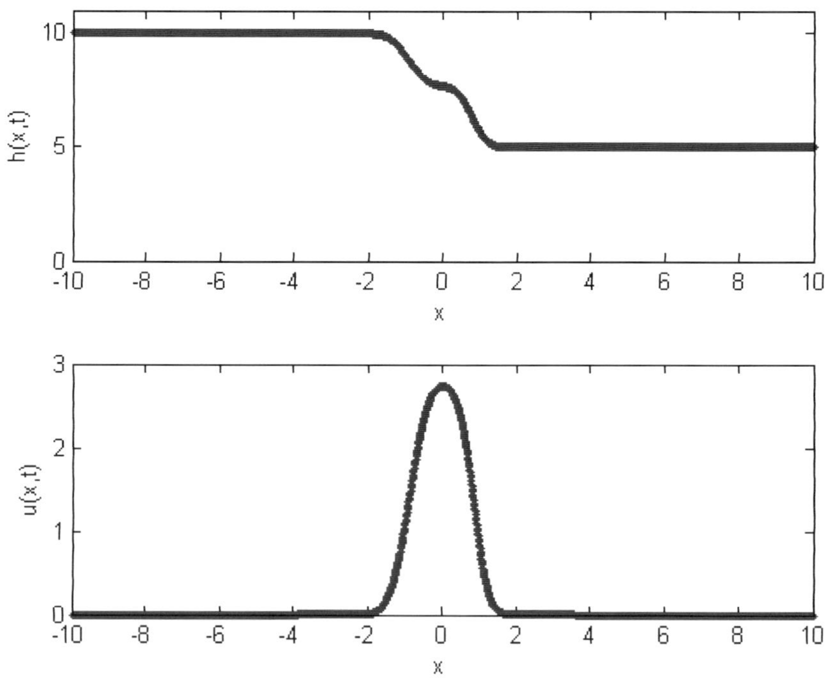

Figure 5. Staggered grid solution with scheme modification as a numerical treatment.

Figure 5 shows an illustration of simulated water depth and velocity at time $t = 0.1$ using the scheme (9) and (10). This simulation uses the same numerical setting as before, that is, 2001 spatial points, $\Delta x = 0.01$, and $\Delta t = 0.005\Delta x$. We observe that artificial oscillations do not occur in these results, as we expect. The solution is slightly diffusive.

4. Conclusion

We have solved the gravity wave-model equations using a staggered grid finite difference method. Our numerical tests involve smooth and nonsmooth solutions. A numerical treatment has been proposed, so that artificial oscillations do not occur in the solution, especially when we deal with nonsmooth problems. Future direction could be about minimising the diffusion around the discontinuity of the nonsmooth solution.

Acknowledgment

This research was supported financially by Sanata Dharma University. Both authors thank Sanata Dharma University for the support.

References

[1] Martins R, Leandro J, and Djordjevic S 2016 Analytical solution of the classical dam-break problem for the gravity wave–model equations *Journal of Hydraulic Engineering* **142** 06016003

[2] Martins R, Leandro J, and Djordjevic S 2016 Influence of sewer network models on urban flood damage assessment based on coupled 1D/2D models *Journal of Flood Risk Management* (in press) DOI:10.1111/jfr3.12244

[3] Mungkasi S, Supriyadi B and Wiryanto L H 2016 Steady flow over an arbitrary obstruction based on the gravity wave equations *Journal of Physics: Conference Series* **776** 012080

[4] Giannoulis I 2016 Interaction of modulated gravity water waves of finite depth *Journal of Differential Equations* **261** 3864

[5] van Oers A M, Maas L R M, Bokhove O 2017 Hamiltonian discontinuous Galerkin FEM for linear, stratified (in)compressible Euler equations: internal gravity waves *Journal of Computational Physics* **330** 770

[6] Zeng Y Q and Liu Q H 2001 A staggered-grid finite-difference method with perfectly matched layers for poroelastic wave equations *The Journal of the Acoustical Society of America* **109** 1.1369783

[7] Stelling G S and Duinmeijer S P A 2003 A staggered conservative scheme for every Froude number in rapidly varied shallow water flows *International Journal for Numerical Methods in Fluids* **43** 1329

[8] Liu Y and Sen M K 2009 An implicit staggered-grid finite-difference method for seismic modelling *Geophysical Journal International* **179** 459

[9] Yang L, Yan H, and Liu H 2016 Optimal implicit staggered-grid finite-difference schemes based on the sampling approximation method for seismic modelling *Geophysical Prospecting* **64** 595

[10] Yan H, Yang L, and Li X Y 2016 Optimal staggered-grid finite-difference schemes by combining Taylor-series expansion and sampling approximation for wave equation modeling *Journal of Computational Physics* **326** 913

[11] Ancey C, Iverson R M, Rentschler M, and Denlinger R P 2008 An exact solution for ideal dam-break floods on steep slopes *Water Resources Research* **44** W01430

[12] Mungkasi S, Li Z, and Roberts S G 2014 Weak local residuals as smoothness indicators for the shallow water equations *Applied Mathematics Letters* **30** 51

A staggered grid finite difference method for solving the elastic wave equations

Maria Yuliani Danggo[1] and Sudi Mungkasi[2]

[1] Department of Mathematics Education, Faculty of Teacher Training and Education, Sanata Dharma University, Mrican, Tromol Pos 29, Yogyakarta 55002, INDONESIA
[2] Department of Mathematics, Faculty of Science and Technology, Sanata Dharma University, Mrican, Tromol Pos 29, Yogyakarta 55002, INDONESIA

E-mail: yanidanggo@gmail.com, sudi@usd.ac.id

Abstract. We consider elasticity (elastic wave) equations for solving acoustics problems. We use the finite difference method on staggered grids to solve the elastic wave equations. On staggered grids, the pressure is computed at a set of spatial points, and the velocity is computed at another set of spatial points. An advantage of using staggered grids is that fluxes do not need to be approximated in the solving process, as the flux values are known. Numerical results confirm that our numerical method is reliable.

1. Introduction

Elastic wave is a wave that its propagation is dependent on the nature of the wave propagation medium itself. It has been an attention for a number of researchers (for example, see [1-9]) due to its importance in physics and engineering. One special example of elastic waves that occurs in everyday life is the propagation of sound, called an acoustic wave [10].

Staggered grid numerical methods provide a way to solve wave equations. These methods have a high accuracy for a large number of problems [11-14]. In this paper, we use the finite difference method with staggered grids to solve the elastic wave equation for acoustic problems. The finite difference method is chosen with staggered grids, because the forms of the equations match with the nature of staggered grids. An advantage of using staggered grids is that we do not need to compute an approximate flux at a certain interface at the domain. In particular, in this paper, we propose a numerical scheme to deal with the elastic wave equations when we use the staggered grid method to solve acoustic problems.

The paper is organised simply as follows. We present the mathematical model in Section 2. The numerical method is explained in Section 3. Numerical results are presented in Section 4. We provide some concluding remarks in Section 5.

2. Mathematical model

The general one-dimensional elastic wave equations are [1, 9]

$$\varepsilon_t(x,t) - u_x(x,t) = 0, \tag{1}$$

$$(\rho(x)u(x,t))_t - \sigma(\varepsilon(x,t),x)_x = 0. \tag{2}$$

Here x represents the space domain and t denotes the time. In addition, ε is the strain, u the velocity, ρ the density, and σ the stress. Note that $\varepsilon_t(x,t) = \frac{\partial \varepsilon(x,t)}{\partial t}$. Other derivative notations are analogous.

In equations (1) and (2), the stress σ and the strain ε are assumed to have the relation

$$\sigma(\varepsilon, x) = K(x)\varepsilon. \tag{3}$$

Here $K(x)$ is the bulk compressibility modulus. The pressure $p(x,t)$ is given by $p = -\sigma$, so equations (1) and (2) can be expressed in linear partial differential equations [15]

$$p_t + K(x)u_x = 0, \tag{4}$$

$$\rho(x)u_t + p_x = 0. \tag{5}$$

Assuming that $\rho(x) = 1$ dan $K(x) = 1$, we rewrite equations (4) and (5) as

$$p_t + u_x = 0, \tag{6}$$

$$u_t + p_x = 0. \tag{7}$$

The system of equations (6) and (7) is a hyperbolic system of partial differential equations. Without loss of generality, in the next sections, we shall focus on solving the system of equations (6) and (7), which is a model for acoustic problems.

3. Numerical method

We note that

$$p_t = \frac{\partial p}{\partial t} \quad \text{and} \quad u_x = \frac{\partial u}{\partial x}, \tag{8}$$

$$u_t = \frac{\partial u}{\partial t} \quad \text{and} \quad p_x = \frac{\partial p}{\partial x}. \tag{9}$$

We take the following discretisations of the derivatives

$$\frac{\partial p}{\partial t}\Big|_{t=t^n}^{x=x_j} \approx \frac{p_j^{n+1} - p_j^n}{\Delta t}, \tag{10}$$

$$\frac{\partial u}{\partial x}\Big|_{t=t^n}^{x=x_j} \approx \frac{u_{j+1/2}^n - u_{j-1/2}^n}{\Delta x}. \tag{11}$$

Using approximations (10) and (11), equation (6) can be written in a fully-discrete form as

$$p_j^{n+1} = p_j^n - \frac{\Delta t}{\Delta x}\left(u_{j+1}^n - u_{j-1}^n\right). \tag{12}$$

Furthermore, we take the following discretisations of the derivatives

$$\frac{\partial u}{\partial t}\Big|_{t=t^n}^{x=x_{j+1/2}} \approx \frac{u_{j+1/2}^{n+1} - u_{j+1/2}^n}{\Delta t}, \tag{13}$$

$$\frac{\partial p}{\partial x}\Big|_{t=t^n}^{x=x_{j+1/2}} \approx \frac{p_{j+1}^n - p_j^n}{\Delta x}. \tag{14}$$

Using approximations (13) and (14), the fully-discrete form of equation (7) is

$$u_{j+1/2}^{n+1} = u_{j+1/2}^{n} - \frac{\Delta t}{\Delta x}\left(p_{j+1}^n - p_j^n\right). \tag{15}$$

The system of equations (12) and (15) is the numerical scheme for solving acoustics problems governed by equations (6) and (7) on staggered grids.

4. Numerical results

This section is devoted to present our numerical results.

We assume that, initially, an acoustics problem has the pressure function

$$p(x,0) = \begin{cases} 1 - \cos(x+\pi), & \text{if} \quad |x| \le \pi, \\ 0, & \text{if} \quad \pi < |x| \le 10, \end{cases} \tag{16}$$

for $-10 \le x \le 10$. This initial pressure is shown in Figure 1. The velocity is $u(x,0) = 0$ for the whole space domain.

Using the given data, the space domain is $[-L, L]$, where $L = 10$. We shall compute the pressure and velocity in the system at every time step. We take that the number of space points is $N = 1000$, the space step is $\Delta x = 2L/(N-1)$, and the time step is $\Delta t = 0.05\Delta x$.

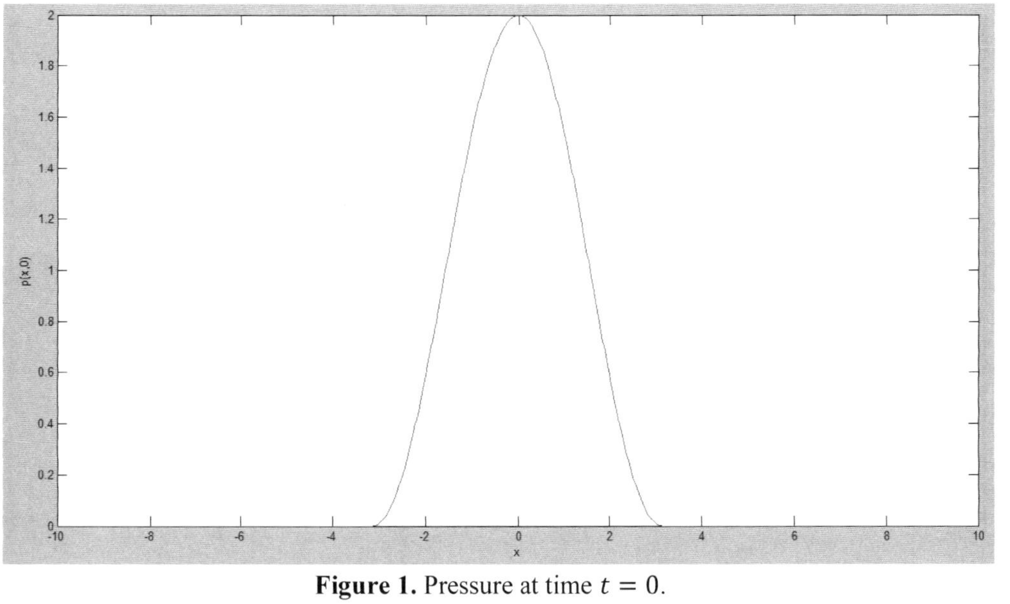

Figure 1. Pressure at time $t = 0$.

As initially, the pressure is centred around $x = 0$, this pressure separates into two waves propagating to the left and right directions. This is illustrated in Figures 2-3. Figure 2 shows the pressure and velocity at time $t = 2.5$, whereas Figure 3 shows the pressure and velocity at time $t = 5$.

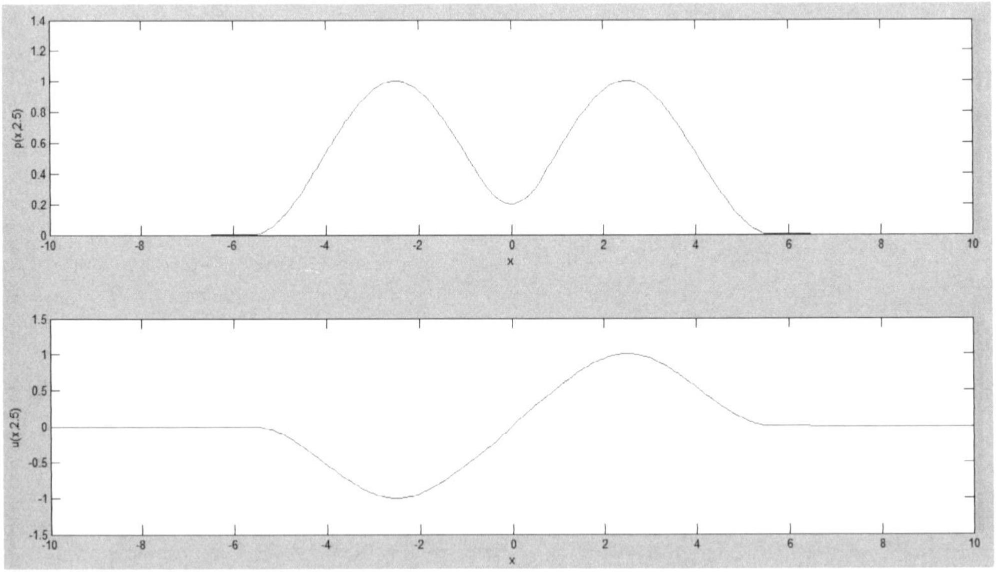

Figure 2. Pressure $p(x, 2.5)$ at time $t = 2.5$.

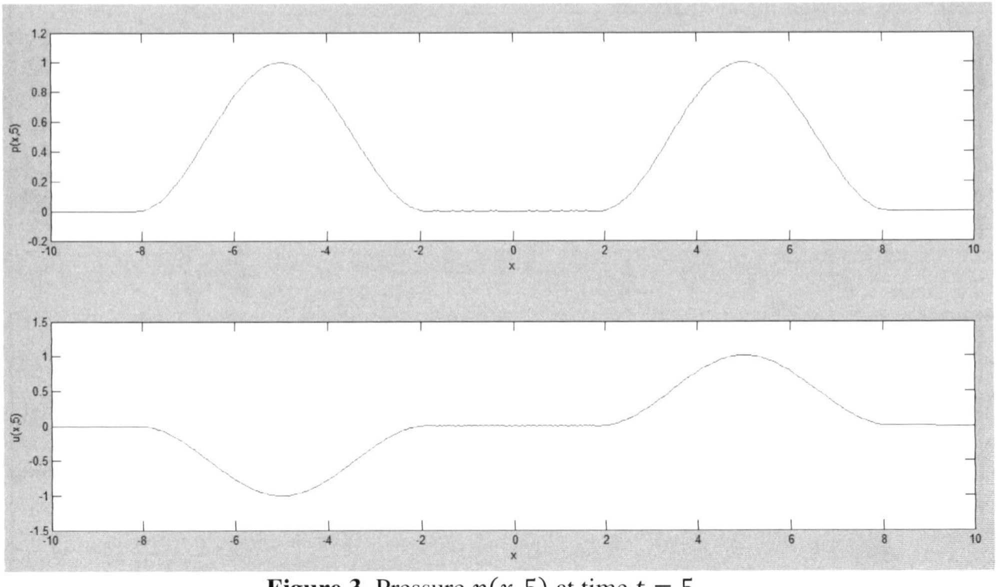

Figure 3. Pressure $p(x, 5)$ at time $t = 5$.

5. Conclusion

The staggered grid finite difference method has been derived and used to solve the elastic wave equations relating to acoustics problems. In this paper, the test uses smooth initial values. We obtain the correct physical behaviour in the numerical solution, as there is a single bump of pressure with zero velocity, the pressure separates into two moving in opposite directions. Future work shall investigate the performance of the numerical method when it is used to solve nonsmooth problems.

Acknowledgment

Expenses of this research were covered by Sanata Dharma University. Both authors are very grateful to Sanata Dharma University for the support.

References

[1] LeVeque R J 2002 Finite-volume methods for non-linear elasticity in heterogeneous media *International Journal for Numerical Methods in Fluids* **40** 93
[2] Nilsson S, Petersson N A, Sjögreen B and Kreiss H O 2007 Stable difference approximations for the elastic wave equation in second order formulation *SIAM Journal on Numerical Analysis* **45** 1902
[3] Hall F and Wang Y 2009 Elastic wave modelling by an integrated finite difference method *Geophysical Journal International* **177** 104
[4] Huang C and Dong L-G 2009 Staggered-grid high-order finite-difference method in elastic wave simulation with variable grids and local time-steps *Chinese Journal of Geophysics* **52** 1324
[5] Sjögreen B and Petersson N A 2012 A fourth order accurate finite difference scheme for the elastic wave equation in second order formulation *Journal of Scientific Computing* **52** 17
[6] Yan H-Y and Liu Y 2012 High-order finite-difference numerical modeling of wave propagation in viscoelastic TTI media using rotated staggered grid *Chinese Journal of Geophysics* **55** 252
[7] Lee J J and Kim H H 2016 Analysis of a staggered discontinuous Galerkin method for linear elasticity *Journal of Scientific Computing* **66** 625
[8] Schneider M, Ospald F and Kabel M 2016 Computational homogenization of elasticity on a staggered grid *International Journal for Numerical Methods in Engineering* **105** 693
[9] Mungkasi S and Supriyadi B 2016 Finite volume numerical solvers for non-linear elasticity in heterogeneous media *International Journal for Multiscale Computational Engineering* **14** 479
[10] Mungkasi S and Ningrum G I J 2016 Numerical solution to the linear acoustics equations *AIP Conference Proceedings* **1746** 020056
[11] Dettmer W G and Perić B 2013 A new staggered scheme for fluid-structure interaction *International Journal for Numerical Methods in Engineering* **93** 1
[12] Bartolo L D, Dors C and Mansur W J 2015 Theory of equivalent staggered-grid schemes: application to rotated and standard grids in anisotropic media *Geophysical Prospecting* **63** 1097
[13] Yang L, Yan H and Liu H 2016 Optimal implicit staggered-grid finite-difference schemes based on the sampling approximation method for seismic modelling *Geophysical Prospecting* **64** 595
[14] Yang L, Yan H and Liu H 2017 An optimal implicit staggered-grid finite-difference scheme based on the modified Taylor-series expansion with minimax approximation method for elastic modeling *Journal of Applied Geophysics* **138** 161
[15] LeVeque R J 2002 *Finite Volume Methods for Hyperbolic Problems* (Cambridge: Cambridge University Press)

The role of visual representation in physics learning: dynamic versus static visualization

Agus Suyatna[1], Dian Anggraini[2], Dina Agustina[2], Dini Widyastuti [2]

[1] Lecturer of Physics Education Study Program, Lampung University,
Jl. Prof. Dr. Sumantri Brojonegoro No. 1 Bandar Lampung, INDONESIA

[2] Student of Physics Education Study Program, Lampung University,
Jl. Prof. Dr. Sumantri Brojonegoro No. 1 Bandar Lampung, INDONESIA

E-mail: asuyatna@yahoo.com

Abstract. This study aims to examine the role of visual representation in physics learning and to compare the learning outcomes of using dynamic and static visualization media. The study was conducted using quasi-experiment with Pretest-Posttest Control Group Design. The samples of this research are students of six classes at State Senior High School in Lampung Province. The experimental class received a learning using dynamic visualization and control class using static visualization media. Both classes are given pre-test and post-test with the same instruments. Data were tested with N-gain analysis, normality test, homogeneity test and mean difference test. The results showed that there was a significant increase of mean (N-Gain) learning outcomes ($p < 0.05$) in both experimental and control classes. The averages of students' learning outcomes who are using dynamic visualization media are significantly higher than the class that obtains learning by using static visualization media. It can be seen from the characteristics of visual representation; each visualization provides different understanding support for the students. Dynamic visual media is more suitable for explaining material related to movement or describing a process, whereas static visual media is appropriately used for non-moving physical phenomena and requires long-term observation.

1. Introduction

Physics studies the physical objects that are abstract and concrete. The ability to solve the physical problems of each person varies depending on the daily experience. A person who has the ability to solve physics problems is often labeled physical intuition. Such people, in general, can solve physics problems more quickly [1]. To grow physical intuition is not an easy thing, but it can be helped by visualizing the phenomenon that is the problem. Visualization of a phenomenon can eliminate the abstract nature of a physical phenomenon. Thus the problem on the phenomenon will become easier to solve. Visual representation is widely used in communication for information in a recognizable form that can be easily interpreted without the use of a adequate calculation. Visual representations show in six types: graphs, tables, maps, diagrams, networks, and icons. Variations in this type can also be classified according to the way they convey the information and the efficiency at which they provide information [2]. Visual representation is very important to communicate ideas in the science classroom; However, such representational designs are not always beneficial to learners [3]. This depends on the nature of the phenomenon to be visualized. Science concepts are often abstract and difficult to understand for students. Visualizations have been shown to help students develop scientific conceptions [4, 5]

Dynamic fluid, impulse & momentum, and rotate motion are topics of physics learned in high school that is very important to be understood by students because the phenomenon is always experienced in everyday life. The ideal and true fluid phenomena, the principle of continuity, the principle of Bernoulli and its application, such as the flow of water in the pipe, the flow of water in the leaking tank, the flight of an airplane, the occurrence of the wind-fired roof, and the simulation of the difference in air pressure when a rider motorcycles passing on the highway overtake large cars are often encountered in daily life, but the fluid stream could not be seen directly so need visualization. Likewise in the material impulse and momentum, on the event of the collision where the force works in a short time and difficult to dismiss it, thus making students difficult to observe. To understand this, learners can not only imagine when the process of impulse and momentum occur. Therefore, the use of visual media will greatly assist students in understanding the concept. Rotational motion events, though easily observable in everyday life, but still difficult to imagine the centripetal style or complex relationship on the gear of bicycle. Everything requires visualization of images.

Vision is the construction of efficient symbolic descriptions from images of the world. An important aspect of vision is, therefore, the choice of representations for the different kinds of information in a visual scene [6]. Visualization objects can be pictures, three-dimensional models, schematic diagrams, geometrical illustrations, computer-generated displays, simulations, animations, videos and so on. Objects can be displayed in a variety of media formats, including paper, slides, computer screens, interactive whiteboards or videos, and may be accompanied by sound and other sensory data [7]. Visualization can be done using images that are static or dynamic. Static visualizations are still images such as photos, pictures, graphics, charts. Dynamic visualization of images that can live like silent movies, simulations, animations, that illustrate the process coherently so that messages can be delivered.

This study aims to examine the role of visual representation in physics learning (Dynamic Fluid, Impulse & Momentum, and Rotate Motion) and to compare the learning outcomes that are taught by dynamic visual and static visual media

2. Method

2.1. Research Design

The study was conducted using quasi-experiment with design Pretest-Posttest Control Group Design. Quasi-experiment is a research design involving an experimental approach but where random assignment to treatment and comparison groups has not been used.

2.2. Participants

The population in this research consists of grade XI IPA senior high school students in odd semester academic year 2016/2017 in Lampung Province. The sampling technique used in this research is purposive sampling technique. Sampling is done by selecting three equivalent Senior High School in Lampung Province, and then two equivalent classes from each school are taken as the sample.

2.3. Procedure

Each pair of two equivalent classes obtained learning material Rotation Motion, Dynamic Fluid, and Impulse & Momentum. The first class learns the material using dynamic visual media and the second class using static visual media. Pre-test is done before the implementation of learning and post-test is done after the learning finished. The test type is an essay. Data were analyzed using calculating normalized gain <g> using formula

<g>=(Sposttest-Spretest)/(Smax-Spretest) [8]

Increased category of learning outcomes according to [8] as follows.

<g> < 0.3 Low category
0.3 < <g> < 0.6 Medium category
<g> > 0.6 High category

In each pair of data group, the test of normal distribution, homogeneity test, and average difference test of N-gain were done.

3. Results and Discussion

The data of pre-test, post-test, and the test of N-Gain in different class which learning using dynamic visual media and using static visual media is shown in Table 1. Normal distribution test on N-gain of all data groups obtained that data distribution is not normal, so that the different test of N-gain uses a statistical test of non-parametric Mann-Whitney U.

Table 1. Pre-test, post-test, N-Gain, and p of difference test

Matter	Using Media Visual Dynamic				Using Media Visual Static				p (2-tailed)
	pretest	posttest	n-gain	Gain category	pretest	posttest	n-gain	Gain category	
Rotate Dynamic	31.05	72.5	0.61	High	33.38	61.76	0.43	Medium	0.014
Dynamic Fluid	13.58	69.13	0.64	High	14.57	53.00	0.48	Medium	0.010
Impuls & Momentum	18.97	75.91	0.70	High	18.83	68.38	0.59	Medium	0.010

The result of data analysis showed that both classes using dynamic visual media and static visual media significantly increase in physic learning outcomes, but based on Mann-Whitney U test results, there was a significant difference of N-Gain ($p < 0.05$) between the learning classes using dynamic visual media and using static visual media for all matter, Rotate Dynamic, Dynamic Fluid, and Impulse & Momentum. Learning outcomes using dynamic visual media is higher compared to classes that use static visual media (Figure 1). Learning outcomes using dynamic visual media obtain high category, while those using visual static media obtain medium category.

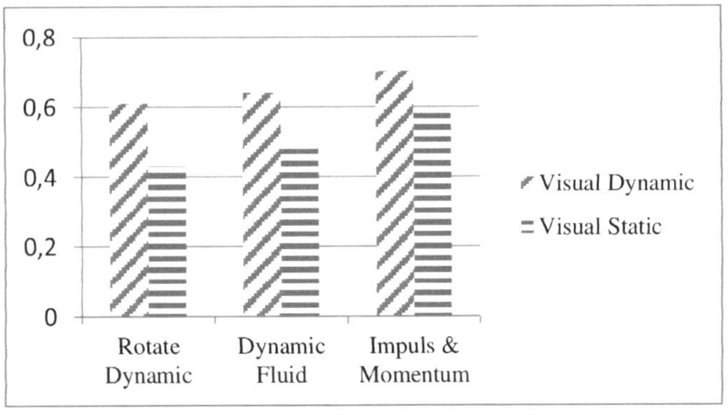

Figure 1. The comparison of N-Gain using dynamic visual media and static visual media

IOP Conf. Series: Journal of Physics: Conf. Series **909** (2017) 012048 doi:10.1088/1742-6596/909/1/012048

The average difference of the learning outcomes in the two group classes is related to the learning process of the two group classes. The teaching methods, which are applied to both groups, are same, what makes it different is in the use of media. The experimental classes use dynamic visuals as a learning media, while the control classes use static visual media. Dynamic visual media as well as static visual media displays the same visualization phenomena, what distinguishes it is only the one moves to describe the actual phenomenon, while the other is a still image. The results showed that the results of physics learning assisted by dynamic visual media are better than using static visual media. This finding is in line with the findings [9], which is the same instructional design methods are effective across different media. The learning results of the classroom that use dynamic visual media are higher because it has the advantages such as it is more realistic in describing a physics event and it can describe the process.

The dynamic visualization of the circular motion in this study is using two-wheeled motion animation linked by chain, centre, and tangent (Figure 2). In this visualization, the rotation of each wheel and the number of turns on each wheel can be observed. In investigating the relationship between period and frequency by using dynamic visuals, it can be seen the time required to spin and the number of rounds that occurred. The use of video and circular motion animation of the gear system of bicycle wheel can show the number of rotations of each wheel, the direction of wheel rotation and the influence of the difference of the radius. The video and animation allow the students to determine the angular velocity, as well as the magnitude and the direction of the linear velocity of the wheels and to predict the effect of radius on the number of wheel rotation (Figure 3). The students who received learning by using dynamic visual media which is 63% can answer the questions about the magnitude and the direction of linear velocity and angular velocity on two tangent wheels, while in the static visual class only 38% can answer correctly.

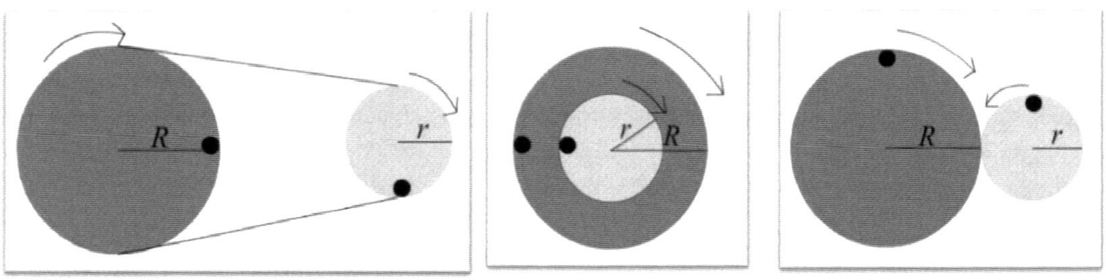

a b c

Figure 2. Wheel system connected by chain (a), Centre (b), and Tangent (c)

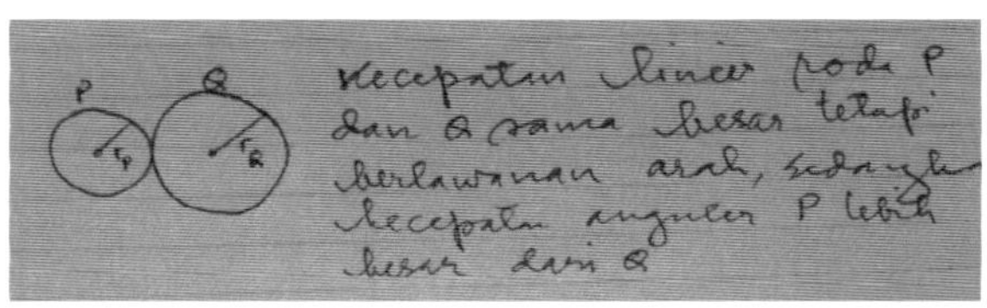

Figure 3. An example of a student's answer to the dynamic visual class of the wheel system

An example of dynamic visualization in learning dynamic fluid matter is the fluid flow on the aircraft wing (Figure 4). By dynamic visualization, it can be described that the streamline flow of the fluid is above and below the wing of the aircraft. Therefore, it can be clearly seen that the fluid that moves above the wing of the plane is faster than the one move under the wings, so that students understand better why the air pressure under the wing is higher compared to the pressure above the wing (Figure 5).

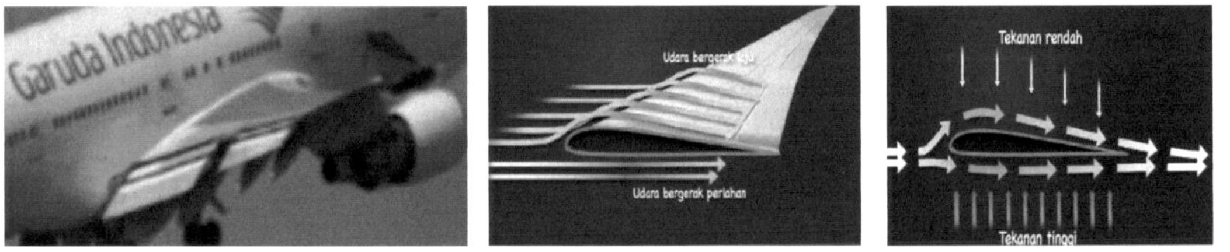

a b c

Figure 4. Wings form when take-off (a), fluid flow (b), and air pressure difference (c)

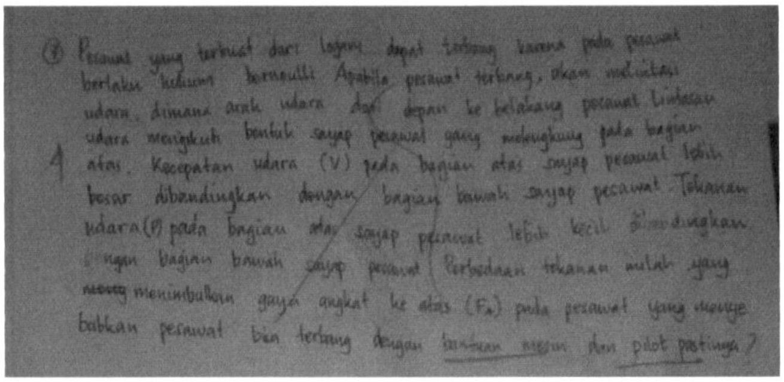

Figure 5. An example of student' answer in the dynamic visual class related to a plane can fly

An example of dynamic visualization in impulse and momentum learning is a video depicting a small car with high speed crashing into a heavy box car but moving at the low speed (Figure 6). This video is followed up by discussing the damage level of the car if the car's speed is lower and the car hit is lighter. Thus the students in dynamic visual media class can explain the greater damage effect of a colliding vehicle based on the principle of momentum (Figure 7). The students in the dynamic visual class obtained an average score of 73.4 for this problem, whereas in the static visual class obtained an average score of 68.5.

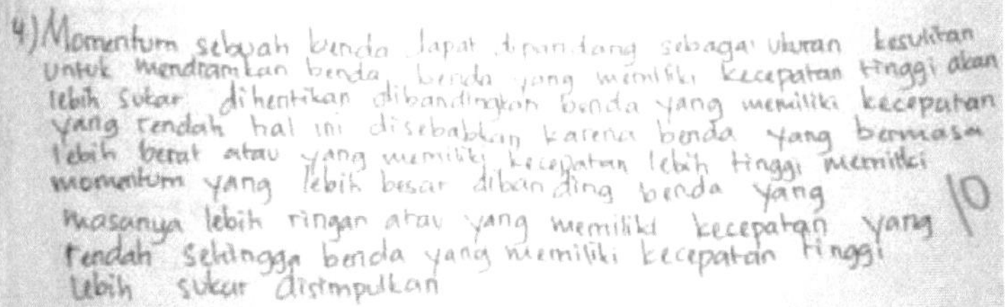

Dynamic visual media is appropriately used to describe material related to movement. This is supported by the results of the study [10] in principle, an animation is created to convey concepts of change and considered effective in expressing processes. Animated graphics seemed superior to static ones, scrutiny of the actual stimuli revealed that the animated graphics conveyed more information, especially about the micro steps between larger steps [12]. Dynamic visualizations can use animated pictures to provide an explicit depiction of abstract concepts and processes [13].

The use of dynamic visual in learning can improve the students' learning outcomes. This is in line with the statement [14] "...with the animation video as a source of learning will facilitate students in the learning process, especially in realizing the concept of material which is abstract." It is also in accordance with the research conducted by [15] that the use of animation media as dynamic visual media can improve the learning outcomes. The learning outcomes which are using dynamic visuals are higher than static visuals also supported by [16] that the students who learn by using dynamic visuals have higher cognitive learning outcomes than the students who learn by using static visuals.

Dynamic visual as a media in learning activities gives the opportunity to the students to observe events/phenomena which looked more realistic because of the motion of the picture. By using dynamic visuals, the students can analyze, prove and conclude by themselves about an event related to the material. In addition, by dynamic visuals, the delivery of information about the subject matter to the students is easier, and the movement of the images provides clear information.

Static visuals are less able to display a dynamic event picture, achieving a coherent understanding is difficult for students using curriculum featuring static pictures in the textbook as these pictures do not convey the dynamic interactions [4]. However, dynamic visual media also has a disadvantage which is it cannot be observed longer because the images are moving and changing rapidly. The experience in this study shows that learning to analyze the direction of centripetal acceleration is more appropriate to use static visual. Visual static is easier to use to indicate the direction of centripetal acceleration that always goes to the same point that is the center of the circle. Thus, the students are more easily understand where the direction of centripetal acceleration of objects at a certain point. The achievement of these indicators, influencing the learning outcomes to reach the indicator calculates the magnitude of centripetal acceleration. The explanation of the direction of centripetal acceleration is an explanation of a concept that is static, so it is better to be explained using static visuals.

A review of research on the design of multimedia explanations conducted in lab at Santa Barbara shows (a) a multimedia effect-in which students learn more from words and based on their own words in both book-based and computer-based environments, (b) a coherence effect-in which students learn more deeply when extraneous material is excluded rather than included in both book-based and computer-based environments, (c) a spatial contiguity effect-in which students learn more deeply when printed words Are in the book-based and computer-based environments, and (d) a personalization effect-in which students learn more deeply when words are presented in conversational rather than formal style-both in computer- Based environments containing spoken words and those using printed words [9].

Differences in scores are obtained larger in classes with dynamic visual than static visuals, because of the advantages of dynamic visual, according to the existing theories, describes a process in a coherent manner, encourage and increase motivation, develop imagination, improve memory, facilitate in understanding the concepts, and give the more realistic picture. If the static visual, the advantages are having a variety of types can be real visual (photography), charts, animation, graphics and so forth. However, static visuals only show the visuals without showing the process coherently so that it gives less understanding of the concept to the students. This causes the use of static visuals gets a lower score compared to the use of dynamic visuals.

4. Conclusion

The use of visual media in Rotation Motion, Dynamic Fluid, and Impulse & Momentum learning can improve student learning outcomes. The average learning outcomes of students who are using dynamic visualization media is significantly higher than the class which obtains learning by using static visualization media. It is seen from the characteristics of visual representation; each visualization provides different understanding support for the students. Dynamic visual media is more suitable for explaining material related to movement or describing a process, whereas static visual media is appropriately used for non-moving physical phenomena and requires long-term observation.

5. Acknowledgment

Thanks delivered to Dr. Abdurrahman, M.Si, Drs. Eko Suyanto, M.Pd., Drs. Feriansyah Sesunan, M.Pd. Ismu Wahyudi, S.Pd., M.Pfis. who have provided input on this research.

References

[1] Larkin J, McDermott J, Simon D P, and Simon H A 1980 Expert and novice performance in solving physics problems *Science* **208(4450)** 1335-1342

[2] Lohse G L, Biolsi K, Walker N, and Rueter H H 1994 A classification of visual representations *Communications of the ACM* **37(12)** 36-50

[3] Cook M P 2006 Visual representations in science education: The influence of prior knowledge and cognitive load theory on instructional design principles *Science education* **90(6)** 1073-1091

[4] Chiu J L and Linn M C 2014 Supporting Knowledge Integration In Chemistry With A Visualization-Enhanced Inquiry Unit *Journal Of Science Education And Technology* **23 (1)** 37-58

[5] Whitworth B A and Chiu J L 2013 Pre-Service Teachers Use of Visualizations in the Science Classroom A Case Study *Virginia Journal of Science Education* **5 (1)** 16-32

[6] Marr D, Lal S, and Barlow H B 1980 Visual Information Processing: The Structure and Creation of Visual Representations [and Discussion]. Philosophical Transactions of the Royal Society of London *Biological Sciences* **290(1038)** 199-218

[7] Vavra K L, Vera J W, Karen L, Linda M P, Stephen P N, and John M 2011 Visualization in Science Education *ASEJ* **41 (1)** 22-30

[8] Hake R 1998 Interactive-Engagement Versus Traditional Methods: A Six-Thousand-Student Survey of Mechanics Test Data for Introductory Physics Courses. *American Journal of Physics* **66(1)** 64-74

[9] Mayer R E 2003 The promise of multimedia learning: using the same instructional design methods across different media. *Learning and instruction* **13(2)** pp 125-139

[10] Manalu K 2013 Gambar Diam Versus Animasi: Visualisasi Dalam Pembelajaran Biologi *Jurnal Al-Irsyad* **3** 46-51

[11] Lin C 2009 Learning Action Verbs With Animation *The Jalt Call Journal* **5 (3)** 23-40

[12] Morrison J B, Tversky B, and Betrancourt M 2000 March. Animation: Does it facilitate learning *In AAAI spring symposium on smart graphics* pp 53-59

[13] Ryoo K & Linn M C 2014 Designing Guidance For Interpreting Dynamic Visualizations: Generating Versus Reading Explanations *Journal of Research in Science Teaching* **51(2)** 147-174

[14] Ismawati D A, Danang T 2016 Pengembangan Media Video Animasi Untuk Meningkatkan Hasil Belajar Siswa Dalam Pelajaran Matematika Sub Pokok Bahasan Hubungan Antar Sudut Kelas VII SMP Negeri 1 Krembung Sidoarjo *Jurnal Mahasiswa Teknologi Pendidikan* **10 (1)**

[15] Sari, N W dan Samawi A 2014 Pengaruh Penggunaan Media Animasi Terhadap Hasil Belajar IPA Siswa Slow Learner. *Jurnal Penelitian dan Pengembangan Pendidikan Luar Biasa* **1(2)** 140-144

[16] Ardiyansyah M 2013 Keefektifan Gambar Statis, Gambar Dinamis Ball-and-Stick, dan Model Molekul Sederhana Dibuat dari Jarum Pentul pada Pembelajaran Bentuk dan Kepolaran Molekul. *Jurnal Pendidikan Sains* **1 (3)** 307-314

Development of Smartphone e-Modul by Problem Solving Method for Biot-Savart Theory

Ihtiari Prastyaningrum, Jeffry Handhika
Universitas PGRI Madiun

ihtiari.prastya@unipma.ac.id
jhandhika@unipma.ac.id

Abstract. Biot-Savart law is an equation that describes the magnetic field created by a current-carrying wire and allows you to calculate its strength at various points. Biot-Savart law is too difficult to be understood, especially about the mathematics concept. Based on the situation, developed an interactive media that's an Electronic Module. This module based on the problem-solving method and can be accessed by smartphone. This research by using a development method, where is, an electronic module is created by Adobe Flash software. By the development of this module is expected that can improve the ability of mathematics concept analytical.

1. Introduction

Students of Electrical Engineering Education Study Program should be equipped with a mature understanding of the concepts of electromagnetic fields. It is intended that later if they become electro teachers, they can pass the understanding of the concept to the students [1]. The selection of the chapter on the law of Biot-Savart is intended to make the theory which is the basis for determining magnetic field strength can be understood first so that it will not lead to errors in conceptual understanding (misconception).

Some educational experts have undertaken research on misconceptions. For example, in 2014 Jeffry Handhika et al, developed a cognitive-charged medium of learning to reduce the alleged misconception of the elementary physics course [2]. Aydin and Balim in 2010 implemented a concept change strategy to eliminate misconceptions in students [3]. Previously, Baser explains the effect of using concept change steps on students' understanding of heat and heat. Some of these studies have reinforced the fact that misconceptions are a serious problem [4].

The terms of misconception are expressed differently by the researchers. According to David Hammer (1996) "*misconceptions to refer only to the phenomenology of patterns in students' responses that are inconsistent with expert understanding*"[5]. While Giuseppe and Fraser 2012 say that "The *ideas and Explanations That vary significantly from accepted knowledge are typically called misconceptions or alternate conceptions*[6].

Misconceptions can occur because of factors such as intuition [7], some things that are abstract and elusive, and the error of explanations from teachers, friends, or teaching resources. In order to reduce misconceptions not only improve methods and learning models but also consider other factors such as teachers, teaching resources such as modules, books, and media.

From the observations that have been done previously obtained the fact that in the Electrical Engineering Education Study Program the field theory course is quite difficult to understand. After further observation, it was found that one of the elusive material is the Biot-Savart law. Both in terms of calculating the magnetic field mathematically, as well as the influence of each variable against the magnitude of the magnetic field magnitude.

Discussion about the law of Biot Savart is studied by students of Electrical Engineering Education who take the field theory course. The field theory course deals with the various theories relating to the electromagnetic field. As already known that the Ampere theorem and the Biot–Savart law is well-known tools used to calculate magnetic fields created by current distributions [8]. The determination of the magnetic field created by a rectilinear *infinite* wire is a classical problem that can be found in every electromagnetism book. The calculation is generally carried out by applying the Ampere theorem to a current circulation on a circle Γ of radius r. Simple symmetry considerations indicate that the magnetic field is orthoradial with $B = B(r)u\theta$, and the use of the Ampere law leads to the well-known result

$$\mu_o I = \oint_{\Gamma} \vec{B}\,.d\vec{l} = 2\pi R\left|\vec{B}(r)\right| \Rightarrow \vec{B}(C) = \frac{\mu_0 I}{2\pi R} \quad (1)$$

The equation model as seen in Equation (1) is an equation that is difficult to comprehend and solved for example by students in the electrical engineering education program. It is based on data in the form of final grades of students Electrical Engineering Education Study Program Semester 3 academic year 2015/2016 which shows the average value of only 67.85. Besides data, student difficulties can also be known from some opinions expressed by students at the time of the interview. Approximately 85% of students find it difficult to understand the equations in field theory, Biot Savart's law for example. Moreover, if you have to connect the equation with real electronics devices.

In the previous semester lectures, lecturers only use Slide Presentations, learning media using only whiteboards. Existing modules Only gives some similarities and concepts in a line Great about field theory. Some examples of problems and solutions are available In the module only refers to the completion of the mathematical case. Some references Mandatory owned by students also difficult to be understood, because according to the student There are some difficulties to connect the equation with the device Often used electronics. And also, the choice of words used is also more complicated. It is based on interviews that have been done with some Students who have taken courses in Medan Theory.

Based on these conditions, an interactive learning media is required. Learning media more communicative and can facilitate students in learning about the theory of electromagnetic fields, especially about the law of Biot-Savart. One of the media that can be used is media e-Module based smartphone. Where the e-Module media can be accessed via smartphone, so it can be used anywhere. This media uses problem-solving method because the interactive model of e-Module is solving some mathematical calculation problem related to Biot-Savart law

2. Biot-Savart Law

The currents which arise due to the motion of charges are the source of magnetic fields. When the charges move in a wire at a certain velocity can produce a current I, the magnetic field at any point P due to the current can be calculated by adding a small part of magnetic field $d\mathbf{B}$, from small segments of the wire $d\mathbf{s}$, (Figure 1.1)

These segments can be assumed as a vector quantity. It has a magnitude of the length of the segment and pointing in the direction of the current flow in the wire. The source of current is very small and be written as $\mathbf{I}d\mathbf{s}$ [9].

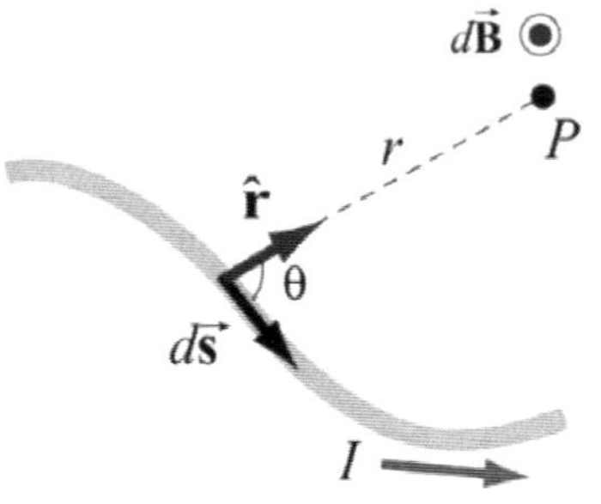

Figure 1. Magnetic field $d\mathbf{B}$ at point P because the effect of the current on the wire

Let r denote as the distance from the current source (in this case is a wire) to the field point P, and $\hat{r}$ the corresponding unit vector. The Biot-Savart law gives an expression for calculating the magnetic field contribution, $d\mathbf{B}$, from the current source, $Id\mathbf{s}$, and the equation of Biot-Savart Law is

$$d\boldsymbol{B} = \frac{\mu_0}{4\pi}\frac{Ids \times r}{r^2}$$

Where μ_0 is a constant called the permeability of free space, with the value $4\pi \times 10^{-7} T.m/A$

Biot-Savart law is used to determine the magnetic field strength either on a straight wire, circular, solenoids or toroid. By an integral part, we can find the magnetic field strength either on a straight wire, circular, solenoids or toroid [10]. In addition to previous research, the Biot-Savart law can also be used as a theoretical basis for determining the electromagnetic energy of the solenoid in the form of special relativity [11].

3. Problem-Solving Method for Mathematics Learning

Problem-solving problem-solving in mathematics is divided into three interpretations, namely problem solving as a goal as a goal, problem-solving as a process, and problem-solving as a basic skill [12]. In mathematics, first, problem-solving is used as one of the learning objectives. In this case problem solving does not depend on specific problems or problems, procedures, or methods, and also the content of mathematics. Second is problem-solving as a dynamic process. In this aspect, problem-solving can be interpreted as a process of applying all knowledge possessed to new and unusual situations. In this interpretation, what needs to be considered are methods, procedures, and strategies used by students in solving a problem. Last is problem-solving as a basic skill. In this interpretation more emphasize problem-solving as the basic skill of counting.

4. Method

This research method is research and development. Subjects in this study were students at the PGRI Madiun University who take a course in Electromagnetic Field (17 students). Data were collected through the e-Module media test. Media that has been created using Adobe Flash has tested eligibility and validated first by some expert lecturers related to the field of science of Electromagnetic Field and lecturer of media learning experts. Media that has been declared feasible, tested to students with the problem-solving method. Where students are given initial knowledge about the concept of Biot-Savart law theory. The following concepts have been described in a mathematical equation model. In turn related to the method of problem-solving, students are given feedback in the form of influence given by each variable that affects the magnetic field value based on Biot-Savart law equation. By using the media, students must be able to present the influence of each of these variables. After that stage is reached, the student is given a conceptual case concerning the Biot-Savart law that must be completed

IOP Conf. Series: Journal of Physics: Conf. Series **909** (2017) 012049 doi:10.1088/1742-6596/909/1/012049

using e-Module media simulation. But previously the students are guided to be able to declare language problem in the mathematical language.

5. Result and Discussion

This result is obtained from the pre test and post test conducted on 15 respondents (students). The first problem is understanding the concept of the variables that affect the magnetic field strength. And the second about the completion of mathematical calculations to determine the strength of the magnetic field. The result showed by Table 1.

Table 1. Percentage of mathematical concepts understanding for Biot-Savart Law

Test	Aspect of Research	Result		
		Before	After	N-Gain
1.	**Magnetic Field Strength of Straight Wire** $B = \frac{\mu_0 I}{2\pi r}$			
	The change of variable I (current) influences the value of the magnetic field strength (B), where if the value of the current flowing large (the amount of electric charge (q) flowing a lot) then the field strength increases, and vice versa. However, if the distance (r) between the electrical current wire with a point around the wire farther, then the magnetic field strength is reduced, and vice versa	13,3 %	66,7 %	0,4
	Can calculate the value of magnetic field strength	13,3 %	46,7 %	0,3
2.	**Magnetic Field Strength of Circular Wire** $B = \frac{\mu_0 I}{2r}$			
	• Changes in variables I and a give effect such as the application of Biot-Savart Law on straight wire, • The explanation of the difference between the first and second equations can be explained mathematically.	20,0 %	73,3 %	0,4
	Can calculate the value of magnetic field strength and the influence of variable change, but without explanation about the differences with the first equation (for straight wire)	20,0 %	53,3 %	0,4
3	**Magnetic Field Strength of the Axis of Solenoid** $B = \frac{\mu_0 I}{L} N$			
	• The change in the value of variables I and N is proportional to the magnetic field strength value (B), whereas L is inversely proportional. • It is an equation for determining the strength of the magnetic field on the solenoid axis, which is twice as large as the solenoid end	26,7 %	80,0 %	0,5
	Can calculate the value of magnetic field strength and the influence of variable change, but without explanation about the differences with the equation of magnetic field strength as the axis of the solenoid.	26,7 %	66,7 %	0,4
4.	**Magnetic Field Strength of the End of Solenoid** $B = \frac{\mu_0 I}{2L} N$			
	• The change in the value of variables I and N is proportional to the magnetic field strength value (B), whereas L is inversely proportional.	26,7 %	80,0 %	0,5

	• It is an equation for determining the strength of the magnetic field on the axis of the solenoid, which the value is half of the value in the solenoid axis			
	Can calculate the value of magnetic field strength and the influence of variable change, but without explanation about the differences with the equation of magnetic field strength as the axis of the solenoid.	40,0 %	80,0 %	0,4
5.	**Magnetic Field Strength of Toroid** $B = \frac{\mu_0 I}{2\pi a} N$			
	• The change in the value of variables I and N is proportional to the magnetic field strength value (B), whereas a is inversely proportional. • The value of a is the average radius of the toroid obtained by the equation $(a_1 + a_2) / 2$	20,0 %	86,0 %	0,6
	Can calculate the value of magnetic field strength	46,7 %	80,0 %	0,5

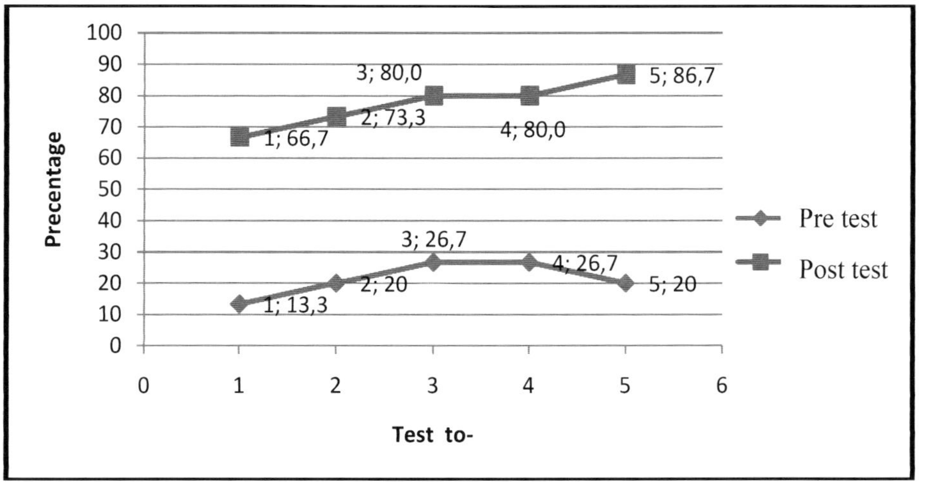

Graph 1. Graph of percentage of pre test and post-test values for the first question

The results in Table 1 are outlined in Graph 1 and Graph 2. Graph 1 shows the student's pre-test and post-test results from the first test (magnetic field strength on a straight wire) to five (magnetic field strength in toroid) for the first type of question. For the first pre-test to post-test to four percentage of students who are able to solve the problem has increased, but when the pretest to five percentage decreased. After analyzing, it is because the type of problem on the fifth pre-test (magnetic field strength in toroid) students must first understand the effect of the radius of toroidal to the area, then connect it with magnetic field strength.

As for the posttest score, the graph shows that the percentage of the number of students who understand the concept of the Biot-Savart law increases with the medium Gain category (0.5 to 0.6). The Gain score is calculated following the equation [13]. After analyzing the students' value, it is concluded that the students of Electrical Engineering Education Study Program need more practice until they can understand the concept of the influence of certain variables on magnetic field strength.

Graph 2 shows an increase in the pretest and posttest results of students for the second type of question (mathematical calculation). In the pretest for the second question, it increases from the first to the fourth but decreases in the fifth question. This is because on the fifth question (field strength in toroid) students must first determine the area of the toroid, then proceed to determine the strength of the field. As for the post, the test has increased, starting from the first test until the fifth test, with the category of medium gain (0.5 to 0.6). Based on the results of research that has been developed, it can be seen that with the e-Module Smartphone module students Electrical Engineering Education Studies University PGRI Madiun can more easily accept the material about the concept of mathematical calculations on Electromagnetic Field Theory focus material Biot -Savart Law.

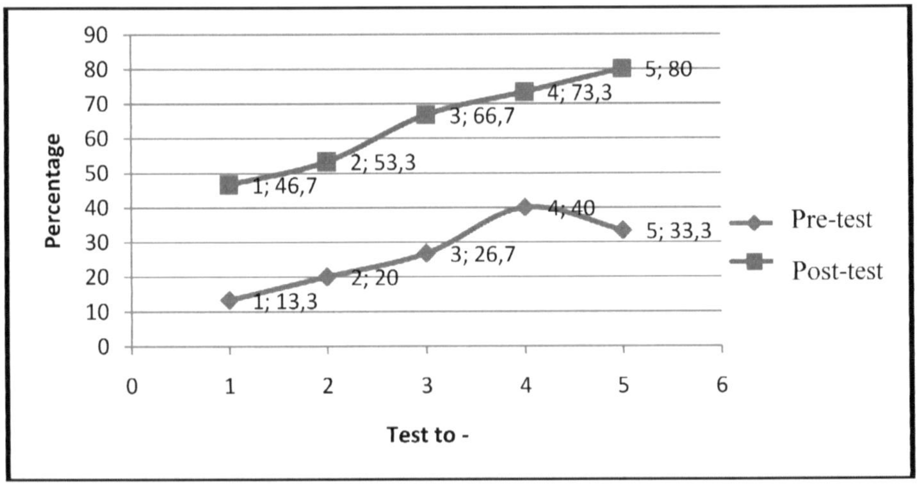

Graph 2. Graph of percentage of pre test and post-test values for the second question

Based on the all of the result, by using smartphone e-Modul understanding of the five derived equations of the Biot-Savart Law equation increases, with the medium gain category.

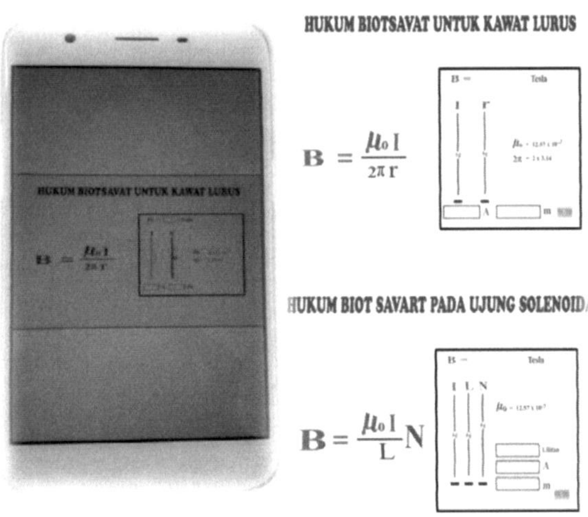

Figure 2. Sample media display on smartphone

Improved understanding focuses on the understanding of the influence of each variable of the principal quantities on magnetic field strength and mathematical arithmetic to determine the strength of the magnetic field. This is due to the e-Module media for the simulation part set for both fields. To calculate the magnetic field strength using e-Module, students only enter the value of the required variables, then tested, then the program will automatically calculate the strength of the magnetic field.

6. Conclusion

The mathematical concept of the law of Biot-Savart can be understood more easily using e-module based smartphone media. The application of this media can improve students' understanding of mathematical concepts, as evidenced by the increase in the percentage of the number of students who can solve problems related to mathematical concepts of Biot-Savart Law. Increasing the percentage with the category of medium gain (0.5 to 0.6). For the future can be developed again similar media but supported by its real equipment. So in addition to understanding, students are also able to apply the concept directly.

References

[1] Kurniadi. 2008. *Mengurangi Miskonsepsi Dinamika Dengan Konflik Kognitif Melalui Metode Demonstrasi*. Jurnal Pendidikan IKIP PGRI Madiun Vol.14, No.1 Juni 2008.

[2] Handhika *et al*. 2014. *Pengembangan Media Pembelajaran Bermuatan Konflik Kodnitif untuk Mengurangi Dugaan Miskonsepsi pada Mata Kuliah Fisika Dasar*. Jurnal Materi dan Pembelajaran Fisika. Vol. 4 No. 2.

[3] Aydin G. & Balim A. G. 2010. *The Activities Based on Conceptual Change Strategies Prepared by Science Teacher Candidates*. Western Anatolia of Education Sciences (WAJES), Dokuz Eylul University Institute, Izmir, Turket ISSN 1308-8971.

[4] Baser M. 2006. *Effect of Conceptual Change Oriented Instruction on Student' Understanding of Heat and Temperature Concepts*. Journal of Maltase Education Research. ISSN 1726-9725.

[5] Hammer D. 1996. *Misconceptions or P-Prims. How May Alternative Perspective of Cognitive Structure Influence Instructional Perception and Intention?*. The Journal of Learning Sciences 5(2), 97-127. Lawrence Erlbaum Associates, Inc.

[6] Giuseppe M. D. & Fraser D. *Myths and Misconceptions in Science Education*. Crucible online www.stao.ca. Volume 43 No. 4. March 2012

[7] Clement J., Brown E. D., & Zietsman A. 1989. *Not All preconceptions are Misconceptions: finding 'anchoring conceptions' for grounding instruction on students' intuitions*. Int. J. Sci. EDUC., vol 11, special issue,554-565.

[8] T. Charitat and F. Graner. 2003. *About the Magnetic Field of a Finite Wire*. European Journal of Physics. 24 (2003) 267-270.

[9] Herman A. Haus and James L. Melscher. 1998 *Electromagnetic Field and Energy*. Massachusetts Institute of Technology.

[10] Rustam Efendi *et al*. 2007. *Medan Elektromagnetika Terapan*. Jakarta: Erlangga.

[11] I. Prastyaningrum & S. Kartikaningsih. 2016. *Approximate theory the Electromagnetic Energy of Solenoid in Special Relativity*. J. Phys.: Conf. Ser. 795 012034.

[12] Branca, N. A. "*Problem solving as a goal, process, and basic skill*" dalam Krulik, S. & Reys, R. E. (editor). 1980. *Problem-solving in school mathematics*. New York: The National Council of Teachers of Mathematics, Inc.

[13] Meltzer, E. D. 2012. *The relationship between mathematics preparation and conceptual learning gains in physics: A possible ''hidden variable'' in diagnostic pretest scores*. Am. J. Phys. 70 –(12). http://ojps.aip.org/ajp/.

IOP Conf. Series: Journal of Physics: Conf. Series **909** (2017) 012050 doi:10.1088/1742-6596/909/1/012050

The influence of intuition and communication language in generating student conceptions.

[1]J. Handhika, [2]C. Cari, A. [3]Suparmi, [4]W. Sunarno

[1]Universitas PGRI Madiun
[1,2,3,4] Sebelas Maret University
jhandhika@unipma.ac.id

Abstract. This research aims to describe the influence of intuition and communication language in generating student conceptions. The conception diagnostic test is used to reveal student conception. The diagnostic test results described and communication language profiled by giving instruction to students to make sentences using physics quantities. Sentences expressed by students are reduced and profiled potential effects. Obtained information that (1) Students generalize non-scientific experience (based on feeling) into the physics problem. This process caused misconception. Communication language can make the students difficult to understand the concept because of the difference meaning of communication and physics language.

1. Introduction

Conception can be seen as an idea of a fact or personal beliefs and develop through experience [1][2]. Conception she or he has, based on experiences in everyday life, instruction, media and so on[1]. conception can develop when making predictions about the situation ("What would happen if ...?") or to explain a phenomenon[3]. Idea helds before instruction is called preconception [4]. Physics concept has been recognized by humans since they first interact with nature. The five senses of human beings made observations of nature. The process of human communication with nature producing experience which is the early foundation of knowledge. The resulting experience is very powerful in influencing the human conception of reading a physical event. Conception arises from experience, or other interaction that involves feelings known by intuition.

Other paragraphs are indented Intuition is explained simply with the rules of "More A-More B" dan "Same A-Same B" [5]. On motion of free fall, for example, children and many adults claim that full Matchbox will hit the ground first in comparison with Matchbox partially filled Matchbox. However, when experiments do, it is clear that the box reaches the ground at the same time, this example is the application of the rules "More A-More B" [5] . Examples of applications "“Same A-Same B" is comparing two water tube, the tube is heated, while the other is not heated [5]. Children and many adults claim that because of the volume of the tube A increases, the weight of A is also increasing. Students equate the amount of weight with volume. Intuition is something that is clear and obvious, or easily accessible, and does not require further evidence, that argument is very simple and not challenging [6]. The term intuition is defined as define intuitions as mental representations of facts that appear to be self-evident [7]. Intuition as form of perception [6]. Intuition is a form of a clear perception of an event, without the need for further evidence. Generalizing an event led to the strengthening of intuition. Intuition can produce the right decision or not. Intuition tend to promote feelings of the decision making process. Intuition is the instinct of human beings coming from within him so that every individual has the potential to use intuition to make decisions. In physics, intuition was first used by philosophers to read natural phenomena.

Intuitions that arise from the students became the subject of the research in this paper. The student experience (within the family and the community) would provide a major influence on the intuition that is raised by the students. By knowing the student experience that led to intuition, the misconceptions can be reduced. In addition through the experience of everyday life, conception can also arise from the learning experience at school and learning resources. The learning experience at school must involve interaction with the component (1) source of teaching, (2) lecturer, (3) friends, and (4) family and community. The interaction of two or more components can also occur and can lead to the conception of the student.

Interaction is certainly involve passive and active communication involving language. Sources of teaching has an important role in generating student conception. The resulting conception can be correct or incorrect, depending on the validity of the source and the ability of students to represent concepts that come from various sources. The development of technology, was able to create a new environment, which is a social networking environment. These developments have an impact on the interaction with teaching resources, which was previously one-way turn out to be two-way, with the involvement of a friend, even lecturers.

The use of language is important, both written and verbal. By using the language, the interaction between the source and the recipient can be established. Understanding the language between the source and the recipient of the key to successful transfer of knowledge and information. In physics, the mathematical language used to represent the concept. The use of the language of mathematics must be consistent and provide valid information, since it has a clear meaning and definition.

To represent the language of mathematics, the language of communication played a role in the birth of the conception of physics. Communication language must be obtained from interaction with professors, teachers, friends, family, and community. Language communications obtained from interaction with the community become important observation in this paper. Physics, like other natural sciences, involves observing nature, measure observation, and synthesize the results with theory, all activities inevitably involves oral and written communication in everyday language [8].

Communities and families is one of learning resources directly obtained by the student. The proportion of students interact with the family and community using the everyday language, of course, greater than the interactions that occur in the classroom. The use of everyday language a bit much also involve physical quantities such as speed, distance and force. Clarification of the use of Physical quantities in the communication can provide information about the causes of conception. based on exposure that have been presented, it is necessary to described the influence of intuition and communication in generating student conception.

2. Methods

in this research, the steps are as follows: (1) the student is given a diagnostic test that can reveal student conception. The tests used are added level of certainty and argumentation to reveal the involvement of intuition in generating conception, this test used to reveal "More A-More B" misconception (2) To determine the effect of the language of communication, students are asked to make sentences using physical quantities, and provide an explanation of the meaning of physical quantities used. Physical quantities that are used in a sentence, among others; position, distance, displacement, speed, velocity, gravity, the speed of gravity, gravity, mass, and weight. Students are also asked to write down the difficulties found at the time of composing sentences. Unstructured interviews were also conducted to multiply the depth of student conceptions. Research conducted on students of sixth semester 2016/2017 fulfilled. research sample number is 16 students. The sentence made by the students profiled as a starting material interview. Interviews conducted only on students who made the unusual sentence. Reduction is carried out to obtain a profound conclusion.

3. Results and discussion

The results presented in two parts; the first part describes the results of the diagnostic test and the description of the conception that gives rise to intuition, the second part describes the communication language written by the students. The concepts discussed in the second part are the concept of speed and velocity, acceleration, and force.

3.1. Diagnostic Test Description

The selected diagnostic test used the FCI test [9] number 10, 23 and test questions from [5] modified with a degree of certainty under the following categories: Tests designed to diagnose misconceptions caused by intuition use. The student's answer profile on each question is as follows: (7) certainly, (6) Almost certainly, (5) Partly certain, (4) Most guess the answer, (1) Overall guess the answer. Students who chose the answer did not agree with the scientist's agreement with the level of certainty 7, concluded experiencing conception. The students are asked to write an argument for their choice of answers. Profile of student answers and level of certainty are described in table 3.1.

Table 3.1 Profile of student answers and level of certainty

Diagnostic Test 1

A ball enter the chanel at "1" and exit at "2". Which of the path representations would most nearly correspond to the path of the ball as it exits the channel at "2" adn roll across the table top.

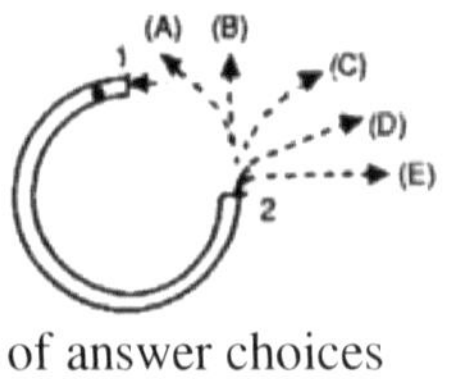

Percentage of answer choices

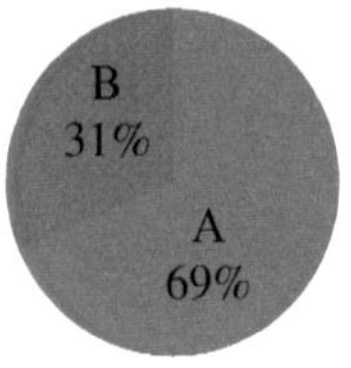

Quantitiy of the level certainty

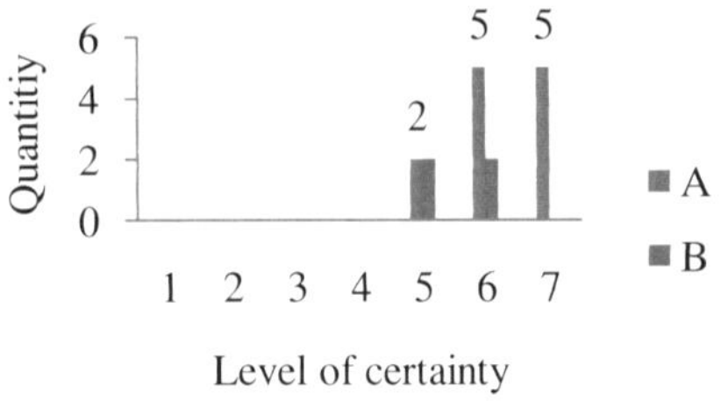

Diagnostic Test 2

Two bottles, one full of water, and the other Half of it, are held at the same hight (15 m) above the ground. In the same manner. They are both dropped at a certain point in time. Based on this condition:

A. The full-filled two times faster than the half-filled bottle.
B. The half-filled two times faster than the full-filled.
C. The full bottle faster than the half-filled bottle but less than two times.
D. The half-filled bottle faster than The full-filled but less than two times.
E. The full-filled bottle has the same time as the half-filled bottle.

IOP Conf. Series: Journal of Physics: Conf. Series **909** (2017) 012050 doi:10.1088/1742-6596/909/1/012050

Percentage of answer choices

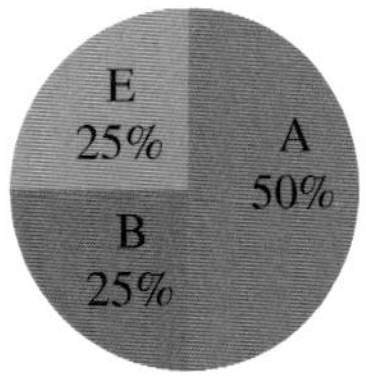

Quantitiy of the level certainty

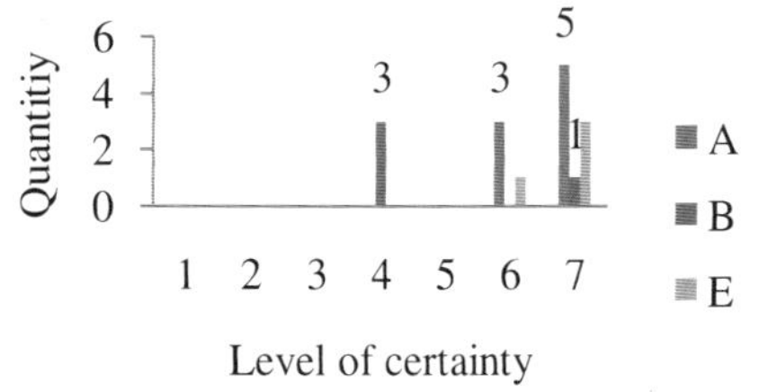

Diagnostic Test 3

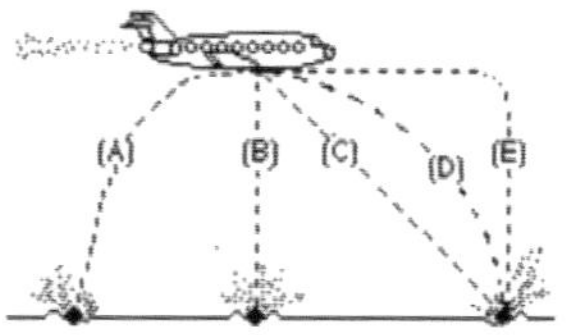

A bowling ball drops accidentally from the cargo plane when the plane flies in a horizontal direction (Notice Pictures).
Which option best describes the ball path after leaving the ship viewed from an observer dwelling on the surface of the earth?

Percentage of answer choices

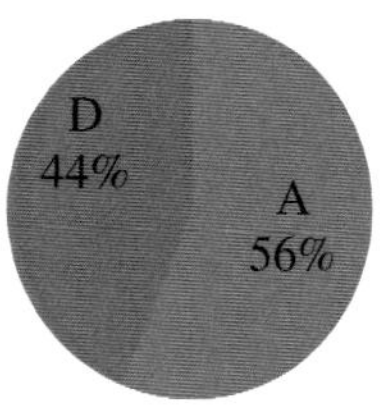

Quantitiy of the level certainty

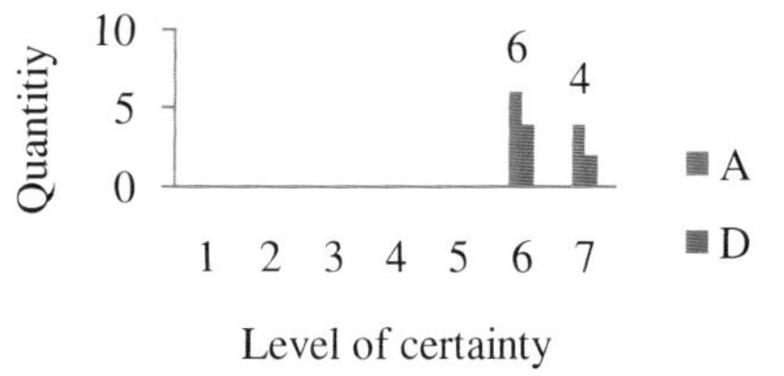

Based on the indicators set out in the methods section, we describe the findings of misconceptions. Diagnostic test 1 shows that 69% of students have chosen answer A with five students choose level certainty 7. Diagnostic test 2, there is 50% have chosen answer A with five students with level certainty 7, and 25% students have chosen answer B, one student with the level of certainty 7. In the diagnostic test 3, found 56% of students have chosen answer A, four students with level of certainty 7 (Table 3.1). Students' argumentation profile presented. A sample student argumentation profile described in Table 3.2.

Table3.2 Examples of student arguments

<table>
<tr><th>No.</th><th>Student argumentation</th><th>Description</th></tr>
<tr><td>1.</td><td>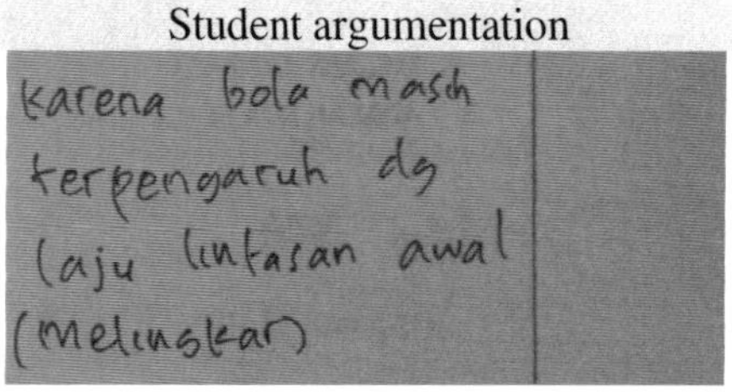
</td><td>"Because the ball is still affected by the initial (circular) trajectory."
Explanation after the clarification:
The student has a conception that the initial trajectory (circular shape) affects the ball path after that.</td></tr>
</table>

IOP Conf. Series: Journal of Physics: Conf. Series **909** (2017) 012050 doi:10.1088/1742-6596/909/1/012050

No.	Student argumentation	Description
2.	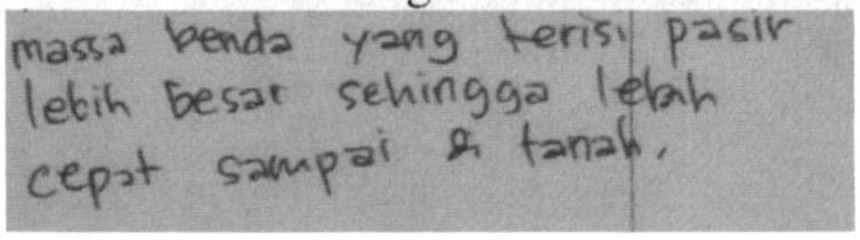 massa benda yang terisi pasir lebih besar sehingga lebih cepat sampai di tanah. 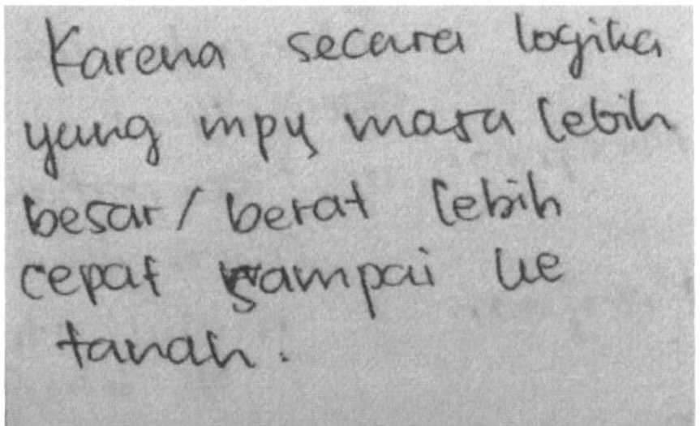 Karena secara logika yang mpy massa lebih besar / berat lebih cepat sampai ke tanah.	"The mass of objects (bottles) filled with sand is bigger, so more quickly to the ground." "Because logically having a bigger / heavier period gets to the ground." Explanation after the clarification: Students have a conception of objects that have a larger mass will be faster.
3.	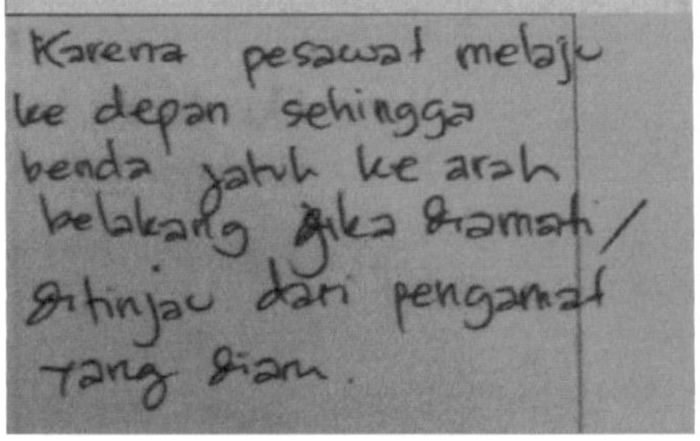 Karena pesawat melaju ke depan sehingga benda jatuh ke arah belakang jika diamati / ditinjau dari pengamat yang diam. 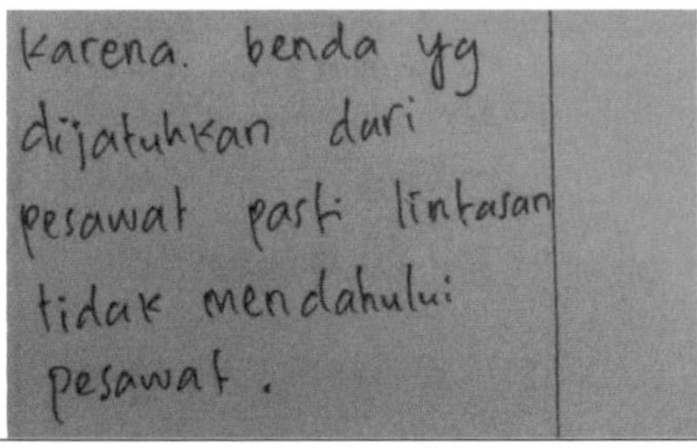 Karena benda yg dijatuhkan dari pesawat pasti lintasan tidak mendahului pesawat.	"Because the plane is moving forward so that the object falls backward if observed from a silent observer." "Because the object dropped from the plane must pass the path does not precede it." Explanation after the clarification: Students have a conception with full confidence that objects that fall from the plane will not fall in front of the plane, would fall behind it.

Student argument for the diagnostic test 1 (Table 3.2) gives information that the initial track conditions affect the motion of the object. Intuition arises because the students are less appropriate in generalizing. For example, if we are accustomed to riding a motorcycle when leaving from home to campus it will feel heavy when leaving from home to campus by foot. The role "Same A-Same B" appears in this condition. The lack of precise in generalizing becomes the forerunner to the emergence of misconceptions. The students' argument for diagnostic test 2 gives information that students have the conception that a bigger mass of the object more faster than a small mass. The rules of "More A-More B" appear in this case, the experience becomes one of cause misconceptions. The interview results provide information that students have more painful experiences of falling objects with a bigger mass than a small mass.

To overcome this misconception, The mathematical language of Newton's law of gravity gives equation:

$$g = G\frac{M}{R^2} \quad (1)$$

If the object is at the height (r) of the earth's surface , equation (1) becomes:

$$g = G\frac{M}{(R+r)^2} \quad (2)$$

Given the values of $G = 6.674 \times 10^{-11}$ Nm^2/kg^2, $M = 5.9742 \times 10^{24}$ kg, and R = 6.378,1 km obtained gravity acceleration as shown in Table 3.3

Table 3.3 Acceleration of gravity for r = 10 - 100000 m

r (m)	*g (m/s²)*

r (m)	*g (m/s²)*
10	9.80
100	9.80
1000	9.80
5000	9.79
10000	9.77
50000	9.65
100000	9.50

Up to 5000 m, gravitational acceleration tends to be constant (9.8 m / s2). Based on equation (2) the mass of the bottle does not affect, so the travel time of both bottles is the same t.

$$t = \sqrt{\frac{2h}{g}}. \qquad (3)$$

The student argument about diagnostic test 3 gives information that the falling object from the plane is unlikely to fall in front of the plane, must fall behind it. This conception arises because the student positioned the reference point as himself, not the observer on the earth. Experience also plays a role in the rise of this conception. For example one of the students mentioned that once dropped a book while riding a motorcycle that was speeding, the book fell behind the rider. This experience is used as a basis by students to strengthen the concept. Limitations Students in illustrating the events in the test matter correctly become the initial problem, then generalizing the experience that is less in line with the problems posed causing misconceptions.

3.2. Section Create sentences

In this section will be described some sample sentences made by students, and used in everyday life. Students are required to make sentences with basic words displacement & distance, speed & speed, acceleration, and style. Examples of sentences made by students are described in Table 3.4.

Table 3.4 Sentences made by the students

Communication language (sentences made by students)	Description
Bahasa: “Saya pindah kosan satu bulan yang lalu” In English: "I moved away one month ago" Bahasa: Jarak dari kampus ke rumah saya 75 km In English: Distance from campus to my house is 75 km	The word "move" and "distance" (In "bahasa" displacement have the basic word “pindah” and other form “perpindahan”) in physics have the same meaning, ie change of position and length of track. The use of language communication can be used to strengthen the concept of physics. Misconceptions are less arise from it
Bahasa: Dari rumah ke kampus saya mengendarai sepeda motor dengan kecepatan 60 km / jam In English: From home to campus I ride a motorcycle with a speed of 60 km / hour Bahasa: saya “ngelaju” dari ponorogo ke kampus In English:	The concept of velocity commonly used in everyday communication, but in its use, the term speed is seldom used. In the communication language, the terms of speed and velocity are equated. In the local language, there is a word that is close to the speed "ngelaju" (local language) but has a different meaning to the speed. When viewed from its application, the language of communication will make it difficult for students to understand the concept of speed and velocity. If students feel the concept of speed and velocity are

I went from ponorogo to campus	the same, the misconception can arise.
Bahasa: Waktu di sma saya tidak masuk kelas akselerasi In English: When I was in high school I was not in an accelerated class	Akselerasi dalam bahasa komunikasi tidak berbeda dengan bahasa fisika.
Bahasa: Gaya berpakaian orang barat berbeda dengan orang indonesia In English: Western dressing style is different from the Indonesian people	In the communication language, the force associated with the style, design of certain clothing and maybe culture.

Not every language used in communication has the same meaning as the language of physics. Student difficulties in understanding the concept of physics can arise because of this factor. Introducing the language of physics needs to be emphasized in every lesson. This description is preliminary information and will be researched and explored in more detail. Misconceptions may arise when less appropriated generalizations are made, such as equalizing speed with speed.

4. Conclusion

It can be concluded that not all incorrect conception are misconceptions. Misconceptions arise because students generalize experiences that involve feelings with the reviewed physical events. The rules of "More A-More B" and "Same A-Same B" provide clear limits on misconceptions. In identifying misconceptions not enough by looking at the incorrect answers, the use of appropriate diagnostic tests, deeper information is needed to identify the misconceptions. Interviews are the most accurate method for identifying misconceptions, but they take a long time.The diagnostic test of two tiers (Table 1.1) comes with a brief argument to be an alternative solution.It is found the meaning of words in different communication languages with a potentially inhibiting physics language In understanding concepts. Misconceptions may arise when students generalize a language of communication that has a different meaning from the language of physics.Increasing research required on this topic.Instructional development to uncover misconceptions has been widely practiced [10-14]. This study is expected to help other researchers in categorizing misconceptions.

References

[1] Aretz S, Borowski A and Schmeling S 2016 A fairytale creation or the beginning of everything: Students' pre-instructional conceptions about the Big Bang theory *Perspect. Sci.* 46–58
[2] Swinkels M F J, Koopman M and Beijaard D 2013 Student teachers' development of learning-focused conceptions *Teach. Teach. Educ.* **34** 26–37
[3] Weil-Barais A and Vergnaud G 1990 Students' Conceptions in Physics and Mathematics: Biases and Helps *Adv. Psychol.* **68** 69–84
[4] John C, David E B and Aletta Z 1989 Not all preconceptions are misconceptions: finding "anchoring conceptions" for grounding instruction on student's intuitions **11** 554–65
[5] Stavy R and Tirosh D 2000 *How Students (Mis-)understand Science and Mathematics : Intuitive Rules Ways of Knowing in Science Series* (New York and London: Teachers College Press)
[6] Isenman L D 1997 Toward an understanding of intuition and its importance in scientific endeavor *Perspect. Biol. Med.* **40** 395
[7] Dreyfus T and Eisenberg T 1982 Intuitive Functional Concepts: A Baseline Study on

Intuitions *J. Res. Math. Educ.* **13** 360–80
[8] Williams H T 1999 Semantics in teaching introductory physics *Am. J. Phys.* **67** 670–80
[9] Hestenes D, Wells M and Swackhamer G 1992 Force concept inventory *Phys. Teach.* **30** 141
[10] Kaltakci D and Eryilmaz A 2008 Identifying Pre-service Physics Teachers' Misconceptions With Three-Tier Tests 1–8
[11] Kaltakci-gurel D, Eryilmaz A and Mcdermott L C 2017 Development and application of a four-tier test to assess pre-service physics teachers ' misconceptions about geometrical optics *Res. Sci. Technol. Educ.* **5143** 1–23
[12] Kanli U 2015 Using a two-tier test to analyse students' and teachers' alternative concepts in astronomy *Sci. Educ. Int.* **26** 148–65
[13] Urban H 2017 Sequential Reasoning in Electricity : Developing and Using a Three-Tier Multiple Choice Test Aim and Research Question **8** 285–92
[14] Siswaningsih W, Firman H, Zackiyah Z and Khoirunnisa A 2016 Development of Two-Tier Diagnostic Test Pictorial-Based for Identifying High School Students Misconceptions on the Mole Concept *J. Phys. Conf. Ser.* **755** 11001

Students conception and perception of simple electrical circuit

ND Setyani[1*], Suparmi[2], Sarwanto[3], J Handhika[4]

[1]Science Education, Postgraduate Program, Universitas Sebelas Maret, Surakarta, Indonesia
[2,3]Lectures of Universitas Sebelas Maret, Surakarta, Indonesia
[4]Lectures of IKIP PGRI Madiun, Indonesia

*Email: nitadepitsetyani@gmail.com

Abstract. This research aims to describe the profile of the students' conception and perception on the simple electrical circuit. The results of this research suppose to be used as a reference by teachers to use learning models or strategies to improve understanding the physics concept. The research method used is descriptive qualitative. Research subjects are the students of physics education program, Universitas Sebelas Maret, Surakarta, Indonesia (49 students). The results showed that students have alternative conceptions. Their conceptions are (1) a high-voltage wire has an electric current and can cause electric shock, (2) the potential difference and the value of resistance used in a circuit is influenced by electric current, (3) the value of resistance of a lamp is proportional to the filament thickness, (4) the amount of electric current that coming out from the positive pole battery is the same for all type of circuit, in series or parallel (battery is constant current sources), (5) the current at any resistor in the series circuit is influenced by the resistor used, (6) the resistor consume the current through it. This incorrect conception can cause misconceptions.

Keywords: Conception, Perception, Simple Electrical Circuit

1. Introduction

Physics is believed to be hard science, so it is a difficult subject for the students [1,2]. The reasons why the concept of physics is difficult to learn are (1) an abstract concept, (2) complex systems, (3) students' prior knowledge or experience is limited, (4) an understanding a finite symbol, (5) misconceptions that occur in students [3]. Research on the physics concept understanding is continuously conducted by researchers for a long time. The results of this research suppose to be used as a reference by teachers to use learning models or strategies to improve understanding the physics concept of the student. The research on students' understanding of the simple electrical circuit concept has been conducted by some experts [4, 5, 6, 7, 8, 9, 10, 11]. Some of the research results about electrical concept could be seen in Table 1 below.

Tabel 1. Research result related to electrical concept

No	Year	Research Results
1.	2004	The students tend to hold multiple misconceptions, even after instruction using DIRECT instrument [5].
2.	2007	a. Bulb give more light when the number of batteries increase (independent from the type of connection). b. Bulb becomes brighter when batteries are connected in parallel. c. Batteries are constant current sources. d. Current is consumed by circuit component. e. Bulb in series always give more brightness. [6] [7]
3.	2009	Describe experiences of using the DIRECT to evaluate and analyze the level of conceptual knowledge [4]
4.	2012	Students held alternative conception on the open and short circuit concept [9]
5.	2012	Teaching with analogy has a significantly positive effect on elimination of misconception [11]
6.	2014	Many students still had difficulties with voltage and potential. The most students were unable to link that two concepts [8]
7.	2014	The students have some common misconceptions in electrical concept [10]

From the result of that research, we can conclude that: (1) Students have alternative conception of electric current, potential difference, and simple circuit, (2) Students have some incorrect conception and misconception on the electrical concept. In this research, we have to know the relationship between conception and perceptions that the students have.

Electrical is an important concept in physics because it is applied our daily life. In our house, there are many electrical types of equipment that we often use, i.e. television, lamp, refrigerator, air condition, etc. Those equipment is an arrangement in complex circuit with the voltage source from the power station. The electrical circuit in our house is the more complex system, so the students need to understand the simple circuits first. Electric currents and simple circuits are fundamental concepts that have to be understood by the students in order easier to learn the electrical concept. A simple circuit is a circuit with one or more batteries, lamps, and short wires. To learn the simple circuit, the students need to understand electric current, voltage, and resistance concept. For the ohmic material, current, voltage, and resistance are described by Ohm's law. And so the students need to understand about the open circuit. At the open circuit, the current is zero.

The students' concept understanding is influenced by the initial concept, their perception on the concept, and the deepening of accepted concepts. Conception is the initial information that students have, before the learning activities. Perception is the basic for building knowledge. If the student's perception is incorrect, it can cause a misconception [12]. Perception occurs when students receive information, and that information will be processed in the brain based on students' knowledge [13]. The student's perceptions are important to building knowledge. It can cause incorrect conception [14]. So we can conclude that if the concepts and perceptions that students have is incorrect, it will cause incorrect conception and even misconception. So it is necessary to choose the right learning model or strategies in deepening the concepts. By knowing the conceptions and perceptions that students have, teachers can choose the right method to improve learning outcomes [14].

Student conceptions can be caused by factors, (1) the environment, (2) learning resources (3) experiences [15]. Conceptual knowledge is built on the experience and contextual knowledge (from the environment) in building intuition [16]. The sources of learning are the books and teachers also the learning process of deepening the concept understanding.

Electrical is an abstract concept. The students will have difficulty understanding simple electrical circuit because of their finite experience about electricity. This issue reinforces the researcher's assumption to conduct research on students' conceptions and perceptions on simple electrical circuit. The research aims to describe the profile of the students' conceptions and perceptions of simple electrical circuit, so the teachers can choose learning strategies or models to improve concept understanding.

2. Methodology
The research method used is descriptive qualitative. Research subjects are the students of physics education program, Universitas Sebelas Maret, Surakarta, Indonesia, as many as 49 students. Data were collected using conceptual tests of simple electrical circuits accompanied by the students' argumentation of their answer and interview.

3. Results and Discussion
Conceptual tests are given to the 2nd students of physics education program, who will take electrical magnetic (basic physics II) course. This conceptual test is done to know the students' conception and perception of simple electrical circuit before the learning activities. However, the students have understood this concept when they were studying in high school. The conceptual test of the simple electrical circuit consists of concepts (1) the requirements of electric current, (2) current, resistance, and voltage in Ohm's law, (3) the factors that affect the value of resistance, (4) series and parallel circuit.
The students' conception and perception profile are described in Table 2. The percentage of students' conceptions is presented in Table 3.

Table 2. Profile of the Students' Conception and Perception

Correct concept	Incorrect concept	Students' Perception
An electrical wire have the same voltage at any point along the wire so it does not has an electric current and does not cause electric shock	1. A high-voltage wire has an electrical current 2. A high-voltage wire, can cause electric shock	1. A high-voltage wire has a potential difference 2. Conductor who has a voltage, it also has electric current in according Ohm's law $V = IR$
1. The potential difference in a circuit is not influenced by the electric current, but it's influenced by the potential difference between the positive and negative poles of the voltage source. 2. The value of resistance of a material not influenced by electric current, but is influenced by the type of conductor (material's resistivity), conductor length and cross-sectional area.	1. The potential difference in a circuit is influenced by electric current. If the circuit is open, there is no electric current; then there is no potential difference. 2. The value of resistance used in the circuit is influenced by electric current. If the circuit is open, there is no electric current; then there is no resistance.	According to Ohm's law $V = IR$, if $I = 0$, then $V = 0$, and $R = 0$. An open circuit has no current, so the potential difference and the resistance in the circuit are zero
The value of resistance is inversely proportional to the cross-sectional area (the thickness of the lamp filament), so the lamp with thick filament will be brighter.	The value of resistance of a lamp is proportional to the filament thickness.	The lamp with thick filament will have bigger resistance and lighter then the thin filament
The amount of electric current that coming out from the positive pole of a circuit is determined by the arrangement of the circuit	The amount of electric current that coming out of the positive pole is the same for all circuit, in series or parallel circuit.	The amount of electric current coming out from the positive pole of a circuit is the same for all circuits because it has not yet entered a resistors or junction

1. At any points on the series circuit, it has a same value of the current. 2. The resistor does not consume the current in a circuit.	1. At any points on the series circuit, it has a different value of the current, and it's influenced by the resistor used. 2. The resistor does consume the current in a circuit.	1. The current in the series circuit is influenced by the resistor used. The difference value of the resistor affects the current used by that resistor. 2. The resistor consumes the current through it, so that the current passing through the resistor, it will be decreased.

Table 3. Percentage of Students' Conception

Students' Conception	Percentage
A high-voltage wire has the same voltage at any point along the wire	
No conception	2.04%
Incorrect conception	42.86%
Correct conception	55.1%
The potential difference and the resistance in a circuit are not influenced by the electric current	
No conception	4.08%
Incorrect conception	93.88%
Correct conception	2.04%
Resistance is inversely proportional to the thickness of the lamp filament (cross-sectional area)	
No conception	2.04%
Incorrect conception	53.06%
Correct conception	44.9%
The amount of electric current that was coming out from the positive pole of battery in a circuit is determined by the arrangement of the resistors in circuit	
No conception	12.24%
Incorrect conception	73.47%
Correct conception	14.29%
The electric current of series and parallel circuit	
No conception	4,08%
Incorrect conception	64,29%
Correct conception	31,63%

Based on Table 2, it can be said that the students have the incorrect conception of the simple electrical circuit. The students' conceptions are rated by their answers and argumentations of the problems.

3.1. The requirements of electric current

The problem that was given in this case, students' conception and perception is presented below.

Case 1:

A bird that flew suddenly hopped on a peeling wire and both of it's legs touch that wire. What will happen to the bird?

A. Bird will be shock because the high-voltage wire
B. Bird will be shock because there is a potential difference on both of bird's legs
C. Bird will be shock because there is a current on the wire
D. Bird will not be shock because there is no potential difference on both of bird's legs
E. Bird will not be shock because the electric current in the circuit is small.

27 students choose the correct answer (d) and the true conception. 21 students had incorrect conceptions, one student choose answer (a), 19 students choose answer (b), 1 student choose answer (c).

IOP Conf. Series: Journal of Physics: Conf. Series **909** (2017) 012051 doi:10.1088/1742-6596/909/1/012051

One student did not choose an answer.
The perception of students that if the wire has high-voltage, then the wire has a current and can cause electric shock.

The student' argumentation of the case 1 is presented in figure 1.

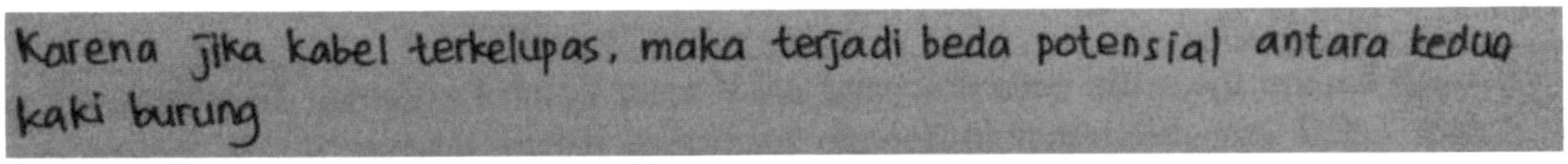

"because it's peeling wire, so it has a potential difference at both of bird legs."

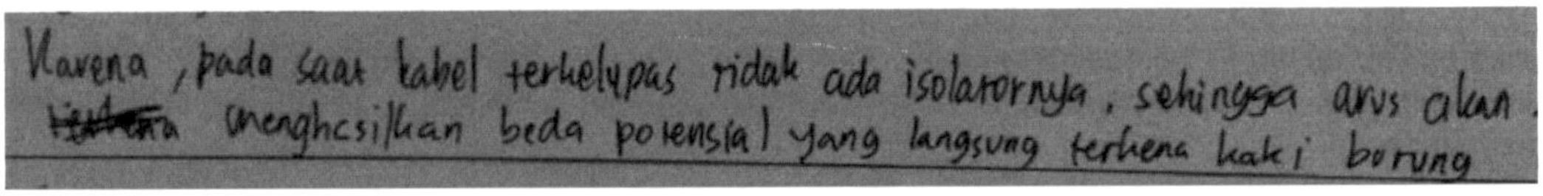

"because the peeling wire hasn't isolator, so electric current occur and cause potential difference on bird legs."

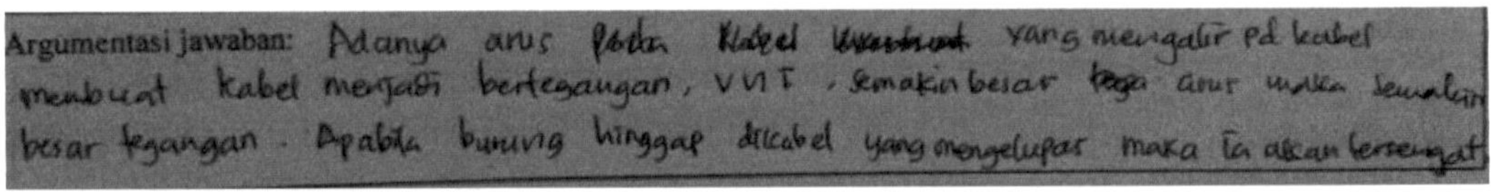

"An electric current can cause wire has voltage, V~I, a higher current cause higher voltage. Bird alight on the peeling wire will get electric shock."

Figure 1. The students' perception of the electric current

Based on figure 1, there is two information. First, students have the perception that the peeling wire will has a potential difference and can cause electric shock. Another perception indicates that the wire has a current that causes a potential difference. The correct concept is a wire with the high-voltage, has not currents because at any point as long as the wire has the same voltage. Electric current may occur if a circuit has the potential difference at two points. Both of bird legs in case 1 are on the same wire that has a same potential difference, so the bird does not get an electric shock. Electrical shock occurs when there is a potential difference on one part of the body with another part that causes an electric current [17].

The result of interview shows that the students can mention the condition of electric current, that is the potential difference and closed circuit. The students' perceptions in this case are due to the experience that they have. When we were touching a high-voltage wire, we can get an electric shock. This students' perception can cause incorrect conception. This result is consistant with the other research result which suggest that the concepts of the students and his or her cognitive processes influence what he or she sees [18].

3.2. Current, resistance, and voltage in Ohm law

The problem that was given in this case, students' conception and perception is presented below:

Case 2:

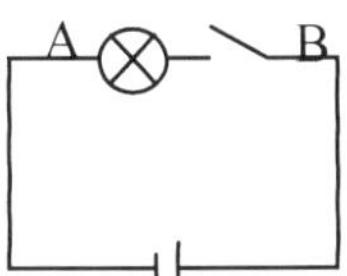

The simple circuit on the side consists of a battery with a voltage V and a lamp with a resistance R. When the switch is open, the following correct statement based on the figure is...

i. The bulb is not light up because there is no current in the circuit
ii. The potential difference between point A and B is zero because there is no current in the circuit
iii. The potential difference between point A and B is V because the potential difference in the circuit is not influenced by the electric current.
iv. The lamp has no resistance because there is no current in the circuit.
v. The lamp has a resistance of R since the value of the resistance is not influenced by the electric current.

Only one student chose the correct answer (c) and given the correct conception. 46 students had an incorrect conception, 31 students choose answers (a), one student choose answer (b), 13 students choose answer (d), and one student choose answer (e).

Two students did not choose an answer.

The perception of students about the open circuit is that there is no current, there is no resistance, and therefore there is also no voltage difference. According to Ohm's law $V = IR$, if $I = 0$, then $V = 0$, and $R = 0$.

The student' argumentation of the case 2 is presented in figure 2.

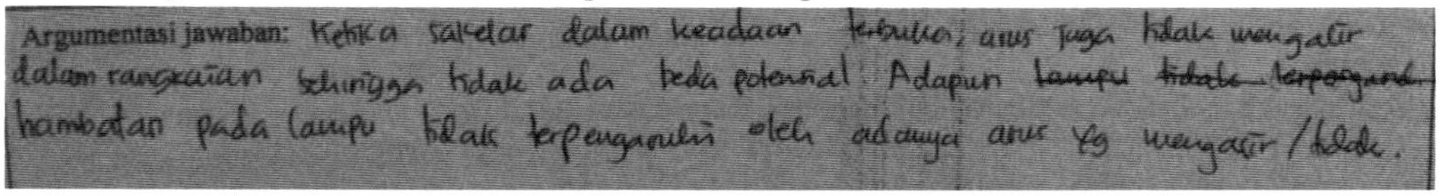

"since the switch is open, no current in the circuit, so there is no potential difference. The lamp resistance is not influenced by electric current."

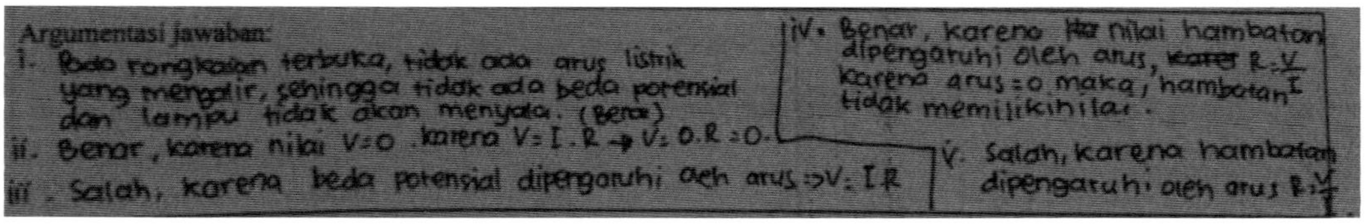

"i. There is no current at opened circuit, so there is no potential difference, and the lamp is off (true).
ii. True, because $V = 0$. $V = IR$, so $V = 0, R = 0$
iii. False, because the potential difference is influenced by current $V = IR$
iv. True, because the resistance is influenced by current, $R = {}^{V}/_{I}$, $I = 0$, so there is no resistance
v. False, because the resistance is influenced by current, $R = {}^{V}/_{I}$"

Figure 2. The students' perception of the Ohm's law

Based on the student' argumentation who have incorrect conceptions in figure 2, there are two groups of the students with different perceptions. That is, the students that assume, since the electric current is zero, the potential difference is zero, but the resistance is constant, and the other students give the argument since the electric current is zero, then the potential difference and the resistance are also zero.

There are the results of the interview with the students:

Researcher : Explain your answer in this case!
Student : The circuit is open, so there is no electric current in the circuit.
Researcher : What about the potential difference and the obstacles?
Student : By Ohm's law, if the electric current is zero, then the potential difference and the resistance is also zero.

From the student' argumentations and interview, it is known that these incorrect perception because they assume according to Ohm's law equation, $R = \frac{V}{I}$. In mathematic language, if one value is zero, it will result in another value is also zero. But in physics language, the current is influenced by potential difference and resistance. So if there is no current, it does not mean there is no potential difference and resistance in the circuit. The potential difference of a circuit is influenced by the potential difference between the positive and negative poles of the voltage source (batteries). The resistance of a wire is to be proportional to the length, inversely proportional to its cross-sectional area, and proportionality constant ρ is called the resistivity of the conducting material. The unit of resistivity is the Ohm-meter $(\Omega.m)$. It's described by the equation $R = \rho\frac{L}{A}$.

The students assume that the electric current is the main concept, which will affect to other quantities, (potential difference and resistance). This is also corresponding with the other research by [9] conducted in Malaysia, that students experience an alternative conception that when there is no current, there is no potential difference and resistance. Understanding mathematic and physics language meaning is necessary to solve this problem. Understanding of mathematic language in the

IOP Conf. Series: Journal of Physics: Conf. Series **909** (2017) 012051 doi:10.1088/1742-6596/909/1/012051

physics language needs to be emphasized to the students and the mathematical information affects student conception. Integration of the language of mathematics and physics language is necessary to reduce incorrect conception [14]. The incorrect perception of the students in interpreting the mathematical language and the physics language is causing incorrect conceptions.

3.3. Factors that influence the resistance

The problem that was given in this case, the students' conception and perception is presented below.

Case 3:
Light bulb A and B are identical in all ways, except that bulb A's filament is thinner than bulb B's. They are connected to 120 V voltage source, so...

a. A will be the brighter because it has the most resistance
b. B will be the brighter because it has the most resistance
c. A will be the brighter because it has the least resistance
d. B will be the brighter because it has the least resistance
e. A and B will have same brightness

A total of 21 students have choosen the correct answer (d) and the correct conception. 27 other students have incorrect conceptions, one student chooses answer (a), five students choose answer (b), 21 students choose answer (c)
One student did not choose an answer.
The perception of students, the resistance of material is proportional to the thickness of the filament. The thin filaments cause the lamp to have a small resistance, so it has more current.

The student' argumentation of the case 3 are presented in figure 3.

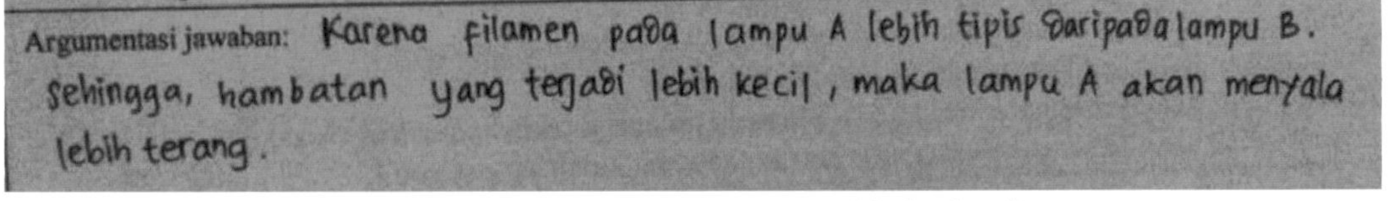

"A's filament is thinner than B's, so A's has the least resistance. A will be brighter."

Figure 3. The students' perception of the resistance

Based on the student argumentation, the students have an incorrect conception about the thickness of the filament to the resistance of material. According to the student's perception, the resistance is proportional to the filament thickness. Conductors that have thin filaments will have the less resistance so it has most current. The true concept holds that the resistance is inversely proportional to the cross-sectional area of the conductor, in this case, it's expressed by the thickness/thinness of the lamp filament. The results of the interviews show that students can explain the factors that influence the resistance correctly.
This students' perception is influenced by their intuition in drawing conclusion, about the relationship of resistance and filament thickness. "More concrete conception and intuition are dominant in student's conclusion" [14].

3.4. Series and Parallel Electric Circuit

The problems that were given in this case, the students' conceptions and perceptions are presented below

Case 4:
Circuit A and B are consists of 2 identical lamps and a battery. The following correct statement is...

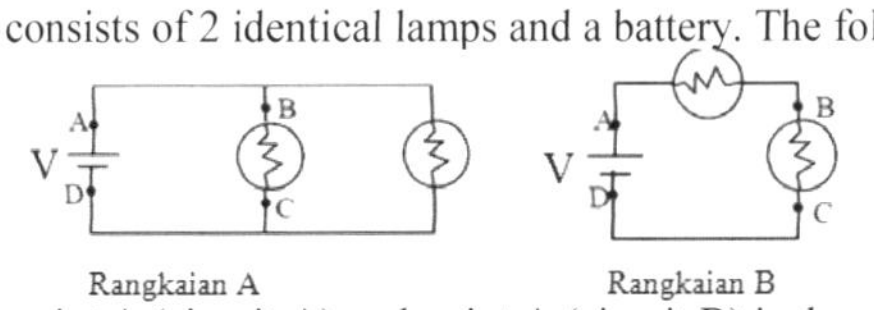

Rangkaian A Rangkaian B

a. The current at point A (circuit A) and point A (circuit B) is the same
b. The current at point B (circuit A) and point B (circuit B) is the same
c. The current at point A (circuit A) is four than the current at point A (circuit B)

d. The current at point A (circuit A) is twice than the current at point A (circuit B)

e. The current at point A (circuit A) is a half than the current at point A (circuit B)

7 students choose the correct answer (c) and the correct conception. As many as 36 other students had incorrect conception, 22 students choose the answer (a), 4 students choose the answer (d), 10 students choose the answer (e)

6 students did not choose the answer.

The perception of students about the current that coming out from the positive pole of the same voltage source, independent from the arrangement of the resistors in the circuit.

Case 5:

A. The correct answer about electric current at the points A, B, C, D, E in the following figure is...

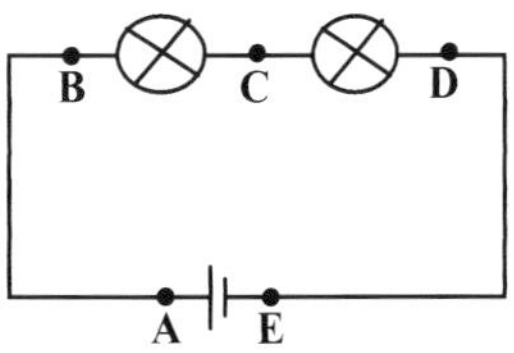

a. A>B>C>D>E
b. A=B>C>D=E
c. A=E>B>C>D
d. A=E>B=C=D
e. A=B=C=D=E

B. The correct answer about electric current at the points A, B, C, D, E in the following figure is...

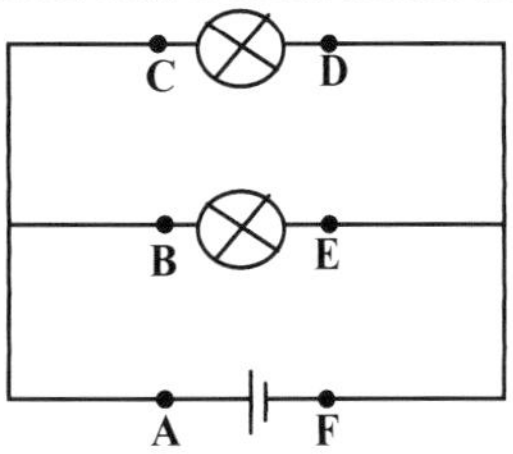

a. A>B>C>D>E>F
b. A=B=C>D=E=F
c. A=F>B=C=D=E
d. A=F>B=E>D=F
e. A=B=C=D=E =F

For case (5.A) 19 students choose the correct answer (e) and the correct conception. As many as 28 other students had incorrect conception, 5 students choose the answer (a), 25 students choose the answer (b)

2 students did not choose the answer.

For case (5.B) 12 students choose the correct answer (c) and the correct conception. As many as 35 other students had incorrect conception, 25 students choose the answer (b), 10 students choose the answer (d)

2 students did not choose the answer.

The perception of students about the current in the series circuit is it's influenced by the resistor used. The difference value of the resistor affects the current on that resistor. The resistor consumes the current through it, so that the current passing through the resistor, it will be decreased.

The student' argumentation of the case 4 is presented in figure 4

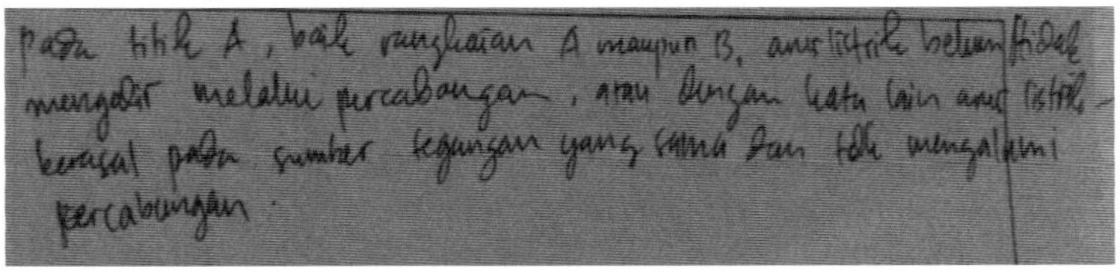

"At point A, in circuit A and B, electric current not yet flow in the junction. It means electric current came from the same voltage source and no junction has the same value."

Figure 4. The students' perception of simple circuit

Based on the student' argumentation in figure 4, the perception of the students about electric current is the current that comes out from the positive pole of the voltage source for the series, or parallel arrangement has the same value. Based on the interview, it is known that the students' perception is due to their concept understanding, the same voltage source will produce the same electric current. In the correct concept, to calculate the current that comes out from the positive pole of the battery is done by calculating the equivalent resistance first. This finding is corresponding with the other research that the students determine an electric current independent of the circuit type. Batteries are constant current sources [6,7].

The student' argumentation of the case 5 is presented in figure 5.

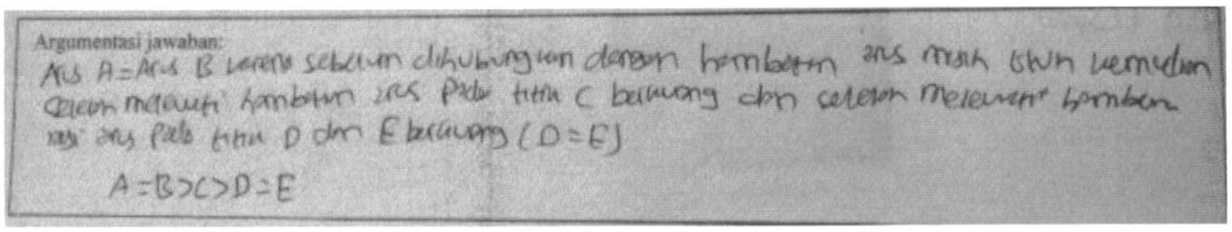

"The current at point A = the current at point B because before it passing through the resistor it was still full, after passing the resistor the current at point C decreases and after passing the resistor at point D and E it decreases too (D = E). A = B> C> D = E"

Figure 5. The student' perception of electric circuit

Based on figure 5, we know that the perception of the students is the resistor consume the electric current through it, so that the current passing through the resistor, it will be decreased. From the interview, the research gets information if the students also assume that the difference value of the resistor affects the current on that resistor. The students forget that the current of the series circuit has a same value. According [6] the misconception of "consumption of the current by circuit components" is frequently encountered in the literature.

So the results showed that the students still have alternative conceptions of current, voltage, and resistance in a simple electrical circuit. Their conceptions are (1) a high-voltage wire has an electric current and can cause electric shock, (2) the potential difference and the value of resistance used in a circuit is influenced by electric current, (3) the value of resistance of a lamp is proportional to the filament thickness, (4) the amount of electric current that coming out from the positive pole battery is the same for all type of circuit, in series or parallel (battery is constant current sources), (5) the current at any resistor in the series circuit is influenced by the resistor used, (6) the resistor consume the current through it, so that the current passing through the resistor, it will be decrease. Incorrect conception is caused by students' perceptions, it's supposed by (1) experience, (2) the understanding of mathematics language and physics language, (3) the intuition to get the conclusions, and (4) the finite concepts understanding. This incorrect conception can cause to misconceptions if the teachers did not apply appropriate learning strategies/models for the deepening of concepts. The teachers can choose a discovery learning where students do themselves directly an activity to find the concept and supported by the explanation of the teacher to provide the correct concept. Also getting students to solve conceptual problems will help students to solve more complex problems. The presentation of the concept in multirepresentation form will also support the deepening of the concept [2].

4. Conclusion

Based on the results and the discussion, it can be conclude that the students conception of simple circuit is influenced by their perception and supposed by (1) their experience, (2) the understanding of mathematics language and physics language, (3) the intuition to get the conclusions, and (4) the finite concepts understanding. The incorrect concept found in this research are:

1. A high-voltage wire has a potential difference, and it can cause electric shock.
2. Conducting wire has a current that produces voltage and causes an electric shock
3. The potential difference and resistance in a circuit are influenced by electric current. If the circuit is open, there is no electric current; then there are no potential difference and resistance.
4. The thin filament causes the lamp to have less resistance, so it has more current.

5. The current that coming out from the positive pole of the voltage source is constant, independent of the type of circuit arrangement (series or parallel circuits).
6. The resistor consumes the current through it, and the current at any resistor in the series circuit is influenced by the resistor used.

This incorrect conception can lead to misconceptions, so it needs to be overcome by learning involving students to discover their concepts through scientific activities. Conceptual test and presented it in multirepresention form can also be used to support conceptual understanding.

Acknowledge
The writer said thanks to all of the students of physics education program, Universitas Sebelas Maret Surakarta.

References

[1] Handhika J, Cari, Suparmi, Sunarno W. (2015). External Representation to Overcome Misconception in Physics. *International Conference on Mathematics, Science, and Education* (2015) 34-37.
[2] Setyani ND, Cari, Suparmi, Handhika J. (2017) Student's Concept Ability of Newton's Law Based on Verbal and Visual Test. *International Journal of Science and Applied Science: Conference Series,* 1(2) p. 162-169
[3] Koba S, Mitchell TC. (2011). Hard-to-Teach Science Concepts: A Framework to Support Learners, Grades 3–5. *NSTA PRESS, 2011,* pp. 5-6
[4] O'Dwyer A. (2009). Prior Understanding of Basic Electrical Circuit Concept by First Year Engineering Students. *All-Ireland Society for Higher Education (AISHE) Conference, NUI Maynooth.*
[5] Engerhardt PV, Beichner RJ. (2004). Students' Understanding of Direct Current Resistive Electrical Circuits. *Am. J. Phys.* 72 (1) 98-115.
[6] Kocukozer H, Kocakulah S. (2008a) Secondary School Students' Misconceptions about Simple Electric Circuits. *Journal of Turkish Science Education.* 4 (1) 101-115
[7] Kocukozer H, Kocakulah S. (2008b) Effect of simple Circuits Teaching on Conceptual Change in Grade Physics Course. *Journal of Turkish Science Education.* 5 (1) 59-74
[8] Timmermann, D and Kautz, C. (2014). Investigating Student Learning of the Voltage and Potential Concepts in Intruductory Electrical Engineering. *Frontiers in Education Conference (FIE) Proceedings, Madrid,* pp 1-4.
[9] Hussain NH, Latiff LA, Yahaya N. (2012). Alternative Conception about Open and short Circuit Concept. *Procedia – Social and Behavior Science 00 (2012 000-000.*
[10] Bilal E, Erol M. (2009). Investigating Students' Conceptions of Some Electricity Concepts. Lat. *Am. J. Phys. Educ.* 3 (2) 193-201.
[11] Ugur G, Dilber R, Senpolar Y, Duzgun B. (2012). The Effects Analogy on Students' Understanding of Direct Current Circuits and Attitudes toward Physics Lessons. *European Journal of Educational Research.* 1 (2) 2011-223
[12] Goldstone, *RL*, Medin, LD & Schyns, PG. (1997). *Perceptual Learning: Advances in Reasech and Theory.* Academic Press. San Diego, California, pp. 1-14.
[13] Slavin, RE. (2006). *Educational psychology.* Pearson Education Inc, USA, pp 201-202

[14] Handhika J, Cari, Suparmi, Sunarno W. (2016). Student Conception and Perception of Newton's Law. *Proceedings of International Seminar on Mathematics, Science, and Computer Science Education.* (MSCEIS 2015)

[15] Handhika J, Purwandari, Cari, Suparmi, Sunarno W. (2015). Profil Konsepsi Mahasiswa Pada Materi Kinematika. *Prosiding Seminar Nasional Pendidikan Sains.* 2015

[16] Lee G & Yi J. (2013). Where Cognitive Conflict arises from?: The Sructure of creating cognitive conflict. *International Journal of Science and Mathematics Education.* 11:601Y623

[17] Thaker, MN, Phadke, BN, Patel, PD. (2013). The Effects of Electrical Hazards. *International Journal of Emerging Technology and Advanced Engineering.* 3(9). 569-574.

[18] Sternberg, JR and Sternberg, K. (2011). *Cognitive Psychology, six edition.* Cengage Learning, Canada , Wadsworth, pp. 107-108

"Alms of the Sea" at Teleng Ria Beach Pacitan: Alternative Literacy Ethnoscience for Junior High School

Linda Novitasari[1], Puput Astya Agustina[2], Ria Sukesti[3], Muhammad Faizal Nazri[4], Jeffry Handhika[5]
[1,2,3,4]Student of physics education program, Universitas PGRI Madiun
[5]Lecturer of physics education program, Universitas PGRI Madiun

vitalinda9@gmail.com, puputas.ag10@gmail.com, yayahbisma@yahoo.com, gerrard08.sg53@gmail.com, jhandhika@unipma.ac.id,
Abstract. The purpose of this research is to explore the potential of ethnoscience and the value of local wisdom contained in the tradition of "sea alms" in Teleng Ria Pacitan Beach, to profile as a science literacy in junior high school. The qualitative descriptive method used, where the researcher conducts the study on group culture under natural conditions through observation, interview, and documentation. The profile obtained in a booklet and socialized in students of Junior High School (SMPN 3 Tegalombo and SMPN 1 Pacitan). From this activity obtained information that the ability of students in interpreting ethnoscience in the tradition of "alms sea" in scientific science knowledge increases. It can conclude that the ethnoscience profile and the value of local wisdom can be used as an alternative science literacy in junior high school and has the potential to be used as a reference provider of data related traditions "Sea alms" and its integration with scientific knowledge.
Key word: *Ethnosciences, Local Wisdom, Sea Alms, Science Literacy*

1. Introduction

"Alms of the sea" is another term for the popular offering tradition in Teleng Ria Pacitan. Many other terms from different offerings traditionini each region. In Yogyakarta for example, it's known as the "*larung samodra*," in Prigi known by the "*larung semboyo*," in Ngebel known as the "*larung sesaji*," in Telaga Sarangan known as "labuhan." In general the purpose of the tradition is same, that is to asking the Gods for fisherman to catch abundant catches in the coming year and be spared from the calamity [1].

In Pacitan, at first the sea alms was just a belief of the traditional fisherman community of Teleng Ria beach to keep peace and balance. Along with the development of sea alms civilization became a tradition of ritual ceremonies that are routinely performed every year. Even today it has become a tourism object to attract foreign tourists [2]. A series of sea charity events include prayer together, "*larungan,*" *ruwatan,*" and a leather puppet shadow title all night. As one of the local cultural heritages, the tradition of alms needs to be preserved and the values of local wisdom and ethnosciences enclosed within them need to be explored.

Ethnoscience is knowledge derived from certain indigenous norms and beliefs that influence the interpretation and understanding of members of nature [3]. Ethnoscience is an activity of transforming the original science which consists of all knowledge of the facts of society derived from hereditary beliefs. The scope of mechanochemistry is infinite, even in the fields of science, agriculture, ecology, medicine and the benefits of flora and fauna [4].

Local wisdom is the view of life and science as well as various life strategies that manifest the activities undertaken by local communities in answering various problems in the fulfillment of their needs [5,14]. Current conditions, many youth people don't know about local wisdom in the area. Whereas local wisdom is a source of dynamic knowledge, developed and passed on by certain populations that are integrated with their understanding of nature and the surrounding culture [6].

Currently, science learning in schools still focuses on the subject matter in general and there is no activity to transform between the original science with scientific science. Though the original science knowledge is closely related to the life of society [7], which is reflected in culture [8]. Although science learning has been relatedwith daily activities, but has not yet related with local wisdom (ethnoscience). Teachers usedit limited as apperception, not until the discussion of the material on deeper local wisdom [9]. The implementation of character education in science education is still done administratively [10].

The result of observation about the understanding of the surrounding community and the junior high school children about the charity 57 respondents from 74 students have not known the tradition of charity of sea alms and the values contained in it. In general, societyjust knowing it as a culture, not understanding scientific science contained. Science learning based approach ethnoscience which related the culture of society [11], so that through the process of transforming the surrounding phenomenon with science knowledge will increase the understanding of the relationship between science in society and scientific science as new insights.

2. Methods

This research used qualitative descriptive methods. Data collection is done from primary and secondary sources. The data source in this research are: (1) primary data source from interview results with informant and questionnaire with students of junior high school (SMPN 1 Pacitan and SMPN 3 Tegalombo); (2) secondary sources fromany documents include photos, videos, and articles about sea alms.

The techniques of collecting data include: (1) interviews with cultural's Pacitan region, fisherman, elderly fisherman, chairman of the traditional fishermen community, (2) the literature review from books, journals, online sites, and etc; (3) looking for any docomentation include photos, videos, and articles about sea alms; (4) observation, conducted by giving questionnaires to junior high school students, namely Junior High School (SMPN 3 Tegalombo and SMPN 1 Pacitan) to know the comparison of sea alms knowledge, and knowing the changes after studying ethnosciences.

Data analysis is done through: (1) data reduction to facilitate the researcher to collect further data; (2) presentation of data to present reduced data for further verification; (3) data verification for conclusion drawing.

Interpretation of the data in this study was conducted by confirming, linking, comparing and reviewing existing data [12]. The conclusion of the results of research conducted based on the results of data analysis has been verified.

3. Results and Discussion

Some literature explains that basically sea alms aims to beg for the salvation of the fishermen and their families and as gratitude for the presence of God over the abundance of blessings given [13].

The results of interviews and documentation obtained, indicates that sea alms symbolizes the fishermens's gratitude for God, asking safety at sea, and as a medium of expectation for the next year to get abundant sea catch. Here's (Table 1) the value of ethnosciences and local wisdom value on the sea alms.

Table 1. The value of local wisdom of sea alms

Value	description
Character value	Fishermen in Teleng Ria routinely organize a crush program from the catch to be included in

	cash. This program is done voluntarily self-awareness of each. It contains character values on how to manage sustenance and teaches the meaning of sincerity and sincerity.
The value of solidarity	The spirit of togetherness and kindship has always been the main thing in the implementation of the tradition of alms sea. With high dsolidaritas the fishermen shoulder to shoulder and work together since the preparation stage until the implementation.
The value of gratitude	Alms of the sea is done as an expression of gratitude to God who has given salvation, protection, and tranquility to the fisherman during fishing in the middle of the sea. With the existence of offering larung is expected the fishermen always get salvation and get the catch of the abundant.
The value of compliance	Fishermen organize sea alms as rasai respect to the ancestors to continue the ritual that has been going on for generations.
Behavior of preserving nature	The materials used on this sea alms is the materials obtained from the environment that is environmentally friendly and easy to decompose, one of which is banana stem. It contains the lesson that if we nourish nature, it will take care of us for a long time.
Culture value	Implementation of sea alms is packaged in the form of a festive cultural carnival with the accompaniment of wayang kulit culture and other entertainment. The existence of a combination of cultural values with rituals indicates the side of cultural preservation that is maintained.
Religious value	The joint prayer on the stage songgo budhoyo in a series of sea alms events vertically contains the meaning of pleading for salvation and asking for sustenance to God. Various tools such as "*tumpeng*" also symbolize the majesty of God.
The soul of nationalism	Decorations in the form of red and white flags on boats used for melar symbolize the attitude of nationalism. It teaches that whatever culture is held, wherever it is locally, it remains all of Indonesia.
Symbolic value	Various tools and components of sea alms can not be separated from certain symbols and meanings. This shows the existence of the symbolic value contained therein. For example, the cone-shaped cone symbolizes God's majesty and the hope that the welfare of life is higher [14].

Aesthetic value	Beauty becomes a value that can not be separated from the sea alms. For example, giving decoration in the form of young coconut leaves (*janur*) on the boat.

Table 2. Profiles of sea alms ethnosciences

Culture value	Ethnoscience
"Tumpeng" at sea alms using a tool called *"gethek,"* which is a square-shaped board composed of banana stems. The selection of banana stem is none other because it is easy to reach and good floating ability.	**Material of hydrostatic pressure** Banana stems can float well. Why is that? The structure of hollow banana stems, so the volume tends to be large, but the mass is lighter because the constituents are dominated by air. This is one of the reasons why banana trunks can float on the surface of the water. This case can be investigated by density's theory. **(1) Density** Based the density's equation: $\rho = \frac{m}{V}$ ....(i) with: ρ = density (kg/m^3) m = mass (kg) V = volume (m^3) Using the equation can be known why the banana stem can float, by looking at the ratio of mass and volume greater volume. This case has a causal relationship. The shape of a hollow banana stem, then the larger volume is bigger →because the volume (*V*) bigger than mass (m), then the ratio of m and v or ($\frac{m}{V}$) less than 1. Meanwhile, the water mass density = 1. Hence, by the equation (i) it is known that its mass value is $<\rho_{air}$. **(2) Hydrostatic pressure** Next look for hydrostatic pressure (ρ_H), hydrostatic pressure is the liquid pressure is calculated from the surface of the liquid to a point in water through equation (ii). Equation: $\rho_H = \rho g h$ ...... (ii) with: ρ = hydrostatic pressure (Pa) g = acceleration of gravity (m/s^2) h = depth (m)

The table above shows a picture of ethnoscience *"gethek"* profile made from banana stems. After being transformed into science knowledge, it turns out there is a great potential of science subject matter in junior high school. That is, science-based science learning by integrating science and students knowledge sciences can gain more understanding of what is encountered. Not only know the ins and outs of theirculture but science knowledge also play a role and vice versa. Thus, in addition to learning and recognizing culture, students can take the important side of exploring the potential of science contained in the culture.

After the value of local wisdom and ethnosciences profiled, post test conducted to the respondent.. The results show that students are able to understand ethnosciences and the value of local wisdom in

IOP Conf. Series: Journal of Physics: Conf. Series **909** (2017) 012052 doi:10.1088/1742-6596/909/1/012052

the tradition of sea alms than before given the material ethnoscience. The following sample of students answers in Junior High School (SMPN 1 Pacitan and SMPN 3 Tegalombo).

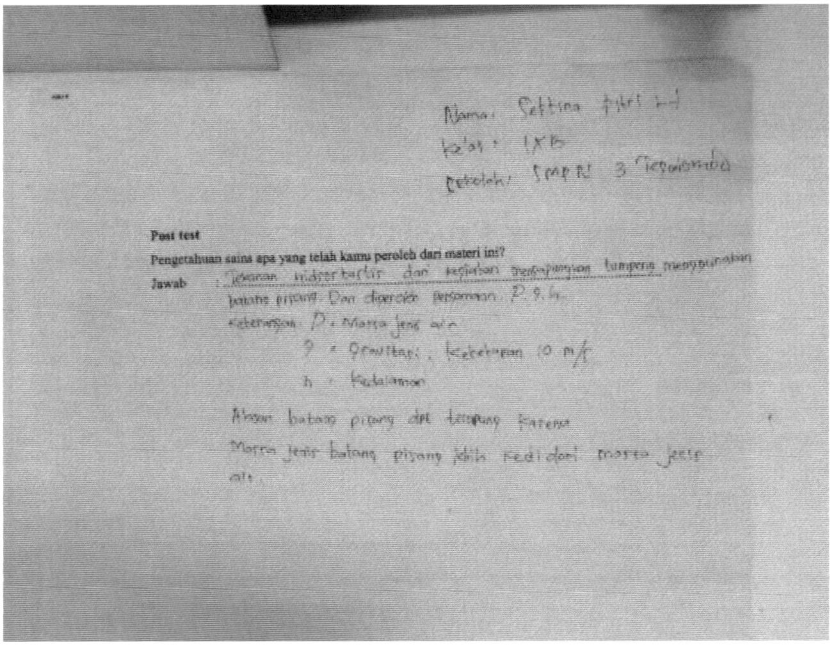

Figure 1. Post test results of science literacy at Junior High School in SMPN 3 Tegalombo

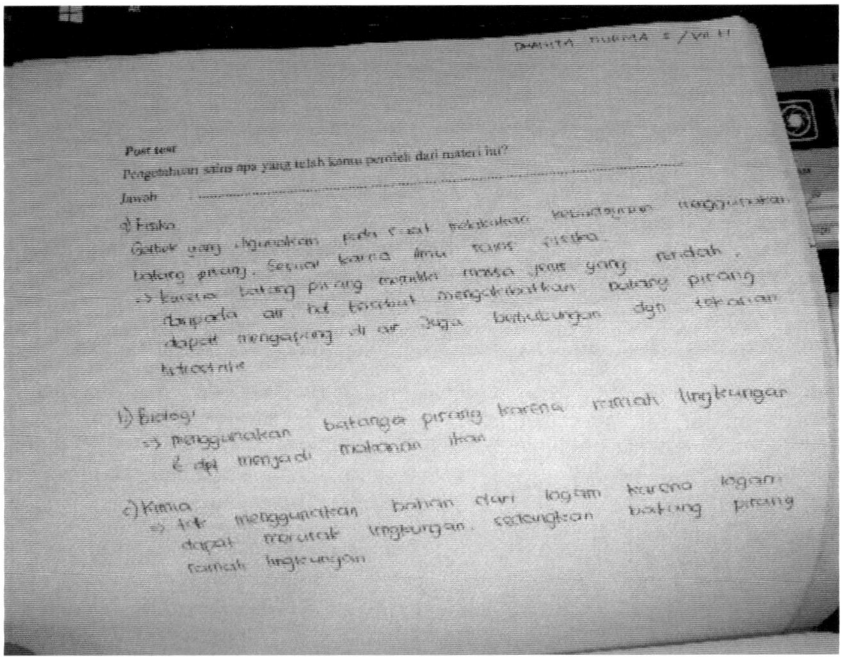

Figure 2. Post test results of science literacy at Junior High School in SMPN 1 Pacitan

Thus, it is evident that science literacy based ethnoscience is essential for the advancement of science knowledge. Through the process of transforming real science in real life, effective for spur the students to think critically. The demand to explore and understand the potential of science knowledge in the culture will also increase student's curiosity, in line with the government's mission which be oriented curriculum 2013 to train students' critical thinking minds.

4. Conclusion

Based on these results, can be concluded that the transformation of real science with science knowledge at sea alms tradition through the efforts of ethnosciences profile and the value of local wisdom effectively become an alternative science literacy. There is a potential for ethnosciences and the value of local wisdom in the tradition of sea alms. The various components in the sea alms that reflected by society can be studied scientifically. science literacy based ethnoscience is essential for the advancement of science knowledge. Through the process of transforming real science in real life, effective for motivate the students to think critically.

References

[1] Permadi, Danang. (2015). Budaya Larung Semboyo dalam Perspektif Hukum Islam. *Skripsi*. IAIN Tulungagung.

[2] UPT PP Tamperan. (2017). *Meriahnya Perayaan Mataraman di Pacitan, dari Larung Sesaji Hingga Wayang Suluk.*

[3] Rasheed Adekunle Fasasi. (2017). Effects of ethnoscience instruction, school location, and parental educational status on learners' attitude towards science. *International Journal of Science Education*, DOI: 10.1080/09500693.2017.1296599.

[4] Rahayu, W. E., & Sudarmin. (2015). Pengembangan Modul IPA Terpadu Berbasis Etnosains Tema Energi dalam Kehidupan untuk Menanamkan Jiwa Konservasi Siswa. *Unnes Science Education Journal, 2*(4), 920-926.

[5] Fajarini, U. (2014). Peranan Kearifan Lokal dalam Pendidikan Karakter. *Sosio Didaktika, 1*(2).

[6] Rusilowati, A., Supriyadi, & Widyatmoko, A. (2015). Pembelajaran Kebencanaan Bervisi Sets Terintegrasi dalam Mata Pelajaran Fisika Berbasis Kearifan Lokal. *Jurnal Pendidikan Fisika Indonesia, 1*(11), 42-48.

[7] Har, E. (2013). Karakter Budaya Sains Asli dan Karakter Budaya Sains Modern pada Pelajar Sekolah Menengah Atas di Sumatera Barat, Indonesia. *Jurnal Pendidikan Sains Sosial dan Kemanusiaan*, 13-26.

[8] Purwadi. (2005). *Upacara Tradisional Jawa.* Yogyakarta: Pustaka Pelajar.

[9] Atmojo, S. (2012). Profil Keterampilan Proses Sains dan Apresiasi Siswa Terhadap Profesi Pengrajin Tempe dalam Pembelajaran IPA Berpendekatan Etnosains. *Jurnal Pendidikan IPA Indonesia, 2*(1), 115-122.

[10] Handhika, J. (2014). Pendidikan Karakter Berbasis Kearifan Lokal dalam Perspektif Keilmuan Fisika. *Jurnal Edukasi Matematika dan Sains.* 2 (1), 34-40.

[11] Nisa', A., Sudarmin, & Samini. (2015). Efektivitas Penggunaan Modul Terintegrasi Etnosains dalam Pembelajaran Berbasis Masalah Meningkatkan Literasi Sains Siswa. *Unnes Science Education Journal, 3*(4), 1049-1056.

[12] Aryadi, M. (2010). *Metode Menafsir Data Kualitatif.* (hal 1-9). Makalah disajikan dalam Semiloka Penelitian dan Penulisan Tesis pada Program Magister Sains Administrasi Pembangunan, Universitas Lampung Mangkurat, Banjarbaru

[13] Purwahida, R.,dkk. 2008. Bahasa dalam Upacara Larung, Sedekah Laut di Laut Bonag, Kecamatan Lasem, Kabupaten Rembang, Jawa Tengah. *Jurnal Pelita*, 3 (1).

[14] http://www.tribunnews.com/lifestyle/2015/03/13/makna-di-balik-bentuk-kerucut-tumpeng (diakses 15 Mei 2017)

IOP Conf. Series: Journal of Physics: Conf. Series **909** (2017) 012053 doi:10.1088/1742-6596/909/1/012053

Development of soft scaffolding strategy to improve student's creative thinking ability in physics

Novinta Nurulsari[1], Abdurrahman and Agus Suyatna[2]

[1,2]Physics Education Department of Graduate Program Lampung University
Jl. Sumantri Brojonegoro No. 01 Gedong Meneng, Rajabasa Kota Bandar Lampung, 35141, INDONESIA

E-mail: [1]novintanurulsari@ymail.com, [2]abdurrahman.1968@fkip.unila.ac.id, [2]asuyatna@yahoo.com.

Abstract. Student's creative thinking ability in physics learning can be developed through a learning experience. However, many students fail to gain a learning experience because of the lack of teacher roles in providing assistance to students when they face learning difficulties. In this study, a soft scaffolding strategy developed to improve student's creative thinking ability in physics, especially in optical instruments. The methods used were qualitative and quantitative. The soft scaffolding strategy developed was called the 6E Soft Scaffolding Strategy where 6E stands for *Explore* real-life problems, *Engage* students with web technology, *Enable* experiment using analogies, *Elaborate* data through multiple representations, *Encourage* questioning, and *Ensure* the feedback. The strategy was applied to 60 students in secondary school through cooperative learning. As a comparison, conventional strategies were also applied to 60 students in the same school and grade. The result of the study showed that the soft scaffolding strategy was effective in improving student's creative thinking ability.

1. Introduction

Each can think creatively, but the ability is still potential. Creative thinking ability in Physics means the ability to think in creating an original idea to solve a problem related to Physics, where the resulting idea is new is the surface capability of a student [1]. Student's creative thinking ability is important to be improved in Physics learning because it can assist students to solve problems in everyday life [2].

The low ability of student's creative thinking was described by some international research results such as TIMMS and PISA scores in recent years. The empirical results of preliminary research also showed that students were not guided to find material-related issues independently. Even students were rarely asked to write down or mention some alternative solutions to one problem that students face. Teachers only played a role in directing students to determine the most effective solution to solving a problem or testing a hypothesis. It was also not supported by the evaluation process of the solution taken as one form of follow-up of decision-making by students.

Many studies had led to the improvement of creative thinking ability, one of them by developing learning process through cooperative learning [3], because life experiences in learning process may have stronger impacts on creative thinking [4]. The implementation of cooperative learning in Physics will not obtain optimal and meaningful results in improving student's creative thinking ability if there are still many students who experience learning difficulties especially in optical instruments material that requires a good conceptual understanding [5]. One strategy that can be combined in cooperative learning is soft scaffolding. Soft scaffolds are dynamic, situation-specific aid provided by a peer teacher to help the learning process. Soft scaffolding requires teachers to continuously diagnose the understanding of learners and provide timely support based on student responses [6]. Scaffolding tool can be designed to help students reflect on their thoughts on different levels of cognitive processes when working on or with their ideas [7]. It means soft scaffolding can be designed as a tool to help students in Physics through cooperative learning.

This study has two objectives, firstly, we developed a soft scaffolding strategy in physics learning especially in optical instruments material through cooperative learning. Secondly, we applied the soft scaffolding strategy to describe the enhancement of student's creative thinking in optical instruments.

2. Method

The methods of this research were qualitative and quantitative. This research was conducted in two phases of development and measurement. In the first phase, we developed the soft scaffolding strategy and the second phase we applied the soft scaffolding to measure student's creative thinking in optical instruments. According to Borg & Gall, research and development is a process used to develop and validate the products that are used in education and learning. The procedures in this study were using the Research and Development (R & D) by Borg and Gall.

Finally, the strategy developed was called the 6E Soft Scaffolding Strategy. Content and construct validity were done by three experts using questionnaires with Likert scale and analyzed by descriptive analysis. The cognitive test instrument to measure student's creative thinking ability after the implementation of the strategy was analyzed using Product Moment Pearson equation to describe the validity and using Alpha Cronbach equation to describe the reliability of the instrument.

The second phase was a phase of measurement. This stage involves 120 students in secondary school, of which 60 students were taught using soft scaffolding strategy and as a comparison, conventional strategies were also applied to 60 students in the same school and grade. The creative thinking ability aspects assessed were flexibility, fluency, originality, and elaboration. Analysis of the improvement and comparison of student's creative thinking ability were done through the pretest and posttest scores by using inferential statistical analysis of t test, The analysis was also supported by the result of the average gain analysis to describe the level of effectiveness of soft scaffolding strategy. The effectiveness of soft scaffolding strategy was also supported by the descriptive analysis results of the implementation of learning, teachers' performances, student learning activities, and student responses.

3. Result and discussion

In the first step, we develop soft scaffolding strategy. In the second step, the effects of the soft scaffolding were discussed. Results of the expert's validity analysis showed the soft scaffolding strategy was feasible to be used. The content validity analysis showed the percentage of 83% which means very high and the construct validity analysis showed the percentage of 87% which also means very high. Finally, the strategy developed was called the 6E Soft Scaffolding Strategy where E6 stands for E1 (*Explore* real-life problems), E2 (*Engage* students with web technology), E3 (*Enable* experiment using analogies), E4 (*Elaborate* data through multiple representations), E5 (*Encourage* questioning), and E6 (*Ensure* the feedback). Specific soft scaffolding strategy can be seen in figure 1.

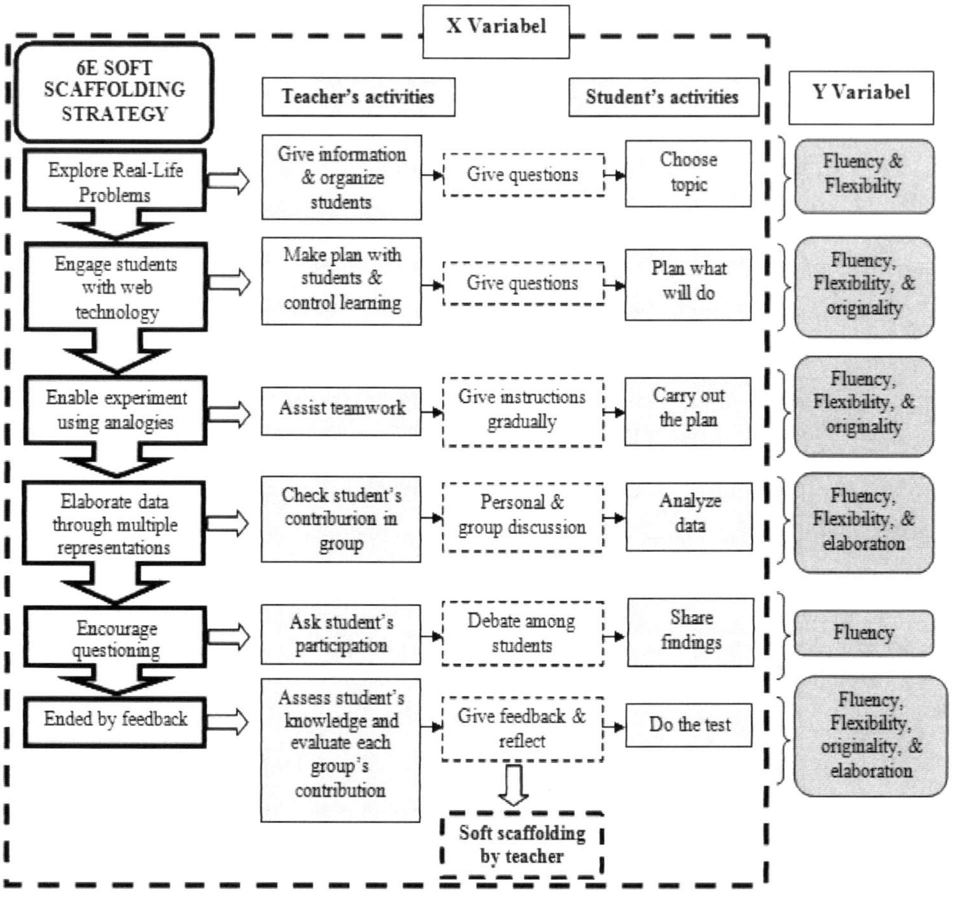

Figure 1. The soft scaffolding strategy developed through cooperative learning of group investigation.

We also examined the instrument used to measure student's creative thinking. The result of validity and reliability of creative thinking ability instrument which consisted of 10 items showed that the instrument was valid and reliable to be used.

The result of this study showed that the soft scaffolding strategy was efficient to improve student's creative thinking in physics, especially in optical instruments material. The effectiveness of the soft scaffolding strategy in improving student's creative thinking ability in Physics learning was known from some indicators, they were (a) the achievement of learning objectives, (b) the achievement of student's learning activity, (c) the achievement of teacher's performances, and (d) students' positive responses [8].

The achievement of learning objectives in experimental class was described by figure 2 which showed the improvement of students' creative thinking ability significantly with the sig value $< 0{,}05$ at the real level of 5%, n-gain analysis also show that strategies such as Figure 1 were effective to be used with mean value of gain were 0,7. The control class that implemented the conventional strategy shows a gain value of 0.2 which means less effective. The result of effect size test using Cohen's showed the result that soft scaffolding strategy gives effect in the medium category.

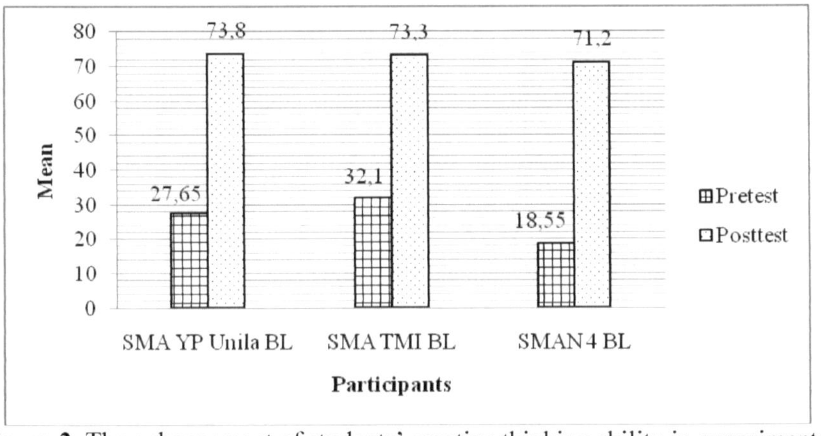

Figure 2. The enhancement of students' creative thinking ability in experimental class

Figure 3 showed the highest percentage of all aspects of creative thinking ability is fluency. Fluency was the ability to think smoothly indicated by several indicators such as the student replied with some answers if there were a question, smoothly expressed his ideas, and could quickly see the errors and weaknesses of an object or situation.

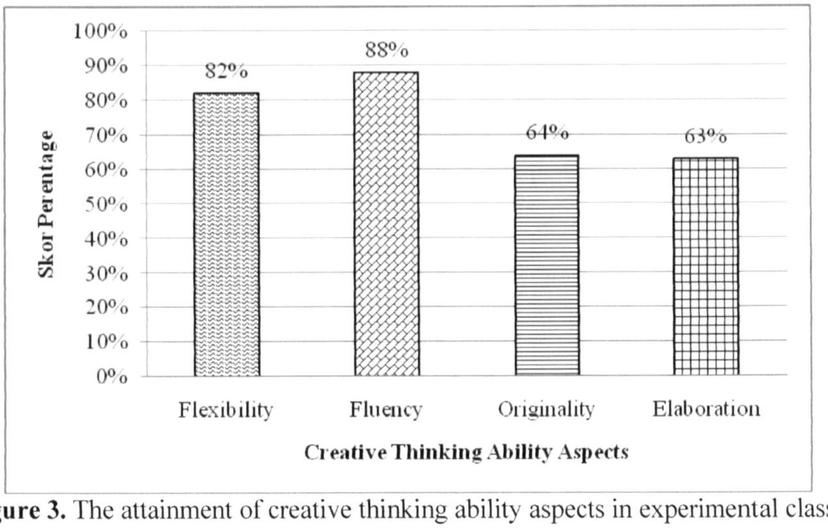

Figure 3. The attainment of creative thinking ability aspects in experimental class

But overall the achievement of students' creative thinking ability based on posttest results was in high category. The achievement of student's learning activity was described by observation score percentage of 72% which means in high category, and teacher's performances got the percentage of 89% which means in a very high category.

Perhatikan gambar dan bacalah ilustrasi berikut.

(a) (b)

Seorang siswa melakukan sebuah eksperimen yaitu memasukkan pensil ke dalam gelas berisi air. Setelah diamati, pensil terlihat patah seperti Gambar (a). Kemudian, anak tersebut mengeluarkan pensil dari gelas berisi air tersebut, lalu memasukkan sepotong kertas HVS ke dalam gelas tersebut. Setelah diamati, ternyata hasil yang ditunjukkan kertas pada Gambar (b) tersebut tidak terlihat patah seperti pensil dalam Gambar (a). Apa yang menyebabkan kondisi pada Gambar (a) dan (b) berbeda? Buatlah dugaan atau hipotesis untuk menjelaskan peristiwa tersebut.

Figure 4. The example question of fluency aspect.

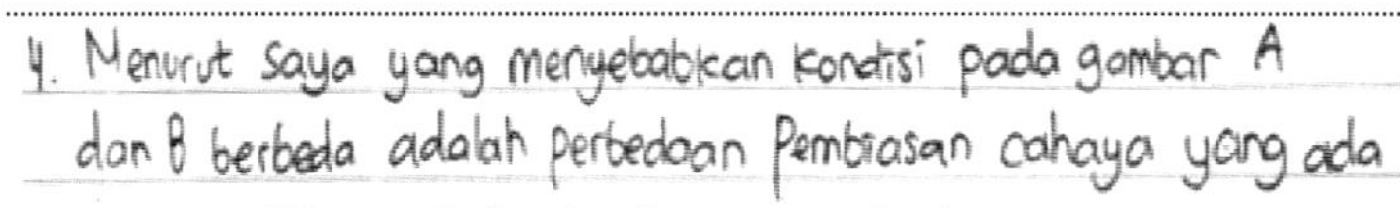

4. Menurut saya yang menyebabkan kondisi pada gambar A dan B berbeda adalah perbedaan pembiasan cahaya yang ada

Figure 5. Student's answers in the pretest

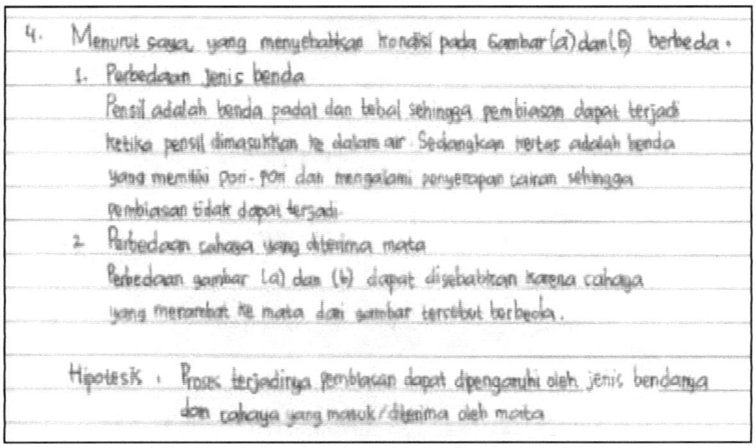

4. Menurut saya, yang menyebabkan kondisi pada Gambar (a) dan (b) berbeda.
1. Perbedaan jenis benda
Pensil adalah benda padat dan tebal sehingga pembiasan dapat terjadi ketika pensil dimasukkan ke dalam air. Sedangkan kertas adalah benda yang memiliki pori-pori dan mengalami penyerapan cairan sehingga pembiasan tidak dapat terjadi.
2. Perbedaan cahaya yang diterima mata
Perbedaan gambar (a) dan (b) dapat disebabkan karena cahaya yang merambat ke mata dari gambar tersebut berbeda.

Hipotesis : Proses terjadinya pembiasan dapat dipengaruhi oleh jenis bendanya dan cahaya yang masuk/diterima oleh mata

Figure 6. Student's answers in the posttest

The general finding of the improvement of student's creative thinking in fluency described through student's answers like a figure 4, 5 and 6. Optical instruments are closely related to the concept of light refraction. When students were asked about the different appearance of two phenomena related to light refraction in [the pretest], the students only answered that the two phenomena were different because of different refraction processes like a figure 5. The student answered without describing his idea through the reason why he answered so and the student has not been able to reveal more than one answer. He just mentioned that phenomena could happen because of the different light refraction without giving a reason why the factor was the cause of the differences between two phenomena. But after learning through the developed strategy, students' answers become more accurate and diverse. Students were also able to give two ideas like a figure 6. He could explain that light refraction can only occur in solids rather than porous objects, then he was able to mention that the difference of light travels to the eye could also affect the refraction of light. The answers given by the student were correct and logical. The student could express his ideas through the reasons of the answer given, even the student found the concept that the refraction was determined by the object.

The other general finding was found when researcher interviewed students after the posttest related to the question number 4 like a figure 4. Students stated that they were able to answer that question like a figure 4 because during the learning process the teacher allowed the students to access information using the web or the internet. Students stated that they could explore new information to develop their ideas through the web. The teacher gave instructions on what information was relevant related to the students' activities, including the design of the project, the phenomena of light reflection and refraction, optical tools in everyday life, and so on. Applying web-based learning evidently could improve students' ability to think fluently through instruction given by teachers [9].

The overall improvement of student's creative thinking ability was caused because students were involved in exploring real-world context physics problems [10], by students discovering problems related to tools to see the solar eclipse and the spearfishing systems through observation method. Web technology also played a role in learning [11], this was shown when students made a plan to create the pinhole camera and the spearfishing system using laptops and the internet in groups through discussion. Students are also stimulated to develop analogy skill [12], they attempt to apply forced analogy in processing tools and materials and linking the concepts of lasers and the spearfishing system. Students make a presentation of data and analysis in the form of multiple representations through diagrams and figures [13]. Students' creative thinking ability also increase when teachers implement the strategy of encouraging questioning [14], through debate activity. The description of soft scaffolding that teachers have successfully implemented can be seen in table 1.

Table 1. Soft scaffolding summary given in learning process

Learning phase	Stage of student's difficulties	Soft scaffolding form	Creative thinking ability successfully improved
Topic selection by using E1 strategy	Describe observation results	Give 5W questions	Fluency & flexibility
	State physics fundamental principles	Give key questions	Fluency
	Ask or answer quantitative questions	Give key questions	Fluency
	Propose an assumption	Give signal questions	Flexibility
	Choose an assumption	Give signal questions	Fluency
Cooperative planning by using E2 strategy	Find keywords to get information on the web or internet	Give signal questions	Flexibility
	Collect some important information sources	Give signal questions	Originality
	Bring together ideas from some sources into meaningful information	Give key questions	Elaboration
	Find new ideas	Give key questions	Originality

	Design procedures	Give key questions	Fluency
Implementation by using E3 strategy	Do the experiment	Give instructions gradually	Fluency
	Collect data	Give instructions gradually	Fluency
	Make an analogy	Give instructions gradually	Flexibility
Analysis & synthesis by using E4 strategy	Make a report	Have personal and group discussion	Elaboration
Presentation the final product by using E5 strategy	Evaluate and ask questions or responses to the topics presented.	Have debate activity	Fluency
Evaluation by using E6 strategy	In this phase, the teacher evaluated by holding a written test to measure the achievement of students' creative thinking ability. Soft scaffolding that teachers gave as feedback is by sharing in the form of weaknesses and strengths of each group.		

Based on table 1, the strategy was implemented through cooperative learning of group investigation and teachers provided soft scaffolding intensively and continuously based on student's difficulty. When students had difficulty in the topic selection stage and cooperative planning that implement the E1 and E2 strategies, the teacher gave soft scaffolding through giving questions for students [15], for example in finding the same idea of two articles and in making procedures of a pinhole camera and a spearfishing system. The questions consisted of 5W questions, key questions, and signal questions. In the implementation stage that implement the E3 strategy, teacher gave soft scaffolding through giving instruction gradually [16], especially how to collect the data related to the hypothesis and find the analogies. In the analysis and synthesis stage that implements the E4 strategy, the teacher gave soft scaffolding through giving discussion [17], especially how to interpret the data in diagrams and figures. In the presentation of final product stage that implements the E5 strategy, the teacher guided students in debate activity [18]. In the presentation of final product stage that implement the E5 strategy, teacher only gave feedback to students as the evaluation of their contributions in learning.

Researcher also found that conversation and interaction between students and teachers were very important to this strategy. Creative thinking ability indicator could directly indicated when student and teacher had discussion like a table 2.

Table 2. The example of interaction between students and teachers using soft scaffolding

Learning phase	Stage of student's difficulties	Interaction between students and teachers using giving signal questions
Cooperative planning by using E2 strategy	Bring together ideas from some sources into meaningful information	The questions the teacher asks are as follows. Look at source A, what did the author mean when writing that ....? Now consider the other source, what was the meaning of the idea written by the author in the paragraph ....? What conclusions were derived from these sources? Is there a common opinion with that written by some sources? Try to describe the similarity of opinion.

	These are the students state: "Pinhole cameras from source A were able to take pictures in cloud even at night, but pinhole cameras from source B were only capable of capturing images when the weather was sunny or there was a light source. It turns out that two forms of pinhole cameras produced were different and made of different tools and materials. It means that the working principle of pinhole cameras depend on the shape and the tools and materials used. So it could be wrong that the pinhole camera can not be used when there was no light. "

Based on table 2, through soft scaffolding in the form of signal questioning, students were able to develop elaboration skills. This was shown based on the findings of the researcher that students are able to develop an idea from source A by adding information from source B to become a unity.

4. Conclusion
From the analysis and discussion, it can be concluded that the soft scaffolding strategy can be used to improve student's creative thinking in physics especially in optical instruments. The soft scaffolding strategy can help the teacher to help students who have difficulties in learning physics. Students also learn better because the strategy not only develops their creative thinking but also their mathematic skill and conceptual understanding in physics so that the soft scaffolding strategy can promote higher order thinking skills.

Acknowledgement
We thank Lampung University (Unila). This project would not be possible without the help from members of our team who are not authors on this paper.

Reference
[1] Runco MA & Jaeger GJ 2012 *Creativity Research Journal* **24** 92.
[2] Tan OS 2009 *Problem Based Learning and Creativity* (Singapore: Cengage Learning).
[3] Siew NM, Chin MK, & Sombuling A 2017 *Journal of Baltic Science Education* **16** 100.
[4] Hong E & Milgram RM 2010 *Creativity Research Journal 22* 272.
[5] Hettmannsperger R, Mueller A, Scheid J, & Schnotz W 2016 *Zeitschrift für Erziehungswissenschaft* **19** 235.
[6] Brush T & Saye J 2002 In *annual meeting of the Association for Educational Communications and Technology, Denver, CO.*
[7] Lin PY, Chang YH, Lin HT, & Hong HY 2017 *Journal of Computers in Education* **4** 43.
[8] Sunyono 2015 *Model Pembelajaran Multipel Representasi* (Yogyakarta: Media Akademi).
[9] Lin CS & Wu RYW 2016 *Eurasia Journal of Mathematics, Science & Technology Education* **6** 1675.
[10] Vijayaratnam P 2009 *Proceedings of the 2nd International Conference on Teaching and Learning, Nilai, Negri Sembilan* 16.
[11] Paulus PB and Nijstad BA 2003 Oxford University Press.
[12] Awang H and Ramly I 2008 *International Journal of Human and Social Sciences* **3** 18.
[13] Rostika D and Juanita H 2017 *EDUHUMANIORA: Jurnal Pendidikan Dasar* **9** 35.
[14] Gall MD 1970) *Review of educational research* **40** 707.
[15] Chen CH & Law V 2016 *Computers in Human Behavior* **55** 1201.
[16] Thompson I 2009 *Written Communication* **26** 417.
[17] Maloch B 2002 *Reading Research Quarterly* **37** 94.
[18] Hmelo-Silver CE 2004 *Educational psychology review* **16** 235.

IOP Conf. Series: Journal of Physics: Conf. Series **909** (2017) 012054 doi:10.1088/1742-6596/909/1/012054

Effect of problem type toward students' conceptual understanding level on heat and temperature

D Ratnasari[1, a)], S Sukarmin[1], and S Suparmi[1]

[1] Science Education Department of Graduate Program Sebelas Maret University, Jl. Ir. Sutami 36A Kentingan, Jebres, Surakarta 57126, Indonesia

[a)]E-mail: dewi_ratnasari@student.uns.ac.id

Abstract. The aim of this research is to analyze the level of students' understanding of heat and temperature concept and effect of problem type toward students' conceptual understanding of heat and temperature. This research is descriptive research with the subjects of the research are 96 students from high, medium, and low categorized school in Surakarta. Data of level of students' conceptual understanding is from students' test result using essay instrument (arranged by researcher and arranged by the teacher) and interview. Before being tested in the samples, essay instrument is validated by the experts. Based on the result and the data analysis, students' conceptual understanding level of 10th grade students on heat and temperature is as follows: (1) Most students have conceptual understanding level at Partial Understanding with a Specific Misconception (PUSM) with percentage 28,85%; (2) Most students are able to solve mathematic problem from teacher, but don't understand the underlying concept.

1. Introduction

The aims of physics learning process at school are to develop students' thinking skill, to provide knowledge, understanding and to develop science and technology [1]. One of an important subject in the curriculum is physics, because of its contribution to the development of science and technology in society [2]. The right Physics learning process should develop conceptual change. Both changes in the form of expanding the concept, as well as change the wrong concept to be true, so that it can apply the concept to solving the problem [3]. Therefore, conceptual understanding is the most definitive thing towards the achievement of Physics learning goals.

The concept in physics learning become an important key to understanding and applying physics in daily life. The concept allows us to understand new experiences by connecting with what we already know. Scientific concepts are media that help to understand its application in everyday life. Scientific concepts are expressed on many levels, from very extensive labels to biology, physics, earth, etc., where the concept is presented in schools and generalized in standard reports or curricula [4].

The physics learning process guides students to be able in finding concept with its way. Therefore, teachers must be able to explore and to develop students' concept. Students should get many opportunities to use their logical reasoning skill, training, formulating a concept and participating in solving complex problems that need their hard effort. Then, students are encouraged to be able in creating accurate conclusion [5].

A good understanding of the physics concept leads students to be able to construct new concepts in other ways in different representations. A good understanding of concepts can make students be skillful in such as remembering, explaining, finding facts, mentioning examples, generalizing, applying, analogizing, and expressing new concepts in other ways [6]. These skills are necessary to physics learning in schools.

Conceptual understanding in physics teaching learning process has been the concern of many researchers. Other studies have revealed a conceptual understanding of physics on various topics [7, 8, 9, 10, 11]. Conceptual understanding makes students can transform knowledge become multiple representations and apply in daily life. Conceptual understanding of various physics concepts that have been obtained at schools can use as adequate basis for the advancement and development of science and technology in society [12]

Profile of students' conceptual understanding level known with appropriate problem type. Based on an interview with physics teachers, most of the teachers focused on mathematics problem than a conceptual problem. Science teachers more value on mathematic learning (algorithms learning) than conceptual understanding giving students who studied in science have the impression that science is "math in disguise" [13]. Students' failure to mixed science and math leads students to transform science problem solving to mathematics problem solving [14]. So, to know conceptual understanding in science need appropriate problem type that not confused with mathematics understanding or algorithms understanding.

Heat and temperature are one of physics materials that have the low conceptual understanding. Many students still difficulty in understanding heat concept and thermodynamic [15, 16, 17, 18]. The most difficult concepts are differentiating between heat and temperature [16, 19]. Therefore, it needs analysis on students' conceptual understanding level on that material so that teachers can recognize how far students' conceptual understanding and teachers can carry out the appropriate step. By knowing the students' conceptual understanding level, teachers can determine the appropriate learning model so that learning indicators can be achieved [11].

Based on the theoretical above, the researcher has conducted studies to investigate the students' conceptual understanding level on 10th-grade students in Surakarta on heat and temperature and effect of problem type arranged by the teacher. An analysis of the level of students' conceptual understanding uses essay test as an assessment instrument.

2. Experimental Method

This research is a descriptive research aimed to describe the level profile of students' conceptual understanding. Subjects of the research are 10th-grade students in the academic year of 2016/2017. It consists of 96 students from one class in every school (high, medium, and low categorized school) in Surakarta. The category of the school is based on the four latest score of a national exam. Data of level of students' conceptual understanding is from students' test result using essay instrument (arranged by researcher and arranged by the teacher). The data will be used to analyze the students' conceptual understanding level to get students' profile that shows the level of conceptual understanding for each student and to know the effect of problem type toward students' conceptual understanding. Data is also collected through interview. The interview is used to confirm and to explore students' conceptual understanding.

3. Result and Discussion

The instrument used to analyze the level of students' conceptual understanding is essay questions that consist of 10 numbers. The aim of free essay test is to avoid guessing answer or trial from students. Essay questions are considered as an appropriate method or a tool to assess the result of complex learning activity. It is because the writing of essay answer will involve the skill of remembering, organizing, expressing, and integrating students' idea [20]. Essay questions will always be chosen by teachers in evaluating skill grade of students even it is not easy to give an objective assessment on each student's answer. This form is usually used by many researchers because of its effectiveness to assess learning achievement result. It is also good to observe high-level thinking ability, such as synthesis and analysis [21]. Students' test result is analyzed by using assessment guidance. The assessment guidance on this question can be seen Table 1 [5].

Table 1. The assessment guidance for conceptual test item [5]

Score	Level of understanding	Criteria for scoring
0	No understanding (NU)	The students did not answer, the answer did not match the question, just repeated the question, irrelevant or unclear response, no explanation given for the answer.
1	Specific Misconception (SM)	The students replied, but scientifically incorrect responses
2	Partial Understanding with a Specific Misconception (PUSM)	The students' answer show understanding of the concept, but there are misconceptions
3	Partial Understanding (PU)	The students' answer contains a part of scientifically accepted concept
4	Sound Understanding (SU)	The students' answer contains all parts of the scientifically accepted concept.

Table 1 is used as a guide in assessing students' answers, therefore teachers can be described students' conceptual understanding level on heat and temperature. The analysis result of students' conceptual understanding level is listed in Table 2.

Table 2. Level of students' conceptual understanding of heat and temperature

Level of understanding	Percentage (%)
No understanding (NU)	10,94
Specific Misconception (SM)	17,40
Partial Understanding with a Specific Misconception (PUSM)	28,85
Partial Understanding (PU)	21,67
Sound Understanding (SU)	21,14

Table 2 shows an overview of students' conceptual understanding level of 10th-grade students on heat and temperature. Based on Table 2, the biggest percentage of students' conceptual understanding is on the level of PUSM or partial understanding with a specific misconception. It shows that students can only understand a part of scientific concept correctly so that they still face misconception. It is also because students tend to memorize than understand the concept. Besides concept understanding data of 96 students, another data is obtained from conceptual understanding level for each question item. Data can be seen in Table 3.

Table 3. Level of students' conceptual understanding on each question

Number of question	NU (%)	SM (%)	PUSM (%)	PU (%)	SU (%)
1	0.00	11.46	47.92	39.58	1.04
2	28.13	6.25	13.54	10.42	41.67
3	31.25	0.00	5.21	26.04	37.50
4	0.00	1.04	18.75	5.21	75.00
5	2.08	6.25	21.88	23.96	45.83
6	2.08	6.25	75.00	10.42	6.25
7	10.42	69.79	19.79	0.00	0.00
8	14.58	7.29	64.58	12.50	1.04
9	10.42	59.38	7.29	22.92	0.00
10	10.42	6.25	14.58	65.63	3.13

Table 3 shows the level of students' conceptual understanding on each question. Based on the data from Table 3, it can be seen that the biggest percentage of students' conceptual understanding of each number is various. However, most of them are dominated on the level SM (specific misconception) and level PUSM (partial understanding with a specific misconception), teachers must concern on this matter because many students still face misconception. The graphic of level students' conceptual understanding of each question is shown in Figure 1.

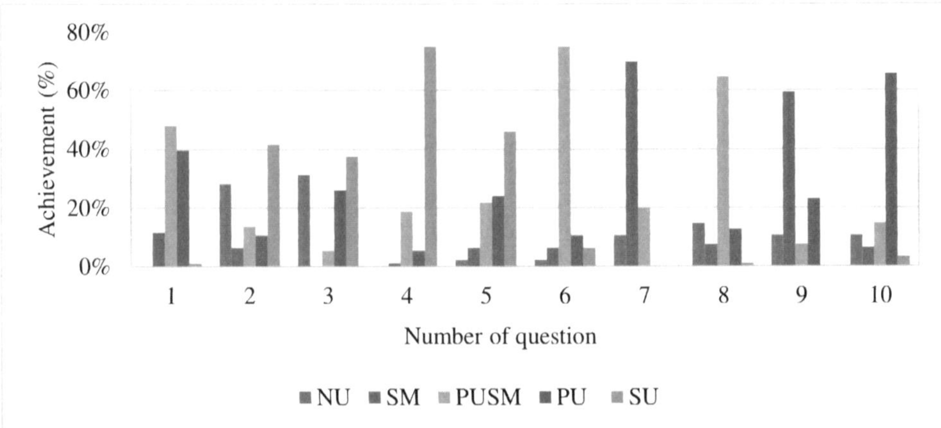

Figure 1. Level of students' conceptual understanding on each question

Based on figure 1, level NU (no understanding) is the biggest at number 3. Number 3 contains expansion concept in the making of the thermometer. In this question, students are asked to analyze if thermometer glass has the same coefficient as alcohol on the expanded space. Expansion concept is one of the concepts that are often used in daily life. However, many students can not answer it at all.

On level SM (Specific misconception), the biggest percentage at number 7. The question is about the influence of thermal conductivity value of the material. In this question, students are asked to determine which objects are cooler if it is touched. The three objects (glass, wood, and iron) have the same temperature at the beginning. Most of the students answer that the condition is because of the influence of specific heat. The example of students' answer can be seen in Figure 2.

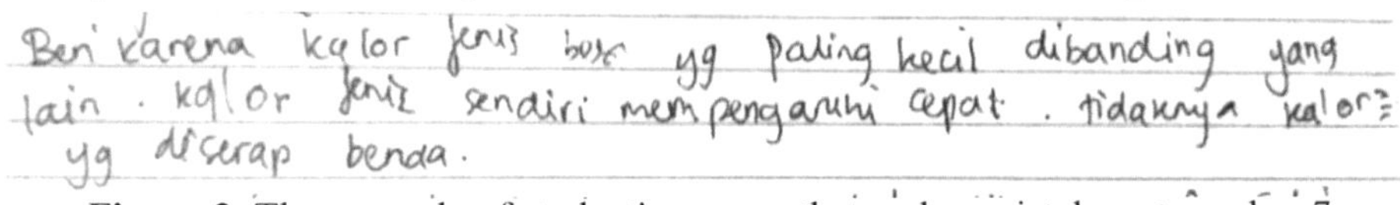

Figure 2. The example of student's answer that makes mistakes at number 7.

Besides students' mistake in answering the question at picture 4, the students were still carried on with the previous concept in junior high school, that the objects were distinguished into conductors and isolators regardless of the magnitude of the thermal conductivity value of each object. Students still assume if the isolators object has a zero thermal conductivity value or no thermal conductivity value. Students' answer can be seen in Figure 3.

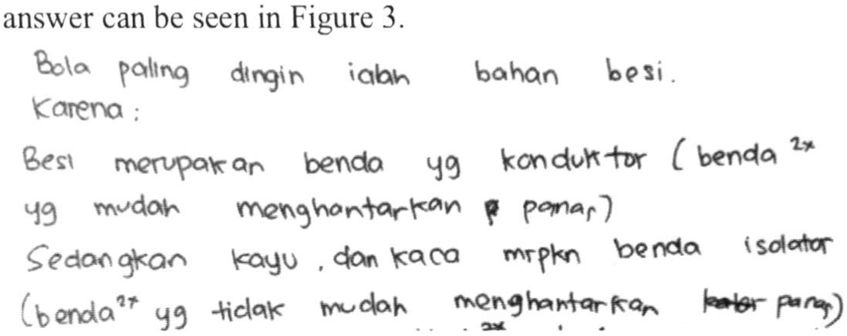

Figure 3. The example of student's answer that makes mistakes at number 7.

Students' answers are not entirely wrong, but wood and glass also have thermal conductivity values even though they are not as big as the thermal conductivity of iron. The object is cold when it is touched because heat transfer from hand to object. The ability of objects to transferring heat is influenced by thermal conductivity value of the material. Students still relate the concept of thermal conductivity with cold/hot objects. Students also find difficulty in differentiating thermal conductivity concept and specific heat concept [22].

The low of students' conceptual understanding because of problem type from teachers only about the mathematic problem. Exercises and example from teachers focused on the mathematic problem, so students only have the algorithmic understanding. Algorithmic understanding is students' understanding of the mathematic problem. The example of students' answer on the mathematic problem about thermal conductivity concept can be seen in Figure 4.

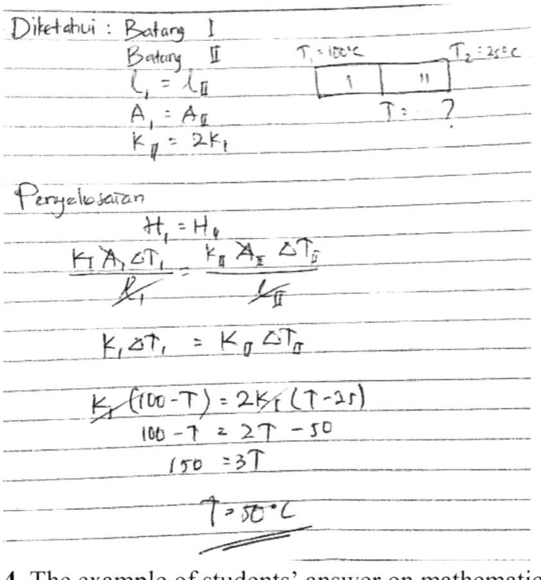

Figure 4. The example of students' answer on mathematic problem

Conventional teaching learning process stressed in the mathematic problem, therefore when students are given conceptual problem, students can not be able to answer it as seen in figure 3. Sometimes, many students can answer the question correctly, but they cannot provide the reasons that support the answer. The result showed that students were able to use formulas and solve mathematical problems, but they did not understand the underlying concepts of the formula [23].

On level PUSM (partial understanding with a specific misconception) shows students' answer that understands a scientific concept, but there is still a misconception. On level PUSM, the biggest percentage at number 6. In this question, students are asked to compare two conditions and to determine which one that has bigger heat energy. In this question, most of the students confuse to answer because students' habit of solving the problem of physics with mathematic calculation. The mathematic calculation can create answer or conclusion, but students' skill to analyze is still low. Students' answer can be seen in Figure 5.

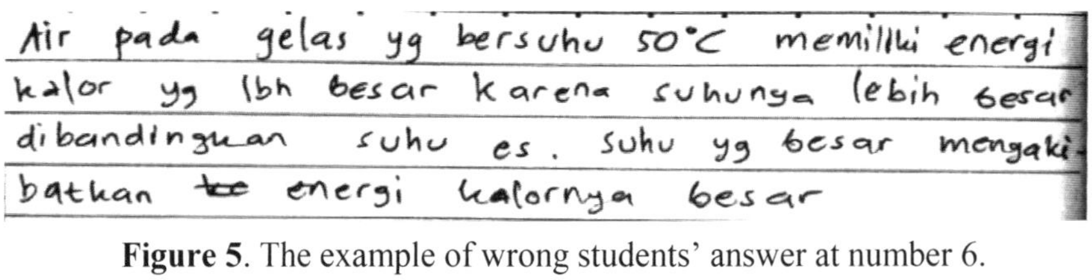

Figure 5. The example of wrong students' answer at number 6.

Based on Figure 5, it can be seen that students still find difficulty in analyzing the relation among variables. Many students have conceptions that amount of heat energy depend on the temperature of the object because students viewed that higher temperature objects would have more heat energy [24]. It is appropriate with the result of students' interview. Students state that there is difficulty in analyzing the relation among the number of heat, mass, specific heat and temperature addition. Nevertheless, there are students who understand the concept of the relationship between mass, specific heat and temperature addition. The example of students' answer can be seen in Figure 6.

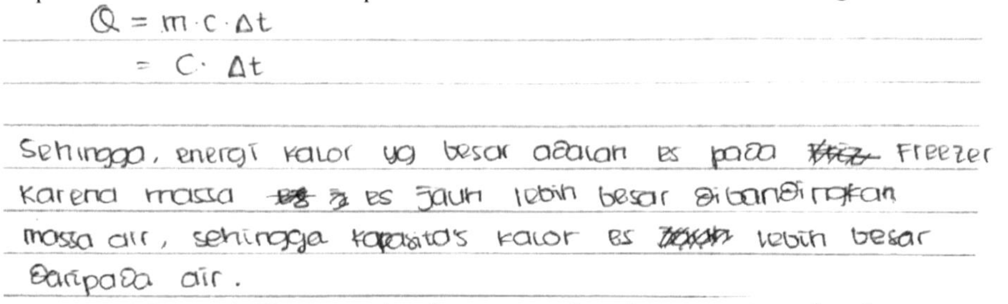

Figure 6. The example of correct students' answer at number 6.

Based on the interview, almost of the student said that difficulty solves number 6 because mass and heat specific not listed in the question. Student prefers counting than applying the concept to solve the problem. The ability of students to complete a mathematic problem (algorithmic) does not depend on the students' conceptual understanding. Based on that statement, students may be able to complete algorithmic question correctly, but they don't understand the basic concept correctly [25]. Problem type from the teacher only about the mathematic problem. The example of students' answer about heat energy based on problem type from the teacher can be seen in Figure 7.

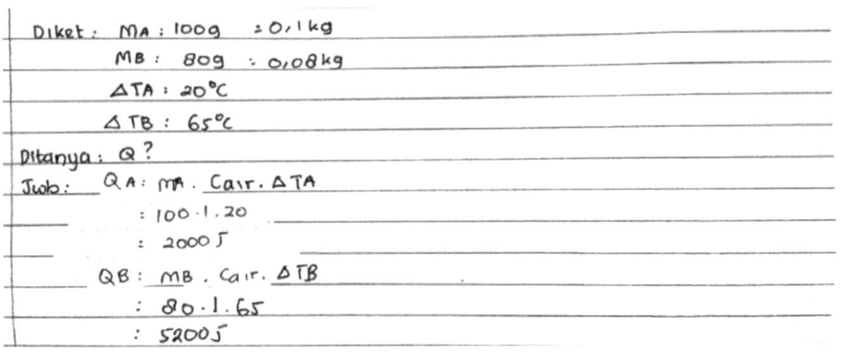

Figure 7. The example of students' answer about the mathematic problem.

Based on Figure 7 students can solve the mathematic problem and calculate heat energy, but based on Figure 5, students still confuse to explain the relation of mass and temperature. The low of students' conceptual understanding in analyzing the relationship between mass and heat is also revealed by the previous researchers. Students think that temperature of an object depends on the mass, in which if there is a big mass, so it also has big temperature and vice versa [26, 27]. If there is small mass, the heat to be absorbed is greater so that the temperature will increase quickly [28]. Students assume that the heat of an object is related to the heat received by the object, so students spontaneously answer that objects with the big specific heat will become hot quickly [29]. Students can use the equation $Q = mc\Delta T$ to determine the amount of heat, but students are unable to connect the four quantity in the equation [23].

The difference of students' conception is influenced by many factors. One of them because the different prior knowledge owned by students. The difference of students' conception is because students come to the class with different knowledge, a knowledge built from daily life. Students' knowledge of the concept of heat and temperature comes from daily life. The concept of heat and temperature directly related to the environment and daily life. Therefore, heat and temperature cannot be directly observed [30]. Students come with various impression and concept that comes from their life. Many impressions and concepts might be wrong [31].

On level PU (partial understanding), the biggest percentage is at question number 10. This question is about the concept application of the Black principle to determine the specific heat of the material. Students are required to design experimental procedure with tools and materials mentioned in the question for the experiment purpose to be achieved. In this question generally, students have not been able to answer the steps in order to determine the specific heat of the metal. Besides that, many students have not understood the usefulness of calorimeters and steps to determine specific heat. Students only know Black principle, but they are unable to apply it to measure specific heat of the metal. Based on the interview result, students are difficult to answer the question because they have never conducted a lab work.

Students have difficulties in distinguishing between Q released and Q received. Many students are confused about the concept of heat and temperature. Temperature perception is only about hot and cold and temperature can be transferred. Students memorize this concept and cannot make any relation between knowledge and physics phenomenon in their daily life [32].

On level SU (sound understanding), shows students' answer that understands a true and complete scientific concept. The lowest percentage is at number 7 and 9. Number 7 has been discussed on level SM (specific misconception), while number 9 concerns about the concept of the heat capacity of the material. In this question, students are asked to analyze the cause of how an object cools fastly if there are two objects that cooled with the same source and at the same time.

At number 7 and 9, students are difficult in differentiating specific heat concept, heat capacity, and thermal conductivity. Most of the students still consider that the cause of how objects cold fastly at number 9 is the influence of specific heat of an object. At number 7, students consider that specific heat is a factor that influences the fast on how heat is absorbed by an object. The example of students' answer that finds difficulty at number 9 can be seen in Figure 8.

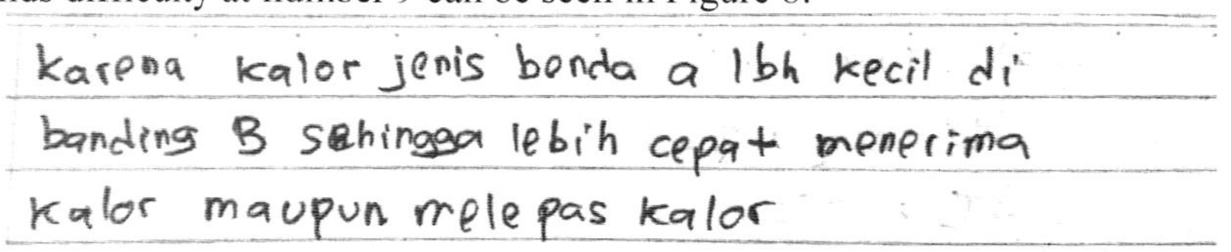

Figure 8. The example of student's answer that makes mistakes at number 9

Heat capacity becomes one of important concept on heat and temperature. Students can calculate the heat capacity, but they don't know the effect of the heat capacity of the material. Based on students' answer in the mathematic problem from the teacher, almost all students can calculate the heat capacity, but not understand the basic concept of heat capacity. The students' answer to the mathematic problem on heat capacity can be seen in Figure 9.

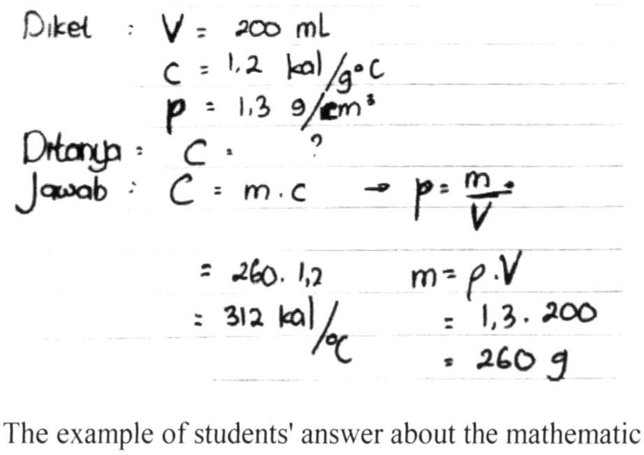

Figure 9. The example of students' answer about the mathematic problem.

The low of the conceptual understanding is because teachers' learning method focused on giving formula or mathematic formulation than the conceptual problem, therefore makes students didn't understand the concept. Students' mathematics ability more dominant than students' perception, so it caused the wrong conception. The integration of mathematics ability and students' perception in physics can decrease the wrong conception [33]. The low of the understanding of heat capacity concept is revealed by the previous researcher that states if the heat capacity is big, the temperature of an object will increase quickly [34]. Another study revealed that students do not consider the value of specific heat and heat capacity as a factor that influences temperature changing [23].

Conceptual understanding of heat and temperature is still low due to the concept of heat and temperature closely related to the daily life of students. One of the main discussions in science education is conceptual understanding and students' misconception of heat and temperature. The main reason for this concern is a fact that heat and temperature are two common words in students' life. This conceptual understanding is a key to understanding other scientific concepts. It is also important for science teachers to understand students' knowledge about this concept and to develop new curriculum and teaching methods for science class [35].

Based on the results of interviews with students, teachers have not used a scientific approach to physics learning. Teachers still use lecture methods with PowerPoint presentation media. So, the concept of physics is less embedded. Teaching with PowerPoint media presentation make teacher tend to omit to ask questions or discussion with students, this is noticed that many science students are relatively shy. In such situations, they would move their eyes from screen to textbook. Thus the instruction would become a monolog of the teacher. Although the PowerPoint presentation has disadvantages, it also has advantages in producing better visual effects and deeper impression [36].

The low of students' conceptual understanding can emerge misconception. A study aimed at identifying students' misunderstanding that emerges misconception on heat and temperature concept. The result of the study indicates that students have difficulty in connecting knowledge gained during at school with experience in daily life. Students' misconception occurs because there is no relation between daily life with school experience [37].

One of the efforts to decrease students' misconception is that teachers must know the level of students' conceptual understanding. Therefore, it needs profile analysis of the level of students' conceptual understanding of heat and temperature. Students' profile can support information media owned by teachers to know how far the difficulty and indicators that have been achieved by students [38]. Teachers must also be able to conduct an analysis of learning indicator achievement. Profile analysis of students' conceptual understanding level is made assisted by software *Microsoft Excel*. Analysis of students' profile using *Microsoft Excel* is developed by research from Wulandari, *et al.* (2015).

Profile analysis of students' conceptual understanding level with *Microsoft Excel* only needs a simple formula to get the profile for each student. Answer keys, students' answer and indicator on each question are processed into *Microsoft Excel* to get students' conceptual understanding level. The example of the profile of students' conceptual understanding level can be seen in Figure 10.

Figure 10 shows the profile of students' conceptual understanding level. At Figure 10 shows the indicators that have been achieved and have not. Besides that, it also shows the students' conceptual understanding level on each number and the percentage achievement of conceptual understanding level on each student. Students' profile can also be used to know students' learning achievement and to find students' difficulty on physics material.

STUDENTS' PROFILE

Student Identity

Name	Fachry Nur H.
Number	12

Number	Score	Indicator		Concept understanding level
		Achieved	Not yet achieved	
1	3	Being able to determine the thermometer exactly, but do not give the reason for the selection of a thermometer.	Determine the reason for selecting the right thermometer based on the properties of the thermometer filler liquid.	PU
2	0	-	Determine the reason for the use of capillary pipe in thermometers accordance with the correct concept completely.	NU
3	0	-	Explain the consequences if the coefficient of expansion of the glass chamber of the thermometer is the same as the thermometer filler according to the correct concept completely.	NU
4	4	Being able to explain the concept of expansion based on problems in daily life according to the correct concept completely.	-	SU
5	2	Being able to explain the cause of the phase chage in daily life, but still contain misconceptions.	Explain the cause of the phase change in everyday life according to the correct concept.	PUSM
6	2	Being able to calculate and explain the heat energy based on the experiment, but there is still a concept error (misconceptions).	Calculates and explains the heat energy based on the experiment according to the correct concept.	PUSM
7	1	Being able to answer, but not in accordance with the correct scientific concept (misconception).	Apply the concept of the effect of thermal conductivity of materials on heat conductivity according to the correct concept based on experimental data.	SM
8	1	Being able to answer, but not in accordance with the correct scientific concept (misconception).	Compare two experiments that absorb more heat energy according the correct concept.	SM
9	0	-	Determine the effect of specific heat and heat capacity of the object on the amount of heat absorbed according to the correct concept completely.	NU
10	2	Being able to design an experiment to determine the specific heat of a material according to the correct concept, but there is a misconception.	Designing an experiment to determine the specific heat of object according to the correct concept.	PUSM

The percentage of concept understanding level

Concept understanding level	Percentage
NU	30%
SM	20%
PUSM	30%
PU	10%
SU	10%

Figure 10. The display of students' conceptual understanding profile

4. Conclusion

Based on the result and the data analysis, students' conceptual understanding level of 10th grade students on heat and temperature is as follows: (1) Most students have conceptual understanding level at partial understanding with a specific misconception (PUSM) with percentage 28,85%; (2) Most students are able to solve mathematic problem from teacher, but don't understand the underlying concept.

Acknowledgements

The writer would like to say her gratitude to her parents for their continuous prayers. For Mr. Sukarmin S.Pd, M.Si, Ph.D. and Mrs. Prof. Dra. Suparmi, M.A., Ph.D. who have guided in completing this article, and for all who have given their help to finish this article.

References

[1] Depdiknas 2006 *Kurikulum Tingkat Satuan Pendidikan (KTSP)* (Jakarta: Depdiknas).
[2] Eraikhuemen, L and Ogumogu, A.E 2014 *AR Int.* **5** 253-262.
[3] Suparno, P 2013 *Miskonsepsi dan Perubahan Konsep dalam Pendidikan Fisika (*Jakarta: Grasindo)
[4] OECD 1999 *Measuring Student Knowledge and Skills* (France: OECD Publication Service)
[5] Nurhuda, T., Rusdiana D., and Setiawan, W. 2017 *IOP Conf. Series: Journal of Physics: Conf. Series* **812** 012105
[6] Eggen, P., and Kauchak, D 2012 *Strategi dan Model Pembelajaran (edisi keenam) (*Jakarta Barat: PT Indeks).
[7] Ponimin, Suparmi, Sarwanto and Sunarno, W. 2017 *IOP Conf. Series: Journal of Physics: Conf. Series* **795** 012053.
[8] Handika, J., Cari, Suparmi 2017 *IOP Conf. Series: Journal of Physics: Conf. Series* **795** 012057.
[9] Handika, J., Cari, Suparmi and Sunarno, W. 2017 *IOP Conf. Series: Journal of Physics: Conf. Series* **795** 012058.
[10] Budiarti, I.S., Suparmi, Sarwanto, Harjana 2017 *IOP Conf. Series: Journal of Physics: Conf. Series* **795** 012051.
[11] Handika, J., Huriawati, F., Fitriani, N. 2017 *J. Phys.: Theor. Appl.* **1** 29-34
[12] Omosewo, E.O 2009 *Afr. Res. Rev.* **3** 314-325.
[13] Pushkin, D. B 1998 *J. Chem. Ed.* **75** 809-810
[14] Redish E. F 2005 *Proceeding of World View on Physics Education in 2005: Focusing on Change*
[15] Driver, R 1989 *Int. J. Sci. Educ.* **11** 481-490.
[16] Linn, M. and N. B. Songer 1991 *J. Res. Sci. Tech.* **28** 885-915.
[17] Lewis, E. and M. Linn 1994 *J. Res. Sci. Tech.* **31** 657-677.
[18] Harrison, A. G., D. J. Grayson, and D. F. Treagust 1999 *J. Res. Sci. Tech.* **36** 55-87.
[19] Erickson, G. L 1979 *Sci. Educ.* **63** 221–230.
[20] Valenti, S., Neri, F. and Cucchiarelli, A. 2003 *JITE: Research* **2** 319-330.
[21] Foltz, P.W., Gilliam S., and Kendall, S.A. 2000 *ILE* **8** 111-129.
[22] Quan, G 2001 *Improvements of Student Understanding of Heat and Temperature* (Washington DC: University of Washington)
[23] Alwan, A.A 2011 *Procedia Soc. Behav.* Sci. **12** 600–614.
[24] Elwan, A 2007 *J. Arabization* **33** 77-103.
[25] Stamovlasis, D., Tsaparlis, G., Kamilatos, C., Papaoikonomou, D., Zarotiadou, E 2005 *Chem. Educ. Res. Pract.* **6** 104-118.
[26] Suparno, P 2013 *Miskonsepsi dan Perubahan Konsep Dalam Pendidikan Fisika* (Jakarta: Grasindo).
[27] Maunah, Nailul & Wasis 2014 *JIPF* **3** 195-200.
[28] Silung, S.N.W. 2015 *Prosiding Seminar Nasional Jurusan Fisika FMIPA UNESA* 180-185.
[29] Silung, S.N.W., Kusairi, S., Zulaikah, S 2016 *JPFT* **2** 95-105
[30] Arnold, M., and R. Millar 1994 *Int. J. Sci. Educ.***16** 131-144.
[31] Kartal, T., Ozturk, N., & Yalvac, H.G 2011 *Procedia Soc. Behav.* Sci. **15** 2758–2763.
[32] Winarti, Cari, Suparmi, Sunarno, W., Istiyono, E 2016 *Proceeding of Asean Education Symposium UPI.*
[33] Handhika, J., Cari, Suparmi and Sunarno, W 2015b *AIP Conf. Proc.* 1-5.
[34] Hafizah, D., Haris, V., & Eliwatis 2014 *Edusainstik Jurnal Pendidikan MIPA* **1** 100-103.
[35] Sözbilir, M 2003 *BUJED* **20** 25-41.
[36] Xingeng, D and Jianxiang, L 2012 *IJEME* **9** 61-65
[37] Gönen, S. and Kocakaya, S.2010 *Eurasian J. Phys. Chem.* **2** 1-15.
[38] Wulandari, R.R.A., Yamtinah, S., Saputro, S 2015 *JPK*, **4** 147-155.

IOP Conf. Series: Journal of Physics: Conf. Series **909** (2017) 012055 doi:10.1088/1742-6596/909/1/012055

Analyzes of students' higher-order thinking skills of heat and temperature concept

Indah Slamet Budiarti[1], A. Suparmi[2], Sarwanto[3], Harjana[4]

Graduate Student's Sebelas Maret University[1],

Lecturers of Sebelas Maret University[2,3,4]
Jl. Ir. Sutami 36A Kentingan Jebres Surakarta 57126, INDONESIA

E-mail: indah_budiarti@student.uns.ac.id, indah_budiarti@yahoo.com

Abstract. High order thinking skills refer to three highest domains of the revised Bloom Taxonomy. The aims of the research were to analyze the student's higher-order thinking skills of heat and temperature concept. The samples were taken by purposive random sampling technique consisted of 85 high school students from 3 senior high schools in Jayapura city. The descriptive qualitative method was employed in this study. The data were collected by using tests and interviews regarding the subject matters of heat and temperature. Based on the results of data analysis, it was concluded that 68.24% of the students have a high order thinking skills in the analysis, 3.53% of the students have a high order thinking skills in evaluating, and 0% of the students have a high order thinking skills in creation.

1. Introduction

Students' capability to be successful in their learning counts a lot on thinking skills. These skills are major in supporting them to solve problems during the learning process. Possessing learning skills, students can develop their cognitive intelligence and ability to relate facts or information by using their knowledge to hence a prediction on a formulated final result.

Thinking skills are not only to define. Thinking is a high order of intellectual or cognitive process [1]. Thinking ability is skill and power to be pervaded in all subjects to improve students' performance and reduce weakness [2]. Teaching and learning activities should involve explicit thinking skills which contribute ease to categorize thinking skills based on the existing framework [3]. High order of thinking has a characteristic to solve special problems and the solutions are results of thinking and reasoning [4]. Characteristics of mental activity derived from the high order of thinking process often involve complex thinking, non-algorithmic, independence in the thinking process and can result in applicative solutions [5], [6], [7]. When one is in any unusual situations, faces plenty of problems, among alternatives and has to decide [8].

The concept of heat and temperature is studied by students from elementary school to college. Temperature is related to the amount of heat. Students had difficulties in Black Principle as two different objects of temperature touch each other, and finally, experience thermal equilibrium, so the temperature will eventually be the same [9]. Most students can not differentiate between heat and temperature [10]. Concepts related to heat and temperature are directly related to living organisms, so that heat and temperature cannot be observed directly in quantity. Concepts developed by students, derived from the interpretation of student ideas in everyday life [11].

High order thinking skills in this research refer to three highest domains of the revised Bloom Taxonomy. Three highest cognitive domains of Bloom Taxonomy (analysis, evaluation, and creation) require high order thinking skills [12]. Based on underlying theories and research objectives in the field, the researcher had performed researchers to analyze high order thinking skills on Heat and Temperature. Meanwhile, 'how are students' high order thinking skills of Heat and Temperature Concept?' is stated as problem formulation in this research.

2. Research Method

The method used in this research is descriptive, to analyze students' high order thinking skills of heat and temperature concept. Samples were taken by using purposive random sampling from 3 Senior High School (SHS) in Jayapura city. Data were collected by applying purposive multiple choice tests and interviews. The tests were adapted from HTCE (Heat and Temperature Concept Evaluation) [13]. The tests consisted of 15 numbers. Five numbers were problems of temperature, five numbers of heat and the rest are of heat transfer. Interviews were taken to collect data about reasons having been written by students.

3. Result and Discussion

Percentage of student's conceptual understanding consistency is based on students' answers as seen on table 1

Table 1. Percentage of Students' High Order Thinking Skills

High Order Thinking Skills	Percentage (%)			
	SMAN 1 Jayapura	SMA YPK	SMA Pembangunan V Yappis	Average
Analysis	75	66.67	53.33	68.24
Evaluation	5	3.33	0	3.53
Creation	0	0	0	0

Based on Table 1, it can be seen that the highest domain of Bloom Taxonomy which is the highest order of high order thinking skills shows average 0%, this indicates that the students had not achieved creation stage, especially on physics concepts of temperature and heat. Table 1 also shows that students' high order thinking skills percentage did not reach 100%, this indicates that 28.24% of students had not achieved high order thinking skills, means that 28.24% of the students were still in low order thinking skills, particularly recognized as LOTS (Lower Order Thinking Skills). Thus, there were still 28.24% of students who achieved knowledge, understanding, and application domain. After further interviews were taken, it was manifested that those three SHS applied learning method dominated by lecture method. Therefore the students were not trained to practice their high order thinking skills. High order thinking skills are achieved from student's learning and teaching process [3].

Related to the percentage data of high order thinking skills collected from 3 SHS becoming samples of this research, the students' high order thinking skills of temperature and heat is presented as follows:

Temperature concept

This question was the same as the problem of previous research [14]. There are glass A and B, of the same material. Glass A contains 150 gram of 60 ^{0}C water, glass B contains 150 gram of 75 ^{0}C water. Both glass A and B, are set at a room temperature of 25 ^{0}C. Which graphic best shows relation between temperature and time on glass A? (Note, the starting point does not always show 0 ^{0}C)

IOP Conf. Series: Journal of Physics: Conf. Series **909** (2017) 012055 doi:10.1088/1742-6596/909/1/012055

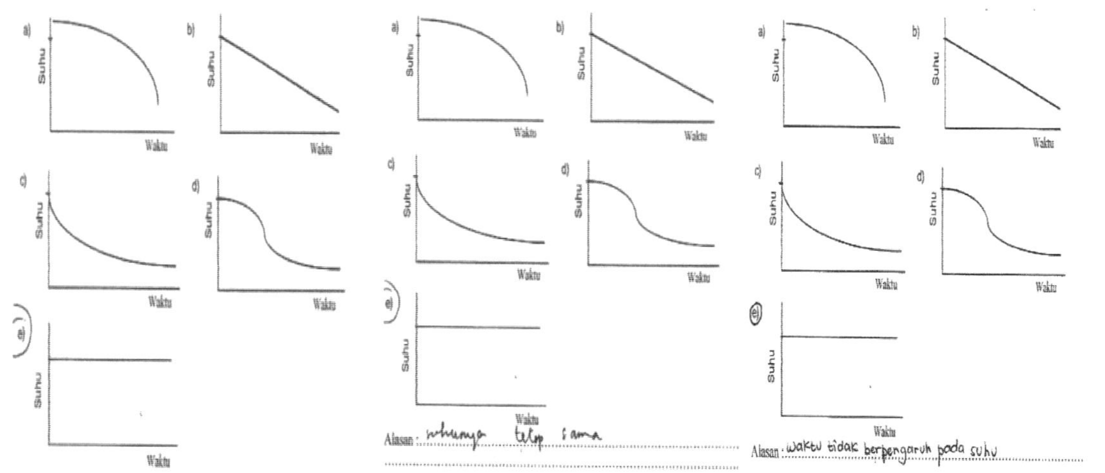

(a) (b) (c)

Figure 1. Student's answer on temperature

The problem about temperature concept above requires cognitive thinking stage from knowledge, understanding, application, analysis, evaluation, and creation. Thus the problem covers creative thinking stage. Students would be able to draw a graphic of the relation between temperature and time if the students had knowledge and understanding about temperature and heat concept. Moreover, students would also be able to apply quantities related to heat and analyze them. If the students were able to analyze the intertwined concept between temperature and heat, they would be able to evaluate it by seeking a relation between temperature and time. Heat (Q) is influenced by substance mass (m), specific heat (c), and temperature change (Δt). Heat rate (H) is influenced by heat and time (t). If $Q = m\ x\ c\ x\ \Delta t$, and $H = \frac{Q}{t}$, thus $H = \frac{m\ x\ c\ x\ \Delta t}{t}$. Thereby temperature change is inversely proportional to time ($\Delta t \infty \frac{1}{t}$), and students can make graphic of the relation between temperature and time, a hyperbola-shaped graph. Based on figure 1, students chose option (e), in which temperature is fixed over time, the student's answer indicated that they had not understood the temperature concept. This conclusion is also supported by the absence of reason on the answer chosen (figure 1a). After a depth interview had performed to the student, it was acknowledged that the student only recognized graphic at option (e). Thus he chose option (e). Based on figure 1a and 1b, students have the same reason. They think that the temperature remains the same at any time.

Heat Concept

If 150 gram of ice and 150 gram of water within the same temperature that is 0 ^{0}C are to be stored in a freezer, of under 0 ^{0}C. Left in storage until they set to the freezer temperature, which substance will lose more heat?

a. 150 gram of ice
b. 150 gram of water
c. Both will lose the same amount of heat as the initial temperature was the same
d. None is incorrect as ice doesn't have heat quality
e. None is incorrect as you can't find water at 0 ^{0}C

Jika 150 gram es dan 150 gram air pada suhu yang sama yaitu 0 °C diletakkan dalam *freezer*, yang bersuhu dibawah 0 °C. Setelah menunggu suhunya sama dengan *freezer*, benda mana yang kehilangan kalor lebih banyak?
a. 150 gram es
b. 150 gram air
c. Keduanya kehilangan sejumlah kalor yang sama karena suhu awalnya sama
d. Tidak ada jawaban karena es tidak mengandung kalor
e. Tidak ada jawaban karena anda tidak dapat memperoleh air pada suhu 0 °C
Alasan : jika suhu awalnya sama maka kalor es = kalor air

Jika 150 gram es dan 150 gram air pada suhu yang sama yaitu 0 °C diletakkan dalam *freezer*, yang bersuhu dibawah 0 °C. Setelah menunggu suhunya sama dengan *freezer*, benda mana yang kehilangan kalor lebih banyak?
a. 150 gram es
b. 150 gram air
c. Keduanya kehilangan sejumlah kalor yang sama karena suhu awalnya sama
d. Tidak ada jawaban karena es tidak mengandung kalor
e. Tidak ada jawaban karena anda tidak dapat memperoleh air pada suhu 0 °C
Alasan : Karena sama-sama dimasukkan ke dalam freezer

Jika 150 gram es dan 150 gram air pada suhu yang sama yaitu 0 °C diletakkan dalam *freezer*, yang bersuhu dibawah 0 °C. Setelah menunggu suhunya sama dengan *freezer*, benda mana yang kehilangan kalor lebih banyak?
a. 150 gram es
b. 150 gram air
c. Keduanya kehilangan sejumlah kalor yang sama karena suhu awalnya sama
d. Tidak ada jawaban karena es tidak mengandung kalor
e. Tidak ada jawaban karena anda tidak dapat memperoleh air pada suhu 0 °C

(a) (b) (c)

Figure 2. Student's answer on heat

The problem about heat concept above requires cognitive thinking stage from knowledge, understanding, application, analysis, and evaluation, thus the problem covers evaluative thinking stage. Students would be able to answer the question above if they had knowledge, understanding, and analysis of the form of matter, the process of the transformation of matter form, and heat. Students would be able to analyze two substances of different kinds and forms. If 150 gram of water within 0 ^{0}C is stored in a freezer, the water will undergo two processes, freezing (temperature 0 ^{0}C and transformation of liquid to solid takes place) and no transformation of matter form in temperature (below 0 ^{0}C) the substance will remain solid (ice). If 150 gram of ice within 0 ^{0}C is stored in a freezer, the ice will undergo one process that is temperature change (below 0 ^{0}C) with the form remains solid (ice). Heat (Q_1) released by water to freeze is determined by water mass (m_a) and melting heat (L). Heat (Q_2) released by water to undergo temperature change, is determined by water mass(m), ice-typed heat (c_{ice}), and temperature change (Δt). Thus total heat (Q_{tot}) released by water is $Q_1 + Q_2 = m\,x\,L + m\,x\,c_{es}\,x\,\Delta t$. Heat released by ice, is determined by ice mass (m_{ice}), ice-typed heat (c_{ice}), and temperature change (Δt). Therefore heat (Q_3) released by ice is $m_{ice}\,x\,c_{es}\,x\,\Delta t$. Water and ice masses are equal, thus $m = m_{ice}$. It is derived that heat released by water (Q_{tot}) is greater than heat released by ice (Q_3), or water releases more heat than ice. Based on figure 2, students did not understand the concept of temperature. Thus students could not relate the concept of temperature, the transformation of matter form and heat. The student answered that ice and water released an equal heat as the ice and water took the same initial temperature (option c). After a depth interview was taken, it was found that students did not understand the concept of form and transformation of matter form, therefore the student assumed that ice and water masses are the same, as well as their initial temperature, leading to his conclusion that ice and water released the same amount of heat (figure 2a and 2b). Based on figure 2c, the student have no reason. After a depth interview was taken, it was found that the student did not understand the concept of heat. He think that ice and water have the same of mass and initial temperature, so ice and water have the same amount of heat.

Heat Transfer Concept.

Florentina took two cartons of chocolate milk quickly, one carton was from the refrigerator, and the other was from the dining table. Why do you think the milk taken from the refrigerator was colder than the one taken from the dining table? Compare to the warm carton of milk, the cold carton of milk is.........

a. colder
b. contains more heat
c. a bad conductor
d. transfer more cold to Florentina's hand
e. transfer more heat to Florentina's hand

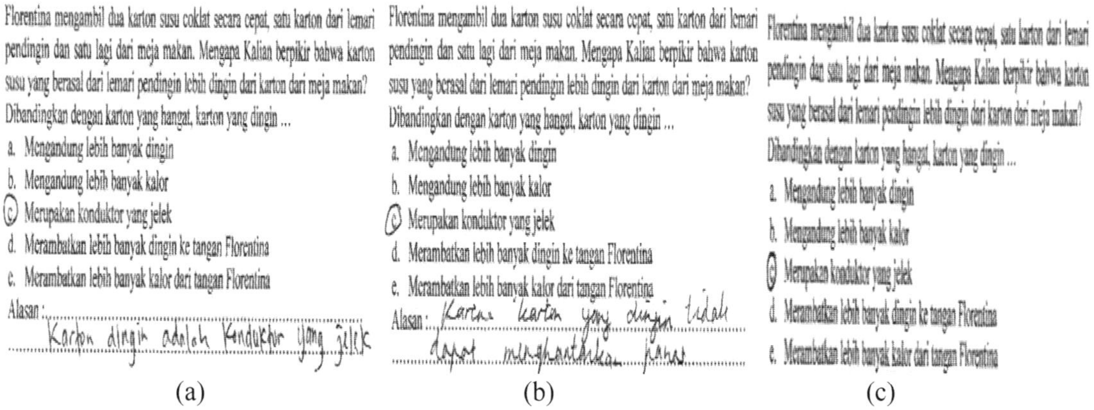
Florentina mengambil dua karton susu coklat secara cepat, satu karton dari lemari pendingin dan satu lagi dari meja makan. Mengapa Kalian berpikir bahwa karton susu yang berasal dari lemari pendingin lebih dingin dari karton dari meja makan? Dibandingkan dengan karton yang hangat, karton yang dingin ...
a. Mengandung lebih banyak dingin
b. Mengandung lebih banyak kalor
c. Merupakan konduktor yang jelek
d. Merambatkan lebih banyak dingin ke tangan Florentina
e. Merambatkan lebih banyak kalor dari tangan Florentina
Alasan : Karton dingin adalah Konduktor yang jelek

Florentina mengambil dua karton susu coklat secara cepat, satu karton dari lemari pendingin dan satu lagi dari meja makan. Mengapa Kalian berpikir bahwa karton susu yang berasal dari lemari pendingin lebih dingin dari karton dari meja makan? Dibandingkan dengan karton yang hangat, karton yang dingin ...
a. Mengandung lebih banyak dingin
b. Mengandung lebih banyak kalor
c. Merupakan konduktor yang jelek
d. Merambatkan lebih banyak dingin ke tangan Florentina
e. Merambatkan lebih banyak kalor dari tangan Florentina
Alasan : Karton karton yang dingin tidak dapat menghantarkan panas

Florentina mengambil dua karton susu coklat secara cepat, satu karton dari lemari pendingin dan satu lagi dari meja makan. Mengapa Kalian berpikir bahwa karton susu yang berasal dari lemari pendingin lebih dingin dari karton dari meja makan? Dibandingkan dengan karton yang hangat, karton yang dingin ...
a. Mengandung lebih banyak dingin
b. Mengandung lebih banyak kalor
c. Merupakan konduktor yang jelek
d. Merambatkan lebih banyak dingin ke tangan Florentina
e. Merambatkan lebih banyak kalor dari tangan Florentina

(a) (b) (c)

Figure 3. Student's answer of Heat Transfer Concept

The problem about heat concept above requires cognitive thinking stage from knowledge, understanding, application, and analysis. Thus the problem covers analytical thinking stage. Students would be able to answer the question above if they had the knowledge, understanding, applicative capability, and analysis on heat concept, heat rate, and transfer. Two similar cartons of milk are in different surrounding temperature, thus the two cartons have different temperature as well. When the two cartons are quickly taken, Florentina will feel the cold milk carton taken from the refrigerator when she holds it. This happens as Florentina's hand touches the milk carton taken from the refrigerator, heat energy from Florentina's hand transfers to the milk carton as a result of temperature differences between hands and the milk carton. Heat is an energy flows from one substance to another which comes in contact and has different temperature. The more heat transfers from Florentina's hand to the milk carton, the colder Florentina's hand feels. Based on figure 3, it can be seen that students did not have an understanding of the concept of heat, and heat rate which became a basic thinking for heat transfer concept. Students answering the question only noticed from where the milk cartons were taken. After a depth interview had been taken, students revealed his argument that substance with heat is a good conductor, while the cold substance is not a good conductor (figure 3a and 3b). This finding shows that the student did not understand the flow of heat energy. Based on figure 3c**,** the student has no reason. After a depth interview was taken, it was found that the student did not understand the concept of heat transfer. He thinks that the cold substance is not a good conductor, so the cold substance cannot transfer of heat.

Count on table 1, it is perceivable that students' high order thinking skills are in the analysis stage. There are 28.23% of students who are in the low order thinking skills. Referring to the depth interviews to twelve students from three SHS, eight students from two SHS testified that the teacher applied a lecture method in the process of teaching temperature and heat. Thus, the teaching model and method applied by the teacher could not support the high order thinking skills, since the students were not personally involved in the learning process. Students' high order thinking skills can be developed or improved by model and method of teaching which support high order thinking skills. Students' high order thinking skills will develop or improve if the students have a good understanding of any concepts.

4. Conclusion

Based on the data analysis, it is concluded that 68.24% of students achieve high order thinking skills in the analysis stage. 28.24% of students show low order thinking skills. Referring to depth interviews to twelve students from three SHS, eight students from two SHS testified that the teacher applied a lecture method in the process of teaching temperature and heat. Students' high order thinking skills can be developed or improved by model and method of teaching which support high order thinking skills.

References

[1] Hamidah, Ida, ed. 2017. *Ideas for 21st Century Education: Proceedings of the Asian Education Symposium (AES 2016), November 22-23, 2016, Bandung*, Indonesia. Routledge,
[2] Y. M. Heong, W. B. Othman, J. B. M. Yunos, T. T. Kiong, R. B. Hassan, and M. M. B. Mohamad, 2011 *International Journal of Social Science and Humanity*, pp. 121–125,
[3] Tee et.al. 2012 JRPPTTE 2 212
[4] S. L. Senk, C. E. Beckmann, and D. R. Thompson, 1997, "Assessment and Grading in High School Mathematics Classrooms," *Journal for Research in Mathematics Education*, vol. 28, no. 2, p. 187
[5] Resnick L. 1987. *Educational and Learning to Think,* Washington D.C.: National Academy Press.
[6] Miri B, David B-C, and Uri Z, 2007, *Research in Science Education*, vol. 37, no. 4, pp. 353–369, Jan.
[7] A. Zohar, 2004 *Higher Order Thinking in Science Classrooms: Students' Learning and Teachers' Professional Development*, Springer Science+Business Media Dordrecht
[8] King F.J., Goodson, L., and Rohani, F. 1998. *Higher- Order Thinking Skills: Definitions, Strategies, and Assessment.* URL:http://www.cala.fsu.edu/files/higher_order_thinking_skills.pdf.
[9] Thomas, M. F. Marquis et all. 1995. Physics Education 30 19
[10] Harrison A. G., Grayson D. J., & Treagust D. F, 1999. *Journal of Research in Science Teaching*, *36*(1), 55-87.
[11] Luera, G. R., Otto, C. A. & Zitzewitz, P. W 2005 J. Phys. Tchr. Educ 2 43
[12] U. Zoller, 1993 *Journal of Chemical Education*, vol. 70, no. 3, p. 195
[13] Pornrat. Et.al. 2013 *International Journal of Innovation in Science and Mathematics Education* 2 29
[14] I. S. Budiarti, Suparmi, Sarwanto, and Harjana, 2017 *Journal of Physics: Conference Series*, vol. 795, p. 012051

IOP Conf. Series: Journal of Physics: Conf. Series **909** (2017) 012056 doi:10.1088/1742-6596/909/1/012056

Integrating numerical computation into the undergraduate education physics curriculum using spreadsheet excel

Ahmad Fauzi[1]

Physics Education Department of Sebelas Maret University,
Jl. Ir. Sutami 36A Kentingan Jebres Surakarta 57126, INDONESIA

E-mail: fauziuns@gmail.com

Abstract. Numerical computation has many pedagogical advantages: it develops analytical skills and problem-solving skills, helps to learn through visualization, and enhances physics education. Unfortunately, numerical computation is not taught to undergraduate education physics students in Indonesia. Incorporate numerical computation into the undergraduate education physics curriculum presents many challenges. The main challenges are the dense curriculum that makes difficult to put new numerical computation course and most students have no programming experience. In this research, we used case study to review how to integrate numerical computation into undergraduate education physics curriculum. The participants of this research were 54 students of the fourth semester of physics education department. As a result, we concluded that numerical computation could be integrated into undergraduate education physics curriculum using spreadsheet excel combined with another course. The results of this research become complements of the study on how to integrate numerical computation in learning physics using spreadsheet excel.

1. Introduction

Physics as a product of science is formed based on a series of observations of natural laws which are generally presented in a mathematical equation. In introductory physics courses, students are usually hampered by their lack of mathematical tools such as in constructing a mathematical equation of physical system and interpreting a graph that may cause difficulties for students to understand the physical system [1,2].

The method used to solve a problem of physics is an analytic method, but the more complex problem that must be solved required a certain numerical method. Along with the development of science and technology, in the arrangement of mathematical models (modeling) and numerical techniques usually, use computer (numerical computation) help to analyze and solve various problems by using visualization and simulation techniques. Numerical computation gives pedagogical advantages, including the development of numerical problem-solving skills, modeling, and visualization of physical systems [3]. Numerical computation is needed by physics graduates as they work, but many types of research have shown that the students are unprepared with computational skills [4,5]. In addition to being required in workplace, numerical computation also required to be taught in the classroom learning process because numerical computation can provide various pedagogy advantages such as being able to develop analytical skills, facilitate learning through simulation and visualization of various systems and physical phenomena and improve the quality of physics education [6,5].`

One of the excellence of numerical computational is to demonstrate more realistic natural phenomena by analyzing the various factors that must be analyzed, where these factors are usually ignored in analytical methods with consideration to make analysis easier. Numerical computation can provide students with opportunities to learn based on the paradigm of solving problems more effectively and efficiently than traditional learning [3]. The advantages of using a computer in physics education should not be realized by simply giving computational problems, besides students achieve the correct numerical solution, their understanding about the physical system on which the problem is based must increase [7]. The students should begin to practice developing models of physical phenomena (modeling skill) and using computers to simulate their physical problems in studying physics process (simulation skill). The modeling and simulating skills consist of several categories: choosing a computational model, assessing computational models, using computational models to test solution, and using computational models to understand a concept [8].

Teaching numerical computations to undergraduate physics students is still a challenge [9,2].Various researches were conducted to find the best way of teaching numerical computation to undergraduate physics students. In reality, however, in many cases that occur in many universities, due to the dense curricula, it is almost impossible to teach numerical computation as a course with its own credit [2]. Various efforts have been done to overcome this problem, such as by integrating numerical computation in basic physics courses combined with lab activities in the laboratory equipped with material resources on the web [2] and providing additional training in mechanics courses and thermodynamics [7]. Doing numerical computation, students need programming experience. Unfortunately, many studies found that most of the students have no programming experience [9]. To overcome this problem, some researchers chose to use easy and free programming languages like Vpython or paid programming languages like Origin and IGOR. However, the use of paid programming languages such as Origin and IGOR also cause problems because students just using ready-made programs, students may assume that numerical computation as black boxes that they do not understand. Therefore, the use of programs that require students to actively make their own computation is the best choice [3].

One of the main obstacles in teaching numerical computation is the language of certain programmers that must be mastered by students. One effort that can be done to help students learn numerical computation without having programming language is by using spreadsheet excel. Spreadsheet excel is selected based on the fact that almost all students have been studied it and the program is also installed on their owned computer. Spreadsheet excel can be used to simulate and visualize various physical phenomena either with analytic or numerical methods without the need for difficult mastery of programming languages [10,11,1].

There were many types of research about numerical computation on high school students or collage students who have not mastered programming languages. As a result, numerical modeling used spreadsheet excel can be used to improve interest and the students' knowledge [10,12]. In that research, the students guided to train their computational skill by the proctored program that will give students a clear instruction. The main trained skills are to recite problem of words into programming tasks, identifying, and updating the input variables.

The case of nonharmonic oscillation such as damped oscillation shows dynamic system usually described by differential equations. Students must master in calculus to solve the differential equation. In order to make the system easier, the differential equation can be replaced by finite difference method [13]. The damped oscillation can be taught easily use an analytical method by exploring graph facilities of spreadsheet excel without using numerical method [14]. Based on that former researchers, in this research, we developed a study that allows students to create their own numerical modeling using spreadsheets excel to solve various physical phenomena to improve students' modeling and simulations skills.

In simple harmonic oscillation, the period does not depend on the amplitude. When the period of oscillation depends on the amplitude, the oscillation is nonharmonic oscillation. The real-world oscillation systems always have some dissipative forces, and oscillation will decrease in amplitude and stop. This phenomenon is called damped oscillation. The assumption of damped oscillation is the fractional damping force proportional to the velocity of the oscillating body. This behavior occurs in friction involving viscous fluid flow, such as sliding between oil and lubricated surfaces. The force on the body due to friction is $F_x = - b\ v_x$ where v_x is the velocity and b is a constant that describes the strength of the damping force. Using Newton's second law, the net force of the body is

$$\Sigma F_x = -bv_x - kx \tag{1}$$

$$m\,\frac{d^2x}{dt^2} = -b\frac{dx}{dt} - kx$$

or

$$\frac{d^2x}{dt^2} = \frac{-b}{m}\frac{dx}{dt} - \frac{k}{m}x \tag{2}$$

equation (2) can be written as

$$\frac{d^2x}{dt^2} + \frac{b}{m}\frac{dx}{dt} + \frac{k}{m}x = 0 \tag{3}$$

equation (3) is a differential equation for x. The general solution of equation (3) is

$$x = Ae^{-\left(\frac{b}{2m}\right)t}cos(\omega' t + \phi) \tag{4}$$

where ω' refers to

$$\omega' = \sqrt{\frac{k}{m} - \left(\frac{b}{2m}\right)^2} \tag{5}$$

the value of ω' is not constant, but depend on b. If the $\frac{k}{m} = \left(\frac{b}{2m}\right)^2$ its refers to critical damping, if $\frac{k}{m} > \left(\frac{b}{2m}\right)^2$ its refers to under damping, and if $\frac{k}{m} < \left(\frac{b}{2m}\right)^2$ its refers to over damping.

The equation (3) can be solved numerically by using a Euler-Cromer method. The equation (3) can be rearranged to

$$\frac{d^2x}{dt^2} = \frac{-b}{m}\frac{dx}{dt} - \frac{k}{m}x \tag{6}$$

based on the definition $\frac{d^2x}{dt^2} = \frac{dv}{dt}$ and $\frac{dx}{dt} = v$ then

$$\frac{dv}{dt} = \frac{-b}{m}\frac{dx}{dt} - \frac{k}{m}x \tag{7}$$

the numerical solution of equation (3) then

$$v_{i+1} = v_i - \frac{b}{m}\,v_i\,\Delta t - \frac{k}{m}x_i\,\Delta t \tag{8}$$

and

$$x_{i+1} = x_i + v_{i+1}\,\Delta t \tag{9}$$

Based on equation (8) and (9) the characteristic of damped oscillation can be investigated by creating a simulation and then choosing the right value of v, b, k and m.

2. Methodology

The participants of this research were 54 students of the fourth semester of physics education department of Sebelas Maret University in the 2015/2016 academic year. In the second semester, the students have studied how to make physics learning media using spreadsheet excel. Therefore in preliminary this course, we focus to train students using spreadsheet excel to solve various physics problem using the analytical method. In this course, the students tried to solve problems of one or two-dimensional motion used spreadsheet excel with analytical analyze. We also asked students to visualize beat phenomena using spreadsheet excel. In this lecture, the students must define the dependent and independent variable of beat phenomena; then they must explore the variables relationship. We found that the majority of students were able to visualize beat phenomena correctly. Students wrote simple programs to compare the motion of a ball in free fall and motion of ball under gravity and air drag, projectile motion by ignored air drag and projectile motion under air drag. The students also learned oscillation/periodic motion with the analytical approach. It was followed by teaching computational modeling using Euler and Euler-Cromer. In this course, students also had to work on the same problem with the analytical and numerical analysis, then students were asked to compare the accuracy of numerical with the analytical solution. Thus, students would be able to determine the most appropriate numerical method to solve various cases.

In this research, we focused numerical computation in modeling and simulations skills. We divide modeling and simulations skills into several categories: choosing a computational model, assessing computational models, using computational models to test solution, and using computational models to understand a concept. Students' numerical computational skills were assessed using performance test, essay question and interview. First, student developed a computational model by using a spreadsheet excel. In this assessment, students must create a computational model and write it correctly in a spreadsheet. Students continued to do a second assessment by answering several essay questions. The essay question was intended to determine how students correlate the computational model to assess computational models, use computational models to test solution, and use computational models to understand a concept.

3. Result and Discussion

3.1. Result

Based on previous research, teaching numerical computations should use a computer program which is familiar and easy to understand by students. Therefore, the first step taken in teaching numerical computing to students is to know the students' ability to use spreadsheet excel. The students' ability in using spreadsheet program is presented in the following table.

Table 1. Students' ability to use spreadsheets excel

No	Indicators	Average Score (maximum score 4)
1	Operating spreadsheet excel program	3,7
2	Defining independent and dependent variable	3,7
3	Defining cell reference	3,3
4	Writing simple mathematics equation	3,7
5	Writing more complex mathematics equation	3,7
6	Making a table	3,7
7	Copying data	3,7
8	Making a graph	3,7
9	Defining *slope*	3,7
10	Using trend line	3,3
11	Converting various physics units	3,7
12	Using basic statistical function	3,3

Based on table 1, we concluded that the students will able follow this course. Before students start the course, we gave additional course to give them a deeper understanding about defining cell reference, trendline, and some basic statistical function.

In this research, we focused numerical computation in modeling and simulation on oscillatory motion phenomena especially nonharmonic oscillation.

Problem 1:
For simple harmonic motion, the general form of the equation is
$$\frac{d^2x}{dt^2} = -kx^{\alpha}$$
When α=k =1 the equation refers to simple harmonic motion, but when α = 3 and k =1 the equation refers to the non harmonic oscillator. Find the numerical solution of nonharmonic oscillation, then simulate the motion and show that the period of the motion depends on the amplitude.
Problem 2:
A nonharmonic oscillation consist of a spring mounted horizontally, attached a 1 kg body and pull of distance 20 cm and release it. If the spring constant is 8 N/m and b= 0,23. a) Analyze the oscillation used a precise numerical method to show critical damping, overdamping, and underdamping oscillation, b) Investigate the effect of b to the period and amplitude the oscillation. You may take b = 0,1 and b = 0,25.

According to performance test and essay questions, the students' modeling and simulations skills can be presented in figure 1.

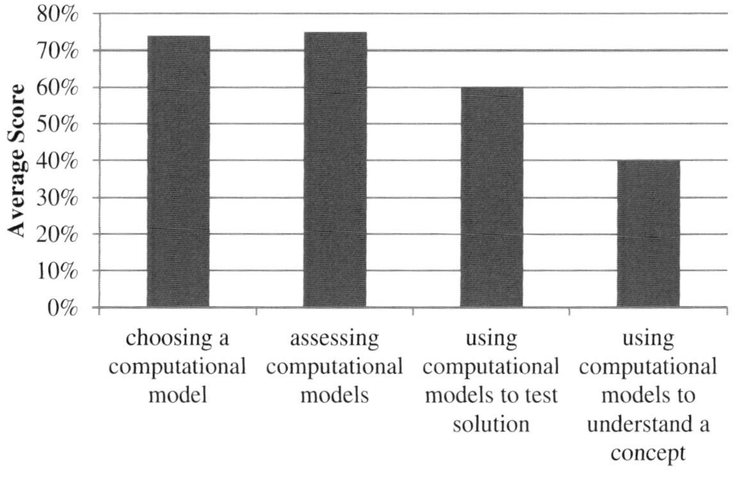

Figure 1. The average score of modelling and simulations skills

There were 62% of students have the ability to create a precise computational model. There were 75 % students able to choose a precise computational model, 75% students were able to asses computational models correctly, 60% students were able to use computational models to test solution correctly, and 40% students were able to use computational models to understand a concept correctly. The observation of students' work showed that most students were able to choose computational model appropriately. Only 25% students choose a wrong computational model, consequently, the students will not be able to assess computational model and test the solution. The examples of the wrong and right computational model done by students shown in figure 2 and 3.

IOP Conf. Series: Journal of Physics: Conf. Series **909** (2017) 012056 doi:10.1088/1742-6596/909/1/012056

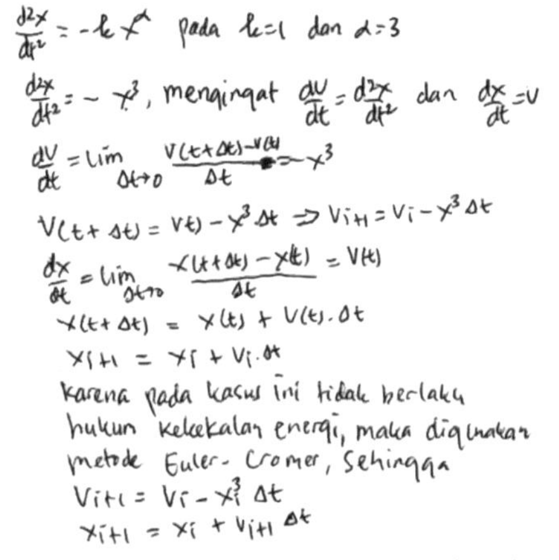

Figure 2. Snapshot of right computational model

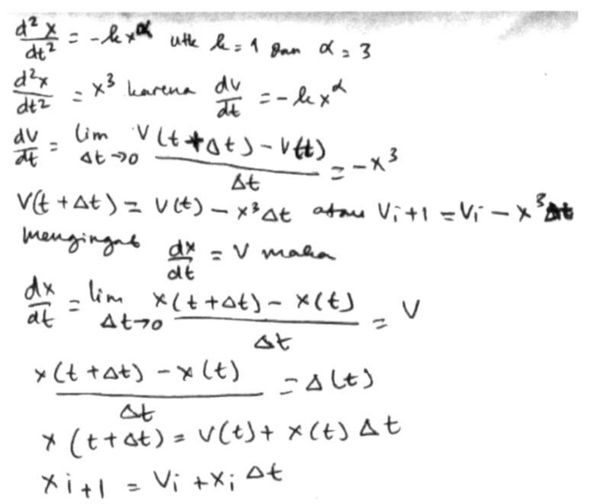

Figure 3. Snapshot of wrong computational model

Based on figure 2 and 3, the students actually understood how to choose a computational modeling. In figure 2, the students chose Euler-Cromer method as the numerical solution of the nonharmonic oscillator. They understood that for nonharmonic oscillator the Euler method is not a good choice because the Euler method yielded a solution that did not quite conserve energy. In figure 3, the students chose Euler method as the numerical solution of the nonharmonic oscillator. They did not understand that for nonharmonic oscillator the Euler method is not a good choice. Based on the interview, we conclude that many students still confused the difference between Euler method and Euler-Cromer method.

According to interview, the most difficulties occurred when students must perform the computational model into spreadsheet environment, performing iterations to assess computational model, and interpreting the results as a graph to find the concept. The difficulties are shown in figure 4.

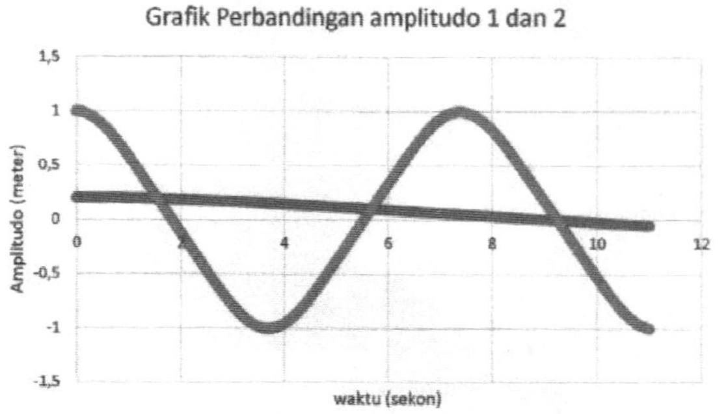

Figure 4. Snapshot the simulation of inaccurate time step (Δt) cause incomplete one-period oscillation

The students are incapable to test the computational models they have made because they couldn't choose the right time step value. It showed on their simulation. Most of the students chose too small time step value so that the results of iterations performed based on graphs can not show the value of one-period oscillation, consequently, they have unable to show that in the non harmonic oscillator, the period of motion depends on amplitude.

Problem number 2 was tested to students after lecturer and students discussed the problem number 1. Based on the mistake done by students in doing problem number 1, problem number 2 were completed with an instruction such as the value of mass, spring constant, amplitude, and time step value. Based on the student's answer, it can be concluded that all students chose Euler-Cromer method to solve the problem. Over than 80% students were able to simulate damping oscillation using spreadsheet excel. The wrong and right simulation of over damping shown in figure 5 and 6.

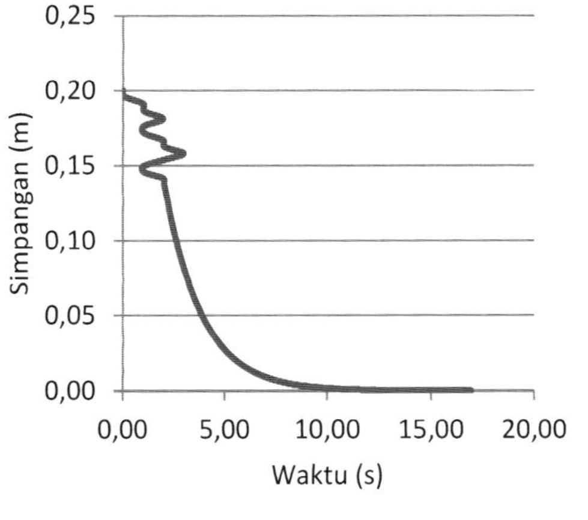

Figure 5. Snapshot the wrong simulation of overdamping

IOP Conf. Series: Journal of Physics: Conf. Series **909** (2017) 012056 doi:10.1088/1742-6596/909/1/012056

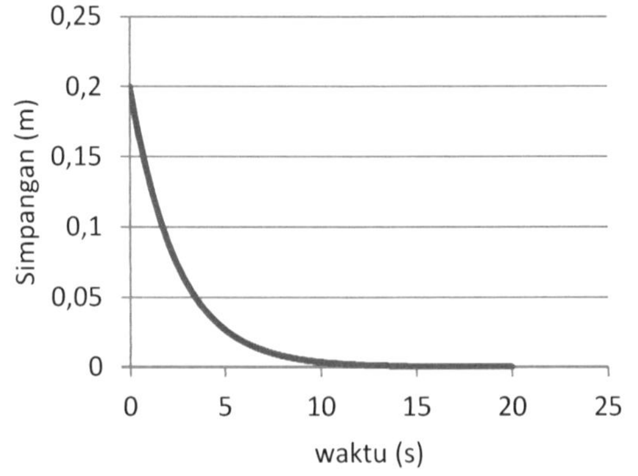

Figure 6. Snapshot the right simulation of overdamping

According to the performance test review, the students made mistake in calculating time, after t = 3 s, the students wrote t = 2 s, 1 s, 1,05 s nor t = 3,05s, 3,1s, 3,15, consequently the simulation was wrong. Based on the interview, it can be concluded that the mistake caused by mistyping the value of time. The students type the number manually, they forgot to use a formula of cell reference in this case.

Numerical computation uses spreadsheet excel gave students chance to investigate the characteristic of nonharmonic oscillation. Most of the students were able to investigate the effect of b to the period and amplitude the oscillation. They made computation that compared the underdamping oscillation at b =0,1 and 0,25 then represent the computation into a graph. In spite of almost students made the simulation correctly, only 40 % students were able to describe the graph correctly. The simulation and the analysis of students are shown in figure 7 and 8.

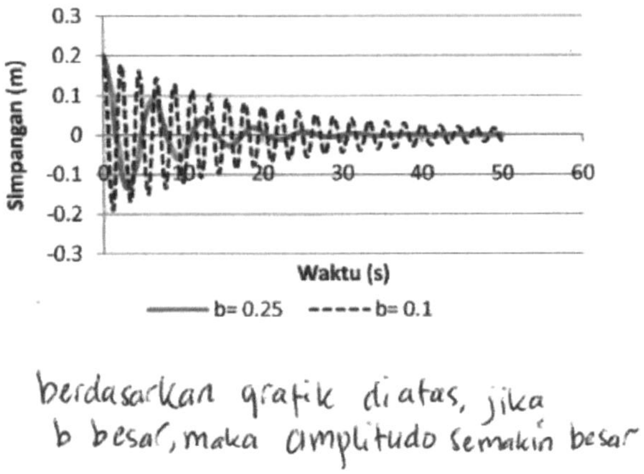

Figure 7. Snapshot of wrong analyze

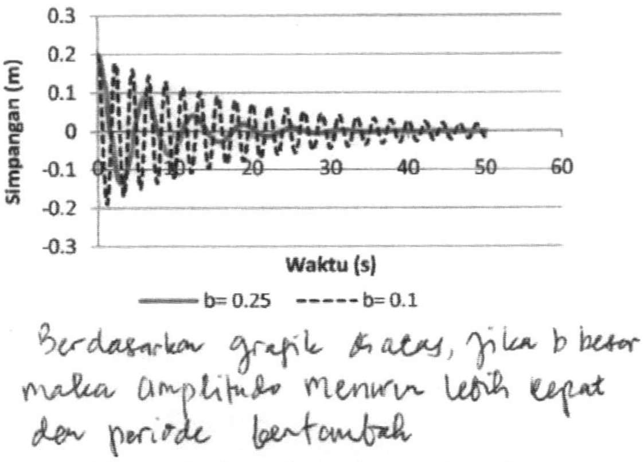

Figure 8. Snapshot of correct analyze

About 40% students said that larger b caused greater damping where the period decrease, 20 % students said that larger b caused the amplitude increase, and only 40% students said that larger b caused larger damping where the amplitude decrease more rapidly and the period increases.

After following this course, some students were interviewed. The students said that this course very interested. About 80% of students said that after following this course, they more interesting to physics and mathematic and only 20% students said that they more interesting to physics. They said that study to solve damped oscillation use numerical method is very interesting because they usually just memorize the general equation of damped oscillation and its general solution used an analytical method. According to students, general equation of damped oscillation is very complicated, in spite of, they know that the general equation of damped oscillation is a homogeneous differential equation of the second order but they doubted able to solve the equation correctly. Most of the students said that use numerical method is very simple and accurate to find the solution of a damped oscillation. Most of the students suggested that the problem involved numerical computation should be completed by an instruction such time step and the value of each variable the problem.

3.2. Discussion
Based on the interview, we conclude that most students very interested to follow the course. They said that they got new knowledge how to solve a complex physics problem such as nonharmonic oscillator using a simple method. The students said that they get the experience how to simulate nonharmonic oscillator to investigate the characteristic of the system. Our finding is consistent with previous research [2,3,5]. Most of the students that chose wrong computation model said that they still confuse about the importance of numerical computation. They said that they do not get a new concept of this course, they just know the concept of analytical method. Based on test performance, we found that most of the students that chose wrong computation model also make a mistake in creating a graph because they couldn't choose the accurate time step. For example, when they asked to investigate the effect of amplitude on the period of nonharmonic oscillation they make a graph of amplitude versus period, nor make a graph of amplitude versus time then calculate the period. Our finding is still consistent with previous research [5,11]. Therefore, for the next research, we advise investigating the motivation of students after following numerical computation course. It is also very important to investigate how to train students to choose the precise computational model.

3.3. Conclusion
We concluded that numerical computation could be integrated into undergraduate education physics curriculum using spreadsheet excel combined with another course. The results of this research become

a complement of the study on how to integrate numerical computational in learning physics for students who have not mastered programming languages.

References

[1] Fauzi A 2016 Proceeding of International Conference on Teacher Training and Education (ICTTE) FKIP UNS 2015. Vol 1 (1) p 290
[2] Serbanescu R, Kushner PJ, and Stanley S 2011 *American Journal of Physics* **79(9)** 919-924
[3] Landau R 2006 *Computing In Science & Engineering* **8(5)** 22-30
[4] Fuller R 2006 *Comput. Sci. Eng* **8(5)** 16–21
[5] Landau R 2007 *Comput. Phys. Commun* **177** 191–196
[6] Chabay R and Sherwood B 2008 *Am. J. Phys* **76(4/5)** 307–313
[7] Buffler A, Pillay S, Lubben F, and Fearick R 2007 *American Journal of Physics* **76 (4&5)** 431—437
[8] Weintrop D, Beheshti E, Horn M, Orton K, Jona K, Trouille L, & Wilensky U 2016 *Journal of Science Education and Technology* **25(1)** 127-147
[9] Ross KR 2006 *Comput. Sci. Eng* **8(5)** 44–50
[10] Benacka J 2016 *Eurasia Journal of Mathematics, Science & Technology Education* **12(7)** 947–964
[11] Lingard M 2003 *Physics Education* **38(5)** 418–422
[12] Caballero M D, Kohlmyer M A, & Schatz M F 2012 AIP Conference Proceedings 1413 p 15
[13] Benacka J 2018 *Spreadsheet in Education (ejsie)* **2 (3)** 289-298
[14] Grigore I, Mironc S, and Barna,ES 2016 *Romanian Report in Physics* **68 (2)** 891-904

The development rubrics skill argued as alternative assessment floating and sinking materials

Viyanti[1,a)], Cari[2], Widha Sunarno[2], and Zuhdan Kun Prasetyo[3]

[1]Program Studi Doktor Pendidikan IPA, UniversitasSebelas Maret, Jl. Ir. Sutami 36A Jebres Surakarta, Indonesia
[2]Program Studi Pendidikan IPA, UniversitasSebelas Maret, Jl. Ir. Sutami 36A Jebres Surakarta, Indonesia
[3]Program Studi Pendidikan IPA, UniversitasNegeri Yogyakarta, Jl. Colombo No 1, Yogyakarta, Daerah Istimewa Yogyakarta 55281,Indonesia

[a)]E-mail: viyanti@students.uns.ac.id

Abstract. The quality of arguing to learners of floating and sinking material can be assessed by using the rubric of an argumentation assessment skill as an alternative assessment. The quality of the argument is measured by the ability of learners to express the claim in a structured manner in order to maintain the claim with supporting data. The purpose of this study was to develop an argument skill rubric based on the preliminary study results which showed a gap between demands and reality related to the students 'floating and sinking students' argument skills. This research was conducted in one of State Senior High School Bandar Lampung. The study population is all students of senior high scholl class XI. Research sample was taken by randomly obtained by 20 students. The research used descriptive survey method. Data were obtained through a multiple choice test both grounded and interview. The results were analyzed based on the level of students' argumentation skills that had met the criteria which developed in the assessment rubric. The results of the data analysis found that the learners are in the range of levels 1 through 3. Based on the data the average learner is at the level of quality argument "high" for component I and the quality of "low" argument for component 2. This indicates learners experience difficulty which making alternative statement supported by reference in accordance with the initial statement submitted. This fact is supported by interviews that learners need a structured strategy to design alternative statements from shared reading sources to support the preliminary statements presented.

1. Introduction

The skill of argument is assumed to be one of the benchmarks of the achievement of the objectives of science learning because learners are able to produce new knowledge through the presentation of preliminary statements as the initial provision of learning. Some scholars have argued, with the reform of learning arguing skills to be a new framework of learning goals that reflect high-level thinking skills [1];[2];[3]. In this regard, learning that focuses on arguing skills requires not only shifting learning goals but also requires teachers to have new and different roles. Teachers need to design learning and assessment to develop students' argumentation skills. This is supported by an expert statement: teachers need to develop instructional designs that include argumentation skills and engage students in developing them [4]; [5]. However, previous research has shown that: teachers have difficulty and have limited ability to integrate argumentation skills in learning [6];[7]. As a result, learners only focus on empowering the claim elements and the lack of constructive knowledge of learners related to the skill of arguing this has an impact on the weak understanding of the concept of floating and sinking matter[8];[9]. The skill patterns of arguments produced by participants students in the preliminary study which demonstrated a lesser diversity of knowledge than students' in terms of the disclosure of a pronouncement (statement), in line with the results of [10] research, that the quality of poor students' argument is attributed to the ignorance of learners in terms of skill elements argue.

Learning that explores alternative concepts of learners and tested with tests designed to empower argumentation skills becomes a challenge for students' to maximize their ability. Students' of science from the point of view of "alternative concepts" are able to understand the learning patterns that involve the internalization of concepts developed as a conflict of knowledge process, where the introduction of such knowledge conflicts is contained in tests containing statements that trigger learners to think. [11] stated that conflicts of knowledge in students' enable learners to make structured statements by classifying and categorizing various problems. Structured knowledge conflicts can be solved by the learner into a picture of the students' able to put the conceptual knowledge at the heart of understanding of science. This is supported by the results of a study of [12], conceptual knowledge can be related to the development of students' argumentation skills. In this regard, assessment instruments with assessment rubrics can facilitate the development of the quality of arguing students' to trigger the emergence of conflicts of knowledge in students'.

An assessment rubric can measure the quality of arguing students' requires clear and measurable criteria. The arranged rubric is focused on the emergence of new knowledge of students' until the emergence of a rebuttal statement that produces an alternative statement. The compiled criteria contain coding schemes to assess the quality of arguing students'. Some experts reveal that: the quality of arguing students' can be judged by a set of criteria including: statements that consider the reasons, the reasons in the statement are acceptable and relevant to alternative statements, the reasons for supporting alternative statements, and the quality of the reasons given in the statements [13]; [14]. The criterion of quality argumentation that can measure the quality of arguing students' has a positive effect in fostering argumentation skills in the development of science learning. Therefore, researchers conduct development research in order to explore the operational and measurable criteria that can measure the quality of the process of arguing students'. This study begins by analyzing elements of argumentative skills from [15] argumentation scheme to map out the criteria of arguing skills for floating and drowning materials. The problem formulation in this research is how to develop the structure of argument skill rubric that is able to measure the quality of arguing students' of floating and sinking material as an alternative assessment?

2. Experimental Method

This research is part of the research of developing an argumentation skill assessment instrument that aims to produce an assessment instrument capable of measuring the skills of high school students' argument floating and sinking material. The development of a skill assessment rubric for floating and sinking material begins with developing a framework of assessing needs oriented to argument skills. The broader objectives resulted in a useful and practical argument scoring tool rubric for the teacher. The study population is senior high school students. The sample of research is class XI amounted to 20 people. Data analysis using qualitative descriptive is based on the structure of the component supporting the argumentation skill. The developed sections were divided into 3 main sections developed using key components of the modified Toulmin argumentation scheme to measure: 1) the process of analyzing data (claim, data, warrant and backing); 2) the ability to analyze alternative statements using various references; 3) the ability to determine the condition that claims can be true (qualifiers) and / or alternative claims (rebuttal).

3. Result and Discussion

This study has identified the quality of arguing students' related to its relevance to the concept of floating and sinking. The conceptual connectivity with argument skills begins by analyzing the difficult concepts of floating and sinking material when applied to [15] argumentation scheme. In addition, analysis of the basic needs of constructing rubrics for assessing students' argumentation skills

IOP Conf. Series: Journal of Physics: Conf. Series **909** (2017) 012057 doi:10.1088/1742-6596/909/1/012057

has been analyzed. The quality analysis of students' arguments is made against the component skill knowledge test of argument for floating and sinking material. The total quality score argued based on the test results will be correlated to levels 1 through 4 for the quality of the students' skills of the skill

The process of cultivating the skill of arguing in judgments is inseparable from the relationship between thinking and perspective that: 1) argues as an important act of building knowledge; 2) argue to build science learning; 3) explicitly combine arguments in judgment as one of the competencies to be developed within the science class; And 4) argue as a vehicle for promoting students' ability to justify the relationship between data and claims as well.

3.1. Development Structure Of Rubric Argumentation Skills
Argumenting skill is a skill that facilitates explanations of internal representation of the phenomenon, scientific principles and cognitive activity of learners when building knowledge of science. The perspective skill perspective that facilitates students' to understand the pattern of science knowledge and understand the phenomenon requires rubrics to determine the quality of arguing students'. The step in developing the instruments of argumentation skills in this development study refers to the perspective of producing students' capable of: 1) uncovering scientific proof in constructing good arguments; 2) make key components argue; 3) analyze and evaluate knowledge; 4) prepare for further learning. Based on that perspective, students' argumentation skills can grow effectively with guidance and support. This is supported by several expert statements, guiding students to build arguments within the framework of writing and assessed by applying TAP analysis makes it easy to distinguish the quality of students' arguments [10];[16]. In another paper, [17] revealed that assessing the written arguments of learners by seeking valid justification in terms of content knowledge. Based on the theoretical study the researchers used the coding for each component to argue in the item in order to facilitate the analysis of the quality of arguing the students'.

3.2. Process Analyzing Statements (Claim, Data, Warrant and Backing)
That is, simple argument structures such as in the form of data, warrant, backing, and claim can be observed in statements which is presented by students' in context [18]. The use of Toulmin's argumentation scheme in this study concentrated on the description of the ability of students' to write statements on the given problem. The resulting statements are analyzed through rubrics containing components arguing claims, data, warrants and backing; Sample statement produced by Students A and B presented in Figure 1 and 2.

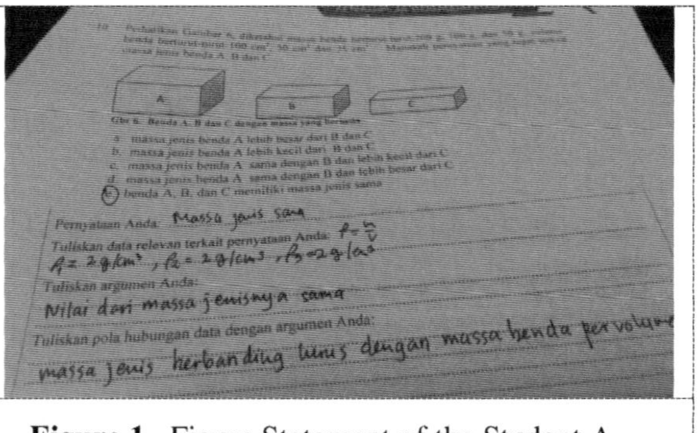

Figure 1. Figure Statement of the Student A

The statements presented by Students A and B generally contain conceptual knowledge related to classification, category, generalization. The mastery of the factual knowledge of each Student A and B statement is characterized by the ability to classify data, classify data based on similarity features, or on the basis of differences; Showing the strength or weakness of a statement, recognizing principles, concluding, mastering theory, showing examples, and recognizing structures.

IOP Conf. Series: Journal of Physics: Conf. Series **909** (2017) 012057 doi:10.1088/1742-6596/909/1/012057

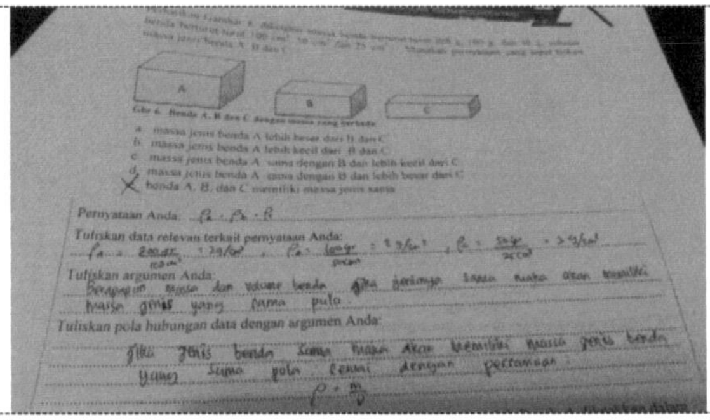

Figure 2. Figure Statement of the StudentsB

"The mass of the type is straight to the mass of pervolume objects (alternative claims supported by reference)". Structurally students A has been able to analyze the process of formation of statements with the support of simple reasons. The quality of statements presented by students' A is analyzed by using the first component of development results included in categories capable of selecting statements and writing down the reasons as an alternative statement supported by scientific knowledge (claims, data and warrant) and obtaining a score of 3.

Furthermore the statement presented by Student B "$\rho_A=\rho_B=\rho_C$ (by selecting the answer option e: objects A, B, and C have the same type of mass)" $\rho_A=200g/100cm^3=2g/cm^3$, $\rho_B=100g/50cm^3=2g/\ cm^3$, $\rho_c=50g/25cm^3=2g/cm^3$ (statements containing data); "regardless of mass and volume of objects if the same type would have the same mass of dose (alternative claim based on data); The same will have the same mass of the same kind of thing in accordance with the equation $\rho=$ m/v (alternative claim supported by reference). "Structurally students B has been able to analyze the process of formation of statements with the support of a simple reason.The quality of statements presented by Student B is analyzed By using the rubric of component I of the development results included in the category capable of choosing the right conclusion statements and provide the reasons and provide more information about the relationship of alternative statements and choice of reasons as well as contacting N concepts and theories of the given problem (claim, data, warrant and backing) and get a score of 4.

The statement of students' A and B indicates that the quality of argument has been structured in which: the resulting statement is supported by the data and some references; the disclosure of data contained in the problem is written clearly and in detail and able to write the physical meaning of each data statement written ; an alternative statement is constructed through the activity of identifying the provided statement, looking at the mass equations of the object and the volume of objects presented in the drawing and capable of uncovering the constituent material; and student B can build connection between claim, data, warrant and backing even at weak level. The quality structure arguing students' A and B illustrate there is a connection between the elements of arguing skills so that learners A and B can be categorized as having good knowledge. The results of the data analysis provide an illustration of the extent to which learners A and B are able to empower their arguing skills, and make it easier for teachers to provide feedback to other undeveloped argument components.

Assessment strategies using rubrics that contain in detail each argumentary skill that the students'have to master can overcome the difficulties of learners making reasonable statements based on reasoning. Teachers can use specific strategies to prepare the task of argumentation [4]. The developed rubric provides an opportunity for teachers to conduct feedback in order to encourage learners to consider the preliminary statements which is presented in the preparation of alternative statements. As a result, students' not only focus from one side of the argument but also one statement during the problem. This is supported by the results of [6] that an alternative statement using the reasoning structure considers the initial statement to encourage a simple alternative statement.

Feedback based on data analysis is important in order to improve students' argumentation skills. It is known that the skill of argument does not come by itself but should always be given training and feedback in the learning process. This statement is supported by the results of [17] research that the quality of arguments generated by learners as a learning material in order to integrate the skills to argue in learning. Another study suggests that feedback activities help students develop epistemological perfection of science [19]. Thus, the effort to integrate argumentation skills and meaningful feedback activities in the learning process of science has many potential benefits for learners to improve their documented skills. This is consistent with the expert's assertion that argumentation skills can be empowered if teachers can perceive each idea and respond to any statement given by students' [20].

3.3. Process Analyzing Alternative Statements Using Various References

The results of further research focused on the ability of learners in describing the ideas set forth in an alternative statement. An example of the best alternative statement from learners to be samples in this study are Students C and D. Alternative statements produced by Students C and D are presented in Figure 3 and 4.

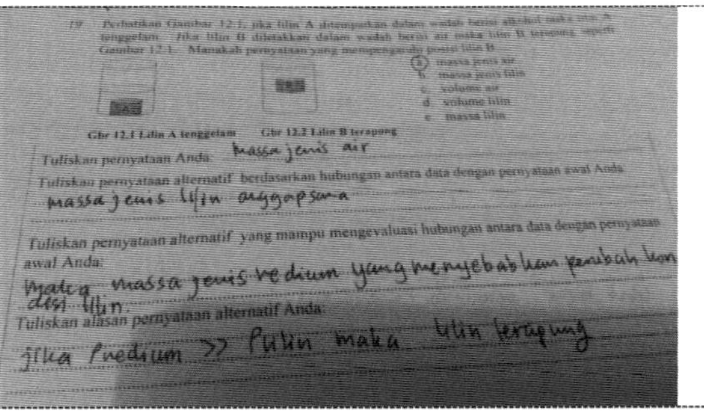

Figure 3. Figure Statement of the Students C

For the problems presented above, learners should direct their initial statement on "water density". Based on the initial statement will bring up an alternative statement "objects whose density is smaller than water will float. An object whose mass is larger than the density of the water type will sink. Based on this the wax will float in water because the wax has a density smaller than water and the volume of water has a mass greater than the volume of wax. Students' can prove this by placing the volume of wax and water volume that has an arbitrary balance. This balance will show that water has a mass greater than a candle ". One of the reasons for the emergence of an alternative statement: "floating objects at different levels in the water because most of the regular objects immersed in fluid have an upward force of fluid until they are in balance with the gravity of the object, and then the object will be at the equilibrium level The same between the two forces acting on the body ".

The statement presented Student C in figure 3 "mass water type (simple claim)"; The wax mass is considered equal (statements containing data); "Then the mass of the type of medium that causes changes in wax conditions (alternative claims based on data); "If $\rho_{medium}>>\rho_{wax}$ then floating wax (alternative claim supported by reference)". Structurally students A has been able to analyze the process of formation of alternative statements contained in the problem. Student C in figure 3 has been able to read data that there are wax-type mass, mass of media type and mass of water type in the problem. Data that can be analyzed by student C as material to write alternative statement. Alternative statements compiled by student C are still categorized as weak or at level 2 where the alternative statement is appropriate and gives a reason (referring to the initial statement (claim and data) with score 2.

Furthermore, the statement presented by Student D in figure 4"wax mass is less than the density of the alcohol (by choosing option answer b: the mass of the wax type (simple claim)": when the waxy mass is less than the density of the alcohol the candle will float as it is inserted in the container contains alcohol (a statement containing data); "if the waxy mass is larger than the mass of the alcohol type, then the wax will sink (alternative claim supported by reference)"; based on buoyancy, drowning and archimedes law (alternative claiming grounds) Structurally students D have been able to analyze the process of formation of alternative statements with the support of simple reasons.

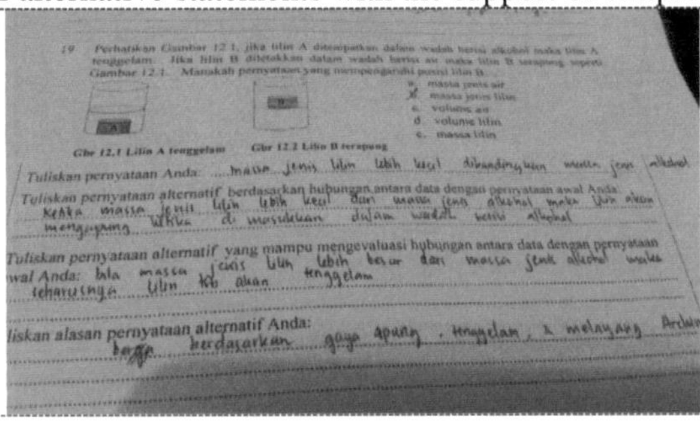

Figure 4. Figur Statement of the Student D

The quality of statements presented by Student D analyzed using the rubric of development results included in the category of appropriate alternative statements and provide reasons for the conclusions that describe relationship between alternative statements and choice of reasons (Claim, dat A, and warrant) and given a score of 3.

As one unity of alternative statement presented by students C and D describes the quality of argument is higher than the statement that only supported data and warrant. Alternative statements by explaining the reasons show that learners have been referring to the correct scientific theory behind the concept. An alternative statement concerns new knowledge that learners have as a result of a conflict of knowledge in the minds of learners. Alternative statements that include cognitive conflict in the thinking of learners indicate that the quality of arguing learners has been structured and illustrates more information in mind. This is consistent with the results of [21], that the quality of arguing learners is possible from the initial statement with the reasons underlying the identification of ideas from the appropriate references. Toulmin's argumentation scheme used as a reference in designing the quality assessment rubric in this research has four levels which are described as the main elements: claim, data, warrant and backing. The findings obtained in this study indicate the need to facilitate the ability of learners to attend opposite positions in order to maintain a contentious position in the matter.

4. Conclusion

Based on data analysis of research results, the average learner for component I was able to achieve score 3 with high category. Meanwhile, for component II the average of students able to achieve score 2 with low category. it is caused by: lack of knowledge about the problems presented, lack of experience in formulating preliminary statements having an impact on the formulation of alternative statements, and reference limitation. Conceptually the cause of the achievement of the score of students' on components I and II resolved by activating and providing an overview of knowledge related to the problems tested.

5. Acknowledgments

Authors wolud like to thank the student of Senior High School XI who are willing to cooperate in this research and provide support for taking of the data.

6. References

[1]Duschl, R. A., Schweingruber, H. A., &Shouse, A. W. (Eds.). (2007). Taking science to school: Learning and teaching science in grades k-8. Washington, DC: National Academy Press.

[2]National Research Council. (2012). A framework for K-12 science education: Practices, crosscutting concepts and core ideas. Washington, DC: The National Academies Press.

[3] Osborne, J., MacPherson, A., Patterson, A., &Szu, E. (2012). Introduction. In M. S. Khine (Ed.): Perspectives on scientific argumentation: Theory, practice and research. (pp. 3 – 15). Dordrecht, The Netherlands: Springer.

[4] Berland, L. K., & Hammer, D. (2012). Framing for scientific argumentation. Journal of Research in Science Teaching, 49(1), 68 – 94.
[5] McNeill, K. L., & Pimentel, D. S. (2010). Scientific discourse in three urban classrooms: The role of the teacher in engaging high school students in argumentation. Science Education, 94(2), 203 – 229.
[6] Simon, S., Erduran, S., & Osborne, J. (2006). Learning to teach argumentation: Research and development in the science classroom. International Journal of Science Education, 28(2 – 3), 235 – 260.
[7] Sampson,V.,&Blanchard, M. (2012). Science teachers and scientific argumentation: Trends in views and practice. Journal of Research in Science Teaching, 49, 1122 – 1148.
[8] Viyanti, Cari, Sunarno. W, Prasetyo, Z.K. (2016). PemberdayaanKeterampilanArgumentasiMendorongPemahamanKonsepSiswa. JurnalPenelitian PembelajaranFisika.Volume 7Nomor 1 April 2016. ISSN 2086-2407.
[9] Viyanti, Cari, Suparmi, Winarti, Budiarti. I.S., Handika. J., et al (2017).Consintency Argued Students Of Fluid. Journal of physics:conference
[10]Osborne, J., Erduran, S., & Simon, S. (2004). Enhancing the quality of argumentation in school science. Journal of Research in Science Teaching, 41(10), 994-1020.
[11] Anderson, L. W. and Krathwohl, D. R., et al (Eds..) (2001) A Taxonomy for Learning, Teaching, and Assessing: A Revision of Bloom's Taxonomy of Educational Objectives. Allyn& Bacon. Boston, MA (Pearson Education Group)
[12]Acar, Omar (2008). argumentation skills and conceptual knowledge of undergraduate students in a physics by inquiry class. Dissertation
[13]Sadler, T. D., &Zeidler, D. L. (2005). The significance of content knowledge for informal reasoningregarding socioscientific issues: Applying genetics knowledge to genetics engineering issues. ScienceEducation, 89, 71-93.
[14]Schwarz, B. B., Neuman, Y., Gil, J., &Ilya, M. (2003). Construction of collective and individual knowledge in argumentative activity. The Journal of Learning Sciences,12(2), 219-256.
[15]Toulmin, S. (1958).The uses of argument. New York: Cambridge University Press.
[16]McNeill, K. L., Lizotte, D. J., &Krajcik, J. (2006). Supporting students' construction of scientific explanations by fading scaffolds in instructional materials. The Journal of the Learning Sciences, 15(2), 153-191.
[17]Zohar, A., &Nemet, F. (2002). Fostering students' knowledge and argumentation skills through dilemmas in human genetics. Journal of Research in Science Teaching, 39(1), 35-62.
[18]Driver, R., Newton, P., & Osborne, J. (2000).Establishing the norms of scientific argumentation in classrooms. Science Education, 84, 287-312
[19]Kenyon, L., Kuhn, L., &Reiser, B. J. (2006). Using students' epistemologies of science to guide the practice of argumentation. In S. A. Barab, K. E. Hay, & D. T. Hickey (Eds.), Proceedings of the seventhinternational conference of the learning sciences (pp 321 – 327). Mahwah, NJ: Erlbaum
[20]Zohar, A. (2008). Science teacher education and professional development in argumentation. In S. Erduran& M. P. Jimenez-Aleixandre (Eds.), Argumentation in science education: Perspectives from classroom-basedresearch (pp. 245 – 268). Dordrecht, The Netherlands: Springer
[21]Sadler, T.D. (2004). Informal reasoning regarding socioscientific issues: A critical review of research. Journalof Research in Science Teaching, 41(5), 513-536.

IOP Conf. Series: Journal of Physics: Conf. Series **909** (2017) 012058 doi:10.1088/1742-6596/909/1/012058

Attainment of students' conception in magnetic fields by using of direct observation and symbolic language ability

Siska Desy Fatmaryanti[1], Suparmi[2], Sarwanto[2] and Ashadi[2]

[1] Student of Doctorate Program on Science Education, School of Postgraduate, Sebelas Maret University, Jl. Ir. Sutami 36A Kentingan Jebres Surakarta 57126, INDONESIA

[2] Lecture of Doctorate Program on Science Education, School of Postgraduate, Sebelas Maret University, Jl. Ir. Sutami 36A Kentingan Jebres Surakarta 57126, INDONESIA

E-mail: siskadesy@student.uns.ac.id

Abstract. This study focuses on description attainment of students' conception in the magnetic field. The conception was based by using of direct observation and symbolic language ability.The method used is descriptive quantitative research. The subject of study was about 86 students from 3 senior high school at Purworejo. The learning process was done by guided inquiry model. During the learning, students were required to actively investigate the concept of a magnetic field around a straight wire electrical current Data retrieval was performed using an instrument in the form of a multiple choice test reasoned and observation during the learning process. There was four indicator of direct observation ability and four indicators of symbolic language ability to grouping category of students conception. The results of average score showed that students conception about the magnitude more better than the direction of magnetic fields in view of symbolic language. From the observation, we found that students could draw the magnetic fields line not from a text book but their direct observation results. They used various way to get a good accuracy of observation results. Explicit recommendations are presented in the discussion section at the end of this paper.

1. Introduction

Scientific theories are characterized by the ability to predict and to observe of natural phenomena. The concept of magnetic field, used in a relation of electric current and magnetic phenomena, is a good example of these characteristics. In this paper, we look in detail at the conception of senior high schools students by using of direct observation and symbolic language ability. The conception encounter with the concept of magnetic field in the context of the relation electric current with magnetic phenomena.

The field concept is theoretically more challenging than the force concept. It also requires a change in the theoretical thinking of students. This change takes place when the surrounding space around the source is being described with a field—without having any other object with which to interact and thus exert a force so that the 'action at a distance' model is replaced with a model describing the action with field source [1]. According to earlier studies, the students fail to described direction of magnetic field around a straight current carrying wire[2]. Another learning, on previous research by Thong & Gustone [3] which looked at the need to help students come to an understanding of both the nature of the construct "field" and the purpose physicists has in creating this construct.

Magnetic phenomena have been experienced by many students from a young age. As these students enter high school and encounter more abstract and mathematically structured magnetic field concepts. When learning process, usually teachers require students to use sophisticated ideas they are not familiar with, but the experiences in their daily lives are not often taken into account in formal education [3].

A growing body of research in physics education indicates that a majority of students have difficulty learning basic physical concepts in a course built around traditional lectures, textbook problems, and verification experiments [4], [5]. These studies indicate that for improved learning, students must engage actively in close interaction of students with their peers. And the laboratory can be an excellent environment [5][6]. The laboratory should help students to understand the role of direct observation in physics and to distinguish between inferences based on theory and the outcomes of experiments.

Physics consists of concepts, hypotheses, theories, and observations. All of them are connected. On laboratory experiments, it is often difficult to separate inferences based on theory from direct observations. Students should understand that experimental evidence is the basis of our knowledge of the laws of physics and that physics is not merely a collection of equations and textbook problems. Many students have no concrete experience with everyday phenomena. Whereas it is important to understand the relationship between observations and concepts in physics. The process of observing phenomena, analyzing data, and explaining observations can develop verbal and mathematical abilities. This is because the process provides an opportunity for students to connect concrete experiences with scientific theory.

In the classroom, objective interpretation raises different interpretations between the formation and meaning of symbols. The difference in symbolic language used in a single concept makes it difficult for students to understand the meaning and relationship between symbols. This difficulty relates to problems in semiotic processes [7]

2. Research Method

The study was conducted at about 86 students from 3 senior high school at Purworejo. The data were gathered by using a paper and pencil test and observation during learning.

In this research, we observe students' activities in doing the experiment. Steps of learning are:

a. Student doing the investigation in group
b. in the process of learning, we observe their direct observation ability
c. Student make a report from their investigation
d. The researcher assessed students' report based on the conception by using symbolic language ability. The range of score is between 1 and 10.

Guided by assessment and observation were using indicators in Table 1. It has been arranged based on some early studies [4], [7],[8].

Table 1. Categories of explanations given by students and corresponding typical answers

Students' Ability	Indicator
Direct observation	describe objects according to observations
	select the information by the purpose of the investigation

	apply the principles of physics in the reported observations
	use the scientific method to obtain the accuracy of the observations
Symbolic Language	use symbols appropriately
	use mathematical rules to solve problems
	read draw and mathematical signs of the magnetic field direction and electric current
	understand the quantitative meaning of units and magnitudes of magnetic field equations

3. Results and Discussion

3.1 Paper and Pencil Test's Results

The test consists of eight problems (four problems about the direction of magnetic fields and four problems about the magnitude of magnetic fields). Students finished the test about 30 minutes. And then the results were analyzed in average score depend on a group of the problem), as seen in Table 2.

Table 2 Average score of Students' conception of magnetic field around current electric wire

Problems	**Symbolic Language (Average score and Standart of Deviation, *N*=86)**							
	1		**2**		**3**		**4**	
Direction of magnetic fields	5,83	1,31	6,01	1,16	6,06	1,28	6,20	1,33
Magnitude of magnetic fields	6,36	1,17	6,35	1,20	6,46	1,12	6,65	1,03

From Table 2, we can see that students conception about the magnitude more better than the direction of magnetic fields in view of symbolic language. For example, in problem 1 (as seen in Figure 1), There were four sketches that describe the relation electric current with magnetic field surrounding. The student should select the direction of the magnetic field from 4 sketches and give their reason.

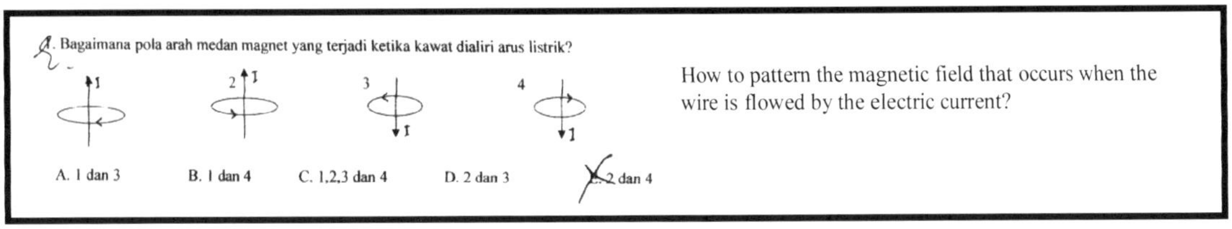

Figure 1. Example of direction of magnetic field problem

In symbolic language, they should read the figure of the magnetic field direction and electric current. After that, they should write the reason of their answer based on their observation. In this section, many students get failed to describe and explain the reason based on their observation.

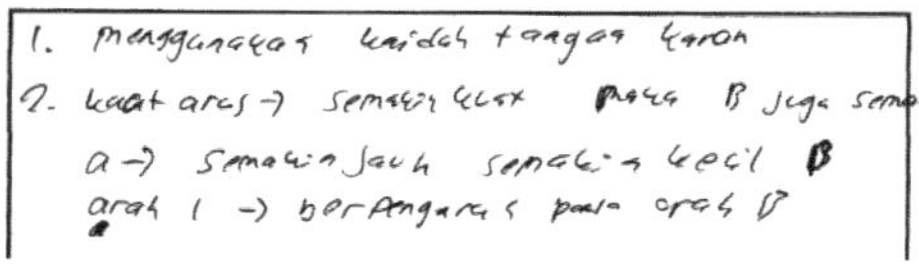

1. Using right-hand rules
2. Increasing electric current *i* makes **B** stronger.
The farther the distance **a**, makes **B** smaller.
i direction influences on **v** direction

Figure 2. Example of the reason of students' answer from problem number 1

As seen in figure 2, they didn't explain about the direction of compass needle-like their observation. They said it because of right-hand rules. They should be written their analysis about the change of electric current direction with the deviation of a compass needle. In another problem, students were asked to give reasons on how to get the magnetic field value. Answers given by students tend to use verbal language. The use of symbols and mathematical signs is still very small.

In indicator using symbolic and mathematics rules to solve the problems, the average score was much better than two indicators before. Students can solve problems with proving in mathematics rules and using notation (as seen in Figure 3).

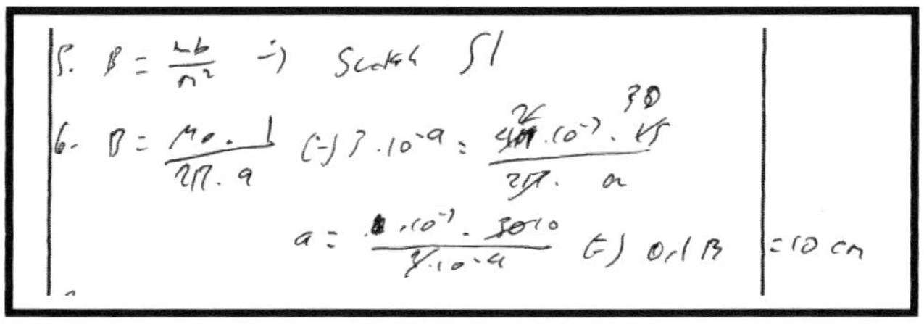

Figure 3. Example of students' answer in a view symbolic language ability

Some students have used the units correctly and know the relationship between units of magnetic field equations. Other students still confused with a unit of magnetic field. They said weber/meter^2 (Wb/m^2) is different with tesla (T). From the previous study [9], also found that symbolic language ability was one of the abilities that needed in learning the magnetic concept. The direction of magnetic field is sometimes stated in a direction to paper view or observer. For example, into the paper or go away from the observer and out of paper or near observer. This state was confusing for the student when applicate to solve the problem.

3.2 Observation during learning process

Learning the concept of the magnetic field has been done with the inquiry through experimentation. Students are required to actively carry out an investigation in groups and conduct their investigation. Nevertheless, the teacher still provides guidance in making conclusions from the results of the investigation.

Observations were made by two observers. The process of focusing on observation is the ability of students to perform direct observation and symbolic language. There are two concepts that are investigated in this study is determining the direction and magnitude of the magnetic field. The first investigation is the magnitude of the magnetic field. Determination of magnitude is done using iron powder and plotting compass. Students made observations on the pattern of iron powder and compass needle deviations on the electric current changes strongly. The second investigation is the direction of magnetic field. Determination of the magnetic field direction is done using some plotting compass that is placed around the electric current carrying straight wire.

Some findings during observation are student can better understand the magnitude and direction of magnetic field. But they have difficulty in describing the deviation of a compass needle in paperwork. The first indicator of direct observation is described object based on observation. The pattern of iron powder indicates the magnitude of magnetic field. From the observation can be observed that if the closer to the wire hence the pattern of iron powder will be more tightly. To observe the relation of the magnetic field and electric current, the students vary the magnitude of electric current. And they draw any pattern changes that occur. The ability of selected information according to the purpose of the investigation is needed. The direction of the magnetic field is the same as the direction of the magnetic force line. The magnetic field magnitude is proportional to the magnetic force line density. The next activity was observed against deviations compass needle around the current-carrying wire. The problems encountered were students' difficulties in reading the compass. In the previous study, also found that pre-service teachers have difficulty interpreting the pattern and direction of the magnetic field with the direction of the compass needle deviation [2]. It was indicated that experiments illustrated how the learning path provides the students with a set of experiences and observations that allow them to explore experimentally the phenomenology of the magnetic field. [5].

In discussion session, found that some student couldn't apply the principles of physics in the reported observations well. They didn't use principles of north and south in magnet when reading compass needle deviation. And at the beginning of observation, principles of flowing electric current still forgotten. Therefore, teacher guidance is necessary during learning. In the inquiry, the assistance provided by the teacher aims to train students in conducting the process of independent inquiry process [10].

4. Conclusion

The results from of the proposed learning illustrate the necessity to introduce a specific experiment. The aim was to allow students to explore magnet property of the object, and giving them a "practical" idea of the magnetic field through the simple observation done by the needle of the compass.

Experiments illustrated how the learning path provides the students with a set of experiences and observations that allow them to explore experimentally the phenomenology of the magnetic field around the electric current wire. Providing a first explanatory model based on an idea that even though it does not have the whole structuration required by the theory of direction of the magnetic field, it has its main conceptual cores, and even if the pupils use the magnetic field in a non-quantitative way, it has its main phenomenological characteristics.

Acknowledgment
This work was supported by the Ministry of Research, Technology, and Higher Education, Republic of Indonesia, through the Doctoral Dissertation Research Grants. This work is also supported by Senior High Schools in Purworejo (SMAN 2, SMAN 4 and SMA Muhammadiyah Kutoarjo) for facilities. The researchers thank to all institution that has been supporting this research and to the reviewers.

References

[1] M. Saarelainen, A. Laaksonen, and P. E. Hirvonen, "Students' initial knowledge of electric and magnetic fields—more profound explanations and reasoning models for undesired conceptions," *Eur. J. Phys.*, vol. 28, no. 1, pp. 51–60, 2007.

[2] S. Fatmaryanti, Suparmi, Sarwanto, and Ashadi, "Student representation of magnetic field concepts in learning by guided inquiry," *J. Phys. Conf.Series*, vol. 795, 2017.

[3] W. M. Thong and R. Gunstone, "Some Student Conceptions of Electromagnetic Induction," *Res. Sci. Educ.*, vol. 38, no. February 2007, pp. 31–44, 2008.

[4] R. Feynman, "Goals of the Introductory Physics Laboratory," *Am. J. Phys.*, vol. 66, no. June, pp. 483–485, 1998.

[5] S. R. C. P. Challapalli, M. Michelini, and S. Vercellati, "Building formal thinking with pupils on magnetic phenomena in conceptual laboratories," *Procedia - Soc. Behav. Sci.*, vol. 93, pp. 946–950, 2013.

[6] J. Dean and D. A. Allwood, "From compass to hard drive — integrated activities for studying magnets," *Phys. Educ.*, vol. 49, no. 6, pp. 663–670, 2014.

[7] S. R. D. E. Lozano and M. Cardenas, "Some Learning Problems Concerning the Use of Symbolic Language in Physics," *Sci. Educ.*, vol. 11, pp. 589–599, 2002.

[8] Brotosiswoyo, "Learning the nature of Mathematics and Natural Sciences and Chemistry Teaching Tips on Higher Education." Ministry of Education, Jakarta, 2001.

[9] S. Fatmaryanti, Suparmi, Sarwanto, and Ashadi, "Analysis of Magnetism Problems in High School Physics National Exam Based on Concept Required and Student's Science Generic Skills," in *International Conference on Science and Science Education*, 2015.

[10] Y. Hsu, T. Lai, and W. Hsu, "A Design Model of Distributed Scaffolding for Inquiry-Based Learning," *Res. Sci. Educ.*, no. 88, 2014.

IOP Conf. Series: Journal of Physics: Conf. Series **909** (2017) 012059 doi:10.1088/1742-6596/909/1/012059

Students' profile of heat and temperature using HTCE in undergraduate physics

Dewanta Arya Nugraha[1], A. Suparmi[2], Retno Winarni[3], Suciati[4]

[1, 4] Science Education, Graduate Program, Faculty of Teacher Training and Education, Universitas Sebelas Maret

[2] Physics Department, Graduate Program, Universitas Sebelas Maret

[3] Primary School Teacher Education Department (PGSD), Faculty of Teacher Training and Education, Universitas Sebelas Maret

Email: dewanta.an@gmail.com

Abstract. Understanding heat and temperature are important to make strong fundamental of physics before understanding the other subject materials. This research aims to describe the students' conception of heat and temperature using *Heat and Temperature Conceptual Evaluation* (HTCE) developed by Thornton and Sokoloff. This research subjects are 10 students of 3rd semester and 24 students of 5th semester Bachelor of Physics. The data collection methods are test and interview. The result are the students' conception of heat and temperature, rate of cooling, calorimetry, rate of heat transfer, perception of hotness, specific heat capacity, change of phase, thermal conductivity. Students are getting difficult on understanding the concept of heat and temperature especially the concept of rate of cooling, change of phase, and rate of heat transfer. The average students' correct answer is 44.88% of 34 students. The lowest mean score is the concept of RHT with the percentage of 17.65%. This research could be used to develop learning media on basic physics course.

1. Introduction

The aim of some physics education research is to study students' understanding or conception in physics using standardized test. Example of the well known standardized tests are Force and Motion Conceptual Evaluation (FMCE) [1], Force Concept Inventory (FCI) [2], Thermal Concept Evaluation (TCS) [3], and Heat and Temperature Conceptual Evaluation (HTCE) [4]. In this study, we focus on the survey using HTCE. Heat and temperature are one of fundamental of physics topic on basic physics course. Heat and temperature are subject matter which contains abstract concept [5]. The studies have done about student understanding show that students get some difficulties in understanding and explaining the heat and temperature concept. The studies about it have been done by Tanahoung, et al.[6], Carlton [7], Yeo and Zadnik [8], Budiarti, et al. [9].

Understanding the concept should be emphasized in the learning because with the understanding of the concept students can understand the whole of the main material on the matter of heat and temperature. Baser states from the point of view of constructivism learning, new knowledge is constructed with existing knowledge [10]. Therefore one of the factors in learning is the initial knowledge, usually an alternative conception of the topic. Thus it can be said that the initial understanding of students before learning also affects the understanding of student concepts after learning. Therefore, the survey conducted to find out the students' understanding heat and temperature.

2. Method

This research method of this preliminary study is descriptive survey method. The data collection are using test and interview. The subjects of this research are 10 students of 3rd semester and 24 students of 5th semester year 2016/2017, Bachelor of Physics, Faculty of Mathematics and Natural Sciences Universitas Sebelas Maret. The test is using HTCE [4], s consist of 27 multiple choice questions with the short reasons. After students doing the test, the researcher gives some interview to get deep information about the answer. The results from the test are divided based on the categories of concept of HTCE as shown in table 1.

Tabel 1. Categories of conceptual areas of HTCE [6]

	Conceptual areas	Question Number
1.	Heat and temperature (H&T)	1, 2, 3, 4
2.	Rate of cooling (COOL)	5, 6, 7
3.	Calorimetry (CAL)	8. 9
4.	Rate of heat transfer (RHT)	10, 11
5.	Perception of hotness (HOT)	12, 13, 14, 15
6.	Specific heat capacity (CAP)	16, 17, 18, 19
7.	Change of phase (PHASE)	20, 21, 22, 23, 25
8.	Thermal conductivity (COND)	26, 27, 28

3. Results

HTCE was tested to students of bachelor physics who have studied about heat and temperature before. The results of the test using HTCE for the students are the truth and wrong answer. 44.88% of 34 students could answer the question correctly. The percentage of the correct answers are shown on figure 1. From figure 1 we can get the three

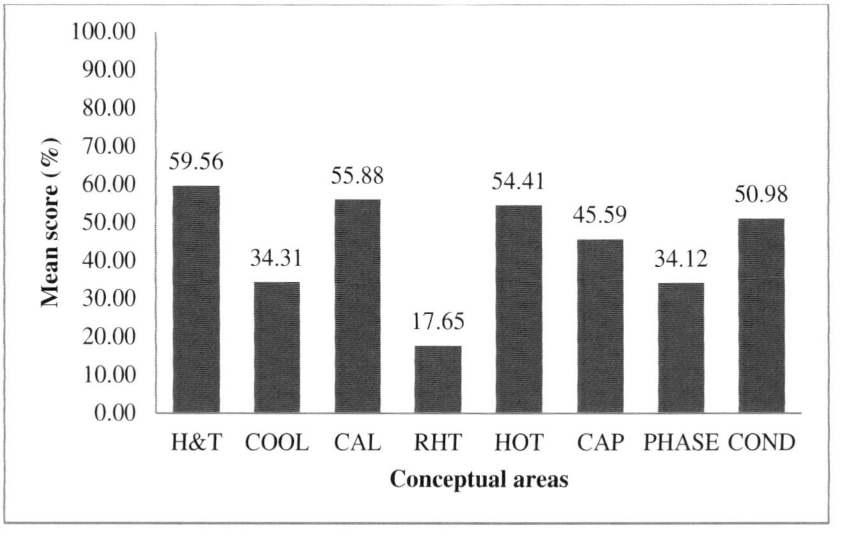

Figure 1. Students mean scores in each conceptual area

Based on Figure 1, The result showed that the students still get some difficulties to understand the concept of rate of cooling, change of phase, and rate of heat transfer. The highest score is in heat and

IOP Conf. Series: Journal of Physics: Conf. Series **909** (2017) 012059 doi:10.1088/1742-6596/909/1/012059

temperature concept with the percentage of 59.56%. The lowest mean score is the concept of RHT (question number 10 and 11) with percentage of 17.65%. More than 50% students get some difficulties on concept of CAP, PHASE, COOL, and RHT.

The 3rd of the most difficult concept is about PHASE. Figure 1 shows that only 34.12% student who can answer the question correctly. The PHASE concept consist of 5 questions, with the question number 20 is the most of student answer false. The students' answer is presented on table 2. Most of the students do not understand the concept of the temperature change during the time (question 20-23). Question 20 – 23 ask about the graph of the change of phase (see figure 2 for question number 20). In these question some students think that when the ice is melting and water is boiling the temperature will change and the energy is used to change the phase and the temperature. It means that the students do not understand the concept. The right answer is the energy only used to change the phase until the ice of water change totally and the energy will change the temperature when the ice or water finished to change the phase. Question number 25 is about the temperature of boiling water on the stove. And the students answer still shows the same perception like the previous question.

Table 2. Students answer about change of phase concept (34 students)

Question	A	B	C	D	E	F	H
20	4	9	1	3	1	9	7
21	4	11	3	1	2	4	9
22	0	0	3	10	6	6	9
23	0	3	2	5	1	15	8
25	1	4	9	13	-	-	7

(shaded) The key answer

Questions 16 -19 refer to a cup which contains water at room temperature. The cup is **perfectly insulated** *so that no heat can transfer into or out* of the cup. A small coffee cup heater inside the cup is used to transfer heat to the water. *The water does not boil.*

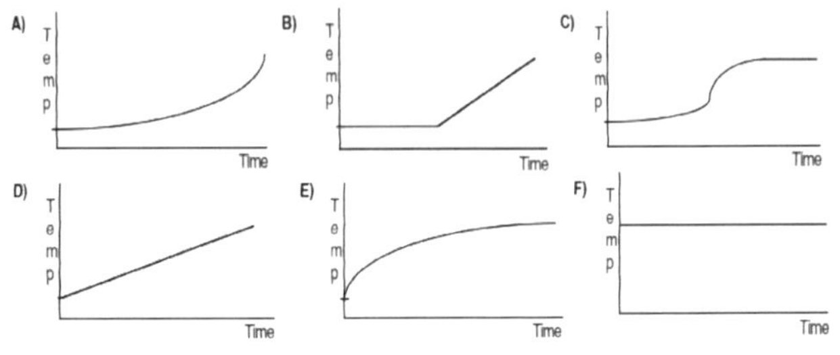

Questions 20-23 refer to the same cup and heater described above but the cup initially contains a mixture of 50 grams of ice and 50 grams of water at 0°C. The cup is **perfectly insulated** so that no heat can transfer in or out. Room temperature is 25°C.

A small coffee cup heater inside the cup allows heat to be transferred to the ice and water mixture. *Heat is transferred at a steady rate* and the mixture is stirred continuously so that the temperature is always uniform throughout the mixture. For each question below, choose the **shape** of the temperature-time graph from the graphs **A** through **F** at the top of the page that best correspond to the temperature of the mixture during the time interval described.

If you think no graph is appropriate, write **H.** (You may choose a graph more than once. The origin of the graphs does not necessarily represent 0°C.)

_____20. Which graph could show a time interval when the ice is melting but there is still some ice in the water?

Figure 2. Question Number 20 [4]

The concept of the rate of cooling is the 2nd most difficult concept answer by students, with only 34.31% can answer it. This concept consists of 3 questions (question number 5, 6, 7). From the question 5 (figure 3), it shows that student can't determine the rate of cooling, but at question 6 (figure 4) most students can answer it. It means that students just answer by their feeling (intuition) without understanding the concept of rate of cooling. Some students said that they choose B on question 5 because the initial temperature is lower than A, but they do not realize that the question is about the rate not about the time to reach the room temperature. Most of the students think that the question 5 and 6 ask the same thing (temperature), so they just think about the time to reach the room temperature.

Table 3. Students answer about rate of cooling concept (34 students)

Question	A	B	C	D
5	6	24	2	2
6	7	20	1	6

The key answer

Questions 5 -7 refer to two identical cups, A and B, each containing 100 grams of water. The cups are in a room where the temperature is 25 °C. The water in cup A is initially at 55°C, while that in cup B is initially at 40°C.

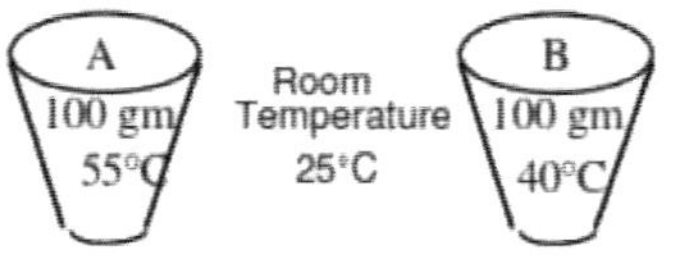

____5. Initially, which cup will cool down at a faster rate?
A) A will cool faster **B)** B will cool faster **C)** they both will cool at the same rate
D) not enough information is given to determine the answer

____6. Which cup will reach its final temperature most quickly?
A) Cup A **B)** Cup B **C)** Both take the same time
D) not enough information is given to determine the answer

Figure 3. Question Number 5-7 [4]

5) B
karena semakin kecil perbedaan suhu, maka energi transfer panas semakin kecil, sehingga semakin kecil perbedaan suhu, maka sistem tersebut akan cepat dingin karena energi yang ditransfer juga kecil.

5) B
Because the smaller the temperature difference, the heat transfer energy is smaller, so the smaller the temperature difference, then the system will quickly cool because the energy transferred is also small

Figure 4. Example of students' incorrect answer

From figure 4 we know that student perception is when the heat transfer energy is smaller, it has the faster rate, but in fact, faster rate means the heat energy transfer during the time is bigger. It shows that students are unable to distinguish the heat transfer and the rate of heat transfer (in this case for cooling). In question 6, most of the students answer the right answer B, but some students still answer A because they think that cup A has the higher temperature with the higher energy. They think that with higher energy, the temperature will cooling fast without thinking that the question focuses on the quickness (time period).

Most students cannot answer correctly on the question about rate of heat transfer concept. In this concept, the percentage of student who can answer the question correctly is 17.65%, this concept is the same concept which is the lowest correct answer on the research done by Tanahoung [6]. In this concept there are 2 questions (question 10 and 11). Actually, the concept is like the concept of rate of cooling but, here focus on heat transfer to maintain the temperature of water. To answer the question students must understand the concept of rate of heat. 12 students answer H (none of the above answers is correct). They answer H because they think that the information given in the question doesn't complete. The said that there's no information mass and the areas of heater that in touch with the water. In this case, it shows that student unable to analyze the rate of heat transfer because there is not enough information. Actually, when they want to calculate they can eliminate the mass and areas because when they find the comparison, they can ignore the mass and the areas (it's the same amount). 7 students answer C because they think that the temperature to maintain in A is higher than the temperature in B, so the rate will be faster. Some students only think that the rate of heat at A is faster than B without calculating the rate of heat because the information doesn't enough (see figure 6).

Table 4. Students answer about rate of heat transfer concept (34 students)

Question	A	B	C	D	E	F	G	H
10	2	8	7	2	0	2	1	12
11	4	4	6	4	0	1	2	13

The key answer

____10. Small coffee cup heaters are placed in cups A and B and heat is transferred to keep the cups at the temperatures shown. The cups contain the same amount of water. Which answer best describes the rate that heat must be transferred to maintain the temperatures shown?

Cup A will require heat at **A**) about five times the rate of B
B) about twice the rate of B
C) a slightly faster rate than B

Both cups will require heat at **D**) the same rate

Cup B will require heat at **E**) about five times the rate of A
F) about twice the rate of A
G) a slightly faster rate than A

H) None of the above answers is correct

Figure 5. Question Number 10 [4]

10. C Alasan : karena temperature yang lebih dipertahankan di A lebih besar
10. C Reason: because the temperature maintained at A is larger
(10) - C - Karena suhu digelas A lebih besar
10. C Because the temperature in A is bigger

Figure 6. Example of students' incorrect answer of question 10

4. Conclusion

From the results, we can conclude that students getting difficult on understanding the concept of heat and temperature especially concept of rate of cooling, change of phase, and rate of heat transfer. The average students' correct answer is 44.88% of 34 students. The lowest mean score is in the concept of RHT students getting difficult on understanding the concept of heat and temperature especially concept of rate of cooling, change of phase, and rate of heat transfer with the percentage of 17.65%. Most of the students still get some difficulties to distinguish rate of heat or rate of cooling with the time needed. Overall of this result, the misconception needed to be corrected. This research could be used to develop learning media on basic physics course.

References

[1] R. K. Thornton and D. R. Sokoloff, 1998 "Assessing student learning of Newton ' s laws : The Force and Motion Conceptual Evaluation and the Evaluation of Active Learning Laboratory and Lecture Curricula Assessing student learning of Newton ' s laws : The Force and Motion Conceptual Evaluation an," vol. 66, no. 4, pp. 338–352.

[2] D. Hestenes, M. Wells, and G. Swackhamer, 1992 "Force Concept Inventory," vol. 30, no. March, pp. 141–158.

[3] P. Wattanakasiwich, P. Taleab, M. D. Sharma, and I. D. Johnston, 2013 "Development and Implementation of a Conceptual Survey in Thermodynamics," *Int. J. Innov. Sci. Math. Educ.*, vol. 21, no. 1, pp. 29–53.

[4] R. K. Thorton and D. K. Sokoloff, "Heat and Temperature Conceptual Evaluation," *physport.org*, 2001. [Online]. Available: https://www.physport.org/assessments/assessment.cfm?I=16&A=HTCE. [Accessed: 28-Dec-2015].

[5] D. A. Nugraha, S. Suparmi, M. Masykuri, and C. Cari, 2016 "Survei Pemahaman Mahasiswa Fisika Pada Materi Kalor Dan Temperatur," *Pros. SNPF (Seminar Nas. Pendidik. Fis.*, pp. 24–28.

[6] C. Tanahoung, R. Chitaree, and C. Soankwan, 2008 "Surveying introductory physics students â€™ understanding of heat and temperature," *THAI J. Phys.*, vol. 165, pp. 165–166.

[7] K. Carlton, 2000 "Teaching about heat and temperature," *Phys. Educ.*, vol. 35, no. 2, p. 101.

[8] S. Yeo and M. Zadnik, 2001 "Introductory thermal concept evaluation: Assessing students' understanding," *Phys. Teach.*, vol. 39, pp. 496–504.

[9] I. S. Budiarti, Suparmi, Sarwanto, and Harjana, 2017 "Students' conceptual understanding consistency of heat and temperature," *J. Phys. Conf. Ser.*, vol. 795, p. 12051.

[10] M. Baser, 2006 "Effect of Conceptual Change Oriented Instruction on Students' Understanding of Heat and Temperature Concepts," *J. Maltese Educ. Res.*, vol. 4, no. 1, pp. 64–79.

IOP Conf. Series: Journal of Physics: Conf. Series **909** (2017) 012060 doi:10.1088/1742-6596/909/1/012060

Profile of student critical thinking ability on static fluid concept

Sulasih[1], A Suparmi[2], Sarwanto[3]

[1]Postgraduate Student's, Sebelas Maret University Surakarta, Teacher at SMA N 1 Sambungmacan, Jl. Raya Timur Km. 15 Sragen 57253, INDONESIA

[2,3]Physics Department of Post Graduate Program Sebelas Maret University, Jl. Ir. Sutami 36A Kentingan Jebres Surakarta 57126, INDONESIA

E-mail:salwa_777@yahoo.co.id

Abstract. Critical thinking ability is an important part of educational goals. It has higher complex processes, such as analyzing, synthesizing and evaluating, drawing conclusion and reflection. This study is aimed to know the critical thinking ability of students in learning static fluids of senior high school students. This research uses the descriptive method which its instruments based on the indicator of critical thinking ability developed according to Ennis. The population of this research is XIthgrade science class Public Senior High School, SMA N 1, Sambungmacan, Sragen , Central Java. The static fluid teaching material is delivered using Problem Based Learning Model through class experiment. The results of this study shows that the average student of XIth science class have high critical thinking skills, particularly in the ability of providing simple explanation, build basic skill, and provide advanced explanation, but they do not have high enough in ability of drawing conclusion and strategic and tactical components of critical thinking ability in the study of static fluid teaching material. The average of students critical thinking ability is 72.94, with27,94% of students are in a low category and 72,22% of students in the high category of critical thinking ability.

1. Introduction

The static fluid concepts are mostly involved in daily life activities. By learning the static fluid concepts teaches the student to think, find problems in everyday life and solve them based on relevant theories and concepts. Static fluid teaching material discusses hydrostatic pressures, Pascal law, Archimedes law, and surface tension. Many students have their own opinions about the concept of the object that is immersed inside of the static fluid that will be sinking, fully immersed, or floating. The behavior of these objects is obeyed to Archimedes law and students still have a misunderstanding of Archimedes law due to their daily lives experience. On the other hand, the ability to think critically is close to the understanding of the concept, therefore by understanding the concept appropriately students are expected to be able to think critically and vice versa.

Fluid is collections of molecules arranged by force (cohesion and adhesion) and by the forces acted by the container wall. Liquid and gas objects are classified as fluid. In studying static fluid we do not need to learn new physics principles to explain pressure and a buoyant force acting on floating and sinking objects [1] since the behavior of floating and sinking objects in static fluid could be explained by using Newton laws [2,3,4,5].

The Archimedes law states that if an object is immersed in the static fluid then it will experience a drag (upward)force equal to the volume of fluid that is removed.Therefore if an object is immersed in the static fluid, its weight will be reduced by the amount of the weight of the liquid that is removed, the object's weight in the static fluid is called as apparent weight. This statement is illustrated in Figure 1. The object that has mass m kg, volume V m^3, height h, area cross section A and density ρ_0 is hung on a spring balance and immersed into the static fluid that has density ρ, schematically shown in Figure 1. Figure 1 shows the derivation of Archimedes law for the apparent weight of the object that is fully immersed by applying the Newton first law where w is the object weight measured in the air.

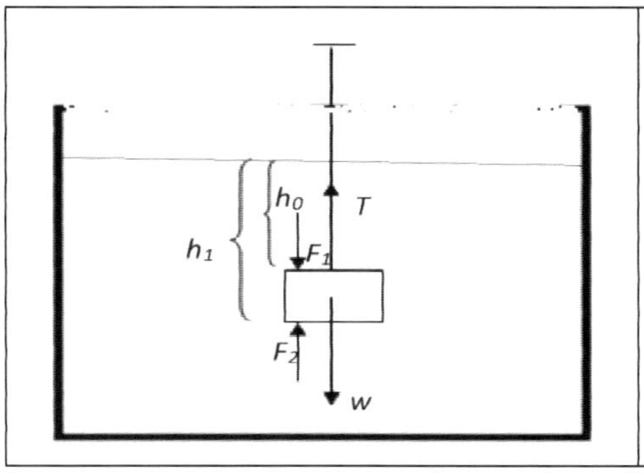

The vertical forces acted on the object are hydrostatic forces F_1 and F_2

$F_1 = P_0A = \rho_0 g h_0 A$, $F_2 = PA = \rho g h_1 A$

The drag force $F_d = F_2 - F_1 = \rho g(h_1 - h_0)A = \rho g h A$

By using Newton first law from figure 1 we have

$T + F_2 = w + F_1$ *then* $T = w - (F_2 - F_1) = w - F_d = w - \rho g h A$

The apparent weight w_a is equal to rope tension T measured by spring balanced is $w_a = T$

Figure 1. Illustration of forces acted on the object hung on the spring balanced scale that is fully immersed in the static fluid

From the observation this law is suitable for floating and fully immersed objects, as shown in Figure 1, for the sink object, on the other hand, the object that falls on the bottom of container will experience normal force acted by the container's bottom, so there is no more drag force from water to the lower part of the block as shown in Figure 2.Figure 2 shows the illustration of forces that acted on an object that sinks in the static fluid in the container. The object that sinks has a size of the object in Figure 1 but the density is different given by ρ_h.

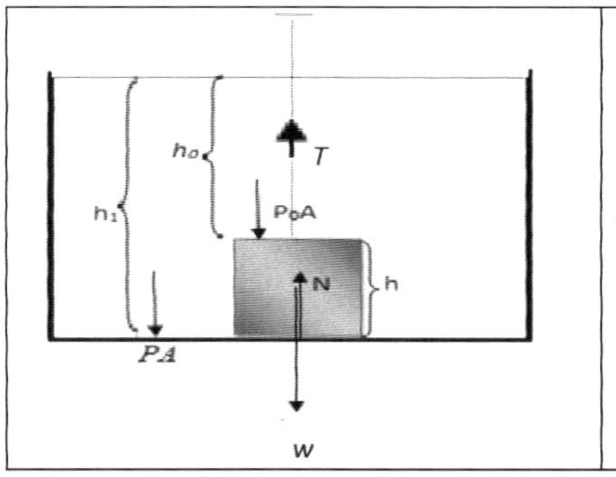

The forces acted on the sink object are F_1, *N, and T*

F_1 *is the hydrostatic force* $h_0 F_1 = \rho g\, h_0 A$

T is the measure apparent weight w_a measured by the spring balance and N is normal forces acted by the bottom of the container, therefore the apparent weight of the object is $T = w_a = w - (N - F_1)$

Figure 2. Illustration of forces acted on the sink objects that are hung on spring balance scale

From the discussion it is expected that the apparent object weight that sinks are smaller than the apparent object weight that is fully immersed, that is $N > \rho g h_1 A$. Therefore it could be said that the Archimedes law is not suitable for the immersed object in the static fluid that is a sink.One of teaching or learning model that should be used to deliver this concept appropriately is problem-based learning model through guided inquiry method.

Critical thinking has been one of the most important human skills in twentieth-century [6]. The thinking process not only reflects to conclude and synthesizes information but also allows individuals to make reasonable thinking in the classroom as well as in everyday life [7]. Students need critical thinking in academic life. All academic tasks require their critical thinking. Finally, however, teachers want them are not only to practice these skills in the classroom but also apply the skills into the real world in their everyday life[8]. Holma [9] suggests that it is not enough for students to have critical thinking skills, but they also need to use these skills effectively. To build the student's critical thinking ability, teachers should often ask some cognitive questions that promote a student to evaluate and synthesis of facts and concepts learned [10].

In psychology view, critical thinking is considered as a skill and focuses on the process of thinking. Some psychologists refer to Bloom's classification to define critical thinking and focus on critical thinking among Bloom's thinking ability [11].

Ennis [6] mentions the critical thinking indicators are as follows: 1) Provide a simple explanation which includes: focus on questions, analyze arguments, and clarify questions; 2) Build basic skills assessing: the credibility of an information source, making observations and judging observation reports; 3) Sum up, including making and assessing deductions, making a decision; 4) Provide further explanations, including defining terms and judging definitions, identifying assumptions; 5) Strategies and tactics, including: deciding actions and interacting with others.

Teachers as educators can use various learning methods to promote the critical thinking. Although educators judge a student who thinks critically about concepts, passion or disposition for critical thinking, and this is not always owned by all students. Many schools expect their students to think critically [13]. Teaching problem-based learning is an effective learning model that encourages students through reinforcing their critical thinking to gain skills to solve problems.

Problem-based learning is a learning model that involves student to solve a problem through the stages of a scientific model [14].So that student can learn the knowledge related to the problem and also have the skills to solve. The results of a study show that students who have difficulty explaining floating, fully immersed and sinking objects and can not identify the force that actions those objects.

2. Research Method

The sample of this research is 11th graders of science class from SMA N 1 Sambungmacan Sragen Central Java Indonesia. The research that has been conducted in January to June 2017 using the descriptive method. The instrument is developed based on the indicator of critical thinking ability of the static fluid that includes the ability of student in:(1) Analyzing the relationship between the depth of fluid and pressure on a fluid, (2) Analyzing the relation of drag force with the depth of the fluid and the area of object, (3) Applying the Archimedes law for floating, fully immersed, and sinking objects phenomenon,(4) Explaining the meaning of surface tension correctly. The instrument of the study has been validated by the expert which consists of 10 essay questions with a rubric score between 0 to 100. The static fluid teaching material is delivered to the students using Problem Based Learning Model through class experiment.

3. Results and Discussion

In this study, students answer sheets of problems that have been developed based on critical thinking ability indicators are analyzed. The indicators of critical thinking ability consist of: (1) provides a simple explanation, (2) build basic skills, (3)Conclude, (4) Provides the further explanation, (5) Set strategies and tactics. Each indicator of student critical thinking ability is represented by the student understanding of static fluid which is provided in the form of student answer sheets.

The results of research that have been conducted shows that, for all indicators of student critical thinking ability, 73.3% of students achieve above the average value of critical thinking ability which is 72.94 with the variance of (16.77), the highest score achieved by the student is 84,54 and the lowest score is 47,73. The student critical thinking ability is classified into two groups, low and high category of each critical thinking ability indicator, shown in Table 1. From Table 1 can be seen that most students have achieved the critical thinking ability in the level of provides a simple explanation, build basic skills and advanced explanation. These results indicate that students' critical thinking ability is good except for Setting strategies and tactics.

Table 1: The average, low, and high score of student for each indicator of critical thinking ability

No	Critical thinking ability	Low	High	Average score
1	Provide simple explanation	26.7%	73.3%	76.7
2	Build basic skills	16.7%	83.3%	75.4
3	Conclude	33%	66.7%	68.5
4	Advanced explanation	20%	80%	74.3
5	Setting strategies and tactics	43.3%	57.7%	69.s8

The following illustration is an example of student answer sheets both for incorrect and correct answers for any indicator of student critical thinking ability. Figure 3 (a) shows the incorrect student answer and Figure 3(b) for the correct student answer for the first indicator of critical thinking ability which provides a simple explanation of hydrostatic pressure concept.

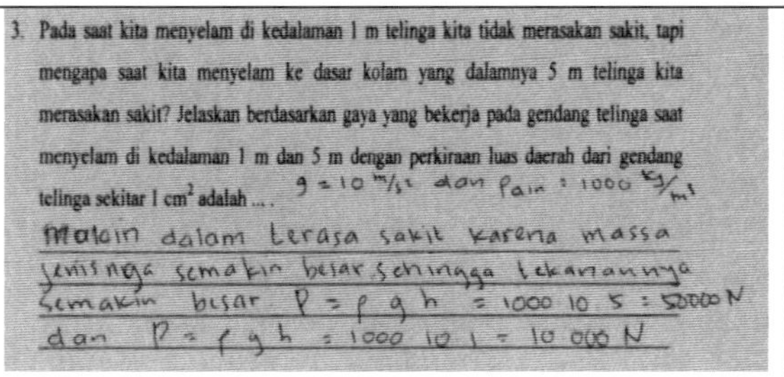
3. Pada saat kita menyelam di kedalaman 1 m telinga kita tidak merasakan sakit, tapi mengapa saat kita menyelam ke dasar kolam yang dalamnya 5 m telinga kita merasakan sakit? Jelaskan berdasarkan gaya yang bekerja pada gendang telinga saat menyelam di kedalaman 1 m dan 5 m dengan perkiraan luas daerah dari gendang telinga sekitar 1 cm² adalah .... g = 10 m/s² dan ρair = 1000 kg/m³
Makin dalam terasa sakit karena massa jenisnya semakin besar sehingga tekanannya semakin besar P = ρ g h = 1000 10 5 = 50000 N
dan P = ρ g h = 1000 10 1 = 10 000 N

When you swim in 1 m depth, your ears do not feel pain, but when you swim at 5m depth why your ears feel pain? Explain the force that acts on the ear drum when you swim in 1m depth and 5 m depth if the area or ear drum is about 1 cm^2!
The deeper is the swim, the more pain is felt since the fluid density is greater for the deeper fluid then the pressure is greater in the deeper.
$P_h = \rho gh = 1000.\ 10.\ 5 = 50000\ N/m^2$
$P_h = \rho gh = 1000.\ 10.\ 1 = 10000\ N/m^2$

Figure 3a: Example of incorrect student answer on the concept of hydrostatic pressure

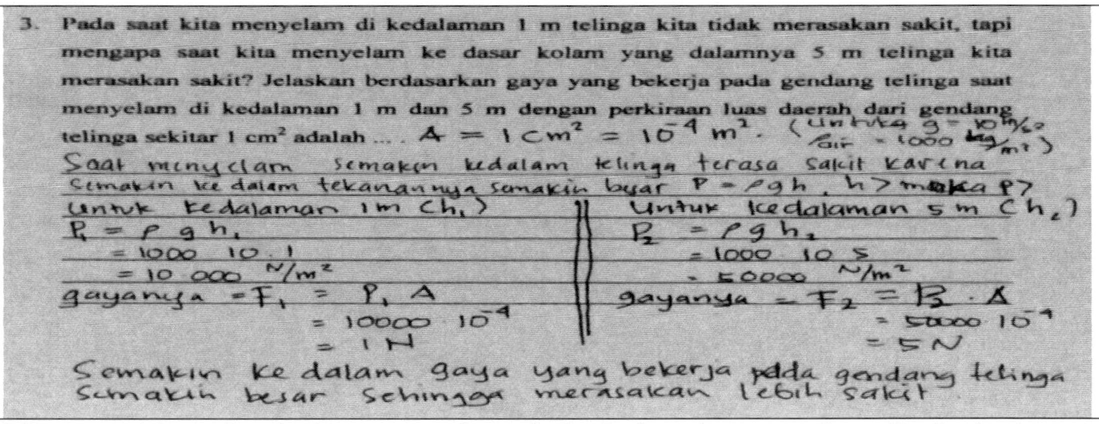
3. Pada saat kita menyelam di kedalaman 1 m telinga kita tidak merasakan sakit, tapi mengapa saat kita menyelam ke dasar kolam yang dalamnya 5 m telinga kita merasakan sakit? Jelaskan berdasarkan gaya yang bekerja pada gendang telinga saat menyelam di kedalaman 1 m dan 5 m dengan perkiraan luas daerah dari gendang telinga sekitar 1 cm² adalah .... A = 1 cm² = 10⁻⁴ m². (untuk g = 10 m/s², ρair = 1000 kg/m³)
Saat menyelam semakin kedalam telinga terasa sakit karena semakin ke dalam tekanannya semakin besar P = ρgh, h > maka P >
untuk kedalaman 1 m (h1): P1 = ρ g h1 = 1000 10 1 = 10 000 N/m²; gayanya = F1 = P1 A = 10000 10⁻⁴ = 1 N
untuk kedalaman 5 m (h2): P2 = ρ g h2 = 1000 10 5 = 50000 N/m²; gayanya = F2 = P2 . A = 50000 10⁻⁴ = 5 N
Semakin ke dalam gaya yang bekerja pada gendang telinga semakin besar sehingga merasakan lebih sakit.

When you swim in 1 m depth, your ears do not feel pain, but when you swim at 5m depth why your ears feel pain? Explain the force that acts on the ear drums when you swim in 1m depth and 5 m depth if the area of each ear drum is about 1 cm^2!
Student correct answer
Forg $= 10\ m/s^2 dan \rho_f = 1000 kg/m^3,\ A = 10^{-4} m^2$

When swim in deeper fluid the ears feel more pain since the hydrostatic pressure P is greater at the deeper fluid, the hydrostatic pressure is given by $P = \rho gh$ and the hydrostatic force is $F= P A$.
At 1m depth, the hydrostatic pressure and force are
$P_1 = \rho gh_1 = 1000.\ 10.\ 1 = 10000\ N/m^2$, *so* $F_1=P_1A = 10000.\ 1.\ 10^{-4} = 1$ *Newton*
At 5 m depth, the hydrostatic pressure and force are
$P_2 = \rho gh_2 = 1000.\ 10.\ 5 = 50000\ N/m^2$, *so* $F_2=P_2A = 50000.\ 1.\ 10^{-4} = 5$ *Newton*
This results show that the hydrostatic pressure and force that act at the ear drums are greater so that it feels more pain.

Figure 3b: Example of correct student answer about the concept of hydrostatic pressure

Based on the sample of students' answers presented in Figure 3 (a) and (b) can be seen that there are students that do not understand correctly about a simple explanation of hydrostatic pressure and they still do not focus on the understanding of the water density which has to be homogenous for the entire swimming pool. In this case, students are expected to understand the relation between force and pressure at any object, particularly force and pressure that resulted by the static fluid. But most students have been able to solve the problem properly according to the concept that the hydrostatic pressure increases by the increase of the fluid depth by using the assumption that whole fluid is homogenous [15]. Even for the incorrect answer, the student has known the definition or concept of the hydrostatic pressure. As shown in Table 1 the percentage of students that be able to provide a simple explanation is more than 70 %.

Figure 4 shows the example of student answer for the fourth indicator of student critical thinking ability which is the level of advanced explanation where the student is asked to analyze the concept of floating, fully immersed, and sinking according to Archimedes law.

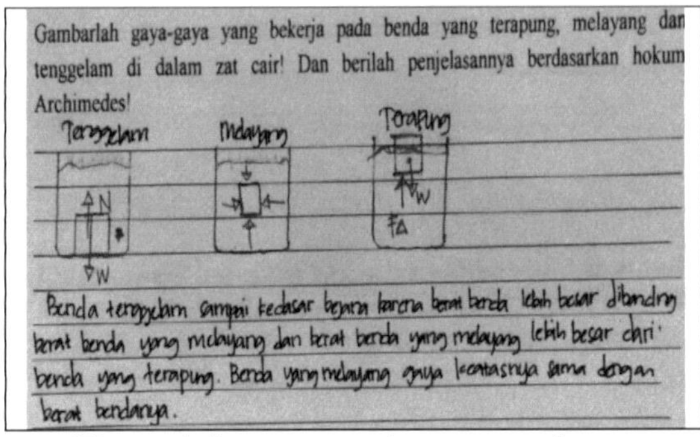
Gambarlah gaya-gaya yang bekerja pada benda yang terapung, melayang dan tenggelam di dalam zat cair! Dan berilah penjelasannya berdasarkan hokum Archimedes!

Benda tenggelam sampai kedasar bejana karena berat benda lebih besar dibanding berat benda yang melayang dan berat benda yang melayang lebih besar dari benda yang terapung. Benda yang melayang gayanya keatasnya sama dengan berat bendanya.

Draw the forces acting on a floating, fully immersed, and sinking objects in a static fluid, and explain it according to Archimedes law!
The weight of the object that sinks to the vessel's bottom is heavier than the object's weight that is fully immersed and object's weight that is fully immersed is heavier than the weight of the floating object. The drag force of an object that fully immersed equals to its weight.

Figure 4. Incorrect student answer for concepts of floating, fully immersed, and sinking objects

Based on the student' s answer sheet shown in Figure 4 can be seen that student still could not understand the concept of floating, fully immersed, and sinking objects if an object is immersed in the static fluid based on the free body diagram of each phenomenon and relation of all forces to Archimedes law. A student can not identify factors that influence why the object is floating, fully immersed, or sinking explicitly. It is seen from the answer sheet that student is still not able to identify all forces that act on the object and then draw it. In addition, they can also not be able to identify the quantities that related with Archimedes law. But Table 1 shows that most students have achieved the advanced explanation indicator (80%), this may be explained that Archimedes law and its application is difficult to part for students in static fluid teaching material.

Figures 5(a) and (b) show the same object that is weighed when it is fully immersed and when it is sinking by using the spring balance scale. The critical thinking ability indicator that will be measured using this problem is the ability of the student to draw a conclusion which is the third indicator of critical thinking ability.

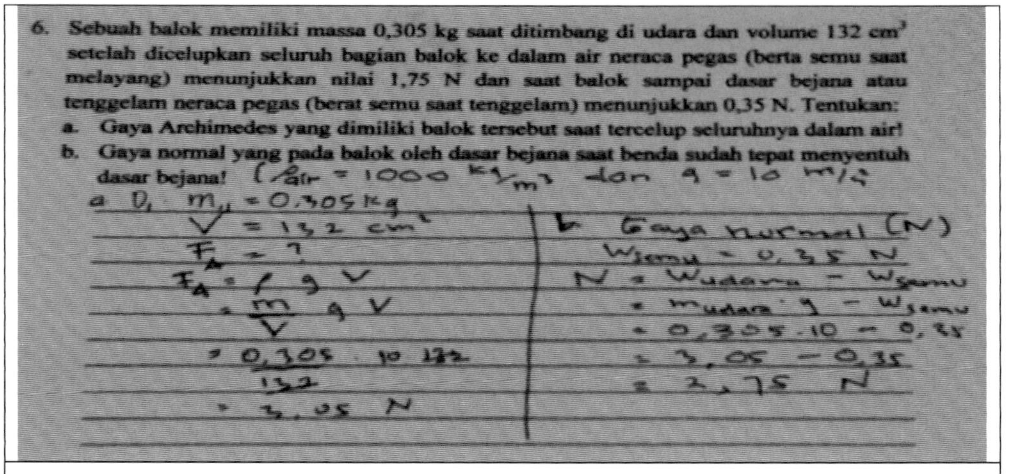
6. Sebuah balok memiliki massa 0,305 kg saat ditimbang di udara dan volume 132 cm^3 setelah dicelupkan seluruh bagian balok ke dalam air neraca pegas (berta semu saat melayang) menunjukkan nilai 1,75 N dan saat balok sampai dasar bejana atau tenggelam neraca pegas (berat semu saat tenggelam) menunjukkan 0,35 N. Tentukan:
a. Gaya Archimedes yang dimiliki balok tersebut saat tercelup seluruhnya dalam air!
b. Gaya normal yang pada balok oleh dasar bejana saat benda sudah tepat menyentuh dasar bejana!

A block that has volume of 132 cm^{3}and mass of 0.305 kg is weighed in the air using spring balance, then the block is fully immersed into the water while it is weighed using the same spring balance and its scale showing 1.75 N. Moreover, then the block is lower until it is sitting at the vessel's bottom (perfectly sink) and the spring balance scale showing 0.35 N. Find:

a. The Archimedes force that acts to the block when it is fully immersed in the water!
b. The normal force that acts to the block by the bottom of the vessel when the object is sink!

Solution

a. $F_A=(m/V)gV=3.05\ N$
b. $N= w\text{-}T= w\text{-}\ w_{semu} = 3.05\ N\text{-}0.35\ N= 2.75N$

Figure 5 (a). Examples of incorrect student's answer on the fully immersed and sinking objects

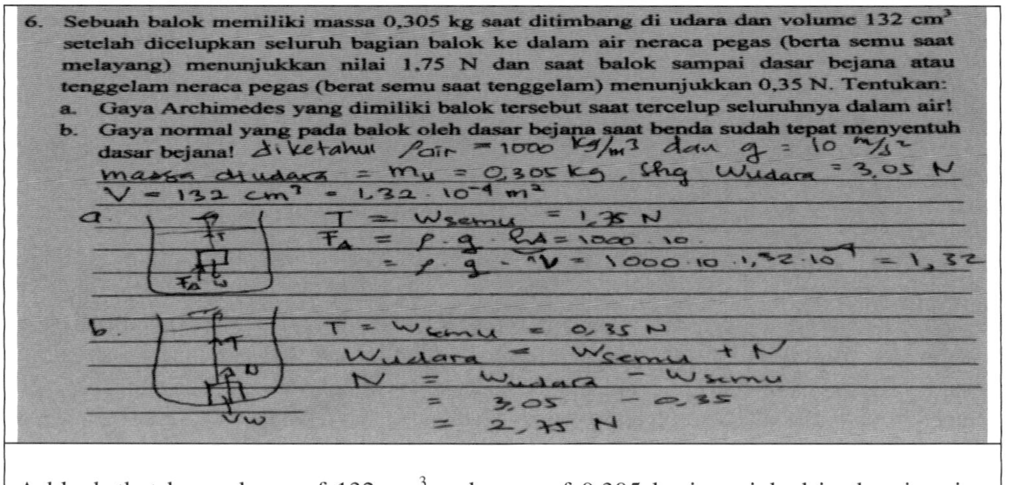
6. Sebuah balok memiliki massa 0,305 kg saat ditimbang di udara dan volume 132 cm^3 setelah dicelupkan seluruh bagian balok ke dalam air neraca pegas (berta semu saat melayang) menunjukkan nilai 1,75 N dan saat balok sampai dasar bejana atau tenggelam neraca pegas (berat semu saat tenggelam) menunjukkan 0,35 N. Tentukan:
a. Gaya Archimedes yang dimiliki balok tersebut saat tercelup seluruhnya dalam air!
b. Gaya normal yang pada balok oleh dasar bejana saat benda sudah tepat menyentuh dasar bejana!

A block that has volume of 132 cm^3 and mass of 0.305 kg is weighed in the air using spring balance, then the block is fully immersed into the water while it is weighed using the same spring balance and its scale showing 1.75 N. Moreover, then the block is lower until it is sitting at the vessel's bottom (perfectly sink) and the spring balance scale showing 0.35 N.

Find

a. The Archimedes force that acts to the block when it is fully immersed in the water!
b. The normal force that acts to the block by the bottom of the vessel when the object is

sink!
Solution
a. $T = w_{semu} = w_{appearent} = 1.75N$
the drag force= $F_A = \rho_{water} g V_{obj} = 1000 \times 10 \times 1.32\ 10^{-4} = 1.32N$
b. $T = w_{semu} = w_{appearent} = 0.35N$
$W_{udara} = w_{semu} + N,\ N = 2.75\ N$

Figure 5 (b). Examples of incorrect student's answer on the fully immersed and sinking objects

The data put into the problem shown in Figure 5 is the real data recorded from the class experiment. It is seen from the data that the apparent weight of the block that calculated from the experimental data using Archimedes law which is equal to $w_{app}=3.05N - 1.32N = 1.73\ N$ and measured data using the spring balanced scale which is equal to $T = w_{app} = 1.75N$. This shows that these two data are in agreement fairly, that means Archimedes law is applicable in that class experiment.

Figure 5 (a) shows the incorrect student answer both parts a) and b) problems. To understand the application of Archimedes law or Newton first law for the fully immersed object, students have to understand how to draw all forces that act on the object as shown in Figure 1. In addition, students have also to understand that the apparent weight can be measured using spring balance scale. From Figure 1 students have to be able to express the relationships of the hydrostatic forces, downward and upward, tension force by spring, and gravitational force by earth on the object. By applying the first Newton law [15] students are able to determine the apparent weight from experimental data. Therefore the apparent can be measured directly using spring balance or can be calculated using the first Newton law with the experiment data so that students can compare the apparent weight from direct measurement and from the calculation.From this discussion can be concluded that to understand the application of Archimedes law for immersed object based on experiment data that related with daily life is very complex that need very high critical thinking skill, not just drawing conclusion ability, therefore there are still many students that experience misunderstanding of application of Archimedes law in this problem.

For problem (b)by using an analogy where the normal force acted on the sink object by the vessel's bottom is similar to the hydrostatic force F_2 (upward direction) in the problem (a). The critical thinking skill used to understand for the sink object is more complex compare to fully immersed object since the Archimedes law is not applicable, but the Newton first law is still applicable, however, both of two student's answers, Figure 5 (a) and (b) are still incorrect. This condition indicates that the student still does not understand the application of Newton law in floating and sinking objects. The experiment data shows that the normal force is much greater than the hydrostatic force F_2.

From the example of student answer sheets show that in a static fluid, the most part that hardly understood is phenomena of floating, fully immersed and sinking objects that its behavior explained using Archimedes law except for sinking object. Furthermore, the application of Archimedes law will be easily understood if the student involved in laboratory activity since students can compare the calculated result obtained from the calculated data using mathematical equation derived theoretically from the first Newton law and measured data from an experiment using spring balance scale. Loverude states that the standard of learning hydrostatics in many sciences class can not predict and explain the behavior of sinking and floating objects on simple objects[1]. The sinking, fully immersed, and floating phenomena are learnt better by using the laboratory with supported by student worksheets and handout and therefore students' critical thinking skills will be improved, particularly for conclusion ability and strategic and tactic ability which is not good enough compare with the other four critical thinking ability indicators, shown in Table 1.As revealed by Newton, PG [16] that in solving the problem related to the critical thinking ability, the student must first know the concept underlying a problem, then visualize the problem schematically to see the relation of the variables or basic concepts, and solve the mathematical equation.

4. Conclusions

The profile of students' critical thinking ability of 11thgrade students in SMA N Sambung Macan, Sragen, Central Java, were mostly high. Students' critical thinking ability is measured through the ability of student in solving static fluid problems, particularly on hydrostatic forces and Archimedes law concepts that learned by problem-based learning model via experiment. For indicators of critical thinking ability that have been set up according to Ennis, it was shown that the critical thinking ability of student are: 27.94% for the low category, 72.22% of high category which means that the critical thinking ability score is above of the average score, where the average score is 72.94. Most of the students have achieved critical thinking ability at the level of providing a simple explanation, build basic skill, and advanced explanation, but they are not high enough for the level in drawing a conclusion and strategic and tactic.

References

[1] Loverude, Kautz, and Heron 2003 *American Journal of Physics* **71**, 1178
[2] Handhika J, Cari C, Suparmi A and Sunarno W 2017 *Journal of Physics: Conference Series* **795** (1), 012058
[3] Handhika J, Cari C and Suparmi 2017 *Journal of Physics: Conference Series* **795** (1), 012057
[4] Setyani N D, Cari C, Soeparmi A and Sunarno W 2017 *International Journal of Science and Applied Science: Conference Series* **1** (2)
[5] Handhika J, Cari C, Suparmi A and Sunarno W 2017 *AIP Conference Proceedings* **1708** (1), 070005
[6] Ennis 1996 *Critical Thinking*. New York: Prentice hall, Upper Saddle River
[7] Utami B, Saputro S, Masykuri and Aminah S N 2016 *International Conference on Educational Research and Evaluation (ICERE)*. ISSN: 2407-1501
[8] Erceg N 2013 *Revista Mexicana de F´ısica E* **59** 65–76
[9] Holma 2014 Under review. ojs.statsbiblioteket.dk/index.php/spf/article/download/18280/20109
[10] Radovanonic 2013 IOPScience
[11] Magno C 2010. *The Role of metacognitive skills in developing critical thinking. Metacognitive learning 5:137-156.* Springer Science Business Media. LLC 2010
[12] Hewitt P 2002 *Conceptual physics: the high school physics program*. Prentice Hall, Upper Saddle River, New Jersey, USA.
[13] Duron R, Limbach B and Waugh W 2006 *International Journal of Teaching and Learning in Higher Education,* **17** (2), 160-166.
[14] Hmelo C E and Ferrari M 1997 *Journal for the Education of the Gifted* **20** 4, 401–422
[15] Halliday and Resnick 1999 Fisika 1. PT. Erlangga Mahameru. Jakarta
[16] Newton P G 1999 *International Journal of Science Education,* **21**(5), 553-576

Analysis of critical thinking ability in direct current electrical problems solving

Hartono[1], Widha Sunarno[2], Sarwanto[2], Dewanta Arya Nugraha[2]

[1,2] Graduate Program in Science Education, Universitas Sebelas Maret, Jl. Ir Sutami Jebres 36A Surakarta;

[1]SMK Negeri 3 Surakarta, Jl. Brigjend Sudiarto 34 Serengan Surakarta;

email: hartonojob@gmail.com

Abstract. This study concern on analyzing the ability of students in critical thinking skills on the subject matter of direct current electricity. Samples were taken using purposive random sampling consisted of 32 students of grade XI, Multimedia 1, SMK Negeri 3 Surakarta in academic year 2016/2017. This study used descriptive quantitative method. The data were collected using tests and interviews regarding the subject matter of direct current electricity. Based on the results, students are getting some difficulties in solving problem in indicator 4. The average of students' correct answer is 62.8%.

1. Pendahuluan

Critical thinking is reasonable and reflective thinking with an emphasis on making decisions about what to believe or do. Based on Ennis (1993) critical thinking indicators can be derived from the critical activity, they are: (1) ability to formulate the problem issues; (2) ability to uncover the facts needed to solve a problem; (3) ability in selecting a logical argument, relevant, and accurate; (4) ability in detecting bias is based on a different angle or point of view; and (5) ability to determine the effect of a statement taken as a decision. Critical thinking is needed by everyone to solve problems in real life [1]. Elder, et al. says that there are six levels of critical thinking, namely unreflective thinking, challenged thinking, beginning thinking, practicing thinking, advanced thinking, accomplished thinking [2]. Yildirim, et al. defines critical thinking as the process of searching, obtaining, evaluating, analyzing, synthesizing and conceptualizing information as a guide for developing one's thinking with self-awareness, and the ability to use this information by adding creativity and taking risks [3].

The characteristics of the direct current electric material are derived from the current, resistance, and voltage. This material tends to be abstract. Direct current electric material is given to the science curriculum in the vocational high school. Learners have difficulty with the concept of electrical resistance, electric current and voltage, energy and electrical power, electrical measuring devices. Learners have a misconception about direct current electricity. The concepts developed by learners derive from interpretations that are evoked from everyday experience [4]. To understand the concepts of physics, learners must be skilled in expressing these concepts in various ways or forms (multiple representations). Representation skills are a must-have ability to interpret and apply concepts to solve problems appropriately. Some representations are representations of concepts in various ways such as verbs, graphs, diagrams, mathematics equations, and so on.

The ability of learners to succeed in learning is determined, among others, with the ability to think [5]. This is a key point in solving problems during the learning process [6]. With the thinking skills of applying learners can train and develop their cognitive intelligence as well as relate the facts or knowledge information acquired earlier to make predictions about the outcome or outcome that was later formulated by Ariska [7]. Teaching and learning activities should involve explicit thinking skills, thus facilitating the ability to categorize thinking based on the available framework [8]. In this study, we concern on investigating the critical thinking ability of learners on the concept of direct current electricity.

2. Method

The research method used in this research is descriptive method to identify the critical thinking ability of learners about direct current electricity. The sample was taken using purposive random sampling consisting of 32 students of class XI Multimedia 1 SMK Negeri 3 Surakarta Lesson Year 2016/2017. The data were collected by using test and interview. The test items consist of 20 multiple choice tests based on the indicator presented by Ennis, et al [1]. Every critical thinking indicator presented with 3 question. The test is including of 2 items about electrical resistance, 8 items about current and voltage, 8 items about energy and electrical power, and 2 items about electrical measuring instruments. After the test, student were interviewed to confirm the student answer and to get the data about their reasons for the answer.

3. Results and Discussion

The data are the percentage of students' answer and the profile of students' answer. The percentage of students answer is presented in table 1.

Tabel 1. Percentage of students' correct answer on critical thinking question

Critical Thinking Skills Indicator	Correct Answer (%)
1	72
2	42
3	64
4	62
5	74

Overall of the result, the average of students' correct answer is 62.8%. The smallest correct answer is in indicator 2 and the most correct answer is about indicator 5. In indicator number 3 the learners have difficulty understanding the problems presented in various representations. A more understanding is achieved when a learner is consistent about what he or she deems to be true [9]. A representation of the problem of various forms of representation demands a scientific perception of the passage. Factors that influence critical thinking processes in solving problem solving problems are as follows: (a) learners are not accustomed to working on multiple choice questions with additional reasons so that learners take longer; (b) learners are less ability to change the story into a mathematical model so that learners difficulties in solving problems; (c) learners tend to often solve problems only by using one rare way to check the results of their work [10].

3.1. Indicator of critical thinking ability in uncovering facts needed to solve a problem on the subject matter of electric current

The electric current is the flow of electric charge or electric charge flowing per unit time. The direction of electric current is from high potential to low potential, so opposite to the direction of electron flow. In order for the electric current to flow from A to B, the positive charge arriving at B must be moved back to A [11]. Thus the potential A is always higher than B. In order for electric current to flow in the wire, then between the two ends of the wire there must be a potential difference. This concept can be seen in Figure 1 below:

IOP Conf. Series: Journal of Physics: Conf. Series **909** (2017) 012061 doi:10.1088/1742-6596/909/1/012061

<table>
<tr><th>ID</th><th>EN</th></tr>
<tr><td>
Nomor 5.
Perhatikan gambar rangkaian (a) berikut:
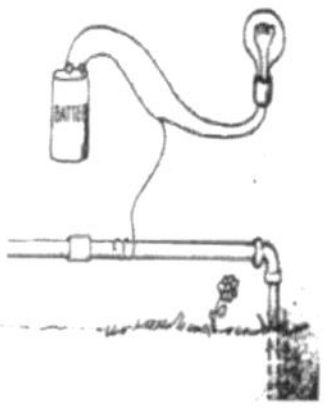
Akankah lampu menyala jika rangkaian dibumikan (grounded) seperti pada gambar ilustrasi?
a. Menyala
b. Mati
c. Redup
d. Berkedip
e. Salah semua
Alasan:
Pembumian rangkaian di titik tertentu tidak berpengaruh pada rangkaian. Elektron bergerak dari kutub – (negatif) ke kutub + (positif) baterai, sedangkan arus listrik bergerak sebaliknya. Aliran elektron ke tanah bumi dapat diabaikan, seakan-akan tidak terjadi aliran. ①
</td><td>
Number 5.
Notice the image below:
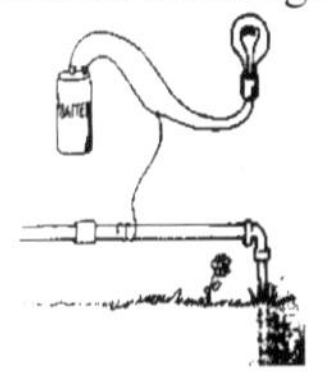
Will the lights turn on when the circuit is grounded as in the illustration?
a. On
b. Die
c. Dim
d. Blinking
e. Anything wrong
Reason:
a. Grounding a circuit in one place has no effect on the circuits. The electronics move from the – to the + side of the battery. Going into the earth will only sidetrack them, so they don't do it.
</td></tr>
</table>

Figure 1. Example of students' answer in electrical resistance material

Based on the answers as in Figure 1, learners already understand the problem by representing the problem in the form of images according to the indicator of critical thinking abilities to choose the argument logical, relevant, and accurate. Learners illustrate a lamp connected to a battery voltage source. One of the connecting cables is connected to a water line (ground). The concept of electron flow from pole - (negative) to pole + (positive). Instead the current flow from the + (positive) pole to the pole - (negative). With ordinary observations, the lamp should not be lit. In fact, the lights are still on. So between the electrons flow with the current flow direction is different. The level of critical thinking of learners at the level is able to uncover the facts needed in solving a problem. Learners are able to think abstract in electricity. So this answer is relevant to the problem.

3.2. The indicator of critical thinking ability in being able to choose logical, relevant, and accurate arguments to solve a problem on the subject matter of current and voltage.

The electric current is the amount of charge flowing through the conductor in each time unit denoted by I. While the electric potential is the sum of the effort each unit of charge is denoted by V. This quantity measures the potential energy of an electric field resulting in the flow of electricity in an electrical conductor. Depending on the difference in electrical potential, an electric voltage may be regarded as extra low, low, high or extra high [12]. By definition, the electric voltage causes a negatively charged electrically attracted object from a low-voltage place to a higher-voltage place. This can be seen in Figure 2 below:

<table>
<tr><th>ID</th><th>EN</th></tr>
<tr><td>
Tegangan Tinggi dan Burung
Misalkan seekor burung berdiri dengan kakinya menjepit bola lampu di rangkaian berikut ini:
Dalam hal ini, burung kemungkinan akan....
a. mengalami kejutan jika saklar dibuka
b. mengalami kejutan jika saklar ditutup
c. mengalami kejutan jika saklar terbuka atau tertutup
d. tidak ada kejutan sama sekali dalam kasus ini
e. tidak semua
Alasan:
a. Pada saat saklar terbuka, burung berada pada satu sisi tegangan shg tidak ada perbedaan tegangan di kakinya
b. Pada saat saklar ditutup, arus mengalir dan melewati hambatan bola lampu. Beberapa arus mengambil jalan memutar mengalir di tubuh burung. Burung mendapat kejutan / KESETRUM Pak ^^ :-)</td>
<td>High Voltage and Birds
Suppose a bird stands with its feet clamped a light bulb in the following series:

In this case, the bird will likely ....
a. experiencing a shock if the switch is opened
b. experiencing a shock if the switch is closed
c. experiencing a shock if the switch is open or closed
d. no surprises at all in this case
e. not all
Reason:
b. When the switch is open the cable on one side of the switch is showing 12 volts, and the wires on the other side are at 0 volts. The bird is on one side so there is no difference in the tension at the foot of the bird. Now close the switch. The current flows through the light bulb. Some currents take a detour and flow through the birds. The bird got a surprise.</td></tr>
</table>

Figure 2. Example of students' answer in electric current and potential difference

Based on the indicator of critical thinking ability learners have not been able to reveal the facts needed in solving a problem of concepts lowering current and voltage electricity. In the circuit, the voltage difference is always in the part of the circuit whose current does not flow. When the switch has opened the cause of the non-flow current is the switch. When closed, the current does not flow because there are obstacles. Could be more clear if it is described there are 2 birds, the first bird parallel with lights, and the second step on the cable only. Then the first bird gets hit by a shock and the second does not have any surprises at all. The electric shock experience explains the second answer because there is a voltage flowing through the lamp.

3.3. Indicator of critical thinking ability in determining the effect of a statement taken as a decision to solve the problem of energy and electrical power.

Electricity is the amount of effort in moving charge per unit of time, denoted by P and expressed in watts. Learners complete by stating that the power delivered can be doubled by doubling the current or voltage to the electrical device. It can be seen in Figure 3 below:

<table>
<tr><th>ID</th><th>EN</th></tr>
<tr><td>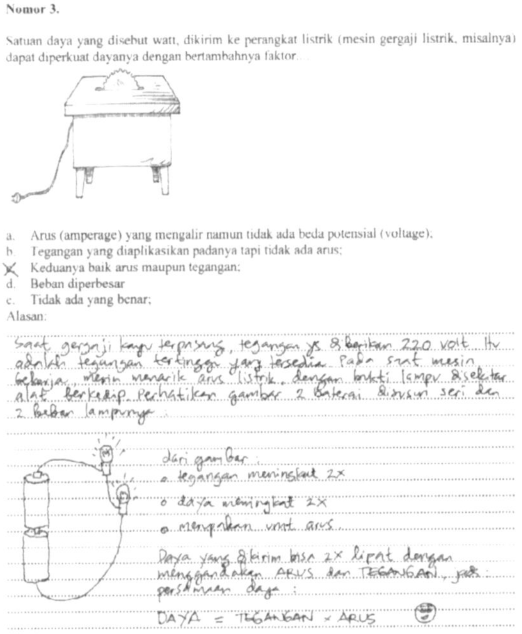
Nomor 3.
Satuan daya yang disebut watt, dikirim ke perangkat listrik (mesin gergaji listrik, misalnya) dapat diperkuat dayanya dengan bertambahnya faktor....
a. Arus (amperage) yang mengalir namun tidak ada beda potensial (voltage);
b. Tegangan yang diaplikasikan padanya tapi tidak ada arus;
X Keduanya baik arus maupun tegangan;
d. Beban diperbesar
e. Tidak ada yang benar;
Alasan:
Saat gergaji kayu terpasang, tegangan yg diberikan 220 volt. Itu adalah tegangan tertinggi yang tersedia. Pada saat mesin bekerja, mesin menarik arus listrik dengan bukti lampu disekitar alat berkedip. Perhatikan gambar 2 baterai disusun seri dan 2 beban lampunya.
dari gambar:
• tegangan meningkat 2x
• daya meningkat 2x
• menyatakan [illegible] arus.
Daya yang dikirim bisa 2x lipat dengan menggandakan ARUS dan TEGANGAN, jadi persamaan daya:
DAYA = TEGANGAN × ARUS</td><td>A unit of power called a watt, delivered to an electrical device (electric saw machine, for example) can be amplified by increasing the ....
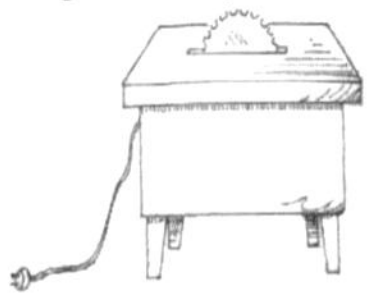
a. A flowing current (amperage) but no potential difference (voltage);
b. Voltage applied to it but no current;
c. Both are currents or voltages;
d. Loads enlarged
e. Nothing is true;
Reason:
c. When the saw is mounted, the applied voltage is 220 volts. That is the highest voltage available for that tool. How to increase output power when a large piece of wood gets into the machine? By increasing the ampere it needs. If you are weighing on the saw and slowing it will draw the surplus current that can be proven by blinking the lights elsewhere in the circuit. It's like dropping home water pressure when someone opens a large tap.
Can you think of a case where the voltage increases without changing the current? Think of a battery set up in series with a light bulb. Then think of two batteries arranged in series with two series bulbs. Voltage and power output multiply but because the load resistance also multiplies, the current in the circuit does not change.</td></tr>
</table>

Figure 3. Example of students' answer in Electrical power material

Based on the answers of students', they are able to uncover the facts needed in solving a problem on the concept of energy and electrical power. Learners are able to think how to increase the voltage without changing the current. By illustrating a battery arranged in series with a light bulb. Then developed two batteries arranged in series with two series bulbs. Voltage and power output multiply but because the load resistance also multiplies, the current in the circuit does not change. So the concept of power = watt = voltage × current, can be applied.

3.4. Indicator of the ability to think critically in the ability to detect bias based on different viewpoints on the material of electrical measuring instrument.

Amperemeter is a tool to measure electric current, either direct current or back and forth in a closed circuit. The most important part of the amperemeter is the galvanometer. Galvanometer works with the principle of the magnetic force between the magnetic field and the coil, the greater the current passing through the coil, the greater the magnetic force generated so that the analog needle deviation increases. There is a limitation of the measurement limit of an amperemeter. This can be seen in Figure 3 below:

IOP Conf. Series: Journal of Physics: Conf. Series **909** (2017) 012061 doi:10.1088/1742-6596/909/1/012061

ID

Nomor 4.

Perhatikan gambar rangkaian (a) berikut:

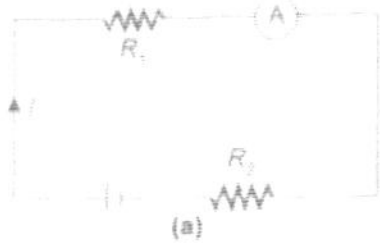

Diketahui R_{sh} = 500 Ω dan R_1 = 1.000 Ω. Dilakukan pengukuran arus dengan amperemeter yang tersedia. Ternyata hasil pengukuran melebihi nilai batas ukur amperemeter (50 mA). Untuk mengatasi masalah tersebut, dengan menambahkan sebuah hambatan yang disusun secara paralel dengan R1, yang disebut dengan hambatan shunt (R_{sh}) seperti pada gambar (b).

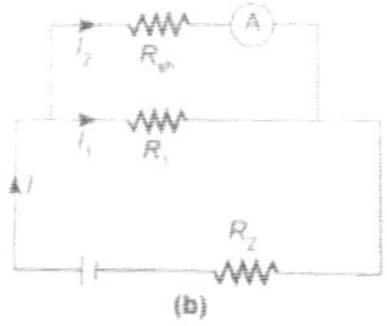

Hasil pengukuran arus pada I_2, menunjukkan skala yang ditunjuk 20 mA, skala maksimum 50 mA, serta batas ukur yang dipilih 25 mA seperti gambar berikut:

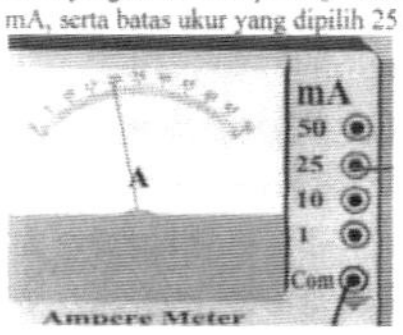

Tentukan besar arus utama (I) yang mengalir dalam rangkaian (b)!

a. 0,015 A
b. 0,150 A (marked X)
c. 1,500 A
d. 5,000 A
e. Tidak semua

Alasan:

Perhitungan Arus Terukur (I_x)

Nilai ukur = $\frac{\text{skala yang ditunjuk}}{\text{skala maksimum}}$ × batas ukur

$= \frac{20}{50} \times 25 = 10$ mA = 0,01 A

Besar arus (I) yang mengalir dalam rangkaian adalah perbandingan R_{sh} dengan R_1 dikalikan dengan I_2 dengan ditambah 1

Fungsi R shunt untuk membagi arus listrik sehingga sebagian mengalir pada R shunt dan sebagian ke R paralel dg R shunt.

sehingga:

$I = (1 + \frac{R_{sh}}{R_1}) I_x = (1 + \frac{500}{1.000}) \cdot 0,01$

$= 1,5 \times 0,01$

$= 0,015$ A

EN

Consider the following circuit images (a):

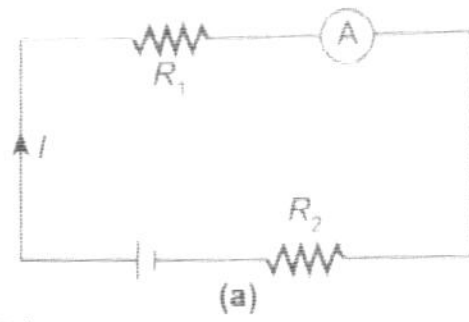

Given R_{sh} = 500 Ω and R_1 = 1,000 Ω. Measurement of current with amperemeter is available. It turns out that the measurements exceed the amperemeter limit value (50 mA). To solve the problem, by adding a constraint arranged in parallel with R_1, called the shunt resistance (R_{sh}) as in figure (b).

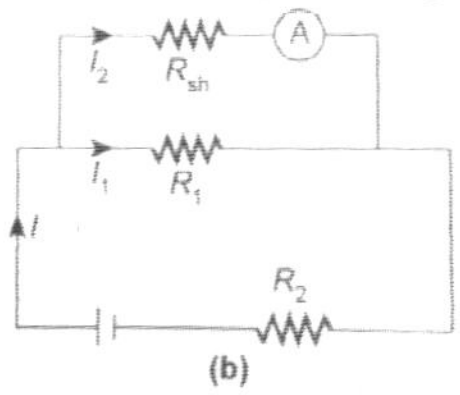

The current measurement results at I_2, showing the designated scale of 20 mA, the maximum scale of 50 mA, as well as the selected 25 mA measurement limit as shown below:

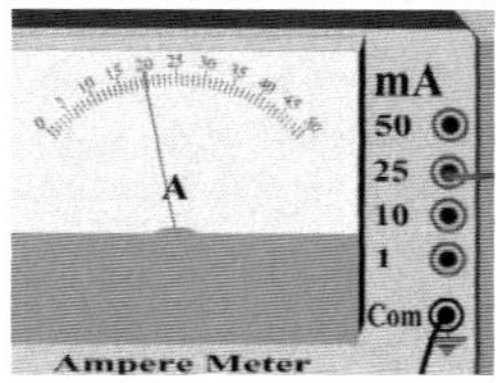

Determine the large main current (I) flowing in the circuit (b)!

a. 0.015 A
b. 0.150 A
c. 1,500 A
d. 5,000 A
e. Not all

Reason:

b. The function of the shunt resistance is to divide the electric current so that it partially flows on the shunt resistance and partly flows to the constraints that are paralleled with the shunt resistance.

And for a solution measuring an electric current that exceeds the measurement limit we can shift the measuring limit if still possible. If not possible we should install shunt resistance in parallel to the amperemeter.

Figure 4. Example of students answer in electrical measuring instrument material

Based on the answer as in figure 4, according to the indicator of critical thinking has not been able to determine a statement taken as its decision on the concept of electrical measuring instrument. The existing problem of learners has not been able to define the concept of shunt resistance. The existence of the measurement of the current value exceeds the measurement limit of an amperemeter. To overcome this problem, we can increase the value of the measurement limits on the amperemeter by adding a parallel-mounted resistance.

4. Conclusion

Based on the results of the discussion, it can be concluded that students are getting some difficulties in solving problem in indicator 4. The average of students' correct answer is 62.8%. Some factor in the school is students prefers studying physics by memorizing without understanding the concept of the materials. Moreover, in physics learning, activities are done by students to solve problems without a deep understanding.

Bibliography

[1] Ennis, R. H. 1993. Critical Thinking Assessment. Theory into Practice. Vol 32, No 3, Hal 179-186
[2] Elder, L & Paul, R. 2008. Critical Thinking development : A Stage Theory with Implications for Instruction. [Online]. http://www.critichalthinking.org
[3] Yildirim, B & Ozkahraman, S. 2011. Critical Thinking in Nursing Process and Education. International Journal of Humanities and Social Science. Vol 1, No 13, Hal 257-262
[4] Budiarti, I.S., Suparmi, Sarwanto, Harjana. 2017. Student's Conceptual Understanding Consistency of Heat and Temperature. J. Phys.: Conf. Ser. 795 012051
[5] Fatmawati, H., Mardiyana, Triyanto. 2014. Jurnal Elektronik Pembelajaran Matematika, Volume 2, Nomor 9, 2014. Jurnal 899-910
[6] Liliawati, W. and Ramalis, T.R. 2009. Semnas IPA UNY. Prociding 790 121
[7] Ariska, M. 2015. Jurnal Inovasi dan Pembelajaran Fisika, Volume 2, Nomor 2, 2015. Jurnal 147-154
[8] Fahim, M & Pezeshki, M. 2012. Manipulating Critical Thinking Skills in Test Taking. International Journal of Education.Vol 4, No 1, Hal 153-160
[9] Stupple, EJN. 2016. Development of the Critical Thinking Toolkit (CriTT): a measure of student attitudes and beliefs about critical thinking. doi:10.1016/j.tsc.2016.11.007
[10] Demirhan, FSUG. 2017. The effect of creative drama on critical thinking in preservice physical education teachers. doi:10.1016/j.tsc.2017.02.018
[11] Forawi, S.A. 2016. Standard-based science education and critical thinking. doi:10.1016/j.tsc.2016.02.005
[12] Demirhan, E., Koklukaya, A.N. 2014. The Critical Thinking Dispositions of Prospective Science Teachers. Procedia-Social and Behavioral Sciences 116 (2014) 1551 – 1555

Using computer simulation to improve high order thinking skills of physics teacher candidate students in Compton effect

Supurwoko[1], Cari[2], Sarwanto[1], Sukarmin[1], Ahmad Fauzi[1], Lisa Faradilla[1], Tiarasita Summa Dewi[1]

[1]Physics Education Department of Graduate Program UNS, 36A. Ir. Sutami street, Kentingan, Jebres, Surakarta 57126, INDONESIA
[2]Physics Department of Graduate Program UNS, 36A. Ir. Sutami street, Kentingan, Jebres, Surakarta 57126, INDONESIA

E-mail: supurwoko@staff.uns.ac.id

Abstract. The process of learning and teaching in Physics is often confronted with abstract concepts. It makes difficulty for students to understand and teachers to teach the concept. One of the materials that has an abstract concept is Compton Effect. The purpose of this research is to evaluate computer simulation model on Compton Effect material which is used to improve high thinking ability of Physics teacher candidate students. This research is a case study. The subject is students at physics educations who have attended Modern Physics lectures. Data were obtained through essay test for measuring students' high-order thinking skills and quisioners for measuring students' responses. The results obtained indicate that computer simulation model can be used to improve students' high order thinking skill and can be used to improve students' responses. With this result it is suggested that the audiences use the simulation media in learning

1. Introduction

One of the important problems in Physics learning is the low quality of learning at various levels of education. The quality of the process and the results of physics learning is determined by some factors one of the availability of laboratory facilities. Laboratory activities are important to be implemented in physics learning, because through laboratory activities aspects of products, processes, and attitudes of learners can be further developed. Physical learning based on laboratory activities can improve the mastery of concepts, generic science skills, and high-order thinking skills of learners. These skills are very important to equip learners in solving various problems faced in society.

Higher-order thinking becomes one of the qualities teachers must have to face the challenges of the 21st century [1]. Although there has been a lot of research on the thinking ability of high-level teachers (Zohar, 2004, Miri, David & Uri, 2007; Dresner, et al, 2013; Ramos, Dolipas, & Villamor, 2013) the fact that prospective teachers cannot explain clearly thinking their high level [2]. This becomes an important foundation for researchers to find out the most appropriate way to improve the current high-level thinking skills of prospective teachers, especially physics teachers, because physics can be categorized as high-level thinking that is difficult to understand and teach discipline [3].

The thinking ability of prospective teachers in understanding the material of physics can influence students' understanding of the material of physics [4]. Students will not be able to deal with issues that demand high-level thinking and problem solving if students are not trained in using their high-level thinking and thinking skills [5].

The selection of this topic is based on the special characters in Compton's own theory of effects, including the high difficulty level and the number of possible experiments so that the developed model can contribute more to the acquisition of the thinking level of physics teacher candidates. In addition, Compton's theory of effects shows that it belongs to the abstract concept. So it needs to be visualized. Abstract concept on Compton effect material cause student have difficulty to understand this concept well. This resulted in low student learning outcomes in the modern physics course in general.

In 1919, Einstein concluded that photons with energy E move in a single direction and carry momentum $E / c = h f / c$. In 1923, Arthur Holly Compton (1892-1962) and Peter Debye (1884-1966) independently proceeded with the idea of Einstein. Before 1992, Compton and his colleagues managed to gather evidence suggesting that the wave-wave theory for light had failed to explain the spread of X-rays to electrons. According to the classical theory, electromagnetic waves with frequencies fo coming to electrons will have two effects (1) the radiation pressure will cause the electrons to accelerate in the direction of propagation or wave propagation, and (2) the electric field from the oscillation of the dating radiation will cause the electron to be isolated at the frequency f ', where f' is the frequency within the framework of the moving electrons. Frequency f 'is different from the frequency f_0 of the dating radiation due to the elopment of the Doppler. Each electron, will first absorb the radiation when the particles move and then radiate back the radiation when the particles begin to move, thus there have been two Doppler shifts in the radiation frequency.

Other electrons will move at different speeds after the interaction, depending on the amount of energy absorbed from the electromagnetic waves, the frequency of the waves scattered at a certain angle from the coming radiation will show the value of the Doppler shift distribution. In contrast to this prediction, Compton's experiments show that, at a certain angle, only one radiation frequency can be observed. Compton and his coworkers realized that they could explain this experiment by treating photons not as waves but as particles with hf energy and $h f / c$ momentum, and assuming that both the energy and momentum of the isolated photoelectron students were eternal . Compton uses a particle model for something called a wave, and is called the Compton effect [6].

Alternative solutions offered include through the utilization of computer technology. Finkelstein says that computers can be used to support the implementation of physics lab [7]. Not only can it be used to collect data, present, and process data, computers can also be used to modify experiments and display complete experiments in virtual form. Garrison states that technology can be adapted into an active learning approach [8]. It needs to be a combination of face to face with online learning. Some things to consider between possible approaches, strategies, techniques, and existing tools. One form of utilization of computer technology on Compton effect material is the development of practical simulations. Simulation is defined as a training method that displays something in an artificial form similar to the actual state. Simulation can be done to identify factors that influence in a system, or prediction of system behavior by taking into account the related factors. Thus, a formula or rules that apply in a system is absolutely necessary for forecasting (prediction / forecasting) in the simulation can run. By doing forecasting, it is expected to understand how factors affect the output of a system; So that proper system control can be done, in order to obtain the desired output [9]. With computer technology the theory of compton effect is realized in the computer program by using software that is easy to learn that is experiment simulation.

In learning that uses the simulation method, students are fostered ability related to interaction skills and communicate in groups. In addition, in the simulation method students are invited to be able to play the role of some behaviors that are considered in accordance with the purpose of learning. The process of teaching and learning with the lab according to Sagala is a learning process where students are given the opportunity to experience themselves, observe the process, observe an object, analyze, prove, and draw its own conclusions about an object, state, or process something[9]. This opinion is supported by Aqib which reveals that practical activities using certain tools can train students' skills in using the tools they have been given and the results they achieve [10].

Gunawan found that interactive multimedia models in basic physics learning can improve generic science skills of prospective teacher students, where indirect observation indicators experienced the highest increase [11]. McKagan found that the use of computer simulations on quantum mechanical materials can improve student learning outcomes and activities in the classroom [12]. The use of computer simulations helps students overcome their learning difficulties in this material. Gunawan finds that the use of virtual laboratory of modern physics can also improve the mastery of student candidates' concepts, especially on the quantum theory of electromagnetic radiation and matter [13].

Based on the above description, in this study has developed a learning media in the form of practical simulation as an alternative solution to limited laboratory facilities and experimental difficulties in the theory of Compton effect abstract. This virtual laboratory model is expected to improve generic science skills of prospective teachers. In this article we will describe the description of the Compton practice simulation of the Compton effect that has been developed and the effect of the use of Compton effect experiment simulation on the improvement of high thinking ability of physics teacher candidate students.

2. Research Methods

This research is a development research with Recursive Reflective Design and Development (R2D2) design developed by Willis [14]. This design is used because it is reflective, recursive, collaborative and evolving so as to give researchers and stakeholders the opportunity to develop the learning product according to the needs continuously until found the product that is considered the most appropriate, effective and efficient.

This study was conducted on 35 physics education students of UNS 2016/2017 academic year selected randomly. Students are given 15 points about high ability thinking test that is in the form of description problem. Problem compiled based on the indicator of high-order thinking ability is the ability to analyze (C4), the ability to synthesize (C5), the ability to create (C6). A limited trial was conducted by applying learning media in the form of a experiment simulation to determine the level of students' ability in high-level thinking. The method used is pre test post test group by looking at comparison of mean value of each indicator in high level thinking. To see the effectiveness of learning outcomes then performed a normalized Gain calculation with the equation:

$$g = \frac{S_f - S_i}{100 - S_i}$$

3. Results And Discussion

The results of the analysis of high-thinking ability of prospective teachers can be increased by utilizing the practical simulation media. This is shown in the comparison of pretest and posttes' average grades of 35 prospective teachers is 40.08 and 68.4. The average analysis of the value of 35 prospective teachers using the equation:

$$\bar{x} = \frac{\sum n}{N}$$

Information : $\bar{x}$ = average value
$\sum n$ = total skor
N = number of prospective teacher

The following is the average value of the teacher's high-level thinking ability for each indicator of high-level thinking ability.

Table 1. The result of calculating the average value of prospective teachers on each indicator of high-level thinking ability.

No.	HOTS Indicators	Question Number	Average	
			Pretest	Posttest
1	The ability to analyze (C4)	2,3,7,9,14 dan 15	37.5	60
2	The ability to synthesize (C5)	1,4,6,8,12, dan 13	42.75	70
3	The ability to create (C6)	5, 10, dan 11	40	75.25

The average value of pretest on the matter of C4, C5 and C6 is almost the same. The value indicates that the high order thinking skills before the treatment. The percentage of pre test score for the C6 question is the highest among other question. This is because the problem of C6 tends to be easier but it can also be done even though it has not received the previous material.

The experiment simulation used using adobe flash professional cs6. Adobe Flash is a very popular program for creating vector-based 2D animations. The latest application program from Adobe Flash is Adobe Flash Professional CS6, this application is the previous version of Adobe Flash Professional CS 5.5. The description of the device developed with a experiment simulation model is as shown in Figure 1.

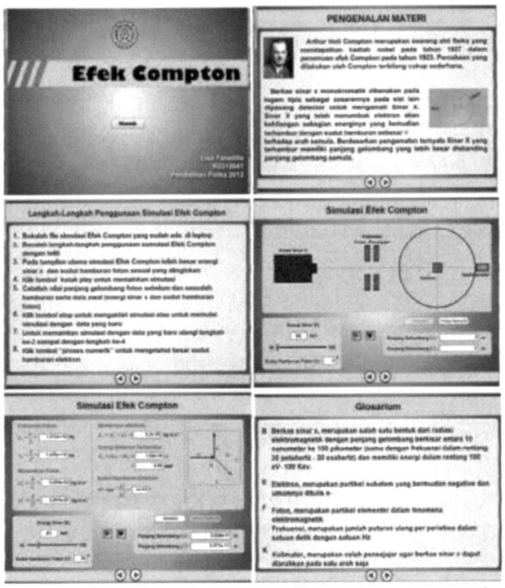

Figure 1. The design of instructional media using a experiment simulation

The learning media consist of 6 pages. The first page shows the title of a experiment simulation Compton effect. Then on the second page there is the introduction of the material, on the third page in the form of the steps of using the simulation media, then the fourth page displays the compact practice experiment simulation media which shows virtual how to experiment the Compton effect by determining the angle the scattering of photons and the enormous energies of x rays fired at the electrons thereby scattering and generating larger scattered lamda values compared with the lamda before scattering. In addition, on the next page, the fifth page is a mathematical process that shows the scattering angle values of the electrons. So student can better understand how to get the scattered electrons angle value. The last page shows the glossary used in Compton effect simulation media.

HOTS prospective teachers can be measured through the three highest cognitive abilities in taxonomy bloom, among others, 1) analyze (C4), 2) evaluate (C5), 3) create (C6) [15]. Assessing high-level skills can improve the grades or achievements of prospective teachers [16]. The following is an explanation of descriptive analysis of candidate answers to the teacher's test based on indicators of high-ability thinking.

IOP Conf. Series: Journal of Physics: Conf. Series **909** (2017) 012062 doi:10.1088/1742-6596/909/1/012062

Indicator 1: The ability to analyze (C4)

Analyzing means breaking down information into sections, determining how parts are related to each other. There are 6 items on the first indicator of question number 2,3,7,9,14 and 15. Indicator about number 3 is analyzing the particle nature of electromagnetic radiation. Here are some examples of answers that show different analytical skills of 2 students.

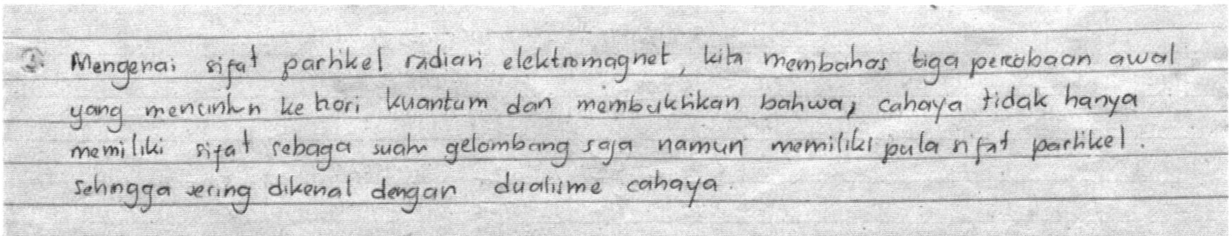
3. Mengenai sifat partikel radiasi elektromagnet, kita membahas tiga percobaan awal yang menuntun ke teori kuantum dan membuktikan bahwa, cahaya tidak hanya memiliki sifat sebagai suatu gelombang saja namun memiliki pula sifat partikel. sehingga sering dikenal dengan dualisme cahaya.

3. Concerning the nature of electromagnetic radiation particles, we discuss early experiments that guide quantum theory and prove that light not only have properties as a wave but also has the nature of particles so known as the dualism of light.

Figure 2. The Student A's Answer Number 3

In the first answer, student A compares three initial experiments that lead to quantum theory and explain the results obtained through the experiment. And based on the interview results show that the answers and interview results accordingly.

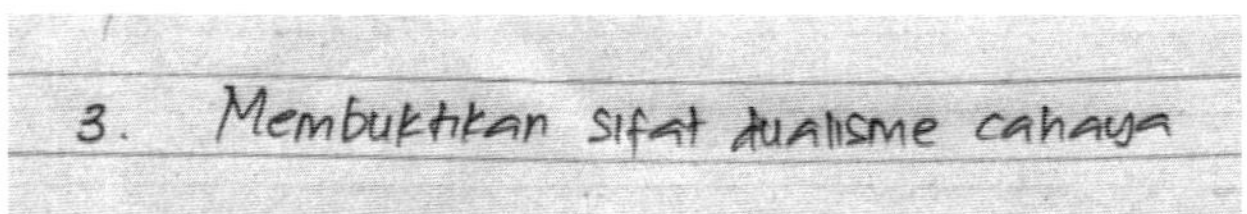
3. Membuktikan sifat dualisme cahaya

3. Prove the nature of light dualism

Figure 3. The Student B's Answer Number 3

Student B also gives the correct answer, just as with student A but student B does not provide the desired explanation. Similarly, the results of the interviews obtained, student B not so understand about further discussion of experiments that prove the nature of light dualism.

Indicator 2: The ability to synthesize (C5)

Synthesis means to justify a statement or plihan with a certain reason..There are 6 items on the first indicator of question number 1,4,6,8,12, and 13. Indicator about number 1 is selecting the most appropriate statement about the Compton effect experiment. Here are some examples of answers that show different analytical skills of 2 students.

IOP Conf. Series: Journal of Physics: Conf. Series **909** (2017) 012062 doi:10.1088/1742-6596/909/1/012062

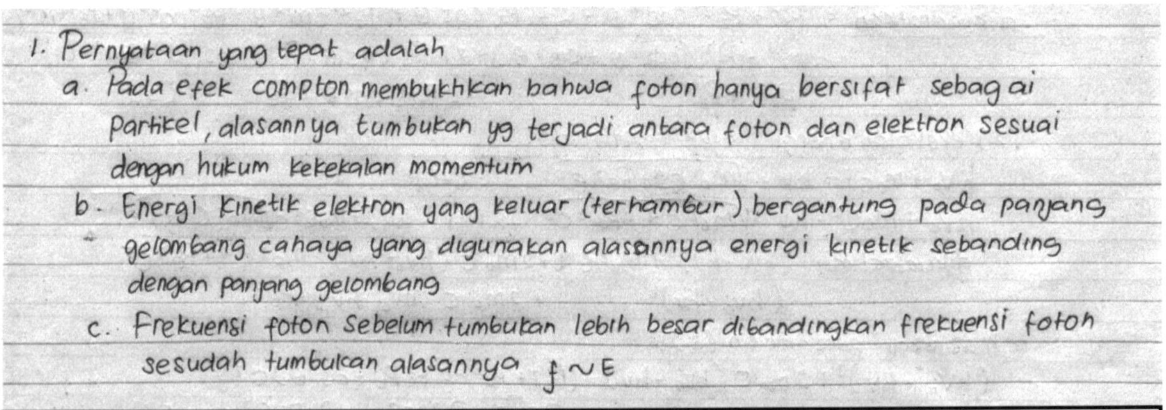

1. Pernyataan yang tepat adalah
a. Pada efek compton membuktikan bahwa foton hanya bersifat sebagai partikel, alasannya tumbukan yg terjadi antara foton dan elektron sesuai dengan hukum kekekalan momentum
b. Energi kinetik elektron yang keluar (terhambur) bergantung pada panjang gelombang cahaya yang digunakan alasannya energi kinetik sebanding dengan panjang gelombang
c. Frekuensi foton sebelum tumbukan lebih besar dibandingkan frekuensi foton sesudah tumbukan alasannya $f \sim E$

1. The exact statement is :
   a. On the Compton effect proven that photons and electrons correspond to the law of conservation of momentum
   b. The kinetic energy of the outgoing (scattering) electrons depend on the wavelength of the light for which reason the kinetic energy is proportional to the wavelength
   c. The frequency of the photons before the collision is greater than the frequency of the photons after the collision of the reason is frequency is proportional to Energy

Figure 4. The Student A's Answer Number 1

In the answer above, student A's answer indicates that the selected statement along with the reason is not yet appropriate. Based on the results of the interview it is because students answer based on the ongoing material and forget the previous material related to Compton effect material.

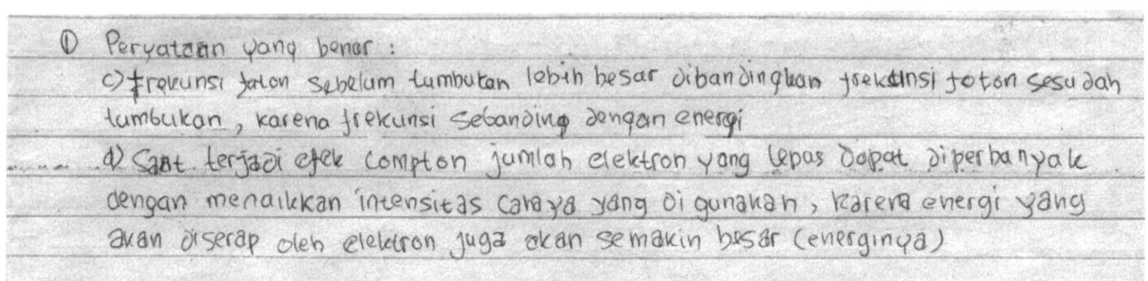

① Peryataan yang benar :
c) frekunsi foton sebelum tumbukan lebih besar dibandingkan frekuensi foton sesudah tumbukan, karena frekunsi sebanding dengan energi
d) Saat terjadi efek compton jumlah elektron yang lepas dapat diperbanyak dengan menaikkan intensitas cahaya yang di gunakan, karena energi yang akan diserap oleh elektron juga akan semakin besar (energinya)

1. Correct statement :
   c. The frequency of the photons before the collision is greater than the frequency of the photons after the collision of the reason is frequency is proportional to Energy

Figure 5. The Student B's Answer Number 1

Whereas in answer student B shows that the statement and the reasons used are appropriate. This is because students are already familiar with Compton effect material.

Indicator 3: The ability to create (C6)

Creating means to create or develop new products, theories, or points of view based on learning. There are 3 items on the first indicator of question number 5, 10, and 11. Indicator about number 5 is formulate the Compton effect equation. Here are some examples of answers that show different analytical skills of 2 students.

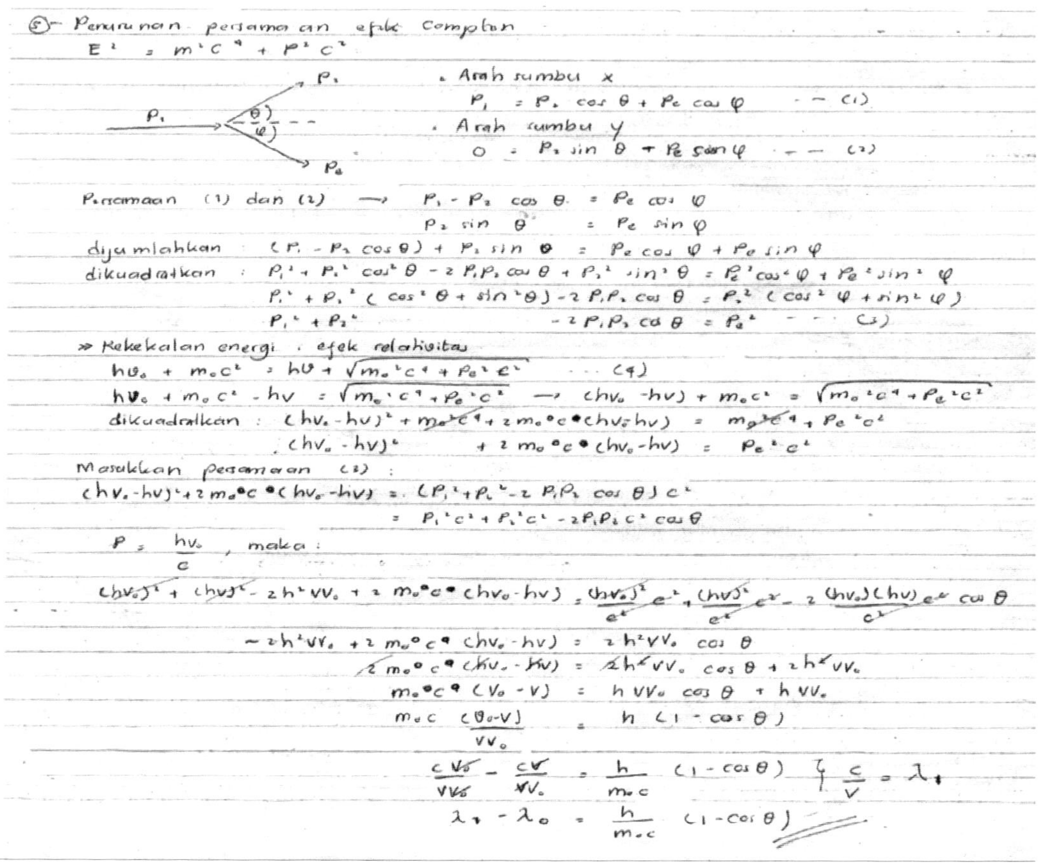

5. Penurunan persamaan efek compton

$E^2 = m^2c^4 + P^2c^2$

• Arah sumbu x

$P_1 = P_2 \cos\theta + P_e \cos\varphi$ --- (1)

• Arah sumbu y

$0 = P_2 \sin\theta - P_e \sin\varphi$ --- (2)

Persamaan (1) dan (2) $\rightarrow$ $P_1 - P_2 \cos\theta = P_e \cos\varphi$

$P_2 \sin\theta = P_e \sin\varphi$

dijumlahkan $(P_1 - P_2\cos\theta) + P_2 \sin\theta = P_e\cos\varphi + P_e \sin\varphi$

dikuadratkan : $P_1^2 + P_2^2\cos^2\theta - 2P_1P_2\cos\theta + P_2^2\sin^2\theta = P_e^2\cos^2\varphi + P_e^2\sin^2\varphi$

$P_1^2 + P_2^2(\cos^2\theta + \sin^2\theta) - 2P_1P_2\cos\theta = P_e^2(\cos^2\varphi + \sin^2\varphi)$

$P_1^2 + P_2^2 - 2P_1P_2\cos\theta = P_e^2$ --- (3)

» Kekekalan energi : efek relativitas

$h\nu_0 + m_0c^2 = h\nu + \sqrt{m_0^2c^4 + P_e^2c^2}$ --- (4)

$h\nu_0 + m_0c^2 - h\nu = \sqrt{m_0^2c^4 + P_e^2c^2}$ $\rightarrow$ $(h\nu_0 - h\nu) + m_0c^2 = \sqrt{m_0^2c^4 + P_e^2c^2}$

dikuadratkan : $(h\nu_0 - h\nu)^2 + m_0^2c^4 + 2m_0c^2(h\nu_0 - h\nu) = m_0^2c^4 + P_e^2c^2$

$(h\nu_0 - h\nu)^2 + 2m_0c^2(h\nu_0 - h\nu) = P_e^2c^2$

Masukkan persamaan (3) :

$(h\nu_0 - h\nu)^2 + 2m_0c^2(h\nu_0 - h\nu) = (P_1^2 + P_2^2 - 2P_1P_2\cos\theta)c^2$

$= P_1^2c^2 + P_2^2c^2 - 2P_1P_2c^2\cos\theta$

$P = \frac{h\nu_0}{c}$, maka :

$(h\nu_0)^2 + (h\nu)^2 - 2h^2\nu\nu_0 + 2m_0c^2(h\nu_0 - h\nu) = \frac{(h\nu_0)^2}{c^2}c^2 + \frac{(h\nu)^2}{c^2}c^2 - 2\frac{(h\nu_0)(h\nu)}{c^2}c^2\cos\theta$

$-2h^2\nu\nu_0 + 2m_0c^2(h\nu_0 - h\nu) = 2h^2\nu\nu_0\cos\theta$

$2m_0c^2(h\nu_0 - h\nu) = 2h^2\nu\nu_0\cos\theta + 2h^2\nu\nu_0$

$m_0c^2(\nu_0 - \nu) = h\nu\nu_0\cos\theta + h\nu\nu_0$

$m_0c\frac{(\nu_0 - \nu)}{\nu\nu_0} = h(1 - \cos\theta)$

$\frac{c\nu_0}{\nu\nu_0} - \frac{c\nu}{\nu\nu_0} = \frac{h}{m_0c}(1 - \cos\theta)$; $\frac{c}{\nu} = \lambda_1$

$\lambda_1 - \lambda_0 = \frac{h}{m_0c}(1 - \cos\theta)$

Figure 6. The Student A's Answer Number 5

A student's answer indicates that his ability to formulate the Compton effect equation is correct. Students explains the coherence of the conservation equation of energy and momentum used, accompanied by an explanation of the treatment of each equation obtained. From the result of the interview show that student A is well-informed about the relationship between conservation of momentum and energy conservation law on the invention of Compton effect formulation.

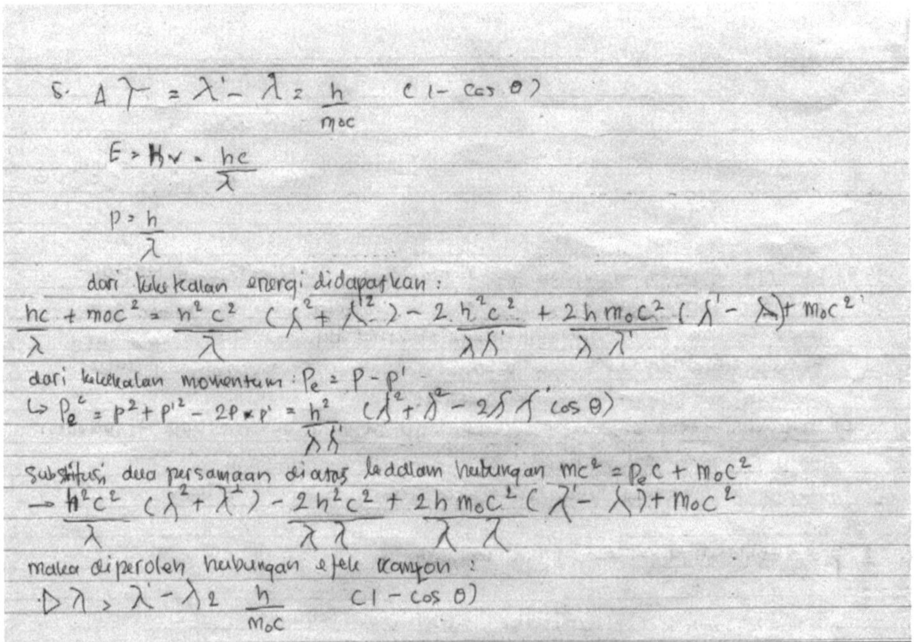

5. $$\Delta\lambda = \lambda' - \lambda = \frac{h}{m_0 c}(1 - \cos\theta)$$

$$E = hv = \frac{hc}{\lambda}$$

$$p = \frac{h}{\lambda}$$

From the conservation of energy we get :

$$\frac{hc}{\lambda} + m_0c^2 = \frac{h^2c^2}{\lambda}(\lambda^2 + \lambda'^2) - \frac{2h^2c^2}{\lambda\lambda'} + \frac{2hm_0c^2}{\lambda\lambda'}(\lambda' - \lambda) + m_0c^2$$

From the conservation of momentum : $p_e = p - p'$

$$p_e{}^2 = p^2 + p'^2 - 2pp' = \frac{h^2}{\lambda\lambda'}(\lambda^2 + \lambda'^2 - 2\lambda\lambda' \cos\theta)$$

Substitution both of the equation above in equation : $mc^2 = p_e c + m_0c^2$

$$\longrightarrow \frac{h^2c^2}{\lambda}(\lambda^2 + \lambda'^2) - \frac{2h^2c^2}{\lambda\lambda'} + \frac{2hm_0c^2}{\lambda\lambda'}(\lambda' - \lambda) + m_0c^2$$

So we get the equation of Compton effect :

$$\Delta\lambda = \lambda' - \lambda = \frac{h}{m_0 c}(1 - \cos\theta)$$

Figure 7. The Student B's Answer Number 5

While student B's answer indicates that the written equation is not wrong, but student B only writes the equation of conservation of momentum and the law of conservation of energy and then comes the equation of Compton effect without any way used to get the formulation. And based on interview result, student B only know that there is equation of Compton effect with both conservation law of energy without knowing process.

The considerable difference is far enough indicates that there are differences in the high order thinking skills of students before and after Compton effect learning by using a practical simulation media.

By utilizing the simulated media students become accustomed to thinking systematically so that in doing the student's questions also become easier and get better results. The PBL model also trains students to be more skilled in solving and finding solutions to problems. To solve a problem requires the high order thinking skill because students must analyze various information to find the right solution.

The result shows that using computer simulation proved superior to improve high order thinking skills of physics teacher candidate students in Compton effect. Students have dared to make decisions that are deemed appropriate based on the evidence they have obtained. This evidence is based on the results of the discussions that they follow in a sustainable manner using the Compton effect simulations. The lowest average score of students is found on the create indicators. In this activity the students seek solutions about new problems encountered by using the concepts already obtained and the completeness of supporting evidence. In this indicator students solve the problem through quiz activity. In order for students to interpret a problem it requires accuracy, clarity, consistency, and jelly in determining the variables. The average comparison of the students' high-order thinking skill based on table 1 obtained the total pretest score is 40 and after implementing the learning process using the experiment simulation the posttest score is 68.3. To see the effectiveness of the product obtained the calculation of the normalized gain value of 0.5 indicates a moderate value. The increase in the average student is obtained because of the intensity of students interact either directly in the classroom through learning media in the form of a experiment simulation. In addition, by using the practical simulation problems presented can also be visualized by students.

Curiosity of students become higher because of the difficulty of the material being studied. High curiosity and high motivation, encourage students to use all their thinking skills to get the perfect solution. In other words, using this experiment simulation can train students to improve high order thinking skills.

4. Conclusions

Based on the results and discussion learning media tools developed through experiment simulations can improve students' higher-order thinking skills. It is proven through validation test result is 3,29 (valid enough). The result of the average score of 35 students in high-order thinking skills after the process using learning media of experiment simulation was 68.5 with the normalized gain value was 0.5. The value is included in the category is. The average value each indicators for pretest and posttest are 37.5 and 60 for ability of analyze (C4), 42.75 and 70 for ability of synthesize (C5), 40 and 75.25 for ability of create. Seen that the considerable difference is far enough indicates that there are differences in the high order thinking skills of students before and after Compton effect learning by using a practical simulation media. By using the practical simulation problems presented can also be visualized by students and can train students to improve high order thinking skills.

References

[1] Collins, Robyn. (2014). Skill For The 21st Century: teaching higher-order thinking. An Electronic Journal for leaders in Educations. 12(4)

[2] Harrison, Neil. (2013) Using The Interactive White Boardto Scaffold a Metalanguage: teaching higher-order thinking skills In Preservice Teacher education. Australasian Journal of Educational Technology, 29(1)

[3] Adeyemo, S.A. (2010). Background and Classroon Correlates of Student's achievement in Physics. International Journal of Educational Research and Technology, 1(1)

[4] Aydin, Suleyman., Keles Pinar Ural., & Hasiloglu, M. Akif. (2005). Establishment for Misconception that Scince Teacher Candidate have About Geometric Optics. The Online Journal of New Horrizons in Education, 2(3)

[5] Viridi, A.S. (2009). Analisis Penyebab Rendahnya Nilai Fisika Siswa di SMA Negeri 6 Bandung. Jurnal Pengajaran Fisika Sekolah Menengah, 1(3): 51-55

[6] Serway. (2010). Fisika untuk Sains dan Teknik Buku 3 Edisi 6. Salemba Teknika

[7] Finkelstein, et.al. (2005). When Learning About the Real World Is Better Done Virtually: A Study of Subtituting Computer Simulations for Laboratory Equipment. Physics Education Research. APS (1) 1 – 8.
[8] Garrison, D.R & Vaughan, N.D., (2008). Blended Learning in Higher Education. San Fransisco : John Willey & Sons, Inc.Serway. J, Fisika untuk Sains dan Teknik, Salemba Teknika, 2009. [3]Aditya, Y.N. 2011. Definisi Simulasi Visualisasi dan Animasi. Jakarta : Ditjen – Dikti
[9] Sagala, Syaiful. (2009). Konsep dan Makna Pembelajaran. Bandung: Alfabeta
[10] Aqib, Z. (2013). Model-Model, Media, Strategi Pembelajaran Kontektual (Inovatif). Bandung: Yrama Widya.
[11] Gunawan. (2009). Pembelajaran Berbasis Multimedia Interaktif untuk Meningkatkan Keterampilan Generik Sains Calon Guru Fisika. Jurnal PIJAR MIPA. Vol 4. No. 2, 46 – 49
[12] McKagan et al. (2008). Developing and Researching PhET simulations for Teaching Quantum Mechanics. American Journal of Physics (76) 406 – 417
[13] Willis.J. (1995). A Recursive, Reflective Instructional Desaign Model Based on Contructivist – Interpretative Theory, Educational Technology

IOP Conf. Series: Journal of Physics: Conf. Series **909** (2017) 012063 doi:10.1088/1742-6596/909/1/012063

Investigating middle school students' difficulties in mathematical literacy problems level 1 and 2

S Setiawati[1], T Herman[2], and A Jupri[2]

[1] Mathematics Education, School of Postgraduate Studies, Universitas Pendidikan Indonesia, Jl. Dr. Setiabudhi No.229 Bandung 40154, INDONESIA
[2] Department of Mathematics Education, Faculty of Mathematics and Science Education, Universitas Pendidikan Indonesia, Jl. Dr. Setiabudhi No.229 Bandung 40154, INDONESIA

E-mail: santysetiawati@student.upi.edu

Abstract. The background of this study is the lack of mathematical literacy skills of students. The proficiency of students' mathematical literacy skills based on the results of the PISA 2015 study shows that Indonesian students at the proficiency level 1. This fact gave rise to this study which aims to investigate middle school students' difficulties in mathematical literacy problems level 1 and 2. Qualitative research was used in this study. An individual written test on mathematical literacy problems was administered, followed by interviews. The subjects of the study were 61 students grade VII in Bandung and 26 of them were interviewed afterward. Data analysis revealed that students' error in performing arithmetic most frequently observed. Other observed difficulties concerned understanding about algebra concept, applying arithmetic operation in algebraic expressions, and interpreting symbols to represent the unknown. In solving mathematical literacy problems, students use their prior knowledge, although sometimes not relevant to the questions. Based on the results, we suggest that mathematics learning in contextual learning and which invites students to participate in the processes of understanding the concepts.

1. Introduction

Mathematics is primarily to assist the human being in understanding and mastering social, economic, a natural problem. In everyday life is often found something which related and needs mathematics to solve daily life problems. The mathematical ability required to solve daily life problems is mathematical literacy. In line with OECD that mathematical literacy helps people to recognize the role of mathematics in the world and to make judgments and decisions needed as citizens [1]. Knowledge and understanding of mathematical concepts are very important, but more important is the ability to solve problems encountered in everyday life [2]. Thus, it is important for students to have the mathematical literacy [3].

Mathematical literacy is defined as the ability of individual's capacity to formulate, apply and interpret mathematics in various contexts involving mathematical reasoning and the use of concepts, procedures, facts to describe, explain and predict phenomena and relate them to daily life [4], [5], [6]. Proficiency in mathematical literacy according to PISA study is divided into 6 levels [1]. This study discusses proficiency of mathematical literacy level 1 and 2. Description of the two levels of mathematical proficiency based on PISA 2015 (OECD, 2016), i.e., students can typically do: (1) at level 1, students can answer questions involving familiar contexts, and able to identify information and to carry out routine procedures according to direct instructions in explicit situations, (2) at level 2, students can interpret, recognise situations in contexts, can employ basic algorithms, formulae, procedures, and capable of making literal interpretations of the results [1]. However, students' mathematical literacy ability is low [7], [8], [9], based on the results of the PISA 2015 study shows that Indonesian students are at the proficiency level 1 [1].

Students experience errors in solving mathematical literacy problems. The interview was done with one of the mathematics teachers in middle school Bandung. Based on the interview, teacher revealed that students are still difficult in representing a problem situation using a variable in algebra. In accordance with Jupri, Drijvers, and Heuvel-Panhuizen [10] in their research that "the most frequently observed difficulty is the ability to translate between problem situations and mathematical situations".

It is an important issue to find out the reason that causing students making errors in solving mathematical literacy problem. This requires attention and need to know the condition in depth what kind of errors that experienced by students in solving mathematical literacy problems. Students who have been able to apply their knowledge to a problem, may not be applied it in different problems. As Ojose points out that student needs to experience problem-solving processes in different situations and contexts in order to use their skills effectively [11]. Therefore, in the current study, we investigate middle school students' difficulties in mathematical literacy level 1 and 2. The research question of this study is as follows: what are the students' difficulties in solving mathematical literacy problems level 1 and 2?

2. Methods

To address the research question, we conducted a qualitative study through an individual written test and interviews. The subjects of the study were 61 students grade VII in Bandung, and 26 of them were interviewed afterward. The instrument consisted of four questions and interview guidelines. The instrument was designed based on the proficiency of mathematical literacy level 1 and 2. Written test is required to collect data or information about the students' work and the whole steps of students' answer in solving the mathematical literacy problems. Furthermore, interviews were conducted with students to strengthen the analysis of mathematical literacy ability.

3. Results and Discussions

3.1. Analysis of students' responses

Students were asked to solve four questions of mathematical literacy problems. After students finish solving the problem, they also interviewed to clarify their answers. Based on the analysis of literacy mathematics test and interview results, the following describes in detail the analysis of students' response.

3.1.1. Analyzing of the first questions. The first question is "Sherly from America is preparing to go to Indonesia for one month to exchange student. She needs to exchange some US dollars (USD) to Indonesian rupiahs (IDR). Sherly knows that the exchange rate between US dollar and Indonesian rupiah is as follows: 1 USD = Rp 13.000 (IDR). Sherly exchange $250 USD into Indonesian rupiah with this exchange rate. How much rupiah (IDR) did Sherly earn?"

The first question is number concept, and the indicator of mathematical literacy is employing mathematical concepts, facts, procedures, and reasoning [1]. The first question belongs to the proficiency level 1 of mathematical literacy due to students can answer the question involving familiar contexts where all the relevant information is present, and the questions are defined [1]. Students' answer will be analyzed to get the information of students' difficulties in solving mathematical literacy problems.

As an illustration, students' answer presented in Figure 1 below:

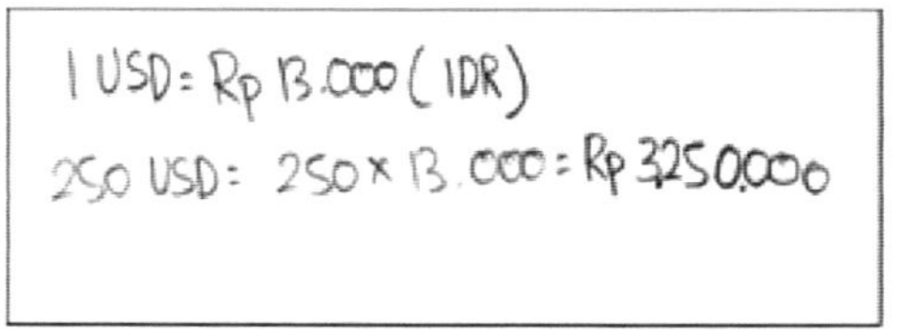

Figure 1. The answer of AK for the first question.

Students' work in Figure 1 shows that students understand the problems and realize to solve this currency exchange with the multiplication of integers. There are students' errors in performing arithmetic operations. Students' answer presented in Figure 2 below:

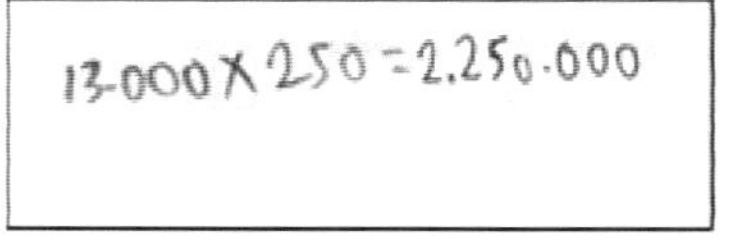

Figure 2. The answer of NT for the first question.

Based on Figure 2, apparently, the student does not add the number 2 with the number 1 at the end of the multiplication, so that it gets 2.250.000, where the results should be 3.250.000. We tried to illustrate the process of multiplication done by students, i.e., in Figure 3.

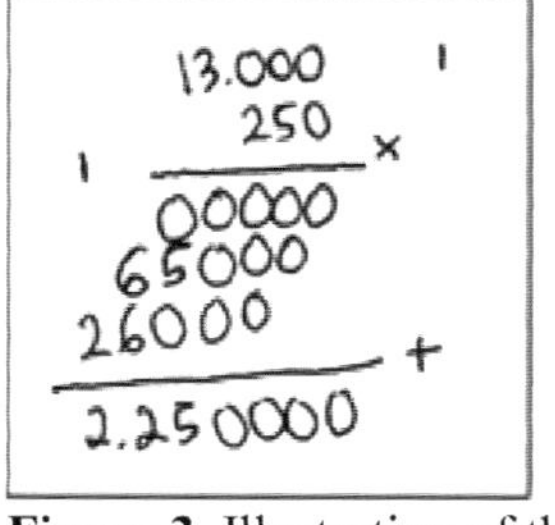

Figure 3. Illustration of the process of multiplication done by students.

Students' error as in Figure 3 caused by inaccuracy of students in thinking to solve the problem.

3.1.2. Analyzing of the second questions. The second question is "Mrs. Dewi has 1 box containing some instant noodles. After instant noodles were cooked, many instant noodles inside the box are 30%. Write a mathematical model with a diagram or symbol stating the number of instant noodles that were cooked!"

The second question is algebra concept, and the indicator of mathematical literacy is formulating situation mathematically [1]. It was designed to represent a situation mathematically, using variables, symbols, and diagrams [1]. Students' work in the second question shows that represent the problem visually, that is a diagram as in Figure 4.

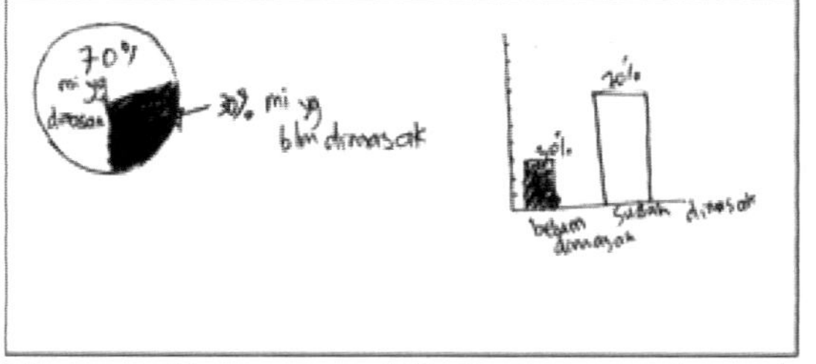

Figure 4. The answer of SL for the second question

To clarify students' answer, here is the result of a conversation with the students in the interview process.

R: How do you solve this problem? Can you explain your answer?

SL: Yes, if one box we judge in percentage is 100%, then noodle which not cook is 30%, so noodle which cook is 70%.
R: What makes you think to make a pie or a bar chart?
SL: I am afraid this pie chart is not correct, so I draw a bar.
R: Actually about this diagram is not yet learned in VII grade, what makes you think that you know that the pie chart and the bar chart look like this?
SL: I see it before and in primary school, bar chart ever learned.

Based on the interview, students revealed that they had studied the diagram before at primary. Thus, students use their prior knowledge to solve the problem [12]. None of the students answer by representing a mathematical symbol. There is one thing that became an intriguing question about representing a situation using appropriate symbols, why none of the students show the mathematical symbol on their work? Whereas, algebra is more focused on the relationship between the number and use of symbolic language or called a variable [13]. The absence of the mathematical symbol in the second question corresponds to a previous study which revealed that obstacles and difficulties in algebra are representing the situation symbolically or representing an expression using symbols [14], [15]. Researcher identifies that students' knowledge of the algebra concepts which newly learned in the first semester is not ready to use and need to be improved by doing lots of exercises.

For the second question, there are students that not answer the question. Here is the result of the conversation with students in the interview process.
R: Why this question was not answered?
FZ: Because I don't know, actually question like this is easy, but determine the process is difficult.
R: Which part makes it hard to answer this question?
FZ: When asked to represent a diagram or symbol.
R: Is that hard to represent it?
FZ: But usually already given how many.

According to students who do not answer the second question, it happens because the students are not familiar with the problem that given unknown number. Students have difficulty interpreting and operating symbols or diagrams to represent the unknown.

3.1.3. Analyzing of the third questions. The third question is "Riko has time from 15.00 WIB up to 18.00 WIB to do some activities, i.e., having an extra lesson, watching television and doing homework. Riko uses $\frac{2}{3}$ of his time to go to place for having extra lesson, $\frac{1}{6}$ of his time to watch television and $\frac{2}{5}$ of his time to do homework. The most time spent by Riko is watching television. Is it true? Explain."

The third question is number concept, and the indicator of mathematical literacy is interpreting, applying, and evaluating mathematical outcomes [1]. This question was included in the proficiency level 2 of mathematical literacy because students can recognize situations, can employ procedures and capable of making interpretations of the results [1]. Here is the students' answer to the third question as in Figure 5, 6, and 7.

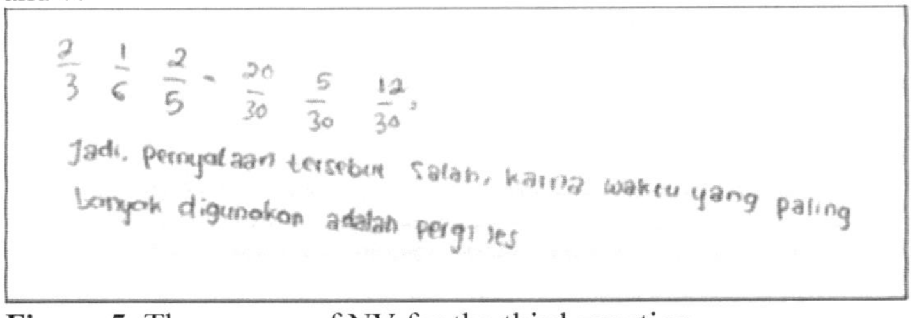

Figure 5. The answer of NV for the third question

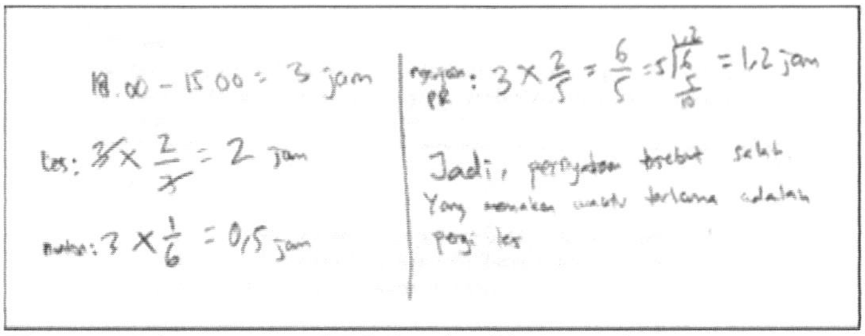

Figure 6. The answer of TG for the third question

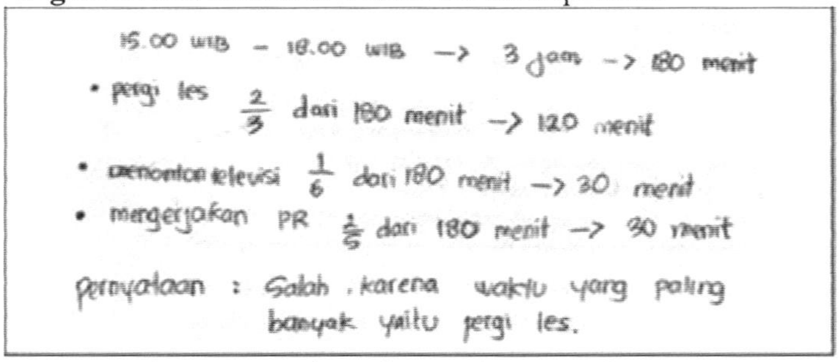

Figure 7. The answer of TK for the third question

Three different steps found in the third question as in Figure 5, 6, and 7. Three students' works fulfill the process of solving mathematical literacy problems. To clarify students' answer as in Figure 5, here is the result of a conversation with students in the interview process.

R: Do you understand this problem?

NV: Understand.

R: How do you solve this problem? Can you explain your answers?

NV: Looking for the denominator. Find the lowest common multiple first, after that sort it.

Students' work as in Figure 5 show that solves this problem by making the same of denominator by looking for the lowest common multiple. Apparently, students sort the fractions from the smallest to the greatest value before writing the conclusions. Students' work shows that the statement in the problem is wrong, and gave the right reason. The conclusion written in Figure 5 was based on the results of the calculations. Students' work as in Figure 6 shows that solving the problem by multiplied the fraction in the problem with 3 hours. Time 3 hours is obtained from the time difference between at 15:00 pm with 18:00 pm. Students' work in Figure 7 show that multiplying fractions by 180 minutes. 180 minutes is obtained from the time difference from 15.00 WIB to 18.00 WIB is 3 hours. Then 3 hours converted into minute, so obtained 180 minutes. Any fractional in the problem multiplied by 3 hours and the students give a conclusion according to the results obtained.

There are also students who made an error in the multiplication of integers with fraction due to students' inaccuracy, as in Figure 8.

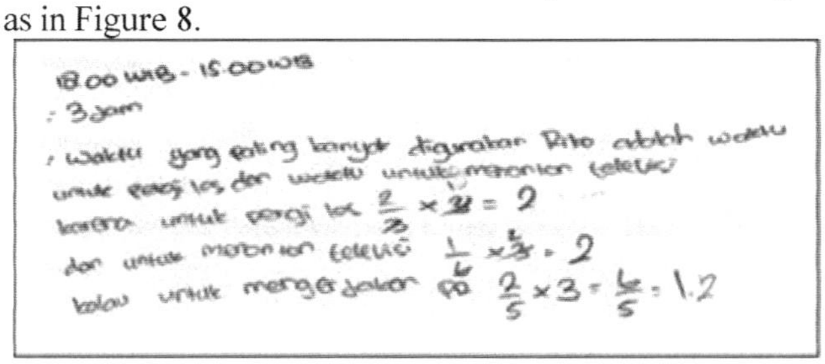

Figure 8. The answer of FR for the third question

Students' work as in Figure 8 shows that made a mistake in the operation of the multiplication of fractions and integers, i.e., $\frac{1}{6}\times 3=2$. The error occurred because the students divide six by three, obtained two, and then two multiplied by one equals two. The error in the arithmetic operation corresponds to Newman's error category, i.e., the error in the process skill, where the student does not perform the mathematical calculation [16]. Students' error because they are inaccuracy in performing the counting operation, that is multiplication between fractional and integer.

3.1.4. Analyzing of the fourth questions. The fourth question is "a rectangle badminton field has a length $(2x+5)$ meters and a width $(2x-3)$ meters. State the perimeter of badminton field in algebraic expressions"

The fourth question is algebra concept and the indicator of mathematical literacy is formulating situations mathematically [1]. It is included proficiency level 2 of mathematical literacy because students can recognize the situation in rectangle context, which perimeter formula of the rectangle, interpretations using algebraic expressions, and employ procedures arithmetic operation in algebraic expressions. Here is the students' answer to the fourth question as in Figure 9, 10, and 11.

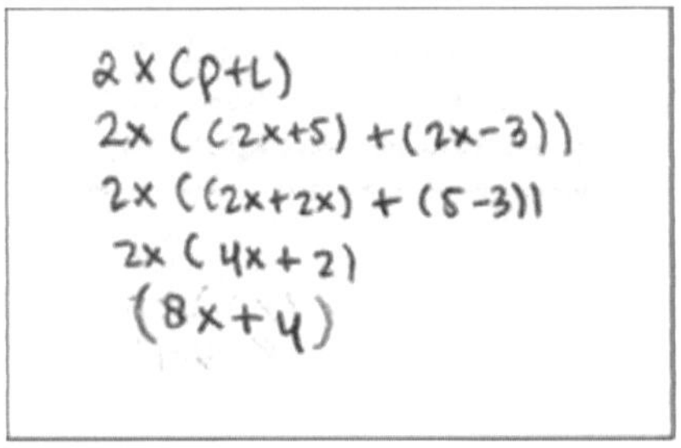

Figure 9. RH's answer for the fourth question

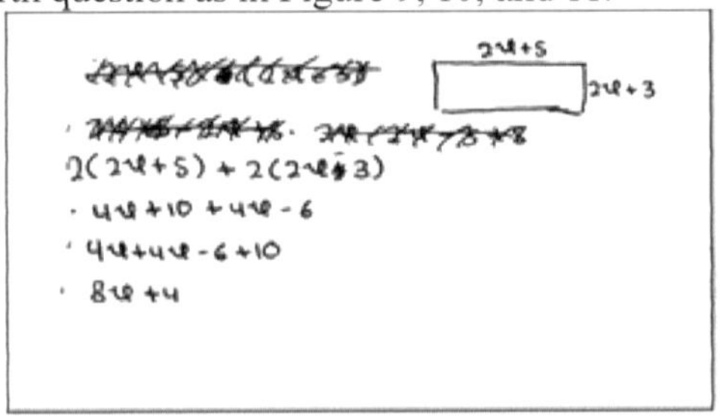

Figure 10. AL's answer for the fourth question

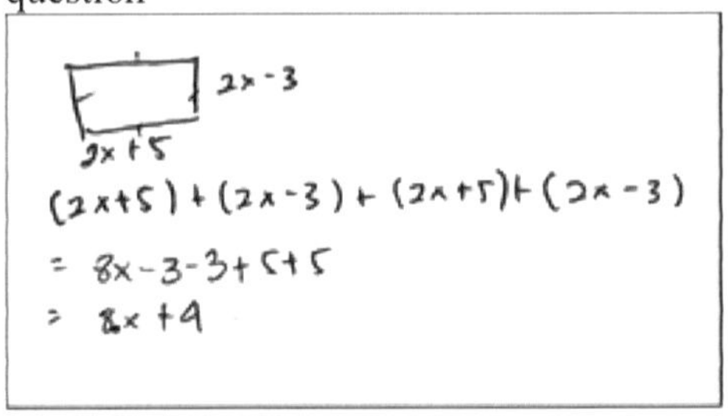

Figure 11. AD's answer for the fourth question

Students' work as in Figure 9, 10, and 11 show the different answer, especially at interpreting the formula. Three different writings the perimeter formula of the rectangle are found for this fourth question. Students' work as in Figure 8 write the perimeter formula with $2\times(p+l)$, Students' work as in Figure 9 write the perimeter formula with $2p+2l$, and students' work as in Figure 10 write the perimeter formula with summing up each side. Students use their prior knowledge to solve the problems, and the purpose of mathematical literacy is to measure students' prior knowledge of mathematics concepts [17].

And also, there are students that make an error. Here is the students' answer as in Figure 12.

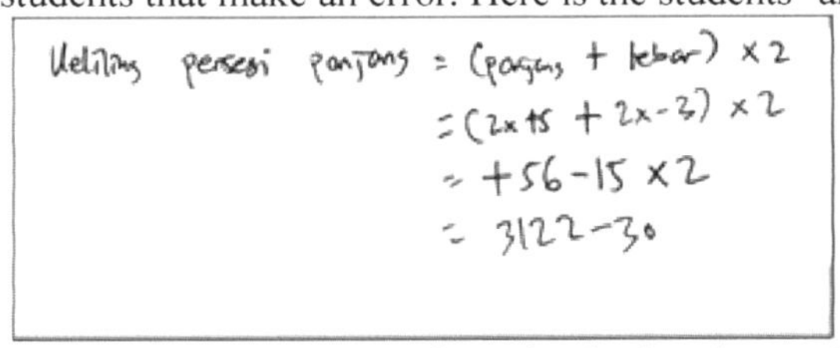

Figure 12. The answer of TG for the fourth question

Figure 12 shows that students' error in applying arithmetic operations in algebraic expressions. To clarify students' answer, here is the result of a conversation with students in the interview process.

R: Do you understand this problem?
TG: No.
R: Why?
TG: Forget.
R: How do you solve this problem? Can you explain your answers?
TG: Anyway the formula is a rectangle if not wrong, length plus width times two. So, $(2x+5+2x-3)\times 2$.
R: What makes you think the algebraic calculation is +56?
TG: I don't know.

The student raises the number $+56-15\times 2$ as the result of the count operation $(2x+5+2x-3)\times 2$ and does not write the variable x in the next step of completion. When interviewed, students said did not know where to get the result +56. For the -15 result, students seem to multiply +5 by -3. Then -15 multiplied by two the result is -30. Although students make mistakes in arithmetic operations with algebra expressions, it remains were seen from the results of the answers that students think to get the results. Observed difficulties concerned applying arithmetic operations in algebraic expressions [10]. Arithmetic operations on algebraic expressions according to Newman's error category, i.e., the category of skill process [18].

Also, there are examples of results of students' answers are wrong in formulating a mathematical situation. The error is in interpreting the questioned. Here is the students' answer as in Figure 13.

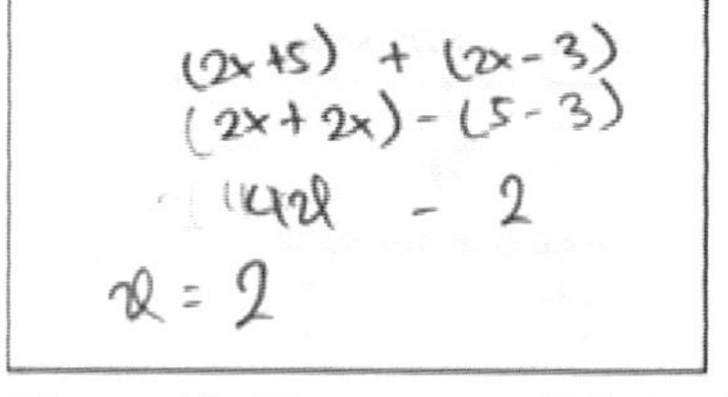

Figure 13. The answer of FR for the fourth question

To clarify students' answer, here is the result of a conversation with the student in the interview process.
R: Do you understand this problem?
FR: Something like that, Miss.
R: How do you solve it? Can you explain your answers?
FR: This problem about algebra, a rectangle badminton field with length $(2x+5)$, and width $(2x-3)$ meter, and asked the perimeter. My answer was $(2x+5)$ summed $(2x-3)$. But I write $2x+2x$ minus $5-3$, because I collect x. Then 5 and 3 is not a variable then there is a minus, so $4x-2$, $x=2$.
R: What makes you think this length summed with width?
FR: I don't know

Figure 13 shows that students' work which misinterprets the question, so they wrong to formulate the perimeter formula of the rectangle. Students understand that this problem about algebraic. However, not understanding the questions, the students simply add up the known of the problem. The sum process operation is no mistake. This student is correct by collecting the variables. For example, $(2x+2x)-(5-3)$. The next step is still correct, i.e., $4x-2$. The error is seen at the time of writing $x=2$. It appears that students get the value of x from result 4 minus 2. For the error as in Figure 13 according to Newman's error category is the transformation problem [18]. Previous research reveals that some students or mistakes most students do on the transformation [15], [19], Things that make students make mistakes that are the inaccuracy of students in thinking.

4. Conclusion
Data analysis revealed that students' error in performing arithmetic most frequently observed. Other observed difficulties concerned understanding about algebra concept, applying arithmetic operations in algebraic expressions and interpreting symbols to represent the unknown. In solving the problem of mathematical literacy problems, students use their prior knowledge, although sometimes not relevant to the questions.

Based on these results, we suggest that in learning process starts with the contextual learning, so that students can understand the concept, and also invite students to participate in the process of understanding the concepts so it is pervasive in students and always remembered for a long period.

Acknowledgments
We thank students and the teacher for their help and their active participation in this research.

References
[1] OECD 2016 PISA 2015 Assessment and analytical framework: science, reading, mathematics and financial literacy (Paris: OECD Publishing)
[2] Johar R 2012 Domain PISA untuk literasi matematika *Jurnal Peluang* **1** 30-41
[3] Vila F and Sanz A 2013 Mathematical literacy in plant physiology undergraduates: results of interventions aimed at improving students' performance *International Journal of Mathematical Education in Science and Technology* **44** 893-904
[4] Dewantara A, Zulkardi, Darmawijoyo 2015 Assesing seventh graders' mathematical literacy in solving pisa-like tasks *IndoMS-JME* **6** 39-49
[5] Wardono, et al 2015 The Realistic scientific humanist learning with character education to improve mathematics literacy based on PISA *International Journal of Education and Research* **3** 349-362
[6] Oktiningrum W, Zulkardi, Hartono Y 2016 Developing PISA-like mathematics task with indonesia natural and cultural heritage as context to assess students' mathematical literacy *Journal on Mathematics Education* **7** 1-8
[7] OECD 2016 PISA 2015 Results: excellence and equity in education (Paris: OECD Publishing) p 192
[8] Mahdiansyah and Rahmawati 2014 Literasi matematika siswa pendidikan menengah: analisis menggunakan desain tes internasional dengan konteks indonesia *Jurnal Pendidikan dan Kebudayaan* **20** 452-469
[9] Wulandari I, Turmudi, Hasanah A 2015 Studi cross-sectional tingkat kemampuan literasi matematis siswa sekolah menengah pertama di bandung berdasarkan pengujian soal PISA *Jurnal Lingkar Widyaiswara* **2** 10-25
[10] Jupri A, Drijvers P, Heuvel-Panhuizen M 2014 Difficulties in initial algebra learning in indonesia *Mathematics Education Research Journal* **26** 683-710
[11] Ojose B 2011 Mathematical Literacy: Are we able to put the mathematics we learn into everyday use? *Journal of Mathematics Education* **4** 89-100
[12] Jumarniati, Pasandaran R, Riady A 2016 The mathematical literacy ability in solving the problem of trigonometric deritative functions *Pedagogy* **1** 66-75
[13] Warren E and Cooper T 2008 Generalizing the pattern rule for visual growth patterns: actions that support 8 years olds' thinking Educ Stud Math **67** 171-185
[14] Yunarni A, Dassa A, Asdar 2015 Profil pemahaman notasi aljabar ditinjau dari kemampuan verbal siswa di kelas v sekolah dasar *Jurnal Daya Matematis* **3** 1-9
[15] Miller A, et al 2017 Preservice teachers' algebraic reasoning and symbol use on a multistep fraction word problem *International Journal for Mathematics Teaching and Learning* **18** 109-131
[16] Jha S K 2012 Mathematics performance of primary school students in assam (india): an analysis using newman procedure *International Journal of Computer Application in Engineering*

IOP Conf. Series: Journal of Physics: Conf. Series **909** (2017) 012063 doi:10.1088/1742-6596/909/1/012063

Sciences **II** 17-21

[17] Breen S, Cleary, O'Shea A 2009 An investigation of the mathematical literacy of first year third-level students in the Republic of Ireland *International Journal of Mathematical Education in Science and Technology* **40** 229-246

[18] Singh P, Rahman A A, Hoon T S 2010 The newman procedure for analyzing primary four pupils errors on written mathematical tasks: a malaysian perspective *Procedia Social and Behavioral Sciences* **8** 264-271

[19] Zakaria E, Ibrahim, Maat S M 2010 Analysis of students' error in learning of quadratic equations *International Education Studies* **3** 105-110

IOP Conf. Series: Journal of Physics: Conf. Series **909** (2017) 012064 doi:10.1088/1742-6596/909/1/012064

The experimentation of learning models viewed from interpersonal intelligence

M T C Gerhana, Mardiyana, and I Pramudya

Mathematics Department of Postgraduate Program Sebelas Maret University, Surakarta, Indonesia

E-mail: Tita.ppsmatematika@gmail.com

Abstract. This research aimed to know experimentation Project Based Learning (PjBL) and Problem Based Learning (PBL) with scientific approach viewed from interpersonal intelligence. The subjects of this research were grade X MIPA students in SMA N 1 Minggir. This research instruments used were test and questionnaire. The result of the research showed that: (1) Students subjected PjBL with scientific approach had a better learning achievement than PBL with scientific approach; (2) students with high interpersonal intelligence had a better learning achievement than low and medium interpersonal intelligence, students with low and medium interpersonal intelligence gave the same learning achievement; (3) In PjBL, students with high interpersonal intelligence had a better learning achievement than low and medium interpersonal intelligence, students with low and medium interpersonal intelligence gave the same learning achievement. In PBL, students with high and medium interpersonal intelligence had a better learning achievement than low interpersonal intelligence, students with high and medium interpersonal intelligence gave the same learning achievement; (4) In high interpersonal intelligence, students subjected PjBL had a better learning achievement than students subjected PBL. In medium and low interpersonal intelligence, students subjected PjBL gave same learning achievement than students subjected PBL.

1. Introduction

The development of education in the current era more rapidly. Students are increasingly required to master all field of learning to be able to ensure a better life. However, education in Indonesia is still low, especially in mathematics education. The low mathematics education in Indonesia were being demonstrated by the international comparison of PISA (Program for International Student Assessment) 2016 in Mathematics subjects that Indonesia is ranked 63 out of 70 countries with an average score of 386 [7]. The low achievement in the mathematics lesson is not only from students but comes from all aspects of education both from government and educators (teachers). Teachers should be more innovative in learning to improve student achievement in the field of mathematics. Innovative learning can be done in various ways, one of them is choosing the right learning model. Learning model is a plan used as a guide in planning the learning in the classroom or learning in the tutorial [10]. The choice of learning model should be by the material discussed and the learning objectives achieved. Learning model such as project-based learning model and problem-based learning model. Project-Based Learning (PjBL) is an innovative learning approach in the 21st century. Through inquiry, students encourage learning reviews, collaborate on research, and create knowledge-based projects that they have [8]. While the Problem-Based Learning (PBL) model is a set of teaching models that use problems as a focus for developing problem-solving skills, materials, and self-organization [4]. The PBL model of teaching is more effective for teaching mathematics. By adopting PBL in teaching mathematics teacher can create some creative thinkers, critical decision makers, problem solvers which is very much needed for the competitive world. And also problem based learning instructional strategy had an effect on content knowledge which provides greater opportunities for the learners to learn a content with more involvement and increase the student's active participation, motivation, and interest among the learners. This leads the learner to have a positive attitude towards mathematics and help them to increase their achievement to a large extent and which will lead to long term memory. It gave a new and desirable kind of experience for the students [5].

Mathematics learning material in this research is trigonometry. Trigonometric material is widely applied in real life, such as counting the height of the building, the height of tower, the height of the aircraft with a certain elevation angle. However, in fact in teaching trigonometry material, teacher less giving a contextual and realitical problem. With the PjBL model and PBL model, teachers can provide contextual and realistic problems. In the PjBL model students can be invited to find real problems, practice the planning, and solve the problem, such as calculating the height of the school flagpole with a clinometer. In the PBL model, teachers can provide contextual problems that can be solved with trigonometric material.

In Indonesia, the ongoing curriculum emphasizes the modern pedagogic dimension of learning, using a scientific approach. The scientific approach is a learning process designed in such a way that learners actively construct concepts or principles through observing stages to identify or find problems, formulate problems, formulate hypotheses, collect data with various techniques, analyze data, draw conclusions and communicate concepts or principles that are found [2]. Therefore, the learning model in this study uses a scientific approach.

Teacher's task besides choosing the right learning model is to pay attention to students' interpersonal intelligence. The chosen learning model must be by students' interpersonal intelligence. Interpersonal intelligence according to Gardner & Checkley is the ability to understand the thoughts, attitudes, and behavior of others [3]. Interpersonal intelligence is the ability to understand and be sensitive to feelings, intentions, motivations, temperaments, temperament of others [9]. Interpersonal intelligence in this study is the ability of a person to understand the thoughts, feelings, attitudes, and motivation of others so that a person can establish good relations and communication with others. PjBL and PBL with a scientific approach is a model of learning that is applied in groups. Students perform tasks assigned by teachers in groups. In the group, students will interact with friends to find the right answer. The accuracy of the answers will affect the mathematics learning achievement. Therefore, interpersonal intelligence also needs to be considered by teachers because it will also affect student learning achievement in the application of mathematics learning model, as in a study conducted by David, et al. that interpersonal intelligence has a positive relationship with learning activities [6]. Some of the problems that will be examined in this research is: (1) Which gives better learning achievement between PjBL model with scientific approach and PBL model with scientific approach; (2) Which had a better mathematics learning achievement between students whose interpersonal intelligence high, medium, and low; (3) In each category learning model, which had a better mathematics learning achievement students whose had the level of interpersonal intelligence; (4) In each level of interpersonal intelligence, which had a better mathematics learning achievement, PjBL model with a scientific approach and PBL model with a scientific approach. Based on the above explanation, the author will conduct research the experimentation of learning models viewed from interpersonal intelligence.

2. Experimental Method

This research employed a quasi-experimental research. In this case to know the difference use of PjBL model with scientific approach and PBL model with scientific approach to students' mathematics learning achievement viewed from student interpersonal intelligence. The method that is used in this research is a quantitative method. Variables that will be observed are learning model as an independent variable, interpersonal intelligence as attribute variable, and student's mathematics learning achievement as a dependent variable. The design used in this research is the 2 × 3 factorial design presented in Table 1.

Table 1. Research design

Learning Model (A)	Interpersonal Intelligence (B)		
	High (B_1)	Medium (B_2)	Low (B_3)
PjBL (A_1)	$(AB)_{11}$	$(AB)_{12}$	$(AB)_{13}$
PBL(A_2)	$(AB)_{21}$	$(AB)_{22}$	$(AB)_{23}$

The population of this research was grade X MIPA students in SMA N 1 Minggir academic year 2016/2017. The used sampling technique is random sampling that is done by selecting 2 classes randomly that is class X MIPA 1 and X MIPA 2. Class X MIPA 1 as an experimental class with PBL model with a scientific approach and class X MIPA 2 as an experimental class with PjBL model with a scientific approach. The research instruments used were tests and questionnaire. Instrument test to get data of student's mathematics learning achievement and instrument questionnaire to get data of student's interpersonal intelligence. The test used in this study as many as 25 questions and problems that were tested as many as 35 questions. A measurement scale that is used for test result is interval scale. Questionnaires used in this study as many as 30 statements and questionnaires tested as many as 50 questions. The measurement scale for questionnaire data uses interval scale which has been converted into nominal scale by categorizing interpersonal intelligence into high, medium, and low categories using following calculation. Interpersonal intelligence questionnaire indicators can be seen in Table 2.

Table 2. Intelligence questionnaire indicators

Interpersonal Intelligence Criteria	Indicators
Feelings of relation in learning	Feelings of learning in groups
Feelings of relation beyond learning	Feelings of interaction on the internet social networking
Ability to empathize	a. Have a sense of caring with others in need b. Paying attention to the latest social issues
Ability to communicate	a. Be able to talk effectively with others b. Pay attention and listen to others while talking
Ability to control emotions	a. Avoiding conflict with others. b. Be able to resolve conflicts in accordance with shared wishes
The breadth of the relation	a. Be able to relate with anyone b. Active follow formal / non-formal organization c. Actively use social media on the internet

Questionnaires given to students are measured using a Likert scale. Questionnaire assessment criteria as presented in Table 3.

Table 3. Questionnaire assessment criteria

Type of Statement	Score of Alternative Answers				
	Strongly Agree	Agree	Neutral	Disagree	Strongly Disagree
Positive statements	5	4	3	2	1
Negative statements	1	2	3	4	5

Calculation of the category of interpersonal intelligence as follows: high interpersonal intelligence if $x_i > \bar{x} + \frac{1}{2}s$; Medium interpersonal intelligence if $\bar{x} - \frac{1}{2}s \leq x_i \leq \bar{x} + \frac{1}{2}s$; Low interpersonal intelligence if $x_i < \bar{x} - \frac{1}{2}s$ [1].

Information:
$\bar{x}$: average questionnaire score
s : standard deviation

The data analysis technique used two-way analysis of variance with a different cell. Before performing the hypothesis test, the prerequisite test is the normality test by Liliefors test and homogeneity test by Bartlet test should be performed. Furthermore, the double comparison test with Scheffe method is done after two-way analysis of variance with a different cell.

3. Result and Discussion

To know the result of experimentation of PjBL model and PBL model viewed from student's interpersonal intelligence, the steps were taken by the author is performing equivalence tests on the student's initial ability, hypothesis test, and double comparison test.

3.1. *Equivalence test on the student's initial ability*

The equivalence test was performed to test the equality of the average population of student's initial ability in experimental class one (a class that will be subject to PjBL model) and experimental class two (a class that will be subject to PBL model). The purpose of the equivalence test is to know that the experimentation result is obtained from the treatment are given, not from other influences. Before performing an equivalence test, the author performs a normality test and homogeneity test of initial ability.

3.1.1. Normality Test. Normality test using Liliefors test. The result of normality test can be seen in Table 4.

Table 4. Summary of normalities data test initial ability of students

Class	n	L_{obs}	$L_{0.05;33}$	Decision	Conclusion
Experimental class one	33	0.1535	0.1542	H_0 is accepted	Normal
Experimental class two	33	0.1526	0.1542	H_0 is accepted	Normal

From Table 4, it appears that for two populations, H_0 is accepted. This suggests that populations for experimental class one and experimental class two come from normally distributed populations.

3.1.2. Homogeneity Test. Homogeneity test is used to determine whether the population comes from the same variances. Homogeneity test using Bartlett test with 5% significance level. Homogeneity test for initial ability showed that ${\chi^2}_{obs} = 0.156$ and $DK = \{\chi^2 | \chi^2 > 3.841\}$, so that ${\chi^2}_{obs} = 0.156 \notin DK$. Thus, the homogeneity test decision is H_0 accepted. From the calculation of homogeneity test, it can be concluded that the population for experimental class one and experimental class two have the same variance.

Normality test and homogeneity test have been fulfilled. The next step is performed one-way analysis of variance with a different cell. Based on calculations obtained $F_{obs} = 3.23$ and $F_{(0.05;1;65)} = 3.99$ so $F_{obs} = 3.23 \notin DK$. Thus, the decision of one way analysis of variance with different cell is H_0 accepted. From the calculation of one way analysis of variance with different cell, it can be concluded that two populations for experimental class one and experimental class two have the same initial ability.

3.2. *Hypothesis Test*

Mathematics learning on trigonometric material applies PjBL model and PBL model. In the PjBL model with a scientific approach, students are given project assignments in groups. The results of the project report can be seen in Figure 1. While in the PBL model with a scientific approach, students are given the task of solving contextual problems related to trigonometry. The results of the student assignments on the model PBL can be seen in Figure 2. The groups in the PjBL model and PBL model are formed heterogeneously.

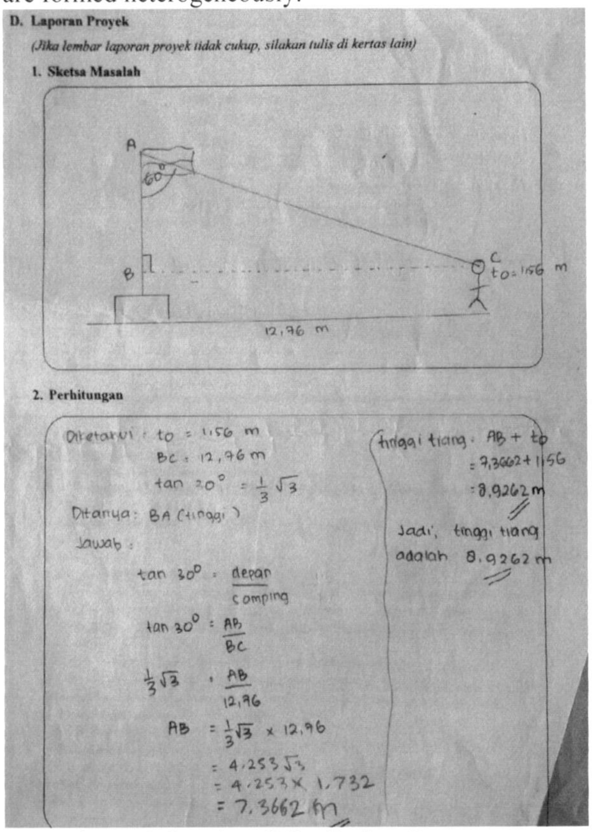

Figure 1. Results of The Project Report

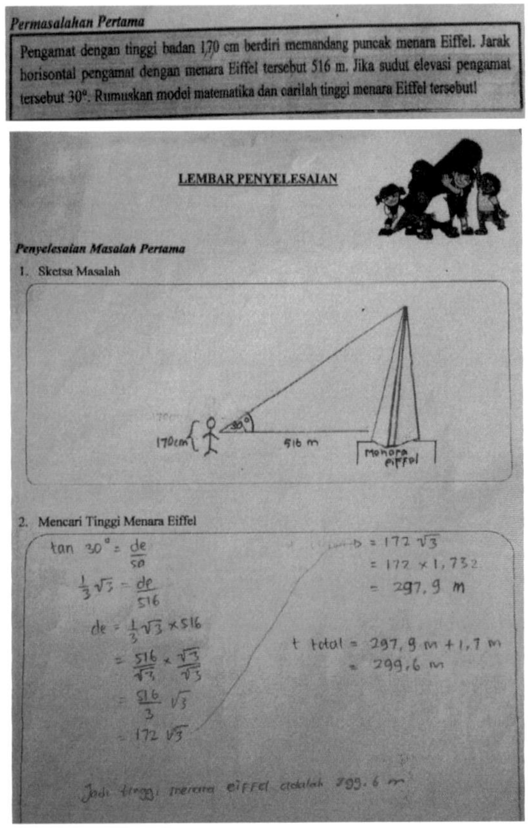

Figure 2. Results of The Student Assignments on The Model PBL

After students complete assignments on PjBL model and PBL model, students are given a test of trigonometry material. In Table 5 presented the results of multiple choice tests.

Table 5. Results of Multiple Choice Tests

	The highest value of multiple choice test	The lowest value of multiple choice test
Students with high interpersonal intelligence level in PjBL model	100	68
Students with medium interpersonal intelligence level in PjBL model	80	36
Students with low interpersonal intelligence level in PjBL model	76	36
Students with high interpersonal intelligence level in PBL model	80	68
Students with medium interpersonal intelligence level in PBL model	80	64
Students with low interpersonal intelligence level in PBL model	68	36

Based on the research result, mathematics learning achievement data based on interpersonal intelligence can be obtained as served in Table 6.

Table 6. Student mathematics learning achievement data based on interpersonal intelligence

Learning Model		Interpersonal Intelligence		
		High	Medium	Low
PjBL	N	9	15	9
	Mean	95.11	62.40	64.89
	Maximum Value	100	80	76
	Minimum Value	80	36	36
	Standard Deviation	6,86	11,69	11,62
PBL	N	7	8	15
	Mean	76.00	71.00	51.47
	Maximum Value	100	80	68
	Minimum Value	68	64	36
	Standard Deviation	11.31	5.55	13.26

3.2.1. Normality Test. Normality test using Liliefors test. The result of normality test based on learning model can be seen in Table 7.

Table 7. Summary of normalities data student mathematics learning achievement

Population	n	L_{obs}	L_{table}	Decision	Conclusion
PjBL	33	0.1364	0.15423	H_0 is accepted	Normal
PBL	30	0.1579	0.16176	H_0 is accepted	Normal
Interpersonal Intelligence High	16	0.1811	0.22150	H_0 is accepted	Normal
Interpersonal Intelligence Medium	23	0.0937	0.18474	H_0 is accepted	Normal
Interpersonal Intelligence Low	24	0.1045	0.18085	H_0 is accepted	Normal

From Table 7, it appears that for four populations, H_0 is accepted. This suggests that populations for PjBL model, PBL model, interpersonal intelligence high, interpersonal intelligence medium, and interpersonal intelligence low come from normally distributed populations.

3.2.2. Homogeneity Test. Homogeneity test using Bartlett test with 5% significance level. Homogeneity test for students mathematics learning achievement based on learning model showed that ${\chi^2}_{obs} = 0.756$ and $DK = \{\chi^2|\chi^2 > 3.841\}$ so that ${\chi^2}_{obs} = 0.756 \notin DK$. Thus, decision is H_0 accepted. From the calculation of homogeneity test, it can be concluded that the population for PjBL model and PBL model have the same variance. Homogeneity test for students mathematics learning achievement based on interpersonal intelligence showed that ${\chi^2}_{obs} = 1.716$ and $DK = \{\chi^2|\chi^2 > 5.991\}$ so that ${\chi^2}_{obs} = 1.716 \notin DK$. Thus, decision is H_0 accepted. From the calculation of homogeneity test, it can be concluded that the population for interpersonal intelligence high, interpersonal intelligence medium, and interpersonal intelligence low have the same variance.

3.2.3. Two-Way Analysis of Variance (ANOVA) with Different Cell. The result of two-way ANOVA with the different cell can be seen in Table 8.

Table 8. Summary of two-way analysis of variance with the different cell

Source	SS	DF	MS	F_{obs}	F_α	Decision
Learning Model (A)	917.71	1	917.71	7.68	4.01	H_{0A} is rejected
Interpersonal Intelligence (B)	7552.43	2	3776.21	31.58	3.16	H_{0B} is rejected
Interaction(AB)	2077.31	2	1038.65	8.69	3.16	H_{0AB} is rejected
Galat	6815.11	57	119.56			
Total	17362.56	62				

Based on the calculation of two way ANOVA with a different cell for inter row effects (learning model) is obtained $F_{obs} = 7.68$ and $F_{(0.05;1;57)} = 4.01$ so $F_{obs} = 7.68 \in DK$. Thus, the decision of two-way ANOVA with a different cell for inter line effects (learning model) is H_0 rejected. Based on the calculation of two-way ANOVA with a different cell for inter line effects (learning model) is obtained $F_{obs} = 31.58$ and $F_{(0.05;2;57)} = 3.16$ so $F_{obs} = 31.58 \in DK$. Thus, the decision of two-way ANOVA with a different cell for inter columns effects (learning model) is H_0 rejected. Based on the calculation of two-way ANOVA with a different cell for interaction effects of rows and columns (interaction of learning models and interpersonal intelligence) is obtained $F_{obs} = 8.69$ and $F_{(0.05;2;57)} = 3.16$ so $F_{obs} = 8.69 \in DK$. Thus, the decision of two way ANOVA with different cell for interaction effects of rows and columns (interaction of learning models and interpersonal intelligence) is H_0 rejected.

According to the result of two-way analysis of variance with different cell, it can be concluded that (a) learning model gives influences toward mathematics learning achievement of students, (b) students' interpersonal intelligence affects to mathematics learning achievement of students, (c) there is an interaction between learning model and students' interpersonal intelligence which influences mathematics learning model of students.

3.2.4. Double Comparison Test. The result of two-way ANOVA with a different cell proves that H_{0A}, H_{0B}, H_{0AB} are rejected, so that double comparison test must be conducted. Double comparison test can be seen below

Table 9. Summary of marginal mean and average of each cell from the student's mathematics achievement test data

Learning Model	Interpersonal Intelligence			Marginal Mean
	High	Medium	Low	
PjBL	95.11	62.40	64.89	72.00
PBL	76.00	71.00	51.47	62.40
Marginal Mean	86.75	65.39	56.50	

The results of two way ANOVA with different cell showing that H_{0A} is rejected. It means that there is an influence of learning model to mathematics learning achievement. The compared learning models provide different learning achievement, so the inter row comparison test needs to be done to find out which learning model provides better learning achievement. The result of average comparison of inter rows can be seen in the following Table 10.

Table 10. Summary of average comparison of inter rows

H_0	F_{obs}	F_{table}	Decision
$\mu_{1.} = \mu_{2.}$	5.35	4.01	H_0 is rejected

Based on Table 10, $F_{obs} = 5.35 > F_{table} = 4.01$ so H_0 is rejected. This indicates that there is a difference in mathematics learning achievement of students who are subjected PjBL model and students who are subjected PBL model. Based on Table 9, average marginal achievement of students with PjBL model is 72.00 and average marginal achievement of students with PBL model is 62.40. Therefore, it can be concluded that the PjBL model gives better mathematics learning achievement than the PBL model.

The results of two way ANOVA with different cell showing that H_{0B} is rejected. It means that there is an influence of interpersonal intelligence to mathematics learning achievement. The compared learning models provide different learning achievement, so the inter columns comparison test needs to be done to find out which interpersonal intelligence provides better learning achievement. The result of average comparison of inter columns is seen in the following table.

Table 11. Summary of average comparison of inter columns

H_0	F_{obs}	F_{table}	Decision
$\mu_{.1} = \mu_{.2}$	28.82	6.32	H_0 is rejected
$\mu_{.2} = \mu_{.3}$	3.80	6.32	H_0 is accepted
$\mu_{.1} = \mu_{.3}$	29.32	6.32	H_0 is rejected

Based on Table 11, for H_0: $\mu_{.1} = \mu_{.2}$ obtained $F_{obs} = 28.82 > F_{table} = 6.32$ so H_0 is rejected. So, there is a difference in mathematics learning achievement of students with high and medium interpersonal intelligence. Based on Table 9, average marginal achievement of students with high interpersonal intelligence is 86.75 and average marginal achievement of students with medium interpersonal intelligence is 65.39. Therefore, it can be concluded that students with high interpersonal intelligence give better mathematics learning achievement than students with medium interpersonal intelligence. H_0: $\mu_{.2} = \mu_{.3}$ obtained $F_{obs} = 3.80 < F_{table} = 6.32$ so H_0 is accepted. This indicates that there is not a difference in mathematics learning achievement of students with medium interpersonal intelligence and low interpersonal intelligence. Therefore, it can be concluded that students who have medium interpersonal intelligence give the same mathematics learning achievement with students who have low interpersonal intelligence. H_0: $\mu_{.1} = \mu_{.3}$ obtained $F_{obs} = 29.32 > F_{table} = 6.32$ so H_0 is rejected. This indicates that there is a difference in mathematics learning achievement of students with high interpersonal intelligence and low interpersonal intelligence. Based on Table 9, average marginal achievement of students with high interpersonal intelligence is 86.75 and average marginal achievement of students with low interpersonal intelligence is 56.50. Therefore, it can be concluded that students with high interpersonal intelligence give better mathematics learning achievement than students with low interpersonal intelligence

The results of two way ANOVA with different cell showing that H_{0AB} is rejected. It means that there is an interaction between learning model and students' interpersonal intelligence which influences mathematics learning model of students. So, double comparative test between cells is required.

Table 12. Summary of comparative test results between cells on the same row

	H_0	F_{obs}	F_{table}	Decision
PjBL model with scientific approach	$\mu_{11} = \mu_{12}$	50.31	11.88	H_0 is rejected
	$\mu_{12} = \mu_{13}$	0.29	11.88	H_0 is accepted
	$\mu_{11} = \mu_{13}$	34.35	11.88	H_0 is rejected
PBL model with scientific approach	$\mu_{21} = \mu_{22}$	0.78	11.88	H_0 is accepted
	$\mu_{22} = \mu_{23}$	16.64	11.88	H_0 is rejected
	$\mu_{21} = \mu_{23}$	24.02	11.88	H_0 is rejected

Table 13. Summary of comparative test results between cells on the same column

	H_0	F_{obs}	F_{table}	Decision
High Interpersonal Intelligence	$\mu_{11} = \mu_{21}$	12.01	11.88	H_0 is accepted
Medium Interpersonal Intelligence	$\mu_{12} = \mu_{22}$	3.23	11.88	H_0 is accepted
Low Interpersonal Intelligence	$\mu_{13} = \mu_{23}$	8.47	11.88	H_0 is accepted

Based on Table 12, for H_0: $\mu_{11} = \mu_{12}$ obtained $F_{obs} = 50.31 > F_{table} = 11.88$ so H_0 is rejected. This indicates in PjBL model; there is a difference in mathematics learning achievement of students with high interpersonal intelligence and students with medium interpersonal intelligence. Based on Table 9, average marginal achievement of students with high interpersonal intelligence in PjBL model is 95.11 and average marginal achievement of students with medium interpersonal intelligence in PjBL model is 62.40. Therefore, it can be concluded in PjBL model with high interpersonal intelligence give better mathematics learning achievement than students with medium interpersonal intelligence in PjBL model. $H_0: \mu_{12} = \mu_{13}$ obtained $F_{obs} = 0.29 < F_{table} = 11.88$ so H_0 is accepted. This indicates in PjBL model there is not a difference in mathematics learning achievement of students with medium interpersonal intelligence and students with low interpersonal intelligence. Therefore, it can be concluded in PjBL model, students who have medium interpersonal intelligence gives the same mathematics learning achievement with students who have low interpersonal intelligence. $H_0: \mu_{11} = \mu_{13}$ obtained $F_{obs} = 34.35 > F_{table} = 11.88$ so H_0 is rejected. This indicates in PjBL model, there is a difference in mathematics learning achievement of students with high interpersonal intelligence and students with low interpersonal intelligence. Based on the Table 9, average marginal achievement of students with high interpersonal intelligence in PjBL model is 95.11 and average marginal achievement of students with low interpersonal intelligence in PjBL model is 64.89. Therefore, it can be concluded in PjBL model, students with high interpersonal intelligence give better mathematics learning achievement than students with low interpersonal intelligence.

Based on Table 12 for $H_0: \mu_{21} = \mu_{22}$ obtained $F_{obs} = 0.78 < F_{table} = 11.88$ so H_0 is accepted. This indicates in PBL model there is not difference in mathematics learning achievement of students with high interpersonal intelligence and students with medium interpersonal intelligence. Therefore, it can be concluded in PBL model, students who have high interpersonal intelligence gives the same mathematics learning achievement with students who have medium interpersonal intelligence. $H_0: \mu_{22} = \mu_{23}$ obtained $F_{obs} = 16.64 > F_{table} = 11.88$ so H_0 is rejected. This indicates in PBL model, there is a difference in mathematics learning achievement of students with medium interpersonal intelligence and students with low interpersonal intelligence. Based on the Table 9, average marginal achievement of students with medium interpersonal intelligence in PBL model is 71.00 and average marginal achievement of students with low interpersonal intelligence in PBL model is 51.47. Therefore, it can be concluded in PBL model, students with medium interpersonal intelligence gives better mathematics learning achievement than students with low interpersonal intelligence. $H_0: \mu_{21} = \mu_{23}$ obtained $F_{obs} = 24.02 > F_{table} = 11.88$ so H_0 is rejected. This indicates in PBL model, there is a difference in mathematics learning achievement of students with high interpersonal intelligence and students with low interpersonal intelligence. Based on Table 9, average marginal achievement of students with high interpersonal intelligence in PBL model is 76.00 and average marginal achievement of students with low interpersonal intelligence in PBL model is 51.47. Therefore, it can be concluded in PBL model, students with high interpersonal intelligence give better mathematics learning achievement than students with low interpersonal intelligence.

Based on Table 13, for $H_0: \mu_{11} = \mu_{21}$ obtained $F_{obs} = 12.01 > F_{table} = 11.88$ so H_0 is accepted. This indicates in high interpersonal intelligence, there is a difference in mathematics learning achievement of students subjected PjBL model and students subjected PBL model. Therefore, it can be concluded in high interpersonal intelligence, students subjected PjBL model give better mathematics learning achievement than students subjected PBL model. $H_0: \mu_{12} = \mu_{22}$ obtained $F_{obs} = 3.23 < F_{table} = 11.88$ so H_0 is accepted. This indicates in medium interpersonal intelligence, there is not a difference in mathematics learning achievement of students subjected PjBL model and students subjected PBL model. Therefore, it can be concluded in medium interpersonal intelligence, students subjected PjBL model give the same mathematics learning achievement with students subjected PBL model. $H_0: \mu_{13} = \mu_{23}$ obtained $F_{obs} = 8.47 < F_{table} = 11.88$ so H_0 is accepted. This indicates in low interpersonal intelligence, there is not a difference in mathematics learning achievement of students subjected PjBL model and students subjected PBL model. Therefore, it can be concluded in low interpersonal intelligence, students subjected PjBL model give the same mathematics learning achievement with students subjected PBL model.

From the calculation, obtained that students subjected PjBL with a scientific approach had a better learning achievement than PBL with a scientific approach. Students with high interpersonal intelligence had a better learning achievement than low interpersonal intelligence and medium interpersonal intelligence, students with low interpersonal intelligence and medium interpersonal intelligence gave the same mathematics learning achievement. In PjBL model with scientific approach, students with high interpersonal intelligence had a better learning achievement than low and medium interpersonal intelligence, students with low and medium interpersonal intelligence gave the same learning achievement. In PBL model with a scientific approach, students with high and medium interpersonal intelligence had a better learning achievement than low interpersonal intelligence, students with high and medium interpersonal intelligence gave the same learning achievement. In high interpersonal intelligence, students subjected PjBL with scientific approach had a better learning achievement than students subjected PBL with a scientific approach. In medium and low interpersonal intelligence, students subjected PjBL with a scientific approach gave same learning achievement than students subjected PBL with a scientific approach.

4. Conclusion

The result of the research showed that: (1) Students subjected PjBL with a scientific approach had a better learning achievement than PBL with a scientific approach; (2) students with high interpersonal intelligence had a better learning achievement than low and medium interpersonal intelligence, students with low and medium interpersonal intelligence gave the same learning achievement; (3) In PjBL with a scientific approach, students with high interpersonal intelligence had a better learning achievement than low and medium interpersonal intelligence, students with low and medium interpersonal intelligence gave the same learning achievement. In PBL with a scientific approach, students with high and medium interpersonal intelligence had a better learning achievement than low interpersonal intelligence, students with high and medium interpersonal intelligence gave the same learning achievement; (4) In high interpersonal intelligence, students subjected PjBL with a scientific approach had a better learning achievement than students subjected PBL with a scientific approach. In medium and low interpersonal intelligence, students subjected PjBL with a scientific approach gave same learning achievement than students subjected PBL with a scientific approach.

References

[1] Budiyono 2015 *Pengantar Penilaian Hasil Belajar* (Surakarta: UNS Press).
[2] Hosnan 2014 *Pendekatan Saintifik dan Kontekstual Dalam Pembelajaran Abad 21* (Bogor: Ghalia Indonesia)
[3] Muhammad Y and Nurdin I 2013 *Pembelajaran Berbasis Kecerdasan Majemuk (Multiple Intelligences)* (Jakarta: Prenadamedia)
[4] Paul E & Don K 2012 Strategies and Models for Teachers : Teaching Content and Thinking Skills (Boston: Allyn & Bacon)
[5] Padmavathy & Mareesh K 2013 Effectiveness of Problem Based Learning In Mathematics *International Multidisciplinary e-Journal 2*(1) 45-51.
[6] David P, Baker, Cocta, Rose, & Cui 2009 Detecting and Understanding The Impact of Cognitive and Interpersonal Conflict in Computer Supported Collaborative Learning Environments *Educational Data Mining 2009*.
[7] Program for International Student Assessment 2016 *PISA Results in Focus* (www.oecd.org/pisa)
[8] Stephanie B 2010 Project-based learning for the 21st century: Skills for the future *The Clearing House* 83(2) p 39-43
[9] Suparno 2004 *Teori Inteligensi Ganda dan Aplikasinya di Sekolah* (Yogyakarta: Kanisius)
[10] Trianto 2012 *Model Pembelajaran Terpadu* (Jakarta: Bumi Aksara)

IOP Conf. Series: Journal of Physics: Conf. Series **909** (2017) 012065 doi:10.1088/1742-6596/909/1/012065

Student's mathematical understanding ability based on self-efficacy

M R Ramdhani[a], B Usodo[b], S Subanti[c]

Magister Pendidikan Matematika, Universitas Sebelas Maret, Jl. Ir. Sutami No. 36-A, Kentingan Surakarta 57126, Indonesia

[a]marlisaramdhani2014@gmail.com
[b]budi_usodo@yahoo.com
[c]sri_subanti@yahoo.co.id

Abstract. Materials in mathematics are provided not only as an ability to memorize, but also to train the ability of mathematical understanding. Students' mathematical understanding ability is influenced by the students' belief in solving the given problems. This research aim to determine the mathematical understanding ability of junior high school students. This research is descriptive qualitative research. Data collection was done through a test, questionnaire, and interview. The result showed that students with high self-efficacy category could master the three indicators of students' mathematical understanding ability well, namely translation, interpretation, and exploration. Students with moderate self-efficacy category can master translation indicator and able to achieve interpretation indicator but they unable to reach exploration indicator. Students with low self-efficacy category only master the translation, but they cannot achieve the interpretation and exploration indicators. So, the students who have high, moderate or low self-efficacy master the indicator of mathematical understanding based on the level of understanding capabilities on each student.

1. Introduction

Mathematics is the science for developing scientific situations, drawing conclusions as well as solving problems. Mathematics also promotes the habit of accuracy, logical, systematic and order arrangements [1]. Currently, not a few students who view mathematics as a subject that is boring and terrifying. Many students find difficulty to understand mathematics. It affects students' low achievement. The absorption capacity of junior high school students in Karanganyar District has the lowest absorption of 42.01% at the district level, 40.26% at the provincial level, and 47.19% at the national level [2]. Mathematical understanding is a very important part in understanding mathematical problems and problems in daily life [3]. A meaningful understanding of learning mathematics should be directed at developing the mathematical connections between different ideas, understand how mathematical ideas are interrelated to one another there by building a thorough understanding and use of mathematics in contexts outside of mathematics [4].

Students' difficulties in proving, most of them were caused by their lack of understanding of mathematical concepts and definitions so that they were unable to construct a mathematical proof and to write mathematical notation or to use mathematical language correctly. It is obvious that students' mathematical proving abilities are associated with their mathematical understanding abilities [5]. Bloom divides the aspects of understanding into three kinds of understanding, namely: translation, interpretation, and exploration [6]. Translation is an ability to understand an idea expressed in another way from the original statement. For example, students can change the story into a mathematical sentence, interpretation is an ability to understand and connect the material or ideas that have been known then explain the problem in detail, and exploration is an understanding that expects a person can make conclusions and provides views on the problem. Knowing the ability of students' mathematical understanding is an indispensable thing because by knowing the ability of the student's, the teacher knows the difficulties faced by students. Based on that case, the teacher can design the right learning so that the subject matter can be delivered well and student achievement can increase. This is by Tambychik's opinion that understanding of difficulties faced by students is crucial and attention on specific mathematics skills may lead to more meaning teaching and learning process [7].

Many students consider that mathematics is a difficult subject so that students do not think deeply in receiving the material. The difficulty of a mathematical problem depends on how students' belief in dealing with it. One important part about self-belief is self-efficacy. Self-efficacy is the reflection of an individual's belief in his or her capabilities to complete tasks [8]. Self-efficacy is the expectation of how far one can perform one behavior in certain situations [9]. Thus, it can be concluded that self-efficacy is a person's belief in certain abilities and how to overcome obstacles. According to Bandura, self-efficacy influences human action through cognitive, motivational and affective processes and are specific to a given domain or context. When people believe that they possess the required capability to accomplish a given task, greater effort and perseverance will be made for goal achievement [10]. Self-efficacy can be grown and studied through four main sources of information, the namely experience of success, vicarious experience, verbal persuasion and physiological state [11]. While Bandura explains that self-efficacy consists of several dimensions, namely Magnitude, Strength and Generality [12]. Magnitude focuses on whether the individual believes that the task begins with a difficulty level of the task, from the moderate up to the most difficult level. Strength reflects an individual's confidence at completing the various components of the task or at various levels of difficulty. Generality is the extent to which self-efficacy on one task generalizes to other tasks or domains. Generally, It is important to understand how self-efficacy estimates on one task generalize to other related tasks, such as academic domains [13].

The results of Lambertus's research shows that the ability of mathematical understanding of students learning to use open-ended approach significantly better improvement of the ability of students' mathematical understanding using conventional learning [3]. Tambychik's research show that respondents lacked in many mathematics skills such as number-fact, visual-spatial and information skills. Information skill was the most critical [7]. Pirie and Martin's show that a teacher must realises that students have different strategies and abilities for finding, collecting, and using their previous understandings, is in a position to consider the best of actions [14]. In addition, Zedan and Bitar's research show that a strong positive correlation between mathematical self efficacy and achievements in mathematics [15]. However, Lilian's research show that effort failed to predict academic achievement, effort could only be regarded as an indirect factor but not a necessary factor in bridging the relationship between attitude, self-efficacy and academic achievement [16]. Based on the explanation above, the purpose of this study is to describe the achievement of each indicator of the mathematical understanding ability of 8th-grade senior high school students based on self-efficacy category, namely high, moderate, and low self-efficacy.

2. Method

Type of this research used is descriptive qualitative in the form of a description explaining on how students' understanding ability based on their self-efficacy. The study was conducted in the even semester of the academic year of 2016/2017 in SMP Negeri 3 Karanganyar. Research subjects taken used purposive random sampling technique based on students' self-efficacy category given in class VIII A which amounted to 32 students. Selected three research subjects each one subject from high, moderate, and low self-efficacy categories. The subject of the study was given a test of students' mathematical understanding ability and conducted an interview to find out how the students' mathematical understanding ability. Subject one is symbolized (S1), subject two is symbolized (S2), and subject three is symbolized (S3), while the test error is symbolized by (F). Table 2.1 is a way to determine high, moderate, and low self-efficacy.

Table 2.1 Measurement Scale of Students' Self-Efficacy Category [17]

No.	Interval	Category
1	$x_i > \bar{x} + 0.5\ s$	High
2	$\bar{x} - 0.5\ s \leq x_i \leq \bar{x} + 0.5\ s$	Moderate
3	$x_i < \bar{x} - 0.5\ s$	Low

Note:
x_i : Students' self-efficacy scores of each respondent
$\bar{x}$: Average self-efficacy score of all sample students
s : Standard deviation of the entire sample

3. Result and Discussion

The self-efficacy category was obtained based on the total score of the self-efficacy measurement scale. Based on that measurement, the score $\boldsymbol{x_i} > 99.57$ is a high self-efficacy, $99.57 \leq \boldsymbol{x_i} \leq 90.31$ is a moderate self-efficacy, and $\boldsymbol{x_i} < 90.31$ is a low self-efficacy. The test of students' mathematical understanding ability in this research is a test used to find out how the students' mathematical understanding ability is based on the aspect of translation, interpretation, and exploration of the material of cube and beam. Interviews were conducted to confirm the answers to students' mathematical understanding test so that the researcher could find out whether between test answers and interviews is consistent. The following is items to measure students' mathematical understanding skills.

1. Father will make a cube box made of wood. If the father has wood which its surface area is 1.350 cm^2, then how long is the length of the cube that should be made by the father?
2. Amira has a rectangular box with a length of 15 cm, height 10 cm, and volume 900 cm^3. If Amira would cover the box using paperboard, then how much paperboard does Amira need?
3. Ibnu has a cube aquarium with a surface area of a cube base 1.6 m. If Ibnu will fill up as much water as ¾ of the entire volume of the aquarium, then how many liters of water is needed by Ibnu?

The students 'mathematical understanding test is based on the students' mathematical comprehension aspects of translation, interpretation, and exploration. Interviews were conducted to confirm and obtain more in-depth information. Following will be presented an answers analysis of students' mathematical understanding and interviews toward research subject based on high, moderate, and low self-efficacy categories.

3.1. High Self-Efficacy Subject Analysis

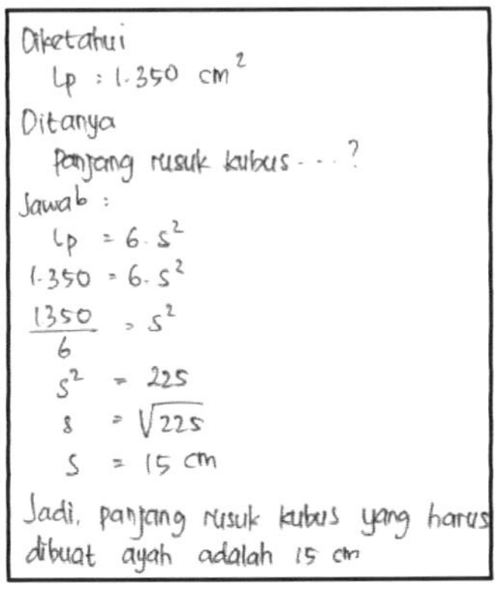

Figure 1. S1N1's answer

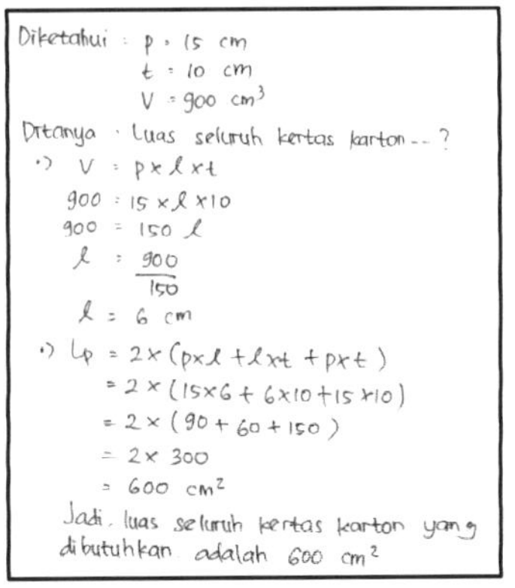

Figure 2. S1N2's answer

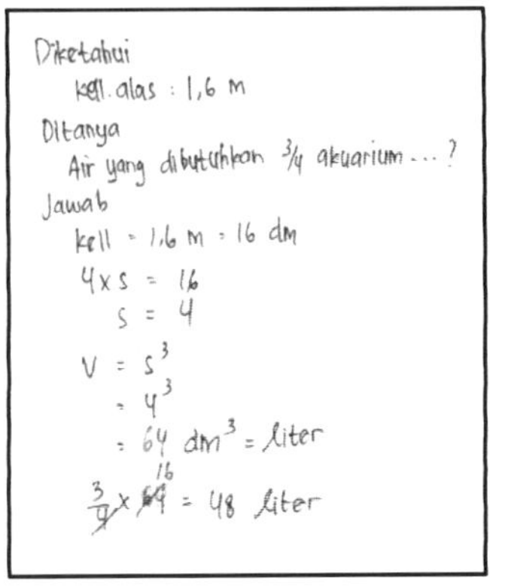

Figure 3. S1N3's answer

IOP Conf. Series: Journal of Physics: Conf. Series **909** (2017) 012065 doi:10.1088/1742-6596/909/1/012065

Figures 1, 2 and 3 show that students write what is known by symbolizing the area of the surface using the symbol Lp, the symbolized length p, the symbolized height t, the symbolized volume p, and the symbolized surface area by the kell. S1 can write down what is being asked correctly. This shows that translating the story into another language to make the problem can be more easily understood by using a certain symbol so that students can determine the right concept to solve the problem correctly [6]. By Patrick's opinion that a variety of internal representations; verbal/syntactic, imagistic, symbolic. We link together these separate representations to create a more complex understanding of that concept [18]. At the time of interview, S1 can explain again the meaning of the symbols and visible to understand the purpose of the three questions.

Students are required to be able to interpret the problem and connect the things that are known with the concept that will be used in finding solutions to the problem. In the beam and cube material, there are several concepts of surface area and volume as well as concepts related to cube beams, for example, the surface area and circumference of square and rectangle. Based on students' answers, students can interpret and define concepts used in solving problems. By Anderson's opinion that interpreting occurs when students can convert information from one representation to another representation. Interpretation involves the conversion of words into words, images into words and so on [3][18]. Answer number 1, S1 uses the concept of surface area to find the length of the cube frame. S1's answer in solving problem number 2 that is to search the width first using the concept of the beam volume. After the width is obtained then look for the surface area of the beam. While the answer number 3, S1 answer by changing the circumference unit from m to dm because it is asked in the form of a liter. Then by using the concept of a rectangular circumference, the length of the base of the cube is obtained. After that S1 looking for the required volume of water is 3/4 times the cube volume. At the time of the interview, S1 can explain again the answer process in order on each answer. Students can explain why they choose to use that concept in solving the problem.

Students calculate by applying the things that are known and the concepts that have been determined. Based on the three questions answered by the students, students are able to determine the exact concept that obtains the correct results on the three questions. As Tambychik's opinion that the ability to create meaningful perceptions for logical thinking and effective use of memory is an important factor in learning skills and problem-solving [7]. At the time of the interview, students can explain the results of the answer; it means that students understand what has been done. When the researcher asks questions about the surface area and the volume of cubes and beam, S1 can answer them correctly and give examples of other real-life problems. Based on the above explanation can be seen that the student persistence in working on the problem so that students with high self-efficacy can master these three indicators, namely translation, interpretation, and extrapolating the research corresponds to Pintrich found that there are positive correlations between self efficacy and persistence [15].

3.2. Moderate Self-Efficacy Subject Analysis

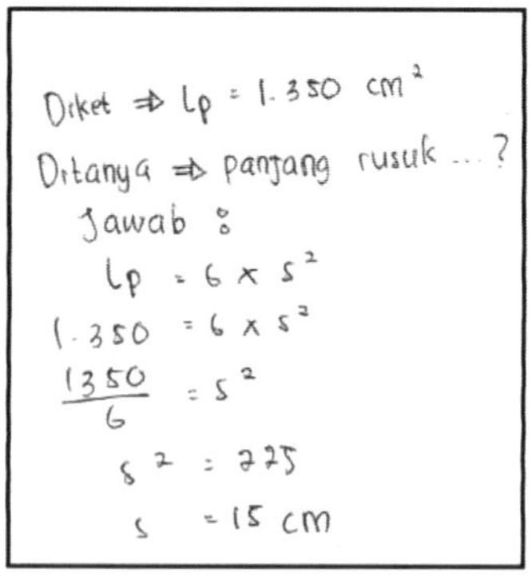

Figure 4. S2N1's answer

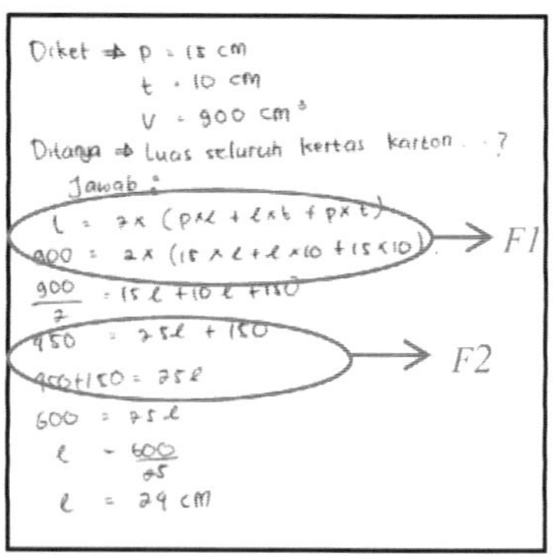

Figure 5. S2N2's answer

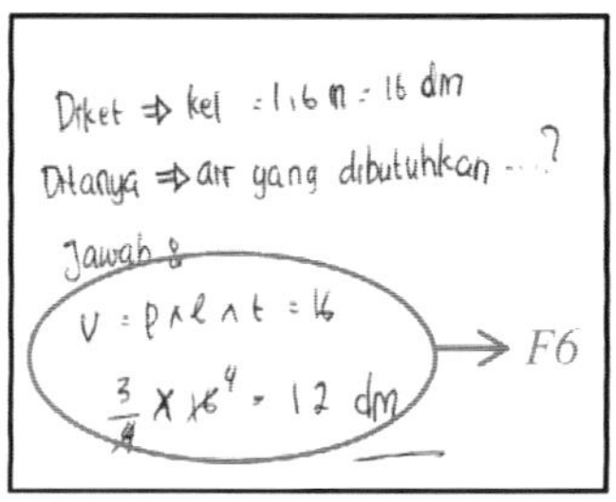

Figure 6. S2N3's answer

Figure 4, Figure 5, and Figure 6 show that students can understand the problem well because S2 can write down what is known, what is questioned by using a particular symbol. It means that S2 can reinterpret the story in a simple language using certain symbols. Kilpatrick stated that representations could be used to understand the mathematics [19]. Similar to S1, S2 also uses Lp for surface area, p for length, t for height, V for volume and *kel* for circumference. At the time of the interview, S2 said that he understands what is being asked. No words are not understood on the item.

Answer number 1, S2 uses the concept of the surface area of the cube to look for the length of the frame. At the time of the S2 interview can also explain again about the test answers. Students also answered that 15 were derived from root 225. This type of problem is an easy item category and usually exemplified by the teacher. The answer number 2, S2 uses the concept of surface area of the beam. Researchers suspect because the question is the surface area, so S2 directly apply the concept of surface area of the beam, meaning S2 does not understand the meaning of each concept on cubes and beams material. Another error is indicated by F2 indicating that the positive 150 which lies on the right side when moving to the left segment remains positive, it should be the switching of the segment changes so that it becomes 450 - 150. At the time of the interview indicates that S2 understands what is being asked. S1 knows that width is unknown, but S2 is confused how to find width using which concept. When the researcher asks the surface area formula, and the volume of the beam S2 can answer correctly. This means that students use memory in memorizing a concept so when they faced with problems that only use one concept, S2 can do correctly, but when faced with problems that use some concept which S2 have not understood. This is in by the Nathan opinion which says that inability to connect conceptual aspects of math, difficulty to make sense of connection between information, incomplete understanding of mathematical language and difficulty in comprehend and visualizing mathematical concept might result in difficulties [7]. When the researcher asks questions that lead to solving problem solutions, students can follow well. This means that students understand the elements of each concept, the students are just lazy to think well. According to the opinion of Pirie as saying that students realize that they have understanding or bills but not immediately accessible, then it should be repeated back to a more basic understanding [14].

Figure 6 shows that S2 writes what is known and questioned but S2 can not provide the right solution in solving the problem number 3. S2 writes $V = p \times l \times t$, meaning that S2 knows the concept of finding the volume is the length multiplied by the width multiplied by the height. S2 has also changed 1.6 m to 16 dm, but S3 looks confused on how to determine the solution so that S2 answer $3/4 \times 16 = 12$. Based on the results of the interview, it shows that S2 understand what is asked but S2 confused to determine the solution in solving the problem. Students cannot determine the right concept in solving the problem, because students do not think well about the elements are known and unknown. So students cannot determine the right conclusions in determining the solution of each problem. When doing an interview, students are not able to make a right conclusion in solving mathematical problems. So the students must do more practice.

3.3. Low Self-Efficacy Subject Analysis

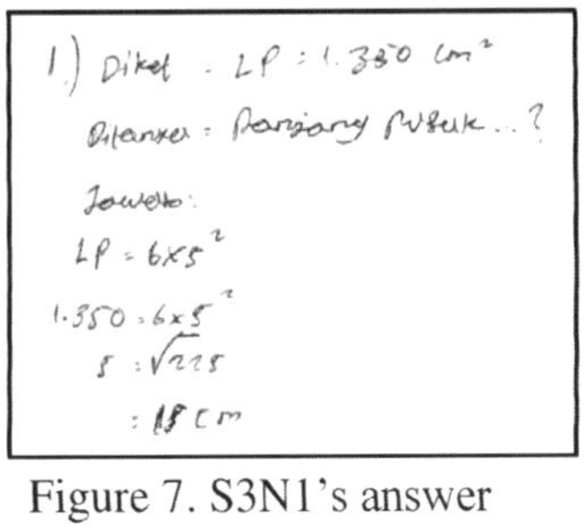

Figure 7. S3N1's answer

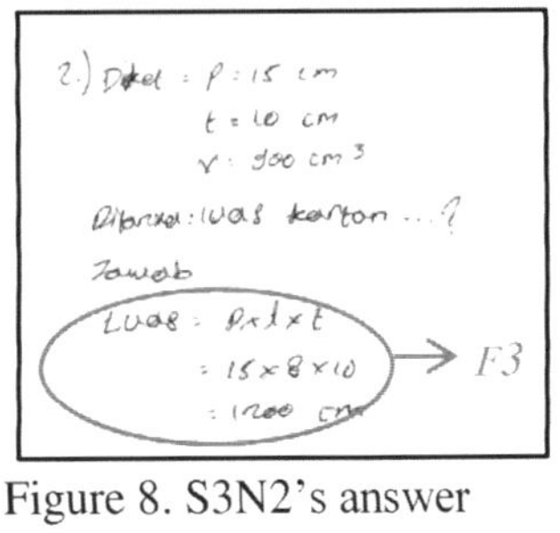

Figure 8. S3N2's answer

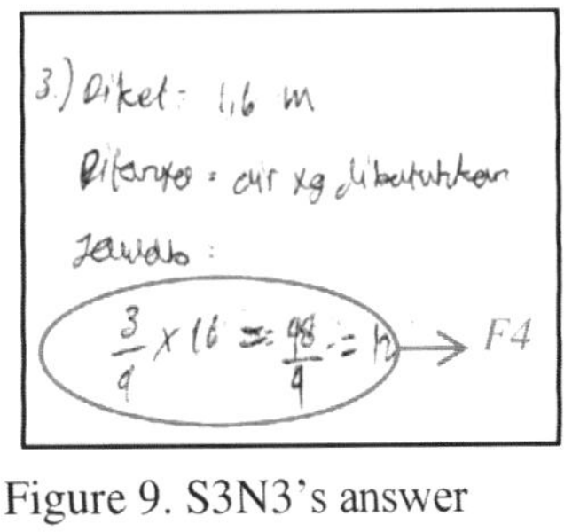

Figure 9. S3N3's answer

Figure 7, Figure 8, and Figure 9 show that students can write down what is known and asked well. S3 write the same symbols with S1 and S2. At the time of the interview, S3 said that he understood the purpose of the three problems and there are not difficult words.

The process of answer number 1, on the third line the researcher is unsure the result of S3 because S3 immediately writes root 225. At the time of the interview, S3 looks difficult in explaining again the process of the S3 answer. S3 cannot answer where the root 225 comes from. Figure 8 shows that S3 cannot solve the problem. S3 writes surface area = $p \times l \times t$, it supposed to the concept of surface area of the beam namely $2 \times (p \times l + l \times t + p \times t)$. Figure 9 shows that S3 writes $3/4 \times 16 = 48/4 = 12$. Students with low self-efficacy categories indicate that students cannot relate the things that are known to be able to determine the concepts used in solving mathematical problems. By Kadir's opinion that students feel difficult in mathematics because they have difficulty understanding and retrieving concepts, formulas, facts, and procedures [20]. When the interview was conducted the students also cannot mention the concepts of the cube beam material. When the researcher leads the students to get the right solution in solving the problem, the students cannot follow it well. S3 just smiled and replied "oh, yes ma'am. I forgot ma'am ", so it is clear that students cannot make conclusions in determining the right solution in solving the problem.

4. Conclusions and Suggestions

Based on the results of research on students' mathematical understanding in class VIII, it can be drawn some conclusions, as follows. Students with high self-efficacy category can master the three indicators of students' mathematical understanding ability well, namely translation, interpretation, and exploration. Students with moderate self-efficacy category can master translation indicator and able to achieve interpretation indicator but they unable to reach exploration indicator. Students with low self-efficacy category only master the translation, but they cannot achieve the interpretation and exploration indicators. The results of this study are expected to give information to the school, especially for the teachers about the ability of students' mathematical connections. Related to the findings that exist, students, mathematical connection ability is very important to have. Therefore, with this research, the teacher can design a good learning process to improve students' mathematical connection ability so that students' achievement will be achieved well.

5. References

[1] Eguawon A O 2002 *Two Problem based learning strategies: Quantitative Ability and Gender as Determinants of Students Achievement in Economic.* Unpublished Ph.D. These, University of Ibadan, Nigeria.BSNP. Laporan Hasil Ujian Nasional Tahun Pelajaran 2015/2016

[2] BSNP 2016 *Laporan Hasil Ujian Nasional Tahun Pelajaran* 2015/2016

[3] Lambertus 2016 *Developing Skills Understanding Of Mathematical High School Student* International Journal of Education and Research Vol. 4 No. **7** July 2016 hal 315-326

[4] NCTM 2000 *Principle and Standards for School Mathematics* Reston VA NCTM

[5] Mukhtar A, Isma 2015 *A Structural Equation Model Examining The Relationship among Remembering, Understanding, Applying and Students'Achievement in Mathematics IIUM Journal of Educational Studies*, **3**:1 (2015) 70-91

[6] David R K 2002 *A Revision of Bloom's Taxonomy: An Overview* Theory Into Practice, College of Education, The Ohio State University Volume 41, Number **4**, 212-218

[7] Tambychik T, Subahan T 2010 *Student's Difficulties in Mathematics Problem-Solving: What do they say?* International Conference on Mathematics Education Research 2010 Procedia Social and Behavioral Sciences 8 (2010) 142–151

[8] Bandura A 1997 *Self-efficacy: The exercise of control* (New York, NY: W. H. Freeman)

[9] Howards F S, Miriam S 2006 *Kepribadian (Teori Klasik dan Riset Modern* (Jakarta: Erlangga)

[10] Marion E B, William G M & Steven E B 2015 *Sources and Influence of Mathematics Self-Efficacy in Jamaican College Students* Caribbean Journal of Psychology: Vol. 7, No. **1**, 2015, 24-42
[11] Ghufron, M N & Rini R 2010 *Teori-teori Psikologi* Jogjakarta Ar-Ruzz Media
[12] Aimee D, Steven R H, Richard D J 2014 *Exploring the Dimensions of Self-Efficacy in Virtual World Learning: Environment, Task, and Content* MERLOT Journal of Online Learning and Teaching Vol. 10, No. **2**
[13] Bong M 1997 *Generality of academic self-efficacy judgments: Evidence of hierarchical relations*. Journal of Educational Psychology, **89**(4), 696-709
[14] Pirie and Martin M L 2000 *The Role of Collecting in the Growth of Mathematical Understanding* Mathematics Education Research Journal 2000, Vol. 12, No.2, 127-146
[15] Zedan and Bitar 2014 *Environment Learning as a Predictor of Mathematics Self - Efficacy and Math Achievement* American International Journal of Social Science Vol. 3, No. **6**
[16] Lilian 2012 *A Study of the Attitude, Self-efficacy, Effort and Academic Achievement of CityU Students towards Research Methods and Statistics* SS Student E-Journal Vol. 1, **154-183**
[17] Budiyono 2015 *Pengantar Penilaian Hasil Belajar* Surakarta UNS Press
[18] Patrick B, Tony H, Steve H, and Jennifer 2007 *How Can We Assess Mathematical Understanding?* Proceedings of the 31st Conference of the International Group for the Psychology of Mathematics Education, Vol. 2, pp. 41-48
[19] Ani M, Elvis N, Rahmad H 2016 *Mathematical Understanding And Representation Ability Of Public Junior High School In North Sumatra* Journal on Mathematics Education Volume 7, No. **1**, January 2016, pp. 43-56
[20] Kadir Z A, Bejor J, Nathiman R, Moamed B, and Khamis J 2003 *Meningkatkan Kemahiran Membentuk Ungkapan Kuadratik* Jurnal Penyelidikan Pendidikan, Jld 7: 12- 19.

IOP Conf. Series: Journal of Physics: Conf. Series **909** (2017) 012066 doi:10.1088/1742-6596/909/1/012066

The application of multiple intelligence approach to the learning of human circulatory system

Lita Kumalasari[1], Yusuf Hilmi A, and Didik Priyandoko

School of Postgraduate Education, Universitas Pendidikan Indonesia (UPI). Jl. Dr. Setiabudi 229, Bandung, 40154, Indonesia

E-mail: litakumalasari21@gmail.com

Abstract. The purpose of this study is to offer an alternative teaching approach or strategies which able to accommodate students' different ability, intelligence and learning style. Also can gives a new idea for the teacher as a facilitator for exploring how to teach the student in creative ways and more student-center activities, for a lesson such as circulatory system. This study was carried out at one private school in Bandung involved eight students to see their responses toward the lesson that delivered by using Multiple Intelligence approach which is include Linguistic, Logical-Mathematical, Visual-Spatial, Musical, Bodily-Kinesthetic, Interpersonal, Intrapersonal, and Naturalistic. Students were test by using MI test based on Howard Gardner's MI model to see their dominant intelligence. The result showed the percentage of top three ranks of intelligence are Bodily-Kinesthetic (73%), Visual-Spatial (68%), and Logical-Mathematical (61%). The learning process is given by using some different multimedia and activities to engaged their learning style and intelligence such as mini experiment, short clip, and questions. Student response is given by using self-assessment and the result is all students said the lesson gives them a knowledge and skills that useful for their life, they are clear with the explanation given, they didn't find difficulties to understand the lesson and can complete the assignment given. At the end of the study, it is reveal that the students who are learned by Multiple Intelligence instructional approach have more enhance to the lesson given. It's also found out that the students participated in the learning process which Multiple Intelligence approach was applied enjoyed the activities and have great fun.

1. Introduction

Howard Gardner with his book Frames of Mind in 1983 proposed the theory of Multiple Intelligence to critics educational system which always focuses on Logical-Mathematical and linguistic capabilities [1]. He proposed every human has their intelligence and ability as their strength and also their weakness. Multiple Intelligence theory is different than intelligence quotient (IQ) in some ways [2]. Multiple intelligence never working alone means every person has not only one intelligence but some dominant intelligence will work together, intelligence will influencing person's performance and products and not by a test score and also influence by the culture around them [2]. The theory of Multiple Intelligence (MI theory) claiming that as each person has a different personality and emotional level, they also have different profiles of intelligence [3]. We can find all intelligence in one person but only some potential in particular intelligence. On the other hand, all intelligence are dynamic and can be developed and also changed. Especially, culture, beliefs, social environment and personal properties are the factors developing the intelligence [4].

According to the Multiple Intelligence theory from Gardner, there are eight different types of intelligence named as logical-mathematical, linguistic, musical, visual-spatial, bodily-kinesthetic, interpersonal, intrapersonal, and naturalist. Logical-mathematical intelligence is the ability to perceive relationship and connections and to use symbolic thought, abstract and sequential reasoning skills [5]. Linguistic intelligence is the ability to think in words and to use language to express effectively both orally and in writing [6]. Musical intelligence is the ability to be sensitive to pitch, melody, rhythm, and tone [4]. This intelligence enables person to recognize, create, reproduce, and reflect on music. Visual-spatial intelligence is the ability to think with pictures, images, shapes, and lines, perceiving and comprehending three-dimensional objects [7]. Bodily-kinesthetic intelligence is the ability to expressing her/his self with movements, gestures and facial expressions, using the effective coordination of brain and body, creating a product using the whole body or a part of the whole body [7]. Interpersonal intelligence is the ability to understand and interact effectively with others. Intrapersonal intelligence is the ability to understand oneself and one's thoughts and feelings and take responsibility for his/her own life [6]. Naturalist intelligence is the ability of recognize and researching all living things in nature and thinking of their creation [7].

One of the main issues in the teaching of Biology is that students sometimes didn't catch the facts and missing the big ideas of concept in Biology learning. This make students become lose interest in learning Biology and also disconnected with their experiences. Teachers in the classroom play an important role to create a conducive and effective learning environment [8]. It is require to planning varied teaching styles which are can encourage students in the learning process through their preferred learning styles. The change and modification of teaching approaches is predominantly to enhance students' performances in Biology learning. Hence one of the approaches are to change the emphasis of teacherr-directed teaching which is also known as the teacher-centred approach of lecturing to more students-centred approaches that involve a higher degree of students' active participation in the learning process [8]. According to this fact, it is quite important to apply multiple intelligence theory in Biology teaching [8].

Teachers can apply the multiple intelligence approaches in their classroom to present their lesson using a wide variety of instructional techniques involving the use of words, numbers and logic, music, group activities, physical activities, pictures and self-reflection to cater for diverse learning styles of the students [9]. Multiple intelligence theory can give a "framework" that help teachers make decisions about the way to deliver the lesson and learning experiences for students. The multiple intelligence approaches to teaching allows teachers to merge their creativity in their instructional planning where student-centred activities could be conduct in eight different forms based on the eight types of multiple intelligence [8].

The final goal of multiple intelligence theory is to increase student understanding of Biology lesson, especially for the difficult concept. Classroom activities often more than one of the multiple intelligence [10]. Below are the following classroom activities examples:

a. Writing a report – linguistic intelligence
b. Composing a song – musical intelligence
c. Group work – interpersonal intelligence
d. Hands-on experiment – bodily-kinesthetic and logical-mathematical intelligence
e. Drawing and ordering pictures – visual-spatial intelligence
f. Self-reflection – intrapersonal intelligence
g. Observing the living things – naturalist intelligence

The purpose of this study is to determine the multiple intelligence profiles from the students and also apply the multiple intelligence approach in Biology learning process to deliver human circulatory system lesson by using some activities based on eight types of multiple intelligence theory.

2. Methodology

This study was carried out as a descriptive research which is to describe the events systematic and factual about the facts and the relationship among the research phenomenon [11]. Describing the facts from the problem as it is without manipulation.

The population of this study was eight types of multiple intelligence from all middle school students in one private school in Bandung area. The samples of this study were all grade 8 students using purposive random sampling technique who enrolled in the academic year 2016-2017. There were eight students in one class, 2 (25%) female and 6 (75%) male students.

The instrument used in this study is Multiple Intelligence Test for young people based on Howard Gardner's MI model to find the potential intelligence from the students as a reference to plan the way for deliver the lesson and also activities as students' experiences. The MI test showed as the percentages of intelligence owned by students.

2.1 Procedures of The Study

To determine the multiple intelligence profile, Multiple Inteligence test for young people based on the Howard Gardner's MI model was given at the beginning of the lesson. The test is using scale test 1-4, and the result is compile as a percentage. The result becomes a reference to plan the lesson and also the activities in the class. The lesson and activities are about the human circulatory system with sub topic are about the parts of the heart. It was crucial to develop appropriate techniques and provide necessary materials that reflect the principles of Multiple Intelligence Theory.

In the next week, the learning process has been given by using some multimedia to cover their intelligence, for example, short clip for visual-spatial and musical intelligence, a slide presentation for linguistic intelligence, discussion for interpersonal and logical-mathematical intelligence. They discussed the parts of the heart and their function including the pathway of blood.

The activities were planned based on the eight types of multiple intelligence and they did all the activities in four rotation. The activities process of the study was as below:

Table 1. Activities process rotation applied in the study

Interpersonal Intelligence	All the activities carried out in a group work
Naturalist and Visual-spatial Intelligence	The procedure started with an observation of mammal heart. They observed the heart, drawing, coloring, and labeling the parts. The students get an experience to see and hold the real heart. This makes them more interest especially the student who has dominant naturalist.
Bodily-kinesthetic and Logical-mathematical Intelligence	The students had a small investigation to measure and compare their resting and heart rate after doing the activity. They make their prediction or hypothesis. They were walking and running to have different pulse rate and make a comparison using a graph and symbol.

Linguistic and Visual-spatial Intelligence	The procedure started with a reading session about the path way of the blood. They gave some random pictures and based on the text they asked to sort the pictures in the right order. The linguistic student loves this activity so much
Table 1. (continued)	
Musical and Linguistic Intelligence	Student created a lyric about the parts of the heart and their function and combined it with the song their choose. They sing their song.
Intrapersonal Intelligence	They were given a self-reflection to get their response to the teaching and learning process in the classroom

3. Results and Discussion

The study results describe the students' Multiple Intelligence profiles from the test given. The data description of the Multiple Intelligence profile of the students is in the form of the rank and mean based on the Multiple Intelligence Test. Table 2 presents these results as follows.

Table 2. Multiple Intelligence Test result for students' profile

Intelligence	Mean	S.D.	Rank
Kinesthetic	14.625	2.56	1
Visual	13.625	2.5	2
Interpersonal	13.125	2.47	3
Logical	12.125	2.29	4
Musical	11.625	2.07	5
Naturalis	11.25	2.31	6
Linguistic	11	3.58	7
Intrapersonal	10.125	2.03	8

Table 2 showed that students' multiple intelligence profile. Of the eight multiple intelligences, students most perceived the bodily-kinesthetic (M = 14,625) and visual-spatial (M = 13,625) intelligences followed by interpersonal (M = 13,625) intelligence. Linguistic (M = 11) and intrapersonal (M = 10,125) intelligences were perceived least by the grade 8 students as shown in Table 2. In figure 1, present the percentage for each intelligence owned by students in grade 8.

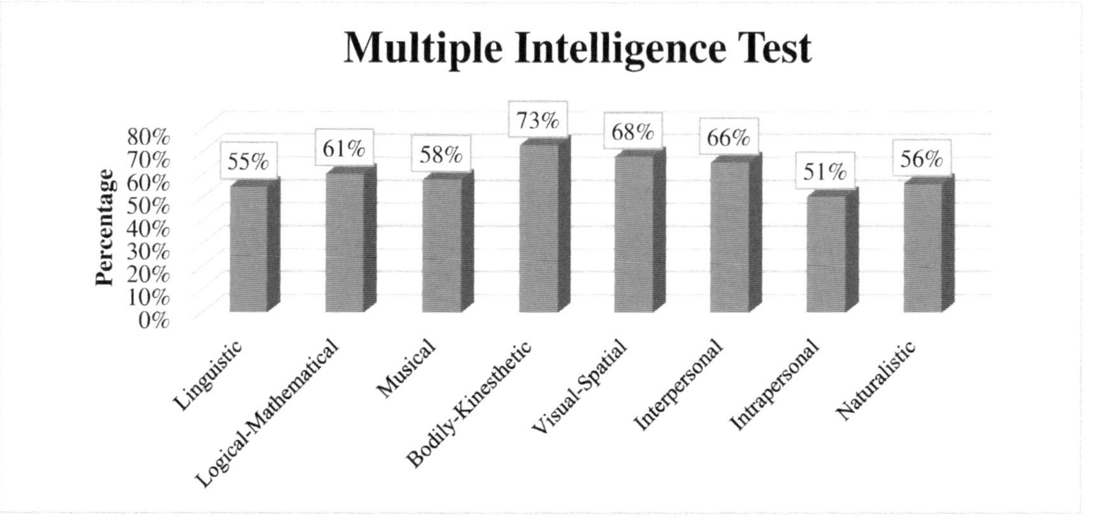

Figure 1. Multiple Intelligence inventory based on Howard Gardner's MI model

From figure 1, it showed the percentage of multiple intelligence profile from the students grade 8. The most predominant are bodily-kinesthetic (73%) intelligence followed by visual-spatial (68%) and interpersonal (66%) intelligence. The least intelligence showed by linguistic (55%) and intrapersonal (51%) intelligence.

The students are good in body movement involved activity, and they loved to move. Also they stronger in visual and interpersonal, it means they are good in group work and the activity that involves the spatial sense. Teacher than planned some activities that cater their predominant intelligence to develop their skills and qualification for understanding the concept of the human circulatory system.

3.1 The learning process results of the students

They did five activities alternately in five rotation in a group work. In the first station, they observed the mammals heart, draw it, coloring and labeling on the worksheet was given. This activity is to cover their visual-spatial intelligence. Based on the analysis of rubric to their worksheet, almost all of them did this very well. They able to drew the mammals heart based on their observation. Figure 2 below show one of their drawing result.

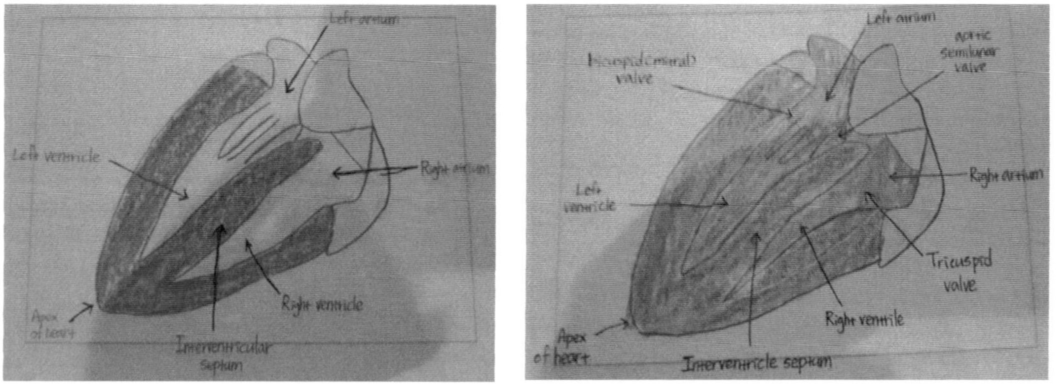

Figure 2. The mammals heart observation result

IOP Conf. Series: Journal of Physics: Conf. Series **909** (2017) 012066 doi:10.1088/1742-6596/909/1/012066

The figure 2 above shown students' drawing the mammals heart based on their observation. He able to describes what he saw very well and also labeling the parts of the heart. It means, by accommodating their visual intelligence they can develop their skill and understand the concept more easier.

In the second station, they used their kinesthetic intelligence to measure and to compare their resting pulse rate and after they did walking and running. They did this in group work. They also used their logical-mathematical intelligence to make a prediction and determine they make the correct prediction or not. They measure their resting pulse rate and collect the data. Next, they were walking and running and measure their pulse rate and collect the data. They create a graph to represent their data and also answers the questions given. Figure 3 below show one of the students result from their worksheet.

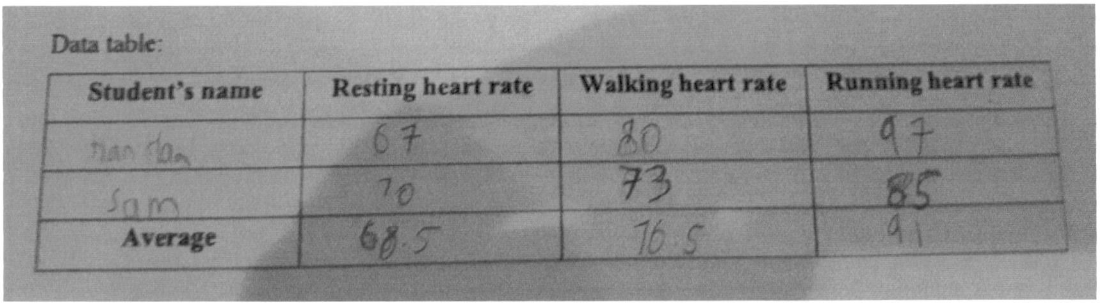
Data table:

Student's name	Resting heart rate	Walking heart rate	Running heart rate
nan dan	67	80	97
Sam	70	73	85
Average	68.5	76.5	91

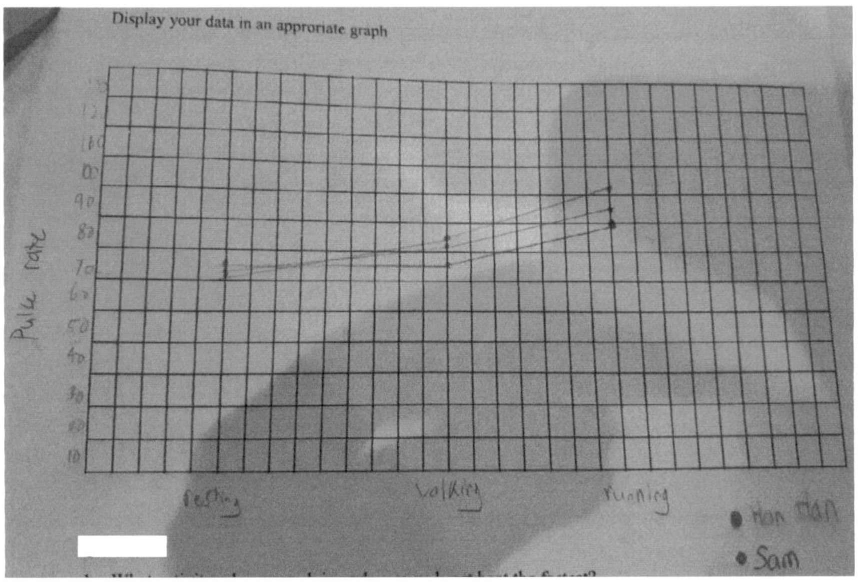

Figure 3. Data table and graph of the comparing pulse rate activity

Figure 3 shown students' result from the activity they were done. They were able to make a graph based on the data they have to represent their data collection. They were also able to conclude that their pulse rate will increase according to their activities. They understand their pulse rate influenced by some factors and that related with how their heart work to pump the blood.

In the third station, they started with the reading text about the path way of blood in the heart. They were given some random pictures and need to sort them in the correct order. Linguistic and visual-spatial intelligence are cover in this activity. Same with the oher station, they work in group work so they can discuss with the team. Figure 4 shown one of the students' work.

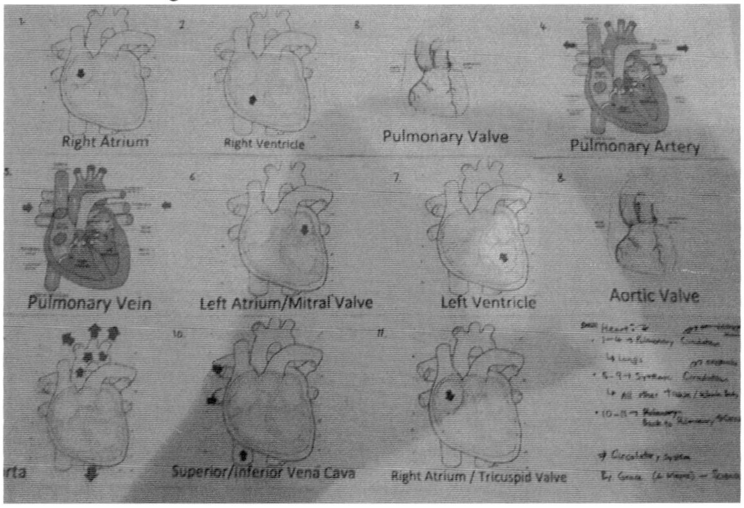

Figure 4. Sort the pictures activity to describe the pathway of blood

Figure 4 shown they did well in sorted the unscrambled pictures. They were able to understand the context of the reading text given about how heart works to pump the blood. They also able to describe it using their own words by creating some sentences from the pictures. Their linguistic intelligence was developed in this activity.

In the fourth station, they created the lyric related to the parts of the heart and how the heart works. They combine it with one song and harmonize the lyrics with the rhythm of the song. The musical intelligence is require in this activity, besides the linguistic intelligence. Figure 5 below show one of the result.

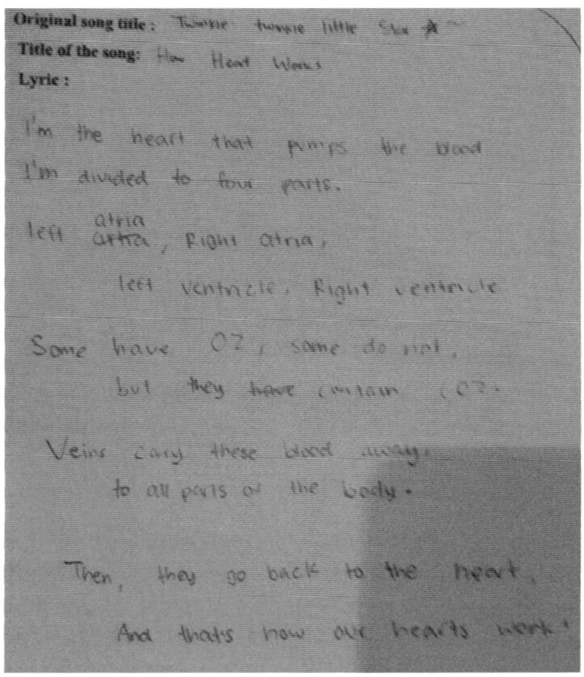

Figure 5. Create the lyric of the heart' parts

Figure 5 shown they did well to make a lyric and also put it in one appropriate song. They able to develop their musical skills and also understand how the heart works. They perform the song in front of the class and singing together with their class mate. Some of the students find difficulties to make a lyric because their musical intelligence is weak, but they can push them self to try making a lyric so without they realized, they improve and develop their skill by finished the task given.

Self-reflection has been given at the end of the learning process. The students reflect their self toward the teaching and learning process. They were given questionnaire with 1-5 scale and eight statements. Students feel this learning process is more interesting and challenging; they loved to join and involved in learning process. They understand the lesson and more conceive of the concept of the heart. They feel happy to interact directly with observed the mammals heart and this gives them a new experience. The students find some difficulties but it does not make them desperate and distract instead makes them more encouraged to solving the problem.

Every student has strength and weak sides no matter what his/her intelligence area is [7]. On condition that teacher can use more intelligence in the learning process, the students also learn how to develop their skill become their strengths. Multiple intelligence approaches make the learning process in the classroom more effective and success.

4. Conclusion

The findings provide the following conclusion: 1) There are three top rank percentage of multiple intelligence profiles from grade 8 students i.e., bodily kinesthetic (73%), visual-spatial (68%) and interpersonal (66%) intelligence. 2) The learning process is planned using multiple intelligence approaches by creating some different activities based on eight types of multiple intelligence theory. That gives the students chance to involved directly in the learning process to understand the concept; they become more connected to their learning. The result shown they able did all the activities very well, and that makes the learning process more effective. 3) Multiple Intelligence approaches helps teacher engage students through their natural curiosity and increase students participation through their excitement. "If students become engaged in the learning of science and develop positive attitudes toward science, there is a greater probability that they will develop high levels of scientific literacy" [12].

References

[1] Mariana Dumitru, Jelea Stela Gabriela. (2010). *The North University of Baia Mare, Faculty of Science, Department of Chemistry-Biology*, **XVII (1)** 100

[2] F.E Gouws. (2008). *Africa Education Review*. **4(2)** 60

[3] H. Gardner. (1999). *Frames of Mind: The Theory of Multiple Intelligence*. New York: Basic Books.

[4] Gonul Gunes, Tuba Gokcek. (2010). *The First Int. Con. Of Living Theorist*, **2(3)** 120

[5] Fang-Mei Tai. (2014). *J. of Human Res. and Adult Learning*. **10(1)** 11

[6] Ali Abdi, Aoosan Laei, Hamze Ahmadyan. (2013). *Uni. J. of Edu. Re*, **1(4)** 281

[7] Sibel G. Yalmanci, Ali Ibrahim C. Gozum. (2013). In. J. on New Trends in Edu, **4** 27

[8] Tajularipin Sulaiman, Abdul R. Abdurahman, Suzieleez Syrene A. Rahim. (2010). *ICMER*, **8** 512

[9] Okoli Josephine Nwanneka, Okoli S. Obianuju. (2014). *In. J. of Sci and Eng Re*, **5** 118

[10] Fred C. Lunenburg, Melody R. Lunenburg. (2014). *In. J. of Schol. Acad Int Div*, **16** 1

[11] Sugiyono. (2016). *Metode Penelitian Pendidikan*. Bandung: Alfabeta

[12] K. Goodnough. (2001). *School Sci and Math*, **10** 180

The Development of Scientific Literacy through Nature of Science (NoS) within Inquiry Based Learning Approach

A. Widowati[1,a], E.Widodo[1], P.Anjarsari[1], Setuju[2]

[1] Science Education Department, Yogyakarta State University,

[2] Mechanical Engineering Education, Universitas Sarjana Tamansiswa Yogyakarta

Email: [a] asri_widowati@uny.ac.id

Abstract. Understanding of science instructional leading to the formation of student scientific literacy, seems not yet fully understood well by science teachers. Because of this, certainly needs to be reformed because science literacy is a major goal in science education for science education reform. Efforts of development science literacy can be done by help students develop an information conception of the Nature of Science (NoS) and apply inquiry approach. It is expected that students' science literacy can develop more optimal by combining NoS within inquiry approach .The purpose of this research is to produce scientific literacy development model of NoS within inquiry-based learning. The preparation of learning tools will be maked through Research and Development (R & D) following the 4-D model (Define, Design, Develop, and Disseminate) and Borg & Gall. This study is a follow-up of preliminary research results about the inquiry profile of junior high school students indicating that most categories are quite good. The design of the model NoS within inquiry approach for developing scientific literacy is using MER Model in development educational reconstruction. This research will still proceed to the next stage that is Develop.

1. Introduction

The implementation of the free economic system at the ASEAN level or known as the ASEAN Economic Community (MEA) has been started in December of 2015. Thus, the people of Indonesia must prepare as well as possible so as to compete in the MEA system. Regarding anticipating the implementation of the MEA, education is important element that should be given top priority. Education is expected to contribute to the formation of the whole person for including soul, body, intelligence, sensitivity, aesthetics, responsibility, and spiritual values. Through education, everyone should be empowered to think independently and critically. In a world that is constantly changing and colored by social and economic innovation, education appears to be one of the driving forces to improve the quality of imagination and creativity as an expression of human freedom and the standardization of individual behavior. Opportunities need to be given to the younger generation to experiment and discover something new.

However, the results of the study indicate that almost half of Indonesian senior high school students (41%) have only limited knowledge about science. It is further revealed that no student can consistently identify, explain, and apply the concept of science to more complex life issues. There are also Indonesian students (6.9%) who do not have science literacy (Anna Permanasari, 2011: U19-U20). In fact, the era of the MEA demands the preparation of Indonesian human resources quality superior. The great challenge faced by Indonesia is how to strive for productive age human resources, reaching its peak in 2020 to 2035 as much as 70% of the total population of Indonesia can be transformed into human resources that have the competence and skills through education in order not to become a burden.

Therefore, students should be equipped with the ability to care and respond to the challenges and problems that develop in society, critical thinking, creative, problem solving and have a good understanding to apply the concept of science in problem solving. This ability can be achieved if students have scientific literacy. Scientific literacy is very important ability to solve various problems due to rapid changes in the field of science and technology, both related to ethics, morals and global issues. Literacy assessment is not solely on the measurement of the level of understanding of science, but also the understanding of various of scientific processes and the ability to apply knowledge and the process of science in real situations. At this time, the nation of Indonesia is facing the MEA era is full of competition in various fields and demanding qualified human resources in order not to lose compete.

Some countries, including Indonesia, have incorporated implicit science literacy in the curriculum. Unfortunately, not all teachers understand how to teach science literacy. It also has an impact on the low quality of science learning outcomes which still indicates that science learning process in Indonesian schools still ignores the acquisition of scientific literacy students. In the 2012 Program for International Student Achievement (PISA) survey, out of a total of 65 countries from the surveyed countries, Indonesia was ranked 64th. Also, there is a lack of availability of teaching materials containing Nature of Science (NoS) explicitly as research conducted by Maharani Savitri and Anita Marina Maryati (2005: 406).

Because of the importance of science literacy as the vital goal of science education and as an effort to realize the quality of human resources of Indonesia is superior as the role of education, including science education. In this case, it is important to be held science lesson that is can develop student science literacy. One of them is the implementation of the Nature of Science (NoS) inquiry approach. It is as Anna Permanasari (2011: U17) states that learning that can build scientific literacy is a lesson that focuses on active student learning, based on scientific inquiry, and constructivism.

Teachers need to comprehend comprehensively about the Nature of Science content and be able to communicate this understanding effectively to students through various strategies or learning approaches. One of the learning approaches that are based on constructivism and can provide an authentic learning experience is inquiry. The results of this study indicate that the application of inquiry on science learning has a positive effect on cognitive outcomes, process ability, and attitudes towards of Science (Ergul, et al., 2011: 62). Based on this, the student's inquiry profile must be known to determine which the inquiry approach appropriate will be implemented in the class.

Based on that reason, the problem of this research is how to produce scientific literacy development through NoS within inquiry-based science learning design? This learning design as the guide for the development of Subject Specific Pedagogic (SSP) for NoS within inquiry-based in science learning.

This research is very important because it will produce the science learning model designed to develop science literacy, that is with NoS within inquiry approach to solve some of the causes of low scientific literacy. First, the student's inquiry profile have not yet comprehensively describe, so it is important to study about it as a part of need assessment. It is assumed that low student's inquiry skill will make neglected and low in science literacy.

2. Method of Research

This research design is Research and Development (R&D). The development model used in this study refers to the model Four D Models and Borg and Gall. Procedure development consists of four main phases (phase define, design, develop, and disseminate) and an additional phase (preliminary testing fields, main product revision, playing field testing, and operational product revision) taken from Borg and Gall procedure. This research has been conducting in June s.d November 2017. Research's subject is the students of the junior high school in Yogyakarta City.

The preliminary testing field has been done at 2013 for collecting information about student's inquiry skill profile. The sample is chosen with purposive sampling technique. The criteria of the school as the sample are scientific activities have been held in its science learning. The number of the students are 211 students from three junior high school of their favorite level. The instrument in the preliminary testing of research is observation sheet for observing the students activities when doing the germination project.

3. Result and Discussion

3.1. Define Phase

In define phase, we must know the need assessment. Need assessment contribute to provide of knowledge on more efficient means of learning science. It is relevant to know what aspects need be improve in learning practice and teacher development program.

The preliminary testing field has been done by using student worksheet in theme "Germination" that apply guided inquiry approach in science learning. In this scientific activities, students investigate relationship between water and seed germination. Students was did the germination observation for several days. Amount of water was as the treatment. The observation has been done to aspects of student inquiry skill, including: make observation note, make question-based on the observation result, make a scientific problem. The observation result of students inquiry profile (N=211 students) as Figure 1.

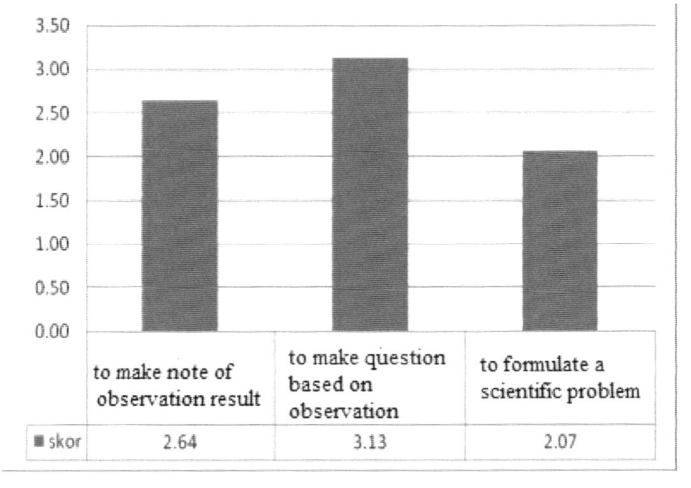

Figure 1. Average score each aspect of students inquiry skill

Figure 1 shows the highest average score of student inquiry skills reaching 3,13 on the skill of making questions based on the observation result. For the other inquiry skills, as formulate scientific problem and make a note of observation result reaching enough category (less than 3). In general, the attainment of inquiry skills are still in enough good category based on Sukarni, et.al research result (2013). As for the percentage of scores on each aspect of inquiry skill aspects input as Table 2.

Table 1. The percentage of scores on each aspect of inquiry skill

No	Score	Aspect of inquiry skill					
		Making observation note		Making question-based on the observation result		Formulating a scientific problem	
		f	%	f	%	f	%
1	Score 1	8	3,79	11	5,21	102	48,34
2	Score 2	67	31,8	46	21,8	32	15,17
3	Score 3	130	61,6	59	27,96	37	17,54
4	Score 4	6	2,84	95	45,02	4	1,90

Note: f= frequency; score 4=very good; 3= good; score 2=enough good; score 1=bad/not good

Table 1 shows that most of the student (more than 60%) have good skill on making observation note, even 2,84% has been very good. Similarly, for the skill to make question-based on observation, most students (more than 60%) also achieved good criteria, even as much as 45,02% had been very good. However, for the ability to determine the formulation of problems that can be investigated is still apprehensive because most (48,34%) lack the ability. These results indicate that junior high school students have not been trained to do inquiry.

The low science literacy of Indonesian students is believed to be due to a lack of learning involving the process of science, for example in terms of formulating scientific questions, using the knowledge it possesses to explain natural phenomena, and drawing conclusions based on facts obtained through inquiry. It shows that learning with inquiry is important to do so that science literacy develops.

3.2. Design Phase

This research uses The Model of Educational Reconstruction (MER) as mean to make an instructional design. The key concern of this model is that science subject matter issues as well as student learning needs, and capabilities have to be given equal attention in efforts to improve the quality of science learning. There are three major steps that are intimately connected as Figure 2.

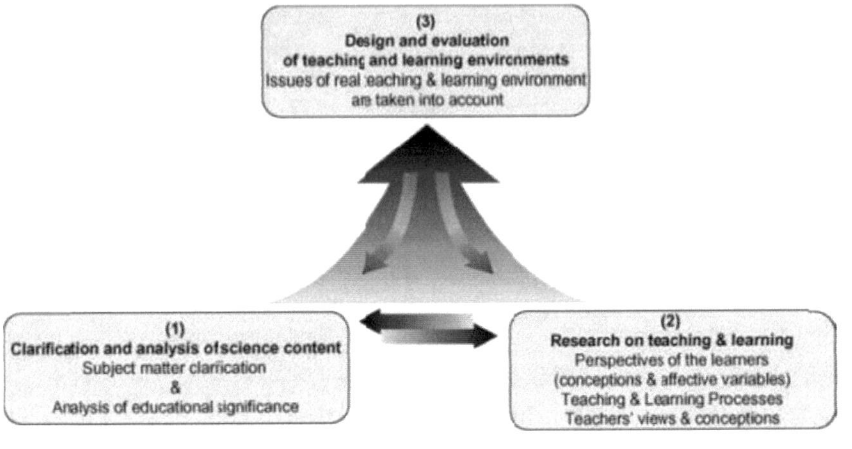

Figure 2. The three components of Model of Educational Reconstruction (MER)

1) To clarify and to analyze the science subject matter. The purpose of this component is to clarify the specific science conception and content structure from an educational point of view. Based on the curriculum of science learning in the junior high school, it is still less oriented to the development of science literacy. Other than that, NoS is still very rarely

delivered in schools. Besides because teachers do not already have an understanding of the importance of NoS and how to learn it, the teaching materials used in schools (including curriculum material 2013) have not explicitly inserted the NoS aspect. The curriculum of education in Indonesia still prioritizes aspects of the content and forgets the context and process dimensions as required by TIMSS.

2) To investigate into student and teacher perspectives regarding the chosen subject. Based on the preliminary testing field show that students' inquiry skills are enough good. The results of the 2015 survey released today show a significant increase in educational attainment in Indonesia by 22.1 points. These results put Indonesia fourth regarding student achievement improvement compared to previous survey results in 2012, from 72 countries that took the PISA test. Indonesia must struggle to improve the student scientific literacy.
3) To design and to evaluate the learning environments (e.g learning materials, learning activities, learning sequences).
   It is important to make an instructional design that is oriented as the effort to develop of scientific literacy through develop an information conception of the Nature of Science (NoS) and apply inquiry approach. By combining both, it is expected that students' science literacy can develop more optimal.

Scientific literacy is closely related to inquiry and Nature of Science (NoS) As Holbrook & Rannikmae (2009: 281) point out that "An understanding of the Nature of Science (NoS) plays an important role in the development of scientific literacy". Based on this, it is important to develop a scientific literacy development model through NoS within inquiry-based learning in science learning as an effort to improve the nation's competitiveness in facing the era of MEA. The integration of the inquiry approach and NoS is believed to further optimize the science literacy (Lederman, Lederman & Antink, 2013: 138).

Table 2. Analysis for scientific activities as integrating between inquiry and NoS

NoS Aspects	Inquiry Aspects
Scientific knowledge is empirically based.	1. Observation is an important way to learn about nature as the object of science. Through observation objects or phenomenon can learn to compare, contrast, and note similarities and differences. 2. Accurate observations and evidence are necessary to draw realistic and plausible conclusions. 3. The analysis of evidence and data is essential to make sense of the content of science.
Scientific knowledge is tentative.	1. The analysis of data from a systematic investigation may provide the student with a basis to reach a reasonable conclusion. Conclusions should not go beyond the evidence that supports them. Additional scientific research may yield new information that affects previous conclusions. 2. The scientific establishment sometimes rejects new ideas, and discoveries or innovation often spring from unexpected findings. 3. Constant reevaluation in the light of new data is essential to keeping scientific knowledge current. In this fashion, all forms of scientific knowledge remain flexible and may be revised as new data and new methods for looking at existing data become available.
Scientific knowledge is the product of	1. An *inference* is a conclusion based on evidence about events that have already occurred. Accurate observation and evidence are necessary to draw realistic and plausible conclusions.

NoS Aspects	Inquiry Aspects
observation and inference.	2. Communicating an observation accurately, one must provide a clear description of exactly what is observed and nothing more. Those conducting investigations need to understand the difference between *what is seen* and what inferences, conclusions, or interpretations can be drawn from the observation. 3. Scientific conclusions are based both on verifiable observations and on inferences. It is because science must be empiric.
Scientific knowledge is the product of creative thinking.	1. Scientists rely on creativity and imagination during all phases of their investigations. 2. Science is a human endeavor relying on human qualities, such as reasoning, insight, skill, energy, and creativity as well as intellectual honesty, tolerance of ambiguity, skepticism, and open mind to new ideas.
Scientific laws and theories are different kinds of scientific knowledge.	1. *Scientific laws* are generalizations of observation data that describe patterns and relationships. Laws may be changed if the new data become available. 2. *Scientific theories* are systematic sets of concepts that offer explanations for observed patterns in nature. Theories provide frameworks for relating data and guiding future research. Theories may be changed if new data become available.
Scientists use many methods to develop scientific knowledge.	1. Investigations can be classified as *observational* (descriptive) *studies* (intended to generate hypotheses), or *experimental studies* (intended to test hypotheses). 2. Experimental studies sometimes follow a sequence of steps known as the Scientific Method: stating the problem, forming a hypothesis, testing the hypothesis, recording and analyzing data, stating a conclusion. However, there is no single scientific method. Science requires different abilities and procedures depending on such factors as the field of study and type of investigation. 3. Different kinds of problems and questions require differing approaches and research. Scientific methodology almost always begins with a question, is based on observation and evidence, and requires logic and reasoning. Not all systematic investigations are experimental.
Scientific knowledge is subjective and culturally influenced.	1. Investigation not only involves the careful application of systematic (scientific) methodology but also includes the review and analysis of prior research related to the topic. Numerous sources of information are available from print and electronic sources, and the researcher needs to judge the authority and credibility of the sources. 2. It is typical for scientists to disagree with one another about the interpretation of the evidence or the theory being considered. Because of this, it is make partly a result of the unique background (social, educational, etc.) that individual scientists bring to their research. Because of this inherent subjectivity, scientific inquiry involves evaluating the results and conclusions proposed by other scientists.

To organize a Nature of Science (NoS) within inquiry approach, relevant learning tools are required. Specific learning tool or Subject Specific Pedagogic (SSP) which not only contains material content, but also invites students to inquiry and contains knowledge about science related to the nature of science (Nature of Science / NoS). The analysis for scientific activities as integrate between inquiry and NoS result as in Table 3 can be use as a guidance to make SSP NoS within inquiry-based learning approach, especially for the scientific activities as the part of learning process.

This research will still proceed to the next stage that is Develop. At this stage, the data obtained from the previous stage results are used as the basis of the development of Subject Specific Pedagogic in Science Learning through NoS within inquiry-based learning.

4. Conclusion

Based on the results and the above discussion, it can be concluded that: (1) In general for each aspect inquiry skill still in the category good enough. The inability of students to do inquiry is still not optimal; (2) An instructional design Nature of Science (NoS) within inquiry-based learning approach that is expected that students' science literacy can develop using The Model of Educational Reconstruction (MER) as means to make an instructional design.

Reference

[1] Abd-El- Khalick, F 2005 Developing deeper understandings of nature of science: The impact of the philosophy of science course on pre-service science teachers' views and instructional planning. *International Journal of Science Education*. **27**(1) pp 15-42

[2] Anna Permanasari 2011 Pembelajaran Sains: Wahana Potensial untuk Membelajarkan Softskill dan Karakter. *Prosiding Seminar Nasional Pendidikan MIPA.* Lampung: FKIP UNILA

[3] Duit, R., Gropengießer, H., Kaufmann, U., Komorek, M. dan Parchmann, I. 2012 "The Model Of Educational Reconstruction – A Framework For Improving Teaching And Learning Science. Sci. Educ. Res. and Pract. in Europe": *Retrospective and Prospective,* ***5,*** p13–37. Retrieved from www.ejmste.org/v3n1/EJMSTEv3n1_Duit.pdf.

[4] Ergul, R., Yeter S.., Sevgül Çalu, Zehra Özd Leku G., Meral A. 2011 The effect of inquiry-based science teaching on elementary school students' science process skills and science attitudes [Electronic version]. B*ulgarian Journal of Science and Education Policy* (BJSEP), **5** 1 p48 Retrieved from http://bjsep.org/getfile.php?id=88, at 11th November 2014

[5] Holbrook, *J.* & Rannikmäe, M. *2009* The Meaning of Scientific Literacy. *International Journal of Environmental and Science Education*, 4, p275 - 288

[6] Lederman, N.G., Lederman, J.S., & Antink, A. 2013 Nature of science and scientific inquiry as contexts for the learning of science and achievement of scientific literacy. *International Journal of Education in Mathematics, Science, and Technology*, **1**(3) p138-147.

[7] Maharani Savitri dan Anita Marina Maryati 2015 Rekonstruksi Bahan Ajar Bermuatan View of Nature of Science untuk Pembelajaran IPA di SMP *Prosiding Simposium Nasional Inovasi dan Pembelajaran Sains 2015*. Bandung: UPI

[8] OECD 2015 *Draft Science Framework.* Retrieved from https://www.oecd.org/pisa/pisaproducts/Draft%20PISA%202015%20Science%20Framework%20.pdf.

[9] Sukarni Hidayati, Surachman, Asri Widowati 2013 Profil Keterampilan Berinkuiri Siswa SMP Kota Yogyakarta dalam Pembelajaran IPA-Biologi. *Research Report.* Yogyakarta: FMIPA, UNY.

Acknowledgments

The authors thank the junior high school in Yogyakarta city, Faculty of Mathematics and Science, YSU, due to the support for this research. The Ministries of Research, Technology, and Universities for the funding of this research.

IOP Conf. Series: Journal of Physics: Conf. Series **909** (2017) 012068 doi:10.1088/1742-6596/909/1/012068

Enhancing the engineering properties of expansive soil using bagasse ash

Niken Silmi Surjandari[1], Noegroho Djarwanti[1], Nafisah Umri Ukoi[1]

[1]Civil Engineering Department, Engineering Faculty, Sebelas Maret University, Jl. Ir. Sutami 36A Surakarta 57126, Central Java, INDONESIA

E-mail: nikensilmisurjandari@gmail.com

Abstract. This paper deals with stabilization of expansive soil on a laboratory experimental basis. The aim of the research was to evaluate the enhancement of the engineering properties of expansive soil using bagasse ash. The soil is treated with bagasse ash by weight (0, 5, 10, 15, and 20%) based on dry mass. The performance of bagasse ash stabilized soil was evaluated using physical and strength performance tests, namely the plasticity index, standard Proctor compaction, and percentage swelling. An X-ray diffraction (XRD) test was conducted to evaluate the clay mineral, whereas an X-ray fluorescence (XRF) was to the chemical composition of bagasse ash. From the results, it was observed that the basic tests carried out proved some soil properties after the addition of bagasse ash. Furthermore, the plasticity index decreased from 53.18 to 47.70%. The maximum dry density of the specimen increased from 1.13 to 1.24 gr/cm3. The percentage swelling decreased from 5.48 to 3.29%. The outcomes of these tests demonstrate that stabilization of expansive soils using bagasse ash can improve the strength.

1. Introduction

Expansive soils cause several problems for civil engineering infrastructure. Various methods are applied to improve the characteristics of expansive soils. Soil stabilization is one of the construction techniques because it improves the engineering properties of soil such as the strength and volume stability. When the soil strength properties cannot be improved by mechanical stabilization, the chemical admixture technique is used to achieve the desired strength. Chemical stabilization is the most popular method utilized to enhance the physical and mechanical properties of problematic soils consisting of soft soil and expansive soil.

In recent years, a considerable number of laboratory experiments have been carried out, and extensive studies have been conducted on reactive soil using various additives such as lime [see references:1, 2, 3, 4]; sand, and cement [see reference 5]. Several waste by products including fly ash [see reference 6], sugar cane straw ash [see reference 7], eggshell powder [see reference 8], blast furnace slag [see reference 9], marble dust [see references 10], stone dust [see reference 11], and used lubrication oil [see reference 12] have been investigated by employing each material alone or in combination with other additives. Many researchers [see reference: 13, 14, 15, 16, and 17] have performed several studies on bagasse ash to evaluate the stabilization properties of expansive soil. Based on the test results, they indicated that bagasse ash admixture caused significant modification and improvement of the engineering properties of expansive soil. Bagasse is the fibrous residue generated after the juice has been extracted from the sugar cane plant. It is deposited as waste and it pollutes the environment [4]. The resulting ash is deposited in stockpiles, which are dumped in waste landfills and constitute environmental problems for society.

In this paper, the expansive soil is embedded with different proportions of bagasse ash (0, 5, 10, 15, and 20% based on dry soil mass). Various tests are carried out on the modified expansive soil to find the index properties, maximum dry density, and percentage swelling.

2. Methodology

2.1 Clay Soil

The soil used in this study was obtained from the district of Jono, in Sragen, Central Jawa from a depth of 0.5 to 1.0 m below the natural ground level. Samples were kept safe and dry in jute bags. They were air dried to allow elimination of water, which may affect sieve analysis. X-ray diffraction (XRD) was conducted to evaluate the soil mineralogy.

2.2 Bagasse Ash

Bagasse ash is a residue obtained from the burning of bagasse in sugar-producing factories. This material contains amorphous silica, which is an indication of cementing properties, which can develop good bonding between grains in the case of weak soil. To stabilize expansive soil, the waste product bagasse ash is collected from Mojo Sugar Company, situated in Sragen Central Java. Bagasse ash is used as a stabilizer to deal with environmental concerns through the reduction of sugar industry waste material. An X-ray fluorescence (XRF) test was conducted to evaluate the chemical composition of bagasse ash material. Figure 1 shows a photograph of bagasse ash material.

Figure 1. Bagasse ash material.

2.3 Experimental Investigation

In this study, the additive contents are defined by the ratio of the weight of a particular additive to the dry weight of the soil, expressed as a percentage. Experimental oedometer tests were conducted by the authors on selected expansive soil samples with different bagasse contents, that is, 0, 5, 10, 15, and 20%. The soil samples were examined to determine the percentage swelling using Equation 1. Swelling tests were conducted on samples prepared at a water content similar to the optimum moisture content for untreated soil with various dry densities depending on the standard compaction (ASTM D 698-91). In the first step, a series of tests were conducted on untreated expansive soil samples. The second set of tests was associated with bagasse-ash-treated soil. The sample used for the swelling test is illustrated in Figure 2.

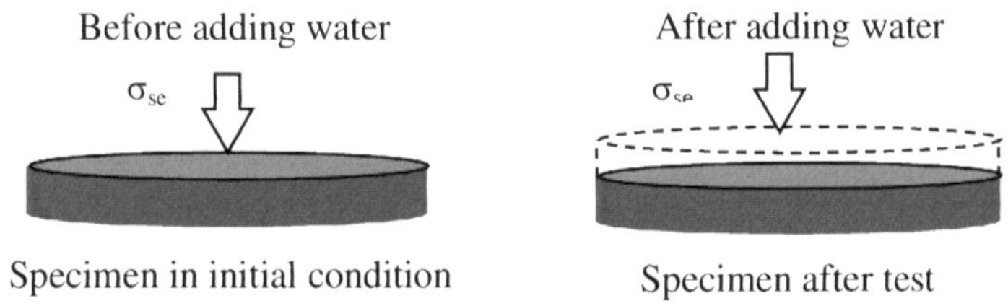

Figure 2. Specimen behavior under swelling test.

An oedometer is one of apparatuses used to measure the degree of swelling. The swelling process is the opposite of the consolidation process; that is, the volume of the soil increases gradually because of excess negative pore pressure [19]. The percentage swelling according to SNI 6424:2008 is calculated using

$$\text{Percentage swelling} = \frac{\Delta h}{h_0} \times 100\% \tag{1}$$

3. Results and discussion

3.1 Characteristics of natural soil

The soil samples used in this paper were collected from the district of Jono in Sragen, Central Java, Indonesia. Table 1 shows the physical properties of the soil used in this investigation. In term of the sizes of particles, the soil was classified as clay of high plasticity (CH) according to the Unified Soil Classification System (USCS). The specific gravity of solids (Gs) was 2.62. The grain size distribution showed that 96.23% was fine-grained material (silt/clay). The Atterberg limits of the fine portion of material were a liquid limit (LL) of about 87.35% and a plastic limit (PL) of 33.37%, which resulted in a plasticity index (PI) of 53.98%. Based on the high plasticity index, the soil can be classified as highly expansive soil.

Table 1. Index properties of expansive soil.

Properties	Value
Natural water content (%)	53.87
Specific gravity (Gs)	2.62
Unit weight (kN/m3)	14.2
Percentage passing 75-µm sieve (%)	96.23
Liquid limit (%)	87.35
Plastic limit (%)	33.37
Plasticity index (%)	53.98
USCS	CH

3.2 XRD test of clay soil from Jono district

XRD was conducted to evaluate the mineralogy of the clay soil. Table 2 shows the XRD results.

Table 2. Major peaks found by XRD test.

No.	Peak no.	2Theta (°)	d (A)
1	22	20.12	4.40979
2	30	26.8194	3.32151
3	3	5.8283	15.1516

3.3 XRF test of Mojo Sugar Company bagasse ash
The chemical composition analysis of the bagasse ash was carried out by XRF test, and the results were as shown in Table 3. The bagasse ash was found to contain mainly silicon, calcium, potassium, and aluminum.

Table 3. Chemical composition analysis of the bagasse ash.

Abbreviation	Ash (%)	Abbreviation	Ash (%)
SiO2	67.33%	MnO	0.27%
CaO	12.51%	ZnO	0.05%
K2O	4.11%	Nd2O3	0.05%
Al2O3	3.45%	CuO	0.05%
Fe2O3	3.39%	V2O5	0.02%
P2O5	3.10%	ZrO2	0.02%
SO3	2.81%	Rb2O	0.02%
MgO	2.04%	Cr2O3	0.01%
Cl	0.46%	Bi2O3	0.01%
TiO2	0.28%		

(Source: XRF test, Lab. MIPA Terpadu UNS)

3.4 Result of Atterberg Limit for Treated Soil
Atterberg limits were found, and grain size tests were conducted on the natural and treated soil in the investigation to determine LL, PL, and PI. Figure 3 shows the variation of LL, PL, and PI with the percentage of bagasse ash.

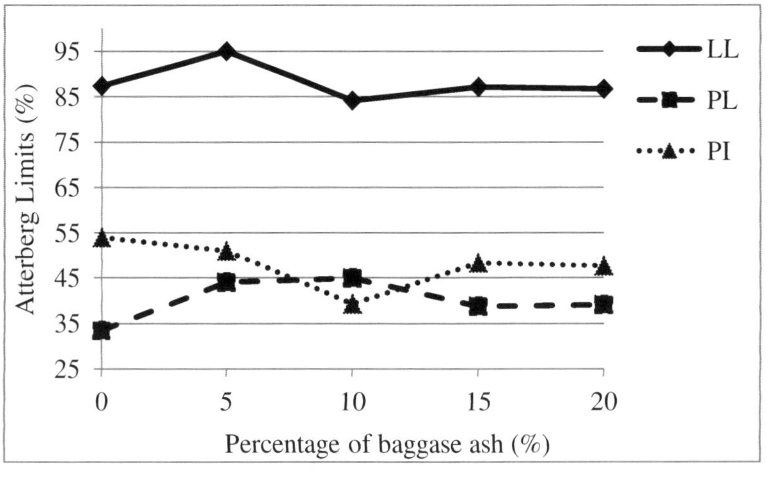

Figure 3. Effect of bagasse ash on Atterberg limit of soil-bagasse ash mixes.

The PI of the clay sample treated by bagasse ash showed a decrease with increases in the quantity of bagasse ash added, as illustrated in Figure 3. The addition of bagasse ash has a remarkable effect on the plasticity of cohesive soils. The general decrease in the liquid limit for the combination of bagasse ash is attributed to the fact that the compounds formed possess cementitious properties due to calcium silicate with soil particles.

3.5 Compaction Behavior

Light compaction tests were conducted on the natural and treated soil in the investigation to determine the maximum dry density (MDD) and optimum moisture content (OMC) values. Figures 4 and 5 show the variation of OMC and MDD with the percentage of bagasse ash, respectively.

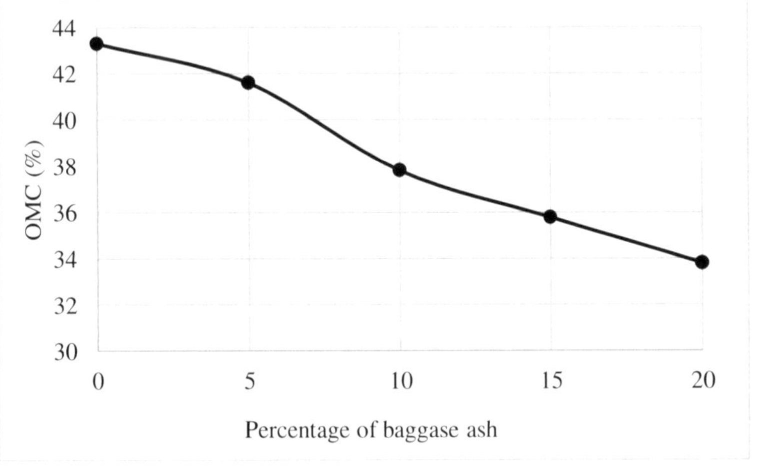

Figure 4. Effect of bagasse ash the optimum moisture content of soil/bagasse-ash mixes.

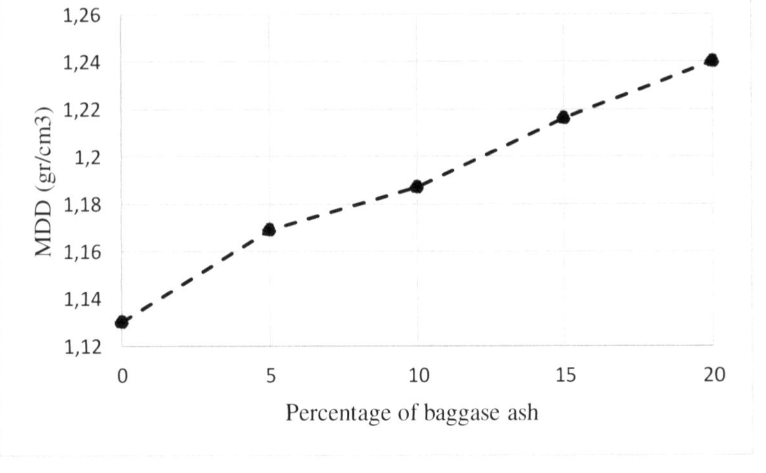

Figure 5. Effect of bagasse ash the maximum dry density of soil/bagasse-ash mixes.

Figure 4 shows that the OMC decreased with increases in the percentage of bagasse ash mixture, and Figure 5 shows that the MDD increased with increases in the percentage of bagasse ash mixture. Bagasse ash causes the soil particles to flocculate. Thus the soil becomes more easily workable and its strength and stiffness increase.

3.6 Percentage Swelling of Treated Soil

A consolidation test according to the ASTM D 2435-02 setup was used to determine the percentage swelling of the soil. The percentage swelling of the soil was calculated with Equation 1. ,Table 4 and Figure 6 show the results of the treated soil.

Table 4. Test results of swelling behavior of treated soil.

Percentage of bagasse ash (%)	H_0 (cm)	H_1 (cm)	Percentage of swelling (%)
0	1.8	1.90	5.48
5	1.8	1.90	5.28
10	1.9	1.97	3.54
15	1.9	1.96	3.08
20	1.9	1.96	3.29

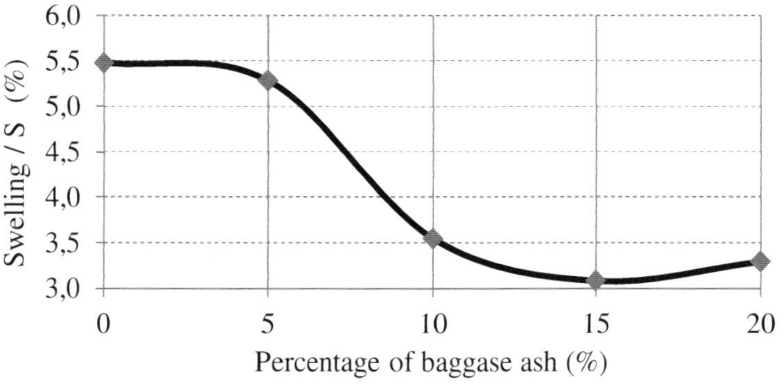

Figure 6. Variation of percentage swelling value with the percentage of bagasse ash.

The addition of 5, 10, and 15% bagasse ash led to a reduction in the percentage swelling. A bagasse ash dose of 15% led to the smallest percentage swelling, so we can say that this was the optimum dose. At a bagasse ash dose of 20%, the percentage swelling tended to increase by a small value.

4. Conclusion

Use of the soil stabilization method by applying waste product bagasse ash was able to successfully improve the expansive soil. The addition of bagasse ash reduces the plasticity index and the percentage swelling and increases the maximum dry density. Bagasse ash is cost-free and available locally; hence it proved economical too. Further research needs to be conducted to find the optimum percentage of bagasse ash that should be applied as a stabilizer and used in combination with other additives, that is, slag furnace, plastic waste, and so on.

References

[1] Barasa, PK. Too, D. Jonah, K. and Mulei, SM.. 2013. Stabilization of Expansive Clay Using Lime and Sugarcane Bagasse Ash. *International Journal of Science and Research (IJSR)*. **4**(4): 2112-2117.

[2] Barasa, PK. Jonah, K. and Mulei. 2015. Stabilization of Expansive Clay Using Lime and Sugarcane Bagasse Ash. *Mathematical Theory and Modeling*. **5** (4): 124-134.

[3] Dang, LC. Hasan, H. Fatahi, B. Jones, R. and Khabbaz, H.2016. Enhancing the engineering properties of expansive soil using bagasse ash and hydrated lime. *International Journal of GEOMATE*. **11**(25): 2447-2454.

[4] Hasan, H. Dang, L. Khabbaz1, H. Fatahi, B. and Terzaghi, S. 2016. Remediation of Expansive Soils Using Agricultural Waste Bagasse Ash. *Procedia Engineering.* **14**(3):1368–1375.

[5] Ramteke, N.B.; Saxena, A.; Arora, TR.2014. Stabilization of Black Cotton Soil With Sand and Cement as a Subgrade for Pavement, *International Journal of Engineering Sciences & Research Technology.* **3**(6):688-692.

[6] Rajakumaran, K.2015. An experimental analysis on stabilization of expansive soil with steel slag and fly ash. *International Journal of Advances in Engineering & Technology.* **7**(6):1745-1752.

[7] Chakraborty, A. Borah, A. and Sharmah, D. 2016. Stabilization of Expansive Soil using Sugarcane Straw Ash (SCSA). *ADBU-Journal of Engineering Technology, AJET.* **4**(1): 2348-7305.

[8] Ahmed, B. Rahman, A. Das, J. 2015. Improvement of subgrade CBR value by using Bagasse ash and Eggshell powder. *International Journal of Advanced Structures and Geotechnical Engineering* **4**(2): 86-91.

[9] Chavan, P. and Nagakumar, MS. 2014. Studies on soil stabilization by using bagasse ash. *International Journal of Scientific Research Engineering & Technology (IJSRET).* 89-94.

[10] Ali, RH. Khan. Shah, AA. 2014. Expansive Soil Stabilization Using Marble Dust and Bagasse Ash. *International Journal of Science and Research (IJSR).* **3** (6) : 2812-2816.

[11] Mudgal, A. Sarkar, R. Sahu, AK.2014. Effect of Lime and Stone dust in the Geotechnical Properties of Black Cotton Soil. *International Journal of GEOMATE.* **7**(2):1033-1039.

[12] Sadeeq, JA; Ochepo,J.; and Salahudeen, BA.2016. Effect of used oil contamination and bagasse ash on some geotechnical properties of lateritic soil. *Leonardo Electronic Journal of Practices and Technologies.* (**28**): 119-136.

[13] Chhachhia, A. and Mital, A. 2015 . Review on improvement of clayey soil Stabilized with bagasse ash. Ijrrest international journal of research review in engineering science & technology. **4** (1): 238-241.

[14] Gandhi, KS. Expansive Soil Stabilization Using Bagasse Ash. 2012. *International Journal of Engineering Research & Technology (IJERT).* **1**(5): 1-3.

[15] Hatmoko, JT. Luli, J. 2007. UCS tanah lempung ekspansif yang distabilisasi dengan abu ampas tebu dan kapur. *Jurnal Teknik Sipil.* **8**(1): 64-77.

[16] Kharade, AS. Suryavanshi, VV. Gujar, BS. and Deshmukh, RR.2014. Waste product 'bagasse ash' from sugar industry can be used as stabilizing material for expansive soil. *IJRET: International Journal of Research in Engineering and Technology.* **3**(3):506-512.

[17] Subhacini, C. Ranjitha, M. Dhanapal, S. Prakash, KA and Shankar, KU.2015. Expansive Soil Stabilization Using Waste from Sugarcane Industry. *Journal for Studies in Management and Planning*, **1**(3):245-252.

[18] Chen,F.H.1975. Foundation on Expansive Soils. Elsevier Scientific Publishing Company. New York.

[19] Craig, R.F.1987. Mekanika Tanah, (Alih Bahasa Soepandji, B.S.) Ed.4. PT Erlangga. Jakarta. Indonesia.

Self-supporting method; an alternative method for steel truss bridge element replacement

Muhammad Arsyad[1], Senot Sangadji[1] and Sholihin As'ad[1]

[1] Civil Engineering Department of Post Graduate Program Universitas Sebelas Maret Surakarta, Jl. Ir. Sutami 36A Kentingan Jebres Surakarta 57126, INDONESIA

E-mail: arsyad@student.uns.ac.id, s.sangadji@ft.uns.ac.id, sholihinasad@staff.uns.ac.id

Abstract. Steel truss bridge often requires replacement of its element due to serious damage caused by traffic accidents. This replacement is carried out using temporary supporting structure. It would be difficult when the available space for the temporary structure is quite limited and or the position of work is at a high elevation. The self-supporting method is proposed instead of temporary supporting structure. This paper will discuss an innovative method of bridge rehabilitation by utilizing the existing bridge structure. It requires such temporary connecting structure that installed on the existing bridge element, therefore, the forces during replacement process could be transferred to the bridge foundation directly. By taking the case on a steel truss bridge Jetis Salatiga which requires element replacement due to its damages on two main diagonals, a modeling is carried out to get a proper repair method. Structural analysis is conducted for three temporary connecting structure models: "I," "V," and triangular model. Stresses and translations that occur in the structure are used as constraints. Bridge bearings are modeled in two different modes: fixed-fixed system and fixed-free one. Temperature load is given in each condition to obtain the appropriate time for execution. The triangular model is chosen as the best one. In the fixed-fixed mode, this method can be carried out in a temperature range 27-28.8° C, while in fixed-free one, the temperature it is allowed between 27-43.4 °C. The D4 is dismantled first by cutting the D4 leaving an area of 1140.2 mm^2 or 127 mm web length to enable plastic condition until the D4 collapses. At the beginning of elongation occurs, immediately performed a slowly jacking on a temporary connecting structure so that the force on D4 is gradually transferred to the temporary connecting structure then the D4 and D5 are set in their place.

1. Introduction

Bridges are essential parts of the road system. The failure or impaired function of the bridges may limit the traffic, restrict the mobility in the network of people and goods, and leads to the economic loss for the community [1]. It is essential that particular attention is given to inspection of bridges as part of the management of road system.

According to Indonesian Bridge Management System (IBMS), the bridge maintenance includes routine maintenance, periodic maintenance, and rehabilitation. Rehabilitation includes repairing or replacement of upper structures, substructures and water flow arrangement. It is important to detect damage on bridge start from visual checking, defect detailing to computer modeling to ascertain the proper repair works and its method [2]. In the case of steel bridges, some treatments are the following:

crack repair, cleaning, cathodic protection, protective coating, strenghtening, and member/element replacement [3].

Traffic accidents may cause damages on steel truss bridge structural element as the position of the main structure is commonly beside traffic lanes. In this case, truss elements often require replacement due to a serious defect. This replacement is carried out by using temporary supporting structure which is placed below the replaced element to support forces that occur at the time of replacement process. It would be difficult to execute when the available space for the temporary structure is quite limited and or the position of the replaced element is at a high elevation. Self-supporting method is proposed instead of temporary supporting structure. It does not require a temporary supporting structure while the forces that occur are accommodated by the existing structure until the replacement is successfully done.

The Jetis Bridge is an A40 steel truss bridge with 9 meters width. Caused by a traffic accident, two main diagonal truss elements (D4 and D5 on the right side) became ripped, twisted, and bent and needed to be replaced with the new ones [4]. Considering limited space around the bridge, the temporary supporting structure as the common method is not applicable. It needs another innovative method by utilizing the existing bridge structure as a self-supporting system. It requires such temporary connecting structure that installed on the existing bridge element, therefore, the forces during replacement process could be transferred to the bridge foundation directly. The method enables the new elements to connect well with minimum intervention to the structure.

2. Methodology

The study of the self-supporting method for steel truss bridge element replacement can be achieved by field measurement and structural analysis modeling. It requires dimension and field temperature data especially the temperature on work hours considering the time of replacement process. The detailed model may be developed taking into account actual dimension of the bridge.

Firstly, the structure was modeled as the actual condition before element replacements where the working load was dead load (upper structure self-weight: steel, concrete, and asphalt). The bearings were set as fixed-fixed at left abutment and fixed-free at right one. The outputs of this modeling were element stresses and node displacements. These data were then used as benchmarks on the next modeling.

Secondly, three (3) alternative shapes of connecting structure were designed and studied. The shapes are "I," "V," and triangular models. In structural analysis modeling, these models were attached to bridge structure to transfer forces to the foundation while the diagonal (D4) on the right side was dismantled. Each model would represent different behavior to steel structure, therefore, it can be determined which model would be chosen for replacement process.

Thirdly, the best connecting structure model was attached on bridge structure in modeling and analyzed to obtain the bridge element stresses and node displacements. Bridge bearings were modeled in two different conditions: fixed-fixed system and fixed-free one. The temperature load was applied from 27°C to certain value that results in xz-displacement as 2 mm.

2.1. Steel Truss Bridge

Truss bridge has been popular for hundreds of years. Steel became a common construction material for bridges, replacing timber and masonry construction to achieve longer span crossings. These structures are still functioning well, although the majority of them are now considered to have sub-standard capacity in the 21st century [5]. Truss elements have evolved from rods, bars, eyebars to box and H-shaped elements. Generally speaking, the box elements are more structurally efficient better than H-shapes, whereas H-shapes are perceived as being more economical regarding of fabrication and easier to connect to the gusset plates because of open access to bolts and easier to maintain because all surfaces are accessible [6].

Steel truss bridge usually contains top chords, bottom chords and diagonals, floor beams (cross girders), and stringers. Bearings support all load and transfer them to foundations. Lateral elements in the plane of top and bottom chords resist wind loads and brace the compression chords and sway elements are thought to square the truss and increase its torsional rigidity.

Generally steel truss bridge in Indonesia uses steel with Fy=350 MPa (i.e Fe510C, ASTM A-572 grade 50, S355JO, HISTAR S460, or SM490) [7]. The connection among bridge elements is usually using bolts. This connection is considered to ease the transportation and erection process [8], usually uses MS (mild steel), HYS (high yield steel), ASTM A325 grade 8.8, or 10.9 HV bolts with tensile strength = 1000 MPa [7]. In more modern steel bridge, prefabricated welded gusset plates connect all elements and eliminate the use of bolts [9].

Jetis Bridge is a Warren type truss without vertical which was built in 1998. With 39.972 m long and 9 m wide including pedestrian, it uses H-shaped elements for its top chords, bottom chords, diagonals, cross girders, stringers, wind bracings and sway elements. The actual dimensions of main elements are summarized in Table 1.

Table 1. Dimension of main element of bridge

No.	Element	Dimension (mm)	No.	Element	Dimension (mm)
1	Top chord	H.340.400.10.13	5	Diagonal 4-13	H.336.249.9.12
2	Bottom chord 1,3,4,5,6,8	H.340.350.11.13	6	Cross girder	H. 947.225.12.21
3	Bottom chord 2	H.340.250.11.14	7	Stringer	H.447.152.8.10
4	Diagonal 1,2,3,14,15,16	H.340.350.12.13			

2.1.1. Load. Generally, Bridges are designed due to dead load, live load and environmental action i.e. temperature, the wind, flood, earthquake, and others [10]. Vehicle load is one of the most important factors to influence the structural safety and applicability of the highway bridge [11]. Design load in RSNI T-02-2005 is calculated according to statistical analysis of common occurrence. Especially in bridge rehabilitation, it needs temporary supports, load restrictions, or diverting traffic, typical courses of action exercised by road authorities during the remediation process [3]. In this case, only dead load, temperature load, and construction load are applied. The dead load consists of self weight of steel structure, concrete slab, concrete pedestrian, and asphalt layer. The temperature load is set to the ambient condition in the bridge location.

2.1.2. Boundary. Generally, The truss bridge boundaries in service condition are set to a fixed-free mode to provide longitudinal movement of the bridge, even in wide steel bridge, sometimes lateral movements of bridge are accommodated. The bridge behavior depends on moving mechanism of overall structure including its bearings [12]. Steel truss bridge usually uses elastomeric bearing which contains rubber at the top of bearing which is vulcanized to a steel plate [13]. In modeling of this element replacement process, these bearings are examined in two (2) different modes: fixed-fixed and fixed free to get the suitable method.

2.1.3. Temperature and displacement. Steel is quite sensitive to temperature change with its lengthening coefficient due to temperature is 12×10^{-6} per°C [14]. The structural responses of bridge are deeply affected by the variation and distribution of temperatures on bridges [15]. Loads and temperature which work on structure results in displacements. Displacement/deflection characteristics are also used as an indicator to diagnose the presence of local damage [16]. In steel truss bridge, at abutment or pier, it is always provided such space to accommodate longitudinal movement. The Jetis Bridge is located in Salatiga, a city in Central Java Province which stretches along 7°17'00"-7°17'23" S and 110°27'56.81"–110°32'4.64" E [17]. Its average temperature is 23°C [18]. The onsite temperature data on May 20th-21st,2017 were 27°C at 08.00 am, increased to 37°C at 11.00 am, and decreased to 27°C to at 05.00 pm. This data will be used in modeling to determine the temperature range for replacement execution.

IOP Conf. Series: Journal of Physics: Conf. Series **909** (2017) 012069 doi:10.1088/1742-6596/909/1/012069

2.2. Replacement method
False-work systems are frequently used in the construction, rehabilitation, and retrofit works of bridge structures [19], beside the cantilever method, float-ins, and tie-backs. Element replacement method can adopt the principles used in the bridge erection method. The method that commonly applied is false-work, by using temporary support structure below the bridge structure which is rested on the ground. This method is practical if the location is quite wide, the soil bearing capacity is adequate, and the bridge elevation is not high. With the self-supporting method, the use of this support is replaced by a temporary connecting structure that transmits loads to the foundation through the existing structure. With this method, the bridge element replacement at any location on the bridge can be done.

3. Result and Discussion

3.1. Modeling of initial condition
The initial condition was defined as the condition of the bridge at preparation stage of replacement without traffic live load. Dead load included upper structure self-weight, concrete deck slab, the asphalt layer and pedestrian concrete. The boundary condition was fixed-fixed at left abutment and fixed-free at right one. The structural model is shown in Figure 1 while the structural analysis result of this condition is summarized in Table 2 and Table 3.

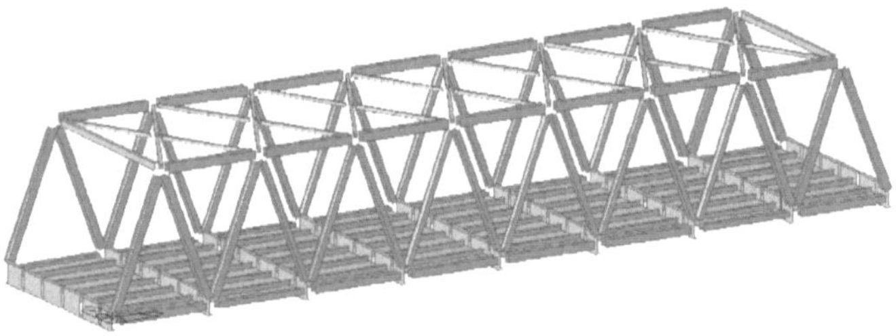

Figure 1. The initial model of bridge structure

Table 2. Element stresses of bridge structure in initial mode

No.	Element	Stress (Mpa)	No.	Element	Stress (Mpa)
1	Diagonal 1	-91.62	17	Top Chord 1	-60.96
2	Diagonal 2	96.24	18	**Top Chord 2**	**-82.93**
3	Diagonal 3	-81.13	19	Top Chord 3	-99.61
4	**Diagonal 4**	**96.76**	20	Top Chord 4	-105.55
5	**Diagonal 5**	**-69.42**	21	Top Chord 5	-99.61
6	Diagonal 6	67.09	22	Top Chord 6	-82.93
7	Diagonal 7	-39.19	23	Top Chord 7	-60.96
8	Diagonal 8	34.71	24	Bottom Chord 1	10.83
9	Diagonal 9	34.71	25	**Bottom Chord 2**	**11.18**
10	Diagonal 10	-39.19	26	**Bottom Chord 3**	**12.31**
11	Diagonal 11	67.09	27	Bottom Chord 4	13.91
12	Diagonal 12	-69.42	28	Bottom Chord 5	13.91
13	Diagonal 13	96.76	29	Bottom Chord 6	12.31
14	Diagonal 14	-81.13	30	Bottom Chord 7	11.18
15	Diagonal 15	96.24	31	Bottom Chord 8	10.83
16	Diagonal 16	-91.62			

The translations especially located around D4-D5 are determined as benchmarks for the next modeling, which the xz-translation differences should not be more than 2 mm to fit the element bolts in their holes. The stresses on elements around replacement location should not be more than 325 MPa to avoid permanent deformation on the elements.

IOP Conf. Series: Journal of Physics: Conf. Series **909** (2017) 012069 doi:10.1088/1742-6596/909/1/012069

Table 3. Node translations of bridge structure in initial mode

No.	Element	xz-Translation (mm)	No.	Element	xz-Translation (mm)
1	D1	0.00	10	TC1	8.29
2	D2-D3	9.66	11	**TC1-TC2**	**14.61**
3	**D4-D5**	**17.89**	12	**TC2-TC3**	**21.12**
4	D6-D7	23.52	13	TC3-TC4	24.60
5	D8-D9	25.45	14	TC4-TC5	24.56
6	D10-D11	23.52	15	TC5-TC6	20.97
7	D12-D13	17.91	16	TC6-TC7	14.28
8	D14-D15	9.70	17	TC7	7.59
9	D16	0.88			

3.2. Modeling of temporary connecting structure

The replacement process starts from dismantling the D4. Due to existing dead load, the forces on D4 is 558.69 kN. A temporary structure is required to transfer this forces to bridge structure when the D4 is dismantled. Three (3) models of the temporary connecting structure are examined starting from the simplest one. H.200.200.8.12 is taken to form the structure. The first is "I" shaped model which connects TC2 to the below node, therefore, the vertical force is handled. The second is "V" shaped model which replaces D4 and D5 roles temporarily. The third is a triangular model which is developed from "V" model with new horizontal beams to hold horizontal forces among the "V" ends.

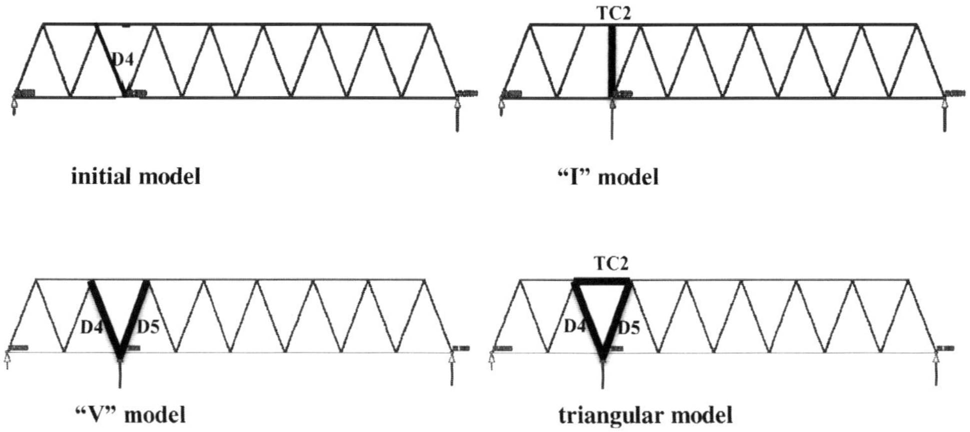

Figure 2. Temporary connecting structure models

From a structural analysis of the models, it can be described the strength and weakness of each one. The "V" and triangular models are applicable while the "I" model does not meet the requirements. Stresses and translations in each model are shown in Table 4 and Table 5 while the strength and weakness of each model are described in Table 6. To avoid slipping of the temporary connecting beam from the top chord while jacking process, the triangular model which has a horizontal lock is selected.

Table 4. Stresses of tested models

No.	Element	Stress (Mpa)			
		initial condition	"I" model	"V" model	triangular model
1	Top Chord 2	-82.93	-232.09	-94.05	-59.24
2	Diagonal 4	96.76	-	-	-
3	Diagonal 5	-69.42	-171.96	-53.83	-55.23
4	Bottom Chord 2	11.18	90.78	11.60	12.56
5	Bottom Chord 3	12.16	85.72	15.73	14.15

Table 5. Translations of tested models

No.	Node	xz-dir. deformation (mm)						
		initial condition	"I" model		"V" model		triangular model	
		Δ	δ	Δ	δ	Δ	δ	Δ
1	TC1-TC2	14.60	8.23	6.37	14.93	-0.33	13.97	0.63
2	TC2-TC3	21.18	36.79	-15.61	21.31	-0.13	20.39	0.79
3	D4-D5	17.89	3.55	14.34	18.39	-0.50	17.32	0.57

Table 6. Strength and weakness of tested model

No.	Element	Strength	Weakness	Remark
1	"I" model	The simplest model, minimize material need	xz-translation >2 mm Potentially endanger TC2	unaccepted
2	"V" model	xz-translation difference < 2 mm and safe to structure	Horizontal forces on V tips are not held well, and more materials need	accepted
3	Triangular model	xz-translation difference < 2 mm and safe to structure	More materials need	accepted

3.3. Modeling the combined influences of temperature and bearing system

The next step is modeling the combined influences of temperature and bearing system to the bridge structure. In the fixed-fixed mode, the longitudinal movement of the bearing is locked. It will resist the node displacements longitudinally, therefore, the node displacements are dominated only by vertical ones. In this mode the vertical displacement is expectedly quite small, therefore, the new elements can be placed easily. This mode will meet a problem when the replacement process is held on a day with high hourly temperature amplitude. The nodes will be higher than the initial condition elevations, and replacement process will be unapplicable.

The node displacement difference at replacement stage to the initial condition should not exceed 2 mm, because the bolt hole was made 2 mm larger than the bolt diameter in fabrication. If the displacement difference is exceeding 2 mm, the bolt can not be installed in its bolt-hole well. There are three important nodes that limit the modeling, the gusset TC2-TC1, TC2-TC3, and D4-D5 with a maximum displacement of 2 mm in one, two, or three of these gussets. The element stresses should not reach 325 MPa.

In the fixed-free mode, the longitudinal movement of the bearing is free on the right side. It will result in more displacements than in the fixed-fixed mode however the vertical displacements are fewer than in fixed-fixed mode in the same temperature amplitude. In this modeling, the combination of boundary modes and temperature load is applied from initial condition 27°C to higher temperature gradually. Stresses and translations of each combination are shown in Tables 7 and Table 8. In fixed-

fixed mode, this method can be carried out in a temperature range 27-28.8°C, while in fixed-free one, the temperature it is allowed between 27-43.4°C.

Table 7. Stresses Due To Boundary Mode and Temperature Load

No	Element	Stress (MPa)			
		fixed-free		fixed-fixed	
		T=27°C	T=43.4°C	T=27°C	T=28.8°C
1	Top Chord 2	-82.93	-232.09	-94.05	-59.24
2	Diagonal 4	96.76	-	-	-
3	Diagonal 5	-69.42	-171.96	-53.83	-55.23
4	Bottom Chord 2	11.18	90.78	11.60	12.56
5	Bottom Chord 3	12.16	85.72	15.73	14.15

Table 8. Translations due to Boundary Mode and Temperature Load

No.	Node	xz-dir. Deformation (mm)								
		Initial condition	boundary: Fixed-fixed				boundary: Fixed-free			
			T=27°C		T=43.4°C		T=27°C		T=28.8°C	
		δ	δ	Δ	δ	Δ	δ	Δ	δ	Δ
1	TC1-TC2	14.60	13.97	0.63	13.02	1.58	13.49	1.11	12.94	1.66
2	TC2-TC3	21.18	20.39	0.79	19.18	**2.00**	19.83	1.35	19.18	**2.00**
3	D4-D5	17.89	17.32	0.57	16.97	0.92	16.91	0.98	16.49	1.40

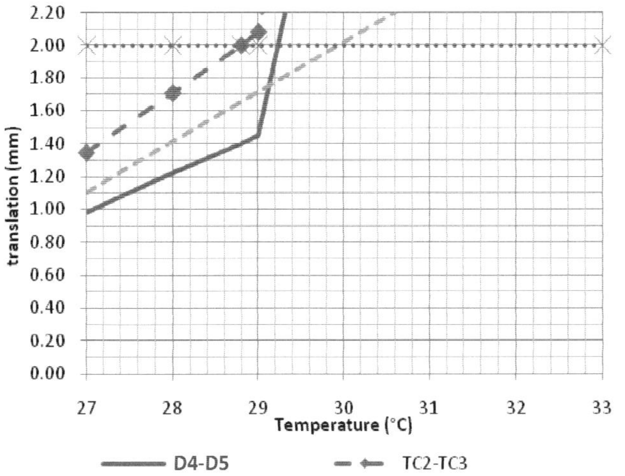

Figure 3. Translations on fixed-fixed mode due to temperature load

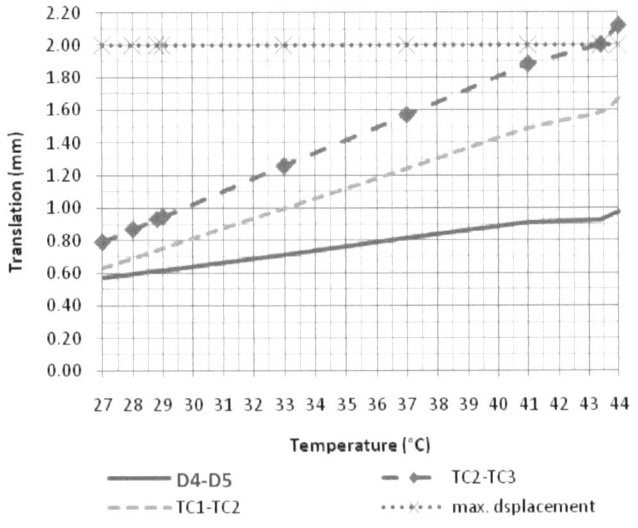

Figure 4. Translations on fixed-free mode due to temperature load

3.4. *Replacement method*

Some considerations are needed to arrange the element replacement method including ambient temperature and its hourly amplitude, boundary support mode, temporary connecting structure dimension, displacements, stresses, and dismantling steps. These will influence the success of the replacement.

3.4.1. Element dismantling steps. To dismantle tension element (D4) from the structure, the temporary connecting structure should be installed first. This structure will receive the load transfer from D4 after it is released. To maintain safety when dismantling D4, it needs special treatment to transfer this force gradually. By knowing the tensile strength of steel material D4 and force that occurs on it, it can be conditioned so that the D4 has collapsed safely while the load is gradually transferred to the temporary connecting structure by slowly-jacking action. With 558.69 kN force and 490 MPa tensile strength steel, the D4 need to be ground/cut symmetrically and leaving 1140.2 mm^2 area or 126.69 mm length of its web to lead the plastic mechanism occurs.

3.4.2. Temporary connecting structure dimension. The temporary connecting structure uses doubled H.200.200.8.12 arranged in a triangular shape at both sides of D4 and D5 by leaving enough space for maneuver while the replacement is held. The total area of this H-arrangement is made slightly larger than the diagonal to provide sufficient safety when the temporary structure works.

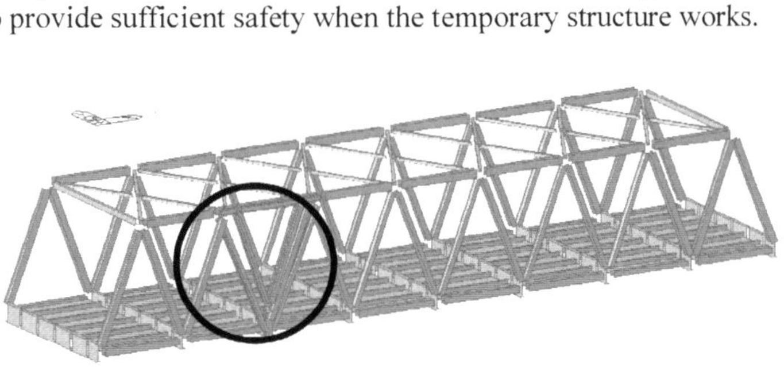

Figure 5. Temporary connecting structure attached to bridge structure

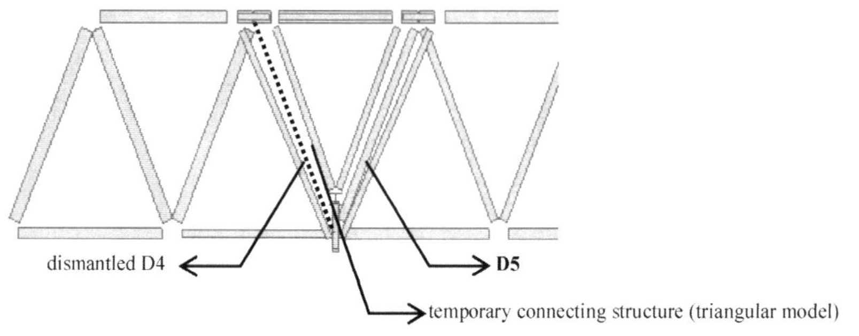

Figure 6. Detail position of temporary connecting structure

3.4.3. Boundary mode selection. The combination of boundary mode and temperature load is very influential in the displacement that occurs. The mode selection in replacement depends on the temperature in the field when replacement takes place.

3.4.4. Jacking and tightening. Jacking is needed while attaching temporary connecting structure to existing bridge and load transfer process from D4. To ensure the force transferred entirely to the temporary connecting structure, the jacking will be followed by tightening using threaded fasteners. The jacking uses a hydraulic jack of 150 tons considering the forces that occur on the elements.

Considering some requirements above, it can be defined the element replacement method of steel truss bridge as shown in Figure 6.

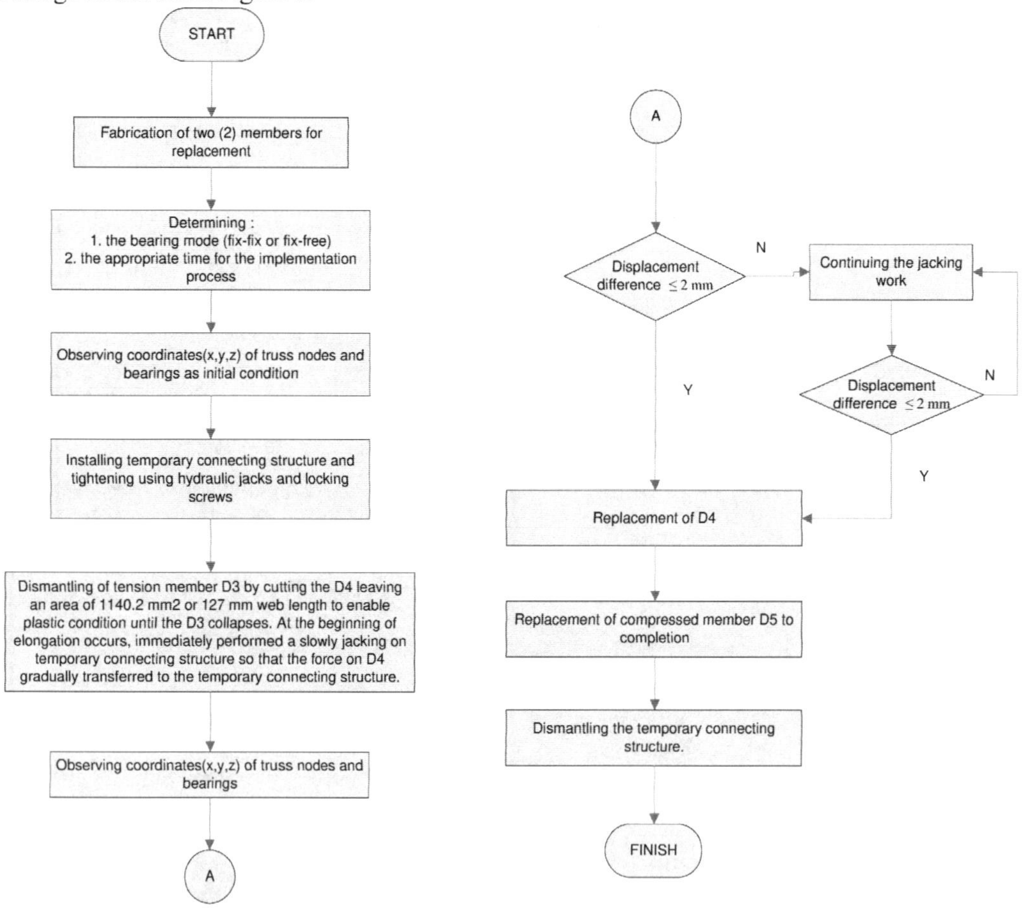

Figure 7. Flowchart of bridge element replacement method

4. Conclusion

Bridge rehabilitation is important in road system to maintain the transportation function to run well. One of rehabilitation work in steel truss bridge is the replacement of its element due to damages. In a condition where the commonly used method is unapplicable caused by limited space of work, high elevation and low bearing capacity of the soil, the self-supporting method is appropriate for steel truss bridge in its element replacement. By taking the case on a steel truss bridge Jetis Salatiga which requires element replacement due to its defects on two (2) diagonals, it can be concluded that the bridge element replacement depends on some factors such as space availability, elevation, soil strength, temperature, and tools.

The method taken should be appropriate to the existing bridge and avoid any damages to the bridge. Element stresses are very significant to be concerned. Overstressed element may cause a new problem for bridge structure. Bolt precision becomes a certain parameter for the success of replacement method and ambient temperature which leads the structure behavior should be considered well.

Acknowledgments

The authors would like to thank Nugroho D (Hutama Karya, PT) for supporting in structural analysis using MIDAS Civil and Itakusuma D for assisting the author in the field measurement. Their support and cooperation are appreciated.

References

[1] Directorate General Of Highways Ministry Of Public Works Republic Of Indonesia 1993 Bridge inspection manual **BMS2**-ME 1-1
[2] Liu J 2015 The detection evaluation and reinforcement research of kankan river bridge *Scientific research publishing:Engineering* 7 462-468
[3] Rashidi M 2017 Remedial modelling of steel bridges through application of analytical hierarchy process (AHP) *MDPI:Applied Sciences* 7(2) 168
[4] Priyanto A 2013 Laporan kunjungan lapangan Jembatan Jetis Salatiga 1-3
[5] Kimpton D 2017 Repair of damage to steel bridges following vehicle impact *Austroads bridge conference, 10th*
[6] Kulicki JM 2000 Highway truss bridges (bridge engineering handbook 1) 16.2
[7] Kementerian Pekerjaan Umum Dirjen Bina Marga 2009 Pemeriksaan jembatan rangka baja 4 82
[8] Dewobroto W 2015 Struktur baja : perilaku analisis dan desain-AISC 2010 8 522
[9] Cai S 2017 Fatigue life assessment of large scale t-jointed steel truss bridge components *Elsevier: Journal of constructional steel research 133* 499–509
[10] Badan Standarisasi Nasional 2005 Standar pembebanan untuk jembatan
[11] Dou YM 2014 Study on the vehicle load of highway bridge based on measured data *Trans tech publications ltd (scientific.net):Applied Mechanics and Materials (Volume 587-589)* 1402
[12] Yarnold M 2014 Identification of bridge movement mechanisms, *Scientific research publishing: Engineering* 6 584-591
[13] Steelman 2013 Shear and friction response of nonseismic laminated elastomeric bridge bearings subject to seismic demands, *ASCE:Journal of bridge engineering* 18-7
[14] Badan Standarisasi Nasional 2005 Perencanaan struktur baja untuk jembatan
[15] Zhou 2016 Thermal correlation analysis of a long-span suspension bridge static responses *SPIE Proceeding Vol. 9804*
[16] Brunnel G 2013 Effect of local damage on the behavior of a laboratory-scale steel truss bridge *Sciencedirect:Engineering structures* Vol.48 p281-291
[17] Badan Pusat Statistik Kota Salatiga 2016 Kota Salatiga dalam angka 7
[18] https://id.climate-data.org/location/976296/
[19] Andre J 2013 Recent advances and existing challenges in the design of bridge falsework systems *Taylor and Francis:Civil Engineering and Environmental Systems* Vol 30

A Fuzzy analytical hierarchy process approach in irrigation networks maintenance

Angga Riza Permana[1,3], Rr. Rintis Hadiani[2], Syafi'i[2]

[1]Student of master program of Civil Engineering, Universitas Sebelas Maret, Jl. Ir. Sutami 36A Kentingan Jebres Surakarta 57126, Indonesia
[2]Lecturer of master program of Civil Engineering, Universitas Sebelas Maret, Jl. Ir. Sutami 36A Kentingan Jebres Surakarta 57126, Indonesia
[3]Dinas Pekerjaan Umum dan Penataan Ruang Kabupaten Ponorogo, Jl. Gajah Mada 67 Ponorogo 63414, Indonesia

Email: anggo26@gmail.com

Abstract. Ponorogo Regency has 440 Irrigation Area with a total area of 17,950 Ha. Due to the limited budget and lack of maintenance cause decreased function on the irrigation. The aim of this study is to make an appropriate system to determine the indices weighted of the rank prioritization criteria for irrigation network maintenance using a fuzzy-based methodology. The criteria that are used such as the physical condition of irrigation networks, area of service, estimated maintenance cost, and efficiency of irrigation water distribution. 26 experts in the field of water resources in the Dinas Pekerjaan Umum were asked to fill out the questionnaire, and the result will be used as a benchmark to determine the rank of irrigation network maintenance priority. The results demonstrate that the physical condition of irrigation networks criterion (W1) = 0,279 has the greatest impact on the assessment process. The area of service (W2) = 0,270, efficiency of irrigation water distribution (W4) = 0,249, and estimated maintenance cost (W3) = 0,202 criteria rank next in effectiveness, respectively. The proposed methodology deals with uncertainty and vague data using triangular fuzzy numbers, and, moreover, it provides a comprehensive decision-making technique to assess maintenance priority on irrigation network.

1. Introduction

Agriculture is an important sector in the development of an country because it plays a role in the provision of basic needs and is also the livelihood of the population in rural areas. In order to support the development of agriculture, the development of irrigation infrastructure and facilities is needed so that the availability of water is sufficient. The number of problems such as breakage to irrigation-networks in the area of Ponorogo regency cause that the performance of irrigation is not optimal.

Ponorogo Regency has 440 Irrigation Areas with a total area of 17,950 Ha and based on Irrigation Infrastructure Baseline Data of 2017 showing 55,32% in good condition, 0,72% in passable condition, 14,42% in mild condition and 29,55% in severe damage. The problems that occur are in addition to limiteds of the budget, years factor and lack of maintenance cause the irrigation-network has decreased function. So far to decide network maintenance of irrigation In Ponorogo District is still using a random choice system. Based on these descriptions, the right system is needed so that the maintenance activities in Ponorogo Regency can run well.

In this study criteria that are used such as physical condition of irrigation network, service area, estimated maintenance cost, and efficiency of irrigation water distribution using fuzzy AHP with reference from previous research [9] which is used criteria, physical condition of irrigation network, service area and estimation Maintenance costs and of [2] that are used criteria of the physical condition of irrigation networks, crop indexes, supporting facilities, personnel organization, documentation and P3A. Both authors use the AHP method as priority maintenance/rehabilitation of irrigation networks.

Fuzzy Logic and Fuzzy AHP methods have been widely used to define infrastructure maintenance priorities such as road maintenance [3,4,12,13], building maintenance [1] and assessment of irrigation network performance [7,13].

Montazar et al. (2012) in his research explains that this combination of fuzzy and AHP methods can improve assessment of the accuracy of procedural assessments and may be a powerful technique for assessing irrigation projects. The purpose of the research is to develop a new methodology (FAHM) and apply techniques to assess the performance of four irrigation projects. According to Jasril et al. (2011) AHP fuzzy covers the weaknesses of AHP, in the fuzzy AHP decision-making system are considered better in describing a vague decision than AHP.

The aim of this study is to make the appropriate system to determine the indices weighted of the rank prioritization criteria for irrigation network maintenance using a fuzzy - based methodology. In the previous research, a method of prioritizing the maintenance of irrigation-networks has been developed using AHP (Analytical Hierarchy Process) methods, such as [2,9] and sequence priority for cost allocation budgeting [8]. It is expected that the system developed by fuzzy AHP method will provide a solution for the maintenance activities of irrigation networks in Ponorogo Regency.

2. Material and methods

2.1. Analytical hierarchy process

The AHP method introduced by Thomas L. Saaty in 1980 was a functional hierarchy with the main input of human perception. This model can process data that are both qualitative and quantitative in the decision-making process. Also, AHP can solve multi-objective and multicriteria problems based on the preference comparison of each element in the hierarchy, making it a comprehensive decision-making model. Linguistic variables are useful for collecting opinions in many cases. Linguistic variables can accept the words from natural language as their values. A linguistic variable differs from a numerical variable in that its value is shown as phrases and expressions rather than numbers [4]. The setting of the quantitative scale of 1 (one) to 9 (nine) is to assess the importance comparison of an element to the others as it is shown in Table 1.

Table 1. The Rating Scale Comparison of AHP

Interest's Intensity	Description	Explanation
1	Both elements are equally important	Two elements have the equally influence on the goal
3	One element is slightly important than the other elements	Experience and judgment support one element more than the other
5	One element is more important than the other elements	Experience and judgment are very in favor of one element over the other
7	One element is clearly more important than the other elements	One strong element is sustained and dominant in practice
9	One absolute element is more important than any other elements	The evidence that supports one element against another has the highest degree of affirmation that might be reinforcing

Interest's Intensity	Description	Explanation
2,4,6,8	Values between two adjacent considerations	This value is given when there are two compromises between two options
Contrary	If **i** activity gets one score compared to **j** activity, then **j** has the opposite value compared to **i**	

The calculation process with the AHP method is as follows:

a. Creates a hierarchical structure of the problem to be solved and determines the matrix comparison in pairs between the criteria. The geometric mean is calculated in each value as the following equation:

$$GM = \sqrt[n]{X_1 \, x \, X_2 \, x \ldots \ldots X_n} \quad (1)$$

b. Performs multiplication of elements in a row and is powered and as shown in the following equation:

$$W_i = \sqrt[n]{a_{11} x \, a_{12} x \ldots x a_{1n}} \quad (2)$$

c. Calculating the priority vector or eigen vector by the weight of each element can be obtained by the following equation:

$$X_i = \frac{W_i}{\Sigma W_i} \quad (3)$$

The result obtained is an eigen vector (X1) as the element weight.

d. Calculating the maximum eigen value (λ_{maks}), by multiplying the reciprocal matrix by the weight obtained, the result of the sum of matrix operations is the maximum eigen value (λ_{maks}) with the following equation:

$$\lambda_{maks} = \sum a_{ij} \, x \, X_i \quad (4)$$

e. Consistency index

This calculation is intended to determine the consistency of answers that will affect the results of the truth. The consistency index calculation can be done using the following equation:

$$CI = \frac{\lambda_{maks} - n}{n-1} \quad (4.1)$$

To find out CI is good enough or not, please note Consistency Ratio (CR). The Consistency Ratio is a parameter to check whether pairwise comparisons have been made consequently or not using the equation below:

$$CR = \frac{CI}{RI} \quad (4.2)$$

The Random Index (RI) value depends on the amount of the matrix as shown in the following table 2. The compilation matrix comparison is acceptable if CR value $< 0{,}1$. If CR $> 0{,}1$ then the comparison assessment should be repeated.

Table 2. Correlation between matrix size and RI value

Random Index Value (RI)															
The amount of the matrix	1	2	3	4	5	6	7	8	9	10	11	12	13	14	15
RI	0.00	0.00	0.58	0.90	1.12	1.24	1.32	1.41	1.45	1.49	1.51	1.48	1.56	1.57	1.59

2.2. Fuzzy analytical hierarchy process

In Multi Criteria Decision Making (MCDM) AHP is developed with a fuzzy approach. This method is known as fuzzy AHP. Fuzzy AHP combines the fuzzy concept with the AHP method. This method is introduced by Chang (1996) using triangular membership (Triangular Fuzzy Number / TFN). Triangle membership function is a combination of two lines. Graphs of triangular membership functions are depicted in figure 1. below:

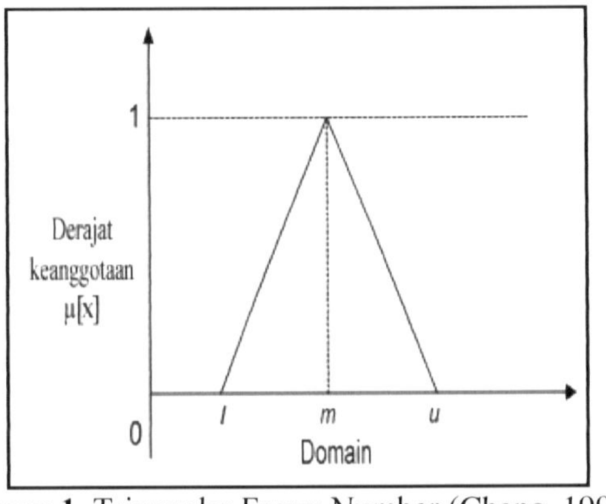

Figure 1. Triangular Fuzzy Number (Chang, 1996)

The intensity value of AHP is defined into the fuzzy triangle scale that is to divide each fuzzy set by two, except for the intensity of the interest of one. The Fuzzy Triangle Scale in [10] can be seen in the table 2. below:

Table 3. Fuzzy Triangle Scale (Chang,1996).

Intensity of Interest AHP	The Set of Linguistics	TFN	Reciprocity
1	Comparison of the same element (equal)	(1,1,1)	(1,1,1)
3	One element is more important than the other (weak)	$(1, \frac{3}{2}, 2)$	$(\frac{1}{2}, \frac{2}{3}, 1)$
5	One element is more important than that other (fairly strong)	$(2, \frac{5}{2}, 3)$	$(\frac{1}{3}, \frac{2}{5}, \frac{1}{2})$
7	One element is much more important than the others (very strong)	$(3, \frac{7}{2}, 4)$	$(\frac{1}{4}, \frac{2}{7}, \frac{1}{3})$
9	One absolute element is more important than the other (absolute)	$(4, \frac{9}{2}, \frac{9}{2})$	$(\frac{2}{9}, \frac{2}{9}, \frac{1}{4})$

Fuzzy Logic completion steps according to [6] are as follows:

a. Creates a hierarchical structure of the problem to be solved and determines the matrix comparison in pairs between the criteria and TFN scale. The geometric mean is calculated in each value.
b. Defines the value of fuzzy synthetic extent (Si) with number i criteria as the following equation:

$$\mathrm{Si} = \sum_{j=1}^{m} M_{gi}^{j} \otimes \left[\sum_{i=1}^{n} \sum_{j=1}^{m} M_{gi}^{j}\right]^{-1} \tag{5}$$

To obtain the equation $\sum_{j=1}^{m} M_{gi}^{j}$, a fuzzy sum operation is performed from the m value of the matrix in the equation below. In the final calculation phase, values (1, m, u) are obtained and used for the next phase.

$$\sum_{j=1}^{m} M_{gi}^{j} = \left(\sum_{j=1}^{m} l_j, \sum_{j=1}^{m} m_j, \sum_{j=1}^{m} u_j\right) \tag{6}$$

l is the lower limit value, m is the expected value and u is the limiting upper value. To obtain the equation $\left[\sum_{i=1}^{n}\sum_{j=1}^{m} M_{gi}^{j}\right]^{-1}$ a fuzzy sum operation is performed from the value M_{gi}^{j} (J = 1, 2, ..., m) to obtain the following equation:

$$\sum_{i=1}^{n}\sum_{j=1}^{m} M_{gi}^{j} = \left(\sum_{i=1}^{n} l_i, \sum_{i=1}^{n} m_i, \sum_{i=1}^{n} u_i\right) \quad (7)$$

Then, calculate the inverse of the vector in the above equation to get the equation below:

$$\left[\sum_{i=1}^{n}\sum_{j=1}^{m} M_{gi}^{j}\right]^{-1} = \frac{1}{\sum_{i=1}^{n} u_i}, \frac{1}{\sum_{i=1}^{n} m_i}, \frac{1}{\sum_{i=1}^{n} l_i} \quad (8)$$

c. Calculates the degree of possibility
To obtain the degree of possibility, $M_2 = (l_2, m_2, u_2) \geq M_1 = (l_1, m_1, u_1)$ is expressed in the equation below:

$$V(M_2 > M_1) = \sup[\min(\mu_{M1}(x), \mu_{M1}(y))] \quad (9)$$

And x and y are values on the axis of the membership function of each criteria. This equation can be defined as follows:

$$V(M_2 \geq M_1) = \begin{cases} 1, \text{ if } & m_2 \geq m_1 \\ 0, \text{ if } & l_2 \geq u_2 \\ \dfrac{l_1 - u_2}{(m_2-u_2)-(m_1-l_1)}, & \textit{Otherwise} \end{cases} \quad (9.I)$$

We need the values of $V(M_2 \geq M_1)$ dan $V(M_1 \geq M_2)$ to compare M_1 and M_2.

d. Compares the degree of possibility among the criteria
The comparison between degrees of possibility between M i (i = 1, 2, 3, 4, 5, ..., k) can be defined by: $V(M \geq M_1, M_2, \ldots M_k) = V[(M \geq M_1), (M \geq M_2), \ldots, (M > M_k)]$ and *min* $V(M > M_i)$, (i=1.2.3,.......,k).
Assume that the equation (10) is:

$$d'(A_i) = \min V(S_i \geq S_k) \quad (10)$$

for k = 1, 2, 3, 4, 5, ..., n; k≠ i. Then the weight vector is given by (11) as follows:

$$W' = (d'(A_1), d'(A_2), \ldots, d'(A_n))^T \quad (11)$$

Where Ai (i=1,2,3,4,5,......, n) is *n* element.

e. Normalization
Normalization of vector weight is given by the following equation:

$$W = (d(A_1), d(A_2), \ldots, d(A_n))^T \quad (12)$$

Where W is a non-fuzzy number of the weight vector.

3. Research Method

The method applied in this study is quantitative descriptive, i.e. research using data from the variables studied. Data for weights between criteria for prioritizing maintenance are obtained by assessing interest in the AHP scale. According to [6] in [1] on the application of AHP method, the primary is the data quality of the respondents and does not depend on the quantity. Therefore, the AHP assessment requires experts as respondents in decision making in alternative selection. The experts here are competent people who have mastered, influence the decision-making or know the information needed.

Data retrieval using a purposive sampling method that is sampled determination technique with a certain consideration. In this study, the data source is the decision makers proposed irrigation network maintenance plan and staff of the Water Resources Sector on Dinas PUPR Ponorogo Regency.

4. Result and Discussion

The problem faced in determining the priority of maintaining irrigation-networks is how to identify the section which should take precedence. Determining the priority of irrigation network element maintenance is effectual because of the limited maintenance fund. In this research, priority maintenance system for irrigation-network is designed to determine the rating on the irrigation-network section that have been ravaged so will generate the priority order of the parts in the irrigation-network is prioritized for maintenance. The phases in preparing this system are as follows:

4.1. Converted comparison between criteria into TFN scale

The first task of the fuzzy AHP method is to decide on the relative importance of each pair of factors in the same hierarchy by using formula (1 - 4.2). The goal is to know the consistency of scores between criteria. After going through the consistency test and stated the importance of interest between the criteria is consistent, then by using triangular fuzzy numbers on table 3., the scale of interest assessment with AHP converted into TFN/ FAHP scale and then calculated in each value the (geometric mean) by applying formula (1). At this stage, the task undertaken is to recapitulate the questionnaire results that have been dispersed to the competent experts in the irrigation-network maintenance field to determine the criteria weight. Total respondents of 26 (twenty-six) persons, the result of the comparison between criteria on TFN/ FAHP scale shown in Figure 2.

Responden	Physical condition of irrigation networks			Physical condition of irrigation networks			Physical condition of irrigation networks			Area of service			Area of service			Estimated maintenance cost		
	Area of service			Estimated maintenance cost			Efficiency of irrigation water distribution			Estimated maintenance cost			Efficiency of irrigation water distribution			Efficiency of irrigation water distribution		
	L	M	U	L	M	U	L	M	U	L	M	U	L	M	U	L	M	U
R1	1/2	2/3	1	1	1 1/2	2	2	2 1/2	3	2	2 1/2	3	2	2 1/2	3	1/3	2/5	1/2
R2	2	2 1/2	3	1/2	2/3	1	1/2	2/3	1	2	2 1/2	3	1/3	2/5	1/2	1/2	2/3	1
R3	1	1	1	2	2 1/2	3	1	1	1	1	1 1/2	2	1	1	1	1/2	2/3	1
R4	3	3 1/2	4	3	3 1/2	4	1	1	1	1	1 1/2	2	1	1	1	3	3 1/2	4
R5	1	1	1	1	1 1/2	2	1/2	2/3	1	1	1	1	1/2	2/3	1	1	1	1
R6	1	1	1	1/3	2/5	1/2	1/4	2/7	1/3	1/3	2/5	1/2	1	1	1	1/3	2/5	1/2
R7	1	1 1/2	2	1/2	2/3	1	2	2 1/2	3	1/2	2/3	1	1	1 1/2	2	1	1 1/2	2
R8	1	1 1/2	2	1/2	2/3	1	2	2 1/2	3	1/2	2/3	1	1	1 1/2	2	1	1 1/2	2
R9	1	1	1	1	1 1/2	2	1	1 1/2	2	1	1	1	1	1	1	1	1	1
R10	2	2 1/2	3	1	1	1	2	2 1/2	3	2	2 1/2	3	1	1	1	1/2	1	1 1/2
R11	1/2	2/3	1	1	1 1/2	2	2	2 1/2	3	1	1 1/2	2	2	2 1/2	3	1	1 1/2	2
R12	1	1	1	1	1	1	1	1 1/2	2	1/3	2/5	1/2	1	1 1/2	2	2	2 1/2	3
R13	1	1 1/2	2	3	3 1/2	4	1	1 1/2	2	1	1 1/2	2	1	1	1	1	1	1
R14	1/2	2/3	1	2	2 1/2	3	1/4	2/7	1/3	2	2 1/2	3	1/3	2/5	1/2	1/4	2/7	1/3
R15	1/4	2/7	1/3	3	3 1/2	4	1/4	2/7	1/3	3	3 1/2	4	3	3 1/2	4	1/4	2/7	1/3
R16	1/4	2/7	1/3	3	3 1/2	4	1/4	2/7	1/3	3	3 1/2	4	3	3 1/2	4	1/4	2/7	1/3
R17	3	3 1/2	4	1	1	1	3	3 1/2	4	1	1	1	1/4	2/7	1/3	3	3 1/2	4
R18	1	1	1	1	1 1/2	2	1	1	1	1	1	1	1/2	2/3	1	1/2	2/3	1
R19	1/4	2/7	1/3	1/3	2/5	1/2	1/4	2/7	1/3	1	1	1	2	2 1/2	3	1/3	2/5	1/2
R20	1	1	1	1	1	1	1/2	2/3	1	1	1	1	1	1	1	1/2	2/3	1
R21	1	1 1/2	2	1	1	1	1/2	1	1 1/2	1	1	1	1	1 1/2	2	1	1	1
R22	1	1 1/2	2	3	3 1/2	4	1	1 1/2	2	1	1 1/2	2	1	1	1	1	1	1
R23	1/2	2/3	1	1	1 1/2	2	2	2 1/2	3	1	1 1/2	2	2	2 1/2	3	1	1 1/2	2
R24	1	1 1/2	2	1/2	2/3	1	1/3	2/5	1/2	1/2	2/3	1	1	1	1	2	2 1/2	3
R25	1	1 1/2	2	1/2	2/3	1	1/3	2/5	1/2	1/2	2/3	1	1/3	2/5	1/2	2	2 1/2	3
R26	1	1 1/2	2	1/2	2/3	1	1/3	2/5	1/2	1/2	2/3	1	1/3	2/5	1/2	2	2 1/2	3
GEOMEAN	**0.879**	**1.088**	**1.320**	**1.020**	**1.272**	**1.581**	**0.742**	**0.931**	**1.148**	**0.974**	**1.200**	**1.466**	**0.919**	**1.083**	**1.263**	**0.795**	**0.987**	**1.211**

Figure 2. Comparison between criteria on TFN/ FAHP scale

4.2. Compiling pairwise comparison

Pairwise comparison of the fuzzy evaluation matrix shown in Table 4. below:

Table 4. Pairwise matrix comparison

	Physical condition of irrigation networks			Area of service			Estimated maintenance cost			Efficiency of irrigation water distribution		
Physical condition of irrigation networks	**1.000**	**1.000**	**1.000**	0.879	1.088	1.320	1.020	1.272	1.581	0.742	0.931	1.148
Area of service	0.758	0.919	1.137	**1.000**	**1.000**	**1.000**	0.974	1.200	1.466	0.919	1.083	1.263
Estimated	0.633	0.786	0.980	0.682	0.833	1.027	**1.000**	**1.000**	**1.000**	0.795	0.987	1.211

maintenance cost												
Efficiency of irrigation water distribution	0.871	1.074	1.347	0.792	0.924	1.088	0.826	1.013	1.257	**1.000**	**1.000**	**1.000**

4.3. Determination of fuzzy synthetic extent value (Si)

In defining the value of fuzzy synthetic extent (Si) then used equation 5. The value of $\sum_{j=i}^{m} M_{gi}^{j}$ is the sum of the fuzzy numbers in each matrix. An example addition of fuzzy value for the physical condition of irrigation networks criterion irrigation can be explained as follows:

$\sum_{j=1}^{m} M_{gi}^{j} = \left(\sum_{j=1}^{m} l_j, \sum_{j=1}^{m} m_{j,} \sum_{j=1}^{m} u_j\right)$

$= \sum$ Physical condition of irrigation networks

$= \{(1 + 0{,}879 + 1{,}020 + 0{,}742), (1 + 1{,}088 + 1{,}272 + 0{,}931), (1 + 1{,}320 + 1{,}581 + 1{,}148)\}$

$= \{3{,}642; 4{,}292; 5{,}048\}$

The value for all matrix-criteria is shown in Table 5.

Table 5. Addition of Fuzzy Value

	$\sum$ L	$\sum$ M	$\sum$ U
Physical condition of irrigation networks	3.642	4.292	5.048
Area of service	3.650	4.202	4.866
Estimated maintenance cost	3.110	3.606	4.218
Efficiency of irrigation water distribution	3.489	4.011	4.692
$\sum$ **Mg**	**13.892**	**16.110**	**18.824**

Next inverse vectors for all matrix-criteria are performed:

$\left[\sum_{i=1}^{n} \sum_{j=1}^{m} M_{gi}^{j}\right]^{-1} = \frac{1}{\sum_{i=1}^{n} u_i}, \frac{1}{\sum_{i=1}^{n} m_i}, \frac{1}{\sum_{i=1}^{n} l_i}$

$= \left[\frac{1}{13{,}892}, \frac{1}{16{,}110}, \frac{1}{18{,}824}\right]$

$= (0{,}0531; 0{,}0621; 0{,}0720)$

Then the value of fuzzy synthetic extent (Si) for the physical condition of irrigation networks criteria can be calculated

$Si = \sum_{j=1}^{m} M_{gi}^{j} \otimes \left[\sum_{i=1}^{n} \sum_{j=1}^{m} M_{gi}^{j}\right]^{-1}$

$= (3{,}642; 4{,}292; 5{,}048) \otimes (0{,}0531; 0{,}0621; 0{,}0720)$

$= (0{,}193; 0{,}266; 0{,}363)$

For the value of fuzzy synthetic extent (Si) on all criteria shown in Table 6.

Table 6. Value of fuzzy synthetic extent (Si)

Value of Fuzzy Synthetic Extent (Si)	L	M	U
S1 (Physical condition of irrigation networks)	0.193	0.266	0.363
S2 (Area of service)	0.194	0.261	0.350
S3 (Estimated maintenance cost)	0.165	0.224	0.304
S4 (Efficiency of irrigation water distribution)	0.185	0.249	0.338

4.4. Calculates the degree of possibility

To get the value of the degree of possibility used equation 9.1. With an explanation as follows:

$$V(M_2 \geq M_1) = \begin{cases} 1 & , m_2 \geq m_1 \\ 0 & , l_1 \geq u_2 \\ \dfrac{l_1 - u_2}{(m_2 - u_2) - (m_1 - l_1)} & . \end{cases}$$

From Table 6. values are obtained:

$V(S1 \geq S2) = 1$; $V(S3 \geq S4) = 0{,}825$
$V(S1 \geq S3) = 1$; $V(S3 \geq S1) = 0{,}721$
$V(S1 \geq S4) = 1$; $V(S3 \geq S2) = 0{,}748$
$V(S2 \geq S3) = 1$; $V(S4 \geq S2) = 0{,}924$
$V(S2 \geq S4) = 1$; $V(S4 \geq S3) = 1$
$V(S2 \geq S1) = \dfrac{0{,}1939 - 0{,}3503}{(0{,}2608 - 0{,}3503) - (0{,}2664 - 0{,}1939)} = 0{,}965$; $V(S4 \geq S1) = 0{,}892$

4.5. Compares the degree of possibility among the criteria

Accomplishment of the degree of possibility follows equation 10. with the following explanation: $d'(A_i) = \min V(S_i \geq S_k)$, $k = 1, 2, 3, 4, 5, \ldots\ldots, n$; $k \neq i$

$d'(1) = V(S1 \geq S2,S3,S4) = \min(1;1;1) = 1$

$d'(2) = V(S2 \geq S3,S4,S1) = \min(1;1;0{,}965) = 0{,}965$

$d'(3) = V(S3 \geq S4,S1,S2) = \min(0{,}825;0{,}721;0{,}748) = 0{,}721$

$d'(4) = V(S4 \geq S2,S3,S1) = \min(0{,}924;1;0{,}892) = 0{,}892$

4.6. Normalization

Normalization of vector weight is given in equation 12. as follows:

$W' = (d'(A_1), d'(A_2), \ldots\ldots, d'(A_n))^T$

$W' = (1;0{,}965;0{,}721;0{,}892)^T$

W = (**0,279 ; 0,270 ; 0,202 ; 0,249**)

The weight vector of the criteria is shown in Figure 3.

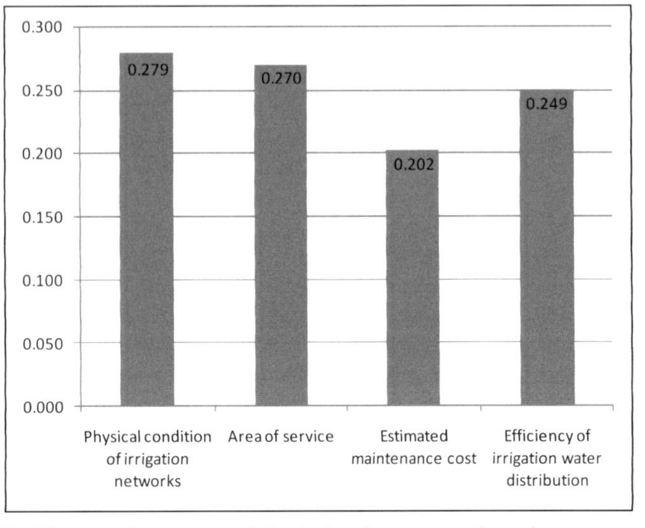

Figure 3. The weight vector of the irrigation network maintenance criteria

5. Conclusion

Based on the analysis using Fuzzy Logic Method, the weight for each criterion is as follows: physical condition of irrigation networks is (W1) = 0.279, the area of service is (W2) = 0.270, the estimated maintenance cost criteria is (W3) = 0.202, and the efficiency of irrigation water distribution is (W4) = 0.249.

Based on the analysis, the area of service becomes the second most important criteria after the physical condition of irrigation networks which became the first criterion to decisive. It shows that both criteria are crucial in the process of decision-making of irrigation-network maintenance.

The proposed methodology deals with uncertainty and vague data using triangular fuzzy numbers, and, moreover, it provides a comprehensive decision-making technique to assess maintenance priority on irrigation network.

References

[1] Agus F 2016 Penentuan Skala Prioritas Pemeliharaa Bangunan Gedung Studi Kasus Gedung Sragen Technopark, Tesis Magister Teknik Sipil: Universitas Sebelas Maret.
[2] Anton Z *et al.* 2016 *Jurnal Teknik Sipil* **UMJ p-ISSN: 2407-1846**
[3] Ashok A *et al.* 2016 *International Journal of Innovative Research in Science Engineering and Technology.* **5** 6858
[4] Babashamsi P *et al.* 2016 *International Journal of Pavement Research and Technology.* **9** 112
[5] Celik M *et al.* 2009 *Expert Systems with Applications.* **36** 190
[6] Chang D-Y 1996 Applications of the extent analysis method on fuzzy AHP *European Journal of Operational Research.* **95** 649
[7] Diah AS *et al.* 2009 *Seminar Nasional Aplikasi Teknologi Prasarana Wilayah.* **ISBN 978-979-18342-1-6** 285
[8] Fauzia M *et al.* 2013 *Konferensi Nasional Teknik Sipil.* **7** 129
[9] I'ied TA 2008 Evaluasi dan peningkatan kinerja jaringan irigasi bapang Kabupaten Sragen Tesis Magister Teknik Sipil: Universitas Sebelas Maret
[10] Jasril *et al.* 2011 *Jurnal Seminar Nasional Aplikasi Teknologi Informasi* **ISSN:1907-5022** 36
[11] Moazami D *et al.* 2016 *Expert System with Applications An International Journal.* **38** 12869
[12] Mahmood M *et al.* 2013 *International Journal of Pavement Research and Technology.* **6** 620
[13] Montazar A *et al.* 2013 *Agricultural Water Management.* **121** 113
[14] Saaty TL 2008 *Decision Making with The Analytic Hierarchy Process*, Int. J. Services Sciences, Vol. 1, No. 1, 2008.

IOP Conf. Series: Journal of Physics: Conf. Series **909** (2017) 012071 doi:10.1088/1742-6596/909/1/012071

Rapid and in-depth analysis for seismic risk evaluation

Novi Dwi Astuti, Meli Anta Alvita, Senot Sangadji, AP Rahmadi and Edy Purwanto

Civil Engineering Department of Post Graduate Program Sebelas Maret University
Jl. Ir. Sutami 36A Kentingan Jebres Surakarta Postal Code 57126, INDONESIA

E-mail: novi87.sukamto@gmail.com, s.sangadji@ft.uns.ac.id

Abstract.The high public demand on housing in urban areas requires the government of Indonesia to adopt a policy of encouraging the development of vertical housing. Cilacap has been allocated Rusunawa (low-income apartment) development in 2006. Evident from some earthquakes occurrence in recent years, however, Cilacap may be seen as an earthquake prone region which posing some risk to this type of vertical structures. The Appropriate strategy should be performed to evaluate the seismic risks of this local government owned four stories low-income apartment. This paper demonstrates two tier evaluation strategy; rapid evaluation and in-depth analysis and compares both results. First evaluation was conducted by means of Building Rapid Visual Screening (RVS) of FEMA 154 of the Rusunawa block A and B. The result was used further to calculate seismic risk score (SR) which exhibit the probability of the building damage given the Maximum Considered Earthquake (MCER) that will occur during the Rusunawa service life. The in-depth analysis was conducted by developing fragility function expressed in the form of fragility curves for the Rusunawa. The fragility shows the probability that certain damage states will be exceeded given the intensity of earthquakes which will occur during building service life. The fragility was developed as lognormal curves in which the building response to earthquake input was analyzed by means of pushover.

1. Introduction

Limited land for urban housing encourages the government to take a policy of vertical housing [1]. The city of Cilacap has been allocated Rusunawa (low-income apartment) development in 2006. It is a government-owned building that functions as a dwelling for people. However, as Cilacap is an area quite prone to earthquakes, therefore, the assessment of the seismic vulnerability of existing buildings, especially Rusunawa, is very important. The appropriate strategy should be taken to evaluate the seismic risks of this local government owned four stories low income-apartment.

The first tier evaluation was conducted by means of Building Rapid Visual Screening (RVS) of FEMA 154. The rapid visual screening (RVS) procedure has been developed to identify, inventory, and screen buildings that are potentially seismically hazardous [2]. RVS will provide a preliminary description of the risk of building against earthquake hazards [3,4]. The result was used further to calculate seismic risk score (SR) which exhibit the probability of the building damage given the Maximum Considered Earthquake (MCER) which will occur during the Rusunawa service life. Meanwhile, to assess the risk more accurately in-depth analysis may be necessary. The analysis should predict the building response over several earthquake intensity in the region. A pushover analysis is performed in this study to thoroughly estimate the maximum force and deformation that occurs and to obtain information on which parts are critical. Many of research indicates that nonlinear static pushover analysis provides sufficient results for low-rise building [5].

2. Rapid Visual Screening

The RVS procedure uses an empirical methodology. It is based on a sidewalk survey of a building and a data collection form completed by the person conducting the survey. The inspector visually observes of the building from the exterior, and if possible, the interior.

There are five kinds of Data Collection Form based on seismicity region namely: Low, Medium, Moderately High, High, and Very High. Determination of Seismicity is based on the Response Spectral Acceleration Area of Maximum Considered Earthquake Risk (MCER) depicted in Table 1

Tabel 1. Seismicity Region Determination from MCER Spectral Acceleration Response

Seismicity Region	Spectral Acceleration Response, SS (short-period, or 0.2 seconds)	Spectral Acceleration Response, S1 (long-period, or 1.0 second)
Low	Low less than 0.250g	less than 0.100g
Moderate	greater than or equal to 0.250g but less than 0.500g	greater than or equal to 0.100g but less than 0.200g
Moderately High	greater than or equal to 0.500g but less than 1.000g	greater than or equal to 0.200g but less than 0.400g
High	greater than or equal to 1.000g but less than 1.500	greater than or equal to 0.400g but less than 0.600g
Very High	greater than or equal to 1.500g	greater than or equal to 0.600g

Data collection form consists of spaces to document building identification information, including the use and size of buildings, building photographs, sketches and documentation of building data related to seismic performance. Based on data collected during the survey, the score can be calculated. This score provides an indication of the seismic performance of buildings. The final score, S is an estimate of the probability of collapse if an earthquake occurs given a maximum-considered earthquake risk, MCER. Mathematically, if P shows the probability of collapse in the MCE_R shaking, then the S score is derived from FEMA 154 which satisfies the equation (1):

$$S = -\log 10(P) \tag{1}$$

For instance, a final score S of 2 implies there is a chance of 1 in 10^2, or 1 in 100, that the building will collapse if such ground motions occur.

3. Risk Score (S_R)

Risk Score, denoted by S_R, which measures building safety regarding how frequently collapse-causing earthquakes occur. It is intended as an estimate of the negative base-10 logarithm of the number of earthquakes that could cause building collapse during the service life of the building, which is commonly taken to be 50 years [6].This is a different measure of performance from S. S has nothing to do with the design life of the building or the number of collapse-causing earthquakes while SR does.

To obtain Risk Score is to add the final score of S with PMFR (Risk Modification Factor). Table 2 shows that, for all seismicity regions between Moderate and Very High, one can add approximately 1.0 to the Final Score to get the Risk Score.

Table 2. Estimated value of PMFR (FEMA 155)

No	Seismicity Region	PMRF
1	Low	0.1
2	Moderate	0.9
3	Moderately High	1.2
4	High	1.1
5	Very High	0.9

4. Pushover analysis

The basic concept of nonlinear static analysis (pushover) is to provide an incremental certain static loading pattern in the lateral direction. This method is simple but the resulting information is able to describe the inelastic response of the building.

Pushover method employs two load cases, i.e. gravitation load and earthquake load, for the non-linear static analysis. A combination of dead load, super imposed dead load and live load was inputted as gravity load case in which each load was multiplied by a factor of 1. As the load was applied to a structural model, the response of the structure was recorded. Then, an around acceleration was inputted as subsequent load case with multiplier factor of 1 as a non-linear static load. This load case was applied with displacement control method. The initial condition for this loading condition was taken from previous gravity load case. The response of structure was then recorded in multiple steps a minimum number of 10 steps and maximum 100 steps. The maximum target was determined when the deviation of 4% times the total building height was reached by the program default [7]. The result of this pushover method is a capacity curve describing base shear force as a function of displacement.

5. Fragility Curve

The fragility curve characterizes the relationship between the probability of a demand parameter will cause a structure to experience a condition of damage that exceeds the limit condition of the prescribed damage states (limit state). This demand parameter can be Peak Ground Acceleration (PGA), Peak Ground Velocity (PGV) or spectral displacement for a certain period. The Federal Emergency Management Agency (FEMA) designs the manual document to explain and simulate the fragility analysis methods of building structures that became the approach method in the assessment of seismic losses.

To simplify damage states, the HAZUS methodology classifies the degree of destruction of buildings into four discrete deterioration conditions, e.g., slight, moderate, extensive, and complete, as illustrated schematically in Figure 1[8]. Each damage condition represents various building damage threshold. For example, slight damage conditions extend from the limit of slight damage to moderate damage limits.

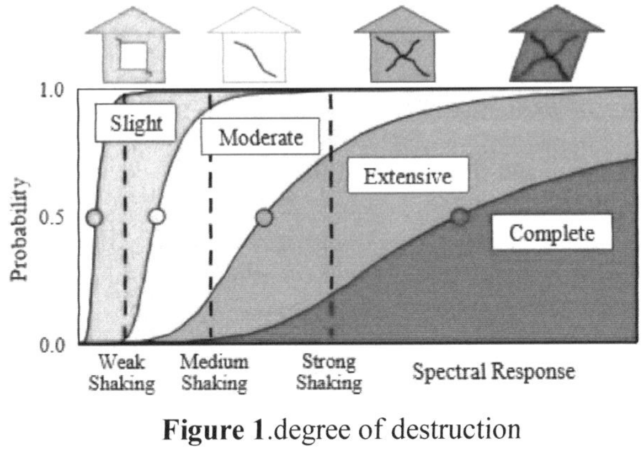

Figure 1.degree of destruction

6. Research methods

This study implements Rapid Visual Screening FEMA 154. Data were obtained from: (1) field survey and (2) direct interview with the technical institution as building manager which has the knowledge of the building condition, documentation, and identification type of damage. Data Collection Form is completed for each building screened through execution of the following steps:

a. Verifying and updating the building identification information;
b. Walking around the building to identify the number of stories and shape, and sketching a plan and elevation view on the Data Collection Form;
c. Photographing the building;
d. Determining and documenting occupancy;
e. Reviewing the soil type and geologic hazards, as identified during the pre-field planning process;
f. Identifying adjacency issues, building irregularities, and any potential exterior falling hazards;
g. Adding any comments about unusual conditions or circumstances that may affect the screening;
h. Identifying the building material, gravity load-carrying system, and seismic force-resisting system to identify the FEMA Building Type (entering the building, if possible, to facilitate this process) and circling the Basic Score on the Data Collection Form;
i. Circling the appropriate seismic performance attribute Score Modifiers (e.g., irregularities, design date, and soil type) on the Data Collection Form;

The dimensions of structural elements, material quality and structural shape are also required for the calculation of loading and modeling of structures.

Based on the as-built drawing and observation results, a structural model will be used as the basis for performance analysis of its structure.

7. Result and discussion

Block A and Block B of Rusunawa Tegalkamulyan at Cilacap was the case study of the investigation. The building is located at 7.89 latitudes and 109.024 longitudes with Spectral Acceleration Response 0.989 for a short period and 0.391 for long-period. It is, therefore, categorized in the moderately high seismicity. Table 3 and 4 show evaluation with RVS.

Table 3. RVS level 1 Rusunawa Tegal Kamulyan Cilacap

No	Data	Description
1	Building Identification Information	address, building name, use, latitude and longitude, and site specific ground motion values, name of the screener(s), and the date and time of the screening. it is desirable to document this information during the pre-field survey.
2	Photo	Photos are taken during the survey and as much as possible to show the entire building to identify compliance with pre-field data.
3	Sketches	Sketches are obtained from as-built drawing documents and verified in the field.
4	Soil type	Soil type is stiff soil obtained from Dinas Cipta Karya Tata Ruang as building manager.
5	Building Characteristics	The building has four stories. The use of the building is for housing with 48 units per block. Total building area of 2010 m2.
6	Fema building type	FEMA building Type Based on construction documents is a Concrete frame buildings with unreinforced masonry infill walls and in accordance with survey observations. The building belongs to the C3 type of building
7	Modifier	Severe vertical irregularity because walls of the building do not stack vertically in plan. (out of plane set back)

		Plan Irragularity: the building is double L shaped, with projections of more than 20 feet.(Reentrant Corner); the exterior beams do not align with the columns in plan.
9	Final Level 1 Score	Basic score C3 = 1,4; V_{L1} = -0,8; P_{L1} = -0,6; Minimum score, S_{MIN}= 0,3 The final level 1score, SL_1= Basic Score + V_{L1}+P_{L1} Because Smin $\geq S_{L1}$ so S_{L1} = 0,3
10	Coment	There are many spalling in concrete beams
11	Action required	the score SL1 less than cut of so detailed structural evaluaation requared. The RVS rating system is made with the assumption that the building is built of quality materials. Damage to structural elements has a significant impact on the expected performance of a building.

Table 4. RVS Level 2 Rusunawa Tegalkamulyan Cilacap

No	Data	Analysis
1	Vertical Irregularity, V_{L2}	Setback: Vertical elements of the lateral system at an upper story are outboard of those at the story below causing the diaphragm to cantilever at the offset. (Score -1) Short column: there are infill walls or adjacent floors that shorten the column. (score -0.5) Split levels: There is a split level at one of the floor levels or at the roof (score -0.5) V_{L2}= -1-0.5-0.5 = -2 (cap at-1.3) V_{L2}= -1.3
2	Plan Irregularity, P_{L2}	Torsional irregularity: Lateral system does not appear relatively well distributed in plan in either or both directions. (Score -0.8) Reentrant corner: Both projections from an interior corner exceed 25% of the overall plan dimension in that direction. Score -0.4 P_{L2} = -0.8 -0.4 = -1.2
3	Redundancy (M)	The building has at least two bays of lateral elements on each side of the building in each direction. Score +0.3 M = +0.3
4	Final Level 2 Score, SL2	S_{L2}= S'+PL2+M $\geq S_{MIN}$ S' = (S_{L1}-V_{L1}-P_{L1}) = 0,3-(-0.8)-(-0.6) = 1.7 S_{L2} = 1.7 + (-1.3) + (-1.2) + 0.3 = -0.5 A negative score implies a probability of collapse greater than 100%, which is not possible. To address this, a Minimum Score, SMIN, is provided. The

	Minimum Score was developed by considering the worst possible combination.

In the first tier analysis using RVS, the Rusunawa building obtained a final score of 0.3. A Final Score, S, of 0.3 implies that there is a chance of 1 in $10^{0.3}$ or 1 in 2 that the building will collapse if MCER ground motions occur. It also implies that the probability of collapse of the building is 50.12% when an earthquake of MCER takes place. Further, the risk score, SR, of the building is 1.3 means $1/10^{1.3}$ collapse-causing earthquakes per 50 years or 1 collapse-causing earthquake per 20 building design lifetimes. It also means the probability of collapse is 5,01%.
The second tier analysis was carried out as the RVS mandated the in-depth-analysis. Pushover analysis was implemented in this level by firstly developing a model of structure (figure 2). The analysis consists of two phases. The first phase of the structure was given a gravity load which is a combination of dead load and reduced live load. The second phase of the structure is given the lateral load of monotonic incremental from the acceleration of the soil. The intensity of lateral loading in the second stage continues to increase until the weakest component of the structure deforms and then continues until the structure collapse.

The computational program repeats the analysis was obtaining several the number of elements reaching the yield. At each loading step, internal force and deformation were calculated and recorded that would produce a capacity curve (figure.3). The structural response curve was then converted into a capacity spectrum in the format of Acceleration-Displacement Response Spectrum (ADRS). This spectrum was used further to develop seismic fragility curve.

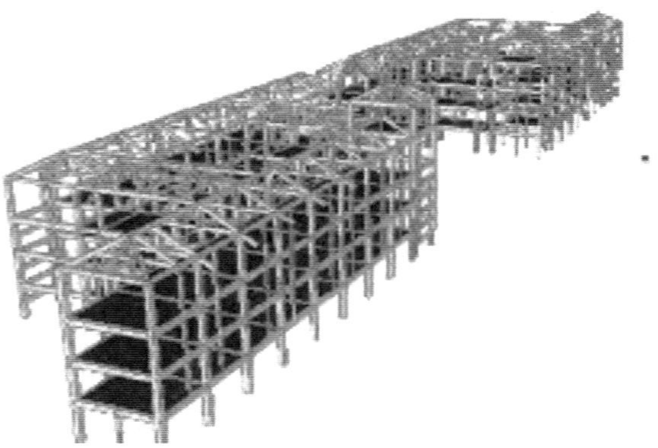

Figure 2. Structural Model

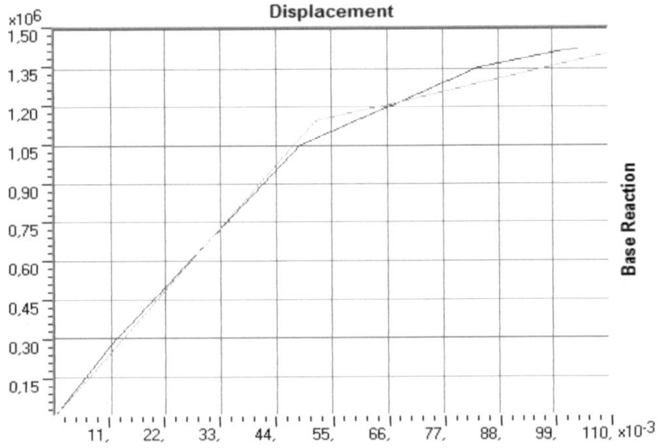

Figure3. Capacity Curve

In addition to estimating the maximum force and deformation that occurs, pushover analysis can also provide information on which building elements are critical when the earthquake occurs. Pushover analysis shows the occurrence of plastic hinges at each stage of the load increase. Plastic hinge prediction that occurs in the structure according to pushover analysis shows the plastic deformation was formed from the end of the beam at the corner of the building as shown in figure 4 and Figure 5.

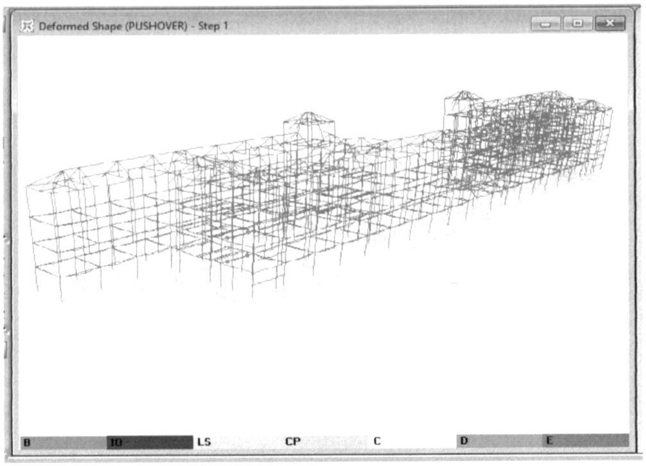

Figure 4. First Step pushover analysis

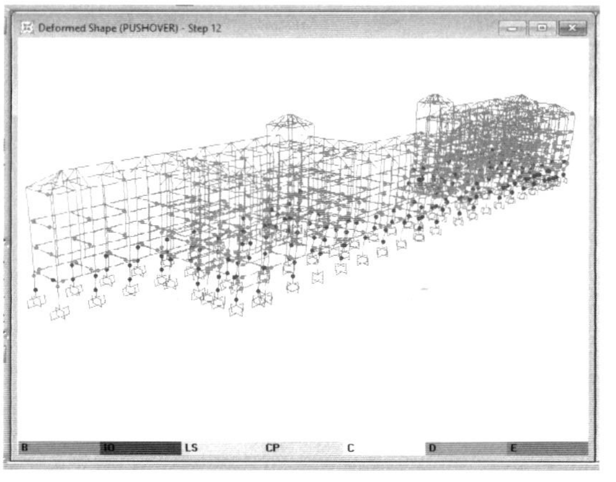

Figure 5. Last Step Pushover Analysis

The color-coded plastic deformation and brief qualitative explanation of the damage level is recapitulated in the Table 5 as follows:

Table 5 The degree of structural damage caused by plastic joints

Hinges colour	Level	explanation
	B	Linear limit, the first melting occurs
	IO	Minor damage, the stiffness of the structure is still the same
	LS	Moderate damage, reduced structure stiffness
	CP	Severe damage, the stiffness of the structure is reduced much
	C	The maximum limit of a capable shear force
	D	Structures in collapse
	E	The structure is unable to resist shear

The fragility curve shows the probability of certain damage to a building when it subjects to a certain intensity of the earthquake. The fragility curves are arranged by the median value and the potential hazard parameters that can be either spectral displacement and spectral acceleration representing certain damage conditions. The method used to determine the median value of the damage is HAZUS-MH MR5 [9]. For a given level of building response, fragility curves distribute damage between four physical damage states: slight, moderate, extensive and complete. Each damage state represents a range of damage to the building with the range starting from threshold.

The procedure to determine the median of spectral displacement ($\overline{S_d}, ds$) based on nonlinier static pushover analysis results. The recapitulation of the damage state is shown in Table 6.

Table 6. Recapitulation of the damage states based on Hazus-MH MR5

Damage states	$\overline{S_d}$, ds (m)	$\overline{S_a}$, ds (g)
Slight	0,006103	0,1112
Moderate	0,009155	0,1571
Extensive	0,039060	0,5004
Complete	0,068966	0,5527

The analysis of fragility is continued by processing four statistically structural data sets to determine the size of the data distribution by computing the standard deviation. Representing uncertainty in the analysis. incorporating standard deviation (Beta) values describe the total variability of fragility curve damage states [10] as shown in the Equation 2.

$$\beta_{ds} = \sqrt{[(CONV\,[\beta_C,\beta_D])]^2 + [(\beta_{M(ds)})]^2} \quad (2)$$

where :

β_C = is the lognormal standard deviation parameter that describes the total variability of damage state,ds,

β_D = is the lognormal standard deviation parameter that describes the variability of the demand spectrum (values of β_d= 0,45 at short periods and β_d= 0,50 at long periods.

$\beta_{M(ds)}$ = is the lognormal standard deviation parameters that describe the variability of the threshold of damage state,ds.

Table.7 Lognormal standard deviation of the damage states

Limit State	Sa, ds (g)	$\beta_{M(ds)}$	β_C	β_D	β_{ds}
Slight	0,11124	0,4000	0,5178	0,45000	0,4765
Moderate	0,1572371	0,4000	0,5178	0,45000	0,4765
Extensive	0,4974554	0,4000	0,5178	0,45000	0,4765
Complete	0,5533506	0,4000	0,5178	0,45000	0,4765

The probability relationship with various demand parameters is shown in the fragility curve (fig.6) by means of the equation 3. The parameter representing the earthquake intensity measures used in the investigation is spectral acceleration [11].

$$P\,(ds|S_a) = \Phi(\frac{1}{\beta_{ds}})\ln\left(\frac{S_a}{\bar{S}a,ds}\right) \quad (3)$$

where :

Φ = cumulative function of probability,

β_{ds} = Lognormal standard deviation of any damage condition,

S_a = spectral acceleration,

$\bar{S}_a, ds$ = median value of spectral acceleration when structure is damaged.

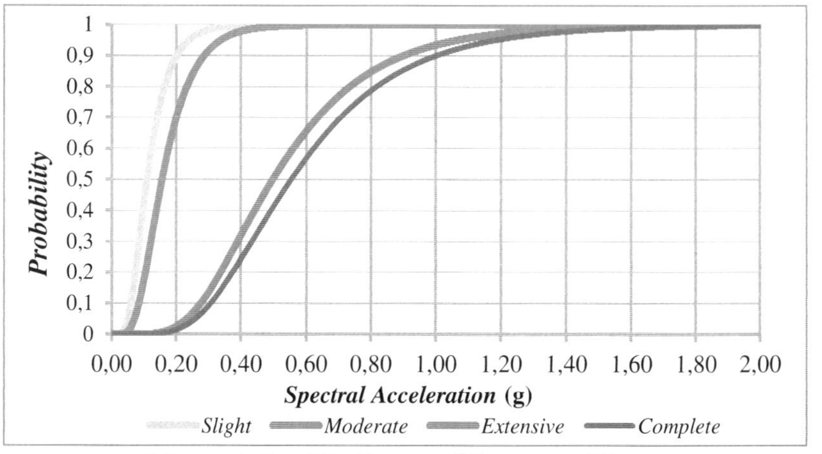

Figure 6. Fragility Curves of Rusunawa Cilacap

8. Conclusion

The fragility curves shown in figure 6 may be used to evaluate a seismic risk of Rusunawa Cilacap rationally by demonstrating how much probability of certain type of damage will occur when a certain level of spectral acceleration takes place in the building occur due to ground shaking. For instance, when the building experience acceleration of 0.3 g, then from the curve it can be inferred that there will be approximately 93% probability the building will suffer moderate and slight damage, but there is only 10% chance that building will be damaged in extensive or even complete manner. Further, if it is expected that the building experience less than 50% probability of extensive damage, then the spectral acceleration of the building should be no more than 0.46 g. By using fragility curves the decision maker and stakeholder then may have a rational basis to take action in minimizing the seismic risk prior the earthquake.

References

[1] Direktorat Jenderal Cipta Karya. (2012). Rusunawa: Komitmen Bersama Penanganan Permukiman Kumuh. Jakarta
[2] Rapid Visual Screening of Buildings for Potential Seismic Hazards: A Handbook (Third Edition), FEMA P-154 / January 2015, Federal Emergency Management Agency, Washington DC
[3] Perrone, D., Aiello, M. A., Pecce, M., & Rossi, F. (2015, August). Rapid visual screening for seismic evaluation of RC hospital buildings. In Structures (Vol. 3, pp. 57-70). Elsevier.
[4] Yakut, A. (2004). Preliminary seismic performance assessment procedure for existing RC buildings. Engineering Structures, 26(10), 1447-1461.
[5] Dewobroto, W. 2006. "Evaluation of Earthquake Resistant Steel Building with SAP2000" Journal of Civil Engineering, Vol. 3, No. January 1, 2006.
[6] Rapid Visual Screening of Buildings for Potential Seismic Hazards: Supporting Documentation (Third Edition) FEMA P-155 / January 2015. Federal Emergency Management Agency, Washington DC
[7] Dewobroto, W. 2007. "Aplikasi Rekayasa Konstruksi dengan SAP2000 Edisi Baru". Jakarta
[8] FEMA. 2010. HAZUS-MH MR5 Advanced Engineering Building Module (AEBM). Washington, DC: Federal Emergency Management Agency
[9] FEMA. 2010. HAZUS-MH MR5 Advanced Engineering Building Module (AEBM). Washington, DC: Federal Emergency Management Agency
[10] HAZUS-MH 2.1 Earthquake Loss Estimation Methodology. Washington, DC: Department of Homeland Security Federal Emergency Management Agency Mitigation Division
[11] Crowley, H., Colombi, M., Silva, V., Ahmad, N., Fardis, M., Tsionis, G., ..& Erberik, M.

(2011). Fragility functions for common RC building types in Europe. *Systemic Seismic Vulnerability and Risk Analysis for Buildings, Lifeline Networks and Infrastructures Safety Gain Project, Deliverable*, 3

The Evaluation of Screening Process and Local Bureaucracy in Determining the Priority of Urban Roads Maintenance and Rehabilitation

Andri Hendhratmoyo[1], Syafi'i[2], and FlorentinaPungky Pramesti[2]
[1]Student of master program of Civil Engineering, UniversitasSebelasMaret, Jl. Ir. Sutami 36AKentinganJebres Surakarta 57126, INDONESIA
[2]Lecturer of master program of Civil Engineering, UniversitasSebelasMaret, Jl. Ir. Sutami 36A KentinganJebres Surakarta 57126, INDONESIA
Email: andri.hmoyo80@gmail.com

Abstract. Due to the limited budget of urban roads maintenance and rehabilitation, its prioritizationis inevitable. Many models have been developed to solve these problems. That is the reason why the purpose of this study was to evaluate the screening process in the decision making of the urban roads maintenance and rehabilitation priority. The prioritization that have to be taken into account on the effect of important criteria are road condition, traffic volume, budget processing and land use. 30 stakeholders were asked to fill in the questionnaires. The object of this case study are 188 urban roads sections at *Ponorogo* in order to examine the priorities. The researchers collected the data from Surface Distress Index (SDI), traffic volume, budget processing and land use of these road sections. Based on analysis, the weights of the criteria were: road condition (W1) = 0,411; traffic volume (W2) = 0,122; budget processing (W3) = 0,363 and land use (W4) = 0,105. The result of this study by the comparison of the index values of the alternatives priorities, *Nyi Ageng Serang* Street, was revealed to have the highest priority over the other streets regarding of maintenance and rehabilitation activities.

1. Introduction

In accordance with its characteristics, roads have a tendency to have a declining condition indicated by the occurrence of roads damage. It is necessary to conduct both routine and periodic maintenance in order to maintain the proper conditions of roads. Because of the local government's budget limitation, it is necessary to formulate the policy from relevant stakeholders in determining the roadsmaintenance and rehabilitation priority with an analytical approach that can integrate with various criteria.

Up to now, many models have been developed to solve various issues related to the decision making, for example by using the methods of Analytical Hierarchy Process (AHP), Fuzzy Logic, VIKOR, and others. [1 - 3] are using the (AHP) method in prioritizing the road maintenance and rehabilitation.[4 - 6] are using fuzzy logic in the priority of road maintenance and rehabilitation. [7] and [8] have first developed research using the fuzzy logic method in determining the priority.

The Fuzzy Logic method, as the prominent component of soft computing builder, has proven to have excellent performance to solve problems that contain uncertainty. Since its introduction by Lotfi Zadeh in 1965, the fuzzy logic sets have an increasing demand by researchers both for the application in certain fields of science,and for developing the concepts given [9].

The budget for road maintenance in *Ponorogo* regency is still much devalued, so it needs various criteria in determining the priority of road maintenance and rehabilitation to reach the right target. Therefore, a screening process is required to decide whether or not a road segment is given the priority of maintenance and rehabilitation. So far the decision on maintenance and rehabilitation of roads still use a random choice system. The impact of the maintenance and rehabilitation of roads, especially in urban areas, becomes improperly targeted.

In this study, the assessment of road conditions uses the Surface Distress Index (SDI) method issued by the Directorate General of Highways Ministryof Public Works [10]. [11] In his research using SDI to determine pavement condition. The value of SDI is used as the basis of screening whether or not the roads get priority, while Fuzzy Logic Method is used to analyze the determining of priority of road maintenance and rehabilitation.

Based on the above description, the right system is necessary so that the activities of urban road maintenance and rehabilitation in *Ponorogo* regency can run well. The purpose of this research was to evaluate the priority of road maintenance and rehabilitation in *Ponorogo* Regency based on the screening process using Surface Distress Index (SDI) method and to evaluate the decision-making system by the policy maker (local bureaucracy).

2. Review of AHP and Fuzzy Logic Method

2.1. Analytical Hierarchy Process (AHP) Method

The assessment of the criteria and sub criteria uses AHP method. The weight calculation starts from sub criteria, criteria, to priority weight of roads maintenance and rehabilitation. This weighting is required to calculate the total weight of the priority of roads maintenance. The researcher used multicriteria methods to weight, by using the appraisal of pairwise comparison matrix based on Analytical Hierarchy Process (AHP) method. [12] The setting of the quantitative scale of 1 (one) to 9 (nine) is to assess the importance comparison of an element to the others as it is shown in Table 1.

Table 1. The Rating Scale Comparison of AHP

Interest's Intensity	Description	Explanation
1	Both elements are equally important	Two elements have the equally large influence on the goal
3	One element is slightly more important than the other elements	Experience and judgment support one element more than the other
5	One element is more important than the other elements	Experience and judgment are very strong in favor of one element over the other
7	One element is more important than the other elements	One strong element is sustained and dominant in practice
9	One absolute element is more important than any other elements	The evidence that supports one element against another has the highest degree of affirmation that might be reinforcing
2,4,6,8	Values between two adjacent considerations	This value is given when there are two compromises between two options
Contrary	If i activity gets one score compared to j activity, then j has the opposite value compared to i	

Source: Saaty, 2008

2.2. Fuzzy Logic Method

The fuzzy logic method introduced by [8] uses Triangular Fuzzy Number (TFN). Triangular Fuzzy Number is a combination of two (linear) lines. Figure 1 describes the graph of Triangular Fuzzy Number (TFN) is illustrated in the form of a triangular curve.

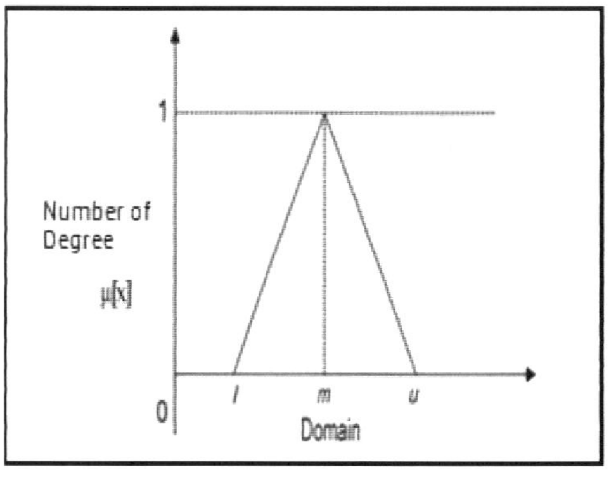

Figure 1. Triangular Fuzzy Number
Source: Chang, 1996

The fuzzy triangle scale defines the intensity value of AHP to divide each fuzzy set by two, except for the strength interest of one. Table 2 explains the triangular fuzzy scale [10].

Table 2. Fuzzy Triangle Scale

Intensity of Interest AHP	The Set of Linguistics	TFN	Reciprocity
1	Comparison of the same element (equal)	(1,1,1)	(1,1,1)
3	One element is more important than the other (weak)	(1, 3/2, 2)	(1/2, 2/3, 1)
5	One element is more important than that other (fairly strong)	(2, 5/2, 3)	(1/3, 2/5, 1/2)
7	One element is much more important than the others (very strong)	(3, 7/2, 4)	(1/4, 2/7, 1/3)
9	One absolute element is more important than the other (absolute)	(4, 9/2, 9/2)	(2/9, 2/9, 1/4)

The fuzzy logic completion steps according to [8] are as follows:

a. Creates a hierarchical structure of the problem to be solved and determines the matrix comparison in pairs between the criteria and TFN scale. The geometric mean is calculated in each value.
b. Defines the value of fuzzy synthetic extent (Si) with number I criterion as the following equation:

$$Si= \sum_{j=1}^{m} M_{gi}^{j} \otimes \left[\sum_{i=1}^{n} \sum_{j=1}^{m} M_{gi}^{j}\right]^{-1} \quad (1)$$

In order to obtain the equation, $\sum_{j=1}^{m} M_{gi}^{j}$, the m value of the matrix in the equation below reveals the fuzzy sum operation. In the final calculation phase, values (1, m, u) are obtained and used for the next phase.

$$\sum_{j=1}^{m} M_{gi}^{j}=\left(\sum_{j=1}^{m} l_j, \sum_{j=1}^{m} m_j, \sum_{j=1}^{m} u_j\right) \quad (2)$$

l is the lower limit value, m is the expected value and u is the limiting upper value. So that the equation $\left[\sum_{i=1}^{n}\sum_{j=1}^{m} M_{gi}^{j}\right]^{-1}$ a fuzzy sum operation is performed from the value M_{gi}^{j} (J = 1, 2,......, m) to obtain the following equation:

$$\sum_{i=1}^{n}\sum_{j=1}^{m} M_{gi}^{j} = \left(\sum_{i=1}^{n} l_i, \sum_{i=1}^{n} m_i, \sum_{i=1}^{n} u_i\right) \quad (3)$$

Then, it calculates the inverse of the vector in the above equation to get the equation below:

$$\left[\sum_{i=1}^{n}\sum_{j=1}^{m} M_{gi}^{j}\right]^{-1} = \frac{1}{\sum_{i=1}^{n} u_i}, \frac{1}{\sum_{i=1}^{n} m_i}, \frac{1}{\sum_{i=1}^{n} l_i} \quad (4)$$

c. Calculates the degree of possibility
To obtain the degree of possibility, $M_2 = (l_2, m_2, u_2) \geq M_1 = (l_1, m_1, u_1)$ is expressed in the equation below:

$$V(M_2 > M_1) = \sup[\min(\mu_{M1}(x), \mu_{M1}(y))] \quad (5)$$

And x and y are values on the axis of the membership function of each criteria. Thisequationcanbedefined as follows:

$$V(M_2 \geq M_1) = \begin{cases} 1 & , if \quad m_2 \geq m_1 \\ 0 & , if \quad l_2 \geq u_2 \\ \dfrac{l_1 - u_2}{(m_2-u_2)-(m_1-l_1)}, & Otherwise \end{cases} \quad (6)$$

We need the values of $V(M_2 \geq M_1)$ dan $V(M_1 \geq M_2)$ to compare M_1 and M_2.
d. Compares the degree of possibility among the criteria
The comparison between degrees of possibility between M i (i = 1, 2, 3, 4, 5, ..., k) can be defined by: $V(M \geq M_1, M_2, \ldots M_k) = V[(M \geq M_1), (M \geq M_2), \ldots, (M > M_k)]$ and *min* $V(M > M_i)$, (i=1.2.3,.......,k).
Assume that the equation (7) is:

$$d'(A_i) = \min V(S_i \geq S_k) \quad (7)$$

for k = 1, 2, 3, 4, 5, ..., n; k≠ i. Then the weight vector is given by (8) as follows:

$$W' = (d'(A_1), d'(A_2), \ldots, d'(A_n))^T \quad (8)$$

Where Ai (i=1,2,3,4,5,......, n) is *n* element.
e. Normalization
Normalization of vector weight is given by the following equation:

$$W = (d(A_1), d(A_2), \ldots, d(A_n))^T \quad (9)$$

Where W is a non-fuzzy number of the weight vector.
f. Calculates the global weight
Weighting the criterion and sub criterion vectors reveals the result, and the multiplication of criterion weight with its sub criterion result in the global weight calculation.

3. Research Method

The study was in *Ponorogo* Regency on urban roads district covering 4 (four) sub districts of *Ponorogo* Regency: *Ponorogo* District, *Jenangan* District, *Siman* District and *Babadan* District.
In this study, the assessment of interests between the criteria and sub criteria determines the basis for the priority of roads maintenance and rehabilitation. In order to get interest assessment, 30 respondents, who are the policy makers of the road maintenance and rehabilitation in *Ponorogo* regency, filled in the questionnaires.

The obtained data are in the form of value/ weight of the degree of importance between the criteria and sub criteria that affect the decision making of the priority of roads maintenance and rehabilitation. The criteria include: road condition, traffic volume, budget processing, and land use.

The researchers used purposive sampling method to determine the respondents and the selected respondents have the knowledge and competence in the field of roads maintenance and rehabilitation. The data of the road segment that became the object of research is 188 road segments. The 188 road segments are ranked based on the Surface Distress Index (SDI), that was SDI values > 50 (above 50) received the priority of roads maintenance and rehabilitation and SDI values< 50 (below 50) did not receive the priority of roads maintenance and rehabilitation, as the roads were categorized in good condition.

After knowing the weights (W) criterion, the next step is to analyze each alternative road segment to get the priority total weight of roads maintenance and rehabilitation from the highest to the lowest. The ranking priority of roads maintenance and rehabilitation used the value of total weight. The calculation of alternative weight starts from the criteria of road conditions on each road segment. Alternative weight assessment determines the sub criteria of road conditions that are total area of cracks, average crack widths, the total number of potholes and the average depth of wheel rutting [9]. Tables 3, 4, 5 and 6 explain the further detailed information.

Table 3. The Assessment of Alternative Weight on Total Area of Cracks

No.	Percentage total area of cracks	Weight Assessment
1	None	1
2	Percentage of cracks: < 10 %	2
3	Percentage of cracks: 10 - 30 %	3
4	Percentage of cracks: > 30 %	4

Table 4. The Assessment of Alternative Weight on Average Crack Widths

No.	Average Crack Width	Weight Assessment
1	None	1
2	Average Crack Width: FINE < 1 MM	2
3	Average Crack Width: MED 1 - 3 MM	3
4	Average Crack Width: WIDE > 3MM	4

Table 5. The Assessment of Alternative Weight on Total Number of Potholes

No.	Total Number of Potholes	Weight Assessment
1	None	1
2	The number of potholes: < 10 / KM	2
3	The number of potholes: 10 - 50 / KM	3
4	The number of potholes: > 50 / KM	4

Table 6. The Assessment of Alternative Weight on Average of Depth Wheel Rutting

No.	Average of Depth Wheel Rutting	Weight Assessment
1	None	1
2	Depth of Rutting: < 1 CM	2
3	Depth of Rutting: 1 - 3 CM	3
4	Depth of Rutting: > 3 CM	4

The next alternative weight assessment is on traffic volume criteria. Table 7 shows the detailed interval division of traffic volume.

Table 7. The Alternative Weight Assessment of traffic volume

Traffic Volume	Weight Assessment
< 2000	1
2000 - 4000	2
4000 - 6000	3
> 6000	4

The following alternative weight assessment is on the criteria of the budget processing. The alternative weight assessment for each road segment can be made as follows:

a. Getting a priority proposal at the District Community Consultations on Development Planningis given a value of 4. If there is no priority proposal the given value is 1.
b. Getting a priority proposal at the Local Government Work Departments Forum is given a value of 4. If it does not get a priority proposal, the value is 1.
c. Getting a priority proposal in the Regency Community Consultations on Development Planningis given a value of 4. If there is no priority proposal, the given value is 1.
d. Listed or entered in the Local Government Budget is given a value of 4. If it is not listed or entered, the given value is 1.
e. Listed or entered in the Revised Local Government Budget is given a value of 4. If it is not listed or entered, the given value is 1.

The last assessment of the alternative weight is on the land use criteria. The amount of value is adjusted to the land or space utilization with the following scoring system:

a. If the road segment can support the social cultural activities, the given value is 4. If it cannot support the activities, the given value is 1.
b. If the road segment can support the trade and industry activities, the given value is 4. If it cannot support the activities, the given value is 1.
c. If the road segment can support the agricultural and plantation activities, the given value is 4. If it cannot support the activities, the given value is 1.
d. If the road segment can support settlement activities, the given value is 4. If it cannot support the activities, the given value is 1.

4. Result and Discussion

Based on the survey of road conditions and the value of Surface Distress Index [10] on 188 urban road segments, the assessment results of road condition ranks are 111 road segments received priority maintenance and rehabilitation (Fair Condition, Poor Condition, and Very Poor Condition). The graph of road conditions as it shown in figure 2.

From the analysis result using the Fuzzy Logic method, the weight for each criterion is as follows: road condition criteria is (W1) = 0.411, traffic volume criteria is (W2) = 0.122, budget processing criteria is (W3) = 0.363 and the land use criteria is (W4) = 0.105. While the priority sequence in roads maintenance and rehabilitation is *Nyi Ageng Serang* Street, the second priority up to the tenth in a row *MT. Haryono Gg. VI Street, Delima Street, Barito Street, Ukel Street, Larasati Street, Truntum Street, MT. Haryono Gg. V Street, Subali Street and Jenar Street.* Figure 3 and 4 shows detailed weight of criterion and sub criteria.

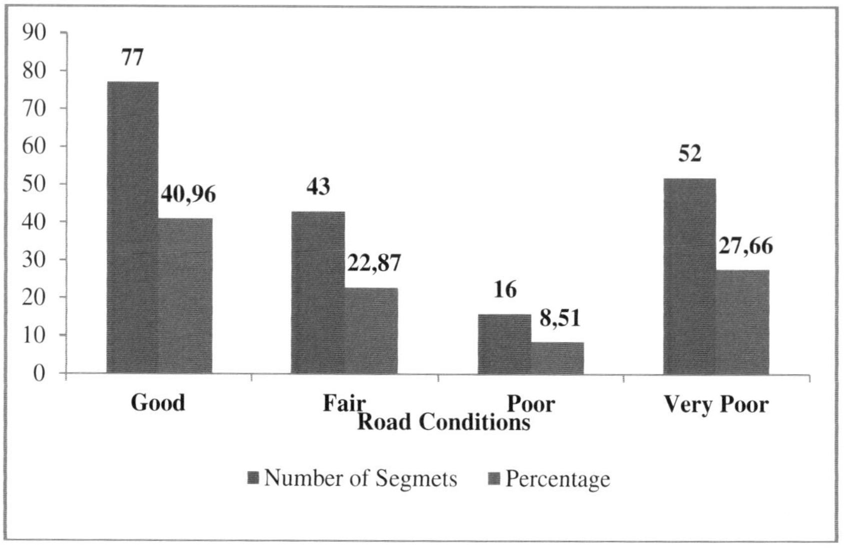

Figure 2. The Graph of Road Conditions

Weight	
Criteria	**Sub Criteria**
Road Condition (W1=0,411)	Total Area of Cracks (0,266)
	Average Cracks Widths (0,197)
	Number of Potholes (0,298)
	Depth Wheel Rutting (0,239)
Traffic Volume (W2=0,122)	Heavy Vehicle (0,284)
	Bus (0,284)
	Light Vehicle (0,216)
	Car (0,141)
	Motorcycle (0,075)
Budget Processing (W3=0,363)	District Community Consultations on Development Planning (0,254)
	Local Government Work Department Forum (0,088)
	Regency Community Consultations on Development Planning (0,218)
	Local Government Budget (0,270)
	Revised Local Government Budget (0,169)
Land Use (W4=0,105)	Social Cultural Activities (0,197)
	Trade and Industry Activities (0,271)
	Agricultural and Plantation Activities (0,274)
	Settlement Activites (0,258)

Figure 3. The Weight of Criteria and Sub Criteria Roads Maintenance and Rehabilitation

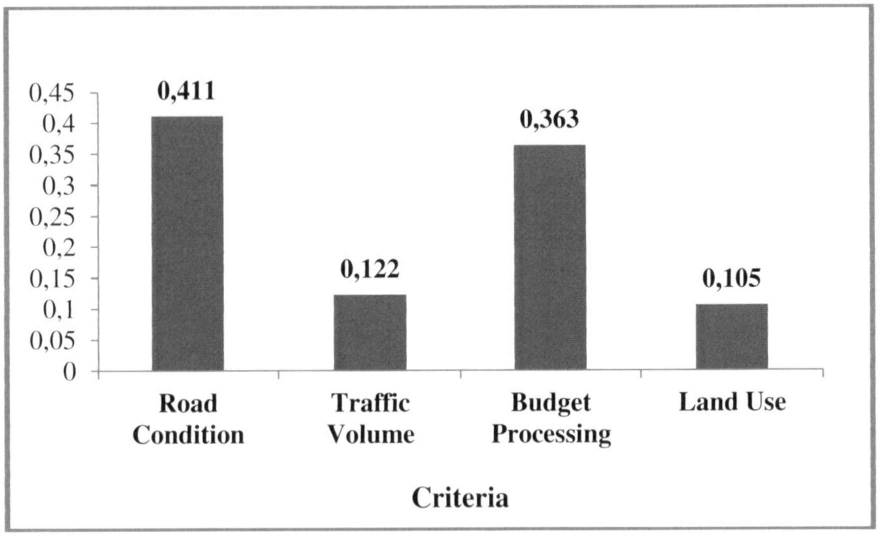

Figure 4. The Graph of Criteria

The important matters based on the result of the study are:

1. Based on the rank of the Surface Distress Index value, from the total 188 road segments, there are 111 road segments that get the priority of maintenance and rehabilitation. It means that 77 road segments do not get the priority (good condition). 111 road segments which get the priority are ranked from a fair condition, poor condition and very poor condition. In this study, Surface Distress Index method is an effective method in the screening process of road condition assessment.
2. The criterion of road condition are the most important criteria of inter interests assessment according to the respondents (policy maker). The next criterion is Budget Processing, and continues to the traffic volume and the land use criteria.
3. Budget Processing criterion become a keyfactor in prioritizing the road maintenance and rehabilitation. The road condition criteria is (W1) = 0.411 and the budget processing criteria is (W3) = 0.363. It shows that the local bureaucracy influence is highly visible in determining the priority of maintenance and road rehabilitation.
4. Budget Processing criterion becomes the most important criterion after the Road Condition. It is because of the most respondents comes from the bureaucracies who understand that the priority of roads maintenance and rehabilitation must go through the correct budget processing and by the applicable of rules.

5. Conclusion

Based on the analysis using the fuzzy logic method, the weight for each criterion is as follows: the road condition criterion is (W1) = 0.411, the traffic volume criterion is (W2) = 0.122, the budget processing criterion is (W3) = 0.363 and the land use criterion is (W4) = 0.105.

Based on the analysis, budget Processing becomes the second most important criterion after the road condition. It shows that the role of local bureaucracy is very important in the process of decision-making of road maintenance and rehabilitation. While the priority order of the roads maintenance and rehabilitation is *Nyi Ageng Serang* Street.

References

[1] Ashok, A., Thasneem, K.M. and Vincent, S. 2016 Pavement Rehabilitation Using Analytical Hierarchy Process (AHP) *International Journal of Innovative Research in Science Engineering and Technology* **5** Issue 5 pages 6858 - 6865.

[2] Subramani, T., 2012Using Expert System Fixing Priority to The Identified Short Term Transport Facility Projection Salem *International Journal of Engineering Trends and Technology* **3** Issue 2 pages 201 - 209.

[3] Zack, J., dan Kruszynski, M., 2015 Application of AHP and ELECTRE III/IV Methods to Multiple Level Multiple Criteria Evaluation of Urban Transportation Projects Transportation Research Procedia *18th Euro Working Group on Transportation* Delft The Nederlands **10** pages 820 – 830.

[4] Babashamsi, P., Goldzafar, A., Yusoff, N.I.M., Ceylan, H. and Nor, N.G.M.2016 Integrated Fuzzy Analytic Hierarchy Process and VIKOR method in the Prioritization of Pavement Maintenance Activities *International Journal of Pavement Research and Technology* **9** pages 112-120.

[5] Mahmood, M., Rahman, M., Nolle, L. and Mathavan, S. 2013 A Fuzzy Logic Approach for Pavement Section Classification *International Journal of Pavement Research and Technology* **6** No. 5 pages 620 - 626.

[6] Moazami, D., Behbahani, H. and Muniandy, R., 2016 Pavement Rehabilitation and Maintenance Prioritization of Urban Roads Using Fuzzy Logic *Expert System with Applications An International Journal* **38** pages 12869 - 12879.

[7] Celik, M., Er, D. and Ozok, A.F. 2009 Application of Fuzzy Extended AHP Methodology on Shipping Registry Selection: The Case of Turkish Maritime Industry *Expert System With Applications* **36** pages 190-198.

[8] Chang, D.Y.,1996 Applications of The Extent Analysis Method on Fuzzy AHP *European Journal of Operational Research* **95** No. 3 pages 649 - 655.

[9] Kusumadewi, S., Hartati, S., Harjoko, A. and Wardoyo, R. 2006 *Fuzzy Multi Attribute Decision Making (Fuzzy-MADM)* (Yogyakarta: Graha Ilmu).

[10] Directorate General of Highways Ministry of Public Works 2011 *Road Condition Survey Guides Number SMD-03/RCS* Directorate General of Highways Ministry of the Public Works Republic of Indonesia.

[11] Ichsan, Saleh, M. Sofyan, danIsya M., 2014 Studi Evaluasi Tingkat Kerusakan Permukaan Jalan untuk Menentukan Jenis Penanganan dengan Sistem Penilaian Menurut Bina Marga (Studi Kasus: Ruas Jalan Bireuen – Takengon), *Jurnal Teknik Sipil Pascasarjana Universitas Syiah Kuala* **3** No. 2 pages 62 – 72.

[12] Saaty, T.L, 1980 *The analytic hierarchy process* (New York: McGraw-Hill).

[13] Jasril, Haerani, E. and Afrianty,I. 2011 *Sistem Pendukung Keputusan (SPK) Pemilihan Karyawan Terbaik Menggunakan Metode Fuzzy AHP (F-AHP)* Jurnal Seminar Nasional Aplikasi Teknologi Informasi, ISSN: **1907-5022** pages 36 - 43.

IOP Conf. Series: Journal of Physics: Conf. Series **909** (2017) 012073 doi:10.1088/1742-6596/909/1/012073

The study on length and diameter ratio of nail as preliminary design for slope stabilization

Indra Gunawan[1], Niken Silmi Surjandari[1], and Yusep Muslih Purwana[1]

[1]Civil Engineering Department, Engineering Faculty, Sebelas Maret University, Jl. Ir. Sutami 36A Surakarta 57126, Central Java, INDONESIA

E-mail: indragunawan@gmail.com
nikensilmisurjandari@gmail.com
ymuslih@yahoo.com

Abstract. Soil nailing technology has been widely applied in practice for reinforced slope. The number of studies for the effective design of nail-reinforced slopes has also increased. However, most of the previous study was focused on a safety factor of the slope; the ratio of length and diameter itself has likely never been studied before. The aim of this study is to relate the length and diameter ratio of the nail with the safety factor of the 20 m height of sand slope in the various angle of friction and steepness of the slope. Simplified Bishop method was utilized to analyze the safety factor of the slope. This study is using data simulation to calculate the safety factor of the slope with soil nailing reinforcement. The results indicate that safety factor of slope stability increases with the increase of length and diameter ratio of the nail. At any angle of friction and steepness of the slope, certain effective length and diameter ratio was obtain. These results may be considered as a preliminary design for slope stabilization.

1. Introduction

Indonesia has a diverse geological structure and topographic condition. One form of the topographic condition is the slopes that can be found in the hilly areas. The common issues in slope stability are the prevention of landslides in slope. The National Board for Disaster Management (BNPB) noted that there were 364 landslides occurred in Indonesia between the period of January – Mei 2017. These disaster resulted not less than 180 casualties of people and 1800 property damage. Based on these data, there should be an alternatives way to reduce the danger of the landslides. Slope reinforcement is one of the alternative to stabilize the slope that is prone to landslides. One method of slope improvement that has been widely applied is soil nailing.

Soil nailing is one of the technique to stabilize the slope with passive inclusions or reinforcements by steel-bar known as soil nails. The inclusions can be fully bonded (grouted nails) or simply driven into the ground. The nails act to limit the soil deformations near the exposed face and transfer the stress to a more stable zone behind the wall.

As soil nailing technology is increasingly applied to reinforce slopes, the number of studies on the effective design of nail-reinforced slope has also increased. The effective design depends on the impact of soil and nail parameter to the stability level of the reinforced slopes, which should be based on a failure behavior and reinforcement mechanisms. The study about the effect of nail parameter such as nail spacing (S_v and S_h) [1], nail length (L) [2], nail inclination angle (^{0}) [3], nail diameter (d) [4], and number of nail (n) [5] to the safety factor have been conducted in recent years. The effect of soil parameters such as slope steepness (β) [1], the angle of internal friction (φ) [6], soil cohesion (c) [7], and unit weight (γ_b) [8] to the safety factor of the slope have also been developed. However, most of the previous study were only focused on the exact parameter impact to the safety factor of the slope; The ratio of this parameter have likely never been studied before.

The purpose of this research was to determine the effective design of soil nailing parameter using the length and diameter ratio (L/d). The properties of the slope were a 20 m height of sand slope in a various angle of friction and steepness of the slope. Data simulation was conducted to determine the impact of length and diameter ratio to the safety factor of the slope.

2. Method

2.1. Soil and nail parameter

The soil and nail parameter that used in this study are based on the previous study and references about nail reinforcement in sand slope [9], [10], [11]. Table 1 shows the soil and nail parameter that used in this study. The length and diameter ratio of nail range from 0 – 700 with the interval of 50 for every diameter that been used in data simulation. The yield stress of the soil nail bar is 517 MPa based on the standard of nail dimensions (ASTM A615 Grade 75).

Table 1. Soil and nail parameter on this study.

Soil		Nail	
Soil parameter	Data input	Nail parameter	Data input
Unit Weight (kN/m^3)	16	Nail diameter (mm)	19, 22, 25, 29, 32
Soil cohesion (kPa)	0	Bor diameter (m)	0,2
Angle of internal friction (^{0})	30, 35, 40	Nail Spacing (m)	1,5 and 2
Slope height (m)	20	Nail inclination (^{0})	15
Angle of the slope (^{0})	45, 60, 90	Bond skin friction (kN/m^2)	140

2.2. Data simulation

Data simulation was conducted to determine the safety factor of the slope based on the impact of the length and diameter ratio. Simplified Bishop Method [12] were utilized in data simulation before and after the soil nailing reinforcement model using Equation 1 and Equation 2 in a computer program. The stability of the slope is determine based on the Bowles safety factor classification [13] that shown in Table 2. The total variations analyzed in this study were 585 variations based on the length and diameter ratio and various condition of soil and nail parameter.

$$FS_b = \frac{\sum[c\,.\,b_i + W_i(1\text{-}r_u)\,tg\,\varphi\,]\left(\frac{1}{Cos\,\alpha_i(1+tg\,\alpha_i\,tg\,\varphi/FS_b')}\right)}{\sum[\,W sin\,\alpha_i]} \tag{1}$$

$$FS_{bT} = \frac{\sum[c\,.\,b_i + W_i(1\text{-}r_u)\,tg\,\varphi]\left(\frac{1}{Cos\alpha_i(1+tg\,\alpha_i\,tg\,\varphi/FS_b')}\right)}{\sum[\,W sin\,\alpha_i\text{-}T_n\,cos\,(\alpha_i+\lambda\,)]} \tag{2}$$

Table 2. Bowles safety factor classification

Safety factor	Slope condition
SF $\geq$ 1,25	Stable
1,07 $\leq$ SF $\leq$ 1,25	Critical
SF $\leq$1,07	Unstable

IOP Conf. Series: Journal of Physics: Conf. Series **909** (2017) 012073 doi:10.1088/1742-6596/909/1/012073

3. Results and discussion

3.1 Safety factor of the slope before reinforcement

The model of the existing slopes (without reinforcement) is described based on the soil parameter in Table 1. Figure 1 shows the example of data simulation on 60^0 existing slopes using a computer program. The result of critical safety factor on this slope variation is 0,349.

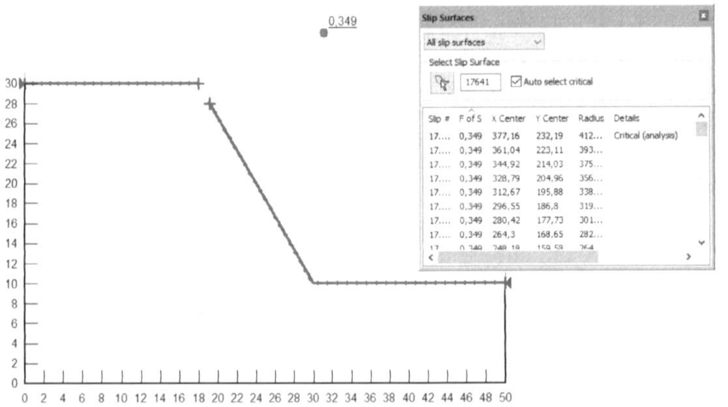

Figure 1. Critical safety factor result on the 226th slope variation ($\beta = 60^0$, $\varphi = 30^0$, no reinforcement)

3.2 Safety factor of the slope after reinforcement

The model of the soil nailing reinforcement on the existing slope is described in the program based on the nail parameter in Table 1 and the length and diameter ratio of nail parameter. Figure 2 shows the example of data simulation on 60^0 reinforced slope using a computer program. The result of safety factor on this slope variation is 1,429. This result indicates that the implementation of soil nails are improved the safety factor of the slope.

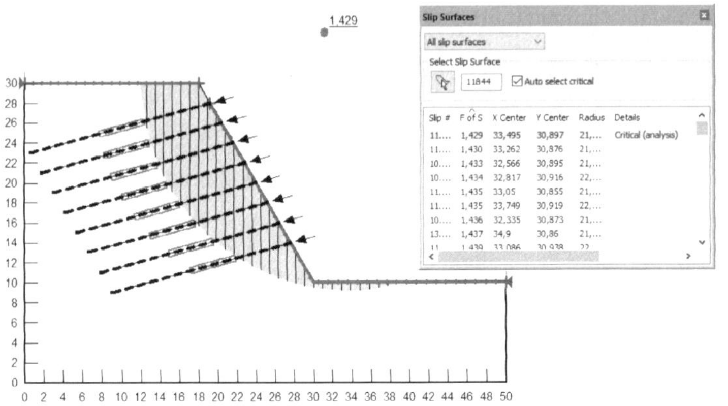

Figure 2. Critical safety factor result on the 290th slope variation ($\beta = 60^0$, $\varphi = 30^0$, d = 32 mm, L/d = 600)

3.3 Impact of length and diameter ratio of nail (L/d) on the safety factor of 45^0 of slope steepness

The safety factor of 45^0 slope reinforcement using soil nailing are calculated using a computer program. The length and diameter ratio of the nail (L/d) that have been used in this variation are 0 – 700 with the interval of 50. The safety factor result based on the length and diameter ratio of the nail from 45^0 of slope steepness is shown in Figure 3 – Figure 5. The average increase of safety factor on the 45^0 slope is shown in table 3. The effective value of L/d from the $45^0$of slope reinforcement using the soil nails are 600 – 650.

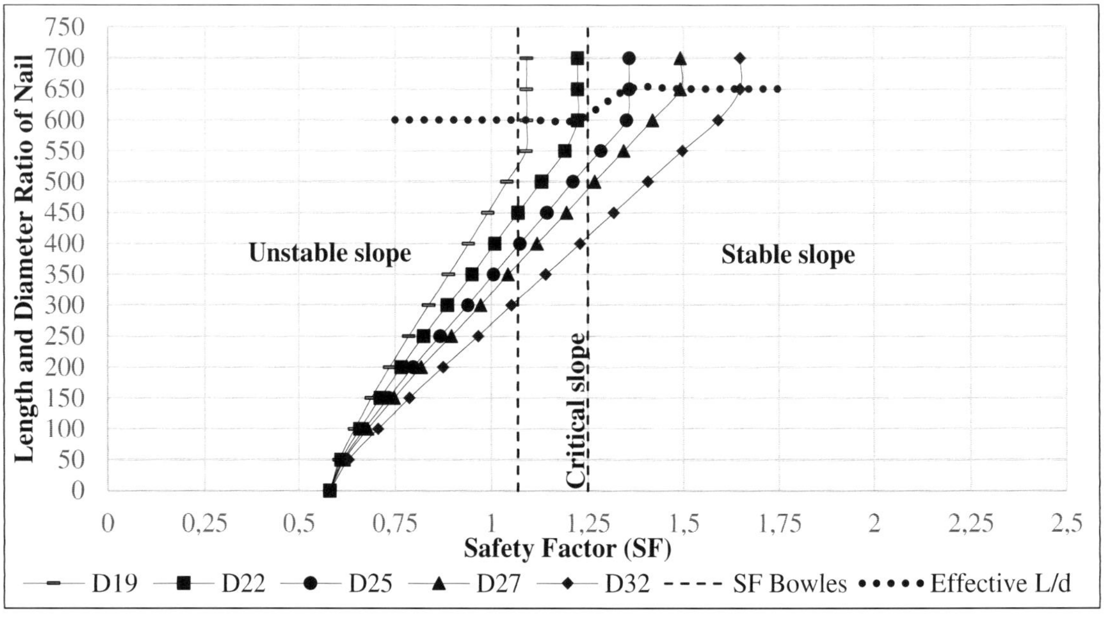

Figure 3. Impact of length and diameter ratio of nail on the safety factor of 45^0 loose sand slope ($\beta = 45^0$, $\varphi = 30^0$)

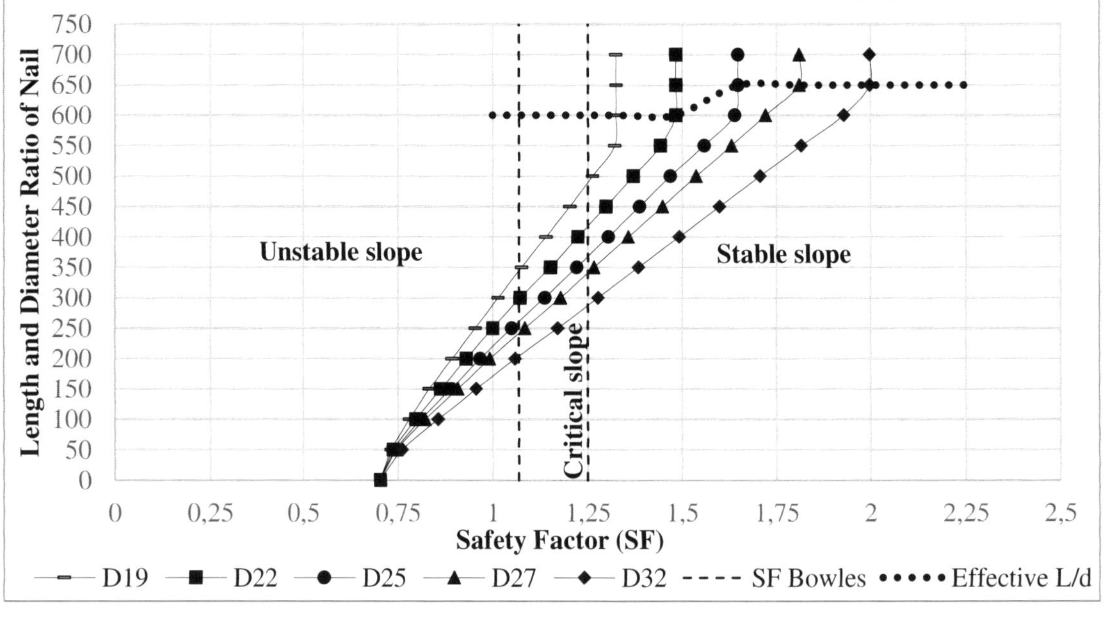

Figure 4. Impact of length and diameter ratio of nail on the safety factor of 45^0 medium sand slope ($\beta = 45^0$, $\varphi = 35^0$)

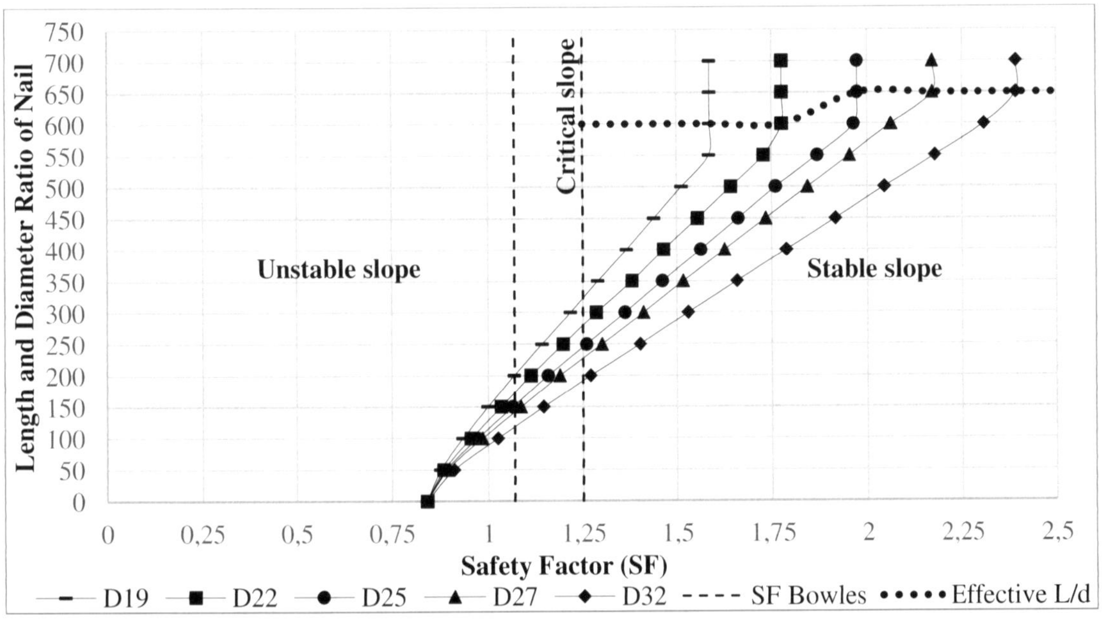

Figure 5. Impact of length and diameter ratio of nail on the safety factor of 45^0 dense sand slope ($\beta = 45^0$, $\varphi = 40^0$)

Table 3. Impact of L/d parameter to average increase of safety factor on the 45^0 slope

Nail diameter	Average increase of safety factor based on L/d
19 mm	± 4,36%
22 mm	± 5,15%
25 mm	± 5,90%
27 mm	± 6,56%
32 mm	± 7,28%

3.4 Impact of length and diameter ratio of nail on the safety factor of 60^{0}of slope steepness

The length and diameter ratio of the nail (L/d) that have been used in 60^0 of slope reinforcement using soil nailing are 0 – 600 with the interval of 50. The safety factor result based on the length and diameter ratio of the nail from 60^0 of slope steepness is shown in Figure 6 – Figure 8. The average increase of safety factor on the 60^0 slope is shown in table 4. The effective value of L/d from the 60^{0}of slope reinforcement using the soil nails are 450 – 550.

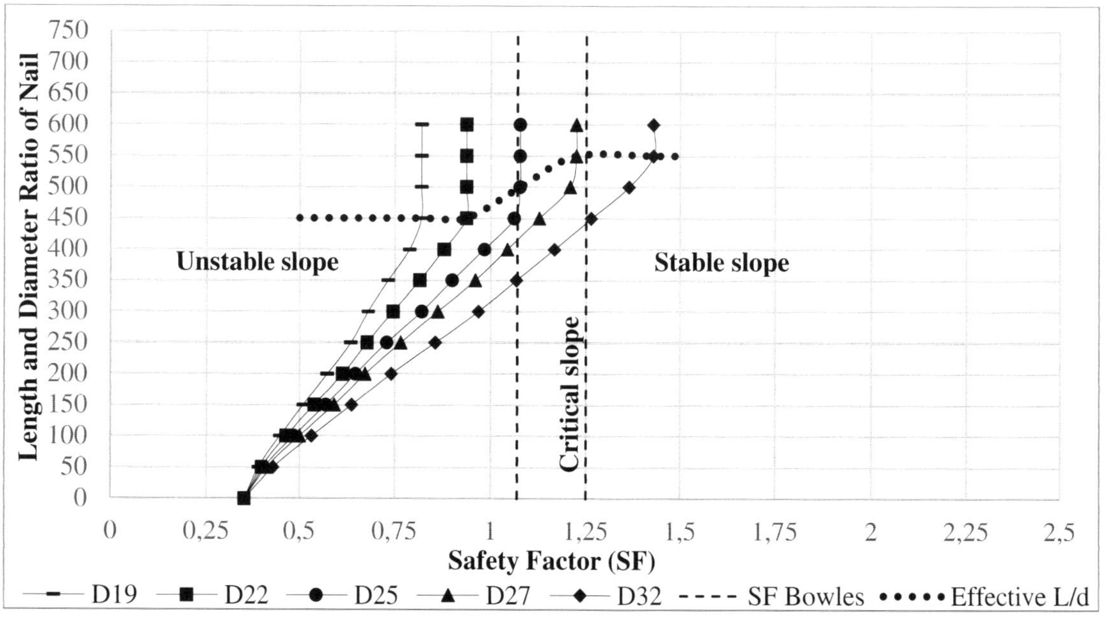

Figure 6. Impact of length and diameter ratio of nail on the safety factor of 60^0 loose sand slope ($\beta = 60^0$, $\varphi = 30^0$)

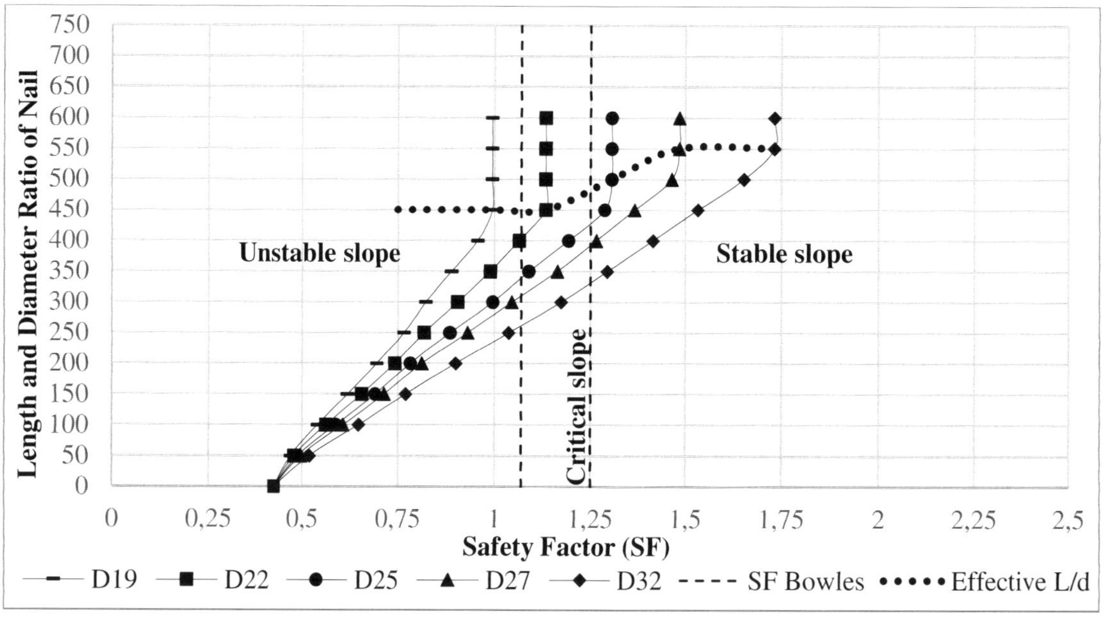

Figure 7. Impact of length and diameter ratio of nail on the safety factor of 60^0 medium sand slope ($\beta = 60^0$, $\varphi = 35^0$)

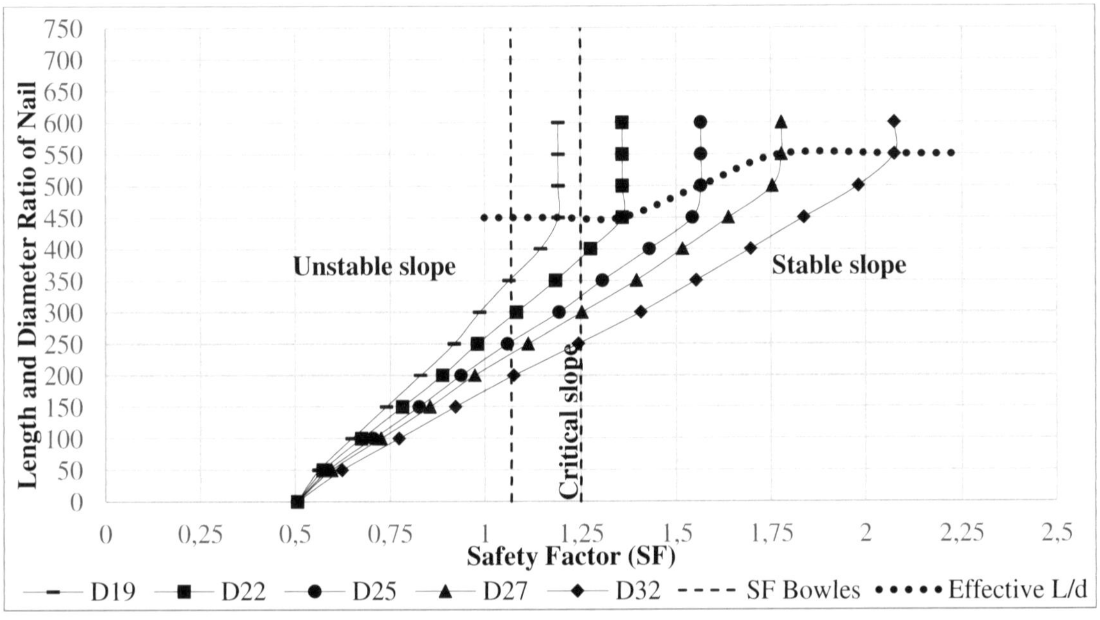

Figure 8. Impact of length and diameter ratio of nail on the safety factor of 60^0 dense sand slope ($\beta = 60^0$, $\varphi = 40^0$)

Table 4. Impact of L/d parameter to average increase of safety factor on the 60^0 slope

Nail diameter	Average increase of safety factor based on L/d
19 mm	± 6,94 %
22 mm	± 8,06 %
25 mm	± 9,27 %
27 mm	± 10,35 %
32 mm	± 11,71 %

3.5 Impact of length and diameter ratio of nail on the safety factor of 90^0 slope

The length and diameter ratio of the nail (L/d) that have been used in 60^0 of slope reinforcement using soil nailing are 0 – 500 with the interval of 50. The safety factor result based on the length and diameter ratio of the nail from 60^0 of slope steepness is shown in Figure 9 – Figure 11. The average increase of safety factor on the 60^0 slope is shown in table 5. The effective value of L/d from the 60^0 of slope reinforcement using the soil nails are 400 – 450.

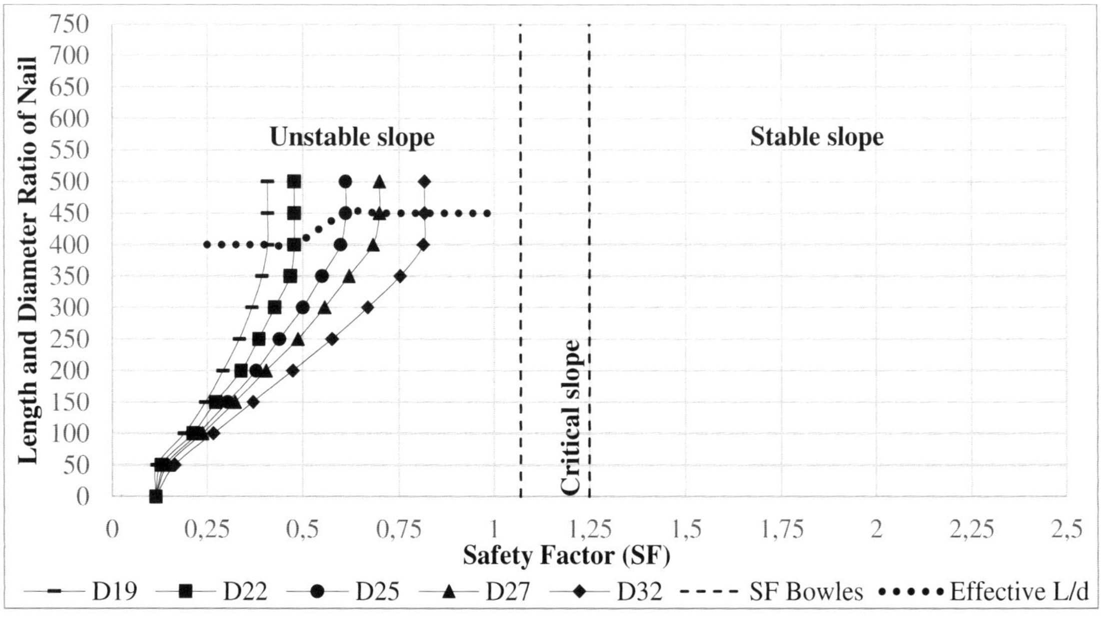

Figure 9. Impact of length and diameter ratio of nail on the safety factor of 90^0 loose sand slope ($\beta = 90^0$, $\varphi = 30^0$)

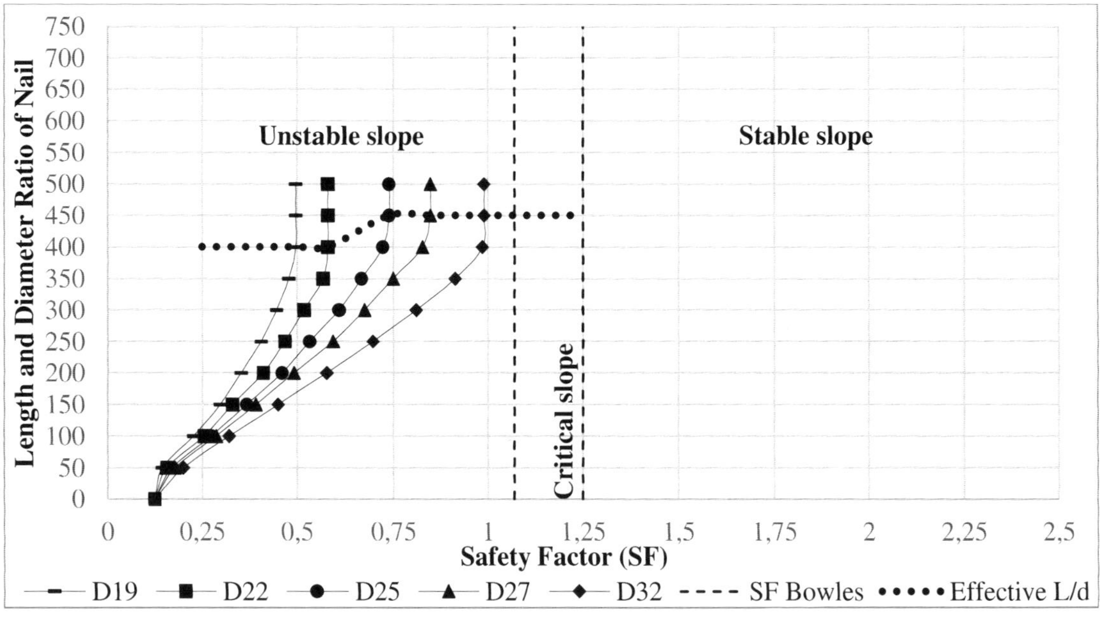

Figure 10. Impact of length and diameter ratio of nail on the safety factor of 90^0 medium sand slope ($\beta = 90^0$, $\varphi = 35^0$)

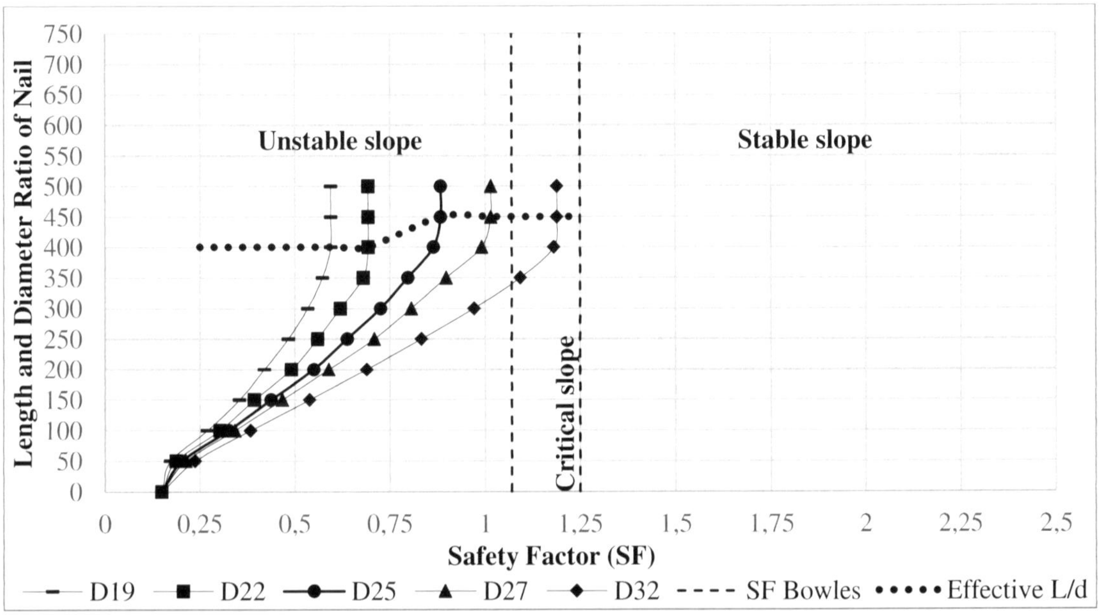

Figure 11. Impact of length and diameter ratio of nail on the safety factor of 90^0 dense sand slope ($\beta = 60^0$, $\varphi = 40^0$)

Table 5. Impact of L/d parameter to average increase of safety factor on the 90^0 slope

Nail diameter	Average increase of safety factor based on L/d
19 mm	± 14,12 %
22 mm	± 15,87 %
25 mm	± 18,41 %
27 mm	± 19,96 %
32 mm	± 22,02 %

4. Conclusion

The result indicates that safety factor of the slope is increases with the increase of L/d value to the effective value of L/d. If the length of the nail passes the effective value of L/d, the safety factor of the slope does not change significantly. The effective value of L/d range from 400 – 650 based on the steepness of the slope. The average increase of slope safety factor also increases with the increase of nail diameter. It indicates that the larger diameter of the nail is more effective to apply in soil nailing reinforcement. However, the use of an efficient nail diameter size is still adjusted to the needs of slope safety factor based on the density and steepness of the slope.

References

[1] Wei, W.B., and Cheng, Y.M. (2010).*"Soil Nailed Slope by Strength Reduction and Limit Equilibrium Methods,"* Elsevier Computers and Geotechnics **37** (2010) 602 – 618.

[2] Hartanto, A. (2012). *"Analisis Stabilitas Lereng dengan Perkuatan Soil Nailing,"* Skripsi S1. Surakarta: Jurusan Teknik Sipil, Fakultas Teknik, Universitas Sebelas Maret.

[3] Hossain, A. Md., dan Islam, A. (2016). *"Numerical Analysis of the Effects of Soil Nail on Slope Stability,"* International Journal of Computer Applications (0975-8887), Volume **141** – No.8, May 2016.

[4] Ling-TangOu. dan Qing-Ming Jiang(2015).*"Stability Analysis of Slope under Different Soil Nailing Parameters Based on the GeoStudio,"* International Journal of Geohazard and Environment 2015 **1** (2): 88-92.

[5] Natalia, Yesi(2016).*"Analisis Transfer Beban Pada Soil Nailing (Studi Kasus : Kawasan Citra Land,"* Manado, Universitas Sam Ratulangi. Jurnal Sipil Statik Vol. **4** No. 8, Agustus 2016 (491 – 497) ISSN: 2337 – 6732.

[6] Octovian C. dan Sartje M. (2014). *"Analisis Kestabilan Lereng dengan Metode Bishop (Studi Kasus: Kawasan Citra Land Sta. 1000 m),"* Jurnal Sipil Statik Vol. **2** No. 3, Maret 2014 (134 – 147) ISSN: 2337 – 6732.

[7] Javia V. et al. (2013).*"An Experimental Study on Horizontal and Inclined Soil Nails in Sand,"* Global Research Analysis, ISSN No 2277-8160.

[8] Ellias, V. And I. Juran, (1991). *"Soil Nailing for Stabilization of Highway Slopes and Excavations,"* Publication FHWA – NHI 00-043, Federal Highway Administration (FHWA), Washington D.C.

[9] FHWA.(2003).*"Geotechnical Circular No. 7. Soil Nail Walls,"* Publication FHWA-IF-03-017, Washington, D.C, U.S. Department of Transportation, Federal Highway Administration.

[10] Prashant A., et. al.(2010), *"Soil Nailing for Stabilization of Steep Slopes Near Railway Tracks,"* Department of Civil Engineering, Indian Institute of Technology Kanpur.

[11] Meyerhoff, G.G. (1963). "*Some Recent Research on The Ultimate Bearing Capacity of Foundation,*" Journal Canadian Geotechnique, Vol.**1**.

[12] Bishop, A.W. (1955). "*The Use of Slip Circle in The Stability if Analysis Slopes,*" Geotechnique, London, Vol. **10**, pp.129.

[13] Bowles, J.E. (1989). *"Sifat-sifat Fisis dan Geoteknis Tanah,"* Erlangga. Jakarta.

IOP Conf. Series: Journal of Physics: Conf. Series **909** (2017) 012074 doi:10.1088/1742-6596/909/1/012074

Utilization of Pb and PbO_2 from lead storage battery waste for batik wastewater treatment using electrochemical method

Riyanto[1,3] **and Siti Wulandari**[2]

[1,2]Department of Chemistry, Faculty of Mathematics and Natural Science, Universitas Islam Indonesia, Jalan Kaliurang KM 14.5 Sleman, Yogyakarta, 55584, Indonesia.

[3]e-mail: riyanto@uii.ac.id

Abstract. In this study, we developed of utilization of Pb and PbO_2 from *lead storage battery* waste for batik wastewater treatment using electrochemical method. This research aimed to degrade batik wastewater with direct oxidation in the electrolysis cell. Degradation has been done by electrolysis of 50 mL batik waste using Na_2SO_4 as electrolyte with various concentrations and voltages. The electrolysis result were analyzed by Spectrophotometer UV-Visible. The result showed that PbO_2/Pb electrode from *lead storage battery* waste has high effect for electrochemical degradation of batik wastewater. Based on UV-Visible spectrum showed that the optimum several of electrolyte concentration and voltage for degradation batik waste was 0.4 M and 9 V, respectively. As conclusions, Pb and PbO_2 from *lead storage battery* waste can be used for electrode for batik wastewater treatment using the electrochemical method.

1. Introduction

Batik production process uses a lot of chemicals that can cause environmental pollution. These chemicals are commonly used in dyeing processes. The chemicals in the process are usually dissolved in water and without removed. In general, pollutants contained in industrial batik waste other than dyes can also be heavy metals, suspended solids or organic substances.

Indigosol dye is often used because it produces bright colors and does not easily fade, but the water used can cause disruption to the environment. Batik waste containing indigosol compounds is dangerous because it can cause some health impacts. These dyes can cause skin diseases and are very harmful can lead to skin cancer. Several techniques have been done to the treatment of batik waste such as coagulation, sedimentation, adsorption and electrochemical (electrolysis). Based on the problem, it is necessary to process liquid waste from batik industry with a method that does not produce other waste, so that electrolysis method is chosen for batik waste treatment. Electrochemical methods have many advantages: not providing additional contamination to the environment, shorter processing time, able to overcome the liquid waste of batik industry with satisfactory results and do not require a lot of complicated analytical process [1-3].

The main factors that support the success of the electrolysis process are electrolytes and electrodes. Several studies have been used NaCl and Na_2SO_4 electrolytes because of good electrical current, and able to degrade batik waste. Research conducted Ariguna et al. [4] electrochemical degradation of Remazol Yellow FG by electrochemical oxidation technique obtained result indicating that the use of NaCl as the electrolyte has high efficiency. Na_2SO_4 is good electrolyte for electrolysis of dye at 8 V using PbO_2/Pb electrode. Sample solution was added 0.71 g Na_2SO_4 able to electrochemical decolorize of indigosol and black B remazol dye of 98.02% and 99.42%, respectively [5].

Widodo et al. [6] have conducted research on the use of PbO_2/Pb and PbO_2/carbon as electrodes for electrodecoloration of black B remazol solution. PbO_2/Pb electrode effective result in electrochemical destruction process of black B remazol dye compared to PbO_2/carbon electrode. At room temperature, the electrochemical decolorization of black B remazol samples using the PbO_2/Pb anode has lasted 100% to CO_2, H_2O and short chain carbon compounds and has no chromophore (color-giving) groups. The electrochemical decolorization using PbO_2/carbon has completely the dyestuff and leaves a benzene-ring compound which can be removed well with the addition of electrolysis time. Lead dioxide as an electrode selected taking into account the nature of this material in electrolysis, stable and electrocatalysis effect is expected to provide synergies in remediation polluted waters with the electrochemical approach.

Batik waste water has treatment by using Pb electrode, as in the research conducted on the degradation of Remazol Brilliant Blue in fabric woven waste by electrocoagulation method starting with the determination of maximum wavelength and calibration curve determination, to determine pH, current strength and contact time. Several studies of the removal of color and organic compounds present in batik industrial wastewater have been carried out, for example by chemical means such as color degradation with oxidation reactions, anaerobic reactions and photocatalysis reactions [7]. In the previous research, the electrolysis method (electrochemical decolorization) was chosen as one of the alternative treatment of dye waste. The advantage of this method compared to other methods is effective and simple [8]. One of the factors supporting the success of the electrolysis process is the electrode used [9].

The use of lead oxide as an anode can overcome all problems in dye waste processing [10]. Lead dioxide has been widely used as an electrode in various studies, among others as electrodes in the oxidation of CN-ions, oxidation of sulfuric acid, and oxidation of organic wastes [11]. Based on the nature of lead dioxide which can transmit current and is inert, this material can be eligible to be electrode so that in this research selected lead dioxide an anode.

Several studies have applied electrochemical methods to treat dyestuff waste with several types of electrolytes such as the Na_2SO_4 electrolyte used in the comparison of the use of two boron-doped diamond anodes and $TiRuO_2$ and as a cathode both using stainless steel [12,13]. Electrochemical degradation of the acid green dye in the electrolyte of NaCl using PbO_2 electrode has been done research by Ghalwa and Latif [14,15]. In this study selected PbO_2 electrode as the anode and Pb as the cathode by utilizing used battery. The PbO_2 electrode is chosen as an anode because it is a stable or non-oxidation electrode so that in the electrolytic cell undergoing a redox reaction is an electrolyte.

2. Experimental Section

2.1. Materials and chemicals

The materials used in this research are batik waste, PbO_2/Pb plate from used battery (lead storage batteries), distilled water, Na_2SO_4 and NaCl. This research used the chemicals were produced by Merck with grade pro analysis. Figure 1 shown used battery and PbO_2/Pb plate from used battery. The electrode used in the electrolysis process is PbO_2 in the anode and Pb in the cathode, the plate used as the electrode is the plate obtained from the motor vehicle battery.

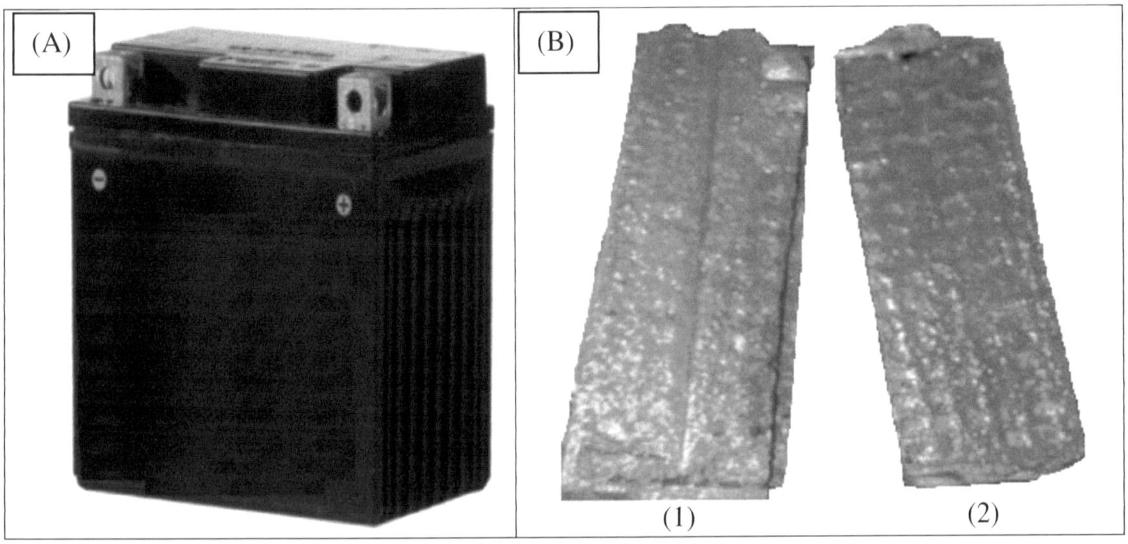

Figure 1. Used battery (A) and (B) PbO_2 plate (1) Pb plate from used battery (2)

2.2. Waste water batik Electrolysis Using PbO_2 and Pb electrode

Waste water batik is used in this research from of batik home industry in Yogyakarta. Added 50 mL sample solutions with the variation of electrolyte are NaCl and Na_2SO_4. The concentrations used in the variations of the two electrolytes were 0.2, 0.3 and 0.4 M. The solution was electrolyzed at 9 and 10 volts for electrolysis time is 3 hours. After the electrolysis of the sample is continued by filtration and transferred into the sample bottle. The analysis included color observation and absorbance measurements with Spectrophotometer UV-Visible from Hitachi U 2010. Figure 2 shows a schematic of the electrolysis cell for batik wastewater treatment using PbO_2 and Pb as anode and cathode, respectively. If the position of the electrode is reversed between the anode and the cathode, the electrolysis process cannot take place, since there is no electric current flowing on the cathode and anode.

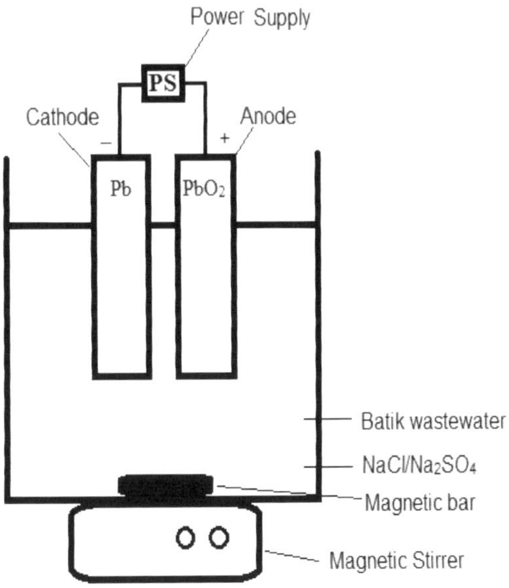

Figure 2. A schematic of the electrolysis cell for batik wastewater treatment

3. Result and Discussion

3.1. The effect of NaCl to electrolysis of batik wastewater treatment

Figure 3 shows the effect of NaCl concentration to electrolysis of batik wastewater using PbO_2 and Pb electrode. Figure 3A shows the effect of NaCl concentration to electrolysis of batik wastewater using PbO_2 and Pb electrode at the potential constant is 9 V with electrolysis time 3 h. Figure 3B shows the effect of NaCl concentration to electrolysis of batik wastewater using PbO_2 and Pb electrode at the potential constant is 10 V with electrolysis time 3 h.

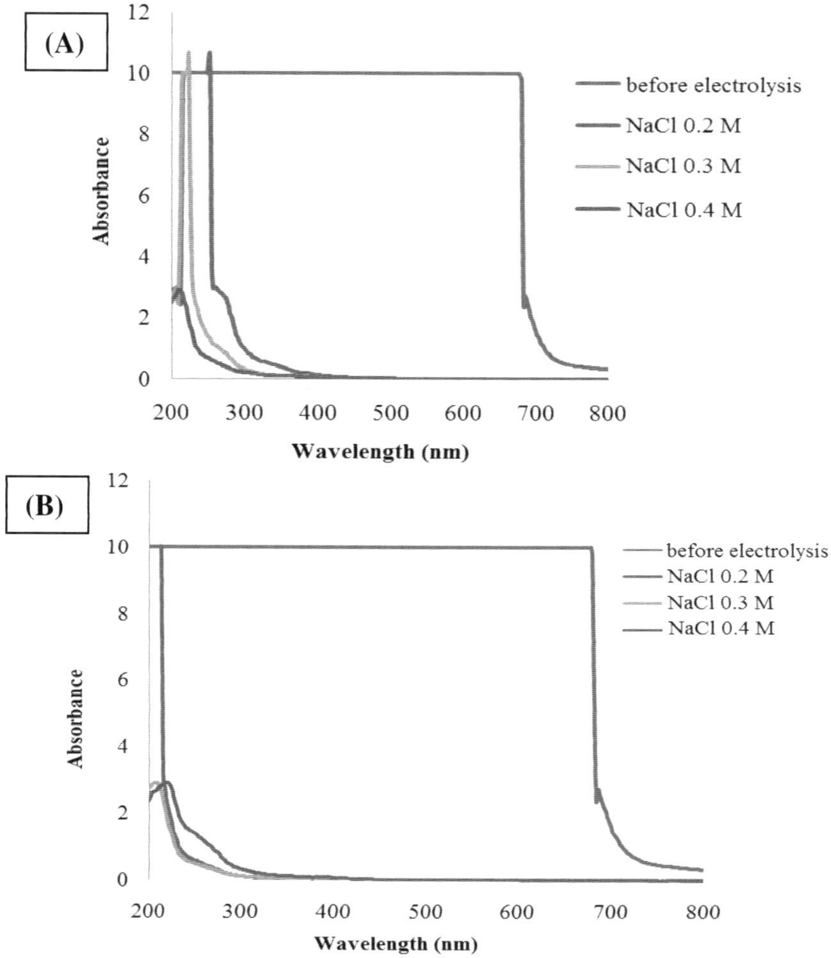

Figure 3. UV-Vis spectra from electrolysis of wastewater batik using PbO_2/Pb electrode electrolysis time 3 h with various concentration of NaCl at potential constant 9 V (A) and 10 V (B)

Figure 3 shows the result of degradation of batik waste before electrolysis and after electrolysis can be seen from UV-Vis spectrum at wavelength 200 nm to 800 nm. The electrolysis with the variation of NaCl concentration added to the solution to be electrolyzed. The electrolysis results show a good variation in NaCl concentration of 0.4 M and there is a peak showing the presence of unconjugated benzene with other double bonds. It proves that batik waste electrolysis has taken place and removed the color perfectly but still leaves a short chain carbon compound and does not have a chromophore (color-giving) group.

The effect of NaCl on electrolysis is in the role of NaCl as a good conductor of electricity and can degrade batik waste during of the electrolysis, so that the result absorbance gets smaller with the added concentration NaCl added to the waste in the electrolysis process. According to Mussa et al. [16] possible reactions to electrolysis waste water batik using NaCl electrolyte:

Anode (+):	$NaCl$	→	$Na^+ + Cl^-$
	$2Cl^-$	→	$Cl_2 + 2e^-$
	$Cl_2 + 2OH^-$	→	$H_2O + OCl^- + Cl^-$
	Dye + Ocl^-	→	$CO_2 + Cl^- + H_2O$
Cathode (-):	$2H_2O + 2e^-$	→	$H_2 + 2OH^-$

3.2 The effect of Na_2SO_4 to electrolysis of batik wastewater treatment

Figure 4 shows the electrolysis result with the variation of Na_2SO_4 concentration added to the solution to be electrolyzed. Electrolysis has been done using Na_2SO_4 as electrolyte solution with electrolysis time and potential constant is 3 hours and 9 V (Figure 4A).

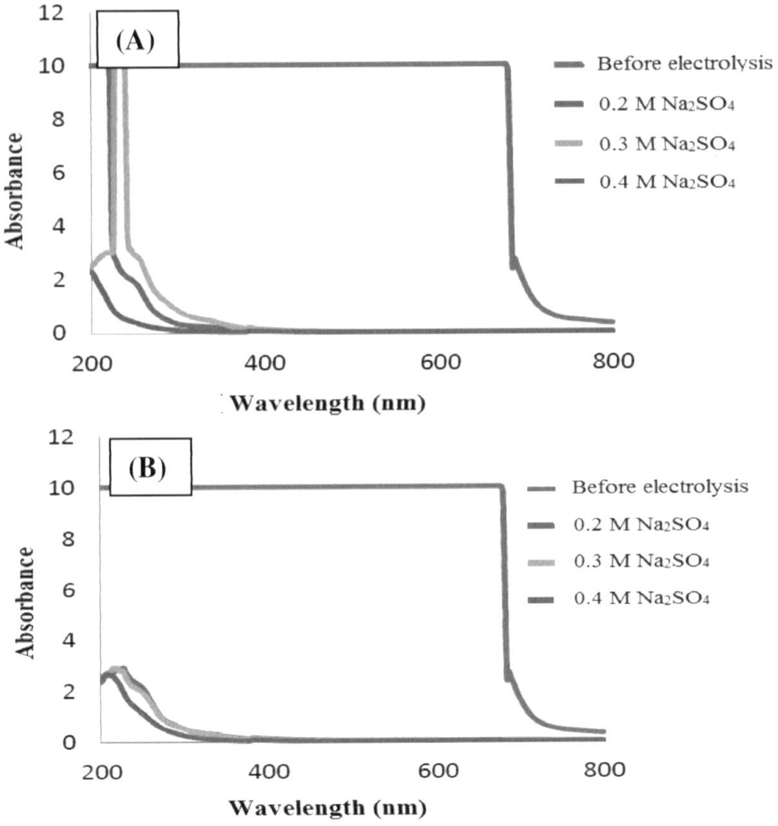

Figure 4. UV-Vis spectra from electrolysis of wastewater batik using PbO_2/Pb electrode electrolysis time 3 h with various concentration of Na_2SO_4 at potential constant are 9 V (A) and 10 V (B)

Figure 4B shows the electrolysis waste water batik using Na_2SO_4 as electrolyte solution with electrolysis time and potential constant are 3 hours and 10 V. Figure 4 shows that there has been a decrease in absorbance of batik waste after electrolysis. It can be seen from the picture that batik waste before electrolysis that the absorption peak at 680 nm identified the chromophore group color character in batik waste with the absorbance of 9.752 and after electrolysis for 3 hours with 9 V in Na_2SO_4 0.2 M its absorbance rises to 10 at a wavelength of 222 nm. The electrolysis results show a good variation of Na_2SO_4 concentration of 0.4 M and there is a peak indicating the presence of unconjugated benzene with other double bonds. It proves that batik waste electrolysis has taken place and removed the color but still leaves a short chain carbon compound and does not have a chromophore (color-giving) group.

Figure 5. Degradation of color of batik waste water (50 mL) after electrolysis at potential constant (A) 9 V and (B) 10 V using NaCl with various concentration (1) 0.2 M, (2) 0.3 M and (3) 0.4 M electrolysis time 3 h

Figure 5 shows the color changes that occur in the degradation process of batik waste. The effect of voltage can be seen in Figure (5A) with 9 V for electrolysis time 3 hours with the variation of NaCl concentration 0.2, 0.3 and 0.4 M experiencing color change significantly. The color of the resulting electrolysis solution is much different from the color of batik waste before electrolysis. The fine powder is probably the residual lead sulfate ($PbSO_4$) that is still attached to the plate. However, the result of electrolysis that has filtered indicates the perfect degradation of batik dye waste. From Figure (5B) with a voltage of 10 V for electrolysis time 3 hours on a variation of NaCl concentration of 0.2, 0.3 and 0.4 M underwent significant color changes. The color of electrolysis solution obtained much different from the color of batik waste before electrolysis, the solution after the electrolysis process must be filtered first because there is still a fine powder marked with a turbid color. However, the color of electrolysis results at 10 V is not a good as the result of electrolysis of 9 V. Base on this data can be concluded, that the higher the voltage the easier the degradation of batik waste.

(A)
(B)

Figure 6. Degradation of color of batik waste water (50 mL) after electrolysis at potential constant (A) 9 V and (B) 10 V using Na_2SO_4 with various concentration (1) 0.2 M, (2) 0.3 M and (3) 0.4 M electrolysis time 3 h

Figure 6 shows the color changes that occur in the degradation process of batik waste. The effect of voltage increase can be seen in Figure (6A) with potential 9 V for electrolysis time 3 hours on the variation of Na_2SO_4 concentration 0.2, 0.3 and 0.4 M experiencing significant color change. The color of the electrolysis solution obtained much different from the color of batik waste before electrolysis, the solution after the electrolysis process must be filtered first because there is still a fine powder but not so visible. The fine powder is probably the residual lead sulfate ($PbSO_4$) that is still attached to the plate. However, the result of electrolysis that has filtered indicates the perfect degradation of batik dye waste. From Figure (6B) with potential 10 V for electrolysis time 3 hours at the variation of Na_2SO_4 concentration 0.2, 0.3 and 0.4 M experienced significant color change. The color of electrolysis solution obtained much different from the color of batik waste before electrolysis, the solution after the electrolysis process must be filtered first because there is still a little fine powder. However, the fine powder of the electrolysis at 10 V is not as much as the electrolysis of the 9 V. Base on this data can be concluded that the higher of potential is more effective for electrochemical degradation of the batik waste. This electrolysis process produces black mud on the use of NaCl electrolyte attached to the PbO_2 anode, it shows a change in the PbO_2 anode with the addition of a black coat attached to the plate. The change of the electrode surface of PbO_2 anode can be seen on Figure 7.

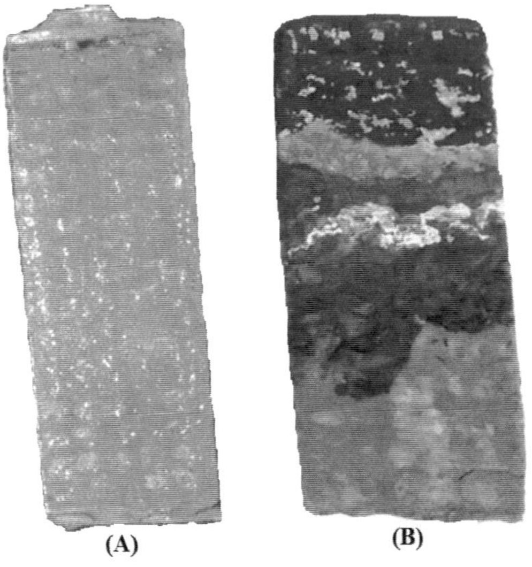

Figure 7. Characterization of electrode surface (A) PbO_2 anode before used electrolysis and (B) after used electrolysis

Figure 7 shows an apparent change in the PbO_2 anode, during which electrolysis occurs attachment of the sludge lining the PbO_2 causing the PbO_2 not used for subsequent electrolysis. The coating attached to the anode causes an inhibition of the flow of electric current, so that the electrolysis process does not take place. The electrolysis process results in sludge when using a NaCl electrolyte in the form of black mud and when using a Na_2SO_4 electrolyte in a purplish blue mud. The weight of the PbO_2 anode after electrolysis increased from the weight before electrolysis. One of the PbO_2 plates used during electrolysis using a 0.2 M NaCl at potential 9 V with an anode weight of PbO_2 before electrolysis 9.2363 g then changed to 9.754 g. On the cathode plate that has used for batik waste water electrolysis, there is a visible reduction of the weight of the eroded electrode but has not changed significantly. The cathode plate is reused for electrolysis batik waste do not show good results, caused by the low current does not even produce a current at all so that the electrolysis process is not the success. The results are not maximal is seen from the solution of the electrolysis is still dark and when filtered to produce a little mud, then the organic degradation process of organic compounds occur is not maximal.

4. Conclusions

The effect of electrolyte on batik waste electrolysis is Na_2SO_4 better than NaCl electrolyte. The effect of potential on batik waste electrolysis is the higher the potential, the more easily decomposed the compound of color substance. PbO_2/Pb electrode from used batteries can degrade batik waste dye compound with the best electrolyte Na_2SO_4 at 0.4 M concentration with potential constant 9 V. The effect of electrolysis on the electrode shows that the mud lining the anode and the cathode plate erode. The erosion properties of the PbO_2/Pb electrode are not significant. The conclusion is that PbO_2/Pb electrode is very suitable for electrochemical degradation of waste water batik.

References

[1] Noorikhlas F 2009 *Product Analysis of the Electrodegradation of Compound Compiler Batik Waste: Electrolysis Remazol Black B Solution* (Indonesia: Diponegoro University) p 12
[2] Chatzisymeon E, Xekoukoulotakis N P, Coz A, Kalogerakis N and Mantzavinos D 2006 *J. Hazard. Mater.* B **137** 998
[3] Dewi P S L, Widodo D S and Haris A 2013 *Chem. Info.* 1 **1** 1
[4] Ariguna I W S P, Wiratini N M and Sastrawidana I D K 2014 *J. Kim. Vis. UPG.* **2** 1
[5] Apipah L, Widodo D S and Hastuti R 2013 *Utilization of Waste Electrode Battery on Electrode Process Colorization of Color Substance Solution* (Indonesia: Diponegoro University) p 56
[6] Widodo D S, Ismiyarto and Noorikhlas F 2009 *J. Sci. Appl. Chem.* 12 **1** 1
[7] Rashed M N and El-Amin A A 2007 *Int. J. Phys. Sci.* 2 **3** 73
[8] Nordin N, Amir S F M, Riyanto and Othman M R 2013 *Int. J. Electrochem. Sci.* **8** 11403
[9] Kong J, Shi S, Kong L, Zhua X and Ni J 2007 *Electrochim. Acta.* **53** 2048
[10] Song Y, Wei G and Xiong R 2007 *Structure and properties of PbO_2-CeO_2 anodes on stainless steel* (China: Beijing University of Chemical Technology) p 15
[11] Li J, Zheng L, Li L, Shi G, Xian Y and Jin L 2006 *Photoelectro-Synergistic Catalysis at Ti/TiO_2/PbO_2 Electrode and Its Application on Determination of Chemical Oxygen Demand* (China: Shanghai) p 102
[12] Panizza M, Barbucci A, Ricotti R and Cerisola G 2006 *Electrochemical Degradation of Methylene Blue* (Genoa: Engineering University of Genoa) p 20
[13] Murniati T, Inayati and Budiastuti S 2015 *J. Ekosains.* 7 **1** 1
[14] Ghalwa A M N and Latif M S A 2005 *J. Iran. Chem. Soc.* 2 **3** 238
[15] Perdana M, Widodo D S and Prasetya N B A 2013 *Chem. Info.* 1 **1** 11
[16] Mussa Z H, Othman M R and Abdullah M P 2015 *J. Braz. Chem. Soc.* 26 **5** 939

Acknowledgements

This research was supported by Ministry of Research, Technology and Higher Education of the Republic of Indonesia, *Directorate General of Research and Development Reinforcement, Directorate of Research and Community Services* through "*Hibah Penelitian Produk Terapan*" Research Grant 2017 for the financial support. We would like to thank the Integrated Instrumentation Laboratory and the Environmental Quality Laboratory at the Islamic University of Indonesia for the ease of use of the equipment to make this research successful.

IOP Conf. Series: Journal of Physics: Conf. Series **909** (2017) 012075 doi:10.1088/1742-6596/909/1/012075

Preparation of carbon-polyvinyl chloride (C-PVC) and its application for electrodes to electrochemical degradation of batik wastewater

Riyanto[1,3] **and Choirun Nisaa**[2]

[1,2]Department of Chemistry, Faculty of Mathematics and Natural Science, Universitas Islam Indonesia, Jalan Kaliurang KM 14.5 Yogyakarta 55584 Indonesia

[3]e-mail: riyanto@uii.ac.id

Abstract. Batik industrial waste contains some hazardous substance, such as Remazol Black B, Remazol Red AB, Remazol golden yellow RNL, naphthol, benzene and other azo dyes class. Batik dye also contains high Pb, COD and TDS which electrochemical degradation is needed. Preparation and application of carbon-PVC (95% w/w: 5%) electrode using the batch system for batik waste degradation with electrolysis method were performed in this study. Variations of voltage and electrolysis time used are 6, 7, 8, and 9 V and 60, 90, 120, and 150 minutes, respectively. Electrolysis was done by the addition of 1.5 g NaCl and the flow of external DC circuit Sanfix SP-303E. The optimum results were obtained from electrolysis with the voltage of 9 V in 150 minutes with 97.68% of waste degraded, and percent reduce TDS until 91.16%, total Pb metal content of 0.636 mg/L and COD figures amounted to 95.76%. This study concludes that electrolysis method using carbon-PVC electrode was used to degrade the batik waste is effective and can be reduce the value of COD, total Pb, and TDS.

1. Introduction

Batik industrial process uses natural and non-natural chemicals. The chemicals in the used dyeing process of batik cloth consist of chemical compounds harmful to health and environment. The use of dyes in high concentrations can cause new problems when processing batik waste. Problems are arising from some batik industry that is difficult to process the coloring waste result from batik industry that is poisonous and dangerous. Batik industrial waste contains some harmful dyes such as black B remazol, remazol red AB, remazol golden yellow RNL, naphthol, and organic compounds [1,2]. Batik waste contains heavy metals such as Pb, Cu, As. In addition, batik waste also has a high COD, and pH is the effect of the pH environment. If the waste is discharged into the environment without first processing, it will cause environmental pollution problems.

Textile waste has been done the treatment using chemical, physical and biological methods. Several researchers has reported methods for the treatment of textile waste by using photodegradation methods [3], absorption using active sludge [4], ozone for textile degradation [5]. The oxidation process using strong oxidizing agents such as ozone and hydrogen peroxide is better [6]. Some of the methods have disadvantages such as generating other waste in the form of mud (sludge) in large quantities, only precipitate; require a long time, so that the operational costs become more expensive and less effective and efficient. Processing was done there are some weaknesses that require expensive chemicals, the process is difficult, not environmentally friendly and generate new waste derived from chemicals used [7].

Batik waste treatment using electrolysis method can be selected to overcome the problem of batik waste which contains organic dye. Electrolysis is one of the electrochemical fields that examines electron transfer on the surface of the electrode. This technique is environmentally friendly, the green technological of the future. Electrochemical techniques are used to treat water waste containing contaminants from organic and inorganic compounds [8]. The PVC-carbon electrodes has used in the electrochemical oxidation process on landfill leachate carried out by Mussa, et al. [9]. Some researcher have been done electrolysis for degradation of phenol [10-11], electrochemical oxidation of alcohol [12], electrochemical degradation of textile wastewater [13-14], olive oil mill wastewaters [15], and tannery wastewater [16]. C-PVC (carbon-polyvinyl chloride) electrodes can used to increase flexibility in the size, shape and the electrode strength. Carbon can be used for electrode because of its conductive and inert properties [17].

Based on this condition, this research focuses on the way of batik waste treatment by the electrochemical method by electrolysis method using PVC-carbon electrode to decrease color degradation, decrease COD number, and total Pb. The objectives of the research are determination of the optimum potential and time required to degrade batik waste using PVC-carbon electrodes. The effectiveness of electrolysis process of batik waste using PVC-carbon electrode can be seen from the color change, percent degradation, TDS, the concentration of Pb, and COD. The characteristics of C-PVC electrodes have been analysis using SEM-EDX.

2. Experimental Section

2.1. Instrumentation and materials

The instrumentation used in this research are DC power supply from Sanfix SP-303E, magnetic stirrer, analytical balance, stopwatch, pH meter, Spectrophotometer UV-Vis Hitachi U 2010, Scanning Electron Microscopy (SEM) EDX series Jeol JSM-6510 LA, and Atomic Absorption Spectroscopy (AAS) Perkin Elmer. The materials used in this research include batik liquid waste from Jaya Kusuma Batik Yogyakarta, NaCl (Merck), silver conductivity point, epoxy glue, silver wire, carbon powder (Merck), polyvinyl chloride (PVC) from Merck, *tetrahydrofuran* (*THF*) analytical grade from Merck and distillate water.

2.2. Preparation of PVC-carbon electrode

PVC-carbon electrodes used for the anode and cathode electrodes. The preparation of the carbon-PVC electrode is 95% carbon from Merck, then added grade 5% PVC powder (w/w) of 5 g total weight. The C-PVC was dissolved in 10 mL of THF solution. The mixture is stirred and allowed to stand for 24 hours. The carbon-PVC was pressed with strength is of 10 tons/cm^2 in the form of a tablet with a diameter of 1.3 cm. The pellet electrode was connected with a silver wire on a hollow glass rod. The PVC-carbon electrode is given a silver conductivity point when paired with silver wire. The outer layer of silver conductivity point was covered with epoxy glue.

2.3. Characterization of C-PVC electrode

C-PVC electrode was analysis using the Scanning Electron Microscopy-Energy Dispersive X-ray (SEM-EDX Jeol type JSM-6510 LA) for determination of morphological structures. The surface morphological was analysis on PVC-carbon electrode before and after electrolysis process with several optimum enlargements. The result of the SEM-EDX analysis is used to know the difference of surface structure of the electrode at the PVC-carbon composite electrode and to know the percentage of elements present in the PVC-carbon electrode.

2.4. Analysis of batik wastewater before and after treatment
Analysis on batik waste samples taken from the washing of batik cloth and wax. Samples of batik waste was stored in plastic bottle according to SNI procedure. Batik waste was measured absorbance with U-Vis Spectrophotometer Hitachi U-2010. Samples was analysis before and after electrolysis with parameter are pH, temperature, COD (mg/L), TDS (mg/L), total Pb. The concentration of Pb was measured by AAS Shimadzu.

Wastewater batik was electrolysis using C-PVC electrode. Wastewater batik is taken as much as 50 mL and put into the electrolysis cells. The electrolysis has been done in the electrolysis cells with a batch system. The electrolysis cells was connecting by power supply is the DC Sanfix SP 303-E potential source. The electrode was used in this study is C-PVC electrode as cathode and anode. The electrode is inserted in the sample then electrolysis is run with potential variations (6, 7, 8, and 9 V). The electrolysis process was carried out with the addition of 1.5 gram NaCl salt with electrolysis time are 60, 90, 120 and 150 minutes. The electrolysis results are stored in a glass erlenmeyer and closed tightly. The electrolysis results are the determination of the waste to known color degradation results.

2.5. COD test of batik waste before and after electrolysis
The sample was taken 2.5 mL and inserted into a closed reaction tube. The samples were added 1.5 mL of digestion solution (the mixture of $K_2Cr_2O_7$ and $HgSO_4$) and added 3.5 mL of H_2SO_4 solution. The sample solution was heated in a 150°C COD reactor for 2 hours then cooled under normal room temperature. Let the suspension precipitate and the clear part of the solution be measured with the UV-Vis Spectrophotometer at the specified wavelength. If the concentration of COD 100-900 mg/L does at a wavelength of 600 nm. If the concentration of COD < 100 mg/L performs at a wavelength of 420 nm. Perform titration by ferroin indicator solution with red titration point of red brown.

3. Result and Discussion

3.1. Characterization of waste water batik before electrolysis
Waste water batik with dark brownish green color taken by using plastic bottle. The plastic bottle was washed with soapy and rinsed clean with distillate water flowing according to SNI 6989.57-2008 standard. The sample was kept in the shade or away from the sun. Table 1 shows the condition of batik waste before electrolysis has alkaline pH, TDS, COD, and large Pb concentration. Waste water batik have alkaline pH, this result same with the statement by Kasam, et al. [18].

Table 1. Result of analysis waste water batik before electrolysis

Parameters	Value
Suhu (°C)	28.5 °C
pH	7.7
COD (mg/L)	11668.8 mg/L
TDS (mg/L)	1443 mg/L
Pb (mg/L)	7.2 mg/L

3.2. Electrolysis of waste water batik using C-PVC electrode
Figure 1 shown the color change occurs after the electrolysis process with the C-PVC electrode with the addition of NaCl 1.5 g. The results are that the color change influenced by the variation factor of the increase of electric potential and the time of electrolysis. Electrolysis at potential 9 V and 150 minutes was the result of colorless waste. The optimum condition for batik waste electrolysis occurred at potential 9 V, time 150 minutes with a decrease of TDS value of 94.73% and final pH value of 9.2.

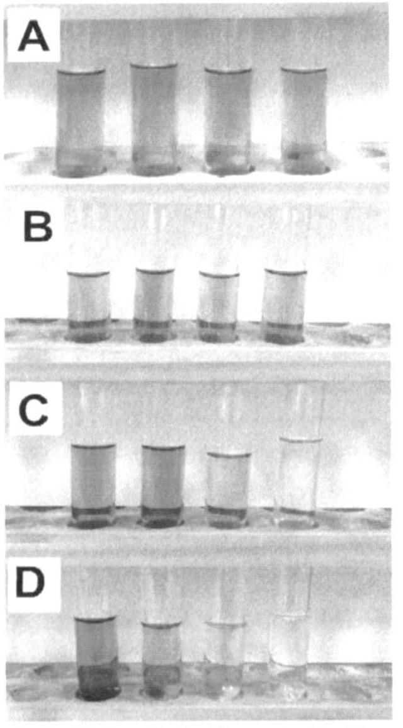

Figure 1. Color change from waste water batik before and after electrolysis using C-PVC electrode in NaCl 1.5 g with electrolysis time 60, 90, 120, 150 minute and potential constant at (A) 6, (B) 7, (C) 8 and (D) 9 V

Measurement of waste degradation quantitatively is determined by Spectrophotometer UV-Vis Hitachi U-2010 to determine the decrease of waste absorbance before and after electrolysis. Colored compounds and organic compounds containing double bonds and chromophore groups can be analyzed using a spectrophotometer UV-Vis. Analysis have been done on wavelength visible (400-800 nm) and UV (200-400 nm). Figure 2 shown UV-Vis spectra from waste water batik before and after electrolysis using C-PVC electrode in NaCl 1.5 g with electrolysis time 60, 90, 120, 150 minutes and potential constant at 6, 7, 8 and 9 V.

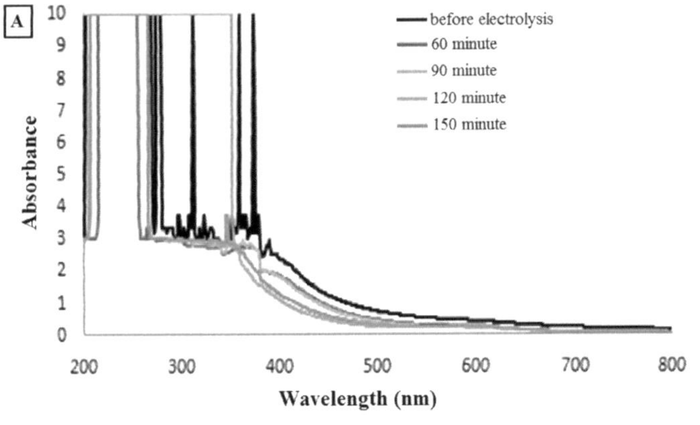

IOP Conf. Series: Journal of Physics: Conf. Series **909** (2017) 012075 doi:10.1088/1742-6596/909/1/012075

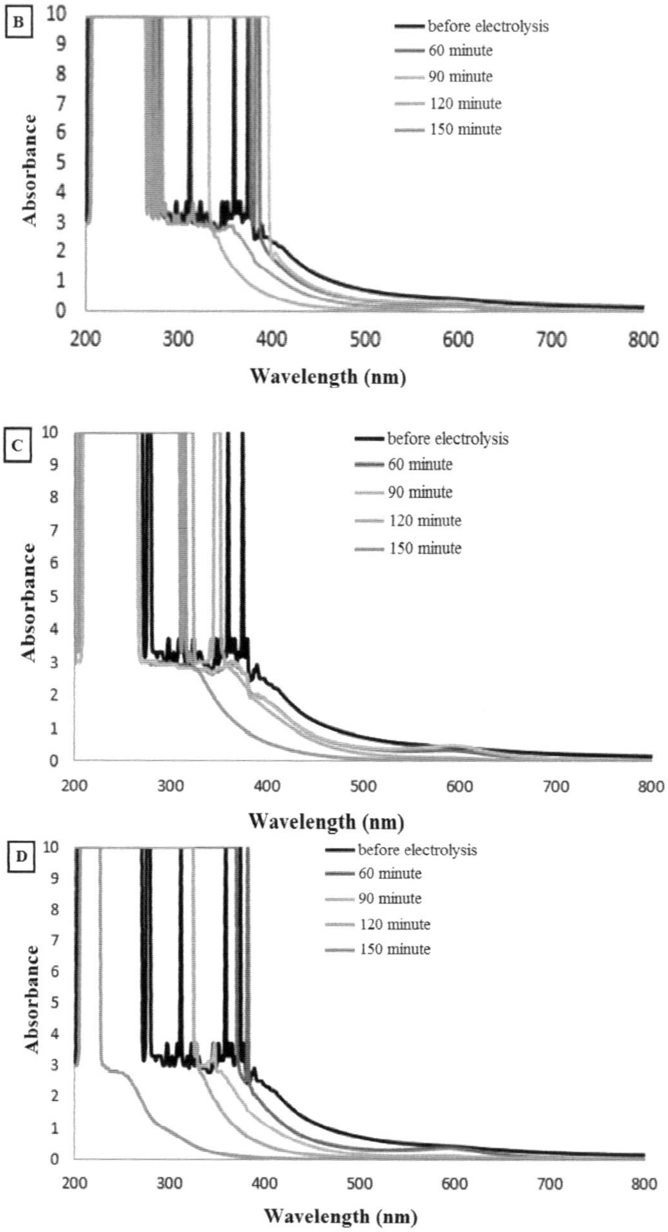

Figure 2. UV-Vis spectra from waste water batik before and after electrolysis using C-PVC electrode in NaCl 1.5 g with electrolysis time 60, 90, 120, 150 minute and potential constant at (A) 6, (B) 7, (C) 8 and (D) 9 V

Figure 3 show the effect of electrolysis time to % degraded waste using C-PVC electrode at potential 9 V with various electrolysis time 60, 90, 120, and 150 minutes. The C-PVC electrode have been used for the electrolysis of waste water batik optimum at potential 9 V. Electrolysis result has been done at 60, 90, 120, 150 minutes is can to degrade waste that is 33.25, 53.34, 78.94 and 97.68%, respectively. Perfectly degraded batik waste occurs at 9 V with a time variance of 150 minutes with a percentage of 97.68% and a non-degraded residual of 2.32%.

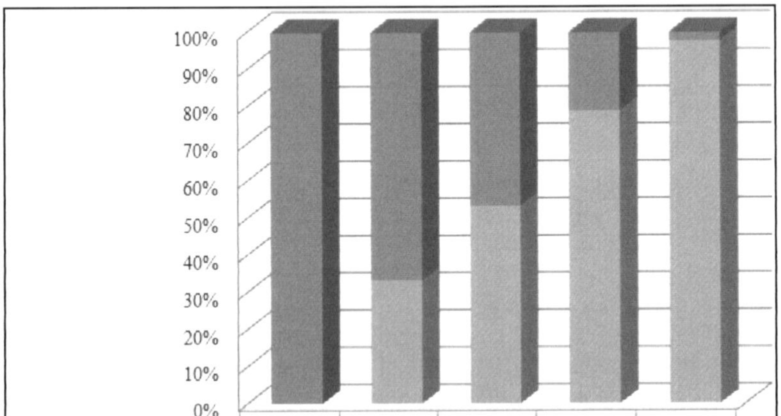

	0	60	90	120	150
% Waste residue	100	66.75	46.66	21.06	2.32
% Degraded waste	0	33.25	53.34	78.94	97.68

Figure 3. The effect of electrolysis time to % degraded waste using C-PVC electrode at potential 9 V with various electrolysis time 60, 90, 120, and 150 minute

3.3. Effect of the electrolysis time and potential to Chemical Oxygen Demand (COD)

According to Government Regulation of the Environment No.5 Year 2014 the maximum limit of textile industry waste for COD that is allowed that is 150-350 mg/L. Waste must be treated with variations time electrolysis and potential up to clear batik waste. Electrolysis was carried out on 50 mL batik waste with electrolysis time 60, 90, 120, and 150 minutes and potentials 6, 7, 8, and 9 V. The NaCl with the concentration of 1.5 mg/L can be used for that helps speed up the decomposition of chemical compounds in batik waste.

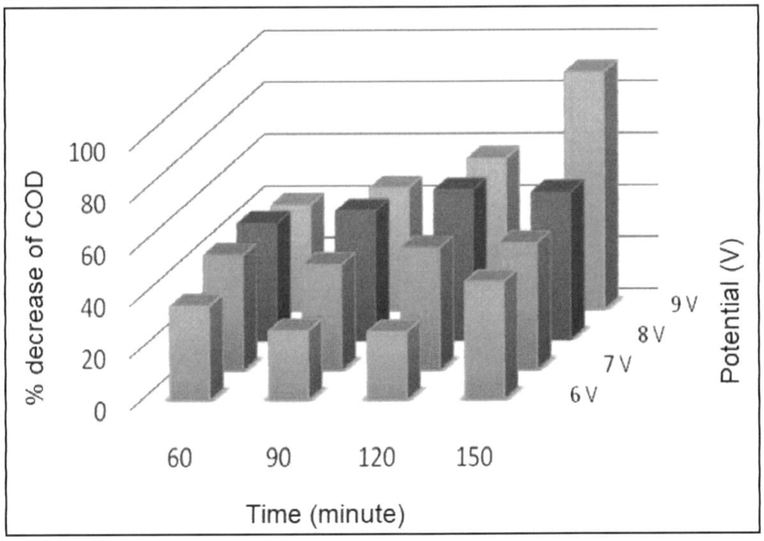

Figure 4. The effect of electrolysis time and potential to decrease of COD at electrolysis batik wastewater using C-PVC electrode

Figure 4 shown effects of electrolysis time and potential to decrease of COD at electrolysis batik wastewater using C-PVC electrode. Electrolysis result shown of batik waste give good enough result to COD number. Based on Figure 4, it can be seen that the optimum waste electrolysis time to decrease COD is 150 minutes and potential 9 V. The longer the electrolysis process and the higher the potential use, the greater the values of COD decrease in batik waste. Electrolysis conducted at 9V potency with electrolysis time for 150 minutes has the best percentage to decrease COD of batik waste amounted to 95.76%. The electrolysis is an effective method for decrease of the COD in batik waste.

3.4. The effect of electrolysis time to concentration of Pb in waste water batik

Figure 5 shown effect of electrolysis time to decrease of concentration of Pb in waste water batik. Analysis of concentration of Pb on batik waste for before and after electrolysis has been done using Atomic Absorption Spectrophotometer (AAS). The result of measurement with the coefficient of determination result R^2 = 0,999 at a wavelength of cathode lamp that is 217 nm. Electrolysis process that flowing electrics causes the migration of Pb^{2+} ions leads to the pole of the cathode. The Pb^{2+} ion was can be reduced to Pb at the cathode. The reduction process on the cathode surface causes Pb to be neutral. These processes cause a decrease of concentration of Pb in waste water batik after electrolysis. Figure 5 shows the optimum time electrolysis is 150 minute can decreasing of Pb up to 0.636 mg/L with the percentage of Pb up to 91.16%.

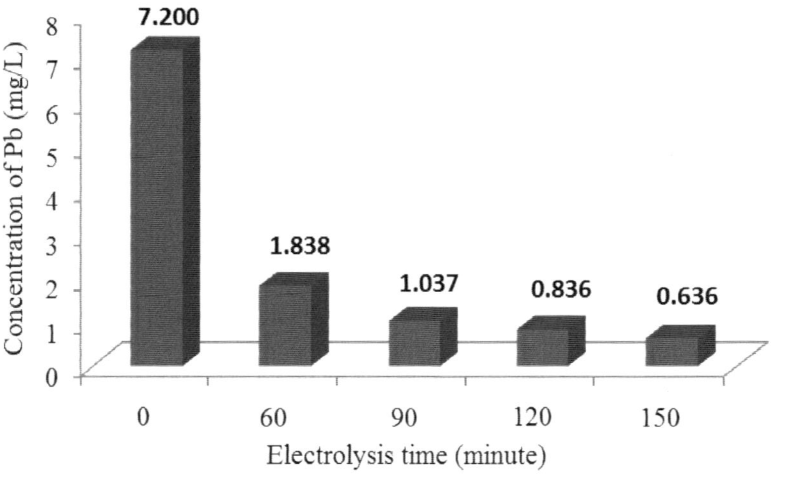

Figure 5. The effect of electrolysis time to decrease of concentration of Pb in waste water batik

3.5. Characterization of C-PVC electrode before and after electrolysis using SEM-EDX

Figure 6A shown of surface morphology of the C-PVC electrode before electrolysis. The surface electrode was covered by vinyl chloride. The carbon was spread evenly with different sizes and coated with chlorine from the bonding with the PVC composite, so that it looks more rigid. Figure 6B shown of surface morphology of the C-PVC electrode after used for electrolysis with same magnification is 5000x. Based on SEM-EDX analysis shown on surface morphology of the C-PVC electrode is cracked and hollow surfaces. This condition is good properties for electrode because of the greater contact between the solution and the electrode. Figure 7A and 7B show the EDX spectra of C-PVC electrode before and after used electrolysis. Based on the EDX spectra shown is not the different chemical composition of C-PVC electrode before and after used electrolysis. C-PVC electrode is good stability for electrolysis of waste water batik in sodium chloride solution.

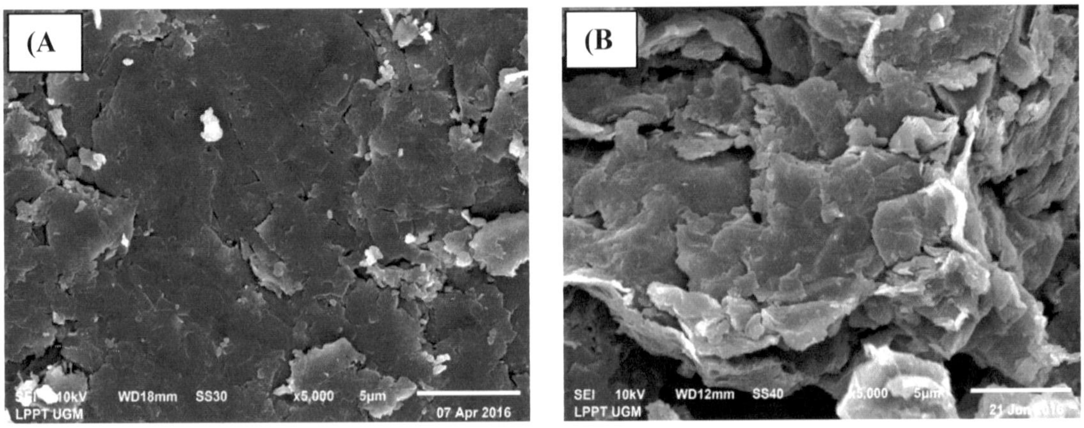

Figure 6. Surface morphology of C-PVC using SEM-EDX with magnification 5000x before used electrolysis (A) and (B) after used electrolysis

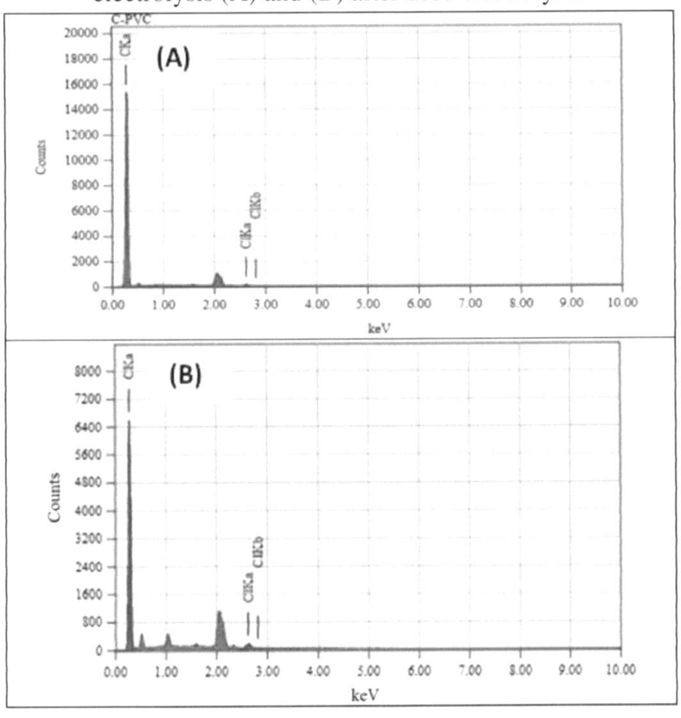

Figure 7. EDX spectra of C-PVC electrode using SEM-EDX before used electrolysis (A) and (B) after used electrolysis

4. Conclusions

Based on the research, it can be concluded are electrolysis method using carbon-PVC electrode effective for degrading waste water batik at the variation of potential 9 V and electrolysis time 150 minutes. Electrolysis using the C-PVC electrode with potential 9 V and electrolysis time 150 minutes can degradation of the waste water batik of 97.68% and can decrease the TDS value to 63 mg/L, Pb of 91.16% and COD of 95.76%. Characteristics of C-PVC electrode after used electrolysis shown not change chemical composition and morphology. C-PVC electrode is the best material for treatment of waste water batik using electrolysis method.

References

[1] Nordin N, Amir S F M, Riyanto and Othman M R 2013 *Int. J. Electrochem. Sci.* **8** 11403
[2] Nordin N, Amir S F M and Yusop M R 2015 *Acta. Chim. Slov.* **62** 642
[3] Prasetya N B A, Haris A and Gunawan 2012 *Molekul.* 7 **2** 143
[4] Yuanita D, Widjajanti E and Sulistyani 2014 *Molekul* 9 **2** 93-100
[5] Ahmet B, Ayfer Y, Doris L, Nese N and Antonius K 2003 *Dyes and Pigments J.* **58** 93
[6] Alkdasi A, Idris A, Saed K and Guan C T 2004 *Global Nest. Int. J.* 6 **3** 222
[7] Riyanto and Julianto T S 2009 *Alternative Processing of Batik Industry Waste By Electrolysis Using Stainless Steel Electrode* (Indonesia: Report of Research Results) p 12
[8] Panizza M and Martinez H 2013 *Elsevier J.* **90** 1455
[9] Mussa Z H, Othman M R and Abdullah M P 2015 *J. Braz. Chem. Soc.* 26 **5** 939
[10] Canizares P, Martinez F, Diaz M, Gomez G J and Rodrigo M A 2002 *J. Electrochem. Soc.* **149** 118
[11] Sathish M and Viswanath R P 2005 *Springer J.* 22 **3** 358
[12] Shen P K and Xu C 2006 *Electrochem. Commun. J.* **8** 184
[13] Chen X, Shen Z, Zhu X, Fan Y and Wang W 2005 *Water Res. J.* **20** 271
[14] Sala M and Bouzan G M C 2012 *Int. J. Photoenergy.* **10** 1
[15] Gotsi M, Kalogerakis N, Psillakis E, Samaras P and Mantzavinos D 2005 *Water Res. J.* **39** 4177
[16] Szpykowicz L, Kaul S N, Neti R N and Satyanarayan S 2005. *Elsevier J.* **39** 1601
[17] Destyorini F, Suhandi A, Indayaningsih N and Suwandi E 2010 *J. Himp. Fisika Ind.* **10** 2
[18] Kasam, Yulianto A and Rahmayanti AE 2009 *Logika J.*, 6 **1** 27

Acknowledgements

This research was supported by Ministry of Research, Technology and Higher Education of the Republic of Indonesia, *Directorate General of Research and Development Reinforcement*, *Directorate of Research and Community Services* through "*Hibah Penelitian Produk Terapan*" Research Grant 2017 for the financial support. We would like to thank the Integrated Instrumentation Laboratory and the Environmental Quality Laboratory at the Islamic University of Indonesia for the ease of use of the equipment to make this research successful.

The Effects of Voltage and Concentration of Sodium Bicarbonate on Electrochemical Synthesis of Ethanol from Carbon Dioxide Using Brass as Cathode

Septian Ramadan[1,a], Sholah Fariduddin[1], Afianti Rizki Aminudin[2], Antisa Kurnia Hayatri[3], Riyanto[1,2]

[1] Program Studi Kimia, Fakultas Matematika dan Ilmu Pengetahuan Alam, Universitas Islam Indonesia, Jalan Kaliurang km. 14,5, Yogyakarta 55581, Indonesia
[2] Program Studi Pendidikan Kimia, Fakultas Matematika dan Ilmu Pengetahuan Alam, Universitas Islam Indonesia, Jalan Kaliurang km. 14,5, Yogyakarta 55581, Indonesia
[3] Program Studi Statistika, Fakultas Matematika dan Ilmu Pengetahuan Alam, Universitas Islam Indonesia, Jalan Kaliurang km. 14,5, Yogyakarta 55581, Indonesia

[a] E-mail: septianramadhan@students.uii.ac.id

Abstract. The effects of voltage and concentration of sodium bicarbonate were investigated to determine the optimum conditions of the electrochemical synthesis process to convert carbon dioxide into ethanol. The conversion process is carried out using a sodium bicarbonate electrolyte solution in an electrochemical synthesis reactor equipped with a cathode and anode. As the cathode was used brass, while as the anode carbon was utilized. Sample of the electrochemical synthesis process was analyzed by gas chromatography to determine the content of the compounds produced. The optimum electrochemical synthesis conditions to convert carbon dioxide into ethanol are voltage and concentration of sodium bicarbonate are 3 volts and 0.4 M with ethanol concentration of 1.33%.

Keywords: carbon dioxide, electrochemical synthesis, ethanol, brass, optimum conditions.

1. Introduction

Energy is one of the major problems faced by almost all countries in the world. Given that energy is one of the main factors of economic growth of a country. Energy problems are becoming increasingly complex as energy demand from all countries in the world continues to increase to sustain economic growth that makes the inventory of conventional energy reserves less and less. The increasing need for energy is in fact also in conflict with the needs of mankind to create a clean environment and free from carbon dioxide (CO_2) pollution.

Global issues about CO_2 are also increasingly widespread talked about and studied along with the increasing number and impact on the world climate. This increase is triggered by the accelerated growth in CO_2-producing energy consumption worldwide, carbon-based energy resource crisis and low energy

IOP Conf. Series: Journal of Physics: Conf. Series **909** (2017) 012076 doi:10.1088/1742-6596/909/1/012076

efficiency in existing technology. According to the Intergovernmental Panel on Climate Change (IPCC), three-quarters of the increase in CO_2 in the air is related to the use of fossil [1].

Concerns over world climate change are contained in the Kyoto Protocol document and the United Nations Framework Convention on Climate Change (UNFCCC) which emphasizes the importance of reducing CO_2 emissions and their absorption from the atmosphere. Similarly, the United Nations Conference on Environment and Development (UNCED) in 1992 in Rio Janeiro, Brazil, which produced two general declarations, one of which also emphasized efforts to reduce global climate change [2].

Due to the high potential of CO_2 hazards, research is needed that can convert CO_2 into other safer forms of compounds, moreover compounds that can be utilized for the welfare of mankind. Currently, there are many studies that try to produce a compound by using CO_2 as its base material. This is one of many ways to overcome CO_2 hazard aside from minimizing its source of production.

The synthesis of other more useful compounds is a great solution for CO_2 processing, as it will provide great and sustainable benefits. Also, it will also provide a dual solution i.e., reducing the amount of CO_2 as well as the production of a more useful compound. Some of the methods that have been done in the synthesis of CO_2-based compounds are chemistry, radiochemistry, thermochemistry, photochemistry, biochemistry, and electrochemistry [3].

The synthesis of alcohol by using CO_2 is also a hot topic. From various studies that have been done, it is known that CO_2 is very potential for the synthesis of renewable energy. The products of the most potent CO_2-based synthesis of the compounds are carbon monoxide (CO), formic acid (HCOOH), formaldehyde (CH_2O), methanol (CH_3OH), oxalic acid ($H_2C_2O_4$), methane (CH_4), ethylene (CH_2CH_2), or ethanol C_2H_6O) [4][5].

Ways can be done to synthesis alcohol from CO_2 is by electrochemistry. The synthesis of alcohol from CO_2 electrochemically has two major advantages over the other way. The first advantage is the selectivity of the product produced on each electrode. Second, the tools and materials used are simple and economical because they do not require high vacuum or temperature conditions [6].

Until now, the optimization process of CO_2-based alcohol synthesis is still very necessary to do. Efficiency in the process and the intended product should still be improved so that a more effective and selective method is obtained. In this study, it is proposed the used brass as cathode and carbon as anode in electrolysis process to get target product, that is ethanol.

2. Experimental Method

2.1. The Series of Electrochemical Synthesis Reactor

The series of electrochemical synthesis reactor used in this study is shown in Figure 1.

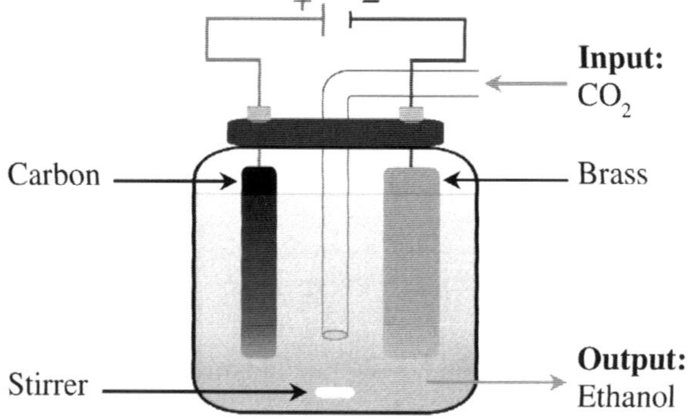

Figure 1. The Series of Electrochemical Synthesis Reactor

2.2. Electrode Preparation
The brass electrode plate is cut to the size of 1.5 × 4 cm, then rubbed using sandpaper until the dirt on the surface of the electrode is lost. The brass electrode plate is then washed with deionized water until clean and dried. The carbon electrode rod from the used battery was washed with deionized water until clean, then washed for 3 hours at 110 °C to remove the water content. The carbon electrode is then cooled in the desiccator.

2.3. Carbon Dioxide Conversion Process to Ethanol
In the variation of the voltage, a total of 50 mL of sodium bicarbonate 0.1 M electrolyte solution was put into an electrochemical synthesis reactor already equipped with a brass electrode plate (cathode) and a carbon (anode) electrode rod. Then the CO_2 gas is fed into the reactor electrochemical synthesis with a flow rate of 0.5 L/minute. The optimum voltage is studied by passing voltage with variations of 1, 3, 5 and 7 volts. Subsequently the sample solution of the electrochemical process was analyzed by gas chromatography to determine the optimum voltage to convert CO_2 into ethanol.

In the variation of the concentration of sodium bicarbonate electrolyte solution, respectively used 50 mL of sodium bicarbonate electrolyte solution 0.05, 0.1, 0.2, 0.4 and 0.6 M and put into an electrochemical synthesis reactor already equipped with a brass electrode plate (cathode) and a carbon electrode (anode) rod. Then the CO_2 gas is fed into an electrochemical synthesis reactor with a flow rate of 1 L/minute. The voltage used is the optimum voltage obtained from the voltage optimization process, flowing for 90 minutes. Subsequently the sample solution of the electrochemical synthesis process was analyzed by gas chromatography to determine the optimum concentration of sodium bicarbonate electrolyte solution to convert CO_2 into ethanol.

The chromatogram obtained from each sample is compared with standard ethanol chromatogram so that it can be known that the compound composition is formed qualitatively and quantitatively. The standard ethanol used is 1% standard ethanol which is also analyzed using gas chromatography.

3. Result and Discussion

3.1. Observations on The Electrode
Before use the brass electrode plate is rubbed using sandpaper paper until the dirt on the surface of the lost electrode is then washed by using the deionized water. This preparation aims to clean the electrode surface of metal impurities or metal oxides covering the active side of the brass electrode. If the active side of the brass electrode is good, it will increase the efficiency of the formation of the compound in the electrochemical synthesis process.

Some studies of converting CO_2 to other compounds using electrochemical synthesis method show that the type of electrode and its preparation techniques greatly affect the results to be obtained. One of the electrodes commonly used to convert CO_2 into ethanol is copper or a mixture of copper with other metals. A mixture of copper and other metals has greater catalytic activity than pure copper in converting CO_2 to alcohol [7]. Therefore, in this study used brass as the cathode which is where the process of CO_2 conversion into ethanol.

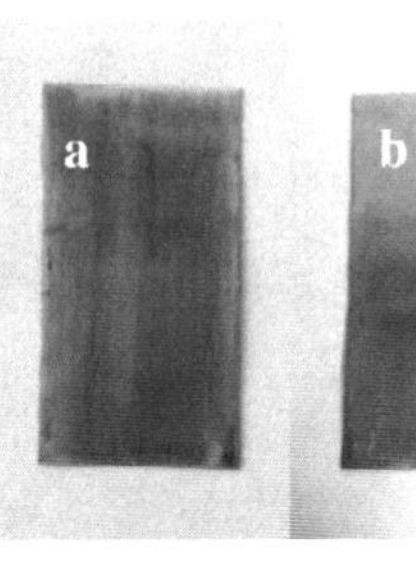

Figure 2. Picture of Brass Electrode (a) Before Use and (b) After Use

IOP Conf. Series: Journal of Physics: Conf. Series **909** (2017) 012076 doi:10.1088/1742-6596/909/1/012076

Figure 2 shows the difference between the brass electrode before and after it is used in the electrochemical synthesis process. Figure 2a is a pre-prepared brass electrode and has not been used for the electrochemical synthesis process. While Figure 2b is a brass electrode that has been used for electrochemical synthesis process for 90 min. The change in the surface of the brass electrode occurs with the formation of a black layer called sponges. The sponges are more clearly visible when the brass electrode is dried in the air.

The formation of the sponge layer is caused by the impurity metal present in the electrolyte solution attached to the brass electrode surface causing the deactivation or poisoning that can poison the electro-catalytic activity of the brass electrode. Also, deactivation of brass electrode can also be caused by the presence of carbon deposits attached to the brass electrode surface. The carbon is a by-product of CO_2 reduction [6].

Carbon is an inert electrode, so it does not react when used as anode in the electrochemical synthesis process. Therefore, in this study used carbon as anode which is where the water oxidation reaction takes place. The inert carbon properties cause carbon not to be damaged during use in the electrochemical synthesis process. Since carbon is inert and the bicarbonate anion (HCO_3^-) does not undergo oxidation in the water solvent, instead the water undergoes oxidation and becomes a source of protons and electrons. This will support the process of ethanol formation at the cathode because the mechanism of reaction of ethanol formation at the cathode involves protons and electron transfer [7].

Activated carbon electrode can increase the quantity of ethanol to be obtained and the optimum time required in the electrochemical synthesis process is also less. The activated carbon electrode has a larger pore size, so the touch surface area is also larger [8]. Therefore, the carbon electrode used in this study is activated first.

3.2. Conversion of Carbon Dioxide to Ethanol

The success of the electrochemical synthesis process is strongly influenced by the voltage flow. The conversion process of CO_2 into ethanol will occur perfectly if the applied voltage is appropriate so that the desired target product will be formed. The reaction that occurs at the cathode (reduction reaction) will produce a different product when the voltage flow is also different. In this study, the applied voltage varied into 1, 3, 5 and 7 volts, then studied the optimum voltage to produce ethanol.

Electrochemical reduction of CO_2 into ethanol is a non-spontaneous reaction (E^o = negative), where an outside source of voltage from the power supply is required. The electrochemical synthesis process carried out in this study uses a fixed voltage, so that the current conditions are stable or remain unnoticed. This will make it easier to determine the optimum voltage to convert CO_2 into ethanol using electrochemical synthesis method.

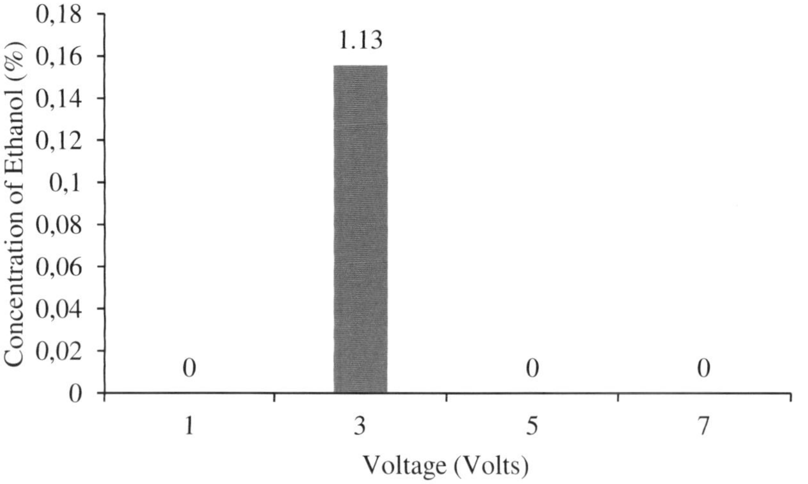

Figure 3. Diagram of Concentration of Ethanol (%) at Voltage 1-7 Volts

Figure 3 shows the concentration of ethanol (%) obtained at voltage of 1-7 volts. It is known that ethanol is only formed at a voltage of 3 volts and no ethanol is detected at other voltage streams. This shows that the voltage required to convert CO_2 into ethanol is very specific. So, it can be determined that the optimum voltage is 3 volts with the resulting ethanol concentration produced is 1.13%.

The composition and concentration of the electrolyte solution greatly influence the results to be obtained in the electrochemical synthesis process. In this study used sodium bicarbonate as an electrolyte solution. The basis of this election is that sodium bicarbonate containing a small alkali metal cation is hydrophilic, so it is not adsorbed on the electrode surface due to its hydration environment and can increase the evolution of hydrogen. This will increase the efficiency of current in the electrochemical synthesis process (CO_2 reduction reaction) in the cathode [9].

In this study used a solution of electrolytes sodium bicarbonate with low concentrations or dilute. The concentration of the sodium bicarbonate electrolyte solution used varied to 0.05, 0.1, 0.2, 0.4 and 0.6 M, then studied the optimum concentration of sodium bicarbonate electrolyte solution to produce ethanol.

Previous studies that have used sodium bicarbonate as an electrolyte solution to convert CO_2 to other compounds have been done [10]. Their results show that if the lower concentration of sodium bicarbonate electrolyte solution is used, the higher the applied voltage. This is due to the conductivity force of an electrolyte solution influenced by the quantity of ions present in the solution. When the electrolyte solution is used, the more concentrated, the more ions are in the solution so that the higher the conductivity and the smaller the voltage difference.

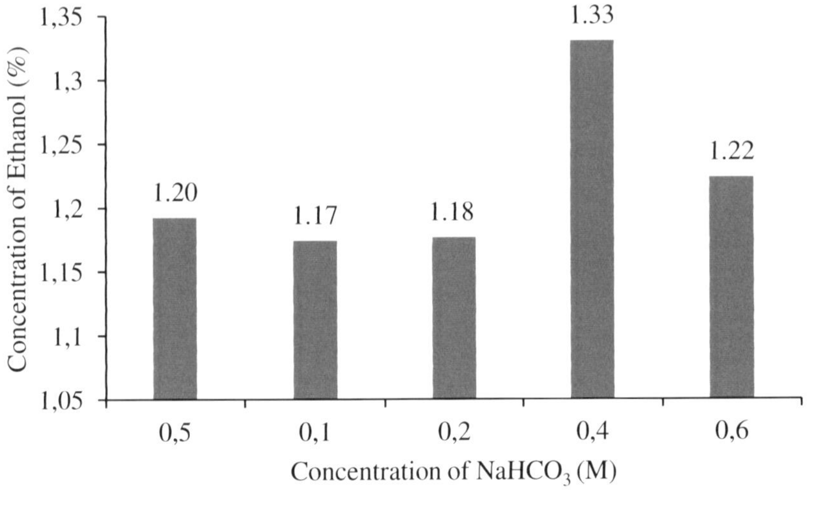

Figure 4. Diagram of Concentration of Ethanol (%) in $NaHCO_3$ 0,005-0,6 M

Figure 4 shows the ethanol concentration (%) obtained at a concentration of electrolyte solution of sodium bicarbonate from 0.05 to 0.6 M. It is known that ethanol is formed at each variation of the concentration of the electrolyte solution of sodium bicarbonate used. The concentration of ethanol in the sample experienced a slight decrease of the electrolyte solution of sodium bicarbonate concentration of 0.05 to 0.1 M, then continued to increase until the concentration of the electrolyte solution of sodium bicarbonate 0.4 M. When the concentration of the electrolyte solution of sodium bicarbonate added to 0.6 M, a decline in the concentration of ethanol in the sample. Figure 4 shows that the optimum concentration of electrolyte solution of sodium bicarbonate is 0.4 M with ethanol concentration produced is 1.33%.

Theoretically, if the more quantity of CO_2 is flowing, the more CO_2 will be converted into ethanol so that the ethanol product that will be formed will also be more numerous. The flow rate of CO_2 given should be noted as it will affect the outcome of electrochemical synthesis products. The solubility of CO_2 in water is only 0.033 mM. In water CO_2 forms equilibrium to H_2CO_3, but only 1% of CO_2 dissolves in the form of H_2CO_3 [9]. This allows the cause of ethanol not to form when CO_2 flow rate is increased. As the CO_2 boost rate is increased it will cause CO_2 bubbles in more and more electrolysis reactors. Excessive CO_2 gas bubbles will affect the stirring factor thus affecting the results of electrochemical synthesis.

Although the quantity of CO_2 that flows more and more, in fact not all CO_2 dissolves in sodium bicarbonate electrolyte solution to be converted to ethanol. Also, the sodium bicarbonate electrolyte solution is also a water-soluble compound and reacts with other compounds to produce CO_2. The reaction of sodium bicarbonate in producing CO_2 gas is as follows [11]:

$$NaHCO_3 + H^+ \rightarrow Na^+ + CO_2 + H_2O$$

4. Conclusion

Ethanol can be synthesis from carbon dioxide by electrochemical synthesis method using brass as cathode. The optimum electrochemical synthesis conditions are voltage and concentration of sodium bicarbonate electrolyte solution are 3 volts and 0.4 M with ethanol concentration of 1.33%.

5. Acknowledgments

This research was conducted in the Laboratory of Chemistry Education, Faculty of Mathematics and Natural Sciences, Islamic University of Indonesia. Thanks to all of the civitas academic of Faculty of Mathematics and Natural Science, Islamic University of Indonesia. Thanks also to Ministry of Research Technology and Higher Education Republic of Indonesia and rector of Islamic University of Indonesia which has also contributed to this research, especially regarding funding. May Allah repay the kindness of you all.

References

[1] Metz B, Davidson O R, Bosch P R, Dave R and Meyer L A 2007 *Climate change 2007: Mitigation Contribution of Working Group III to the Fourth Assessment Report of the Intergovernmental Panel on Climate Change* (Cambridge UK and New York, USA: Cambridge University Press)

[2] Hakim M S 2015 Study of Conversion Carbon Dioxide (CO_2) to Methanol Using Copper Electrode by Electrochemical Reduction Method *Thesis* (Yogyakarta: Islamic University of Indonesia)

[3] Kaneco S, Iiba K, Hiei N H, Ohta K, Mizuno T and Suzuki T 2002 Electrochemical Reduction of Carbon Dioxide to Ethylene with High Faradaic Efficiency at a Cu Electrode in CsOH/Methanol *Electrochimica Acta.* (**44**):26, 4701-4706

[4] Jones C, Robertson E, Arora V, Friedlingstein P, Shevliakova E, Bopp L, Brovkin V, Hajima T, Kato E, Kawamiya M, Liddicoat S, Lindsay K, Reick C H, Roelandt C, Segschneider J and Tjiputra J 2013 Twenty-First-Century Compatible CO_2 Emissions and Airborne Fraction Simulated by CMIP5 Earth System Models under Four Representative Concentration Pathways *DOI*: **10**.1175/JCLI-D-12 00554

[5] Kuhl K P, Hatsukade T, Cave E R, Abram D N, Kibsgaard J and Jaramillo T F 2014 Electrocatalytic Conversion of Carbon Dioxide to Methane and Methanol on Transition Metal Surfaces *J. Am. Chem. Soc.*, 2014, **136**(40), pp 14107–14113

[6] Lee J A and Tak T 2001 Electrocatalytic Activity of Cu Electrode in Electroreduction of CO_2 *Electrochimica.* **46**: 3015 3022

[7] Yang H P, Yue Y N, Qin S, Wang H and Lu J X 2016 Selective Electrochemical Reduction of CO_2 to Different Alcohol Products by an Organically Doped Alloy Catalyst *DOI*: **10**.1039/c6gc00091f

[8] Afiati A A 2016 Electrosynthesis of Ethanol from Carbon Dioxide (CO_2) Gas Using Carbon Electrode *Thesis* (Yogyakarta: Islamic University of Indonesia)

[9] Fitriani S 2012 Study of CO_2 Reduction Reaction by Electrochemical Method Using Cu Electrode *Thesis* (Depok: Indonesia University)
[10] Hori Y, Murata A and Takahashi R 1889 Formation of Hydrocarbons in the Electrochemical Reduction of Carbon Dioxide at a Copper Electrode in Aqueous Solution *J. Chem. SOC., Faraday Trans*. I, 1989, **85**(8), 2309 2326
[11] Ratna 2008 *The Effect of The Developer Material on The Wafer* (Surabaya: Widya Mandala Catholic University)

IOP Conf. Series: Journal of Physics: Conf. Series **909** (2017) 012077 doi:10.1088/1742-6596/909/1/012077

Stabilizer Effect on Thermal Decomposition of Aged Solid Propellant

Dwi Setyaningsih[1*], Ariadne L. Juwono[1], Heru Supriyatno[2]

[1] Department of Physics, Faculty of Mathematics and Natural Science, Universitas Indonesia, Kampus UI, Depok, Indonesia

[2]Indonesian National Institute of Aeronautics and Space, Rumpin, Bogor, Indonesia

*Corresponding Author: dstya81@gmail.com

Abstract. Composite solid propellants experience aging process during storage, and their properties will alter. The stabilizer added was used to reduce the effects of aging. This research studied the influence of the stabilizer on changes of the composite solid propellant thermal decomposition due to the onset of aging during the 40 days of storage at a temperature of 333 K. AO 2246 as a stabilizer added by as much as 0 to 1% by weight of propellant binders. The testing method used is heating the samples at various heating rate using DSC. And the Ozawa method was used to determine the values of the activation energy and frequency factor on the Arrhenius parameters. The optimal composition of stabilizer was 0.5 wt% for those on the value of activation energy and frequency factors for unaged and aged propellants were nearly the same i.e. 196 kJ/mol and 1.22 x10^{12} min^{-1}.

1. Introduction

The composite solid propellant is widely used for missiles rocket and launch vehicles. The composite solid propellants consist of oxidizing agents, binders, and additives. Ammonium Perchlorate (AP) had good characteristics, performances, compatibility, and availability so it most widely used as an oxidizing agent [1]. Hydroxyl Terminated Polybutadiene (HTPB) acts as a binder, also serves as a fuel which would be oxidized in the combustion reaction. The selection of HTPB as fuel binder because of its nature and can allow high solid content up to 90% weight [2,3].

The propellants experience aging during storage, and their properties will alter. An additive that serves as a stabilizer can be added to reduce the effects of aging. For example, amine and hindered phenols act as H donor on free radicals thus preventing the propagation of a chain reaction [4]. Hindered phenol compounds, such as types of AO 2246 is used because it was evidence that AO 2246 was a proper stabilizer for HTPB that used as a propellant binder [5].

Combustion of the propellant is a complex process. At the burning surface area, gas decomposition of Ammonium Perchlorate and HTPB, are inter-diffused and produce diffusion flame streams [6]. A various research was conducted to study the propellant thermal decomposition processes. A variety of mathematical and analytical methods used to determine kinetic parameters of thermal decomposition. Differential Thermal Gravimetry (DTG) and Differential Scanning Calorimetry (DSC) were the most widely used instrument in the study of the thermal decomposition of the material. The value of the kinetic parameters of thermal decomposition of the material such as activation energy (E) and frequency factor (A) can be determined using some methods such as Ozawa, Kissinger, or Flynn Wall Ozawa (FWO).

This research studied the influence of the AO 2246 as a stabilizer on changes of the composite solid propellant thermal decomposition due to the onset of aging during the 40 days of storage at a temperature of 333 K.

2. Experimental

2.1. Material

The composition of the composite solid propellant was HTPB as a fuel binder with curing agent Isophorone Di Isocyanate (IPDI) with a 17.6 wt%, 2.25 wt% of DiOctyl Adipate (DOA) as the plasticizer, and 80 wt% of solid content. Ammonium Perchlorate acts as an oxidizing agent with three-grain size 200 μm, 100μm, and 50 μm, while the Aluminum powder serves as a metal fuel. The type of hindered phenols, AO 2246 as a stabilizer added by as much as 0%; 0.2%; 0.5%; and 1% by weight of the binder. The samples code, S00, S02, S05, and S10 represent the amount of AO 2246 added.

Manufacture of propellant was done in batches planetary mixer for 135 minutes. The mixing stages started with mixing HTPB and DOA, followed by the addition of Aluminum powder, stabilizer and other additives. The addition of Ammonium Perchlorate is done in two stages, and the last stage was the addition of the IPDI. The propellant is cured in an oven at a temperature of 60 C (333 K) for three days.

Testing of thermal decomposition conducted for the unaged propellant and aged propellant that has been stored for 40 days at a temperature of 333 K

2.2. Testing

Thermal decomposition of unaged and aged samples were tested using DCS 60 Shimadzu instrument. They were heating at the range 303 – 823 K of temperature at four heating rates, i.e. 15, 20, 25, and 30 K/min under nitrogen atmosphere. Because of high heating rates and explosivity of propellant, the only small sample size was used about 1.5 – 1.8 mg. Ozawa or iso conversional method was used to determine the activation energy (E) and frequency factor (A).

2.3. Methods

The Ozawa method that used to determine the activation energy (E) and frequency factor (A) is based on basic kinetic equation for heterogeneous chemical processes and not necessary to know the reaction order. Activation energies determined by this method were the sum of activation energies of chemical and physical reactions in the thermal decomposition process. The relationship between reciprocal of an absolute exothermic peak (Tp^{-1}) and a logarithm of heating rates (log β) is a linear equation and described in equation (1) can be used to determine the activation energy and frequency factor.

$$\text{Log}\,\beta = aT_p^{-1} + b \quad (1)$$

Where a and b are the parameters of linier equation, a (slope)= -0.4567E/R and R is gas constant.

With assumed that reaction kinetics followed the Arrhenius law and exothermic reaction was a single step process, the value of frequency factor (A) can be calculated using equation (2)

$$A = \frac{E}{R}e^{b} \qquad (2)$$

3. Result and Discussion

Figure (1) shows the DSC patterns for unaged propellant with 1% stabilizer at various heating rates. There is an endothermic and an exothermic peak on the curve. Endothermic peak has almost the same value for different heating rate, which is about 520 K. This peak represents the phase transition of Ammonium Perchlorate from orthorhombic to cubic [7,8]. While, an exothermic peak is the temperature where thermal decomposition reaction occurred, which is different for various heating rates. The higher of heating rates, the higher of exothermic temperature.

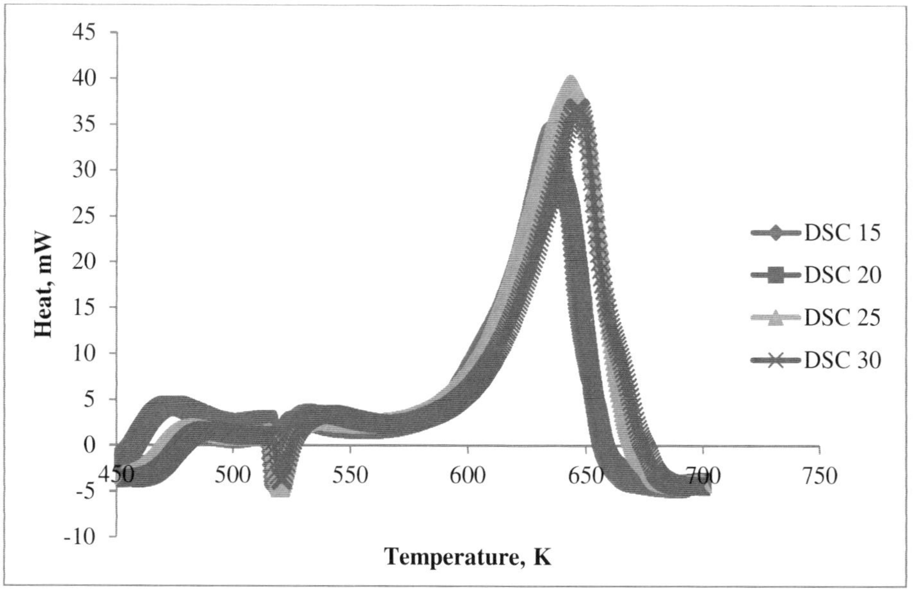

Figure 1. DSC Curve For Unaged Propellant at 1% Stabilizer

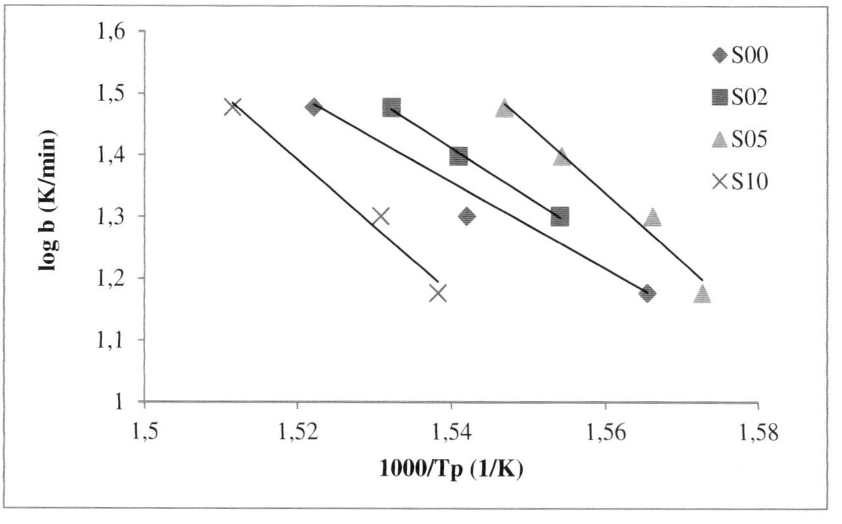

Figure 2. Ozawa Plot For Unaged Propellant

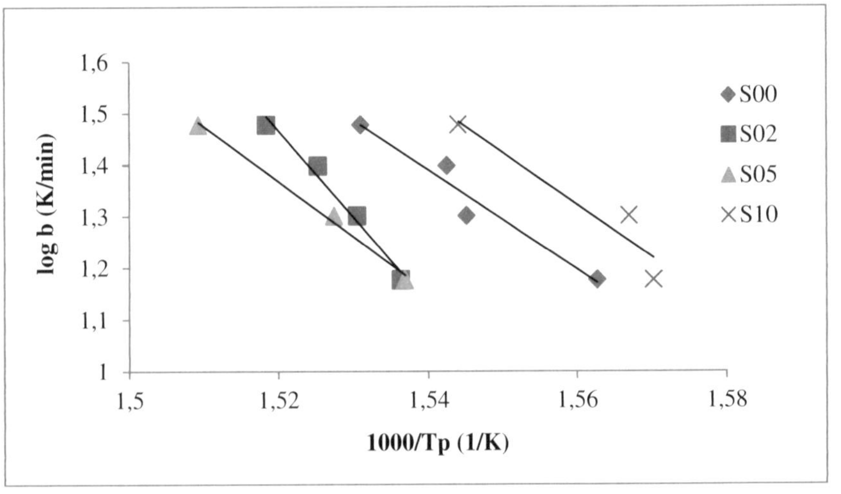

Figure 3. Ozawa Plot for Aged Propellant

Figure (2) and (3) show the relationship between the reciprocal of absolute exothermic peak and logarithm of heating rates for unaged and aged propellant. Based on Ozawa method and using equation (1) and (2) the value of activation energy (E) and frequency factor (A) can be determined as seen in Table 1.

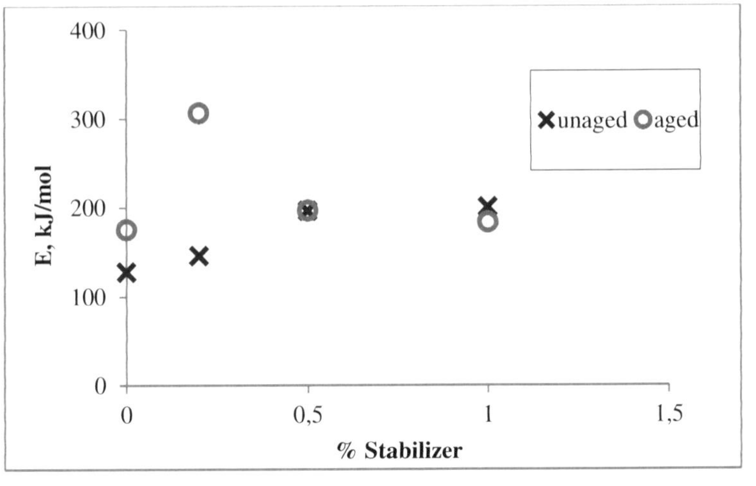

Figure 4. Activation Energy For Unaged and Aged Propellants

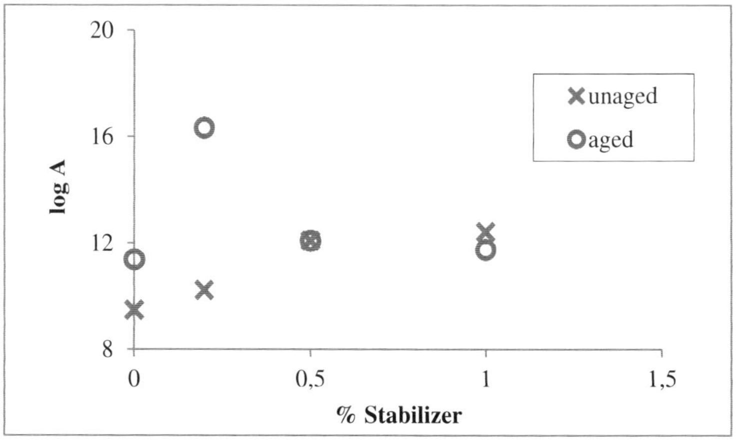

Figure 5. Logarithmic of Frequency Factor for Unaged and Aged Propellants

Figure (4) and (5) show the changes in the value of activation energy and frequency factor respectively. The greater the amount of stabilizer was added, the value of the activation energy and the frequency factor are increased for unaged propellant. However, after experiencing aging, the changes of the values of activation energy and frequency factor were different for various stabilizer composition.

Table 1 shows the results of activation energy and frequency factor for unaged and aged propellant. The value of activation energy is about 127 – 200 kJ/mol for unaged propellant and 175 – 306 for aged propellant. These values are accordance with the literature. Sell et.al studied about the kinetics of AP/HTPB propellant thermal decomposition using thermogravimetry. The activation energies calculated by Ozawa method were between 100 – 230 kJ/mol [9]. Goncalves et.al studied about kinetics of unaged propellant thermal degradation using DSC and Ozawa method and the calculated activation energy and frequency factor were 134 kJ/mol and 2.04 x 10^{10} min^{-1} [10]

Table 1. Thermal Decomposition Parameters For Unaged and Aged Solid Propellants

Code	% Stabilizer	Unaged			Aged		
		E,kJ/mol	A, min^{-1}	Log A	E,kJ/mol	A, min^{-1}	Log A
S00	0	127.58	2.91 x 10^9	9.46	175.24	2.31 x 10^{11}	11.36
S02	0.2	145.75	1.63 x 10^{10}	10.21	306.67	2.10 x 10^{16}	16.32
S05	0.5	195.98	1.22 x 10^{12}	12.08	196.35	1.22 x 10^{12}	12.09
S10	1.0	200.26	2.62 x 10^{12}	12.08	183.37	5.51 x 10^{11}	11.74

On the propellant S00 and S02, the value of activation energy and a frequency factor of aged propellant were increased very significantly. It means, that the aged propellant are less reactive, but the decomposition reaction rate is higher than unaged ones. In the other hand, the propellant S05, had no changes. So, the reactivity and thermal decomposition rate of propellant do not change for aging of 40 days at a temperature of 333 K. The S10 propellant, with an addition of stabilizer 1%, the value of E and A were slightly decline. It said that on the propellant S05 and S10, the AO 2246 stabilizer has an optimal influence by minimizing the changes of the propellant properties due to aging, in this case, the thermal properties. Otherwise, on the addition of stabilizer of less than 0.5% by weight of binders, the changes of propellant thermal properties are still significant.

4. Conclusion

The changes of thermal properties propellant due to aging can be minimize by addition of AO 2246 as a stabilizer. For unaged propellant, the greater stabilizer added, the greater the activation energy and frequency factor. The AO 2246 stabilizer gives an optimal effect for the composite solid propellant at an addition of more than 0.5 wt% due to the changes of thermal decomposition properties after experiencing aging for 40 days at a temperature 333 K are very slightly. The value of activation energy and frequency factors for unaged and aged propellants at 0.5 wt% AO 2246 were nearly the same i.e. 196 kJ/mol and 1.22 $x10^{12}$ min^{-1}.

References

[1] S. Chaturvedi and P. N. Dave 2015 Solid Propellants: AP/HTPB Composite Propellants *Arab. J. Chem.*

[2] L. Meda, G. Marra, L. Galfetti, S. Inchingalo, F. Severini, and L. De Luca 2005 Nano-Composites for Rocket Solid Propellants *Compos. Sci. Technol.* **65**, no. 5 SPEC. ISS. 769–73

[3] L. Galfetti, L. De Luca, F. Severini, S. Cerri, L. Lentini, L. Meda, G. Marra, and V. Babuk 2006 Experimental Investigation And Numerical Modeling of The Condensed Combustion Products of Micro And Nano Aluminized Solid Propellants *57th International Astronautical Congress (IAC)* 1–12.

[4] I. Vulic, G. Vitarelli, and J. M. Zenner 2002 Structure-Property Relationships: Phenolic Antioxidants with High Efficiency and Low Colour Contribution *Polym. Degrad. Stab.* **78** no. 1,27–34.

[5] L. F. C. Sbegue and L. D. Villar 2016 Comparative Assessment of Stabilised Polybutadiene Binder under Accelerated Ageing *J. Aerosp. Technol. Manag.* **8**, no. 2 122–9

[6] J. A. F. F. Rocco, J. E. S. Lima, A. G. Frutuoso, K. Iha, M. Ionashiro, J. R. Matos, and M. E. V Suárez-Iha 2004 TG Studies of A Composite Solid Rocket Propellant Based on HTPB-Binder *J. Therm. Anal. Calorim.* **77**, no. 3, 803–13.

[7] R. F. B. Goncalves, J. a F. F. Rocco, and K. Iha 2008 Thermal Decomposition Kinetics of Aged Solid Propellant Based on Ammonium Perchlorate - AP/HTPB Binder *44th AIAA/ASME/SAE/ASEE Joint Propulsion Conference & Exhibit* 325–42.

[8] J. K. Sharma, P. Srivastava, G. Singh, M. S. Akhtar, and S. Ameen 2015 Catalytic Thermal Decomposition of Ammonium Perchlorate and Combustion of Composite Solid Propellants Over Green Synthesized CuO Nanoparticles *Thermochim. Acta* **614** 110–15

[9] T. Sell, S. Vyazovkin and C. A. Wight 1999 *Combust. Flame*, **119** 174

[10] R. F. B. Goncalves, J. a F. F. Rocco, K. Iha, M. Ionashiro, J. R. Matos, and M. E. V. Suárez-Iha 2004 Thermal Degradation of Composite Solid Propellant Examined by DSC - *J.Therm.Anal.Cal.* **75**

Acknowledgement

The authors gratefully acknowledge support from Indonesian National Institute of Aeronautics and Space for these research.

IOP Conf. Series: Journal of Physics: Conf. Series **909** (2017) 012078 doi:10.1088/1742-6596/909/1/012078

Study of Synthesis Polyethylene glycol oleate Sulfonated as an Anionic Surfactant for Enhanced Oil Recovery (EOR)

Yulianti Sampora[1]*, Ariadne L. Juwono[2], Agus Haryono[1], Yan Irawan[2]

[1]Department of Physics, Faculty of Mathematics and Natural Sciences Universitas Indonesia, Depok 16424, Indonesia

[2]The Research Centre for Chemistry-Indonesian Institute of Science Kawasan Puspiptek Serpong, Banten, Indonesia 15314.

*Corresponding author: *samporayuli@gmail.com*

Abstract. Mechanical Enhanced Oil Recovery (EOR) through chemical injection is using an anionic surfactant to improve the recovery of oil residues, particularly in a reservoir area that has certain characteristics. This case led the authors to conduct research on the synthesis of an anionic surfactant based on oleic acid and polyethylene glycol 400 that could be applied as a chemical injection. In this work, we investigate the sulfonation of Polyethylene glycol oleate (PDO) in a sulfuric acid agent. PDO in this experiment was derived from Indonesian palm oil. Variation of mole reactant and reaction time have been studied. The surfactant has been characterized by measuring the interfacial tension, acid value, ester value, saponification value, iodine value, Fourier Transform Infrared (FTIR), and particle size analyzer. There is a new peak at 1170-1178 cm-1 indicating that S=O bond has formed. PDO sulfonate exhibits good surface activity due to interfacial tension of 0,003 mN/m. Thus, polyethylene glycol oleate sulfonate was successfully synthesized and it could be useful as a novel an anionic surfactant.

1. Introduction

Recently, the enhanced of oil recovery (EOR) technology gains much attention in many countries due to the lack of oil resources and difficulty in finding new oil sources. Application of EOR technology gives an additional chance to get out more oil from the reservoir, possibly about another 20% [1]. This technique is more efficient than other techniques such as gas and thermal engineering. One of the chemical injection techniques is using surfactants. Basically, the surfactant for EOR made from vegetable oils and palm oil derivatives with polyester formed.

The previous study, Rafa, and Favero have conducted a petroleum-based surfactant study [4, 5], whereas biomass-based polymeric surfactants have been obtained from natural resources. It has attracted much attention as a polymeric surfactant and potential for EOR applications due to lower cost and derived from renewable sources. This polymer also has a good viscosity and interface activities. Another study from Elraies (2011) reported the esterification of sulfonated fatty acids from castor oil [6]. Synthesized of a vegetable oil based methyl ester (oleic acid) as a cationic polymeric for EOR applications was performed by Irawan (2015) [7]. Babu (2015) was synthesized an anionic surfactant (sodium methyl ester sulfonate) from castor oil [8]. An anionic surfactant have a negative charge and some properties such as stable, decreasing IFT or interface tension, and low adsorption to rocks in petroleum reservoirs [2]. Therefore, an anionic surfactant could be considered being the effective chemical compound for EOR techniques.

The surfactant used depends on conditions and characteristics of the intended reservoirs. This statement is similar as reported by Hirasaki who investigate an anionic surfactant for adsorbing positive charges on clay surfaces in sandstone and matrix on carbonate surfaces at neutral pH [9]. The reservoir conditions containing sandstone is more suitable using an anionic surfactant which can be obtained by the process using a sulfonation agent such as concentrated sulfuric acid. Sulfuric acid could be hydrolyzed. Neutralization process is needed as soon as the sulfuric group formed [3]. When the crude oil contacts the water that contains the brine, the natural surfactant will be collected on the interface and form an adsorbed film that lowering the interface tension from the crude or water. Decreasing sources can be affected by the adsorption of the surfactant to rocks in the petroleum reservoir. The utilization of sulfonate agents can reduce the adsorption process so that the oil recovery would be increased. The composition of reservoir rock plays a role in determining the interaction between mineral-added reagent (surfactant or polymer) and the effect on liquid interface properties such as surface load and wettability [11]. Stachurski said that the charge value of a solution in the zeta potential of emulsion droplets is a function of pH and independent of the length of aliphatic hydrocarbon chains [12]. Iglauer (2010) has observed the effect of surfactant adsorption on Berea sandstone with a standard size for sandstone of 74-250μm [13]. In surfactant flooding, mechanism of the recovery process was based on decreasing interface tension between two phases (IFT). The role of IFT in oil recovery has been explained regarding capillary (Ca) numbers (Hirasaki). In fact, the presence of hydrophobic parts in water-soluble structures is also known to affect the interface properties (surface tension, IFT, wettability). The value which usually used are 10^{-3} mN / m) to support the application of EOR techniques [14]. However, some studies of surfactant flooding polymers suggest that in some cases there is no significant correlation between low IFT values and the optimum value of IFT is greater than 10^{-3}mN/m or commonly called by ultralow IFT [15]. Therefore, the effectiveness of the surfactant polymer value is very difficult to achieve, and the range of IFT value is 0.1 – 15 mN/m [4]. There is a lack of evidence about an anionic surfactant based on petroleum are more costly and non-renewable so that the current research aimed an anionic surfactant based on palm oil (oleic acid) and polyethylene glycol are renewable, more economical, and have ultralow IFT values.

2. Materials and Methods

2.1. Materials
For an anionic surfactant synthesis, laboratory grade chemicals such as sulfuric acid (H_2SO_4) pro analysis (95-97%) from E.Merck cat. 1.00731.2500, Hydrogen peroxide (H_2O_2) pro analysis (35-37%) from E.Merck Cat. 1.07209.2500, Polyethylene glycol oleate (PDO) technical. For titrimetric analysis using Potassium hydroxide (KOH) pro analysis from E.Merck.Cat. 1.05033.1000, technical ethanol, a reagent for ester value analysis, acid value, iodine value, and saponification value.

2.2. Methods
In this study, the sulfonation reaction was carried out with the polyethylene glycol oleate (PDO) ester as an anionic surfactant. This reaction has been performed by using Li [17], for PEG Dioleate (PDO), the sulfonation process was carried out using 50% of total weight H_2O_2 reactants. Furthermore, the sample is washed using 50% methanol to convert all remaining fatty acids [18]. The following sulfonation process: (1) amount of sulfuric acid as a sulfonation agent, was added with varying mole ratio of 1:0,5; 1:1; and 1:1,5. While the PDO has been stirred in a reactor, sulfuric acid was added gradually. The reaction temperature has been maintained at 80°C for 2, 4 and 6 hours. PDOS (Polyethylene glycol Oleate Sulfonate) was bleached with H_2O_2 50% total reactant weight at 75°C for 2 hours. The product and methanol (50% total reactant weight) were mixed in 2 hours at 50°C; PDOS

was neutralized with NaOH solution at 50°C for 2 hours; then the residue was evaporated and characterized.

2.3. Characterization of PDOS an anionic surfactant

The chemical and physical properties of oleic acid and PEG based an anionic surfactant was analyzed by titration method based on ASTM D4662-03 to obtain the acid value (Av), ester value (Ev), saponification value (Sv), and iodine value (Iv). The functional group present in the PDOS surfactant was determined used FTIR spectrophotometer (Shimadzu-IR Prestige21). Interfacial Tension (IFT) measurement on the surfactant solution was conducted to measure the interfacial stress of the surfactant solution with oil. The result was expected with smaller than 10^{-3} mN/m. IFT measurement using Spinning Drop Interfacial Tensiometer TX 500. The particle size analysis of surfactants was determined using HORIBA Nano Particle Analyzer SZ-100 at 25°C. All samples were prepared in deionized water and sonicated using Elmasonic E 15H for 10 minutes.

3. Result and Discussion

3.1. Synthesis of PDOS based an anionic surfactant

Figure 1 shows the results of iodine value at 1:1-mole ratio for 2 hours was still high, then decreased by 4 hours, but increased again at 6 hours reaction. While the mole ratio of 1:0,5, sulfuric acid as sulfonation agent used is not optimal, so iodine value is still high. The highest number of iodine on TM (without methanol) indicates that there are still many reactants that do not react [6, 8, 18].

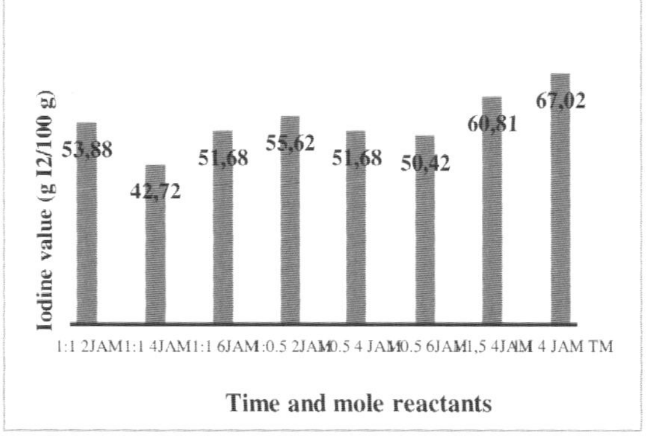

Fig.1. Iodine value graph

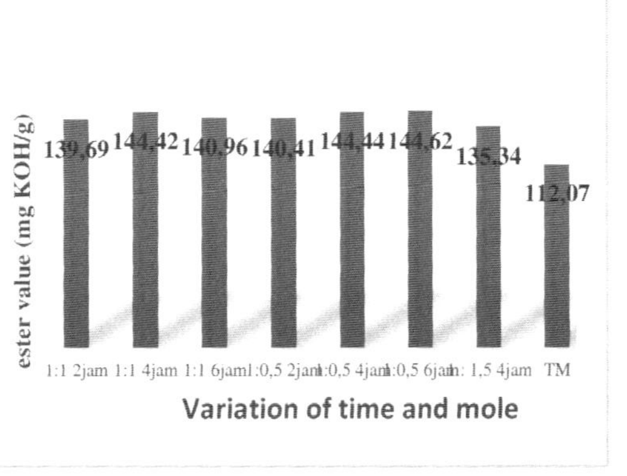

Fig.2. Ester value graph

In Figure 2, ester value is an indicator of the ester compound. From the variation of time and mole ratio, the TM sulfonation (without methanol) has shown the significant number, the smallest ester value among all variables. In figure 3, the optimum value is at a mole ratio of sulfuric acid and PDO of 1: 1 for 4 hours, the smallest value compared to other variables. The odds are already reacting all over the reactant. This also is seen in stark contrast to the saponification value of TM (without methanol). In figure 4, it shows that the smallest acid value which means the optimum result is in the reaction of 1:1-mole ratio for 4 hours. From the four images of the curves, concluded the optimum reaction of sulfonation is at a mole ratio of 1:1 during 4 hours.

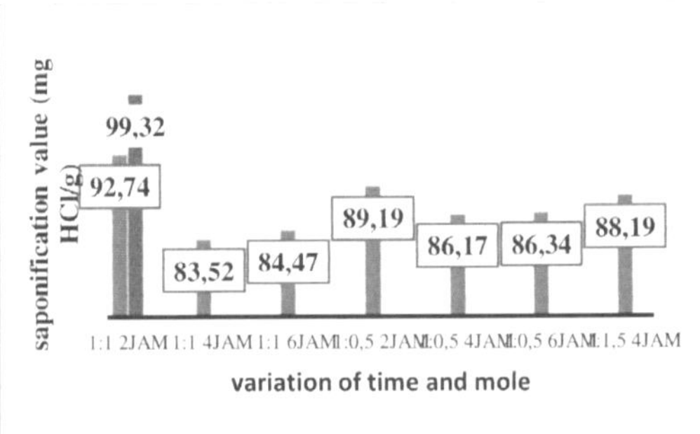

Fig. 3. Saponification value graph

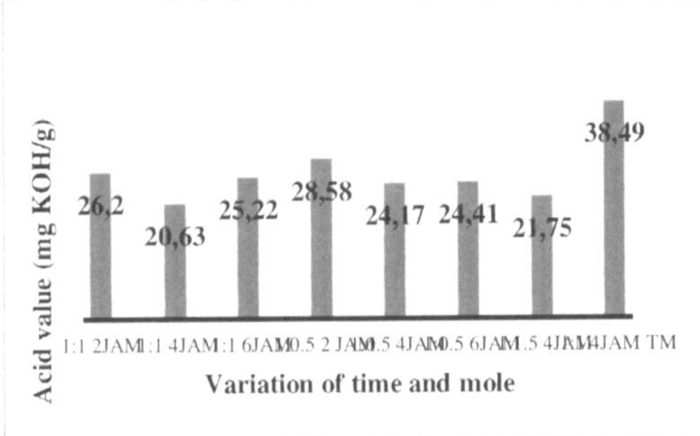

Fig. 4. Acid value graph

3.2. Characterization of PDOS based an anionic surfactant

Characterization of Polyethylene glycol oleate sulfonated based an anionic surfactant was recorded in Table 1. Analysis of acid value was conducted to find out the progress of the synthesis of PDOS. Decreasing acid value from oleic acid into PDO occurred because of extension chain of reactive carboxyl from oleic acid formed an anionic surfactant, whereas the decline acid value from PDO into PDOS occurred due to neutralization process on sulfonation reaction with NaOH and formation of $NaSO_3$ or sodium sulfonate. On the other hand, the ester value increased with sulfonation reactions. Increasing ester value occurred because of extension chain of reactive carboxyl from PDO [17], it will react with the base which means the residual acid in the anionic surfactant product is reduced [6, 8, 18]. Babu in his study made sodium methyl ester sulfonate from ricinoleic acid, with the result of acid value, saponification value and iodine value are 10-12, 185-190, and 80-90 respectively.

Table 1. Character raw materials and sulfonation products

Materials	Physical properties		Av (mg KOH/g)	Ev(mg HCl/g)	Sv(mg HCL/g)	Iv(mg I_2/100g)	HLB
	Shape	Color					
PDO	Liquid	dark yellow	61,78	56,28	105,73	45,63	8,25
PDOS 1:1 4 hours	Liquid	brown	20,63	144,42	83,52	42,72	10,09
Oleic acid	Liquid	yellow	180,04	22,04	117,77	85,59	˜1,0*
PEG 400	Viscous	colorless	2,87	25,24	18,67	-	12,69-13,1*

The saponification number of the nonionic surfactant in the present invention has decreased by 60%. The saponification number and acid number of the nonionic surfactant of the present invention can be used to calculate the HLB value based on the Griffin method [16] with the formulas as follows:

HLB = 20 (1-S/A), where S=the saponification number of the ester
A = acid number of fatty acids

To obtain an HLB value for the 8 to 13 surfactant PDO, as seen in Table 1, an anionic surfactant in the present invention categorized in an O/w emulsion, for example, surfactant application for EOR.

3.3. Mechanism of reaction

Figure 5 shows the complete reaction mechanism:

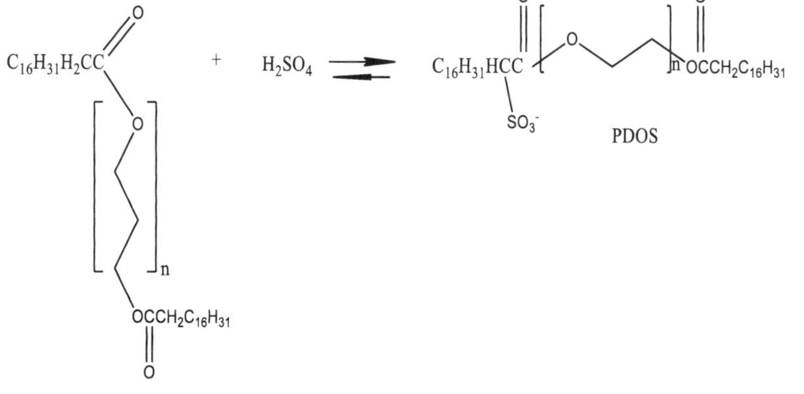

Fig.5. Sulfonation reaction of polyethylene glycol oleate

The scheme of PDO sulfonation is shown in Figure 5. When H_2SO_4 as a sulfonating agent reacted with PDO, SO_3^- ion binds with C alpha atom in the oleic chain in the PDO structure. The possibility of such reactions also describes the reaction of sulfonation in the fatty acid molecule. Indications for determine the formation of sulfonate in the PDO can be seen from the emergence of the S=O stretch peak at wavelengths 1170,79 cm^{-1} (PDOS) which did not appear at the PDO as shown in figure 6. Elraies has stated the sulfonate group is presented in peaks between 1160-1120 cm^{-1}, whereas the peaks at 1410 and 1068 cm^{-1} are another indication of the presence of sulfonate groups due to the S=O stretching vibration. Babu has identified the sulfonate group at wave number 1158 cm^{-1}.

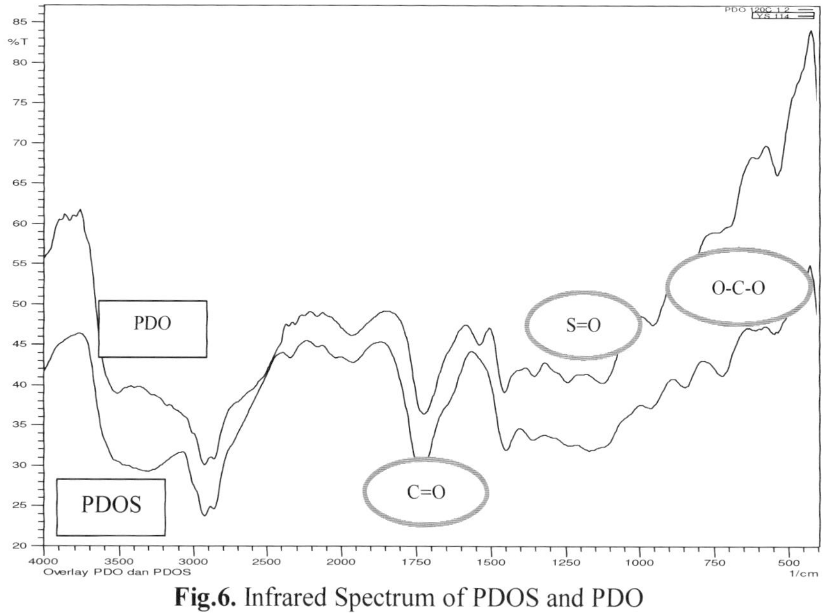

Fig.6. Infrared Spectrum of PDOS and PDO

3.4. Interfacial Tension (IFT) and Particle Size Analyzer (PSA)

Figure 7 shows that the smaller surfactant concentration has been used for the reservoir well formulation X. The IFT value obtained is also high, more concentration of surfactant, experienced an increase in IFT value and reached a minimum number in the range of 0.3% surfactant. IFT scores ranging from 10^{-3} mN/m or so-called ultralow IFT indicated that it could be used as a surfactant for EOR applications.

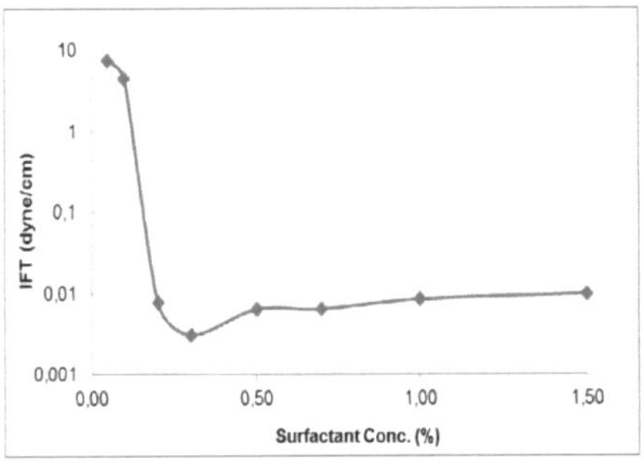

Fig.7. Interfacial Tension (IFT) Graph

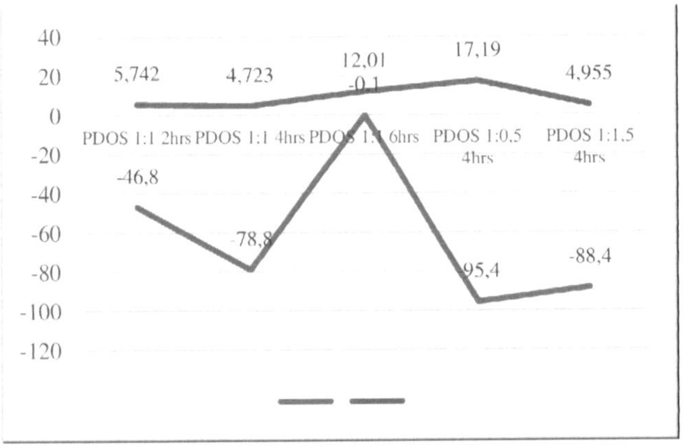

Fig.8. Particle Size and Potential Zeta of PDOS

As previously mentioned, the surfactant size has an effect on the adsorption process by rocks, where the rock size in the maximum sandstone is 250 μm. The particle size of the surfactant is smaller than the standard size so that the surfactant could not be adsorbed by the surface of the rocks. Therefore, particle size analysis is also required. In figure 8, the PDOS with a mole ratio of 1: 1 acid for 4 hours sulfonation has obtained PSA value of 4.723μm. Thus, the particle size of PDOS as an anionic surfactant has required in the EOR technique. Zeta potential value is a parameter that controls electrostatic interaction in dispersion particles [19]. In this case, Zeta potential may confirm that the surfactant solution is anionic because of the negatively charged (-). From the figure 8, PDOS with 1:1-mole ratio at 4 hours has a high zeta potential value, which is about (-)78,8, an anionic.

4. Conclusion

Synthesis of an anionic surfactant has been carried out by sulfonation reaction with time variation of 2 hours; 4 hours and 6 hours and the mole ratio of sulfuric acid to polyethylene glycol oleate (PDO) ester are 1:0,5; 1:1; 1:1,5. Results of PDOS product analysis showed the optimum result at 1: 1-mole ratio for 4 hours. Product characterization using FTIR exhibit a new peak at 1170 cm^{-1} (PDOS) indicating S = O bond for sulfonation process. Acid value, ester value, saponification value, and iodine value was obtained to confirm the synthesis of PDOS. The particle size was observed for PDOS at 4,723μm and zeta potential at -78.8. It has indicated that the anionic surfactant could be used in EOR system.

References

[1] Kokal S and Al-kalbi A 2010 *Enhanced oil recovery: challenges & opportunities* (E. Advanced, and S. Aramco) 64–69
[2] Abadli F 2012 *Simulation Study of Enhanced Oil Recovery by ASP (Alkaline, Surfactant, and Polymer) Flooding for Norne Field C-segment*
[3] Alfonso J and Ortega T 2012 *Sulfonation / Sulfation Processing Technology for An anionic Surfactant Manufacture*
[4] Raffa P, Broekhuis AA and Picchioni F 2016 Polymeric Surfactants for Enhanced Oil Recovery: a Review, *J. Petroleum Sci. Engineering*

[5] Favero C, Gaillard N and SNF SAS 2010 *Process for Enhanced Oil Recovery* United States of America US20100197529 A1
[6] Elraies KA, Tan IM, Awang and Saaid M 2010 The synthesis and performance of sodium methyl ester sulfonate for enhanced oil recovery *J. Pet.Sci. Technol.* 28 1799-1806
[7] Irwan Y, Saraswati and Ghozali M 2015 Sintesis Biopolymer Berbasis Minyak Nabati *Jurnal Sain Materi Indonesia* ISSN 1411-1098
[8] Babu K, Maurya NK, Mandal A and Saxena VK 2015 Synthesis and Characterization of Sodium Methyl Ester Sulfonate for Chemically –Enhanced Oil Recovery *Brazilian .J. Chem. Eng.* 32 no. 3 795–803
[9] Hirasaki GJ, Miller CA and Puerto M 2011 Recent advances in surfactant EOR *SPE J.* 16 889–907
[10] Bera A and Mandal A 2015 Microemulsion: a novel approach to enhanced oil recovery: a review *J. Petrol. Explore. Prod. Technol.* 5 225-268
[11] Kanicky JR et all 2001 *Surface Chemistry in the Petroleum Industry* Chapter 11 Handbook of Applied Surface and Colloid Chemistry
[12] Stachurski J 1996 The effect of the Zetta Potential on The Stability of a Non-Polar Oil-in-water Emulsion *J. of Colloid and Interface Sci.* 184 433-436
[13] Iglauer S et.all 2010 New surfactant classes for enhanced oil recovery and their tertiary oil recovery potential *J. Petrol. Sci. Eng.* 71 23-29
[14] Zhang DL, Lui S, Yan W, Puerto M, Hirasaki J and Miller CA 2006 *Favorable attributes of alkali-surfactant-polymer-flooding* Paper SPE 99744 presented at the PFE symposium on improved oil recovery Tulsa OK
[15] Yu Q. Jiang H and Zhao C 2010 Study of interfacial tension between oil and surfactant polymer flooding *J. Pet. Sci. Technol.* 28 1846–1854
[16] Griffin WC 1949 Classification of Surface-Active Agent by HLB *J. Society of Cosmetic Chemist* 1 311-326
[17] Li H 2013 *Process for Preparing Alpha-Sulfo-Fatty Acid Ester Salt Surfactant* United States of America US8604226 B2
[18] Cohen L, Roberts DW, Pratesi C, Politécnica E, De Algeciras E and De Cádiz U 2010 Φ - Sulfo Fatty Methyl Ester Sulfonates (Φ -MES): A Novel Anionic Surfactant 21 1033–1038
[19] Kaszuba M et.all 2010 High-concentration zeta potential measurements using light-scattering techniques *J.Philosophical Trans. R.Soc.* A. 368, 4439-4451

Study on the colorimetric properties of 2,4,6-triarylpyridine derivative compound for imaging Formaldehyde

D Ovianto[1], F A R Ma'ruf[1], N N Fadilah[1], I B A R Sugiharta[1] and B Purwono[1]

[1] Department of Chemistry, Universitas Gadjah Mada, Sekip Utara, Bulaksumur, Yogyakarta 55281, Indonesia

E-mail: dadang.ovianto@mail.ugm.ac.id

Abstract. The derivative compound of 2,4,6-triarylpyridine was synthesized in two steps. The target compound was tested as colorimetric chemosensor against formaldehyde and the limit of detection was also determined using spectrophotometer UV-vis. The first step of the synthesis was the formation of 4-phenyl-2,6-bis(4-nitrophenyl)pyridine compound (**1**) from 4-nitroacetophenon and benzaldehyde. The second step of the synthesis was the formation of 4-phenyl-2,6-bis-(4-aminophenyl)pyridine (**2**) from reduction of the nitro group in compound **1** by HCl 37% solution and Sn metal. Compound **1** and **2** were characterized by spectrometers FTIR, ^{1}H NMR and direct inlet-mass spectrometry. The results showed that compound **1** was synthesized with a yield of 78.3% and compound **2** with a yield of 68.9%. Compound **2** as chemosensor showed color transition from colorless to yellow in chemosensor test against formaldehyde in ethanol solvent. The limit of detection of formaldehyde was measured as 4.7×10^{-3} M using spectrophotometer UV-vis.

1. Introduction

Formaldehyde (FA) is classified as a known human carcinogen by IARC [1,2,3]. Formaldehyde is commonly found in the environment, since it is formed primarily from natural sources and human activities. Naturally formed by biomass burning process or decomposition and volcanic activity. Meanwhile, direct human activities such as emissions from industrial and fuel combustion vehicles[16]. Formaldehyde is one of the chemicals that in practice is used as a food preservative [4,5]. Early detection is important to ensure that formaldehyde residues have no harmful effects on the health of the body by controlling levels of formaldehyde. Many methods based on spectrophotometric, fluorometric, piezoresistive, amperometric or conductive measurements have been proposed for detecting the concentration of formaldehyde [6]. Other techniques such as liquid and gas chromatography have been shown to be more selective and accurate for determining formaldehyde [7]. However, sensor technology is required that can be used in real time and in-situ detection simultaneously and reuseable [8,9].

Pyridine derivative compounds exhibit various activities such as antimalarial, antioxidant, anticonvulsant, anesthetic, antibacterial and antiparasitic [10,11]. While as a heterocyclic compound, pyridine derivative compounds have conjugated orbitals that can be added chromophore or fluorophore groups lead to optical signal changes in case of interaction with the substrate. Therefore pyridine derivative compounds can be developed as colorimetric sensor compounds [12]. The structure of 2,4,6-triaryl pyridine is a fluorescence material that is compatible with hole-transport capability and high fluorescence quantum yield for analysis [13]. Wang et al. (2014) succeeded in synthesizing 2,4,6-triaryl pyridine by reacting benzaldehyde, acetophenone and sodium hydroxide powder into mortar, yielding

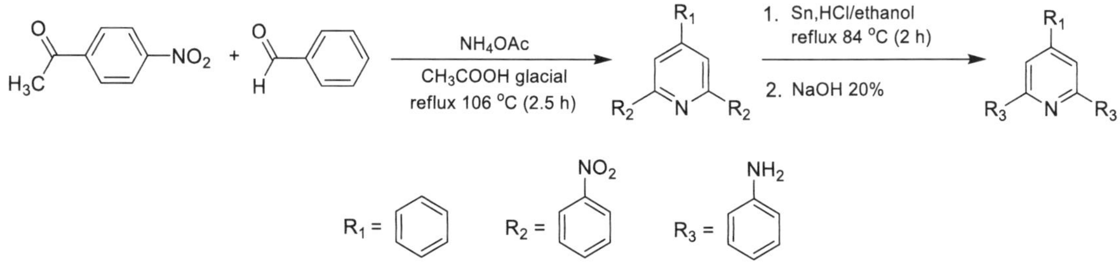

Scheme 1. Synthetic procedures for chemosensor compound.

1,3,5-triarylpentane-1,5-dione in the form of white needle solids with 85% yield. The solids are reacted with ammonium acetate in ethanol, yielding a white solid with a yield of 75% [14]. Type of introduction of formaldehyde substrate which can be utilized as chemosensor that is imine formation. The formation of the covalent bond has the advantage of being strong, reversible and can be used in water or slightly protic solvent [15].

Herein, our research group has designed 4-phenyl-2,6-bis(4-aminophenyl)pyridine that possess an amine group that it can form imine bond with formaldehyde. The pyridine ring was introduced as chromophore and the amine group as recognition moiety. The synthesis of 2,4,6-triaryl pyridine generally resulting high yield and selectivity, while derivative substances have not been widely used as colorimetric chemosensors.

2. Experiment

2.1. *Material and reagents*

All commercial grade chemicals and solvents were purchased and were used without further purification. The 4-phenyl-2,6-bis(4-nitrophenyl)pyridine (**1**) was synthesized according to Scheme 1 and was modified from Tamami and Yeganeh (2001) procedure [16].

2.2. *Synthesis of 4-phenyl-2,6-bis(4-aminophenyl)pyridine (2)*

The synthesis route of chemosensor compound was demonstrated in Scheme 1. The ethanol solution (20 mL) of compound **1** (0.15 g, 0.38 mmol) and tin granular (0.27 g, 2.3 mmol) was stirred to a temperature of 60 °C. After that, the mixed solution was added HCl 37% (5 mL). Then, the reaction of mixture solution was stirred at 84 °C for 2h. The reaction result was kept at room temperature and then stored in a refrigerator for 24h. The reaction product was filtered and put in a solution of sodium hydroxide

(6.0 g/30 mL). The solution was stirred until the yellow crystal precipitate was formed, then filtered and washed with distilled water three times. The crude were purified by recrystallization using ethanol solvent to get slightly yellow crystal of chemosensor in 68.9% yield (m.p. 199.1-199.4 °C, 208 °C) [17], ^{1}H NMR ($CDCl_3$, 500 MHz) δ: 7.41-7.44 (t, 1H), 7.48-7.50 (t, 2H), 7.69-7.71 (d, 2H), 7.68 (s, 2H), 8.02-8.04 (d, 4H), 6.77-6.78 (d, 4H), 3.81 (s, 4H). FTIR (KBr) ν: 1180 cm^{-1} (C–N), 1612 cm^{-1} (C=N), 3371 and 3425 cm^{-1} (NH_2). GC-MS for $C_{23}H_{19}N_3$ m/z 337.0 (100%).

2.3. *Instruments*

FTIR spectra were recorded on a Shimadzu Prestige 21. Mass spectra were recorded on a Shimadzu QP-2010S. ^{1}H NMR spectra were recorded on JEOL JNM-ECZ5000R/S1 spectrometer at 500 MHz. Chemical shifts are reported in ppm downfield from tetramethylsilane as internal standard and chloroform as a solvent. UV-vis spectra were recorded on a Shimadzu D-1800 spectrophotometer. Photoluminescence spectra were performed on a Shimadzu RF-6000 spectro fluorophotometer.

2.4. *General procedure for colorimetric spectra experiments*
A total of 2.5 mg of chemosensor compound was dissolved in 25 mL of solvents (acetonitrile, ethanol, DMSO, acetic acid, ethyl acetate). Then re-made the same solution for all solvents with the addition of 100 μL formaldehyde 37%. The two treatments before and after addition were subjected to spectral readings between 800-200 nm using a UV-vis spectrophotometer and observed. Both treatments before and after the addition of formaldehyde were done between 800 to 200 nm spectral scan using spectrophotometer UV-vis and observed.

2.5. *Limit of detection calculations*
About 3 mL of stock solution included in the 14 vial and then added a 37% formaldehyde (1.09 g / mL) with a volume of 0, 10, 20, 30, 40, 50, 60, 80, 100, 200, 400, 700, 1000 and 1500 μL. The absorbance readings were performed at a certain wavelength at maximum absorption wavelength using a UV-vis spectrophotometer.

3. Results and discussion

3.1. *Synthesis of compound 1*
In this work, compound **1** was prepared in a yield of 78.3% from benzaldehyde, 4-nitroacetophenone and ammonium acetate by a one-pot synthetic procedure. The chemical structure was confirmed by FTIR.

3.2. *Synthesis of compound 2*
The crude was purified by recrystallization using ethanol solvent, the chemosensor compound of solid with a yield was obtained by 68.9% and had melting point between 199.1-199.4 °C.

3.3. *Structure determination of compound 2*
The success of the reduction reaction of compound **1** was shown in the peak loss at 1350 cm^{-1} wavenumber as the stretching asymmetric aryl-NO_2 vibration. This proves that in compound **2** does not have a nitro functional group. The molecular ion has M^+ value equal to the mass of the theoretical molecule of chemosensor molecule. Ingole et al. (2010) reported substituted 2,4,6-triaryl pyridine compounds were analyzed using GC-MS having M^+ molecular ions appearing as base peak [18]. ^1H NMR showed the total hydrogen atoms in each molecule there are a number of 19 hydrogen atoms (6.7-8.0 ppm, C-H arom., 15H; 3.8 ppm, NH_2, 4H).

3.4. *Solvatochromic test of compound 2*
Test of compound **2** was dissolved in acetonitrile, ethanol, DMSO, acetic acid, ethyl acetate solvent and showed a colorless appearance except in acetic acid showed yellow color. Then compared with compound **2** with the addition of saturated formaldehyde in various solvents, it was found that only ethanol solvents can give rise to peaks with observable differences before and after the addition of formaldehyde in the visible light region.

The UV-vis spectrophotometer spectra of Figure 2 showed that the spectra changes with the appearance of peaks in the region 400-600 nm after the addition of formaldehyde occur in acetonitrile, ethanol and acetic acid solvents. Between acetonitrile and ethanol showed the similarity of the new peak shape which indicates the color change of the solution and indicates a positive solvatochromic phenomenon, but the peak on ethanol has a shift approaching the visible light region. It showed that ethanol is more effective as a qualitative colorimetric chemosensor. Ethanol is a protic solvent, including acetic acid, while the other solvents is aprotic solvent. It can be concluded that compound **2** can be used as a chemosensor compound using a protic solvent as ethanol because it qualitatively and quantitatively can provide color change before and after addition of formaldehyde effectively.

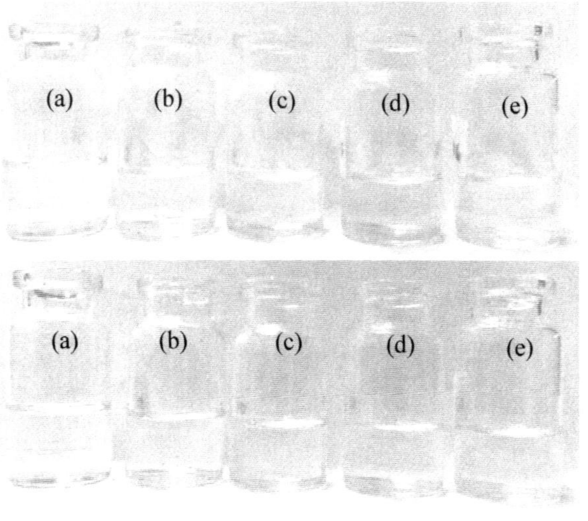

Figure 1. Compound **2** (top) and after addition of formaldehyde (below) in the solvent: (a) acetonitrile, (b) ethanol, (c) DMSO, (d) acetic acid, (e) ethyl acetate.

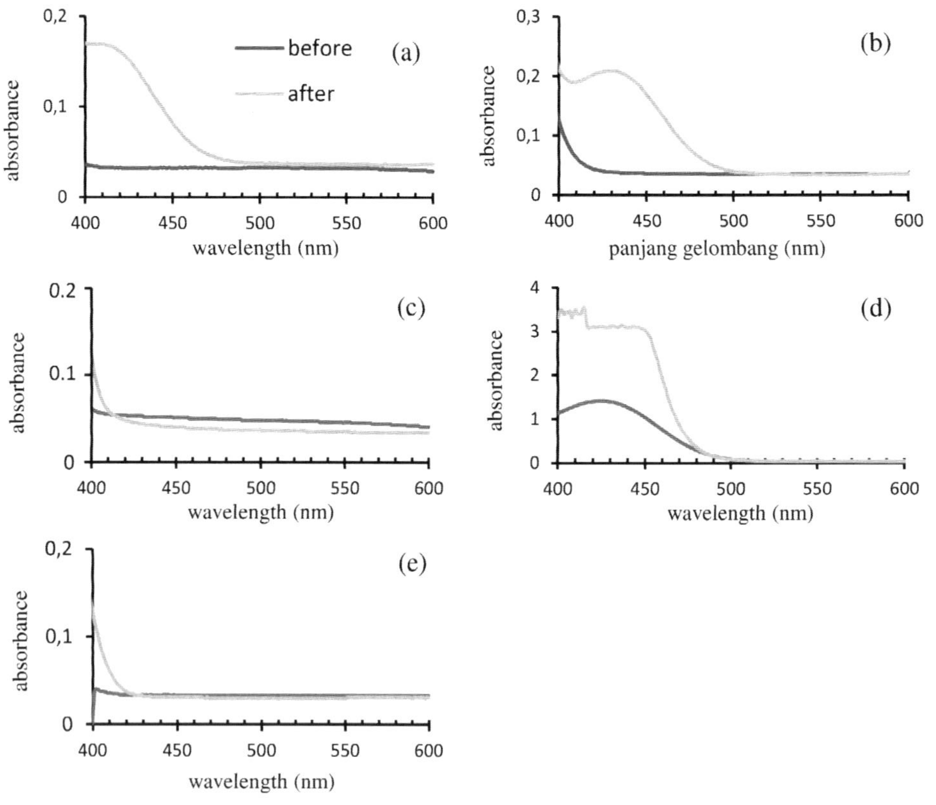

Figure 2. UV-vis spectrophotometer spectra before and after addition of formaldehyde to the solvent: (a) acetonitrile, (b) ethanol, (c) DMSO, (d) acetic acid, (e) ethyl acetate.

3.5. *Test of chemosensor against formaldehyde*

The result of the chemosensor test of compound **2** on formaldehyde as the substrate dissolved in the ethanol solvent, gives positive response in the form of color change. Compound **2** in the ethanol solvent at a concentration of 2.97×10^{-4} M (solution A) and the formaldehyde-added compound **2** (solution B) were read in the 340-600 nm wavelength range, and then obtained the maximum absorption wavelength at 430 nm. The wavelength region of light appears where the blue-purple color is absorbed, complementary to the yellow observation color. The yellow color corresponds to the color of the solution after it has been added formaldehyde (Figure 3).

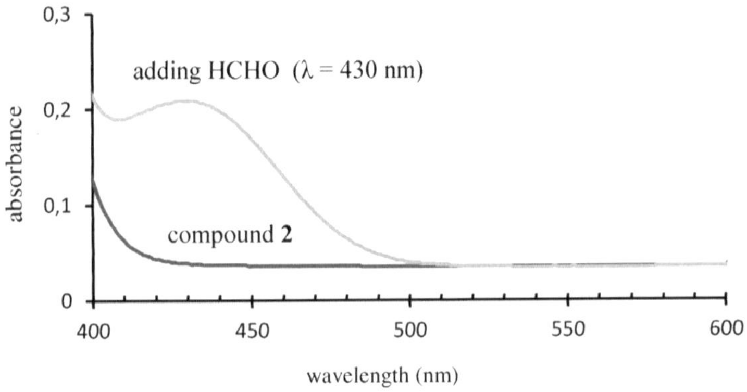

Figure 3. Spectra of spectrophotometer UV-vis of chemosensor compound 2.97×10^{-4} M in ethanol (blue) and after adding 50 μL HCHO 37% (yellow).

3.6. *Limit of detection calculations*

The value of the detection limit is calculated based on the formula (3 × SD) / slope in which SD is the standard deviation of the blank solutions (Table 1).

Tabel 1. The standard deviation of blank solutions of compound **2** (2.97×10^{-4} M).

Blank	Absorbance	Blank	Absorbance
Blank 1	0.048	Blank 5	0.048
Blank 2	0.048	Blank 7	0.050
Blank 3	0.048	Blank 8	0.049
Blank 4	0.048	Blank 9	0.049
Blank 5	0.049	Blank 10	0.049
Standard deviation = 0.000699			

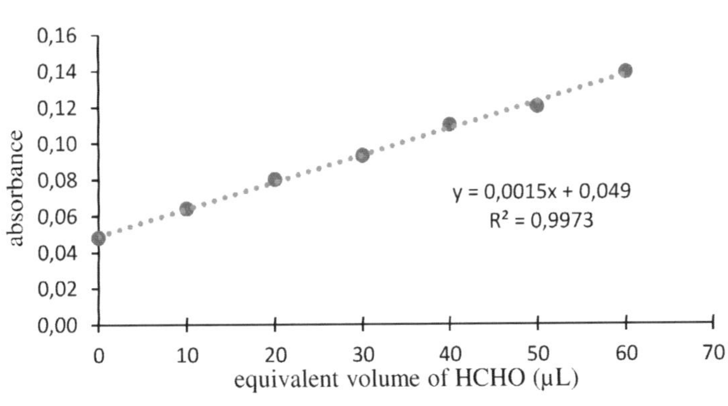

Figure 4. The standard curve of absorbance of compound **2** on formaldehyde using a UV-vis spectrophotometer at λ 430 nm.

Using the absorbance plot versus the equivalent volume of formaldehyde, Figure 4 showed the linearity relationship below the 60 μL equivalent volume. Based on these data, the standard deviation value of 0.000699 and slope value 0.0015 μL^{-1}, obtained detection limit value of 4.7×10^{-3} M.

4. Conclusions

In conclusion, we have presented a rapid and efficient chemosensor for formaldehyde, which showed colorimetric recognition for HCHO in ethanol solutions. The colorimetric spectra showed difference appearance between before and after addition of formaldehyde to chemosensor solution about visual color changes from colorless to yellow at 430 nm. The limit of detection of formaldehyde was measured as 4.7×10^{-3} M. We believe that these characteristics of chemosensor through formation of imine bond, make it attractive for molecular modifications and applications as colorimetric sensor for HCHO.

Acknowledgments

We gratefully acknowledge funding from Ministry of Research, Technology and Higher Education of the Republic of Indonesia with Pekan Kreativitas Mahasiswa 2017.

References

[1] Edrissi B, Taghizadeh K, Moeller B C, Kracko D, Doyle-Eisele M, Swenberg J A and Dedon P C 2013 *Chem. Res. Toxicol.* **26** 1421-23
[2] IARC 2012 *IARC Monogr. Eval. Carcinog. Risks Hum.* **100** 9–562
[3] Swenberg J A, Moeller B C, Lu K, Rager J E, Fry R C and Starr T B 2013 *Toxicol. Pathol.* **41** 181−189
[4] Bianchi F, Careri M, Musci M and Mangia A 2007 *Food Chem.* **100** 1049–53
[5] Chen L, Jin H, Xu H, Sun L, Yu A, Zhang H and Ding L 2009 *J. Agric. Food Chem.* **57** 3989–94.
[6] Chung P, Tzeng C, Ke M and Lee C 2013 *Sensors* **13** 4468-84
[7] Tomkins B A, McMahon J M and Caldwell W M 1989 *J. Assoc. Off. Anal. Chem.* **72** 835-839
[8] Chen E X, Yang H and Zhang J 2014 *Inorg. Chem.* **53** 5411-13
[9] Zeng J B, Fan S G, Zhao C Y, Wang Q R, Zhou T Y, Chen X, Yan Z F, Li YP, Xing W and Wang X D 2014 *Chem. Commun.* **50** 8121-23
[10] Kim B Y, Ahn J B, Lee H W, Kang S K, Lee J H, Shin J S, Ahn S K, Hong C I and Yoon S S 2004 *Eur. J. Med. Chem.* **39** 433-447
[11] Adib M, Tahermansouri H, Koloogani S A, Mohammadi B and Bijanzadeh H R 2006 *Tetrahedron Lett.* **47** 5957-60
[12] Cave G W, Hardie M J, Roberts B A and Raston C L 2001 *Eur. J. Org. Chem.* **17** 3227-31
[13] Wang H Y, Shi J J, Wang C, Zhang X X, Wan Y and Wu H 2012 *Dyes Pigments* **95** 268-274
[14] Wang H Y, Chen L F, Zhu X L, Wang C, Wan Y and Wu H 2014 *Spectrochimica Acta Part A: Molecular and Biomolecular Spectroscopy* **121** 355-362
[15] Zhou W, Dong H, Yan H, Shi C, Yu M, Wei L and Li Z 2015 *Sensors and Actuators* B **209** 664
[16] Tamami B and Yeganeh H 2001 *Polymer* **42** 415
[17] Gavrishova T N, Li V M and Budyka M F 2010 *Russian Journal of Applied Chemistry* **83** 926-928
[18] Ingole P G, Jadhav S V and Bajaj H C 2010 *Int. J. ChemTech Res.* **2** 289-294

IOP Conf. Series: Journal of Physics: Conf. Series **909** (2017) 012080 doi:10.1088/1742-6596/909/1/012080

Synthesis and Elucidation Structure of *Tetrakis*-diphenylaminecopper(II) Chloride Hexahydrate

H Syaima[1], S B Rahardjo[2], and E Suciningrum[2]

[1]Chemistry Graduate Program, Graduate School, Universitas Sebelas Maret, Jl Ir. Sutami 36 A, Kentingan, Jebres, Surakarta, 57126, Indonesia

[2]Department of Chemistry, Faculty of Mathematics and Natural Sciences, Universitas Sebelas Maret, Jl Ir. Sutami 36 A, Kentingan, Jebres, Surakarta, 57126, Indonesia

E-mail: husna_syaima@yahoo.com

Abstract. $CuCl_2{\cdot}2H_2O$ with diphenylamine formed a complex compound in 1:4-mole ratio of metal to the ligand in methanol. Its structural properties were investigated by employing metal content analysis by Atomic Absorption Spectroscopy (AAS), magnetic susceptibility, UV-vis and FTIR spectroscopy. The forming of the complex was indicated by shifting of UV-Vis spectra. The result of analysis Cu(II) in the complex showed empirical formula of the complex were $Cu(diphenylamine)_4Cl_2(H_2O)_6$. The electrical conductivity of complex showed the charge ratio of cation and anion = 2:1. Finally, the proposed formula of the complex was $[Cu(diphenylamine)_4]Cl_2{\cdot}6H_2O$. Based on infrared spectra, it was revealed that diphenylamine existed as monodentate bind to copper(II) through the functional group of N-H. The electronic spectral study of the complex showed three transition peaks on 861, 592, and 419 nm corresponding to the $^2B_{1g} \rightarrow {}^2A_{1g}$, $^2B_{1g} \rightarrow {}^2B_{2g}$ dan $^2B_{1g} \rightarrow {}^2E_g$ transitions. The complex was paramagnetic and indicated that ligands form square planar geometry around the Cu(II).

1. Introduction

Copper complexes are able to form many geometries with different types and numbers of ligands. It is reported that copper complex forming square planar [1-3], tetrahedral [4-5], octahedral [6-8], trigonal planar [9-10], and square pyramidal [11-12] geometry. A large number of complexes have been synthesized. However, the various design of the copper complexes based on different ligands are still commonly interesting to learn among scientists. Each copper complex compound has unique chemical and physical properties.

Such compounds, particularly where the copper ion coordinately binds to organic molecules, play important roles in biological processes [13], e.g., blue copper protein and primarily associated with copper-dependent cellular enzymes [14]. Also, copper compounds are often used as a catalyst in industrial processes [15], in some cases allowing the production of material otherwise difficult to obtain, also being applied in the medicinal and pharmacological field [13],[16-17].

Diphenylamine is a chemical compound used as stabilizers in nitrocellulose-containing explosives and propellants. It is also used in the perfumery, as antioxidants in the rubber and elastomer industry, pharmaceuticals, the production of dyes, photography chemicals, and etc. [18]. Diphenylamine has a lone pair of electron on the nitrogen atom [19] as a donating site. Our attention is focused on ligands

IOP Conf. Series: Journal of Physics: Conf. Series **909** (2017) 012080 doi:10.1088/1742-6596/909/1/012080

containing N-H group during last years due to their coordination chemistry and ability to be an antibacterial agent [20-21].

The purpose of the present work is to report the data of the new complex of copper with diphenylamine ligand and its properties which are characterized by Atomic Absorption Spectroscopy (AAS), molar conductivity, UV-Visible, IR, and magnetic susceptibility measurements.

2. Experimental

2.1. Materials

All chemicals and solvents were of analytical grade and used as received without further purification. The metal salts ($CuCl_2 2H_2O$, $CuSO_4 5H_2O$, $NiCl_2 6H_2O$, $NiSO_4 6H_2O$, and $AlCl_3.6H_2O$) and diphenylamine were purchased from Merck.

2.2. Physical Measurements

The copper content was determined by Atomic Absorption Spectrometer (AAS) Shimadzu AA-6650. Fourier transform infrared (FTIR) spectra of diphenylamine, $CuCl_2 2H_2O$, and its complexes as KBr pellets were recorded in the spectral range 4000–400 cm^{-1} with a Perkin-Elmer Series 2000 apparatus. Molar conductivity (Λ*m) of 1 mM solution in methanol was measured on Jenway CE 4071 conductivity meter at 25 ^oC. Magnetic susceptibilities were measured on AUTO Sherwood Scientific 10169 Magnetic Susceptibility Balance (MSB) at room temperature. Spectra UV-VIS of metal complex and ligand was recorded in methanol solvent using Shimadzu UV-3601 spectrophotometer.

2.3. Synthesis of Cu(II) complex

The complex has been prepared by reaction of $CuCl_2{\cdot}2H_2O$ and diphenylamine in methanol. Methanolic solution (10 mL) of $CuCl_2{\cdot}2H_2O$ (0.341 g; 2 mmol) was added drop wise to methanolic solution (20 mL) of diphenylamine (1.354 g; 8 mmol) with constant stirring for 6 hours. The solution was then concentrated to 15 mL of volume and left to stand for 72 hours. The precipitate was filtered with filter paper and dried in a desiccator for 24 hours.

3. Result and Discussion

3.1. Electronic spectra

Figure 1 shows the shift of $CuCl_2{\cdot}2H_2O$ maximum absorption wavelength towards smaller wavelengths in the complex Cu(II)-diphenylamine. Shifting λ_{max} indicated the formation of the complex. It is due to the substitutions of H_2O molecules with diphenylamine and shows that diphenylamine has greater ligand field energy. Electronic spectra of the complex of Cu(II)-diphenylamine showed a wide absorption at λ_{max} 816, 592, and 419 nm assigned to the $^2B_{1g} \rightarrow {}^2A_{1g}$, $^2B_{1g} \rightarrow {}^2B_{2g}$ dan $^2B_{1g} \rightarrow {}^2E_g$ transition. It was reported that the three transitions attributed to square planar geometry such as in Cu-N-Nicotinoyl-N-o-hydroxythiobenzhydrazide [22]. However, the predicted geometry of the complex needs to confirm by other characterization methods.

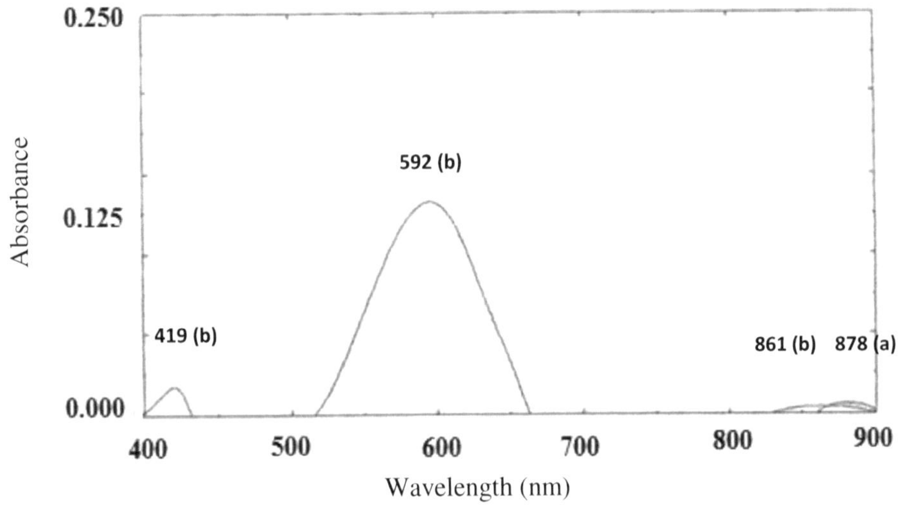

Figure 1. Electronic spectra of $CuCl_2 \cdot 2H_2O$ solution (a) and the complex solution (b)

3.2. Analysis of The Amount of Copper with Atomic Absorption Spectroscopy

The result of measurements of the copper content in complex was 6.91 ± 0.16 %. If the result of the measurement is compared with the copper content theoretically on various possible complex formulas, the formula can be approximated as $Cu(diphenylamine)_4Cl_2 \cdot 6H_2O$.

3.3. Analysis of Electrical Conductivity

The molar conductivity of the complex was determined in methanol (10^{-3} M) at room temperature. The molar conductance values of the complex lied in 33.96 ± 0.16 Scm^2mol^{-1}, thus indicating that is an electrolyte. By comparing the conductivity values of the complex sample with a molar conductivity of the standard solution (shown in Table 1), it is seen that the charge ratio of cation:anion= 2:1. This fact showed that the chloride ions were not coordinated to the central metal ion Cu(II) and only act as an anion. The complex dissociates in solution to be $[Cu(diphenylamine)_4]^{2+}$ and 2 ions of Cl^-. Thus, it strengthened the proposed formula of the complex was probably $[Cu(diphenylamine)_4]Cl_2 \cdot 6H_2O$.

Table 1. Molar Conductivity of Metal Salts and Cu(II) complex in Methanol

Solution	Λ_m (S cm^2 mol^{-1})	Cation : Anion Charge
Methanol	1.04	-
$CuSO_4 \cdot 5H_2O$	8.60	1:1
$NiSO_4 \cdot 6H_2O$	7.42	1:1
$CuCl_2 \cdot 2H_2O$	47.16	2:1
$NiCl_2 \cdot 4H_2O$	53.16	2:1
$AlCl_3 \cdot 6H_2O$	90.96	3:1
The complex	**33.96 ± 0.16**	**2:1**

3.4. Infrared Analysis

IR spectrum of Cu(diphenylamine)$_4$Cl$_2$·6H$_2$O showed a shift in the absorption of the aryl-NH-group toward a smaller wave number of 1319 cm^{-1} (diphenylamine ligand) to 1313 cm^{-1} Cu(diphenylamine)$_4$Cl$_2$·6H$_2$O which indicated that the cluster (aryl-NH-) ligands of diphenylamine are coordinated on Cu^{2+} central ions. In the Cu(diphenylamine)$_4$Cl$_2$·6H$_2$O there was an uptake of the -OH broad group at 3396 cm^{-1} which indicated the absorption of the O-H group of H_2O molecules, as occurs in complex [Zn(L)(L')]·3H$_2$O (L=2,6-pyridinedicarboxylate and L' = 4,4'-dipyridylamine) showing widespread uptake in the 3400 cm^{-1} region which is a water of crystalline dihydrate [23]. The absorption of clusters O-H broad from this H_2O molecule caused the absorption of the N-H group overlap, therefore, it did not appear. This is due to the nearly same region of the O-H and N-H infrared absorption [24].

3.5. *Magnetic Properties* (μ_{eff})

The μ_{eff} value measured for the present Cu(II) complex was 1.71 BM, which showed one unpaired electron, paramagnetic, and discrete magnetically non-coupled spin-only value for copper(II) ion. The magnetic moment value of Cu(II) complex indicated that [Cu(diphenylamine)$_4$]Cl$_2$·6H$_2$O is present in square planar geometry [25]. Based on the above results, the structure in Figure 2 is suggested for the complex.

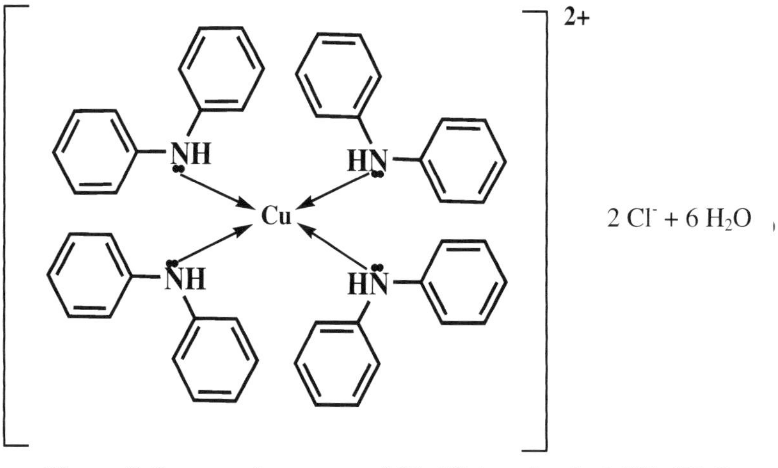

Figure 2. Suggested structure of [Cu(diphenylamine)$_4$]Cl$_2$·6H$_2$O

4. Conclusion

The complex of [Cu(diphenylamine)$_4$]Cl$_2$·6H2O, namely *Tetrakis*diphenylaminecopper(II) chloride hexahydrate, was able to be synthesized by mixing methanolic solution of Cu(II) and methanolic solution of diphenylamine with a mole ratio of 1:4 and stirring for 6 hours. Electronic spectra of the complex showed a wide absorption at λ_{max} 816, 592, and 419 nm assigned to the $^2B_{1g} \rightarrow {}^2A_{1g}$, $^2B_{1g} \rightarrow {}^2B_{2g}$ dan $^2B_{1g} \rightarrow {}^2E_g$ transition. The functional group coordinated to the Cu(II) was possibly N-H, forming square planar geometry. The complex was paramagnetic.

Acknowledgement

The authors gratefully express acknowledgments to Indonesian Ministry of Research Technology and Higher Education (RISTEDIKTI) for financial support.

References

[1] Turkyilmaz M 2017 *Journal of Thermal Analysis and Calorimetry* **129(1)** 451-460.
[2] Said M, Ahmad J, Rehman W, Badshah A, Khan H, Khan M, Rahim F and Spasyuk D

2015 *Inorganica Chimica Acta* **434** 7-13.

[3] Dong Y, Dong X, Dong W, Zhang Y and Zhang L 2017 *Polyhedron* **123** 305-315.

[4] Angel N, Khatib R, Jenkins J, Smith M, Rubalcava J, Le B, Lussier D, Chen Z, Tham F, Wilson E. and Eichler J 2017 *Journal of Inorganic Biochemistry* **166** 12-25.

[5] Huang M, Guo Y, Shi Y, Zhao L, Niu Y, Shi Y and Li X 2017 *Inorganica Chimica Acta* **457** 107-115

[6] Alabdali A *Journal of Applied Chemistry* 3 **2** 5-10.

[7] Bibi S, Mohamad S, Manan N, Ahmad J, Kamboh M, Khor S, Yamin B and Abdul Halim S 2017 *Journal of Molecular Structure* **1141** 31-38.

[8] Olalekan T, Ogunlaja A, VanBrecht B, and Watkins G 2016 *Journal of Molecular Structure* **1122** 72-79.

[9] Alfonso S, González S, Higuera-Padilla A, Vidal A, Fernández M, Taylor P, Urdanibia I, Reiber A, Otero Y and Castro W 2016 *Inorganica Chimica Acta* **453** 538-546.

[10] Khalaji A, Peyghoun S, Akbari A, Feizi N, Dusek M and Eigner V 2017 *Journal of Molecular Structure* **1127** 511-514

[11] Pahonţu E, Julea F, Chumakov Y, Petrenco P, Roşu T, and Gulea A 2017 *Journal of Organometallic Chemistry* **836-837** 44-55.

[12] Jopp M, Becker J, Becker S, Miska A, Gandin V, Marzano C and Schindler S. 2017 *European Journal of Medicinal Chemistry* **132** 274-281.

[13] Ndagi U, Mhlongo N, and Soliman M 2017 *Drug Design, Development and Therapy* **11** 599-616.

[14] Anastassopoulou J and Theophanides T 1995 *Bioinorganic Chemistry* 209-218

[15] Routaray A, Nath N, Maharana T, Sahoo P, Das J and Sutar A 2016 *Journal of Chemical Sciences* **128(6)** 883-891.

[16] Rajalakshmi S, Fathima A, Rao J and Nair B 2014 *RSC Adv* **4(60)** 32004-32012.

[17] Hu K, Li F, Zhang Z. and Liang F 2017 *New J. Chem,* **41(5)** 2062-2072.

[18] Srinivasan K, Vaheethabanu J, Manisankar P and Stalin T 2011 *Journal of Molecular Structure* **987** 214-224

[19] Ejelonu B C and Olagboye S A 2015 *Research Journal of Pharmaceutical, Biological and Chemical Sciences* **6** 2 184-190

[20] Nagesh G, Mahendra Raj K and Mruthyunjayaswamy B 2015 *Journal of Molecular Structure* **1079** 423-432.

[21] Xue L, Deng D. and Wang Q 2017 *Russian Journal of Coordination Chemistry* **43(3)** 181-188.

[22] Shrivastav A N K, Singh, and SM Singh 2003 *BioMetals* **16** 311-320.

[23] Saylor S M R M, Supkowski, and R L LaDuca 2008 *Inorganica Chimica Acta* **361** 317–326.

[24] Rahardjo S B, Lestari W W, Syaima H 2017 *IOP Conf. Series: Earth and Environmental Science* **75** 012025 doi :10.1088/1755-1315/75/1/012025

[25] Turan N, Körkoca H, Adigüzel R, Çolak N and Buldurun K 2015 *Molecules* **20** 9309-9325

Modeling Alkyl p-Methoxy Cinnamate (APMC) as UV absorber based on electronic transition using semiempirical quantum mechanics ZINDO/s calculation

Salmahaminati, Muhlas Abdul Azis, Gani Purwiandono, Muhammad Arsyik Kurniawan, Dwiarso Rubiyanto[1] and Arif Darmawan[2]

[1]Chemistry Departement, Faculty of Mathematics and Natural Sciences, Islamic University of Indonesia, Yogyakarta, 55581, INDONESIA

[2]Agency for the Assessment and Application of Technology (BPPT), Puspiptek Serpong, Tangerang Selatan, 15314, INDONESIA

E-mail: salmahaminati@uii.ac.id

Abstract. In this research, modeling several alkyl p-methoxy cinnamate (APMC) based on electronic transition by using semiempirical mechanical quantum ZINDO/s calculation is performed. Alkyl cinnamates of C1 (methyl) up to C7 (heptyl) homolog with 1-5 example structures of each homolog are used as materials. Quantum chemistry–package software Hyperchem 8.0 is used to simulate the drawing of the structure, geometry optimization by a semiempirical Austin Model 1 algorithm and single point calculation employing a semiempirical ZINDO/s technique. ZINDO/s calculations use a defined criteria that singly excited –Configuration Interaction (CI) where a gap of HOMO-LUMO energy transition and maximum degeneracy level are 7 and 2, respectively. Moreover, analysis of the theoretical spectra is focused on the UV-B (290-320 nm) and UV-C (200-290 nm) area.The results show that modeling of the compound can be used to predict the type of UV protection activity depends on the electronic transition in the UV area. Modification of the alkyl homolog relatively does not change the value of wavelength absorption to indicate the UV protection activity. Alkyl cinnamate compounds are predicted as UV-B and UV-C sunscreen.

1. Introduction

Currently, the development research of sunscreen product has been conducted to find the new compounds with properties such as efficacious, good and does not effect on the skin. The suncreen compound can protect skin from the effects of Ultra Violet (UV) emitted by the sun. The legal mechanism UV of a sunscreen agent is the energy absorption of UV light used for excitation of the electronic state of the compound [1]. One of the sunscreens component is octyl p-methoxy cinnamate[2].

Based on its chemical structure, the interesting part in octyl p-methoxy cinnamate are the alkyl and the benzyl chain part of the derived compound. Both of these compounds can be obtained by performing modifications on either one part of the structure or combination of some parts. The structure of alkyl p-methoxy cinnamate is presented in Figure 1.

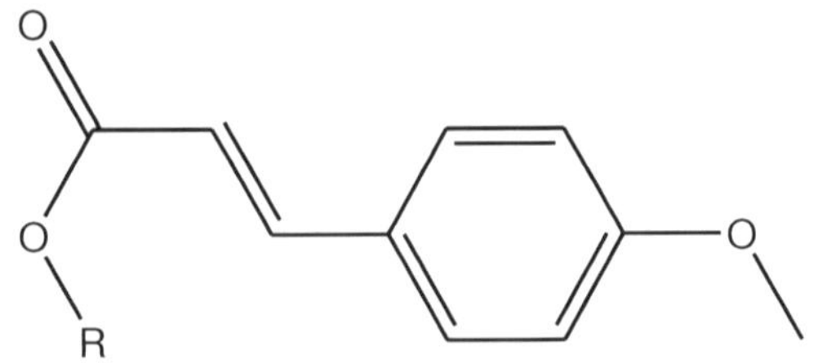

Figure 1. alkyl p-methoxy cinnamate (APMC), (R is alkyl group)

Light absorption by molecules in ultraviolet regions depends on the electronic structure. The UV spectra of the organic compound are closely related to the transitions between electronic levels [3]. The design of sunscreen compound has been investigated in both experiments [4] and computational chemistry [5]. Simulating a sunscreen compound using a computational chemistry method has some advantages. The cost and time for spending the design process of sunscreen compound by using computational chemistry are cheap [6,7].

The use of computational chemistry methods for organic compound can be conducted with qualitative approach [5]. The qualitative approach is performed by reviewing the structure model of chemical compounds on the aspects of transition spectra on the calculations. The previous study was conducted to investigate the transition area especially on the region UV-B and UV-C [8]. Application of computational methods for using in development research was pretty much [9,10]. Modeling of sunscreen compounds has been performed by Walters, et al, [5] and also conducted by Iqmal, et al, [2] using a semiempirical calculation of molecular orbitals ZINDO to examine the activity various isoamic derived from cinnamon compounds. The application of semiempirical approach of molecular orbitals ZINDO/s has also been utilized by previous research. Mudhab [11] used this method to study the aspects of geometry and transition spectra of complex compounds Co (II)-fenantrolin and Co (II)-bipyridine. ZINDO/s method can be selected to learn the design and new model compounds of cinnamate derivatives and salicylates. The purpose of this study is to be able modeling of alkyl p-methoxy cinnamate derivatives using a semiempirical approach of molecular orbitals ZINDO/s computing in the form of spectral calculations of UV B and UV C which are the UV light areas where relatively harmful to the skin.

2. Materials

The computational chemistry equipment used in this study is a computer unit with specifications: Pentium 41.4 GHz processor, 256 MB SDRAM memory, and HD 20 GB, and HyperChem 8.0 Windows based computational chemistry software.

This computational study using the structures of alkyl p-methoxy cinnamates with alkyl type refers to the alkyl acetate material available at the chemical catalog. Type alkyl is limited to homolog C1 (methyl) until C7 (heptyl)

3. Methods

3.1 Geometry Optimization

For each compound that is used as research material, 2D structures are created with Hyperchem packages program [12]. They equipped with hydrogen atoms as each atom to complete the structure and form into a 3D structure. The process continued with the structural geometry optimization. In structural energy minimization, we obtain the structural conformation molecules which are stabilized. The method was AM1 (Austin Model 1) and the convergence limit was determined after reaching the gradient limit which is 0.01 in energy per position change. And the single calculations point energy is performed for recording the data.

3.2 Spectrum Determination

The stabilized structure was calculated by an AM1 approach with single point calculation using semiempirical ZINDO/s for generating the transition. In this calculation, the criterion which is found for the methods configuration is a single excitation (Singly excited - CI) and the gaps HOMO-LUMO energy transition was 7. The maximum degeneration rate determined to the limit of 2. The calculation will be activated in the recording process file so the output of the calculations could be shown in detail. The data calculation results of the discontinuous spectra are in the region UV-visible wavelength. This data is used to predict the type of sunscreen compound activity.

4. Results and Discussion

Study of alkyl p-methoxy cinnamate is done by modifying the substitution of the alkyl chain. This study was to know the influence length of the alkyl chains substituent and the effect of isomer form changes for every homolog. The changes in alkyl chains include the length of it as well as in the shape of isomers. The analysis is done using an alkyl chain homolog C4 (butyl) to C7 (heptyl). The analyzed compounds are grouped and given the code as it is shown in Table 1. From each homolog that might happen, no all of them is done, as an example in homolog Hexyl; there is only four homologous. The derivatives with substantial variations in the benzyl region are not done because of the difficulty level of the synthesis compounds relatively complex. The presence of O-CH_3 groups in the benzyl region would complicate the synthesis procedure of alkyl p-methoxy cinnamate derivatives which is desired to substitute. The wavelength different is found because of the molecule in a pure state, while the outcome experiments analyzed has a relatively low purity [11].

The solvent effect also has the influence on the measurable uptake in more polar solvents, the red shifts uptake $\pi \rightarrow \pi^*$, and blue shifts uptake n (lone pair electron) $\rightarrow \pi^*$. The data is generated from chemical calculations of the wavelength and the intensity.

Table 1. Structure derivative and spectrum data of alkyl p-methoxy cinnamates

Alkyl (R)	Group (C1-C7)	Alkyl (R) structure	Wavelength number (nm) with maximum intensity
Methyl	C1	CH3•	292.5
Propyl	C3	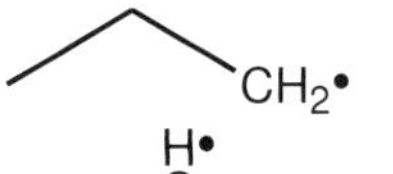	290.8
Isopropyl	C3	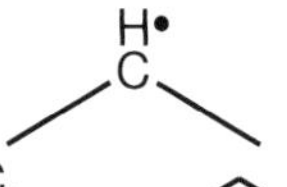	289.9
n-butyl	C4	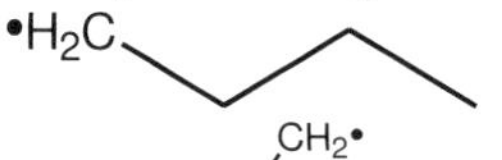	291.8
Isobutyl	C4	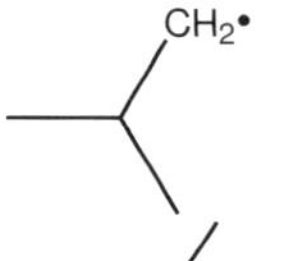	290.1
t-butyl	C4	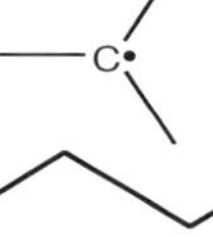	292.2
n-amyl	C5	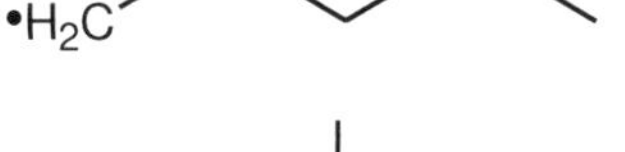	292.2
Isoamyl	C5	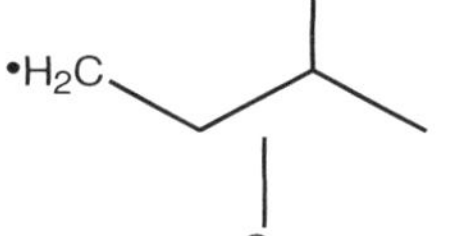	292.2
t-amyl	C5	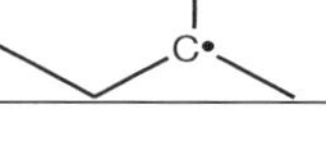	292.2

2,2 dimethyl propyl	C5	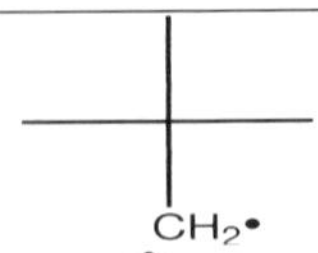	292.2
n-hexyl	C6	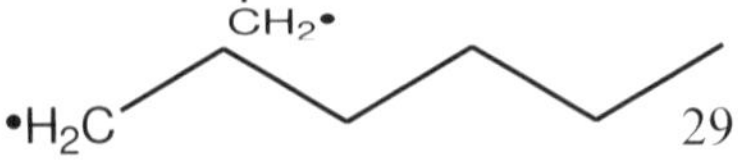	291.5
Isohexyl	C6	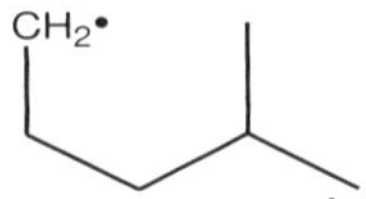	289.7
t-hexyl	C6	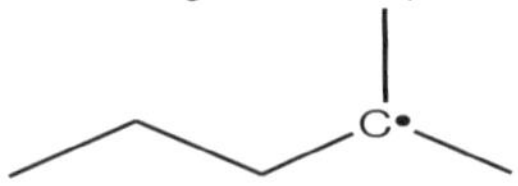	260.6
2,2 dimethyl butyl	C6	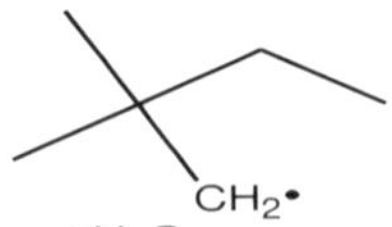	283.9
2,3 dimethyl butyl	C6	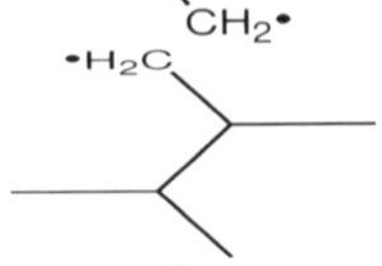	289.5
n-heptyl	C7		290.5
Isoheptyl	C7	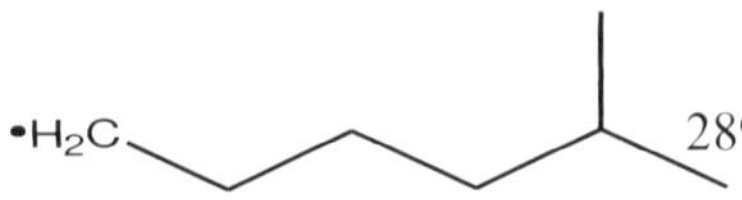	289.3
t-heptyl	C7	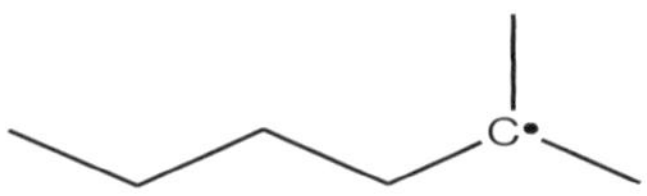	290.2
2,2 dimethyl amyl	C7	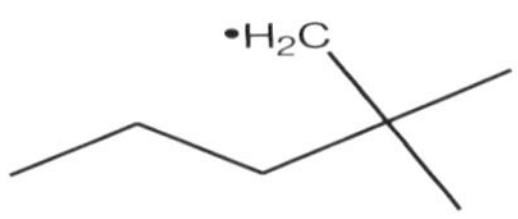	290.0
2,2,3 trimethyl butyl	C7	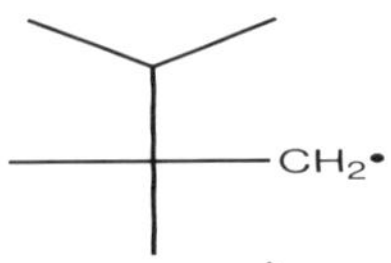	290.3
3,3 dimethyl amyl	C7	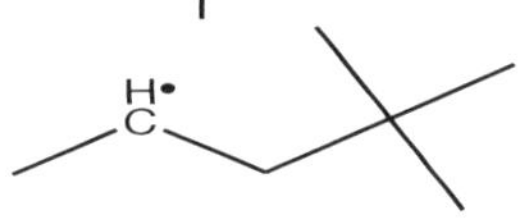	291.3
2,3,3 trimethyl butyl	C7	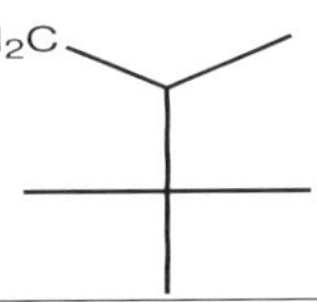	290.2

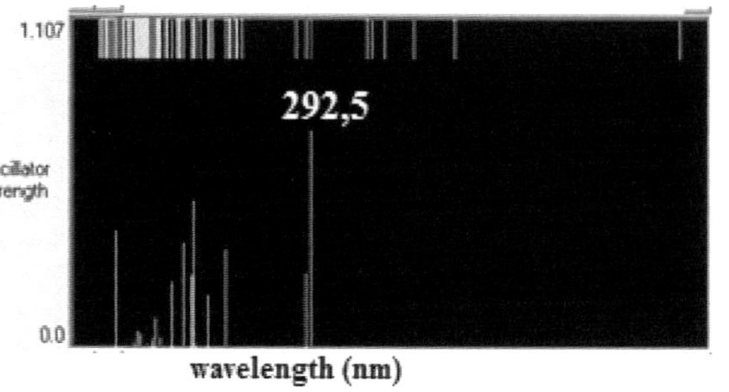

Figure 2. UV spectrum of methyl p-methoxy cinnamate (C1)

The spectrum of methyl p-methoxy cinnamate is in Figure 2. The ribbons which show the intensity of uptake rays are due to the electronic transition. The spectrum (up) represents all of the electronic transitions without showing the energy. The spectrum (bottom) has large intensity. The spectrum (up) and (bottom) are discontinuous because the chemical calculations are assumed that:

1. The compound is in a state of pure gas, so there is no effect from the other compounds.
2. The electronic transitions are from ground state to excited state.

The alkyl p-methoxy cinnamate structure is for some clusters which has simple conjugated chromophores in the benzene ring. Each of them has an electron transition type and different wavelengths. The unconjugated chromophore has the electron transition of [11]:

O which is the transition of $n \rightarrow \sigma^*$ $\lambda = 185$ nm
C = O which is the transition of $n \rightarrow \pi^*$ $\lambda = 300$ nm
C = O which is the transition of $n \rightarrow \sigma^*$ $\lambda = 190$ nm

The chromophore group with no conjugated system has several auxochromes groups. They are a saturated group which bound and change the wavelength with the maximum absorption intensity. Alkyl p-methoxy cinnamate has the auxochromes group which is a methoxy group ($-OCH_3$). This compound would have a relative absorption to the UV light area and also the activity as a UV-B and UV-C sunscreen.

Changes in the form of an isomer to every alkyl homolog by the alkyl chain of the relative substituents do not cause the shifts of wavelength absorption significantly. The size of the alkyl chain is also not relatively having a good effect in the wavelength and intensity. Thus the type of UV light protection is also not influenced by the length of the alkyl chain cinnamate compounds.

The changes of straight and branch chain in alkyl homolog of methyl (C1) to heptyl (C7) show that there is no effect in the difference maximum wavelength. It could be explained because the cluster of alkyl does not contain the presence of chromophores.The chromophore is contained of double or triple bond. In the experiment, the synthesis of alkyl p-methoxy cinnamate could happen from the esterification reaction [13,14]. It would be difficult to synthesize if the chemical starting point of alkyl chain has chromophore group.

5. Conclusion

The derivative of alkyl p-methoxy cinnamate which is the alkyl of methyl, propyl, butyl, amyl, hexyl, and heptyl is a compound of UV B and UV C absorbers. Substitution of alkyl in this cinnamates compounds relatively did not give the changes of the wavelength absorption

References

[1] Kimbrough DR 1997 *J.Chem. Ed* 74. 5
[2] Tahir I, Wijaya K, Subarni T, Wahyuningsih TD 2004 *Indonesian J.Chem.* 2004 3 230
[3] Sastrohamidjojo H 1991 *Spectroscopy* 2nd Ed. Liberty Yogyakarta
[4] Finnen MJ 1987 *J. Pharmacol. Skin* 72 69

[5] Walters C, Keeney A, Wigal CT, Johnston CR, Cornelius RD 1997 *J. Chem.Ed*. 74 99
[6] Grant GH, Richards WG 1995 *Computational Chemistry* Oxford Science Publication Oxford University Press. Oxford
[7] Leach AR 1996 *Molecular Modelling : Principles and Aplications,* Addison Wesley Longman Southampton University London
[8] Davis MR, Quigley MN 1995 *J. Chem. Educ*. 72 279
[9] Darmawan A, Hardi F, Yoshikawa K, Aziz M, Tokimatsu K 2017 *J. Apenergy* 05 058
[10] Darmawan A, Budianto D, Aziz M, Tokimatsu K 2017 *J. Apenergy* 03 122
[11] Tahir I, Wijaya K, Subarni T 2002 *Indonesian J. Chem* 2 64
[12] Anonym 1992 *HyperchemTM Release 3 for Windows: Manual,* Autodesk Inc., Tulsa
[13] Salmahaminati, Jumina 2017 *J.Phys. Conf. Ser* 795 01 20 65
[14] Salmahaminati, Jumina 2017 *J.Egypro* 03 469

IOP Conf. Series: Journal of Physics: Conf. Series **909** (2017) 012082 doi:10.1088/1742-6596/909/1/012082

Low cost heterogenous catalyst from (*Achatina Fulica*) snail shell and its application for biodiesel conversion via microwave irradiation

Is Fatimah, E. A. Kurniastuti, I. A. Basthiani, and A. Fakhri

Chemistry Department, Universitas Islam Indonesia, Kampus Terpadu UII, Jl. Kaliurang KM 14, Sleman, Yogyakartam Indonesia 555584.

E-mail : isfatimah@uii.ac.id

Abstract. Research on preparation of heterogenous catalyst from *Achatina Fulica* snail shell and its application biodiesel conversion has been investigation. Research aimed to obtain low cost and reusable catalyst for biodiesel production. The catalyst was prepared by grinding and calcining the snail shell at 900°C for 2 hours. The obtained solid was analysed by using XRD, SEM-EDX. FTIR, and also basicity measurement. Catalyst was used in the cenvertion of rice bran oil transesterification at varied volume of oil methanol ratio of 20-80 under microwave and reflux methode. The transesterification result were analyzed by using GCMS.

1. Introduction

Today, more than 90% of the world's energy needs are supplied from fossil fuels. The use of fossil energy resources as human needs, especially in the fields of transport, power generation, and industrial causes of fossil energy resources dwindling, especially petroleum. The use of fossil energy sources over the years has led to increased global temperatures, also known as global warming. It is caused by high levels of carbon released as a product and waste gas. This condition lead to observe and search for alternative fuel sources that are more environmentally friendly in order to overcome the negative effects arising from petroleum-based fuels [1].

Biodiesel is the most potential alternative energy resource for developing countries. Within this scheme, Castor oil, *Ricinuscommunis L.*, (castor bean, castor, Jatropha oil, ricin, Higuerilla, Mamona, mamoeira, palmachristi) is widely used since it is a non-food oil can easily be found and cultivated in all tropical countries in the world[2]. *Ricinuscommunis* L grow well in tropical conditions are hot and humid and has a growing period of 4 to 5 months. Just like other vegetable oils, castor oil is a triglyceride of various fatty acids and about 10% glycerin. Fatty acids consist of about 80-90% ricinoleic acid, linoleic acid 3-6%, 2-4% and 1-5% oleic acid saturated fatty acids. The high content of ricinoleic acid is the reason castor oil is versatile in the field of technology.

Refer to the principles of green chemistry, improving effectivity and efficiency of biodiesel production can be optimized by several optimization procedures. Two of variables affecting the condition are the selection of catalysis and mechanism as well as the simpler method in synthesis. Catalysts provide active sites involved in the reaction process. Usually homogeneous catalyst of NaOH or KOH were utilized in the production but the procedure gives ineffectivity since the catalyst can not be reuse and becomes consumable chemicals. As many other organic reactions, replacement of homogeneous catalyst with heterogeneous form is an interesting choice. Some investigations attempted to replace homogeneous catalyst with the heterogenous form such as CaO, $CaCO_3$, $Ca(OH)_2$ or other supported basic catalyst[3–5]. Based on some investigations, exploration on Ca-based catalyst from naturally-occurring materials were reported. For example, the conversion of ostrich egg shell, snail shell, egg shell as CaO source. In this investigation, Ca-based catalyst for biodiesel production was conducted using waste of snail (*Achatina Fulica*) shell. In advance, the simpler method of biodiesel production was achieved over microwave-assisted method. Research was aimed to characterize and test the catalyst for biodiesel production in the microwave-assisted method.

2. Material and Methodes

2.1. Material

Castor Oil was obtained from PT.Sinar Kimia, chemicals consist of methanol, HCl, and n-hexane were purchased from Merck, Germany. Snail (*Achatina Fulica*) shell was obtained from Paddy field area of Sleman District, Yogyakarta.

2.2. Preparation of catalyst

Snail shell was cleaned with water to remove proteins and other substancesby washing using warm water before sun-dryed. The shel was crushed and followed by calcination at 900°C for 2h. The sample was characterized by x-ray diffraction (XRD), Fourier-Tramsform Infra Red (FTIR) and Scanning Electrone Microscope-Energy dispersive x-ray (SEM-EDX). Rigaku Miniflex instrument 600 Benchtop X-Ray Diffraction (XRD) was utilized to determine crystallinity of the sample, FTIR NICOLET AVATAR to determine the functional groups contained in the sample snail calcination results and SEM JEOL microscope instrument was used for surface profile of sample.

2.3. Catalytic Activity in Biodiesel Production

Biodiesel production from castor oil was conducted over two method: reflux and microwave-assisted method. For each reaction the molar ratio of oil to methanol was 20:80. Microwave method was operated at medium power for varied time of irradiation. After each reaction, biodiesel fraction was separated using a separating funnel and the sample was analyzed using Gas Chromatography-Mass Spectra GC-MS type Shimadzu QP-2010.

3. Result and Discution

3.1. Catalyst production from Snail Shells

Sample obtained from snail shell was characterized using XRD and the pattern is presented in Figure 1.

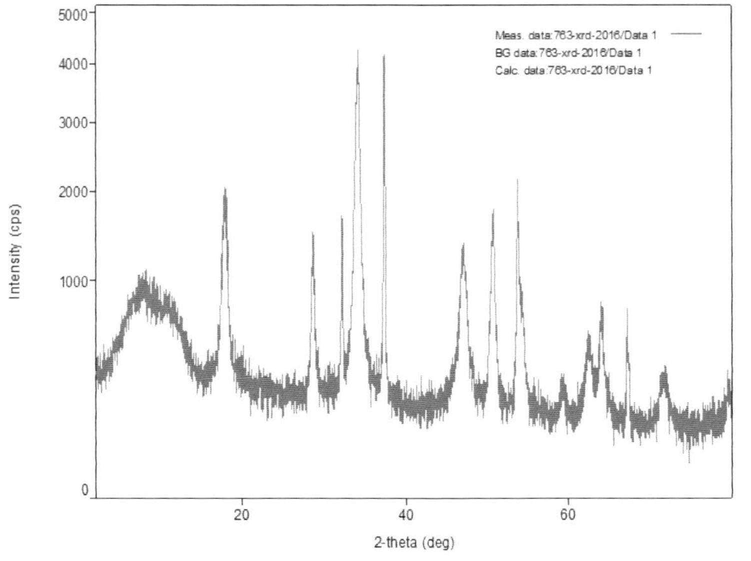

Figure 1. Diffractograms Catalyst Snail Shell

The pattern in Figure 1 shows that the sample contains $Ca(OH)_2$ as showed by reflections at 2θ 28;47, 34;014 and 50.688°.

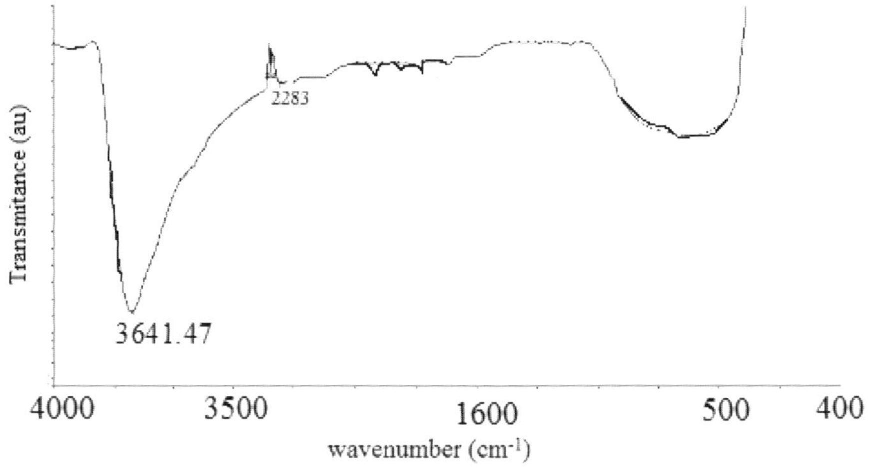

Figure 2. IR Spectra snail shell

FTIR spectra of material (Figure 2) demonstrated the spectra of OH group at 3641.47cm^{-1}. Another important spectra are at 2100-2260cm^{-1} as indication of alkyne groups, 875.4cm^{-1} as indication of (-CH). The presence of Ca-O is identified at the frequencies of 417.99cm^{-1}. These results are consistent with the results of the study reported by Liu et al, (2008) that the infrared spectra of $Ca(OH)_2$ is shown from the OH groups that emerged in the area 3600-4000cm-1. Meanwhile, according to Alba-rubio et al, (2010), CaO group found on the tape under 600cm-1 is on the area around 405cm^{-1}[6].

The change of surface morphology of material is illustrated in Fig 3. From EDX analysis it is found that the O content in $Ca(OH)_2$ sample was about 71.4% Wt, and Ca 27% Wt content in $Ca(OH)_2$ sample.

IOP Conf. Series: Journal of Physics: Conf. Series **909** (2017) 012082 doi:10.1088/1742-6596/909/1/012082

(a)

(b)

(c)

Figure. 3. SEM profile of materials with the magnification of (a) 5000x, (b) 10000x, (c) 20000x

3.2. Catalytic activity in biodiesel production

Result of catalyst testing in biodiesel conversion at varied condition is presented in Table 1.

Tabel 1. Catalyst Activity in Biodiesel Product

Catalyst weight (g)	Method	Reaction time	Volume	Conversion (%)
2	Reflux	2 (h)	58 mL	81.5%
2	Reflux	3 (h)	67 mL	86.05%
2	Reflux	5 (h)	26 mL	92.49%
3	Reflux	5 (h)	18 mL	-
5	Reflux	5 (h)	21 mL	98.96%

2	Microwave	10 minutes	65 mL	91.65%
2	Microwave	20 minutes	56 mL	95.25%
2	Microwave	30 minutes	49 mL	90.13%

Based on data in Table 1 it can be found that both reflux and microwave produce biodiesel. On the reflux method, time of reaction as well as catalyst weight give no effect the production of biodiesel. Compared to reflux method, microwave method give higher volume and conversion with the optimum rection time of 20 mins.

Based on the resulting data, methods reflux produce methyl ester by 98.96% at the optimum transesterification reaction for 5 hours with the use of a catalyst as 5gram, while the microwave method produced methyl ester by 95.25% at the optimum transesterification reaction for 20 minutes. Thus, it can be concluded that the differences were relatively small, the microwave method is more profitable because of the time required is much faster and simpler to use than using reflux.

4. Conclusion

Ca-based catalyst has been successfully derived from Snail shell and the material exhibits catalytic activity in biodiesel production. The material showed the dominant content of $Ca(OH)_2$ and the catalyst is active for reflux and microwave method.

References

[1] Ibrahim N, Kamarudin S K and Minggu L J 2014 Biofuel from biomass via photo-electrochemical reactions: An overview *Journal of Power Sources* **259** 33–42

[2] Sivasamy A, Cheah K Y, Fornasiero P, Kemausuor F, Zinoviev S and Miertus S 2009 Catalytic applications in the production of biodiesel from vegetable oils *ChemSusChem* **2** 278–300

[3] Hatefi H, Mohsennia M and Niknafs H 2014 Transesterification of Corn Oil Using MgO and ZnO as Catalysts *International Conference on Chemistry, Biomedical and Environment Engineering* 37–41

[4] Tan Y H, Abdullah M O, Nolasco-Hipolito C and Taufiq-Yap Y H 2015 Waste ostrich- and chicken-eggshells as heterogeneous base catalyst for biodiesel production from used cooking oil: Catalyst characterization and biodiesel yield performance *Applied Energy* **160** 58–70

[5] Lengyel J, Cvengrošová Z and Cvengroš J 2009 Transesterification of triacylglycerols over calcium oxide as heterogeneous catalyst *Petroleum & Coal* **51** 216–24

[6] Gupta P and Paul S 2014 Solid acids: Green alternatives for acid catalysis *Catalysis Today*

Evaluation of Ultrafiltration Performance for Phospholipid Separation

N Aryanti, D H Wardhani, Z S Maulana and D Roberto

Department of Chemical Engineering, Diponegoro University
Jl. Prof. Soedarto, SH, Kampus Undip, Tembalang, Semarang, 50278, INDONESIA

E-mail: nita.aryanti@che.undip.ac.id

Abstract. Ultrafiltration membrane for degumming of crude palm oil has been applied as an alternative method since the membrane process required less procedure than the conventional degumming. This research focused on the examination of ultrafiltration performance for phospholipid separation from model crude palm oil degumming. Specifically, profile flux and rejection, as well as blocking mechanism, were investigated. Feed consisting of Refined Crude Palm Oil – Isopropanol – Lecithin mixtures were represented as crude palm oil degumming. Lecithin was denoted a phospholipid component, and the concentrations of lecithin in feed were varied to 0.1%, 0.2%, and 0.3%. The concentration of phospholipid was determined as phosphor content. At the concentration of lecithin in feed representing phospholipid concentration of 8,45 mg/kg, 8,45 mg/kg, 24,87 mg/kg and 57,58 mg/kg, respectively. Flux profiles confirmed that there was a flux decline during filtration. In addition, the lecithin concentrations do not significantly effect on further flux decline. Rejection characteristic and phospholipid concentration in the permeate showed that the phospholipid rejections by ultrafiltration were in the range of 23-79,5% representing permeate's phospholipid concentration of 1,73 - 44,25 mg/kg. Evaluation of fouling mechanism by Hermia's blocking model confirmed that the standard blocking is the dominant mechanism in the ultrafiltration of lecithin mixture.

1. Introduction

Application of membrane technology for edible oil refining process has been performed and is able to eliminate undesired components [1-4]. Ultrafiltration combined with solvents such as n-hexane has been known as the widest application degumming technology based on membrane method [5-8]. Phospholipid separation from edible oil by ultrafiltration membrane can be performed by one stage of the process through the separation of phospholipid and solvent micelle. The phospholipid forms a micelle having a molecular weight about 20,000 Da and a molecular size approximately of 20-200 nm. The ultrafiltration membrane separates the micelle from the solvent-oil mixture, and the phospholipid is retained by the ultrafiltration membrane [9]. However, the main challenge in the application of membranes especially ultrafiltration is the phenomena called fouling. Fouling is an irreversible membrane change caused by specific physical and / or chemical interactions between the membrane and the various components present in the process flow. Membrane fouling can be identified by the permeate flux decreased due to the effect of blocking on the surface as well as inside the membrane pore [10].

Fundamental studies of fouling mechanisms on the ultrafiltration membranes have been performed, for example for ultrafiltration of coconut cream [11], organic compounds [12], whey models [13] and

Polyethylene Glycol [14]. Study of fouling mechanism on ultrafiltration for oil degumming or separation of oil components are limited only for degumming of crude sunflower and soybean oil [6] and POME [10].
Separation of phospholipid and triacylglycerol could not be performed based on different sizes because they have almost similar size (900 Da). However, due to a specific characteristic of phospholipid as an amphiphilic molecule forming a micelle with a solvent, then it is possible to be separated by the ultrafiltration membrane. The oil components (fatty acids and phospholipids) and the solvents in the degumming process have complex effect in the formation of fouling. Fouling can be formed by fatty acid components of the oil structure and phospholipids. In addition, the micelle in the solvent-oil system has a specific effect on the formation of fouling. The micelles have an affinity with other polar components and the wetting properties (hydrophilicity and hydrophobicity) of the membrane. Furthermore, the formed micelle is also capable of binding other impurities and contributes to the formation of the membrane fouling. The phospholipid amphiphilic molecule can be strongly adsorbed on the membrane and allows changes in the fouling properties of reversible fouling to irreversible fouling. As a consequence, the solvent-oil ultrafiltration phenomenon is complex and have specific properties. This due to the characteristic of separation is determined not only by the particle size of phospholipids and fatty acids alone. Therefore, the analysis of the fouling mechanism in the ultrafiltration membrane for degumming crude palm oil (CPO) should be studied comprehensively. In order to understand the fouling mechanism and dominant foulant, the study was performed by ultrafiltration of CPO model comprising a mixture of Refined CPO-Solvent-Lecithin.

2. Materials and method

2.1. Materials
For preparing feed solution, a mixture of Refined Crude Palm Oil (CPO) from Indofood, isopropanol (Multi Kimia Raya, Semarang) and Lecithin (Indrasari, Semarang) were used. Polyethersulphone (PES), PolyEthylene Glycol (PEG) and N-Methyl Pyrrolidone (NMP) were materials for membrane preparation. The PES material was Veradel3100P from Solvay.

2.2. Methods

2.2.1. Membrane preparations
In this research, flat sheet ultrafiltration membranes were used. The membranes were prepared by a non-solvent induced phase separation method [15]. PolyEthylene Glycol (PEG) was the additive and and N-Methyl Pyrrolidone (NMP) was the solvent.

2.2.2. Ultrafiltration Cell
Experimental works were carried out by a laboratory-scale Ultrafiltration cell. Schematic diagram of the Ultrafiltration Cell is illustrated in Figure 1.

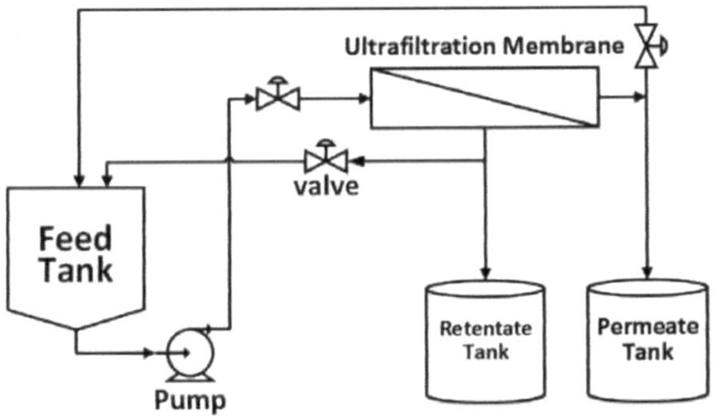

Figure 1. Schematic Illustration of Ultrafiltration Cell System.

All experimental runs were performed at room temperature (29 ± 2°C). In the first step, membranes were compacted by filtering water through the membrane at a pressure of 1 bar for 30 minutes. For every run, a new circular membrane sheet with an area of 13.85 cm^2 was applied. The feed comprising a mixture solution of Refined CPO, isopropanol, and Lecithin at a certain concentration. In this study, the concentrations of Lecithin were varied (0.1%, 0.2% and 0.3% w/w). The ultrafiltration was operated by filtering feed solution at a temperature about 24 - 25°C and 1 bar operating pressure. Permeate fluxes (J) were determined by measuring the volume of permeate collected at 5-minute intervals for 120 minutes and calculated by using equation (1).

$$J = \frac{Q}{(A.t)} \tag{1}$$

where W is representing total weight of permeate, A is the membrane area, and t is a time interval. The filtrations were conducted at total recycle mode where both permeate and retentate were recycled back to the feed tank to maintain same concentration. After each flux determination, the collected permeate was returned to the feed tank. After 120 minute operation, a permeate of 10 ml was collected and analysed.

2.2.3. Membrane Rejection
The filtration efficiency of phospholipid removal from the feed solution was calculated based on the rejection of phospholipid. The rejection (R) was calculated by equation (2).

$$R = \left(1 - \frac{C_p}{C_f}\right) x100 \tag{2}$$

where C_p and C_f are a concentration of phospholipid in the permeate and the feed, respectively. The concentration of phospholipid was determined based on phosphor content by AOAC 1999 method.

2.2.4. Blocking Mechanism
Blocking mechanism of CPO degumming model ultrafiltration was studied based on Hermia's model. This model has been previously applied for fouling mechanism evaluation of konjac glucomannan separation [16], ultrafiltration of batik wastewater [17] and ultrafiltration of dye solution [18]. The

Hermia's model describes the mechanism of membrane fouling based on blocking filtration law, consisting of complete pore blocking, standard pore blocking and intermediate pore blocking and cake filtration. The blocking law filtration is expressed in the term of permeate time and filtration time and developed for dead-end filtration as shown in equation (3) [19]:

$$\frac{d^2t}{dV^2} = k\left(\frac{dt}{dV}\right)^n \tag{3}$$

where t is filtration time, and V is the permeate volume, k is constant, and n is a value illustrating the different fouling mechanism.

The values of n are described as follow: complete blocking having n value of 2, intermediate blocking is represented with n = 1, the standard blocking illustrated with n = 1.5 and the cake layer formation has n value of 0. In the complete blocking model, it is assumed that each solute participated in blocking the entrance of the membrane pores completely. For intermediate blocking, it is assumed that every solute stays on previously deposited solute. Standard blocking considers the deposition of each solute to the internal pore wall. The cake layer formation applied based on the accumulation of the solute on the membrane surface in the cake form [20]. The Hermia's model was then linearized based on the n value for each model using fitting equation (4) to (7) regarding permeate flux versus time as presented in the following.

For Complete Blocking (n = 2) :

$$\ln J = \ln J_0 - k_c t \tag{4}$$

For Intermediate Blocking (n = 1) :

$$\frac{1}{J} = \frac{1}{J_0} + k_i t \tag{5}$$

For Standard Blocking (n = 1.5) :

$$\frac{1}{\sqrt{J}} = \frac{1}{\sqrt{J_0}} + k_s t \tag{6}$$

For Cake/Layer Formation (n = 0) :

$$\frac{1}{J^2} = \frac{1}{J_0^2} + k_{cf} t \tag{7}$$

Where k_c, k_i, k_s and k_{cf} are constants for complete blocking, intermediate blocking and cake/layer formation, respectively.

IOP Conf. Series: Journal of Physics: Conf. Series **909** (2017) 012083 doi:10.1088/1742-6596/909/1/012083

3. Result and discussion

3.1. Effect of lecithin concentration

The performance of the ultrafiltration membrane at various lecithin concentration in the term of flux profile is presented in Figure 2.

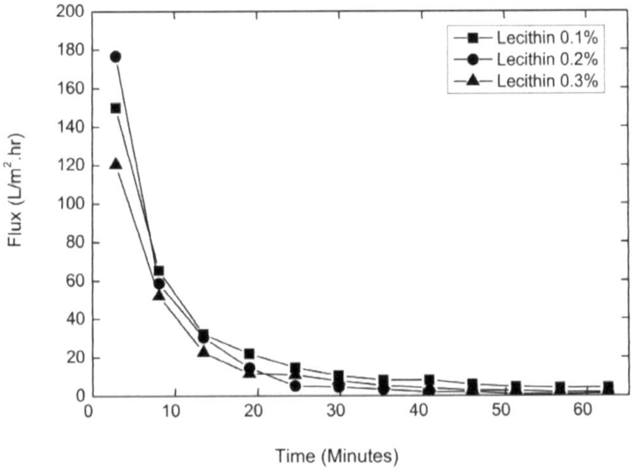

Figure 2. Flux profile during ultrafiltration of Refined Crude Palm Oil-Lecithin-Isopropanol at various Lecithin Concentration

The figure shows flux decline over the time during ultrafiltration then reach the constant flux. The flux decline was more pronounced at the end of filtration than at the beginning of filtration. This was due to gel deposited on the membrane surface [6, 21,22]. The gel layer was formed as a result of concentration polarisation caused by concentration gradient at boundary layer on the membrane surface. The gradient was denoted due to solute accumulation retained by the membrane. The gel layer was a part of the polarisation layer of the concentration from the micelle (phospholipid) and arose when the critical solubility is reached [23]. In addition, the retention of phospholipid agglomeration in the membrane pores had a contribution to the increase of deposited layer on the membrane surface and pore clogging [24].

Scanning Electron Microscope images of fouled membrane as displayed in Figure 3 confirmed that there was a foulant layer on the membrane surface. It was predicted that the foulant layer comprised oil and lecithin (phospholipid). Lecithin is categorised as lipid-containing phosphatidylcholine and other types of lipid. As a phospholipid, lecithin has characteristic of the amphiphilic molecule and hence it was likely to have interaction with the typically hydrophilic Polyethersulphone membrane. Moreover, the phospholipids could form reverse micelles with the isopropanol that contribute to the fouling layer formation [24].

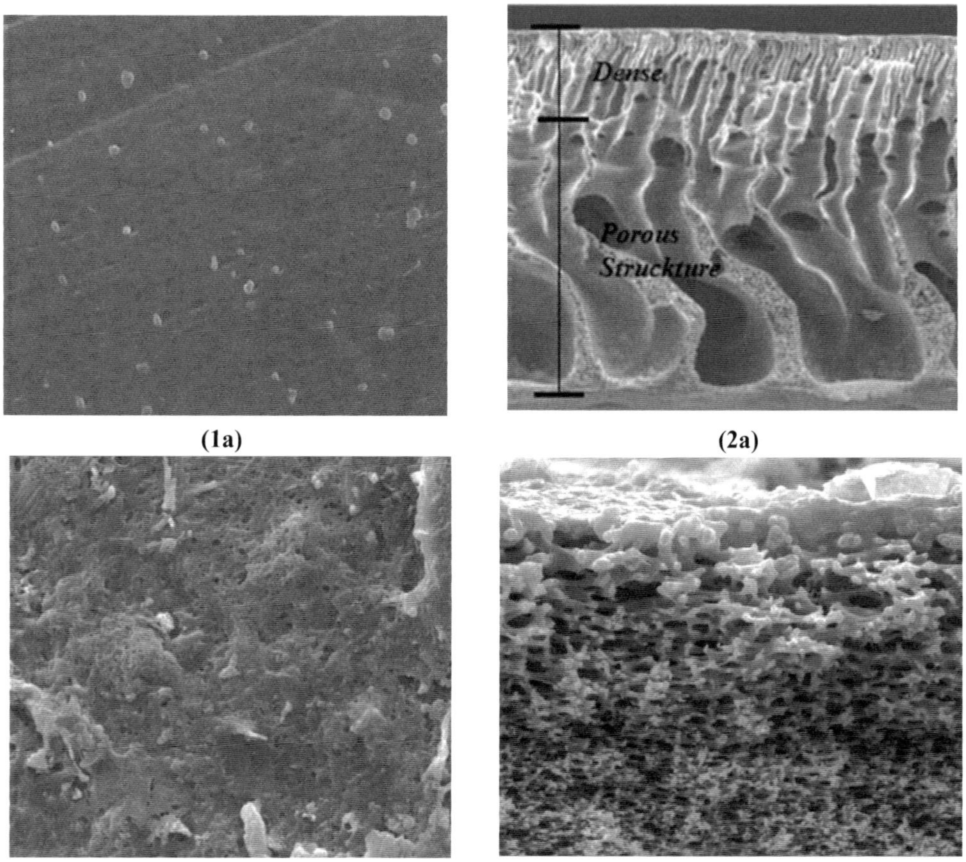

(1b) **(2b)**

Figure 3. Scanning Electron Microscope Images of Clean Membranes (1a) Surface (2a) Cross-Sectional Structure and Fouled Membrane of 0.1% (w/w) Lecithin (1b) Surface (2b) Cross-Sectional Structure at magnification of 10.000x

3.2. Rejection of phospholipid

A rejection indicates membrane selectivity in rejecting or removing a compound in the feed. Phospholipids are compounds that form micelle when dispersed in water. In a non-polar solvent such as isopropanol, the phospholipids forms reverse micelles having the average molecular weight of 20.000 daltons (10-200 nm) [23]. Ultrafiltration can reject compound with a molecular weight in the range of 300-500.000 Dalton. Hence, in this case, the phospholipid was expected could be retained in the retentate and the oil and isopropanol were produced in the permeate. Figure 4a and 4b display rejection of phospholipid and concentration of phospholipid in the permeate, respectively.

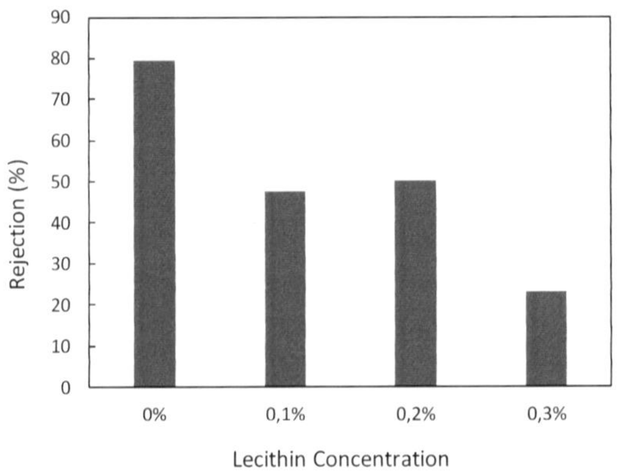

(a)

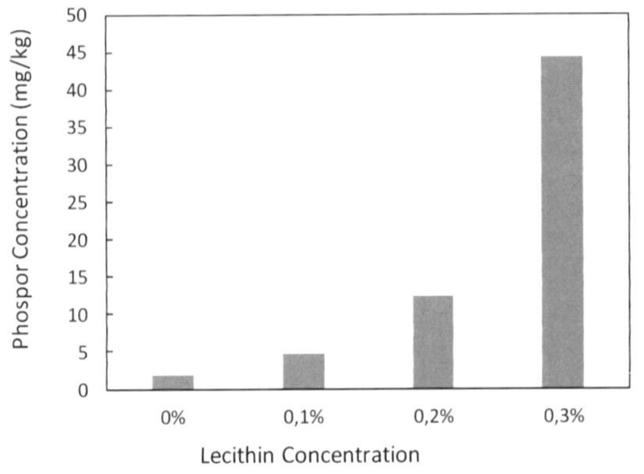

(b)

Figure 4. (a) Rejection of phospolipid (b) Phospor concentration in permeate. Phospor concentrations in the feed were 8,45 mg/kg, 24,87 mg/kg, 45 mg/kg and 57,58 mg/kg for 0%, 0.1%, 0.2% and 0.3% (w/w) lecithin, resepctively

The figure presents that the higher phospholipid concentration, the lower of the rejection. The rejection indicates that performance of the membrane in the term of phospholipid rejection is poor (in the range of 25-80%). Membrane rejection was unpredictable, and the performance depended on the specific solvents. There was the possibility of strong solvent and membrane polymer interaction causing a membrane swelling and affect the membrane pores contraction. Phospholipids have molecular weight approximately 700 Da. When the phospholipids mixed with the solvents, they generated micelles having a size approximately 20.000 Da and hence rejected by the membrane [7].

3.3. Blocking mechanism by Hermia's model

Hermia's model was applied in order to understand the membrane fouling during ultrafiltration of Refined CPO-Isopropanol-Lecithin. By fitting the experimental data into the Hermia's linearized equation (equation (4) to equation (7)), mechanism prevailing the fouling was identified. Figure 5 showed the fitting experimental data to four types of Hermia's model. Table 1 showed all the corresponding correlation coefficients (R^2), and the bold maximal R^2 value indicating the best fitting model.

Table 1. R^2 value of Blocking Mechanism based on Hermia's Model.

	R^2		
	Lecithin 0.1%	Lecithin 0.2 %	Lecithin 0.3%
Complete Blocking	0.9236	0.8891	0.9272
Intermediate Blocking	0.9591	0.9642	0.9934
Standard Blocking	0.9983	0.9909	0.9965
Gel/Cake Formation	0.9945	0.9862	0.9923

According to Table 1, it was confirmed that ultrafiltration of Refined CPO-Isopropanol-Lecithin mixture at all lecithin concentrations was fit with the Standard Blocking model. In this model, it was predicted that there was deposition of solutes to the internal pore wall. This finding indicated that solutes were penetrating inside the membrane pore and deposited on the internal wall. When the solutes were deposited inside the pores, the size of the solutes was smaller than the membrane pores. It was predicted that the droplet of CPO plays a role in standard blocking. Moreover, it was possible that the PES membrane have hydrophobic characteristic and hence the standard blocking was taken place

[25]. Comparing to Figure 3 illustrating SEM of membrane surface shows that there was foulant deposited on the membrane surface. It was analysed that the standard blocking mechanism occurred in the initial filtration. Further, at the end of filtration, the mechanism was cake/gel formation [25].

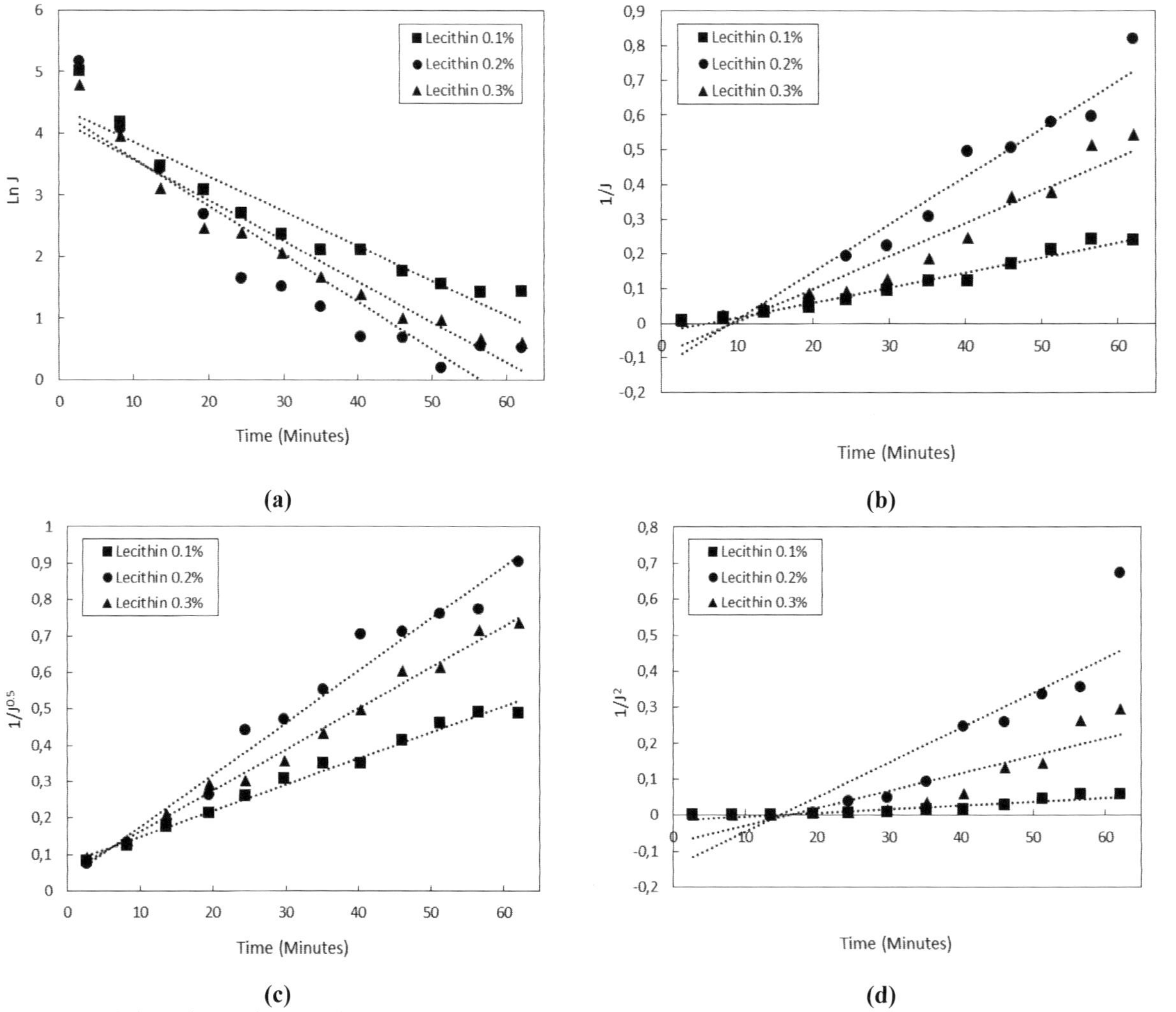

Figure 5. Fitting of experimental data to Hermia's model: (a) complete blocking (b) intermediate blocking (c) standard blocking (d) cake/gel layer formation

4. Conclusion

An ultrafiltration cell with flat sheet ultrafiltration membrane was performed for phospholipids separation of the Refined CPO-Isopropanol-Lecithin mixture. The ultrafiltration membranes were laboratory-made membrane produced through a non-solvent induced phase separation method. The membranes were prepared based on Polyethersulphone (PES) material with the N-Methyl Pyrrolidone (NMP) and Poly Ethylene Glycol (PEG) as the solvent and additive, respectively. Feed consisting of Refined CPO – Isopropanol – Lecithin mixtures were represented as crude palm oil degumming. Lecithin was denoted a phospholipid component, and the concentrations of lecithin in feed were varied to 0.1%, 0.2%, and 0.3%. The concentration of phospholipid was determined as phosphor content. At the concentration of lecithin in feed representing phospholipid concentration of 8,45 mg/kg, 8,45 mg/kg, 24,87 mg/kg and 57,58 mg/kg, respectively. Analysis of blocking mechanism was studied based on Hermia's blocking model. Flux profiles confirmed that there was a flux decline during filtration. In addition, the lecithin concentrations do not significantly effect on further flux decline. Rejection characteristic and phospholipid concentration in the permeate showed that the phospholipid rejections by ultrafiltration were in the range of 23-79,5% representing permeate's phospholipid concentration of 1,73 - 44,25 mg/kg. Analysis of fouling mechanism based on the highest R^2 value confirms that the standard blocking is the dominant mechanism in the ultrafiltration of lecithin mixture.

Acknowledgments
Authors acknowledge to Directorate of Research and Community Service, Ministry of Research and Higher Technology, the Republic of Indonesia for the financial support. The research was funded by Fundamental Research Grant 2017.

References
[1] de Morais Coutinho C, Chiu M C, Basso R C, Ribeiro A PB, Gonçalves L A G and Viotto L A 2009 *Food Res. Int.* **42(5)** 536-50
[2] Subramanian R, Nakajima M, Raghavarao K S, and Kimura T 2004 *J. Am. Oil Chem. Soc.* **81(4)** 313-22
[3] Niazmand R, Farhoosh R, Razavi S M, Mousavi S M and Noghabi M S 2011 *Procedia Food Sci.* **1** 90-94
[4] Arora S, Manjula S, Krishna A G and Subramanian R 2006 *Desalination* **191(1-3)** 454 -66
[5] Ochoa N, Pagliero C, Marchese J and Mattea M 2001 *Sep. Purif. Technol.* **22** 417-22
[6] Pagliero C, Mattea M, Ochoa N and Marchese J 2007 *J. Food Eng.* **78(1)** 194-97
[7] García A, Álvarez S, Riera F, Álvarez R and Coca J 2006 *J. Food Eng.* **74(4)** 516-22
[8] Ribeiro A P, Bei N, Gonçalves L A, Petrus J C and Viotto L A 2008 *J. Food Eng.* **87(4)** 514-21
[9] Cheryan M 2005 *Memb. Technol.* **2005(2)** 5-7
[10] Said M, Ahmad A, Mohammad A W, Nor M T and Abdullah S R 2015 *J. Ind. Eng. Chem.* **21** 182-88
[11] Ng C Y, Mohammad A W, Ng L Y and Jahim J M 2014 *J. Food Eng.* **142** 190-00
[12] Guo X, Gao W, Li J and Hu W 2009 *Water Environ. Res.* **81(6)** 626-32
[13] Corbatón-Báguena M, Álvarez-Blanco S and Vincent-Vela M 2015 *Desalination* **360** 87-96
[14] Vela M C, Blanco S Á, García J L and Rodríguez E B 2009 *Chem. Eng. J.* **149(1)** 232-41
[15] Aryanti N, Sandria F K I, Putriadi R H and Wardhani D H 2017 *Eng. J* **21(3)** 23-35
[16] Aryanti N, Wardhani D H and Supandi S 2016 *Sci. Study Res. Chem. Chem. Eng. Biotechnol. Food Ind.* **17(2)** 125-37
[17] Aryanti N, Pramudono B, Prawira C N P, Rhenardi R and Sandria F K I 2015 *AIP Conf. Proceed.* **1699** 060002-1-9
[18] Aryanti N, Sandria F K and Wardhani D H 2017 *Adv. Sci. Lett.* **23(3)** 2598-600
[19] Hermia, J 1982 *Trans. Ind. Chem. Eng.* **60** 183-87
[20] Amin I N H M, Mohammad A W, Markom M, Peng L C and Hilal N 2010 *Desalination* **261** 313-20
[21] Penha F M, Rezzadori K, Proner M C, Zin G, Fogaça L A, Petrus J C, de Oliveira J V and Di Luccio M 2015 *J. Food Eng.* **155** 79-86
[22] Kim I C, Kim J H, Lee K H and Tak T M 2002 *J. Memb. Sci.* **205(1)** 113-23
[23] Pagliero C, Ochoa N, Marchese J and Mattea M 2001 *J. Am. Oil Chem. Soc.* **78(8)** 793-96
[24] Sehn G A, Gonçalves L A and Ming C C 2016 *Grasas y Aceites* **67(1)** 120-27
[25] Mah S K, Chuah C K, Lee W C and Chai S P 2012 *Sep. Purif. Technol.* **98** 419-31

Microencapsulation of Natural Anthocyanin from Purple Rosella Calyces by Freeze Drying

A Nafiunisa, N Aryanti[1], D H Wardhani, and A C Kumoro

Chemical Engineering Department, Diponegoro University, Semarang, INDONESIA

E-mail: nita.aryanti@che.undip.ac.id

Abstract. Anthocyanin extract in powder form will improve its use since the powder is easier to store and more applicable. Microencapsulation method is introduced as an efficient way for protecting pigment such as anthocyanin. This research was aimed to characterise anthocyanin encapsulated products prepared from purple Roselle calyces by freeze drying. The liquid anthocyanin extracts from ultrasound-assisted extraction were freeze-dried with and without the addition of 10% w/w maltodextrins as a carrier and coating agents. The quality attributes of the powders were characterised by their colour intensity, water content, and solubility. Analysis of encapsulated material was performed for the powder added by maltodextrin. The stability of the microencapsulated pigment in solution form was determined for 11 days. Total anthocyanin content was observed through pH differential method. The results of the colour intensity analysis confirm that the product with maltodextrin addition has more intense colour with L* value of 29.69 a* value of 54.29 and b* value of 8.39. The result with the addition of maltodextrin has less moisture content and more soluble in water. It is verified that better results were obtained for powder with maltodextrin addition. Anthocyanin in the powder form with maltodextrin addition exhibits higher stability even after 11 days. In conclusion, the microencapsulation of anthocyanin with maltodextrin as a carrier and coating agent presented a potential method to produce anthocyanin powder from purple Roselle.

1. Introduction

Anthocyanins are easily found in many kinds of fruits, vegetables, flower, or leaves. Roselle flower is a source of anthocyanin that is easy to find in Indonesia; it could be used as a good source for producing red colourant for many foods. Anthocyanin has broad colour scheme from bright red, purple, red, dark red and dark blue. It is well known as harmless water-soluble pigments that provide red to the purple shade which makes them an important group of natural colourants [1]. Most colours used in the food industry are artificial from synthetic sources. Due to their harmful effects, there is currently an increasing trend towards applying natural dyes [2]. Anthocyanin also has side properties such as antimicrobial and anticancer that make it suitable as an alternative for synthetic dyes [3],[4].

Produce high-quality food colourant from the natural resource such as Roselle flower as an alternative for synthetic colourant is a new scientific challenge. The stability of anthocyanin is very susceptible to colour degradation. The degradation depends on various factors including the structure of anthocyanin, pH, temperature, oxygen, light, concentration, ascorbic acid, sugar, and water activity [5],[6]. The degradation started from the extraction process or storage [7],[8],[9].

Microencapsulation was offered as an efficient way to introduce natural compounds into food products. In microencapsulation process, anthocyanin as the core material is coated with thin film of wall materials that protect it from environmental disturbance. Encapsulated colour is easier to handle and offer better stability. Microencapsulation with the addition of coating materials can be done with various techniques, include spray-drying [10],[11],[12], freeze-drying [3],[13],[14],[15], microwave-assisted technique [16]. Freeze drying is considered as the most suitable method to obtain high-quality dry pigment product, compared to other methods [13]. Various materials are used as coatings agent such as maltodextrin, Arabic gum, disaccharides, starches, chitosan-cellulose and many others [4],[13],[17], of which maltodextrin gave better physicochemical properties as colourant powder. The previous study has considering maltodextrin as an appropriate coating agent for food colourant product [5],[18],[19]. However, specific investigation of microencapsulation product of purple Roselle is limited.

The objective of the present study was to characterise anthocyanin encapsulated products prepared from purple Roselle calyces by freeze drying. Compare it with the character of non-encapsulated powder. Specifically, the effect of maltodextrin addition on the product characteristic was addressed. The analysis of powder stability in solution form was also conducted to predict its behaviour when applied as a colourant.

2. Materials and Methods

The study is consist of several phases which is anthocyanin extraction from Roselle calyxes, preparation and microencapsulation of the liquid extract with maltodextrin using Freeze-dryer, analysis of encapsulation efficiency, physical properties characterization of the powders, and stability analysis.

2.1. Anthocyanin Extraction

Dry purple Roselle flower petals were bought from the Roselle garden in Blitar, East Java, Indonesia. Dried calyxes were ground and sieved until 100 mesh in size then stored in the dry container. Distilled water as a solvent was supplied from Indrasari Chemical Store, Semarang. The ratio of solute and solvent are 1:15 w/w. The anthocyanin from the Roselle calyxes was extracted by ultrasonic assisted extraction method, conducted at a frequency of 40 kHz for 30 minutes at room temperature. After extraction, the liquid extract was then filtered through a filter paper and stored in the refrigerator for not more than 6 hours.

2.2. Microencapsulation of Powders

Maltodextrin obtained from Multikimia Raya Chemical Store, Semarang, were used as wall material. Maltodextrin was dissolved in liquid anthocyanin extract at ambient temperature (25°C) to obtain 10% total solid concentration. The solutions were frozen in the freezer for 24 hours before drying by freeze dryer. The solutions were dried in a freeze dryer (*Heto Powerdry* LL 1500) for the next 24 hours (minus 100°C vacuum pressure). Dried materials were ground using a pestle and mortar stored in dark container in a freezer (−18°C) until usage. A blank sample was also prepared by the liquid extract without any addition of wall materials and then freeze-dried for 24 hours in similar conditions (-100°C vacuum pressure).

2.3. Materials Encapsulation Efficiency

The efficiency of microencapsulating process was determined by the Equation 1 [20],

$$\%\ EE = \frac{Total\ Mass - Surface\ Mass}{Total\ Mass} \times 100\% \qquad (1)$$

Where total mass is the total anthocyanin in sample and surface mass is the mass of anthocyanin in powder surface. Total mass of anthocyanin in the sample was obtained by dissolving 0.1 grammes of microencapsulated powder in 1 mL of aquadest. Ethanol anhydrate was added to extract it for 5 minutes before filtration [16]. Surface mass was calculated by washing 0.1 gramme of sample in 45 mL of ethanol anhydrate. After filtration, one mL aquadest was added into the filtrate and calculate the mass of the surface materials.

As for process yield, the equation state by Battista (2015) was used.

$$\% \text{ EY} = \frac{total\ antocyanins\ in\ solid\ materials}{total\ antocyanins\ extracted} \times 100\% \quad (2)$$

Both total anthocyanin extracted and total anthocyanin in solid materials was analysed by spectrophotometer using pH methods [21]. To find the amount of total anthocyanin Equation 3 was used.

$$\text{Total Antosianin (mg/L)} = \frac{\text{Absorbansi} \times \text{DF} \times 1000}{55{,}9 \times 1} \quad (3)$$

2.4. *Physical Properties Characterization of Powders*

Anthocyanin powder with and without wall material is characterised by its physical properties (moisture content, solubility, and colour intensity) then to be compared. The moisture content of the powders was determined by AOAC (1995) oven method, and the solubility analysis was done by Kainuma method [22].

To investigate the effect of wall material addition on the colour of encapsulated powders, colour intensity analysis was conducted for both samples [23]. Colour intensity was analysed by Chromameter (Color Reader CR-400/410) based on Hunter's Lab Colorimetric System. L* represent lightness (100=white, 0=black), a* represent the color of red (+) until green (-), b* represent the color of yellow (+) until blue (-). The other parameter considered as Chroma, Hue, and ΔE^*_{ab}, and it can be calculated by the equation,

$$H_{ab} = tan^{-1}\frac{b^*}{a^*} \quad (4)$$

$$C_{ab} = \left(a^{*2} + b^{*2}\right)^{\frac{1}{2}} \quad (5)$$

$$\Delta E^*_{ab} = \left(\Delta L^{*2} + \Delta a^{*2} + \Delta b^{*2}\right)^{\frac{1}{2}} \quad (6)$$

By taking the value of L*, a*, and b* the colour can be visualised using Colour Converter from Nixsensor. The value will take a calculation to convert it into RGB (Red Green Blue), and the colour can be visualised to see with bare eyes.

2.5. *Stability analysis*

The stability analysis was done for the solution state of the powder. The powder sample is dissolved into aquadest by 10% of volume. The solution is stored in the closed dark bottle in the same room condition (25-30°C, 1 atm) for 11 days. Total anthocyanin is analysed for a period of time by pH methods using Spectrophotometry UV-Vis (UV Mini 1240 Shimadzu) [21]. KCl buffer solution with pH 1.0 and Na-Acetate buffer solution with pH 4.5 was used to dissolve the Roselle extract. The absorbance test was done at λvis-max and 700 nm wavelength from each solution. The visible-maximum wavelength for cyanidin-3-glucoside is 510 nm. The concentration (mg/l) anthocyanins were calculated according to the following Equation 7:

$$\left[(A_{510} - A_{700})_{pH\ 1,0} - (A_{510} - A_{700})_{pH\ 4,5}\right].MW.\frac{1000}{\varepsilon . l} \quad (7)$$

Where A is the absorbance at each wavelength and pH, MW is the molecule weight (449.2 g/mol for cyanidin-3-glucoside, ε is the extinction coefficient (26900), and l is the path length of the cuvette in cm [24].

3. Results and Discussion

3.1. Encapsulated materials efficiency

Maltodextrin was used as wall material for anthocyanin extract from Roselle calyxes. By adding maltodextrin, it is expected that the pigment material will be protected from damage. Therefore the effectiveness of the micro-encapsulation process needs to be calculated. Encapsulation materials efficiency analysis shows the amount of anthocyanin encapsulated in the powder [11],[17],[20]. Encapsulation methods were said to be successful if it can achieve high retention of the core materials [4]. The physical characteristic of encapsulated materials should be better than the non-encapsulated ones.

The encapsulated powder with 10% of maltodextrin was analysed for its encapsulated materials efficiency and encapsulation process yield. From the calculation, we get that the encapsulated materials efficiency is 33.3% and encapsulation process yield of 56.9%. Showing that there are certainly anthocyanin materials encapsulated in the wall material. The previous research indicates that the addition of wall material will increase its encapsulation efficiency in microencapsulation of procyanidin [15]. Wall materials like Arabic gum and maltodextrin also help to gives high encapsulation efficiency in microencapsulation of phytosterols by spray drying [11]. Other previous study states that encapsulation efficiency will be increased if greater carrier concentration was added, giving additional protection [18],[19].

3.2. The effect of maltodextrin to the water content and solubility of encapsulated anthocyanin

The correlation between the addition of maltodextrin with solubility and water content in the encapsulation material can be seen in the Table 1.

Table 1. The Properties of Freeze-dried Pigment Powder

No	Powder	Properties	
		Water content	Solubility
1.	Rosella Powder with addition of 10% maltodextrin	11.40%	24%
2.	Rosella Powder without addition of maltodextrin	18.20%	15%

Table 1 shows that powder with addition of maltodextrin has lower water content than the powder without any addition of maltodextrin. A prominent characteristic of freeze drying product is that it is very hygroscopic. The powder tends to absorb water from the environment either through absorption or adsorption [25]. This character is the result of drying methods in freeze-drying where the process of water content removal uses the subliming principle. The water in solid form is converted to gas at a vacuum pressure condition [26]. Microencapsulation process can protect sensitive materials such as pigment from moisture, heat, light or oxidation [27]. This is the reason why the encapsulated powder has lower water content. Maltodextrin has been used as wall materials in microencapsulation of anthocyanin from purple corn [8], blackberry by-product [14], saffron petal [28], Jaboticaba pomace [29], and encapsulation of phenolic from grape skin [18].

The powder used as wall materials must exhibit good solubility which is a device factor for the quality of the product. The solubility in water at 25°C was showing that the encapsulate powder has a higher solubility than non-encapsulate powder. This behaviour is because anthocyanin in the extract will show interaction with the carriers during the encapsulation process. The solubility of a freeze drying material is strongly influenced by the wall material. Maltodextrin is a polysaccharide with high polarity, so it is soluble in the water [29]. The interactions between anthocyanins and their carrier make anthocyanins powder also have the same properties as maltodextrin [30].

3.3. The effect of maltodextrin to the colour intensity of encapsulated anthocyanin powder

After freeze-drying, colour intensity of the samples were characterised by using Chromameter. The value of L*, a*, b*, C_{ab}, H_{ab}, and ΔE^*_{ab} from both powder sample are represented in Table 2.

Table 2. L*, a*, b*, C_{ab}, and H_{ab} value of microencapsulated powders

Parameter Value	Powder	
	10% maltodextrin	without maltodextrin
L*	29.696	23.121
a*	54.296	47.879
b*	8.393	12.999
C_{ab}	54.941	49.612
H_{ab}	0.153	0.265

A high value of parameter a* is due to a significant amount of anthocyanin in the extract of Roselle calyxes [3]. Table 2 also shows a higher value of colour parameter a* in encapsulated powder than non-encapsulated powder. This result shows that encapsulated powder is redder than the other one. Maltodextrin as wall materials gives protection to the anthocyanin encapsulated in it. The previous study shows that maltodextrin has great protecting effect for anthocyanin stability [19] and reduce degradation of anthocyanin rather than naked anthocyanin in solution state [16]. The L* notation of the powder shows that powder encapsulated in maltodextrin is lighter than non-encapsulated powder. Maltodextrin is white flourish materials which have L* value of 98.18±0.15 with a really low value of a* and b* [32]. Since the anthocyanin is encapsulated inside the maltodextrin, it is natural that the colour of the powder is slightly lighter.

Chroma, representing colour intensity, which is the distance of a colour from the origin (a*= b*=0) in the a*, b* plane. Hue angle expresses in degrees form (0° to 360°), 0° being a location on the +a* axis which represent red colour. Then the value rotates anti-clockwise to 90° for the +b* axis to indicates yellowness, axis, 180° (green) for -a*, 270° (blue) for -b*, and back to 360° = 0° [12]. Table 2 indicates that the chroma value of encapsulated powder is greater than non-encapsulate powder. On the other hand, the colour of powder is more vivid. As for H_{ab} parameter value shows that the Hue is not significantly different. The H_{ab} values are less than 10° which indicates that the powder is considered red colour [33]. CIE total colour difference between the two powders was also calculated to have the number of the difference value. The calculation gives a result of 10,277 ΔE^*_{ab} this value is considered to be significantly different in colour [23].

The visualisation of the colour was considered by the colour converter from NixColorSensor. The value of L*, a*, and b* are recalculated to obtain the value of RGB (Red Green Blue). The visualisation of colour using RGB value is easy to find by many graphic programs such as CorelDraw, Photoshop, and many others. The value of RGB for the encapsulated powder is 141 0 60, as for non-encapsulated powder are 116 0 39. The visual colour presented in Figure 1

.a. b.

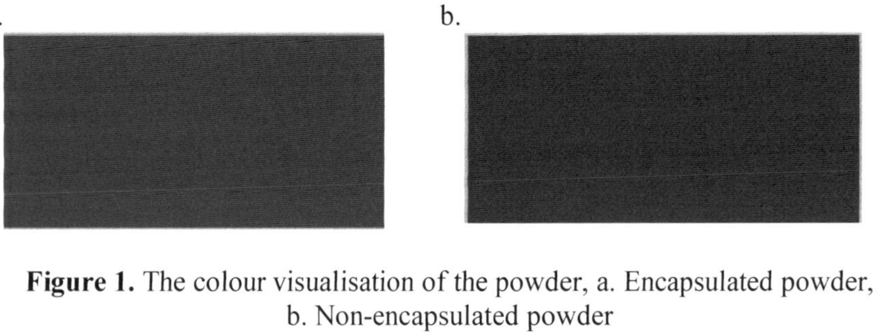

Figure 1. The colour visualisation of the powder, a. Encapsulated powder, b. Non-encapsulated powder

3.4. The effect of maltodextrin to the stability of encapsulated anthocyanin powder

The anthocyanin powder obtained from encapsulation process was dissolved in aquadest to obtained anthocyanin solution. The solution was left for 12 days and being observed in a certain time. The results were provided in Figure 2.

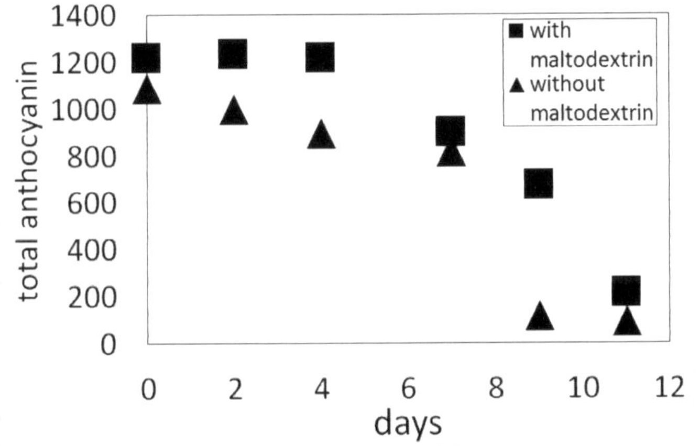

Figure 2. The relation between storage time and total anthocyanin

Figure 2 shows that anthocyanin solution obtained from powder without maltodextrin have fewer anthocyanin from the very beginning. The amount of total anthocyanin from both samples are getting lower every single day. The solution from encapsulated powder shows a stable trend of total anthocyanins degradation each day. In contrast, the solution from non-encapsulated powder shows a dramatic decline after nine days. After being stored for 11 days, both samples showed a significant decrease in a total of anthocyanins. The solution from non-encapsulated powder exhibits 91.61% of anthocyanin lost from day one until day 11, While the solution from encapsulated powder showed a fewer decrease of 82.53%. This result indicates that addition of maltodextrin as wall materials affected the stability of anthocyanin in dissolved powder as the encapsulated anthocyanin gives more stable condition compared to non-encapsulated anthocyanin.

Degradation of anthocyanin pigment causes the phenomena of decline in the total anthocyanin content. Degradation often occurs in almost all kind of pigment, such as anthocyanin. The previous research shows that degradation of pigment also happens in red pigment from Roselle [5],[34], anthocyanins from red onion [35], and colour degradation in vitamin C fortified cranberry juice [36]. The degradation often occurs starts from the extraction and processing of anthocyanin was done [9],[35]. Maltodextrin as the wall materials could indicate to have the capability to protect the structure and the function of molecules activity thru the drying and storing process [37]. That can be the reason why total anthocyanin of solution from non-encapsulated powder is fewer than the encapsulated one from the beginning. In solution state, maltodextrin reduces the degradation of anthocyanin rather than the non-encapsulate one. The same result also shows in previous research in Roselle extract colourant in a drink [23].

4. Conclusion

The addition of maltodextrin in encapsulation process provided better results than encapsulation process without maltodextrin. The encapsulated anthocyanin powder shows better physical properties with lower water content and higher solubility. It also gives more vivid red colour with higher Chroma value. The stability analysis also shows that encapsulated anthocyanin powder is more stable in solution form, indicate that the powder has better capability to apply as a food colourant. Finally, the study shows that microencapsulation of anthocyanin with maltodextrin as a carrier and coating agent presented a potential method to produce anthocyanin powder from purple Roselle.

Acknowledgments
Authors acknowledge to Directorate of Research and Community Service, Ministry of Research and Higher Technology, the Republic of Indonesia for the financial support. The research was funded by Higher Education Excellence Research Grant 2017 (344.74/UN7.5.1/PP/2017).

References
[1] Osorio C, Acevedo B, Hillebrand S, Carriazo J, Winterhalter P and Morales A L 2010 *J. Agric. Food Chem* **58** 6977-6985
[2] Downham A and Collins P 2000 *International Journal of Food Sciences and Technology* ***35*** 5-22.
[3] Jafari S-M, Mahdavi-Khazaei K and Hemmati-Kakhki A 2015 *Carbohydrate Polymers* 1-18
[4] Mahdavi S A, Jafari S M, Assadpour E and Ghorbani M 2016 *Journal of Food Engineering* **181** 59-66
[5] Selim K A, Khalil K E, Abdel-Bary M S and Abdel -Azeim N A 2008 *Alexandria Journal of Food Science and Technology* **Special Volume** 7-20
[6] Tsai P and Huang H 2004 *Food Research International* **37** 313-318
[7] Clifford M N 2000 *J. Sci. Food Agric.* **80** 1063-1072
[8] Lao F and Giusti M 2017 *Food Chemistry* **227** 376-382
[9] Patras A, Brunton N P, O'Donnell C and Tiwari B K 2010 *Trends in Food Science & Technology* **21** 3-11
[10] Araujo-Diaz S B, Leyva-Porras C, Aguirre-Banuelos P and Álvarez-Salas C 2017 *Carbohydrate Polymers* **167** 317-325
[11] Battista C A D, Constenla D, Ramírez-Rigo M V and Piña J 2015 *Powder Technology* **286** 193-201
[12] Ersus S and Yurdagel U 2007 *Journal of Food Engineering* **80** 805–812
[13] Lončarić A, Pichler A, Trtinjak I, Pilizota V and Kopjar M 2016 *LWT - Food Science and Technology* **73** 391-396
[14] Yamashita C, Chung M M S, Santos C d, Roberta C, Mayer M, Moraes I C F M and Branco G 2017 *LWT - Food Science and Technology* 1-27
[15] Zhang L, Mou D and Du Y 2007 *J. Sci. Food Agric.* **87** 2192-2197
[16] Zaidel D N A, Sahat N S, Jusoh Y M M and Muhammad I I 2014 *Agriculture and Agricultural Science Procedia* **2** 82–89
[17] Wang W, Jung J and Zhao Y 2016 *Carbohydrate Polymers* 1-30
[18] Lavelli V, Harsha S C P S and Spigno G 2016 Food *Chemistry* 1-34
[19] Pai D A, Vangala V R, Ng J W, Ng W K and Tan R B H 2015 *Journal of Food Engineering* **161** 68-74
[20] Fang Z and Bandhari B 2011 *Food Chemistry* **129** 1139-1147
[21] Giusti and Wrolstad, 2001 Giusti and Wrolstad, 2011
[22] Daramola B and Osanyinlusi S A 2006 *African Journal of Biotechnology* **5** 917-920.
[23] Duangmal K, Saicheuaa B, and Sueeprasan S 2008 *LWT - Food Science and Technology* **41** 1437-1445
[24] Berg S, Bretz M, Hubberman E M and Schwarz K 2012 *Journal of Food Engineering* **108** 158-165
[25] Mosquera L H, Moraga G and Martínez-Navarrete N 2012 *Food Research International* **47** 201-206
[26] Oberoi, D. P. S., and Sogi, D. S., (2015 *Journal of Food Engineering* **165** 172-178
[27] Jafari S M, Assadpoor E, He Y and Bhandari B 2008 *Drying Technology* **26** 816-835
[28] Khazaei K M, Jafari S M, Ghorbani M and Kakhki A H 2014 *Carbohydrate Polymers* **105** 57-62
[29] Souza A C P, Gurak P D and Marczak L D F 2017 *Food and Bioproducts Processing* **102**

186–194

[30] Zhang H, Yu D, Sun J, Liu X, Jiang L, Guo H and Ren F 2014 *Nutr. Res. Rev.* **29** 1-15

[31] Caliskan G and Dirim S N 2016 *Powder Technology* **287** 308-314.

[32] Jimenez-Agular D, Ortega-Regules A, Lozada-Rambrez J, Peirez-Peirez M, Vernon-Carter E and Welti-Chanes J 2011 *Journal of Food Composition and Analysis* **24** 889-894

[33] Aurelio D-L, Edgardo R G and Navarro-Galindo S 2008 *International Journal of Food Science and Technology* **43** 322–325

[34] Petersson E V, Liu J, Sjöberg P J R, Danielsson R and Turner C 2010 *Analytica Chimica Acta* **663** 27-32

[35] Roidoung S, Dolan K D and Siddiq M 2017 *Food Research International* 1-32

[36] Balasubramani P, Palaniswamy P T, Visvanathan R, Thirupathi V, Subbarayan A and Prakash-Maran J 2015 Int. J. Biol. Macromol. **72** 210–217

The effect of combination of sugar palm fruit, carrageenan,and citric acid on mechanical properties of biodegradable film

S A Rinanda, M Nastabiq,S H Raharjo, S K Hayati, M A Yaqin, and Ratnawati

Chemical Engineering Department, Faculty of Engineering, Diponegoro University, Jl. Prof. Soedarto Semarang 50239, INDONESIA

Email: rinandasafiraaulia@gmail.com

Abstract. Biodegradable film is a type of plastic material that can be degraded naturally and is usually made of organic material. The material commonly used is polysaccharides. The purpose of this study is to observe the effect of the combination of sugar palm fruit, carrageenan, and citric acid (CA) on the mechanical properties of the biodegradable films, such as tensile strength, elongation and film thickness. The experiment begins with dissolving the sugar palm fruit porridge and carrageenan with ratios of 1:0, 3:1, 2:1, 1:1 in water. The mixture was heated using a heater and magnetic stirrer at 80° C for 10 minutes. Glycerol and citric acid (CA) were added to the solution and stirred for 5 minutes. Each film solution was printed on a modified acrylic and, dried for 18 hours in an oven at 55° C. The formed film layer was then removed from the acrylic mold and inserted in a desiccatorsat 23° C for 1 hour. Then the film analyzed for its tensile strength, elongation using Dynamic Mechanical Thermal Analysis (DMTA), and thickness. The optimum result shown by sugar palm fruit and carrageenan ratio of 1:1 with 1% citric acid (CA).

1. Introduction

Biodegradable film is a type of plastic material that can be degraded naturally and usually made of organic material. In previous research, a bio plastic made of nata de cassava has a long stage because through the fermentation process to make nata with conditions that must be sterile because it involves bacteria [1]. Further research biofilm produced by adding lactate acid is not commonly used because of its expensive price [2]. Therefore, this research is a middle way that can utilize sugar palm fruit the availability is abundant and easy to obtain and use, carrageenan as a thickener, glycerol as a plasticizer which is easier and cheaper than collagen and addition of citric acid as a cross linker. The components of biodegradable film are sugar palm fruit and carrageenan as the main material, glycerol as a plasticizer and citric acid as a cross linker. The resulted films are expected to have better mechanical properties, such as tensile strength, elongation, and thickness.

2. Component of biodegradable film

The main components of the biodegradable film are grouped into three classes, namely hydrocolloids, lipids, and composites. Hydrocolloid compounds are very numerous in nature, one of which is the galactomannan polysaccharides commonly found in sugar palm fruit. The most often used hydrocolloid in biodegradable films are carbohydrates, especially starch. Sugar palm fruit has a high content of carbohydrate, so it is a potential raw material in making a biodegradable film. The carbohydrates are mainly starch and galactomannan. The weakness of biodegradable film from carbohydrates is that it is easily torned, so it is necessary to add the plasticizer to strengthen the film produced. Also, it also needs the addition of cross linker, such as citric acid that forms a bond that inhibits recrystallization and retrogradation and increases the tensile strength of the film.

2.1. Sugar palm fruit
Sugar palm fruit has a high content of starch and crude fibber. In addition, the galactomannan polysaccharide compounds are also present. This carbohydrate can be utilized for raw materials in making a biodegradable film.

2.2. Carrageenan
Carrageenan is a polysaccharide extracted from several species of seaweed or red algae (*Rhodophyceae*). These natural polymers have the ability to form thermo-reversible gels or thickened solutions when they are added the salt solution, so they are widely used as gelling, thickening and stabilizers in various industries such as food, pharmaceuticals, cosmetics, printing, and textiles [3]. Carrageenan can also be used as a foodstuff coating or an edible film-forming agent [4].

2.3. Glycerol as plasticizer
The plasticizer is an additive added to the polymer to improve its performance by increasing ionic mobility and interaction between ionic and polar groups in the polymer chain [5]. One type of plasticizer used is glycerol. The addition of glycerol as plasticizer will reduce the density and force between substrate molecules (starch) with glycerol so that the thin layer formed is more flexible and smooth.

2.4. Citric Acid
Citric acid is a weak organic acid found in certain plant leaves and fruits. The chemical formula of citric acid is $C_6H_8O_7$. Citric acid (CA) has three carboxyl groups which will bind to the hydroxyl group (-OH) of starch and form an ester group. The bond inhibits recrystallization and retrogradation of the starch and increases the tensile strength of the formed film.

3. Materials and methods
The research was conducted in Instrumentation Laboratory, Chemical Engineering Operations Laboratory, Chemical Engineering Department, Diponegoro University Semarang.

3.1. Materials
The materials for this research were sugar palm fruit obtained from vegetable traders around Tembalang, carrageenan from CV. Karagenan Indonesia, while aquadest, glycerol, and citric acid from chemical store Indrasari Semarang.

3.2. Equipment
The equipment used in this work were basins, thermometers, ovens, acrylic plate, stopwatch, analytical scales (4 digits behind the comma), beaker glass, funnel, heater, and magnetic stirrer.

3.2.1. Fixed variables. Glycerol concentration: 40% (v/w) glycerol / sugar palm fruit – carrageenan and type of solvent: aquadest (total weight 125 gr)

3.2.2. Free variables. Mass ratio of sugar palm fruit to carrageenan for total weight of 5 gr: 1:0; 1:1; 2:1; 3:1 and citric acid concentration (w/w) (mass of citric acid per overall mass): 1% ; 3% ; 5%

3.3. Methods

3.3.1 Biodegradable film making. The biodegradable film was made according to modified methods of Adzaly [6]. First, a pulp of sugar palm fruit was made by cutting the fruit to small pieces to facilitate its destruction by using a blender. Then sugar palm fruit was blended by adding aquadest with a ratio of 1: 1. A solution of sugar palm fruit-carrageenan was made by dissolving the pulp of sugar palm fruit and carrageenan in aquadest according to variable ratios in a measuring glass up to 125 ml. The mixture was heated using a heater and magnetic stirrer at 80°C for 10 minutes. Glycerol and citric acid were added to the solution and stirred for 5 minutes. Each film solution was caste on an acrylic plate, and dried for 18 hours in an oven at 55°C. The formed film layer was removed from the acrylic plate and inserted in the desiccator and kept at 23°C for 1 hour. The films were analyzed for thickness, tensile strength, and elongation.

3.3.2 The characterizations of a film. First, thickness analysis. It was conducted to determine the thickness of the film. The measurements were performed using a digital micrometer. *Second,* tensile strength analysis. The tensile strength of the film was measured at the time of elongation percentage measurement by the same procedure and equipment. *Third,* percent extension (elongation). The sample was cut into rectangular with the size of 12 cm x 2 cm. Then the sample was placed on the analyzer. The equipment was operated by stretching the sample until the film was broken and the final length of the film before it was broken was recorded.

4. Result and discussion

4.1. Effect of sugar palm fruit: carrageenan ratio and citric acid concentration on tensile strength

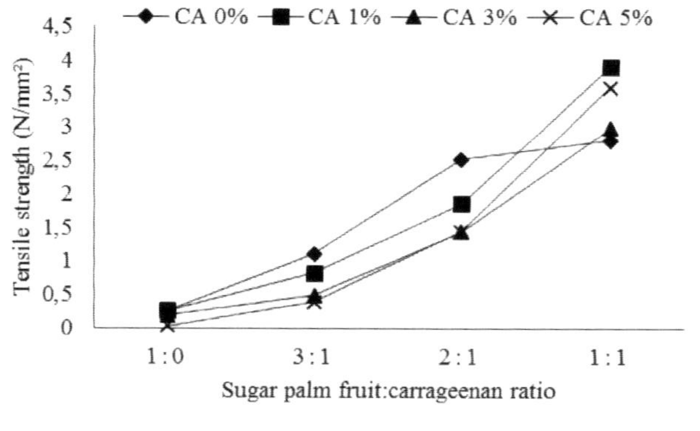

Figure 1. Effect of sugar palm fruit: carrageenan ratio and citric acid (CA) concentration on tensile strength

Figure 1 shows that the smaller the concentration of sugar palm fruit or the greater the concentration of carrageenan, the greater the value of tensile strength. This is due to more hydrogen bond in the film caused by the existence of carrageenan. An increase in carrageenan caused the formation of more hydrogen bonds between starch and carrageenan so the film would have greater bond density and stronger structures [7]. The optimum composition for the tensile strength is the ratio of sugar palm fruit: carrageenan 1: 1 with the use of 1% of citric acid. The use of citric acid increases the strength of the film. This is due to the crosslinking formed between citric acid and carbohydrate that increases the value of tensile strength [8].

4.2. Effect of sugar palm fruit: carrageenan ratio and citric acid concentration on % elongation

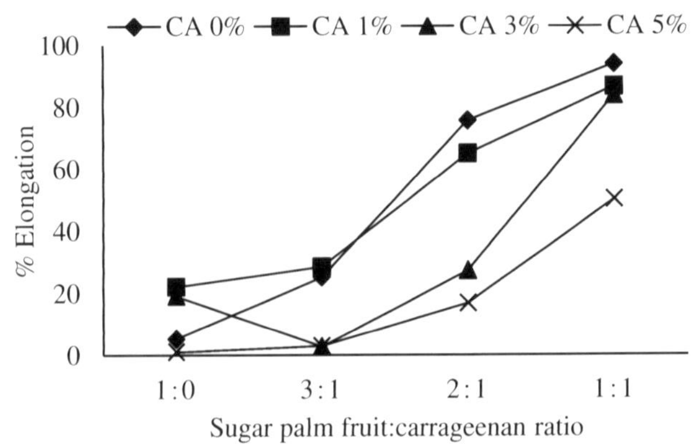

Figure 2. Effect of sugar palm fruit: carrageenan ratio and citric acid (CA) concentration on % elongation

Figure 2 shows that the smaller concentrations of sugar palm fruit or, the greater the concentration of carrageenan increases the value of the elongation of the film. This is due to the increased water holding capacity of the resulted film matrix, so it is better and elastic. This result is similar that the greater the concentration of carrageen increased the elongation of the film [9]. Increasing the concentration of citric acid affects the value of the elongation where the greater the concentration of citric acid decreases the elongation. The value of elongation in the addition of 1% citric acid slightly decreases compared to that without citric acid. This is due to the occurrence of crosslinking causing bonding between one polysaccharide molecule with other polysaccharide molecules, or even with glycerol [10]. Consequently, the matrix of the formed film becomes less elastic.

4.3. Effect of sugar palm fruit: carrageenan ratio and citric acid concentration on thickness

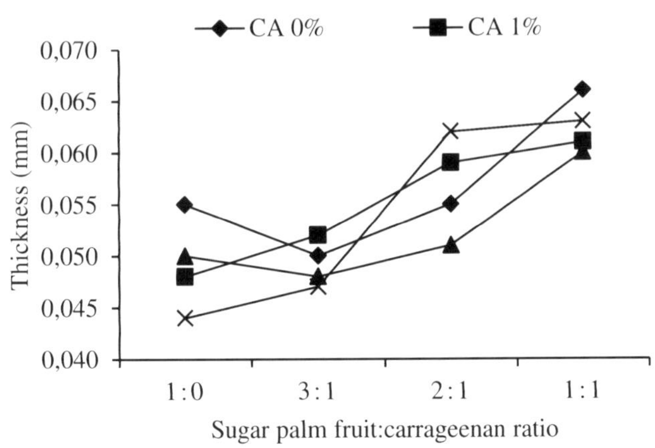

Figure 3. Effectof sugar palm fruit: carrageenan ratio and citric acid (CA) concentration on thickness

The thickness of the films, with or without an addition of citric acid, is almost the same. The thickness tends to increase the amount of sugar palm fruit decreases and a number of carrageenan increases. In the previous study, it was reported that the film thickness of the biopolymer was mainly affected by starch concentration [11]. In this work, although the amount of sugar palm fruit decreases, the increase of the amount of carrageenan and citric acid increases the film thickness due to the total solid. Also, the thickness is also influenced by several other factors including the mixed water content after heating, the depth of the acrylic cast pool, the time and temperature in the heating process [12].

5. Conclusions

Based on itsphysical and mechanical properties, the ratio of carrageenan: starch of 1:1 and citric acid concentration of 1% are the best carbohydrate ratio and citric acid concentration, respectively, because the resulted filmhas the highesttensile strength, i.e.3.91 N / mm^2. Meanwhile the ratioof sugar palm fruit: carrageenan at constant citric acid concentration has no significant effect on the film thickness.

Acknowledgments

The authors acknowledge the financial support from the Directorate General of Higher Education, Ministry of Research Technology, and Higher Education of the Republic of Indonesia through Student Creativity Program with a letter of agreement no. 547/B3.1/KM/2017.

References

[1] Heru P dan Eli R.2011. *Bioplastik Nata De Cassava Sebagai Bahan Edible Film Ramah Lingkungan*. Jurnal Penelitian Saintek, Vol. 16, Nomor 2.
[2] Tuti I.S., Hotman P. Manurung, dan Fery Permadi.2008. *Pembuatan Edibel Film dari Kolang-Kaling*. Jurnal Teknik Kimia, No. 4, Vol. 15.
[3] Campo V.L., Kawano D.F., Junior, S. 2009. *Carrageenans: Biological Properties,Chemical Modifications, and Structural Analysis Carbohydrate Polymers*. pp. 167-18
[4] Meyer R.C., Winter A.R. dan Weiser H.H. 1959. *Edible protective coatings for extending theshelf life of poultry*. Food technology. pp. 146-148
[5] Pradipta, I. M. D dan Johar M. 2012. *Pembuatan Dan Karakterisasi Polimer Ramah Lingkungan Berbahan Dasar Glukomanan Umbi Porang*.Prosiding Pertemuan Ilmiah Ilmu Pengetahuan dan Teknologi Bahan 2012. ISSN 1411-2213
[6] Adzaly, N. Z., Jackson, A., Villalobos-Carvajal, R.,Kang, I., Almenar, E. 2015. *Development of a novel sausage casing*. Journal of Food Engineering. **152**, pp.24-31
[7] Wu, Y., Geng, F., Chang, P.R., Yu, J., and Ma, X. 2009. *Effect of Agar on The Microstructure and Performance of Potato Starch Film*. Carbohydrate Polymers **76**: pp299- 304
[8] Ghanbarzadeh, B, Almasi, H., and Entezami, A. A. 2010. *Improving the barrier and mechanical properties of corn starch-based edible films:Effectofcitricacidandcarboxymethylcellulose.*Industrial Crops and Products p7
[9] Irianto, H.E., Darmawan, M., and Mindarwati, E. 2006. *Manufacture of edible film of the composite carrageenan, tapioca starch, and beeswax*. Jurnal Penelitian Perikanan Indonesia **1(2)**: pp 93-101
[10] Lacroix Monique , Genevieve Delmas-Patterson, Ley Tien Canh. 2002. Crosslinked protein and polysaccharide biofilms. Patent WO 2001037683 A9
[11] Basiak, E., Debeaufort, F., Lenart, A.. 2016. *Effect of oil lamination between plasticizedstarch layers on film properties*. Food Chem. **195** p56
[12] Rahmansyah, H. 2015. *Peningkatan Sifat Fisik dan Mekanis Edible Film Karagenan Melalui Penambahan Penaut Silang Kation Ca^{2+} dan Nanopartikel Karbon*. Departemen Kimia. FakultasMatematikadanIlmuPengetahuanAlamInstitutPertanian Bogor

IOP Conf. Series: Journal of Physics: Conf. Series **909** (2017) 012086 doi:10.1088/1742-6596/909/1/012086

Synthesis of zeolite from rice husk ash waste of brick industries as hydrophobic adsorbent for fuel grade ethanol purification

A Purnomo[1,*], M Alhanif[1], C Khotimah[1], UA Zuhra[1], BR Putri[2] and AC Kumoro[1]

[1]Department of Chemical Engineering, Faculty of Engineering, Diponegoro University, Semarang 50275, Indonesia

[2]Departemnt of Environmental Engineering, Faculty of Engineering, Diponegoro University, Semarang 50275, Indonesia

*E-mail: ariipurnomoo@gmail.com

Abstract. A lot of researchers have devoted on ethanol utilization as renewable energy to substitute petroleum based gasoline. When ethanol is being used as a new fuel candidate, it should have at least of 99.5% purity. Usually produced via sugar fermentation process, further purification of ethanol from other components in fermentation broth to obtain its fuel grade is a crucial step. The purpose of this research is to produce synthetic zeolite as hydrophobic adsorbent from rice husk ash for ethanol-water separation and to investigate the influence of weight, adsorption time and initial ethanol concentration on zeolite adsorption capacity. This research consisted of rice husk silica extraction, preparation of hydrophobic zeolite adsorbent, physical characterization using SEM, EDX and adsorption test for an ethanol-water solution. Zeolite with highest adsorption capacity was obtained with 15: 1 alumina silica composition. The best adsorption condition was achieved when 4-gram hydrophobic zeolite applied for adsorption of 100 mL of 10% (v/v) ethanol-water solution for 120 minutes, which resulted in ethanol with 98.93% (v/v) purity. The hydrophobic zeolite from rice husk ash is a potential candidate as an efficient adsorbent to purify raw ethanol into fuel grade ethanol. Implementation of this new adsorbent for ethanol production in commercial scale may reduce the energy consumption of that usually used for the distillation processes.

1. Introduction

Indonesia has a diverse energy potential, both from fossil energy and non-fossil energy. Fossil energy sources include petroleum, natural gas, coal, and coal bed methane (CBM). Indonesia also has relatively large non-fossil energy sources, which are mainly water and geothermal. But its utilization is still less than the current potential. As clearly stated in the Blueprint of the Ministry of Energy and Mineral Resources of the Republic of Indonesia, the Indonesia's existing petroleum oil reserves is about 9.1 billion barrels while the current annual production of 387 million barrels. With this condition, the current oil reserves will be exhausted in the next 23 years [1]. Also, the currently operated refinery infrastructures can only produce domestic fuels, which meet about 56% of the national fuel needs. Therefore, it is necessary to find alternative energy sources as substitutes to fossil fuel. One of the alternative to gasoline is ethanol, which can produce from biomass in large scale.

The ethanol used as a fuel is called fuel grade ethanol (FGE) should have at least 99.5% purity and an octane number of 118 [2]. The fuel grade ethanol can be used as fuel either in a pure state or its mixture with gasoline at various concentrations [3]. Indeed, the presence of water in fuels, even at the lowest content, is highly undesirable. Therefore, water separation from ethanol mixtures becomes technological problem, especially in taking into account the azeotropic liquid vapors equilibrium relation. The azeotrope composition of the ethanol-water mixture is 95.63% ethanol and 4.37% water (by weight). Ethanol boils at 78.4°C, while water boils at 100°C, but the azeotrope boils at 78.2°C, which is lower than either of its constituents. The conventional method to solve the azeotropic problem is through azeotropic distillation. However, it consumes a large amount of energy. Therefore, it is necessary to develop more efficient and economical separation methods,.which can combine with the conventional distillation or fermentation processes, such pervaporation with porous membranes [4] or adsorptive separations either in vapor phase [5] or liquid phase [6].

Concerning the heterogeneity of their crystal structure and specifically due to their molecular sieving capability, zeolites have been proven to be the most suitable adsorbents for separation of water-ethanol mixtures. The superiority is on the side of the 3A molecular sieves having a crystal structure which allows only the water molecules with molecular size 0.265 nm to penetrate inside the zeolite micropores. The other possibilities are that on the side of other dealuminated (highly hydrophobic silicalites [9]. These zeolites have silica-to-aluminum ratio n = SiO_2 /Al_2O_3 > 100 [11]. In contrast to the 3A molecular sieves, the conventional 5A zeolite does not separate ethanol and water molecules. Both ethanol and water molecules can penetrate inside the zeolite micropore volume (the molecular size of ethanol is 0.446 nm [8]). But ethanol can absorb because high silica content of zeolite and its known as hydrophobic properties. Zeolite hydrophilic/hydrophobic property mainly depends on the Si/Al ratio, i.e., zeolitic hydrophilic property increases as the aluminum content in the zeolite framework increases and vice versa [12]. Hydrophobic/hydrophilic nature of zeolites also appears to depend on their framework structure [13]. Pure siliceous zeolite beta has been reported to be much more hydrophobic than silicalite-1 and the other siliceous 12-numbered ring zeolites, even though they contain almost no aluminum. Regular Si/Al-zeolites have adsorption capacities for alcohols/water in the following order H_2O > methanol > ethanol > propanol, whereas silicalites are expected to give the reversed order of adsorption [14].

Some the crystalline forms of silica encountered in nature and some others exist as synthetic forms which extremely have no natural counterpart. Silicalite is a new hydrophobic crystalline silica molecular sieve, which its crystal structure is similar to zeolite ZSM-5. As a consequence, it can consider as the Al-free analogs of ZSM-5. Due to large surface area and pore volume, silicalite has good adsorption capacity. The competitive adsorption of a binary mixture (e.g. water and alcohols) on silicalite molecular sieve has shown that hydrophobicity increases with decreasing of the Na_2O content of the adsorbent. This fact can consider as an advantage of silicalites compared to ZSM type zeolites [15]. The influence of alkali metal cations on the stability of silicalite is as $Li^+ > Na^+ > K^+ > Rb^+ > Cs^+$, while particle size and morphology increase vice versa [16]. This adsorbent has a very beneficial hydrophobic and organophilic characteristic, which makes it suitable to use in selective adsorption of some organic materials from water, either in liquid or vapor phases [17]. Therefore, the synthesis of this material will be desirable due to its excellent potential for separation of polar organic compounds from gaseous or liquid mixtures (e.g. separation of small amounts of ethanol from fermentation broth). Due to the abundant availability of domestic resources used in this synthesis, the commercial production of this material is considered to be economical. This paper presents the synthesis of zeolite from rice husk white ash obtained from brick manufacturer waste and its capacity to absorb ethanol from the ethanol-water mixture. Rice husk ash highly potential as the raw material for zeolite production due to its high silica content (±94.5%) [18], less undesirable impurities and large surface area [19].

2. Materials and Method

2.1. Materials

The white rice husk ash used in this research was a donation from a brick manufacturer in Lampung-Indonesia. Technical grade ethanol (70% v/v), reagent grade sulfuric acid (98% w/w), sodium hydroxide crystal (75% w/w) and distilled water purchased from Indrasari Semarang. And the aluminum hydroxide as the aluminum source obtains from Multi Kimia Raya Semarang. Zeolite synthesis

The hydrophobic zeolite prepared by mixing sodium silicate and sodium aluminate. White rice husk ash was initially dried in an electric oven at 105°C to reduce the moisture content to no higher than 10% (w/w). The dried's ash was then ground with a mortar and sieved to obtain 300 um ash particles. Ten grams of rice husk ash was vigorously mixed with 10 M aqueous sodium hydroxide in a Pyrex glass reactor at 120°C under reflux for several hours to produce sodium silicate. Further, the slurry was then heated at 350°C for 4 hours until it turns into solid. Then the solid was dissolved in 100 mL of aqua dest and allowed to stand for 12 hours. Preparation of sodium aluminate done by dissolving 20 grams of NaOH crystal into 100 mL of aqua dest with continuous stirring. The NaOH solution heated to 100 °C. After the NaOH solution reaches 100°C, 8.5 grams of $Al(OH)_3$ was gradually added into the solution for 20 minutes until the crystal completely dissolved. Sodium silicate and sodium aluminate mixed with a 5: 1, 10: 1, and 15: 1 Si/Al ratio. The mixture was then dried at 150°C for three days and adjusted its pH with concentrated H_2SO_4 to a pH below 10. The mixture was re-dried and further ground to obtain uniform products.

2.2. Characterization and adsorption test of zeolite

The morphology of the zeolite was characterized using scanning electron microscope (SEM). The adsorption test of ethanol was carried out to determine the ability of zeolite as a hydrophobic adsorbent to produce fuel grade ethanol. The variables studied in this study were the adsorbent weight, adsorption time, and initial ethanol content. As much as 100 ml ethanol solution with 7.5%, 10%, and 12.5% (v/v) concentration was prepared from 70% (v/v) ethanol and each poured into a 500mL beaker glass. Then, 4, 8, and 12 grams of the synthesized zeolites introduced into the ethanol solution. The zeolite-ethanol solution mixtures left for 60 90, and 120 minutes to allow adsorption of ethanol molecules onto zeolites. Then, the zeolite separated from each solution. The zeolite adsorption capacity indirectly determined by measuring the refractive index of the ethanol solution by using the refractometer.

3. Results and Discussion

3.1. The morphology of zeolites

The SEM analysis was used to observe the microstructure of the surface of rice husk ash where main material of zeolite hydrophobic. The observed microstructures of zeolite particle with Si/Al ratio of 5:1, 10:1 and 15:1 depicted in Figure 1, Figure 2 and Figure 3. Also, the elemental analysis of the zeolite from the EDX analysis presented in Table 1.

Figure 1 and figure 3 show the uniformity of shapes of hydrophobic synthetic zeolite with the pore diameters 0.3-1μm and 0.05-0.01μm, respectively. Zeolite with Si/Al ratio of 15:1 (Figure 3) had the smallest pore diameter or largest specific surface area and more porous nature of X-type zeolite so that it is suitable for use as a hydrophobic adsorbent [20]. In contrast to that, the zeolite with Si/Al ratio of 10:1 (Figure 2) is rich in pores 0.01-1.5 um diameter. The particles were non-uniform in size and shape with some crystals apparently fused together to form agglomerate particles. The particle size of isolate crystal from SEM micrograph was approximately 2μm. This larger pore diameter may suggest the zeolite has less adsorption capacity due to the smaller specific surface area as compared to the former zeolites.

The EDX analysis reveals that for zeolites obtained from raw materials with Si/Al ratio of 5:1, 10:1 and 15:1, the actual Si/Al ratio in the zeolites were 5.67, 12.52 and 18.47, respectively. This fact shows that the Si/Al ratios in the zeolites reasonably higher than expectation. The zeolite with Si/Al ratio of 12.52 was likely to be $Na_{6.6}(AlO_2)_{6.6}(SiO_2)_{89.4}$ or ZSM-5 type MFI, while zeolite with Si/Al ratio of 18.47 was expected to be $Na_{4.8}(AlO_2)_{4.8}(SiO_2)_{91.2}$ or another ZSM-5 type MFI [20]. The zeolite with Si/Al ratio of 18.47 will be more suitable for ethanol adsorption from ethanol-water mixture [11]. Although all adsorbents are distinctly hydrophobic, the ethanol enrichment in the adsorbed phase changes drastically. Only those adsorbents with a small channel system (ZSM-5) have practical potential for hydrophobic separation [20]. The sodium content in the zeolite was the lowest for zeolite with Si/Al ratio of 1:15, while no significant different in sodium content observed for zeolite with Si/Al ratio of 5:1 and 10:1. This fact is mainly due to the higher concentration of sodium content in the reactant mixture during the fusion treatment [21]. In the case of a high content of (AlO4)- -Na+ dipoles in the cavities, the first water molecules are strongly localized on these dipoles and are the basis for the formation of the associates. These structures hinder the adsorption of the ethanol molecules on the oxygen walls of the cavities [20].

Figure 1. Image of zeolite with Si/Al 5:1 with 3000× magnification

Figure 2. Image of zeolite with Si/Al 10:1 with 3000× magnification

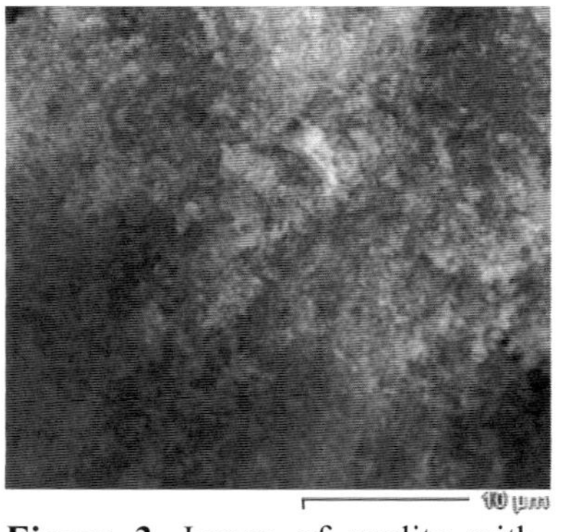

Figure 3. Image of zeolite with Si/Al 15:1 with 3000× magnification

Table 1. EDX analysis of zeolite from rice husk ash waste (atom %).

Elements	Zeolite Si/Al: 5:1	Zeolite Si/Al: 10:1	Zeolite Si/Al: 15:1
O	40.81	41.55	44.45
Si	27.11	19.41	35.83
Al	4.76	1.55	1.94
Na	12.87	12.91	3.72

3.2. Effect of initial ethanol concentration

The adsorption capacity of zeolite can test through dynamic adsorption test. In the industry, fermentation of glucose using *Saccharomyces cerevisiae* may yield fermentation broth with 7 – 12 % (v/v) ethanol content [22]. Therefore, the zeolites obtained in this work were tested their adsorption capacity using 4-gram zeolite and 100 mL ethanol solution with initial ethanol concentrations varied between 7-12 % (v/v) for 120 minutes. The results of ethanol adsorption test on zeolites with various Si/Al ratios and initial ethanol concentrations depicted in Figure 4.

In an adsorption system, the behavior of the adsorbed phase differs significantly from the bulk phase. In the bulk free liquid, the water molecules associated in clathrates. The first ethanol molecules can fill the cavities of the clathrate structure. The positive deviations from Raoult's law for the activity coefficients in the bulk phase caused by repulsive forces between the ethanol molecules and the water clathrates. With higher ethanol content, a pseudo two-phase system of the clathrates and a random ethanol-water mixture exists. However, as the concentration of ethanol increases the degree of association is lower because of the decreasing number of hydroxyl groups in the bulk phase [20].

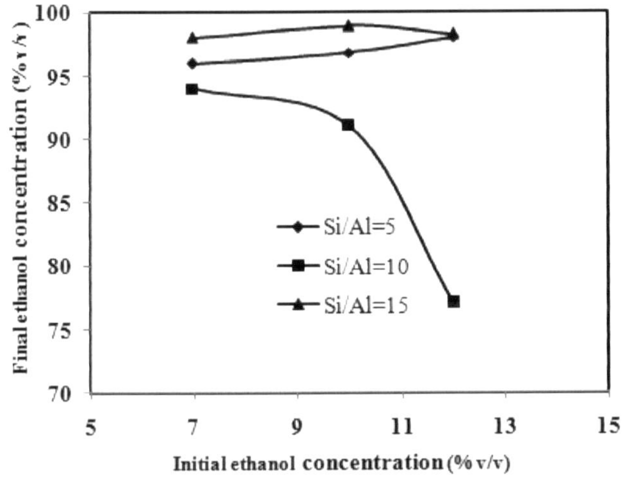

Figure 4. Effect of initial ethanol concentration on ethanol adsorption capacity

Figure 4 confirms that zeolite with Si/Al ratio of 15:1 possesses highest adsorption capacity compared to the other zeolites with less Si/Al ratios for all initial ethanol concentrations studied. With lowest sodium content, the water molecules were not trapped inside the zeolite pores and did not further interact with sodium cations surrounding the zeolite pores [23]. Highest ethanol concentration (98.93% v/v) can obtain when this zeolite was used to absorb ethanol solution with 10% (v/v) initial concentration. The least ethanol adsorption capacity observed for zeolite with Si/Al ratio of 10:1. This zeolite is likely to be hydrophilic as its ethanol adsorption capacity decreases with the increase of initial ethanol concentration. Surprisingly, the adsorption capacity of zeolite with Si/Al ratio of 5:1 increases with initial ethanol concentration. This finding agrees well with previous work reported in the literature that dealuminated [9], or highly hydrophobic silicalites [9] has the higher preference to absorb ethanol rather than water molecules. Gaara and Akporiaye also reported that the activity of Y-zeolite increase in the yield with decreasing aluminum content suggesting the rise of hydrophobicity of dealuminated zeolite [24].

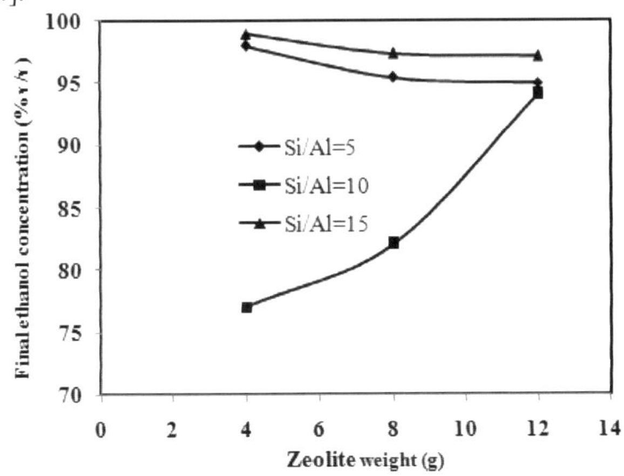

Figure 5. Effect zeolite weight on ethanol adsorption capacity

3.3. *Effect of zeolite weight*
This study was conducted by adsorption of 12% (v/v) ethanol solution onto zeolites for 120 minutes at various zeolite weights. Figure 5 presents the superiority of zeolite with Si/Al ratio of 15:1 to adsorb ethanol compared to the other zeolites. However, no significant adsorption capacity differences between zeolite with Si/Al ratio of 15:1 and that with Si/Al ratio of 5:1 observed. There was also noticeable phenomenon is that, when the zeolite weight reached a certain value, a gradual decline in ethanol concentration in the products appeared. Ivanova and coworkers also found that as the zeolite amount increases, the amount of ethanol adsorbed onto zeolite clinoptilolite pore surface gradually decreases [25]. a similar result was reported by Sumin et al. [26] on the adsorption of carbon dioxide using hydrophobic zeolite. On the contrary, an improvement in ethanol concentrations with the increase of zeolite weight observed for zeolite with Si/Al ratio of 10:1.

3.4. *Effect of adsorption time*
Three different adsorption times were chosen to study the effect of time on the adsorption of ethanol onto zeolites with three different Si/Al ratios. As shown in Figure 6, the ethanol concentration increases linearly with adsorption time on to zeolite with Si/Al ratio of 15:1, which is a high adsorptive capacity zeolite. Adsorption time only slightly affected the ethanol concentration of the product for zeolite with Si/Al ratio of 5:1. However, a different pattern was found for zeolite with zeolite with Si/Al ratio of 10:1.

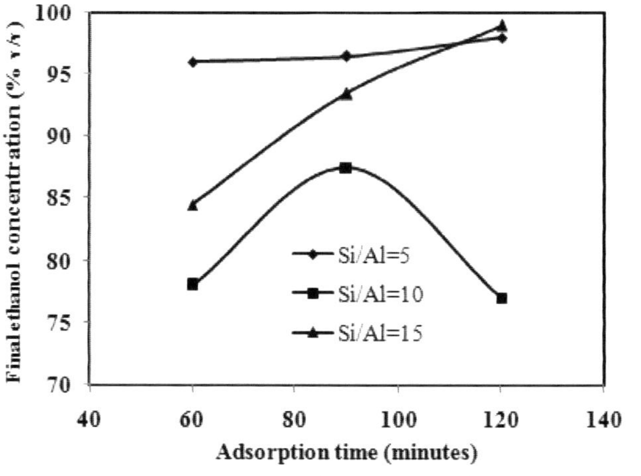

Figure 6. Effect of adsorption time

As the adsorption time increases from zero to 90 minutes, the ethanol concentration in the product also increases. However, further increase of adsorption time using zeolite with Si/Al ratio of 10:1 reduces the ethanol concentration in the product. This phenomenon suggests desorption of ethanol from zeolite pores surface at long adsorption time. For particles with low adsorptive capacity, before the particles get saturated, the adsorption capacity increases with time. As time propagates, however, the zeolite particle gets saturated, and the amount of adsorbed ethanol molecules decreases gradually [26].

4. Conclusion
Based on the experimental data, it can conclude that zeolite with Si/Al ratio of 15:1 is the most suitable adsorbent for ethanol. No significant effect of initial ethanol concentration and zeolite weight on the product ethanol concentration. Longer adsorption time results in higher ethanol concentration of the product for adsorption using zeolite with Si/Al ratio of 5:1 and 15:1. Different trends observed for the effect of initial ethanol concentration, zeolite weight and time during adsorption of ethanol using

zeolite with Si/Al ratio of 10:1. A relatively good adsorption condition was adsorption of 10% (v/v) ethanol solution using 4 grams of zeolite with Si/Al ratio of 15:1 for 120 minutes to obtain the product containing 98.93% (v/v) ethanol.

Acknowledgement
The authors would like to thank the financial support from the Directorate General of Higher Education-Ministry of Research, Technology and Higher Education of the Republic of Indonesia for the Student Creativity Program Grant 2017. The authors also would like to express their gratitude to Diponegoro University for providing necessary facilities for this research.

References
[1] Kementerian EDSM RI 2010 *Rencana Strategies Kementerian Energi Dan Sumber Daya Mineral Tahun 2010 - 2014* (Jakarta: Kementerian Energi dan Sumberdaya Mineral Republik Indonesia)
[2] Prihandana R, Noerwijari K, Gamawati P, Adinurani, Setyaningsih D, Setiadi S and Handoko R 2008 *Bioetanol Ubi Kayu Bahan Bakar Masa Depan* (Jakarta: Agromedia Pustaka)
[3] Badan Standarisasi Nasional 2008 *Bioetanol Terdenaturasi Untuk Gasohol* (Jakarta: Badan Standarisasi Nasional)
[4] Sekulic J, Ten Elshof JE and Blank DHA 2005 *J.Membrane Sci.* **254** 267-274.
[5] Caputo D, Iucolano F, Pepe, Colella C 2007 *Microporous Mesoporous Mater.* **105 (3)** 260-267
[6] Adnadevich B, Mojovic Z and Abu Rabi A 2008 *Adsorption* **14** 123-131.
[7] Pruksathorn P and Vitidsant T 2009 *Am. J. Eng. and Appl. Sci.* **2 (1)** 1-7
[8] Ravel AA and Ponomarev AM 1983 *Kratkii Spravochnik Fiziko khim* (Velichin, Leningrad: Himia)
[9] Einicke WD, Heuchel M, Szombathely MV, Bräuer P, Schöllner R and Rademacher O 1989 *J. Chem. Soc. Faraday Trans.* **85 (1)** 4277-4285
[10] Falamaki C, Sohrabi M, and Telebi G 2001 *Chem. Eng. Technol.* **24(5)** 501-506
[11] Keltsev NV 1984 *Osnovy Adsorbtsionnoi Tekhniki* (Moscow: Khimiya)
[12] Haugh GS 1985 *Stud. Surf. Sci. Catal.* **24** 609-615
[13] Breck DW 1964 *Zeolite Molecule Sieves* (New York: JohnWiley)
[14] Milestone NB and Bibby DM 1981 *J. Chem. Technol. Biotechnol.* **31** 732-736.
[15] Bowen TC, Noble RD and Falconer JL 2004 *J. Membr. Sci.* **245 (1-2)** 1-33
[16] Kungang T and Ruben X 1985 *Stud. Surf. Sci. Catal.* **24** 73-80
[17] Grose WR and Flanigan EM 1977 Crystalline silica US-Patent 4 061 724
[18] Husin AA 2002 *Pemanfaatan Limbah Untuk Bahan Bangunan* (Bandung: Pusat Penelitian dan Pengembangan Pemukiman)
[19] Taherzadeh MJ and Karimi K 2008 *Int. J. Mol. Sci.* **9** 1621-1265
[20] Einicke WD, Reschetilowski W, Higuchi M, Szombathely MV, Brauer P, Schollner R, Schwieger W and Bergk KH 1991 *J. Chem. Soc. Faraday Trans.* **87(8)** 1279-1282
[21] Sakthivel T, Reid DL, Goldstein I, Hench L, and Seal S 2013 *Environ. Sci. Technol.* **47 (11)** 5843–585
[22] Dombek KM and Ingram LO 1985 *Appl. Environ. Microbiol.* **51**197-200
[23] Byrappa K and Yoshimura M 2001 *Handbook of Hydrothermal Technology: A Technology for Crystal Growth and Materials Processing* (New York, USA: William Andrew Publishing LLC)
[24] Gaara K and Akporiaye D 1996 *J. Mol Catal. A: Chem.* **109 (2)** 177-187
[25] Ivanova E, Damgaliev D, and Kostova M 2009 *J. Univ. Chem. Technol. Metal.* **44 (3)** 267-274
[26] Sumin L, Youguang M, Shuhua S and Chunying Z 2010 *Braz. J. Chem. Eng.* **27(2)** 327 - 338

IOP Conf. Series: Journal of Physics: Conf. Series **909** (2017) 012087 doi:10.1088/1742-6596/909/1/012087

Swelling power and solubility of modified breadfruit flour using *Lactobacillus plantarum*

Istiana Norita Rahma[1], Raja Haris Pratama, Alfiyanti, Deo Reynaldo Alwi, Woro Indriani Setyo Tri Astuti and Dyah Hesti Wardhani

Chemical Engineering Department, Faculty of Engineering, Diponegoro University
Jl. Prof. Soedharto, SH Tembalang Semarang 50275, Indonesia

Email: istiananorita@gmail.com

Abstract. Breadfruit has the high nutritional and carbohydrate content which is comparable with wheat, so it is potential to be processed as the breadfruit flour. However, some studies showed the breadfruit flour has low swelling power (SP) and water solubility (WS) compared to wheat. Hence, the flour required modification. The main purpose of this research was to increase the value of SP and WS on breadfruit flour through a fermentation process. Sterile breadfruit flour (50 g) was suspended using sterile distilled water in a 250 ml glass beaker to obtain the flour concentration of 25%, 30%, 35%, 40% and 45% (g/ml). A certain concentration of *L. plantarum* was added (5, 7.5, 10, and 12.5%, ml/ml) then incubated at 30°C for 24, 48, 72, 96 and 120 h. The obtained solid cake was washed before being dried. The effect of cultural concentration on swelling power (SP) and water solubility (WS) is presented in Figure 1. The optimum fermentation condition was achieved by a 35% concentration of breadfruit flour which fermented with 10% *L. plantarum* for 24 h fermentation. At this condition, all parameter showed an increase of carbohydrate, protein, SP and WS into 84.761, 3.776, 8.870% and 0.136%, respectively.

1. Introduction

The change of lifestyles makes people more preferably eat bread for their breakfast and it also happens in Indonesia. Indonesia had to import wheat as much as 5,692,630 MT [1] in the first half in 2016 to keep these needs. On the other hand, Indonesia has natural resources of agricultural products that are carbohydrate-rich content such as breadfruit (*Artocarpus altilis*), yams, taro, and so on. Breadfruit has a similar proximate content with wheat flour. According to Ajani et al. [2] wheat flour contains 97.57% carbohydrate, 0.42% fat, 1.08% protein, 0.93% ash, 0.08425% fiber and 12.755% water; while breadfruit flour has 96.91% carbohydrate, 0.595% fat, 1.345% protein, 1.045% ash, 1.26% fiber and 9.065% water. Although the proximate contents are similar, the breadfruit flour still as a less desirable characteristic such as low SP and low solubility [3]. SP of breadfruit flour is 4.84-6.32 and WS 0.0801 – 0.1129 % [4]. To have comparable properties with the wheat, the breadfruit should be modified to improve its functional properties.

According to Balagopan et al., [5], SP increase in volume and weight, which is the starch in the air. Swelling strength improves quality. The strength of the swelling provides a soft texture on food products [6]. Flour that has high WS will make the flour is easier to be digested by the body [7]. Retnowati et al. [8] reported that SP is a parameter of acceptability of starch as food because it can affect the texture. Flour with high SP values can be applied to the manufacture of products such as noodles, vermicelli, and analog rice. High WS and viscosity flour absorbed water and formed a gel easily. Hence, it can be used as a gelling agent or thickening agent.

One method of flour modification is fermentation which is relatively efficient and low energy requirement but extend shelf life [9]. Research on fermented breadfruit is rarely found; Appiah et al. [10] reported that natural fermentation of breadfruit increases a concentration of protein and fat. Natural fermentation improved SP of breadfruit flour although insignificant, from 7.02 to 8.72 [10].

L. plantarum which found in a natural fermentation process converts the sugar into the organic acid and lowering the pH. According to Putri et al. [11], modification of cassava starch assisted by ultraviolet light as well as fermentation of lactic acid bacteria such as *L.plantarum* and *L. amylophillus* result in changing of physical properties such as SP and solubility. In this study, the highest SP was found in non-fermented starch, 22.28 g/g, while the highest water solubility (WS) was in fermented starch with *L. amylophyllus* at 13.95 g/100g. The starch content of cassava was 6.61 ml/g. However, fermentation using *L. amylophyllus* and *L. plantarum* increased the starch to 8.54 ml/g and 10.6 to 11.25 ml/g, respectively. Gunawan [12] reported that modification of cassava flour fermented with *Lactobacillus plantarum* had the highest protein content (8.58%) compared to that of *Saccharomyces cerevisiae* (2.29%) or *Rhizopus oryzae* (4.72%).

The purpose of this research is to study the effect of concentration of microorganisms, fermentation time, and concentration of flour on functional properties and proximate of breadfruit flour.

2. Material and Method

Breadfruit flour was prepared by peeled the skin of fresh breadfruit, washed with water to remove the dirt and sliced to shrink in size. After dried under the sun, breadfruit was milled and sieved to obtain the flour of 80 meshes and ready to be used. Bacterial cultures *L. plantarum* 3074 was bought from the Food and Nutrition Research Center Gajah Mada University. *L. plantarum* was cultivated in MRS broth according to Anggraini et al. [13] and incubated at 37°C for 24 h. Breadfruit flour fermentation was fermented following the method developed by Putri et al. [11]. Sterile breadfruit flour (50 g) was suspended using sterile distilled water in a 250 ml glass beaker to the concentration of 25%, 30%, 35%, 40% and 45% (w/v). The suspension was inoculated with *L. plantarum* (5, 7.5, 10 and 12.5%, v/v). The glass beaker was then covered with aluminum foil and incubated at 30°C for 24, 48, 72, 96 and 120 hours. Fermented breadfruit flour is then washed three times with sterile distilled water before being dried in an oven. The breadfruit flour was subjected to analysis of the water content [14], the ash content [14], protein [14], fat [14], carbohydrates [14], SP [14] and WS [15].

3. Result And Discussion

3.1. *Effect of Number Culture Bacteria*

The experimental results on the effect of the amount of bacterial culture of the SP and WS modified breadfruit flour can be seen in Figure 1. Figure 1 shows that the SP and WS in modified breadfruit flour increased as the number of bacteria culture was added. The SP increased because the number of bacteria that produced extracellular amylase enzyme that broke down or degraded polymer of breadfruit flour into simpler compounds was higher. This extracellular amylase enzyme broke the starch polymer bonds into shorter oligosaccharides or simple sugar molecules. The degradation process caused the granular become porous after drying and subsequently increased the ability to absorb the water. This absorption led the starch granules to swell and expand so that the SP went up [16].

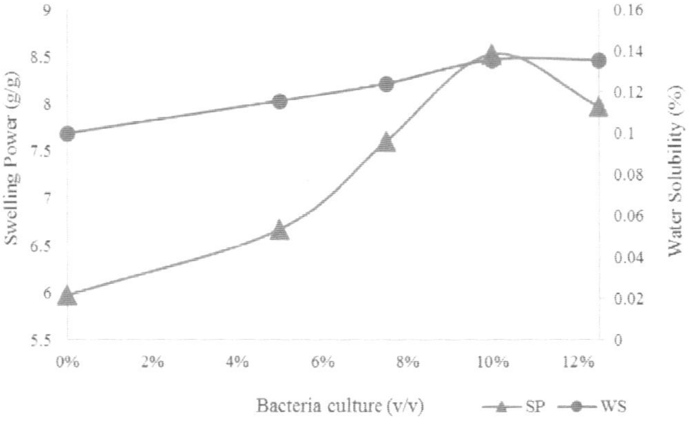

Figure 1. Effect of The Amount of Bacterial Culture of The Swelling power (SP) And Water solubility (WS) Modified Breadfruit Flour

The WS increased because bacteria produce pectinolytic and cellulolytic enzymes that could destroy the cell walls of breadfruit [17]. Tian et al. [18] stated that the smaller the granule size, the higher the starch solubility. However, this increase was not significant could be due to formation of amylose - like fragments caused by enzymatic hydrolysis of amylopectin at the amorphous region of granule. Thus, the fragments could realign and form new hydrogen bonds resulting in greater internal granule stability [19]. This result was consistent with studies conducted by Surojanametakul et al. [20] in the modification of rice with fermentation process using *L. plantarum* which resulted in increasing SP value from 7.6 to 8.9 and WS value from 0.094% to 0.126%.

3.2. Effect of Fermentation Time

The experiment result of the effect of fermentation time against SP and WS in modified breadfruit flour can be seen in Figure 2.

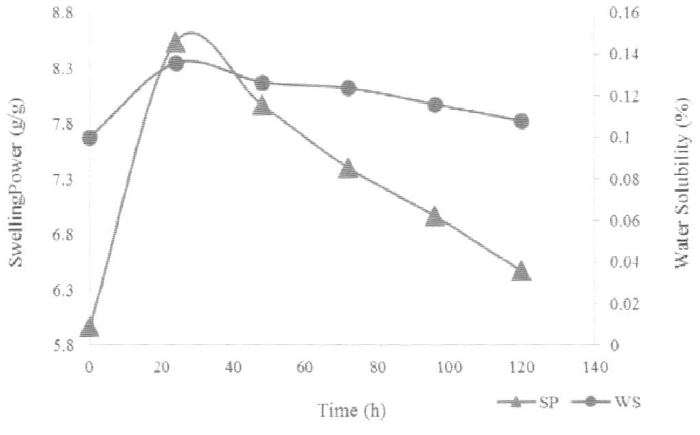

Figure 2. Effect of Fermentation Time Against Swelling Power (SP) and Water Solubility (WS) in Modified Breadfruit Flour

Figure 2 shows that the SP and WS of modified breadfruit flour were optimum after 24 fermentation period, and tended to decrease afterward. The SP increased in fermentation for 24 hours because the bacteria were in the exponential phase where the maximum speed of microbial growth resulted in a high number of bacteria. This was consistent with studies conducted by Zareian et al. [21] who found the exponential phase of *L. plantarum* was 24-36 hours. A higher number of bacteria produced more enzyme, hence digested more flour. Swelling power is strongly influenced by the presence of amylose groups as one of the components of the starch composition and the development of starch granules caused by the substitution of acetyl groups in starch which can weaken the hydrogen bond. The longer the fermentation process leads to reduce amylose, resulting in a decrease of amylose which leads to increase in swelling power [22].

The WS increased in fermentation for 24 h because of the flour degradation by bacteria into simple sugar. Therefore, it was easier to interact with the water due to granule flour bond water and release amylose into the dispersing medium [23].

Swelling power and water solubility decreased along with increasing fermentation time caused by lack of nutrients. The bacteria then enter the stationary phase. Hence, the bacteria activity to break down the starch of breadfruit reduced.

3.3. The Effect of Flour Concentration

The experiment result of the effect of flour concentration against swelling power (SP) and water solubility (WS) in modified breadfruit flour can be seen in Figure 3.

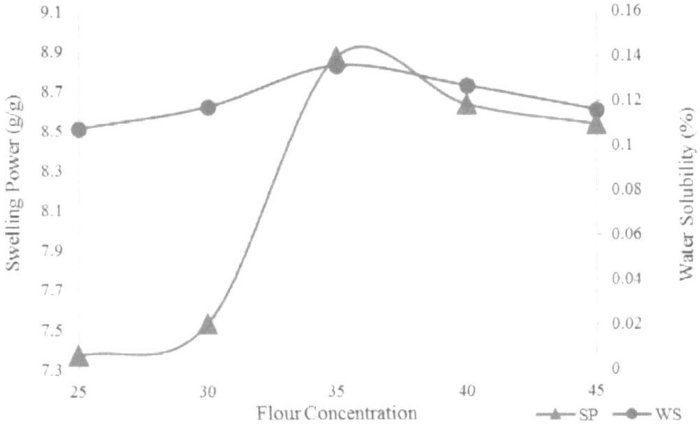

Figure 3. Effect of Flour Concentration Against Swelling power (SP) and Water solubility (WS) In Modified Breadfruit Flour

Figure 3 shows that the SP increased as the rising value of the concentration of flour; however, it declined in the flour concentration of 40% and 45%. Values of SP increased at variable concentrations flour 25%, 30%, and 35% due to the higher concentration of flour provided the more amylose number; therefore, the ability to absorb water was higher. Water was absorbed in each granule cause SP value increased because the granules continued to swell and throng around [24].

Values of the WS did not increase significantly because Lii et al. [25] stated that high amylose content is making the properties of flour granules become more rigid and not broken easily, so the value of SP declined on the flour concentration of 40% and 50%.

3.4. Proximate Analysis and Functional Properties of Modified Breadfruit Flour

From this research, it was found that the composition of modified breadfruit flour that gave the best result of SP and WS was 35% concentration of breadfruit flour using 10% L. Plantarum and fermented for 24 h. Therefore, this composition was analyzed on its proximate. The experiment result of proximate analysis and functional properties of the best-modified breadfruit flour could be seen in Table 1.

Table 1. Proximate Analysis and Functional Properties of 100 Grams the Best Modified Breadfruit Flour

Proximate Analysis	Native	Best Modified	Wheat Flour
Carbohydrate	83.430	84.761	97.570
Protein	2.697	3.776	1.080
Water	12.272	10.136	12.755
Ash	0.632	0.471	0.930
Fat	0.969	0.858	0.420
Swelling power (SP)(g/g)	5.970	8.870	9.350
Water solubility (WS) (%)	0.100	0136	1.350

The addition of *L. plantarum* bacteria in fermentation could increase the protein and carbohydrate content in modified breadfruit flour. Carbohydrate content increased by 1.595 % while protein content 39.95%. The increases in protein content were due to the activity of *L. plantarum* during fermentation which produced the proteinase enzyme as well as peptidoglycan-degrading enzymes. Cell wall was composed of components glycoprotein and lipoprotein [26]. *L. plantarum* was a microorganism that acted as Single Cell Protein (SCP), so it can increase the protein content in the food during fermentation.

The decrease of ash content was caused by the activity of lactic acid bacteria that produced phytase enzymes that hydrolyzed the phytase compounds to protect minerals. Phytase enzymes supported the release of mineral ions. Therefore, the WS increased. This result was supported by Lopez et al. [27] and Panda et al. [28], in which phytic acid (InsP6 - myoinositol hexaphosphat) protected several mineral components. Hence, its availability became limited by lactic acid bacteria fermentation would degrade phytic acid and increase the bioavailability of food.

Syahputra et al. [29] reported that during fermentation, the fat of the material would be hydrolyzed into a simpler substance by lactic acid bacteria for energy sources during fermentation. Taufik [30] reported that the fat content would decrease with fermentation due to the decreasing activity of bacterial fatty acid synthesis along with the decreasing of the bacterial population. Appiah et al. [7] and Gunawan [12] reported that modification would increase the swelling power (SP) and fermentation with *L. plantarum* would increase the protein. Compared with the wheat flour, the modified breadfruit flour with the best formula had comparable proximate contents, the number of SP and WS. Therefore, the breadfruit flour has a potency to be used as a substitute for the wheat flour.

4. Conclusions

The swelling power (SP) increased along with the number of bacteria, extended fermentation time, and flour concentration. Bacteria produced extracellular amylase enzyme that broke down polysaccharides into simpler compounds. Higher concentration of flour provided more amylopectin, hence led ability to absorb more water. However, the solubility did not increase significantly because of the fat degradation. The fat would also be absorbed by the surface of the granules to form a layer that inhibited the water binding by flour granules.

5. Acknowledgment

The author would like to thank the management board and staffs of Laboratory of Public Sevices Chemical Engineering, Faculty of Engineering, Diponegoro University, for the supports during the research as well as to Ministry of Research Technology and the Higher Education Republic of Indonesia for financial support in PKM 2017.

References

[1] APTINDO (Asosiasi Produsen Tepung Terigu Indonesia). 2014. Overview industri tepung terigu nasional indonesia. Retrieved October 20, 2016, from http://www.aptindo.or.id/.

[2] Ajani, Alice., F., Oshundahunsi O., Rahman, Akinoso., A., Arowora Kayode., A., Abiodun Aderibigbe., and O., Pessu Patricia. 2012. Proximate composition and sensory qualities of snacks produced from breadfruit flour. Global Journal of Science Frontier Research Biological Sciences, 12 7.

[3] Mutmainah Fajriyatul., M., Dimas Rahadian A., and Amanto, Bambang Sigit. 2013. Kajian karakteristik fisikokimia tepung sukun (*Artocarpus communis*) by soaking time variation and acetic acid concentration. Jurnal Teknosains Pangan, 2 4 : 46-53.

[4] Appiah, Francais. 2011a. Nutrient composition, functional properties, digestibility and formulation of selected food products from breadfruits (*Artocarpus spp.* and Tre*culia africana*). Thesis. Kwame Nkrumah University of Science and Technology. Kumasi : Ghana.

[5] Balagopalan, C., Padmosa, G., Nanda, S.K. and Moorthy, S.N. 1988. cassava in food, feed and industry, C.R.C. Press Inc. Boca Raton, Florida. pp. 189-196.

[6] Gary G. Hou. 2010. Asian Noodles: Science, Technology, and Processing, ISBN: 978-0-470-17922-2.

[7] Appiah, F., Oduro, I. and Ellis, W.O. 2011b. Fuctional properties of *Artocarpus altilis* pulp flour as affected by fermentation. Agriculure and Biology Journal of North America, 773-779. Doi:10.5251/abjna.2011. 2.5.773.779.

[8] Retnowati, D.S, Kumoro, A.C, Budiyati, S. 2010. Modifikasi pati ketela pohon secara kimia dengan oleoresin dari minyak jahe. Jurnal Rekayasa Proses 4 No 1 : 1-6. Diponegoro University. Semarang.

[9] Juodeikiene, Grazina., Bartkiene, Elena., Viskelis, Pranas., Urbonaviciene, Dalia., Eidukonyte, Dalia., and Bobinas, Ceslovas. 2012. Fermentation processes unsing lactic acid bacteria producing bacteriocins for preservation and improving functional properties of food products, advances in applied biotechnology, Prof. Marian Petre (Ed.). ISBN:978-953-307-820-5, Intech.

[10] Appiah, F., Oduro, I. and Ellis, W.O. 2011C. Proximate and mineral composition of Artocarpus altilis pulp flour as affected by fermentation. Pakistan Journal of Nutrition, 10 7 : 653-657.

[11] Putri, D. W. R., W, Haryadi D and N, Cahyanto M. 2011. Effect of biodegradation by lactic acid bacteria on physical properties of cassava flour. International Food Research Journal, 18 3 : 1149-1154.

[12] Gunawan., T., Widjaja., S., Zullaikah., L., Ernawati., N., Istianah., W., Aparamarta H., and D., Prasetyoko. 2015. Effect of fermenting cassava with Lac*tobacillus plantarum, Saccharimices cereviceae, and Rhizopus oryzae* on the chemical composition of their flour. International Food Research Journal, 22 3 : 1280-1287.

[13] Anggraini et al. 2014. Optimalisasi fermentasi tepung jali (*Coix lacryma-Jobi L.*) termodifikasi ditinjau dari kadar protein terlarut. Proceeding Seminar Nasional Sains dan Pendidikan Sains IX 5 No 1 : 608-611, ISSN : 2087-0922.

[14] AOAC. 1995. Official Method of Analysis. Washington: Association of Official Analytical Chemists.

[15] Kainuma, K., Odat, T., & Cuzuki, S. 1967. Study of flour phosphates monoesters. Journal of Technology Society Flour, 14, 24-28.

[16] Kusumaningrum, Annisa., and Sumardiono, Siswo. 2012. Upaya perbaikan sifat fisikokimia tepung ubi kayu melalui proses fermentasi sawut ubi kayu dengan starter bakteri asam laktat *Lactobacillus casei* dan pemanfaatannya sebagai bahan baku snack tradisional pilus dan roti muffin. Semarang: Diponegoro University.

[17] Hersoelistyorini, Wikanastri., Dewi, Sri Sinto., and Kumoro, Andri Cahyo. 2015. Sifat fisikokimia dan organoleptik tepung mocaf (modified cassava flour) dengan fermentasi menggunakan ekstrak kubis. The 2nd University Reasearch Coloquium, ISSN : 2407-9189. 10-17.

[18] Tian SJ, Rickard JE, Blanshard JM. 1991. Physicochemical properties of sweet potato starch. J Sci Food Agri. 57 4 :451-491. doi:10.1002/jsfa.2740570402.

[19] Numfor, Festus A. 1999. Physicochemical changes in cassava starch and flour associated with fermentation, effect on textural properties. The Journal of Food Technology in Africa. 4 1 : 31-35.
[20] Surojanametakul, Vipa., Satmalee, Prajongwate., Panthavee, Wanchai., and Phonkaivon, Naraporn. 2013. Effect of *Lactobacillus plantarum* fermentation on some properties of immature rice : KDML 105. Proceedings of the 51st Kasetsart University Annual Conference, Bangkok, Thailand, 5-7 February 2013.
[21] Zareian, Mohsen., Ebrahimpour, Afshin., Bakar, Fatimah Abu., Mohamed, Abdul Karim Sabo., Forghani, Bita., Ab-Kadir, Mohd Safuan B., and Saari, Nazamid. 2012. A glutamic acid-producing lactic acid bacteria isolated from malaysian fermented foods. International Journal of Molecular Sciences, 13 : 5482-5497. Doi:10.3390/ijms.13055482.
[22] Sasaki, T. dan Matsuki, J. 1998. Effect of wheat starch on structure on swelling power. Cereal Chemistry. 75: 525 – 529.
[23] Anggraeni, Yeni Puspita., and Yuwono, Sudarminto Setyo. 2014. Pengaruh fermentasi alami pada chips ubi jalar (*ipomea batatas*) terhadap sifat fisik tepung ubi jalar terfermentasi. Jurnal Pangan dan Agroindustri, 2 No 2 : 58 - 69.
[24] Armanda, et al. 2016. Karakteristik fisikokimia tepung sorghum coklat Utuh (whole grain brown sorghum flour) terfermentasi ragi tape. Jurnal Pangan dan Agroindustri 4 No. 2 : 458-467.
[25] Lii, Cheng-Yi., Tsai, Mei-Lin., and Tseng, Kuo-Hsuen. 1996. Effect of amylose content on the rheological property of rice flour. Cereal Chem, 73 4 : 415-420.
[26] Reddy G, Altaf MD, Naveena BJ, Venkateshwar M, Kumar EV. 2008. Amylolytic bacterial lactic acid fermentation-A review. Biotechnology Advance, 26 1 : 22-34. http://doi.org/c6m9ms.
[27] Lopez, H.W., Krespine, V., Guy, C., Messager, A., Demigne, C., dan Remesy, C. 2001. Prolonged fermentation of whole sourdough reduces phytate level and increases soluble magnesium. Journal Agriculture Food Chemistry, 49 : 2657-2662.
[28] Panda, S.H., and Ray, R. C. 2008. Direct conversion of raw starch to lactic acid by Lactob*acillus plantarum* MTCC 1407 In semi solid fermentation using sweet potato (*Ipomea batatas L*) flour. Journal of Science & Industrial Research, 67 : 531-537.
[29] Syahputra A, Pato U, Rossi E. 2015. Variasi penambahan sukrosa terhadap mutu cocoghurt menggunakan *Enterococcus faecalis* UP-11 isolated dari Tempoyak. J Faperta. 2 1.
[30] Taufik E. 2004. Dadih susu sapi hasil fermentasi berbagai starter bakteri probiotik yang disimpan pada suhu rendah: karakteristik kimiawi. J Pet, 27 3 : 88-100.

IOP Conf. Series: Journal of Physics: Conf. Series **909** (2017) 012088 doi:10.1088/1742-6596/909/1/012088

Screening and Characterization of Polygalacturonase as Potential Enzyme for Keprok Garut Orange *(Citrus nobilis var. chrysocarpa)* Juice Clarification

E Widowati, R Utami and K Kalistyatika[1]

[1]Food Science and Technology, Faculty of Agriculture, Universitas Sebelas Maret Surakarta, Indonesia
Jl. Ir. Sutami 36A Kentingan Jebres Surakarta 57126 Indonesia

E-mail: esti_widowati@yahoo.com

Abstract. Use of thermostable enzyme from bacilli for industrial application is significant. This research aimed to isolate thermophilic pectinolytic bacteria from orange peel and vegetable waste which produced thermostable polygalacturonase, to investigate the polygalacturonase ability in clarifying keprok Garut orange juice, and to characterize polygalacturonase based on pH optimum, temperature optimum, enzyme stability, enzyme kinetics K_M, and $V_{max.}$ Obtained, 14 isolates that further selected to 4 best isolates based on highest polygalacturonase activity and keprok Garut orange juice clarification ability. Four selected enzyme isolates were AR 2, AR 4, KK 4, and KK 5 had ability to increase juice transmittance, decrease juice viscosity and also reduce total soluble solid. Furthermore 4 selected isolates were partially purified by ammonium sulphate precipitation and dialysis method. Four partially purified enzymes were known that enzyme character of AR 2 optimum at pH 6; AR 4 optimum at pH 5.5; KK 4 optimum at pH 6; and KK 5 optimum at pH 4.5. Four enzymes were optimum at temperature 60^0C thus stable at temperature 50-60^0C, this characteristic indicate that enzymes were thermostable. AR 2 showed active activity stable at pH 4-7; AR 4 showed active activity stable at pH 6-7; KK 4 showed active activity stable at pH 4-6; however KK 5 stable at pH 4-5. Enzyme AR 2 and KK 4 was getting inactive at pH 11, thus AR 4 and KK 5 inactive at pH 12. K_M value of AR 2, AR 4, KK 4, and KK 5 was 0.0959; 0.0974; 0.0966; and 0.178 mg/ml respectively. V_{max} of AR 2, AR 4, KK 4, and KK 5 was 0.0203; 0.0202; 0.0185; and 0.0229 U/ml respectively. Enzyme AR 2 was the most compatible enzyme to be applied in keprok Garut orange juice clarification for it had the lowest K_M value..

1. Introduction

Keprok Garut orange *(Citrus nobilis var. chrysocarpa)* belongs to keprok orange *(Citrus nobilis)* species which has best quality among other orange species in Indonesia [18;21] particularly cultivated in Garut and it has been recognized as superior variety. Highest pectin percentage in keprok Garut orange is about 3-3.5% [1]. Orange has been commonly processed into orange juice.

Orange juice with a high content of pectin cause turbidity and viscosity effect in juice. This condition allows difficulty of extraction and filtration, as pectin particle blocks filtration machine where the juice must pass. Pectinase (especially polygalacturonase) addition in the extraction process improves fruit juice yield through an easier process, decreases juice viscosity and degrades the gel structure, thus improving juice concentration capacity. Enzymatic clarification can increase yields by more than 90% compared to conventional mechanical juicing, besides improving the organoleptic (color, flavor) and nutritional (vitamins) properties and technological efficiency (ease of filtering) [14]. Industry prefers to higher temperature for juice clarification condition. For high temperature decrease viscosity of the solution, increase substrate solubility and reduce the risk of microbial contamination. Thermostable enzymes isolated from thermophilic organisms are needed to adapt that juice clarification condition in the industry.

Polygalacturonase is one of many well known pectinolytic enzyme that catalyzes hidrolysis of polygalacturonic acid producing D-galacturonic acid, has *thermostable* characteristic, optimum pH 3-6 [17]. Pectin depolimerisation with polygalacturonase can decrease juice viscosity and contribute in improving flavor of juice [14;25]. The most commonly bacteria produced polygalacturonase are *Bacillus* sp. [14] dan *Pseudomonas* sp. [7].

This research was done to screen and characterize polygalacturonase from pectinolytic bacteria found in keprok Garut peel and vegetable waste which had potential in keprok Garut orange juice clarification.

2. Material and Methods

2.1. *Material*

Pectin citrus and D-galacturonic acid used in this research were from Sigma Aldrich. Various isolate sources like keprok Garut peel and vegetable waste (darker colored mustard green and cabbage) were collected from local market (Surakarta, Indonesia). Those samples were rotted for five days then incubated at 55^0C for three days. Keprok Garut orange for polygalacturonase clarification ability test was also obtained from the same local market.

2.2. *Isolation and screening of pectinolytic bacteria*

Isolates of pectinolytic bacteria were isolated from decaying like keprok Garut peel and vegetable waste and repeatedly sub cultured on pectin plates, then purified colonies obtained from quadrant streak method. The medium composition consist of yeast extract 1g/L; Na_2HPO_4 0.9 g/L; KH_2PO_4 1g/L;$MgSO_4.7H_2O$ 0.5 g/L; KCl 0.2 g/L; Pectin citrus 1g/L; and Bacteriological Agar 15g/L.The isolates were maintained in pectin agar slant at 4^0C for further studies. Morphological characteristics of isolates were identified using Gram staining, endospore staining, catalase test and colony morphology. Isolates which were isolated from decayed keprok Garut orange peel designated as KK, then from decayed vegetable waste designated as AR.

2.3. *Enzyme production*

The bacterial isolated were separately cultured at 55^0C for 24 hours in pectin citrus broth (yeast extract 1g/L; Na_2HPO_4 0.9 g/L; KH_2PO_4 1g/L; $MgSO_4.7H_2O$ 0.5 g/L; KCl 0.2 g/L; and Pectin citrus 1g/L). The pH of the medium was adjusted at range 4-6 (must be acid condition to ensure the only pectinolytic enzyme produced was polygalacturonase). The broth from each culture was centrifuged (Hettich) at 6000 rpm for 15 min at 4^0C and the supernatant was collected as crude enzyme extract.

2.4. *Enzyme assay*

Polygalacturonase activity was assayed using D-galacturonic acid as standard. To measure the activity of polygalacturonase, the assay mixture (1 cc) containing 0.1 cc volume of enzyme and 0.9 cc (0.7% w/v pectin dissolved in 0.025 M asetat buffer (pH 4.8) was incubated at 55°C for 30 min. The reducing sugar released was measured by the dinitrosalicylic acid method (Miller, 1959). The reaction in the test tube was stopped by adding 1 cc of DNS, boiled during 10 min at 90^0C, cooled and added 0.5 ml potassium sodium tartrate 40% [15;12]. Optical density (OD) of the solution was measured by spectrophotometer (UV-Vis Shimadzu), at 540 nm. A standard curve was prepared using D-galacturonic acid (0 to 90 μg/ml) and a linear regression analysis was used to determine the total reducing sugar. Blanks were prepared with inactivated enzymes. One unit of enzymatic activity (U) was defined as one μmol of galacturonic acid released per minute.

2.5. Juice clarification
Keprok Garut orange that was already peeled then extracted without additional water. Orange juice was filtered by the manual filter to separate from big orange dreg. 30 cc juice was treated with different enzymes (3 cc) and incubated for 60 min at 55^0C. Enzymes were also given to 1% pectin solution. Physical characteristic was measured by viscosity by relative viscosity method [27], trasmittance (%T) using spectrophotometer (UV-Vis Shimadzu) at 600nm while 1% pectin solution at 1100nm, total soluble solid (^{0}Brix) determined by Hand Refractometer (ATAGO).

2.6. Growth rate and enzyme production curve
Growth rate and enzyme activity was determined by taking samples aseptically from submerged fermentation at the interval 15 minutes for 2 initial hours of fermentation, start from 3rd hour till 13th hour samples were taken every 1 hour .

2.7. Polygalacturonase purification
Polygalacturonase was purified using ammonium sulphate followed by dialysis. Different levels of ammonium sulphate saturation were used (50, 60, 70, and 80%). Known volume and units of crude enzyme filtrate was treated with these levels of ammonium sulfate saturation overnight at 4^0C and then centrifuged (Hettich) at 12.000 rpm for 10 minutes. The obtained precipitate was dialyzed in a selophan membrane against sodium acetate buffer (pH 5.2;0.05 M), overnight at 4^0C in cool chamber under agitation, with change of the buffer every 8 hours.

2.8. Enzyme specific activity determination
Protein concentrations in crude enzyme, and the partially purified one were determined as described by Lowry (1951) using bovine serum albumin as standard. Enzyme specific activity was defined as enzyme units per mg protein.

2.9. Characteristics of partially purified Polygalacturonase:

2.9.1. Optimum pH
0.1 cc volume of enzyme and 0.9 cc (0,7% w/v pectin dissolved in 0.025 M asetat buffer) pH 4.0; 4.5; 5.0; 5.5 and 6.0 was incubated at 55°C for 30 min. Enzyme activity was measured by DNS method.

2.9.2. Optimum temperature
0.1 cc volume of enzyme and 0.9 cc (0,7% w/v pectin dissolved in 0.025 M asetat buffer pH 4.8) was incubated at 35, 40, 45, 50, 55 and 60°C for 30 min. Enzyme activity was measured by DNS method.

2.9.3. Thermal stability
0.1 cc volume of enzyme dissolved in 0.1 cc 0.025 M asetat buffer in optimum pH of each enzyme then incubated at 30, 40, 50, 60, 70, 80, and 90°C for 30 min. Enzyme activity was measured by DNS method.

2.9.4. pH stability
0.1 ml volume of enzyme dissolved in 0.1 ml 0.025 M asetat buffer pH 3, 4, 5, 6, 7, 8, 9, 10, 11, 12 then incubated at optimum temperature of each enzyme for 30 min. Enzyme activity was measured by DNS method.

2.9.5. Enzyme kinetics
The Michaelis constant (K_M) of the purified polygalcturonase (PG) was determined by measuring the reaction velocities (μ mol galacturonic acid /min) at various concentrations of pectin 0.05; 0.1; 0.2; 0.4; 0.8; and 1 (mg.ml^{-1}) dissolved in 0.025 Masetat buffer pH 4.8at 55°C for 30 min. The data were plotted according to Lineweaver-Burk plot [16].

3. Result and Discussion

3.1. Isolation and characterisation of pectinolitic bacteria
Pectinolytic bacteria was isolated from decaying like keprok Garut peel and vegetable waste and incubated at 55^{0}C to obtain thermophilic pectinolytic bacteria. High temperature of incubation was needed to screen potential thermostable polygalacturonase produced by the pectinolitic bacteria so it will be appropriate for industry clarification process which uses high temperature. According to [2] the temperature of clarification of orange juice is 37-55^{0}C. Nine isolates from vegetable waste (designated as AR) and five isolates from keprok Garut orange peel (designated as KK) were already obtained.

Gram staining showed that all isolate was Gram negative except AR 5. Endospore staining with malachite green and safranin solution showed no isolate had endospore. Catalase test showed all isolate positive (found forming CO_2), it means all isolates were aerob as reported by [26] all isolates collected from sweet orange peel also showed had no endospore, positive in catalase test, and had characteristics as Gram negative bacteria.

3.2. Screening potential isolates for juice clarification.
First priority choosing four best-potential isolates was from four highest polygalacturonase activity by DNS method to ensure that dominant enzyme act in juice clarification (decreasing viscosity and increasing transmittance) was polygalacturonase. The activity of PG in crude enzymes was studied (Table 1) the range of PG activity in crude enzime produced was 0.001 Unit/ml to 0.123 Unit/ml. Enzyme isolate AR 2 had highest PG activity 0.123 Unit/ml. AR 4 was second highest PG activity 0.075 Unit/ml, KK 4 as the third with 0.074 Unit/ml and KK 5 as the fourth with0.064 Unit/ml. Screened four isolates from 14 isolates, those are AR 2, AR 4, KK 4 and KK 5.Orange juice that already clarified by pectinolitic enzyme will show clearer and less viscous. [9] pectic enzyme are used to reduce hase or gelling of juice. The addition of pectinases is preferred since it increases the volume of free-run juice and reduces pressing time. Range of juice viscosity with crude enzyme addition was 0.896 cp – 1.211 cp,whereas in control (orange juice without addition of crude enzyme) had viscosity value 0.946 cp, it showed decreasing viscosity. Similar result had been reported by [8] pectinase addition could reduce mango juice viscosity caused by depolimerisation of soluble pectin. Pectinase can hydrolyze protopectins and pectins to smaller chains like galacturonic acid [22], leading to a reduction of water holding capacity. As a consequence, free water was released to the system caused a reduction of viscosity [10]. It promoted the reduction of waste loss because the less viscous pure was easier for the filtration, showing a significant increase in juice yield [19]. Range of orange juice %transmittance with enzyme addition was 2.10 – 2.90 whereas in control %transmittance was 1.90. All crude enzyme could increase transmittance value of orange juice. It had been reported by [26] that enzyme addition could increase % transmittance in both orange juice and 1% pectin solution. All orange juice added with crude enzyme show ^{0}Brix 8, where it was less than control with ^{0}Brix 9 for pectinase degradated pectin substance into simple reduction sugar as stated by [8].

Table 1. Enzyme activity and clarification ability in juice

Isolate	Crude Enzyme Activity (Unit/ml)	Total Cell (celll/ml)	Orange Juice			1% Pectin Solution	
			Viscosity (cp)	% T	TPT (^{0}Brix)	Viscosity (cp)	% T
AR 1	0.059	1.190. 10^6	0.938	2.60	8	1.234	83.10
AR 2	0.123	0.540. 10^6	0.912	2.90	8	1.124	83.80
AR 3	0.036	0.548. 10^6	0.896	2.60	8	1.173	83.80
AR 4	0.075	0.654. 10^6	0.930	2.70	8	1.197	83.20
AR 5	0.031	0.688. 10^6	1.211	2.10	8	1.211	83.00
AR 6	0.039	1.288. 10^6	1.078	2.10	8	1.178	83.00
AR 7	0.019	1.142. 10^6	0.923	2.50	8	1.222	83.40
AR 8	0.046	0.756. 10^6	0.905	2.60	8	1.221	83.40
AR 9	0.027	0.786. 10^6	0.925	2.40	8	1.174	83.80
KK 1	0.053	0.618. 10^6	0.904	2.50	8	1.198	83.40
KK 2	0.001	0.552. 10^6	0.951	2.50	8	1.203	83.20
KK 3	0.058	0.744. 10^6	0.917	2.40	8	1.464	83.30
KK 4	0.074	0.442. 10^6	0.945	2.70	8	1.193	83.30
KK 5	0.064	0.218. 10^6	0.900	2.50	8	1.173	83.00
Control			0.946	1.90	9	1.199	82.90

3.3. Determination of bacterial growth periode and pectinase productivity
Determination of bacterial growth period and pectinase productivity was used to determine time of product recovery. In brief enzyme synthesis period with highest activity was used as time production of partialy purified polygalacturonase. The highest activity was seen at the end of exponential phase, before the initial stationary phase. AR 2 and AR 4 isolate showed log total cell/ml at 10th hour after inoculation, the highest activity of enzyme AR 2 at 9th hour; AR 4 at 10th hour. KK 4 showed log total cell/ml at 11th hour after inoculation, the highest enzyme activity at 12th hour; whereas KK 5 showed log total cell/ml and highest enzyme activity at 12th hour. This result has similar time production with [26] which reported pectic enzyme isolated from sweet orange peel (designated as KJ-9) had time production at 10th hour.

3.4. Enzyme Characterization

3.4.1. Optimum pH
Palmer stated that the changes of enzyme tertiary structure caused hydrophobic group contact with water thus solubility of enzyme decreased. Decreasing enzyme solubility caused decreasing enzyme activity gradually[13]. Polygalacturonase activity found to be in optimum condition at several specific pH. Changes in pH have an effect on the affinity of the enzyme for the substrate [4]. According to [5] polygalacturonases have a relatively higher optimum temperature 50-60^0 C and a high tolerance to low pH 4.5-6.0. These properties could lead to their becoming acceptable in food industry for fruit juice sprocessing. Enzyme isolate AR 2 optimum at pH 6; AR 4 optimum at pH 5.5; KK 4 optimum at pH 6; and KK 5 optimum at pH 4.5 (figure 1).

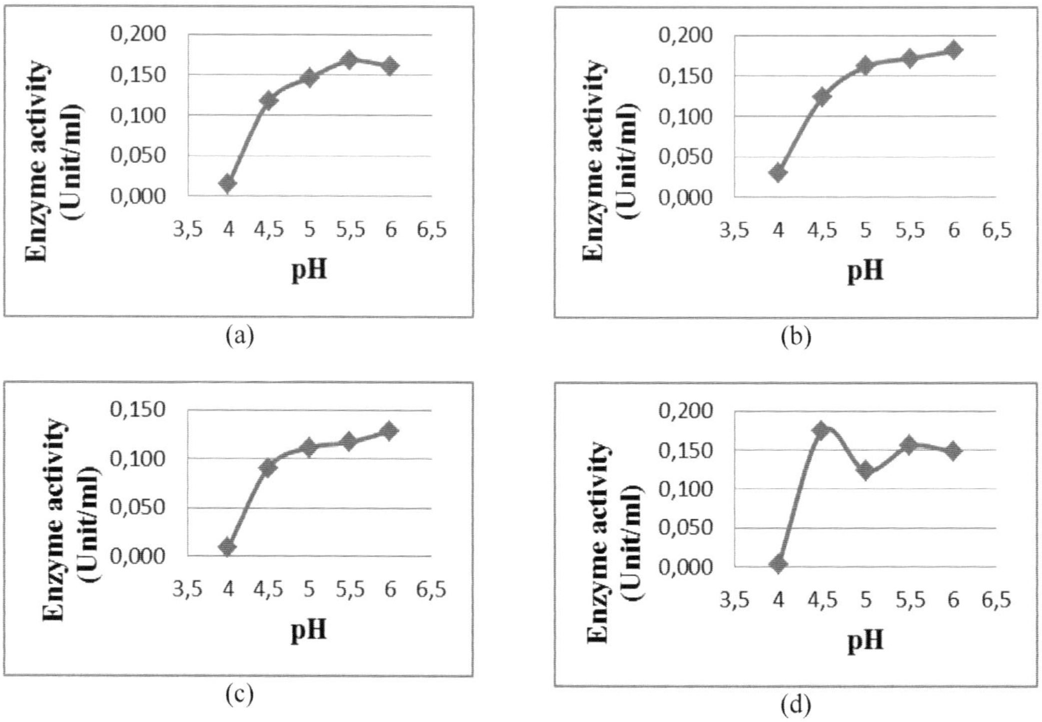

Figure 1. Optimum pH of enzymeAR 2 (a), AR 4 (b), KK 4 (c), KK 5 (d).

3.4.2. Optimum Temperature

Effect of temperature is related with reaction rate that calaysed by enzyme.This could be due to the breakage of secondary, tertiary and quaternary bonds that maintain the three dimensional structure of enzymes at high temperature and there by would lead to conformational changes of the enzyme active site. It is known that enzyme activity increases with increasing temperature up to the optimum temperature as the result of increasing kinetic energy, which can favor rate of collisions between substrate and enzyme during hydrolysis process [20]. Optimum temperature condition of all enzymes was found to be at 60^{0}C (figure 2). This optimum activity was similarly determined in the previous study by [3] which showed highest polygalacturonase activity of *Bacillus* sp. AD 1 was found to be at 60^{0}C. Regarding bacterial pectinases, the optimal temperature for a PGase from *Streptomyces* sp. QG-11-3 is 60°C [23]. Higher optimum temperature offers some advantages for industrial process such lower viscosity of the substrate, reducing the risk of microbial contamination and provides better solubility at high temperature [24].

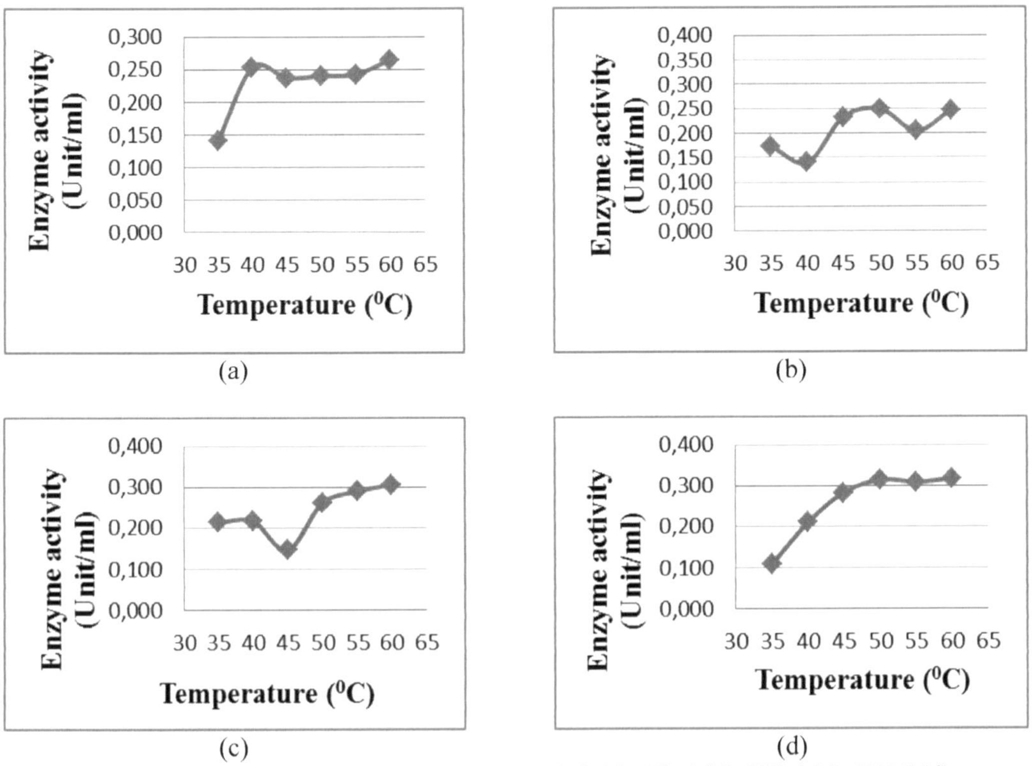

Figure 2. Optimum temperature ofenzymeAR 2 (a), AR 4 (b), KK 4 (c), KK 5 (d).

3.4.3. pH and thermal stability

Enzyme AR 2 isolate showed active activity and constantly stable at pH 4-7; AR 4 showed active activity and constantly stableatpH 6-7; while KK 4 at pH 4-6; and KK 5 showed active activity and constantly stable atpH 4-5. Enzyme AR 2 and KK 4 isolates start being completely inactive at pH 11, while AR 4 and KK 5 start being completely inactive at pH 12 (figure 3). Generally pH showed enzymes still active and stable at pH 4-10. Based on thermal stability result, enzyme AR 2, AR 4, KK 4 andKK 5 isolates were stable at 50-60^{0}C (figure 4). Thermal stability results of PG indicated that all enzymes were thermostable enzyme.The result tended similarly with pevious research. The optimal pH pectic enzymes may act range between 3.5-11, while the optimal temperatures vary between 40-75°C [6; 9].

IOP Conf. Series: Journal of Physics: Conf. Series **909** (2017) 012088 doi:10.1088/1742-6596/909/1/012088

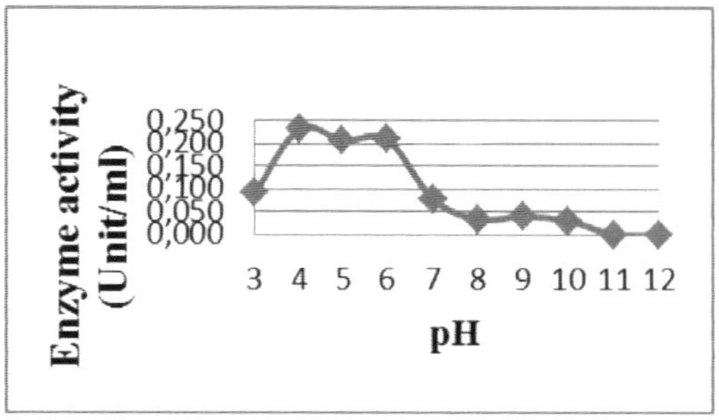

(a)

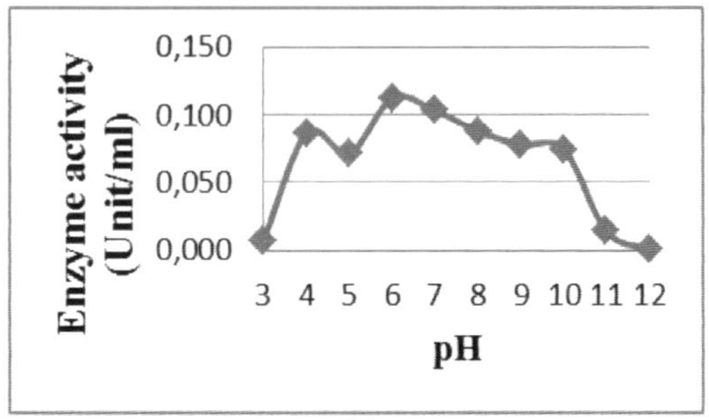

(b)

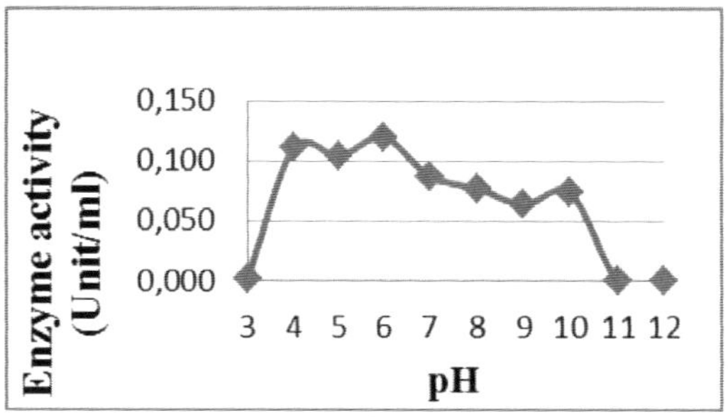

(c)

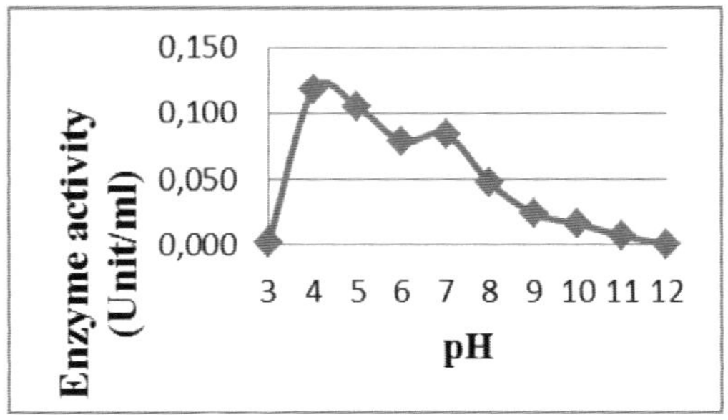

(d)

Figure 3. pH stability of enzyme AR 2 (a), AR 4 (b), KK 4 (c), KK 5 (d) at incubation 60^0C

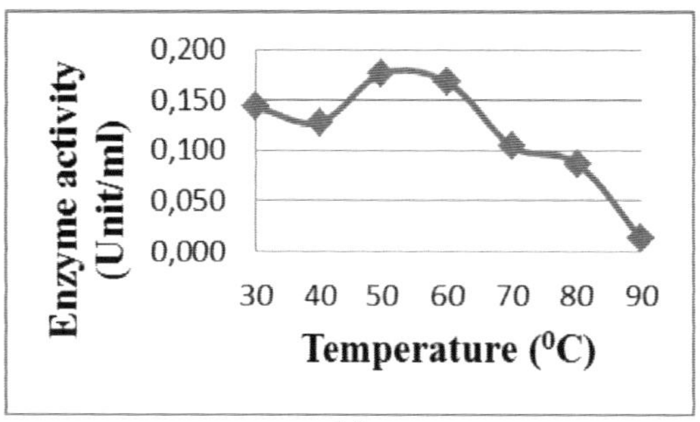

(a)

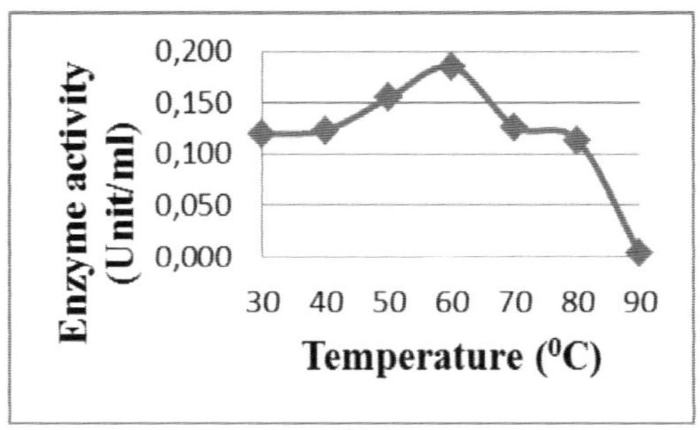

(b)

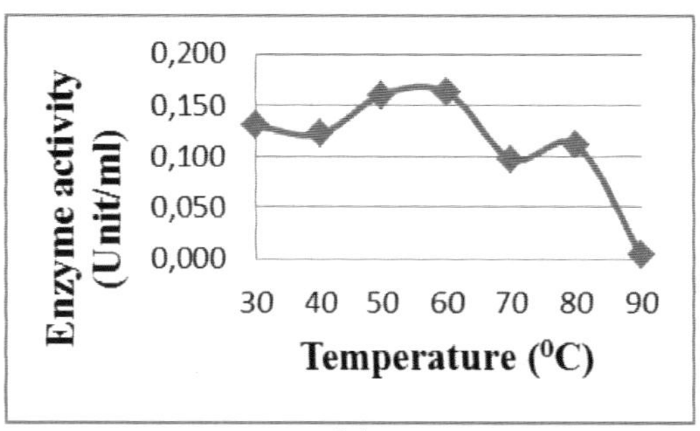

(c)

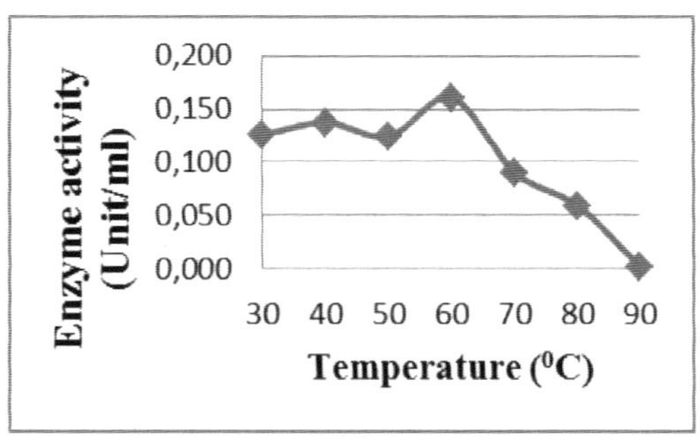

(d)

Figure 4. Thermal stability of enzyme AR 2 at pH 6(a); AR 4 at pH 5,5(b); KK 4 at pH 4(c); KK 5 at pH 4,5(d).

3.4.4. Enzyme kinetics

Table 2. K_M and V_{max} Enzyme AR 2, AR 4, KK 4, and KK 5 Isolate

Isolate	K_M (mg/ml)	V_{max} (U/ml)
AR 2	0.0959	0.0203
AR 4	0.0974	0.0202
KK 4	0.0966	0.0185
KK 5	0.178	0.0229

The K_M and V_{max} of AR 2, AR 4, KK 4, and KK 5 enzymes is shown on Table 2. Enzyme isolate AR 2 had lowest K_M value compared with others, was found to be 0.0959 mg/ml. Therefore, this low K_M value signified strong affinity of this enzyme for the substrate. For the lower K_M value, determines the higher affinity between substrate and enzyme. Higher affinity between polygalacturonase with pectin (pectic acid) as substrate makes higher potential to degrade pectin. [16] stated that the difference of V_{max} and K_M value was related with enzyme purification level.With result above, obtained AR 2 as the most ideal enzyme to be applied in keprok Garut orange juice clarification. The highest K_M value was found in KK 5 to be 0.178 mg/ml; the highest V_{max} was found in KK 5 to be 0.0229U/ml; whereas the lowest V_{max} was found in KK 4 to be 0.0185U/ml. This range of K_M and V_{maks} was in line with range of K_M and V_{max} respectively reported by [26] to be 0.0170-0.025mg/ml and 0.0386-0.040U/ml.

4. Conclusion

14 were screened up to four best isolates based on highest polygalacturonase activity and keprok Garut orange juice clarification ability. Four selected enzyme isolates are AR 2, AR 4, KK 4, and KK 5 had ability to increase juice transmittance, decrease juice viscosity and also reduce total soluble solid. Four partially purified enzymes had each propertiesAR 2 optimum at pH 6; AR 4 optimum at pH 5.5; KK 4 optimum at pH 6; and KK 5 optimum at pH 4.5. Four enzymes were optimum at temperature 60^0C thus stable at temperature 50-60^0C, this characteristic indicates that enzymes were thermostable. All screened enzymes were stable in acidic condition, while enzyme AR 2 and KK 4 was getting inactive at pH 11, thus AR 4 and KK 5 inactive at pH 12. K_M value of AR 2, AR 4, KK 4, and KK 5 was 0.0959; 0.0974; 0.0966; and 0.178 mg/ml respectively. V_{max} of AR 2, AR 4, KK 4, and KK 5 was0.0203; 0.0202; 0.0185; and 0.0229 U/ml respectively. Enzyme AR 2 was the most compatible enzyme to be applied in keprok Garut orange juice clarification for it had the lowest K_M value.

Acknowledgemnet
This research was funded by Start-up insentive research PNBP UNS 2014.

References

[1] Aak 1994 *Cultivation of citrus plants* (Yogyakarta: Kanisius Publisher)
[2] Arthey D and Ashurst P R 2001 *Fruit processing nutrition product and quality management*. 2, ed (Maryland: AM Aspen Publication)
[3] Dey A, Karmakar M, and Ray R R 2011 Extracellular thermostable polygalacturonase from *Bacillus sp. AD 1 .Der Pharmacia Lettre*. 3 2 358-367
[4] Dixon M and Webbs E C 1971 *Enzymes* (Great Britain: Williams Clowes and Sons)
[5] Freitas P M, Martin N, Silva S and Gomes E 2006 Production and partial characterizationof polygalacturo-nases produced by thermophilic Monascus sp. N8 and by thermotolerant *Aspergillus sp. N12* on solid-state fermentation. *Brazil. Brazilian Journal of Microbiology*. 37 302-306
[6] Gummadi T and Panda 2003 Purification and biochemical properties of microbial pectinases – a review. *Process Biochem*. 38 987–996.
[7] Hatanaka C and Imamura T 1974 Production of an exopolygalacturonase liberating

digalacturonic acid by a *Pseudomonas*. *Agriculture Biology Chemistry*. 38 11
[8] Iriani E S, Said EG, Suryani A and Setyadjit. 2005. Effect of difference concentrationof pectinase addition and incubation condition toward rendement and quality of mango kuini juice (*Mangifera odorata griff*). *Jurnal Pasca panen* 2 1 11-17
[9] Kashyap D R, Vohra P K, Chopra S, and Tewari R. 2001 Applications of pectinases in the commercial sector: a review. *BioresourTechnol.* 77 215-227
[10] Lee W C, Yusof S, Hamid N S A and Baharin B S 2006 Optimizing conditions for enzymatic clarification of banana juice using response surface methodology (RSM). *Journal Food Engineering* .73 55-63
[11] Miller G L 1959 Use of dinitrosalicylic acid reagent for determination of reducing sugar.*Analytical Chemistry*. 31 426 – 432
[12] Ordonez R G, Morlon J, Gasparian S, and Guyot J P. 1998.Occurrence of a thermoachidophilic cell-bound exo-pectinase in *Alicyclobacillus acidocaldarius*. *Folia Microbiology 4*. 3 6 657-660
[13] Palmer T 1981 *Understanding enzymes* (England: Elli Horwood)
[14] Pedrolli DB, Monteiro, A.C., Gomes, E., and Carmona, E.C. 2009.Pectin and pectinases: production, characterization and industrial application of microbial pectinolytic enzymes. The Open Biotechnology Journal 3 9-18
[15] Phutela U, Dhuna V, Sandhu S and Chadha BS 2005 Pectinase and polygalacturonase production by a thermophilic *Aspergillus fumigatus* isolated from decomposting orange peels. *Brazil. Brazilian Journal of Microbiology*. 36 63-69
[16] Putra G P G 2007 Determination of endogenous polygalacturonase enzyme kinetics (PG) from kakao seed pulp. *Jurnal Biologi XIII* .1 21-24
[17] Reza S M 2007 Isolation and characterization of polygalacturonase produced by *Tetracoccosporium* sp. *Iran. Iran Journal Chemistry*. 26 1
[18] Sarwono B 1994 *Citrus and citrus family* (Jakarta:Penebar Swadaya)
[19] Sato A C K , Sanjinéz-Argandoña E J and Cunha R L 2006 The effect of addition of calcium and processing temperature on the quality of guava in syrup. *International Journal of Food ScienceTechnology*. 41 417-424
[20] Suhartono 1989 *Enzyme and biotechnology* (Bogor: PAU Bioteknologi IPB)
[21] Suminar E 2007 *Multiplication in vitro of keprokgarutorange (citrus nobilis var. chrysocarpha) to multiplyfree-disease seed.* Reseach Report of Young Researcher Unpad.(Bandung: LembagaPenelitianUniversitasPadjadjaran)
[22] Sun Y S, Wang J, Wu F, Chen X, Liao and Hu X 2006 Optimizing enzymatic maceration in pretreatment of carrot juice concentrate by response surface methodology. *International Journal of Food Science Technology*. 41 1082-1089.
[23] Torres E F, Sepulveda TV, Gonzales GV 2006 Production of hydrolytic depolymerising pectinases. *Journal Food Technology Biotechnology*. 44 2 221–227
[24] Turner P, Mamo G and Karlsson E N 2007 *Potential and utilization of thermophiles and thermostable enzymes in biorefining* (Sweden: Microbial Cell Factories)
25] Walter R 1991 *The chemistry and technology of pectin* (New York: Academic Press, Inc)
[26] Widowati E, Utami R, Nurhartadi E, Andriani M A M and Wigati A W 2013 Production and characterization of pectinase from pectinolitic bacteria in sweet orange juice. *JurnalAplikasiTeknologiPangan*.1
[27] Yuwono S and Susanto T 2001 *Testing of physical properties of food* (Surabaya: UNESA Publisher)

The effect of Taro (*Colocasia esculenta* L.) and Lesser Yam flour (*Dioscorea esculenta* L.) as thickener agent on physical characteristics of frozen wheygurt

E. Nurhartadi, R. Utami, E. Widowati, and B.M. Karunawati[1]

Department of Food Science and Technology Faculty of Agriculture
Sebelas Maret University
Jl. Ir. Sutami 36A Kentingan Jebres Surakarta 57126, INDONESIA

E-mail: edhi.nr@gmail.com

Abstract. The results showed that the addition ratio of taro and lesser yam flour affected to the physical characteristics of frozen wheygurt. The addition of lesser yam flour increased total soluble solids until the addition ratio of 2:2 due to the higher ash content of lesser yam (2.87%) than taro (0.44%). Sample with addition ratio of 1:3 and 0:4 significantly different compared to other samples, due to the starch content difference between taro (70-80%) and lesser yam (51.34%). Addition ratio of taro and lesser yam flour do not have a significant effect on the viscosity of the frozen wheygurt, due to both starch have similar setback viscosity. Lesser yam setback viscosity was 684.8 cP, while taro was 838.3 cP. Setback viscosity showed a high tendency of retrogradation. The addition ratio of taro and lesser yam flour have a significant effect to the overrun of frozen wheygurt. Addition ratio of taro and lesser yam flour have a significant effect to melting rate of frozen wheygurt. This result was caused by higher peak viscosity of taro starch compared to lesser yam, thus produced thicker gel than lesser yam. This lead increased water contents in the mixtures entrapped and slows down water mobility, hence melting rate would decrease.

1. Introduction

Cheese whey is a by-product of the cheese-making process in the form of green-yellowish liquid. The cheese-making industry generated 9 kilos of cheese whey to make a kilogram of cheese[1]. Cheese whey contains 0.006-0.0070% fat, 0.3%-0.9% protein, 3.04-5.00% lactose, 0.5-0.62% ash[2]. Cheese whey can be utilized as various fermented products because of its compounds can be used by bacteria as growth medium source[3].

Frozen yogurt is now a rising trend among other fermented drink products. Frozen yogurt is a dessert which combines ice cream texture with nutritive and health properties of yogurt[4]. The textural quality of frozen yogurt determined by dairy fat and total solids of the used milk. However, cheese whey contains low total solids which make its derived product has low viscosity compared to commercially fermented products[3].

The viscosity of whey based product can be improved by thickener addition. Hydrocolloid often used as thickener. One of the most common hydrocolloid types used is starch. Arrowroots starches are suitable for acidic and frozen product like frozen yoghurt[5]. Taro (*Colocasia esculenta* L.) and lesser yam (*Dioscorea esculenta* L) have potential to be used as thickener because of their high starch content (approximately 51.34-70.92%), and the same amylopectin and amylose fraction ratio (taro

21.44:78.56 and lesser yam 24.30:75.7). However, these two materials have different gelatinization profile. Taro starch has high peak gelatinization viscosity[6] thus makes ice cream made from taro starch melts slowly[7]. Lesser yam starch has low peak gelatinization viscosity[8] thus makes ice cream can melt in an ideal time range[9]. In this study, taro and lesser yam flour are combined to find the best physical characteristics of frozen wheygurt. Furthermore, these two tubers have prebiotic contents which can be utilized as lactic acid bacteria (LAB) growth medium in human digestion system and metabolites formation, such as oligosaccharides in taro[10] and inulin in lesser yam[11].

The aim of this study was to determine the effect of taro and lesser yam flour on frozen wheygurt physical characteristics which includes viscosity, overrun, melting rate and total solids.

2. Materials and Methods

2.1. *Materials*

Mozzarella cheese whey obtained from cheese industry in Boyolali (Indonesia). Lactic acid bacteria used in this study were *Lactobacillus acidophilus* FNCC 0051 and *Lactobacillus plantarum* FNCC 0027 from Food and Nutrition Culture Collection Yogyakarta (Indonesia). Taro flour bought from Naya Tepung Bogor (Indonesia) and lesser yam flour bought from Kusuka Ubiku Bantul (Indonesia). Other materials used in frozen wheygurt preparations were egg yolk, sucrose (Gulaku®), whipping cream (Anchor®, 33.33% fat).

2.2. *Starter Preparation*

Both *L. acidophilus* FNCC 0051 and *L. plantarum* FNCC 0027 were inoculated in separate sterile deMan Rogosa Sharpe (MRS) broth medium and incubated at 37°C for 24 hours. 2% (v/v) of each culture of MRS broth were inoculated to 100 ml pasteurized skim milk and incubated at 37°C for 24 hours. The cultured skim milk contains 10^8 cells/ml of bacterial cells.

2.3. *Frozen wheygurt preparation*

Frozen wheygurt was prepared by mixing the materials (cheese whey, 12% (w/v) sucrose, 13% (w/v) whipping cream, 2% (w/v) egg yolk and 2,5% (w/v) thickener) for 10 minutes. The mixtures were pasteurized at 75°C for 30 min and cooled until 45°C. 2% (v/v) of each *L. acidophilus* and *L. plantarum* culture starters were inoculated into pasteurized mixtures and incubated at 37°C for 18 hours. Frozen wheygurt then refrigerated for 5 hours, frozen for 30 minutes at -18°C for 30 minutes, mixed using mixer for 20 minutes and frozen for 24 hours.

2.4. *Physical characteristics analysis*

The samples total soluble solid was determined using gravimetric method (AOAC 941.08). The viscosity was determined using Brookfield Viscometer[12]. Overrun was evaluated using following equation:

Overrun (%): $\frac{A-B}{B}$ x 100 (Rezaei *et al.*, 2012).

Note: A: Weight of unit mix, B: weight of equal volume (ml) of frozen yogurt

To evaluate melting rate, 2 grams of frozen yogurt was placed in a sealed cup and frozen for two days. Samples then were taken out from the freezer and put in room temperature. The time required for the frozen wheygurt to melt was measured using a stopwatch[14].

3. Results and Discussion

3.1. *Total soluble solid*

The adequacy of total soluble solids in ice cream dough serves to increase the thickness of ice cream dough and to maintain the air bubbles stability[15]. The increasing amount of dissolved solids shows that the amount of water in the dough decreases and reduces the amount of ice crystallization[16]. Moreover, total soluble solids also affect the ice cream texture. If total soluble solids contents were high, the ice cream texture will be flaccid, otherwise if total soluble solids contents were low resulted in ice crystal formation and coarse texture[17].

The result (Table 1) showed that the addition ratio of taro and lesser yam flours have a significant effect on the total soluble solids of wheygurt. The addition of lesser yam flour increased the total soluble solids until addition ratio of 2:2. This is caused by the higher ash content of lesser yam (2.87%) than taro (0.44%) [8; 6]. Sample with addition ratio 1:3 and 0:4 significantly different compared to other samples, due to the starch content difference between taro (70-80%) and lesser yam (51.34%). Total soluble solid is influenced by the composition of the mixtures. Total soluble solid in ice cream is composition materials in solid form[16]. This study used taro and lesser yam flour which have starch contents. Starch composes mix solid content, furthermore, the amylopectin molecule on the starch is very tight causing produced frozen yogurt has high total soluble solid[18]. Total soluble solid of all samples was under required standard (34%)[19] due to water and solid content ratio which is not balanced and syneresis occurred by the gel matrix of taro and lesser yam flour. The produced frozen wheygurt has higher water content and lower solid content.

Table 1. Effect of taro and lesser yam flour as thickener agent on physical characteristics of frozen wheygurt*

Formulation Taro: Lesser Yam	Total Soluble Solids (%)	Viscosity (cP)	*Overrun* (%)	Melting Rate (min)
4:0	18,618^{abc} ± 1,061	9,22^{a} ± 2,220	2,770^{a} ± 0,0270	7,715^{a} ± 0,370
3:1	18,986^{ab} ± 0,737	10,07^{a} ± 1,746	2,477^{b} ± 0,215	7,946^{a} ± 0,135
2:2	19,378^{a} ±0,532	9,51^{a} ± 1,114	2,562^{ab} ± 0,103	7,656^{a} ± 0,382
1:3	17,920^{bc} ± 0,839	9,06^{a} ± 0,540	1,951^{c} ± 0,190	6,388^{b} ± 0,118
0:4	17,709^{c} ± 1,421	9,76^{a} ± 0,890	1,899^{c} ± 0,187	6,645^{b} ± 0,299

*** Results in the same column with the same superscript were not significantly different (α=5%)**

3.2. Viscosity

Frozen yogurt is an oil-in-water emulsion food system. Emulsions formed in frozen yogurt are unstable thermodynamically. Hydrocolloids usually act as emulsion stabilizers. As a material that can serve to maintain longer emulsion stability through adsorption mechanism. The way it works traditionally as a stabilizer by improving the structure, compacting, gelation of the aqueous continuous phase[20].

Taro and lesser yam flours used in this study because of their starch contents, a hydrocolloid usually used as thickener. These flours can increase the viscosity because these flours contain high amylopectin fraction about 56-60%. Starch consists of two fractions which are amylose and amylopectin. Amylose is a water-soluble fraction, whereas amylopectin is a water-insoluble fraction. Starch can not dissolve in cold water and form an impenetrable suspension of light with a strong movement to settle. If the starch solution is further heated up to the point of gelling of the starch wherein the starch granules expand. The expanding starch granules will absorb water up to the maximum size of starch granules. Under these conditions the maximum viscosity increase has been achieved[21]. This resulted in the water content in the dough to be bound, thus increasing the viscosity of frozen wheygurt dough. Taro flour has a high peak viscosity, amounted to 2865.3 cP[6]. Viscosity peak indicates maximum viscosity that can be achieved by starch during gelatinization. The higher the viscosity peak, the more viscous and dense gel produced[8].

The results showed that addition of taro and lesser yam flour do not have a significant effect on the viscosity of the frozen wheygurt (Table 1), due to both starch have similar setback viscosity. Lesser yam setback viscosity is 684.8 cP[22], while taro is 838.3 cP. Setback viscosity showed a high tendency of retrogradation. Starch retrogradation caused by the re-establishment of hydrogen bonds between the molecules of amylose and amylopectin [6].

3.3. Overrun

Overrun shows the amount of air entrapped in the ice cream mixture[23]. Overrun shows the ability of foaming and foam stability associated with a decrease in surface tension caused by protein molecules absorption[12].

The results showed that addition ratio of taro and lesser yam flour have a significant effect to the overrun of frozen wheygurt (Table 1). At the peak of gelatinization, taro has a viscosity of 2865.3 cP[6] while lesser yam has a viscosity of 1030.4 cP[22] thus the gel produced by taro was thicker and caused more air entrapped in frozen wheygurt dough and the overrun was higher. Overrun is influenced by mixtures viscosity[24], total soluble solids, thickeners as well as whipping process[25]. High mixtures viscosity resulted in high overrun value. This is in accordance with Stokes' Law where the higher the fluid viscosity, the slower particles velocity, consequently hinders air bubbles to coagulate. Hence, air bubbles remain dispersed and small-sized thereby increasing ice cream overrun[24].

In this study, whipping process was done using a hand-held mixer to replace ice cream maker, so there is a possibility of inconsistencies in the whipping process. Conventional method does not allow simultaneous stirring and cooling. The process is repeated and temperature change is occurred, so it is very influential to the overrun of ice cream[26].

3.4. Melting rate

Melting rate is the time required for ice cream to melt completely like the initial ice cream dough at room temperature. Good ice cream is ice cream that is resistant to melting at room temperature and within a certain time[27].

The results showed that addition of taro and lesser yam flour had a significant effect to melting rate of frozen wheygurt (Table 1). This result was caused by the higher peak viscosity of taro starch than lesser yam, thus makes taro produced thicker gel than lesser yam. This leads to more water contents in the mixtures being entrapped and slows down water mobility, therefore melting rate will decrease[15].

Melting rate depends on materials used in ice cream dough. Total soluble solids proportion in ice cream affected the ice cream melting rate. The increasing of total soluble solids can lower ice cream dough freezing point, hence more water contents being entrapped and decrease water mobility. The increasing of entrapped water in the mixtures will slow down ice cream melting rate[15]. Dough viscosity and overrun also affected the ice cream melting rate. Ice cream with high viscosity has higher melting resistance[29]. The high overrun rate will have slow melting rate due to lack of heat propagation rate caused by the large volume of air in ice cream[24]. In general, ice cream with low overrun melts faster than ice cream with high overrun[28].

4. Conclusions

Addition ratio of taro and lesser yam flour affected the physical characteristics of frozen wheygurt. Sample with higher taro addition has better physical characteristic compared to sample with higher lesser yam addition. The addition of lesser yam flour increased total soluble solids until the addition ratio of 2:2. Sample with addition ratio of 1:3 and 0:4 significantly different compared to other samples, Addition ratio of taro and lesser yam flour do not have a significant effect to the viscosity of the frozen wheygurt, Addition ratio of taro and lesser yam flour have a significant effect on overrun and melting rate of frozen wheygurt.

References

[1] Siso M I G 1996. *Bioresource Technology* **87** 1-11.
[2] Kumar PA and Bangaraiah P 2014. *International Journal of Pharma and Bio Sciences*. **5** 1101-11

[3] Fatma, Soeparno, Nurliyani, Hidayat C, dan Taufik M 2012. *Agritech* **32** 352-61
[4] Mahrous H and Salam R 2016. *American Journal of Food and Nutrition* Vol.**6** 1-10
[5] Noraziah A Y and Theng C C 2013. *4th International Conference on Biotechnology and Food Science*
[6] Kafah F F S 2012. Fakultas Teknologi Pertanian. Institut Pertanian Bogor
[7] Syahbania N 2012. Jurusan Teknologi Pertanian, Fakultas Pertanian. Universitas Hasanuddin
[8] Richana N and Sunarti T C 2004. *J. Pascapanen* **1** 29-37
[9] Kusumawati A 2013. Program Studi Pendidikan Biologi, Fakultas Keguruan dan Ilmu Pendidikan. Universitas Muhammadiyah Surakarta
[10] Nuraida L Palupi N S Ekasari P D Widayanti N W 2006. *Prosiding Mikrobiologi dan Bioteknologi PATPI*
[11] Prabowo A Y , Estiasih T and Purwantingrum I 2014. *Jurnal Pangan dan Agroindustri* **2** 129-35
[12] Hakim L Purwadi P, and Masdiana 2002. Fakultas Peternakan. Universitas Brawijaya
[13] Rezaei R, Khomeiri M, Aalami M, and Kashaninejad M 2012 *J Food Sci. Technol.* **51** 2809-2814
[14] Hubeis M, Andarwulan N, and Yunita M 1996 *Buletin Teknologi dan Industri Pangan ITB.* **7** 1-1-10
[15] Widiantoko R K and Yunianta 2014. *Jurnal Pangan dan Agroindustri* **2** 54-66
[16] Windhianingrum N, Purwadi I and Thohari 2015. Fakultas Peternakan. Universitas Brawijaya Malang
[17] Abdullah M Rehman S Zubair H Saeed H M Kousar S and Shahid M 2003. *Pakistan Journal of Nutrition* **2** 305-11
[18] Dewi L K Thohari I and Radiati L E 2014. Fakultas Peternakan. Universitas Brawijaya. Malang
[19] Badan Standarisasi Nasional 1995. Standar Nasional Indonesia (SNI). SNI-01-3713-1995. Es Krim. (Jakarta : Dewan Standarisasi Indonesia)
[20] Phillips G O and Williams P A 2009. *Handbook of Hydrocolloids 2nd Edition.* (Cambridge: Woodhead Publishing Limited)
[21] Wuestenberg T 2015. *Cellulose and Cellulose Derivatives in the Food Industry: Fundamentals and Applications*. E-book.. (Wienheim: John Wiley and Sons)
[22] Alsuhendra R 2009. Program Studi Tata Boga, Jurusan Ilmu Kesejahteraan Keluarga, Fakultas Teknik. Universitas Negeri Jakarta
[23] Syahputra E 2008. Departemen Teknologi Pertanian, Fakultas Pertanian, Universitas Sumatera Utara
[24] Praptiningsih Y, Tamtarini R, and Anisa 2012. Fakultas Teknologi Pertanian. Universitas Jember
[25] Handayani N, Sulistyowati M, dan Juni S 2014.*Jurnal Ilmiah Peternakan* **2** 1-7
[26] Hartatie E.S. 2011. *Gamma* **7** 20-6
[27] Susilawati N F and Aditya W N 2014. *Jurnal Teknologi dan Industri Hasil Pertanian* **19**243-56
[28] Bodyfelt F, Clark S, Costello M, and Drake M (ed.) 2008. *The Sensory Evaluation of Dairy Products* 2nd Edition. (New York: Springer)

Acknowledgments
The publication of this paper was supported by Hibah MRG-UNS (PUT-MRG) 2017 from Universitas Sebelas Maret.

IOP Conf. Series: Journal of Physics: Conf. Series **909** (2017) 012090 doi:10.1088/1742-6596/909/1/012090

Diversity of arbuscular mycorrhiza in the rhizosphere of Cajeput in agroforestry system with different fertilizer management of maize.

Parwi, B Pudjiasmanto , D Purnomo and VR Cahyani

Doctoral Study Program of Agricultural Science, Graduate School, Sebelas Maret University

Email: vitaratri@staff.uns.ac.id

Abstract. This study investigated the diversity of arbuscular mycorrhiza in rhizosphere of cajeput with different fertilizer management of maize. This research was conducted by observation on cajeput agroforestry system in Ponorogo that have different fertilizer management of maize: conventional management (CM), universal management (UM) and alternative management (AM1, AM2, and AM3). The result showed that the highest infection of arbuscular mycorrhiza was observed in the plot of AM3, while the lowest colonization was observed in the plot of CM. Infection of arbuscular mycorrhiza in roots cajeput from five fertilizer management , ranging from 32.64% - 63.33%. In all fertilizer management, there were eight species of arbuscular mycorrhiza which five species were *Glomus* genus, one species was *Acaulospora* genus and two species were *Gigaspora* genus. *Glomus constrictum* was the dominant species in all fertilizer management. *Acaulospora favoeta* was found only in the plot of AM3. Spore density varies between 150-594 / 100g of soil. The highest spore density was observed in the plot of AM3, while the lowest spore density was observed in the plot of AM1. The highest diversity index value of arbuscular mycorrhiza (Species richness and Shannon-Wiener) was observed in the plot of AM3.

1. Introduction

Arbuscular mycorrhiza is favorable soil microorganisms to host plants and ecosystem. Arbuscular mycorrhiza obtains carbon from the host plant. On the other hand, arbuscular mycorrhiza has a role for growth and yield crop, namely: increase adsorption nutrient and water [1], can protect the plant from root pathogens [2], preventing stress plants from unfavorable environments [3]. Furthermore, arbuscular mycorrhiza can improve soil fertility on degraded soils [4].

The diversity of arbuscular mycorrhiza depends on soil type, plant type and climate condition. The diversity of arbuscular mycorrhiza is also dependent on fertilization management. High N fertilization has a negative impact on the diversity of arbuscular mycorrhiza [5]. On the other hand, fertilization of manure and straw have a positive effect on the diversity of arbuscular mycorrhiza [6].

Arbuscular mycorrhiza exist in rhizosphere Cajeput in an agroforestry system in Sukun Village, district of Ponorogo has an intercropping system with a permanent cycle of maize as the main plant, soybeans and fallow. Each farmer has differences in fertilizer management for maize cultivation. Suspected that different fertilizer management will affect the diversity of arbuscular mycorrhiza in the rhizosphere of cajeput. This study investigated the diversity of arbuscular mycorrhiza in rhizosphere of cajeput (*Melaleuca leucadendron* LINN) with different fertilizer management of maize (*Zea mays*

L). The information diversity of arbuscular mycorrhiza in cajeput rhizosphere can be used as a consideration in fertilizer management to create a sustainable agroforestry system

2. Materials And Methods

2.1. Location of sampling sites

This research was conducted by observation on cajeput agroforestry system in Ponorogo (7°52' S, 111°35' E, altitude 265 m) that have different fertilizer management of maize: conventional management (CM), universal management (UM) and alternative management (AM1, AM2, and AM3). Conventional management (CM) used 360 kg Urea / ha + 200 kg NPK / ha +2 t of chicken manure / ha. Universal management (UM) used 650 kg Urea / ha + 330 kg NPK / ha + 3.3 t of chicken manure / ha. Alternative management of AM1 (650 kg Urea / ha + 330 kg NPK / ha + 4 t of chicken manure / ha), AM2 (730 kg Urea / ha + 430 kg NPK / ha + Maize straw) and AM3 (650 kg Urea / ha + 330 kg NPK / ha + 3.3 t of chicken manure / ha + Maize straw). In CM, UM, AM1 and AM2 managements used minimum tillage, whereas AM3 used intensive tillage

2.2. Collection of root and soil samples

Soil samples were taken randomly from the rhizosphere of cajeput on an agroforestry systems. Each treatment was done randomly by three replications. In each replication, the soil sample was taken randomly for about five sub samples, so that total samples were 5 x 3 x 5 = 75 sub samples. Each sub sample comprised of 2 kg soil, which was put into a polybag and kept in room temperature before having soil analysis. Before the implementation of soil analysis, the same replication was mixed to obtain soil samples as a composite of sub samples.

Root samples were cleaned with running water to remove soil particles, then inserted into bottles containing 60% alcohol.

2.3. Estimation infection of arbuscular mycorrhiza

The technique of Phillips and Hayman [7] was used to detect infection of arbuscular mycorrhiza in cajeput root. Infection of Arbuscular mycorrhiza was determined from fress root. Root samples were cut into 1 cm and inserted into a glass beaker containing 10% KOH at 70-80 °C for 10 minutes. The KOH solution was removed, and the sample was washed with water. Root samples were treated using 1% HCl and stained with with Staining solutions (glycerol : lactic acid : aquadest = 2 : 2 : 1 + trypan blue 0.05%). Infection of arbuscular mycorrhiza was observed under the microscope. Infection of arbuscular mycorrhiza percentage was calculated from the number root infection segment out of total root segment

$$\text{Infection of Arbuscular mycorrhiza (\%)} = \frac{\text{Number of AM positive segments}}{\text{Total number of segments observed}} \times 100$$

2.4. Isolation of arbuscular mycorrhiza

Mycorrhiza spores were isolated by wet sieving [8] and decanting by sucrose centrifugation [9] procedures. 100 g soil was dissolved into 500 ml of water. The suspension was filtered with 250 µm, 90 µm, and 45 µm sieves filters. Filtrate on sieve 45 µm was added 60% sugar and stirred for 10 minutes. The spores were filtered with a 45µm sieve and placed in a petridish to be observed under a binocular stereomicroscope. Spores were sorted into groups and counted.

2.5. Identification of arbuscular mycorrhiza

Isolation of mycorrhiza spores was based on spore morphology including spore color, spore size, the number of spore walls, bolbours, hyphae and Melzer reagent. The identification is based on the morphological description provided by The international collection of vesicular arbuscular mycorrhizal fungi (*http:// invam.caf.wvu.edu*) and originally published species descriptions.

2.6. Statistical analysis

The diversity of arbuscular mycorrhiza was measured by spore density, Relative abundance, Species richness and Shanon Wiener index [10]. The formula used to calculate these parameters is given in Table 1. Data analysis was done using one-way analysis of variance (ANOVA) by SPSS program. LSD analysis (Students t least significant difference) was used to compare the treatments. The Pearson correlation coefficient was used to determine the relationship between spore density with Species Richness. Regression analysis was calculated to assess the significance of the relationship between spore density and species richness

Table 1. Diversity measures used to describe AM communities

Spore density	The number spores in 100 g soil
Relative abundance (RA)	(Spore number of species / total number of identified spore samples) x 100
Species Richness (SR) was estimated from Margalef Indeks (D_{Mg})	D_{Mg} = S- 1/ln N
Shannon-Wiener index of diversity (H')	H' = - $\sum$pi (ln pi)

Pi is the relative abundance of each identified species per sampling site and calculated by the following formula,
Pi = ni/Ni where n is the spore numbers of a species and N is the total number of identified species per sampling sites

3. Results And Discussion

3.1. Diversity of arbuscular mycorrhiza

Soil samples of cajeput rhizosphere were collected from five location that different of fertilization management revealed the presence of several arbuscular mycorrhiza species. A total of eight species of arbuscular mycorrhizas were wet sieved from the soil sample collected from five different of fertilization management. The identified species were *Glomus claroideum, Glamus lamellosum, Glomus coronatus, Glomus ambisporum, Glomus constrictum, Gigaspora rosea, Gigaspora margarita* and *Acaulospora favoeta* (Table 2). Similarly in this study also eight arbuscular mycorrhiza species were isolated by Pagano and Scotti [11] from two species of *Eucalyptus* plants in semiarid Brazilian. A number of arbuscular mycorrhiza species in this research was highest than in rhizosphere of *Eucalyptus globules* in Portugal (five species) was reported by Silvia *et al*, [12].

Spore abundance arbuscular mycorrhiza was highest *Glomus constrictum* (60.87% - 73.56%), followed by *Glomus claroideum* (13.61% - 26.81%), and lowest *Gigaspora rosea* (2.49% - 5.66%). Spore abundance arbuscular mycorrhiza was dominated genera *Glomus*. A similar observation of spore abundance arbuscular mycorrhiza was reported Pagano and Scotti [11] that the dominant arbuscular mycorrhiza on Eucalyptus is *Glomus sp.* Furthermore, Chen and Dell [13] proposed that spore abundance arbuscular mycorrhiza in Eucalyptus rhizosphere in China was highest *Glomus mossae* (67%), followed by *Glomus geosporum* (15.4%) and lowest *Acaulospora scrobicullata* (2.6%). On the other hand, Silvia *et al.* [12] which states that arbuscular mycorrhiza colonizing in the Eucalyptus rhizosphere was dominated by *Gigaspora margarita* (40%). Spore density in the agroforestry system of Cajeput ranges from 150 - 594/ 100g of soil. Spore density was highest in e plot of AM3 and lowest in the plot of CM (Figure 1).

Diversity indices (H' and SR) of arbuscular mycorrhiza was observed different on five kind fertilizer management. Shannon-wiener index (H') in the plot of CM (1.06 ± 0.11) was higher than in the plot of UM (0.94 ± 0.05), although the number spores in the plot of CM was lower than in the plot of UM. Shannon Wiener index was determined number spores and species distribution [14]. Shannon-Wiener index (H') was lowest (0.90 ± 0.07) in the plot of AM1 and highest (1.07 ± 0.03) in the plot of AM3. Richness species (SR) was highest (7.17 ± 0.57) in the plot of AM3 and lowest (5.46 ± 0.58) in the plot of AM1 (Table 3). The correlation analysis found that the spore density was positively correlated with Richness species (r = 0.848) (Figure 2). A similar observation was reported by Kavitha and Nelson [15] that spore density is positively correlated with Richness species (r=0.618)

Table 2. Relative abundance arbuscular mycorrhiza

Species	CM	UM	AM1	AM2	AM3
Glomus Claroideum	23.34±6.92	13.61±4.93	17.41±5.23	23.00±1.43	26.81±2.21
Glomus lamellosum	0.41±0.71	1.93±0.76	0.42±0.72	1.06±0.64	0.46±0.42
Glomus coronatus	3.70±1.45	1.23±0.76	2.73±3.23	2.19±0.82	2.68±1.64
Glomus ambisporum	2.31±2.46	2.53±0.80	2.78±1.70	3.06±2.05	1.27±0.25
Glomus Constrictum	61.87±9.37	73.56±3.17	71.63±4.14	64.73±2.87	60.87±2.35
Gigaspora Rosea	3.03±2.76	4.35±1.79	3.20±2.29	2.49±2.14	5.66±1.63
Gigaspora Margarita	5.30±6.39	2.79±0.37	1.83±1.63	3.46±1.70	2.13±0.91
Acaulospora favoeta	0	0	0	0	0.02±0.006

CM: Conventional management, UM: Universal management, AM: Alternative management 1, AM2: Alternative management 2, AM3: Alternative management 3.

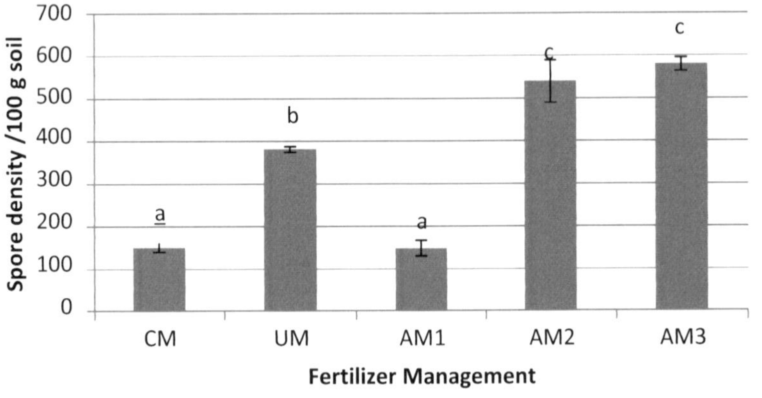

Figure1. Spore density (CM: Conventional management, UM: Universal management, AM: Alternative management 1, AM2: Alternative management 2, AM3: Alternative management 3).

Table 3. Diversity of arbuscular mycorrhiza

Fertilizer Mangement	Shannon –Wiener(H')	Species Richness (SR)
CM	1.06±0.11 c	5.47±0.58 a
UM	0.94±0.05 ab	6.50±0.58 b
AM1	0.90±0.07 a	5.46±0.58 a
AM2	1.05±0.03 bc	6.84±0.00 c
AM3	1.07±0.03 c	7.17±0.57 c

CM: Conventional management, UM: Universal management, AM1: Alternative management 1, AM2: Alternative management 2, AM3: Alternative management 3.

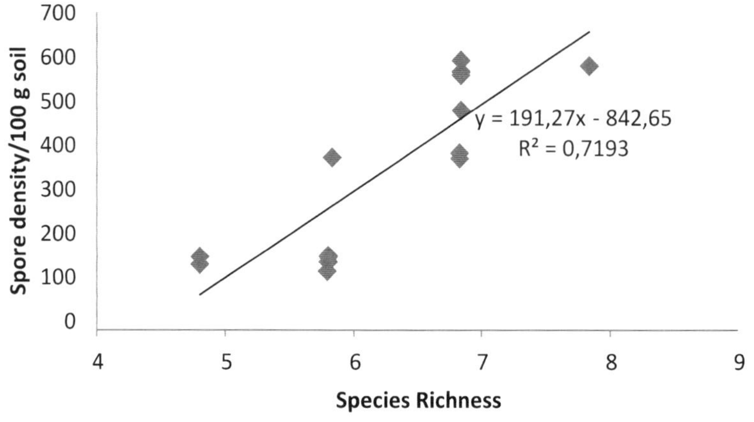

Figure 2. Linear regression between spore density with Species Richness

3.2. Infection of arbuscular mycorrhiza

Infection of arbuscular mycorrhiza was observed in the five plots studied with different infection of arbuscular mycorrhiza. Infection of mycorrhiza was highest in the plot of AM3(63%) and lowest in the plot of CM (36%) (Figure 3). A similar observation infection of arbuscular mycorrhiza (35-55%) in rhizosphere *Eucalyptus camaldulensis* and *E.grandis* in semiarid Brazil was reported Pagano and Scotti [11]. However, infection of arbuscular mycorrhiza only of 16-39% in rhizosphere of *Eucalyptus globules, E. maideni* and *E. sideroxylon* rhizosphere was repored from Northern Algeria [16] and only of 1 – 26% in rhizosphere of *Eucalyptus grandis* in native forest Northern Queensland Australia was reported Adams *et al,* [17]. Infection of arbuscular mycorrhiza in plant roots depends on host plants, soil, climate and land history [13].

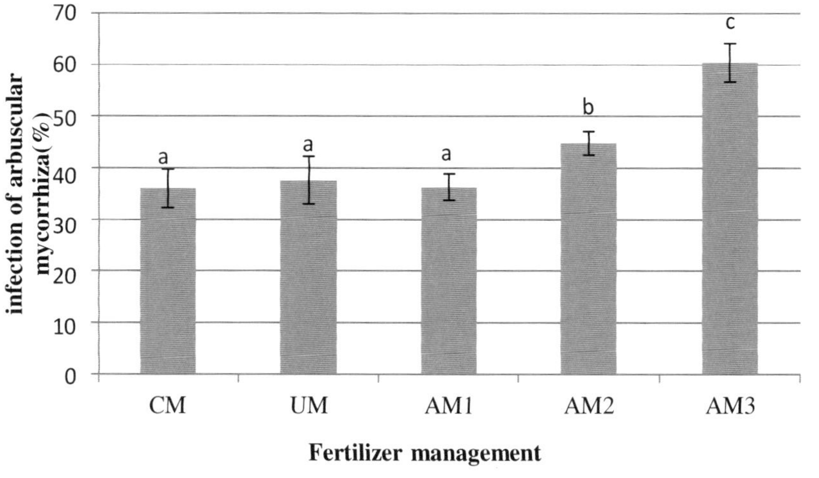

Figure 3. Infection of arbuscular mycorrhiza. (CM: Conventional management, UM: Universal management, AM1: Alternative management 1, AM2: Alternative management 2, AM3: Alternative management 3).

4. Conclusion

Communities of arbuscular mycorrhiza in the rhizosphere of cajeput in agroforestry system showed varies in diversity on different fertilizer management of maize. The highest infection of arbuscular mycorrhiza, highest spore density and highest diversity index value of arbuscular mycorrhiza were found in the plot of AM3. The lowest colonization was observed in plot of CM, and lowest spore density was observed in the plot of AM1. In all fertilizer management, there were eight species of arbuscular mycorrhiza which five species were *Glomus* genus, one species was *Acaulospora* genus and two species were *Gigaspora* genus. Arbuscular mycorrhiza species were dominated by *Glomus constrictum* in all fertilizer management. Especially in the plot of AM3 was found *Acaulospora favoeta.*

References

[1] Cahyani VR 2009. Proceeding of International Seminar of Upland for Food Security. 7-8 November 2009, Faculty of Agriculture Jenderal Soedirman University. ISBN: 978-979-99046-1-4, 254-260
[2] Kamal R, Gusain YS and Kumar V 2014. *International Journal of Current Microbiology and Applied Sciences*, **3(7)**: 564-585
[3] Khalil SE and Yousef RMM 2014. *International Journal of Advanced Research*, **2(6)**: 723-737.
[4] Enkhtua B, Rydlova J and Vosatka M. 2000. *Applied Soil Ecology*, **14**: 201-211
[5] Tiana H, Drijberc RA, Zhang JL and Li XL 2013. *Agriculture, Ecosystems and Environment* **164**: 53– 61.
[6] Wu F, Dong M, Liu Y, Ma X, An L, Peter J, Young W and Feng H 2011. *Plant Soil* **342**: 233–247
[7] Phillips JM AND Hayman DS 1970. *Trans. Br, mycol. Soc.* **55 (I)**:158-161
[8] Gerdemann, JW and Nicolson TH 1963. *Trans. Br, Mycol. Soc.* ***46***: 235-244
[9] Walker C, Mize W, McNabb HS 1982. *Can J Bot* **60**:2518–2529
[10] Jefwa JM, Sinclair R and Maghembe JA 2006. *Agroforestry Systems* **67**:107–114
[11] Pagano MC, Scotti MR 2008. *Mycoscience* **49**:379–384
[12] Silva MCS, Mendes IR, Almeida T, Paula, Luz JMR, Cruz C, Soares DM, Bazzolli, Kasuya MCM 2014. *European Journal of Agriculture and Forestry Research* **(2) 3**: 25-42
[13] Chen YL, Liu S & Dell B 2007. *Mycorrhiza* **17**:527–535
[14] Songachan LS, Kayang H 2011. *Mycosphere* **2(4)**, 497–505.
[15] Kavitha T and Nelson R 2013. *American-Eurasian J. Agric. & Environ. Sci.*, **13 (7)**: 982-987
[16] Adjoud D and Hargas RH 2000. *Mycorrhiza* **9**: 287-290.
[17] Adams F, Reddell P, Webb MJ and Shipton WA 2006. *Australian Journal of Botany* **54**: 271–281

IOP Conf. Series: Journal of Physics: Conf. Series **909** (2017) 012091 doi:10.1088/1742-6596/909/1/012091

Mycorrhizal diversity in the rhizosphere of sugarcane and grass on different soil types

Vita Ratri Cahyani*, **Dewi Rastikawati, Nestri Yuniardi, Jauhari Syamsiyah, Suntoro**

Faculty of Agriculture, Sebelas Maret University
Jl. Ir. Sutami No 36A Kentingan Jebres Surakarta 57126, INDONESIA

E-mail: vitaratri@staff.uns.ac.id

Abstract. Mycorrhiza has been known well as beneficial microbiota for supporting plant growth and production. Understanding of the variability and the consistency of the mycorrhizal diversity on various habitats is important for developing mycorrhizal utilization. Mycorrhizal diversity in the rhizosphere of sugarcane from 4 (four) soil types and the rhizosphere of grass from 3 (three) soil types were investigated in the present study. The results showed that *Glomus* indicated as a versatile genus because it was found as a common and dominant genus in the sugarcane rhizosphere on all of four soil types (Alfisol, Andisol, Inceptisol, Vertisol) and in the grass rhizosphere on all of three soil types (Ultisol, Oxisol, Histosol). In addition, *Acaulospora* was found as a common genus in grass rhizosphere. Statistical analysis indicated that P availability in the rhizosphere of sugarcane had a significantly negative correlation with mycorrhizal spore density, in which decreasing P availability significantly related with increasing spore density.

1. Introduction

Mycorrhiza has been known well as beneficial microbiota for supporting plant growth and production. It also protects plants against abiotic stresses such as drought, acidic or alkalic soil. Mycorrhiza establish broadly distributed potentially symbiotic association with the majority of higher plants and has a role for absorbing nutrient, especially phosphor (P) for plants [1-4]. It also improves drought [5] and freezing tolerance [6], acidic or alkalic soil and salinity tolerance [7-9], protects plant from soil-borne pathogen [10], improves plants phytohormone [11].

The existence of arbuscular mycorrhiza widely found in rhizosphere area, including sugarcane [12-15] and grass [16]. Indigenous mycorrhiza shows better potential than commercial mycorrhiza [17]. Most of the sugarcane field found in agriculture plantation, while grass field found in the nonproductive land. Many factors including host types, soil, and environmental conditions can affect mycorrhizal diversity, infectivity and effectivity [18]. Many studies reported that mycorrhizal diversity determines plant biodiversity, ecosystem variability, and productivity [19-21].

The purpose of the present study was to examine mycorrhizal diversity in the rhizosphere of sugarcane and grass on different soil types. Examining the diversity of mycorrhiza on sugarcane lands and grass lands can extend and deepen the understanding of the variability and the consistency of the existence of mycorrhiza in the rhizosphere of the two types of plants which grown continuously on various habitats. Knowledge of mycorrhizal diversity is expected to be the basis within categorizing various mycorrhiza types based on its ability for supporting plants growth and fitness [22].

2. Materials and Methods

2.1. Materials

Rhizosphere soils of sugarcane were collected from 4 soil types: Alfisol taken from Jumantono (07°37′47″S 110°56′51″E), Andisol taken from Tengaran (07°15′51″ S 110°26′57″ E), Inceptisol taken from Delanggu (07°39′20″ S 110°44′21″E), Vertisol taken from Jatikuwung (07°31′06″S 110°50′44″ E). Rhizosphere soils of grass were collected from 3 soil types: Oxisol taken from Tuntang (07°21′08″S 110°28′43″E), Ultisol taken from Jatinangor (06°55′49″S 107°45′38″ E), and Histosol taken from Rawa Pening (07°15′48″S 110°27′00″ E).

2.2. Sampling of Rhizosphere Soils

Rhizosphere soil of sugarcane and grass were taken in about 500 g from each location with three replications. Nonrhizosphere soils were randomly taken from the bulk soil of nonplant area for the same each soil type as rhizosphere soils at a depth of 0-20 cm with three replications. Soil samples then were analyzed for chemical characteristics of pH H_2O and P availability.

2.3. Isolation and Identification of Mycorrhiza

Isolation of mycorrhiza from rhizosphere and nonrhizosphere soil of sugarcane and grass were conducted using a modified method of wet-sieving and decantation [23] followed by adding sugar solution [24]. A suspension of 100 g of each soil samples and 250 ml aquadest were stirred and allowed a moment for heavier particle settling. Then, the suspension was filtered using a series of sieves with three different diameter sizes of 250 μ, 90 μ, 45 μ. The suspension of filtrate was transferred and divided to several test tube with a balanced weight then added 60% of sugar solution and centrifuged at 2000 rpm for 3 minutes. The sugar solution was pipetted on 45 μ sieves then sprayed with aquadest to petri dish and observed under the microscope. Particle retained in the last sieve 45 μ were collected in the petri dish, and re-suspend in a shallow of standing aquadest then observed using the binocular microscope for spore details (40-400x).

As for mycorrhizal identification, some mycorrhizal spores from petri dish were transferred to slide glass and observed by dropping Melzer solution then covered with cover glass. Identification of spores morphology was using *Species Description of* INVAM (*International Culture Collection (Vesicular) Arbuscular Mycorrhizal*) [25]. Spore density was calculated from total spores found per 100 g soil. Statistical analysis was done to determine the correlation between mycorrhizal spore density and soil chemical characteristics.

3. Results and Discussion

The characteristic of chemical soil content was divided into two parameters, pH and P availability (Figure 1). The effect of soil pH is an important factor toward the development and efficiency of mycorrhiza in various crops that affected mycorrhizal diversity [26-27]. The result of pH measurement was shown in Figure 1 that Alfisol has pH 6,1 which belonged to acidic soil, Andisol has pH 7,1 which belonged to neutral soil, Inceptisol has pH 6,5 which belonged to acidic soil, and Vertisol has pH 7,3 which belonged to alkalic soil.

Figure 1 shows soil type which higher soil acidity or higher soil alkalinity tended to have a lower P availability, while Andisol as a neutral soil in pH had a lowest P availability, this results were from the effect of alophan besides of pH. Figure 1 also showed that all of the soil types on grass rhizosphere were belong acidic soil. P availability of Histosol from grass rhizosphere known as the highest P availability compared to the other soil types both of grass rhizosphere and sugarcane rhizosphere.

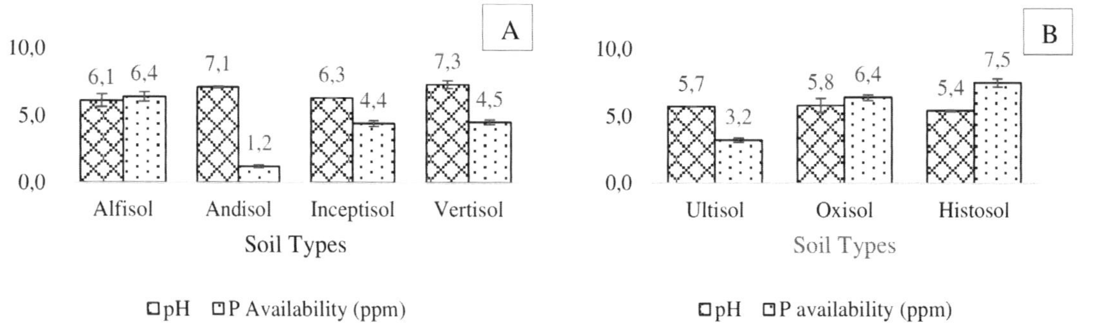

Figure 1. pH H_2O and P availability of sugarcane rhizosphere (A) and grass rhizosphere (B)

Spore density from sugarcane rhizosphere was higher and more diverse than spore density from grass rhizosphere (Table 1) and (Table 2). In sugarcane rhizosphere with acidic and neutral soil types (Alfisol, Inceptisol, and Andisol) had lower spore density than spore density of alkalic soil type (Vertisol). But, spore diversity of acidic and neutral soil were more diverse than spore diversity of alkalic soil type. Sugarcane rhizosphere on Alfisol contained spore density of 109 spores/100 g soil which consisted of *Acaulospora*, *Gigaspora*, *Glomus*, and *Scutellospora*. Sugarcane rhizosphere of Andisol had 142 spores/100 g soil which consisted of *Gigaspora*, *Glomus*, and *Scutellospora*. Sugarcane rhizosphere on Inceptisol had 114 spore/100 g soil which consisted of *Acaulospora*, *Gigaspora*, and *Glomus*. Whereas, sugarcane rhizosphere on Vertisol had 152 spores/100 g soil which consisted of *Acaulospora* and *Glomus*.

Table 1. Mycorrhizal spore density and diversity in the sugarcane rhizosphere

	Alfisol	Andisol	Inceptisol	Vertisol
Spore density (spores 100^{-1}g soil)	109	142	114	152
Spore diversity based on Genus	*Glomus*	*Glomus*	*Glomus*	*Glomus*
	Acaulospora	-	*Acaulospora*	*Acaulospora*
	Gigaspora	*Gigaspora*	*Gigaspora*	-
	Scutellospora	*Scutellospora*	-	-

In the grass rhizosphere on Ultisol had 174 spores/100 g soil which consisted of *Acaulospora*, *Gigaspora*, and *Glomus*. The highest spore density was shown by Oxisol which reached 286 spores/100 g soil than Ultisol that had 174 spores/100 g soil, and the lowest was 68 spores/100 g soil from Histosol. Mycorrhiza spore density and diversity of sugarcane rhizosphere were higher and more diverse than grass rhizosphere. The matter was suspected because soil type of sugarcane rhizosphere had more diverse of pH. The value of soil pH affected an existence of mycorrhiza [28].

Table 2. Mycorrhizal spore density and diversity in the grass rhizosphere

	Ultisol	Oxisol	Histosol
Spore density (spores 100^{-1}g soil)	174	286	68
Spore diversity based on Genus	*Acaulospora*	*Acaulospora*	*Acaulospora*
	Glomus	*Glomus*	*Glomus*
	Gigaspora	-	-

In this study, the correlation was found between spore density with pH and between spore density with P availability. The results of statistical correlation analysis between pH and spore density in the rhizosphere of sugarcane and the rhizosphere of grass were (r=0,478) and (r=0,487). The correlation analysis between P availability and spore density in the rhizosphere of sugarcane and the rhizosphere of grass were (r= -0,917**) and (r= -0,218). The results indicated that decreasing P availability in the sugarcane rhizosphere significantly related with increasing mycorrhizal spore density.

Table 3. Mycorrhizal spore morphology from the rhizosphere of sugarcane (magnification 400x)

	Shape	Color	Spore surface texture	Special characteristics	Reaction with Melzer
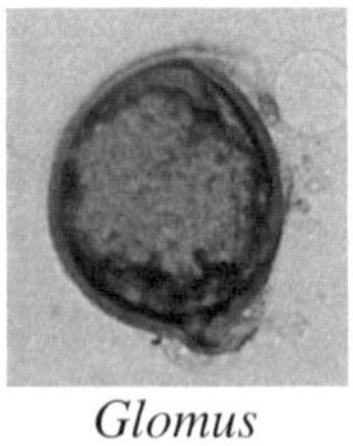 *Glomus*	Round	Brown	Smooth	Subtending hyphae	-
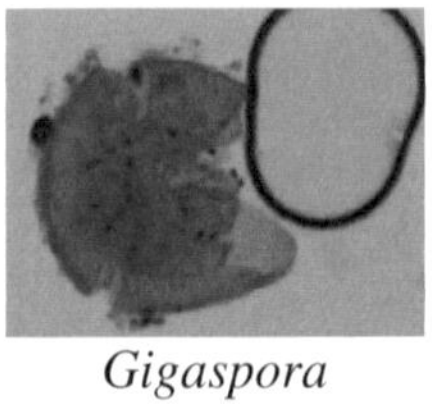 *Gigaspora*	Round	Dark yellow	Smooth	-	Slight change from dark yellow to light yellow
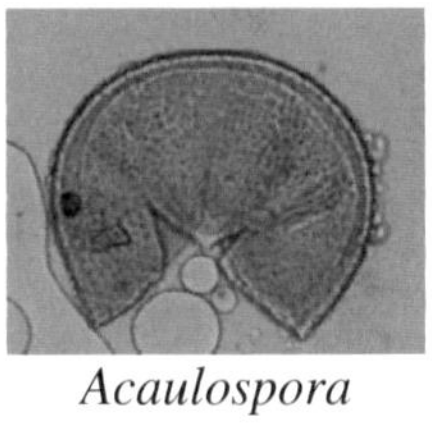 *Acaulospora*	Round	Hyaline	Rough	Have an ornament	A color change inside spore
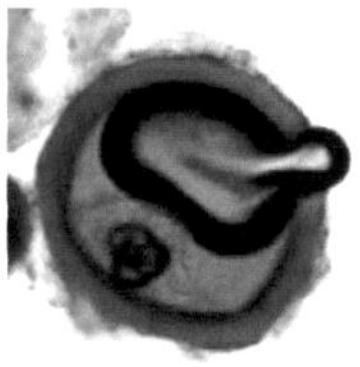 *Scutellospora*	Round	Dark brown	Smooth	-	None

Table 3 shows that spore of sugarcane rhizosphere consisted of four genera; there were *Glomus*, *Gigaspora*, *Acaulospora*, and *Scutellospora*. *Glomus* from sugarcane rhizosphere was round, had a smooth surface, there was subtending hyphae, and did not react with Melzer solution. *Acaulospora* of sugarcane rhizosphere was hyaline, and when it was dropped by Melzer solution the color changed from hyaline into light brown, and in center part of the spore changed darker than side part, it has a rough surface as its ornament. *Scutellospora* of sugarcane was dark brown, had a smooth surface, and no reaction after dropping Melzer solution, it had three layers of the cell wall, and had special characteristic germinal shield within the spore. *Gigaspora* in this study was dark yellow with a smooth surface, had no special characteristic, and had reacted with Melzer then turned from dark yellow to light yellow. *Acaulospora* of sugarcane rhizosphere was hyaline; there was an ornament on its spores surface, it had three layers of the cell wall, after adding a drop of Melzer solution, the color of spore was changed, in the center part of the spore turned into more darker.

The present study shows similar finding as reported by Kumalawati [14] that *Glomus*, *Acaulospora*, *Gigaspora* were found as a common mycorrhiza of the sugarcane rhizosphere in three districts of South Sulawesi province. The difference is that South Sulawesi, especially in Gowa, found *Sclerocystis* besides the three genera, and they did not found *Scutellospora*. The other study that found similar result was reported by Srikumar [12] that *Glomus* and *Gigaspora* were found in four study sites areas in India.

Table 4. Mycorrhizal spore morphology from the rhizosphere of grass (magnification 400x)

	Shape	Color	Spore surface texture	Special characteristics	Reaction with Melzer
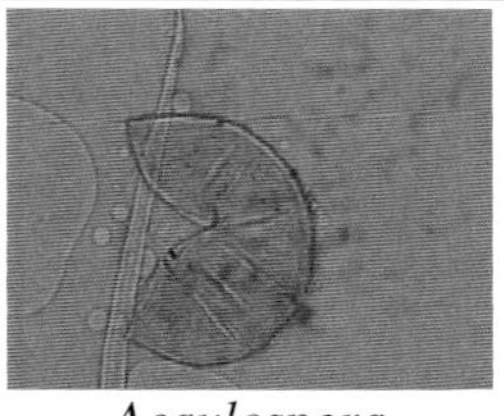 *Acaulospora*	Round	Hyaline	Smooth	-	Slight change from hyaline to pale yellow
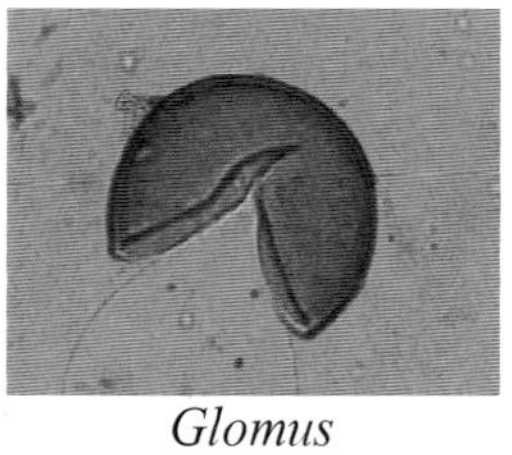 *Glomus*	Round	Brown	Smooth	Subtending hyphae	None
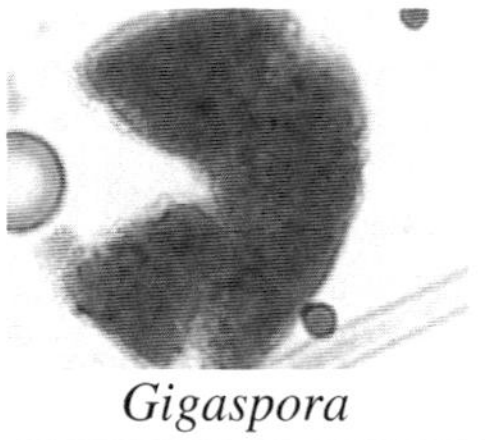 *Gigaspora*	Subglobose	Red	Rough	Have an ornament	React thoroughly

Table 4 shows that on the grass rhizosphere there were only found three genera; there were *Acaulospora*, *Glomus,* and *Gigaspora.* In Ultisol had 174 spores/100 g soil that consisting of *Acaulospora*, *Gigaspora* and *Glomus*. Whereas, grass rhizosphere in Oxisol contained spores density of 286 spores/100 g soil and in Histosol 68 spores/100 g soil, consisting of *Acaulospora*, and *Glomus*. *Acaulospora* in this study was hyaline with the surface structure was smooth, had no special characteristic, and reacted with Melzer from hyaline to pale yellow. *Glomus* in this study was brown with the surface structure was smooth, had subtending hyphae, and had no reacted with Melzer. *Gigaspora* in this study was red with the surface structure was rough, had an ornament, and reacted with Melzer from red to orange. Another study [15] reported the same genera that found in the several rhizosphere of grasses. *Acaulospora*, *Glomus*, and *Gigaspora* were found on the grass rhizosphere in Penelokan village of Bali.

Conclusion

Glomus was only found as a common and dominant genus on all of four soil types (Alfisol, Andisol, Inceptisol, Vertisol) in the sugarcane rhizosphere. The rhizosphere of sugarcane from Vertisol indicated the highest of mycorrhizal spore density of 152 spores/100 g with the lowest of mycorrhizal diversity. The rhizosphere of sugarcane from Alfisol indicated the lowest of mycorrhizal spore density of 109 spores/100 g with the highest of mycorrhizal diversity.

Acaulospora and *Glomus* were found as a common genera on all of three soil types (Ultisol, Oxisol, Histosol) in the grass rhizosphere. The rhizosphere of grass from Oxisol indicated the highest of mycorrhizal spore density of 286 spores/100 g and from Histosol indicated the lowest of mycorrhizal spore density of 68 spores/100 g. In addition, the genus of *Gigaspora* was only found in the rhizosphere of grass on Ultisol.

It was found that P availability in the sugarcane rhizosphere had a negative correlation significantly with mycorrhizal spore density, in which decreasing P availability significantly related with increasing spore density.

Acknowledgement

This research was supported in part by a Grant of "Hibah Penelitian Strategis Nasional 2016" from Directorate General of Higher Education Indonesia.

References

[1] Cahyani VR 2009 Proceeding of International Seminar of Upland for Food Security. 7-8 November 2009, Faculty of Agriculture Jenderal Soedirman University. ISBN: 978-979-99046-1-4, 254-260.
[2] Papilaya PM 2013 *International Journal of Sciences and Research (IJSR)* 4: 1935-1942.
[3] Syafria H, Jamarun N, Zein M and Yani E 2015 *International Journal of Agriculture Science IJAC* 1:47-54.
[4] Al-Karaki GN 2016 *Agrofor International Journal.* 1: 154-161. doi: 10.7251/AGRENG1602154A.
[5] Sulistiono W, Taryono, Yudono P and Irham 2017 *Journal of Agricultural Science* 9: 95-108.
[6] Otgonsuren B, Lee JT and Lee MJ 2015 *International Journal of Advanced Research in Biological Sciences*2:191-204.
[7] Shinde SK, Shinde BP, and Patale SW 2013 *Int. J. Life Sci& Pharma Res.* 3: 11-15.
[8] Cahyani VR, Ariyanto DP, Qodri BM, Pramesti GPA, Purwani RH, Purba SM, Asakawa S and Kimura M 2014 Proceedings of 11th International Conference of the East and Southeast Asia Federation of Soil Science Societies, 21-24 October 2013, IPB, Bogor, ISBN 978-979-19904-1-7, 434-435.
[9] Elhindi KM, El-Din AS and Elgoban AM 2016 *Saudi Journal of Biological Sciences* 24: 170-179. Sjbs.2016.02.010.
[10] Newsham KK, Fitter AH and Watkitason AR 1995 *Journal of Ecology* 83: 991-1000.
[11] Pozo MJ, Lopez-Raez JA, Azcon-Aguilar C and Garcia-Garrido JM 2015 *New Phytologist* 205: 1431–1436. doi: 10.1111/nph.13252.
[12] Srikumar R, Thangaraj R and Murugaian P 2009 *Agriultural Science Digest* 9: 1-4.
[13] Prabudoss V 2011 *International Journal of Current Research* 3: 228-234.
[14] Kumalawati Z, Musa Y, Asrul L and Ridwan I 2014 *International Journal of Scientific & Technology Research IJSTR* 3: 201-203.
[15] Srinivasan M, Kumar K, Kumutha K and Marimuthu P 2014 *Journal of Applied and Natural Sciences* 6: 29-293.
[16] Dewi NKS, Susrama IGK, Sritamin M, Adnyana M and Wirawan IGP 2014 *International Journal of Biosciences and Biotechnology* 2: 26-31.
[17] Berruti A, Lumini E, Balestrini R and Bianciotto V 2016 *Front. Microbiol.* 6:1559. doi: 10.3389/fmicb.2015.01559.

[18] Moreira M, Baretta D, Tsai SM, Gomes-daCosta SM and Cardoso EJBN 2007 *Sci. Agric.* 64 4: 393-399.
[19] Van der Heijden MGA, Klironomos J, Ursic M and Sanders IR 1998 *Nature* 396.6706 69.
[20] Van der Heijden MGA 2002 Mycorrhizal Ecology 157: 243-265.
[21] Klironomos J, McCune J, Hart M, Neville J 2000 *Ecology Letters* 3 2: 137-141
[22] Alarcon A, Hernandez-Cuevas LV, Ferrera-Cerrato R and Franco-Ramirez A 2012 *J. Biofertil Biopestici* 3 1 p 1000115.
[23] Gerdemann JW, Nicolson TH 1963 *Trans. Brit. Mycol. Soc.* 46, 336-338.
[24] Walker C, Mize W and McNabb Jr HS 1982 *Can. J. Bot.* 60, 2518-2529.
[25] INVAM [International Name of Vesicle Arbuscular Mycorrhizae] 2016 http://www.invam.wvu.edu/the-fungi. Accessed on June 2016.
[26] Singh G, Goyne KW and Kabrick JM 2015 *Geoderma Regional* 5: 117-126.
[27] Singh S 2004 *Mycorrhiza News* 16 2: 2-23.
[28] Kluber LA, Carrino-Kyker Sr, Coyle KP, DeForest JL, Hewins CR, Shaw AN, Smemo KA, Burke DJ 2012 *PLoS ONE* 7(11): e48946. doi:10.1371/journal.pone.0048946.

IOP Conf. Series: Journal of Physics: Conf. Series **909** (2017) 012092 doi:10.1088/1742-6596/909/1/012092

Histology of Epiphyseal Plate of Adolescent Rat Stimulated by Laserpuncture

Selfi Handayani[1*], Ari H Ramelan[2], Bambang Purwanto[3], Koosnadi Saputra[4], Didik G Tamtomo[5]

[1]Doctoral Program in Medical Sciences, Postgraduate Universitas Sebelas Maret Surakarta Indonesia.

[2]Department of Physics, Faculty of Mathematics and Natural Sciences, Universitas Sebelas Maret Surakarta Indonesia

[3]Department of Internal Medicine, Faculty of Medicine, Universitas Sebelas Maret Surakarta Indonesia.

[4]Acupuncture Reseach Laboratory for Health Services, National Institute of Health of Republic Indonesia /Academy Acupuncture of Surabaya.

[5]Department of Anatomy, Faculty of Medicine, Universitas Sebelas Maret Surakarta Indonesia

E-mail: selfi_handayanii@yahoo.com

Abstract. Epiphyseal plate was used for determining longitudinal bone growth. Laserpuncture was believed to stimulate height growth. We used 40 male Wistar rats which aged three weeks old and weighed more than 40 g as subjects. They randomly divided into group A or B and each group evenly divided into four subgroups which were a negative control and others applied with laser on GV20, ST 36 or combination of GV 20+ST 36 respectively. These acupoints were then stimulated using the laser. After treatment, mice were sacrificed, then tibias were taken for histology preparation processes. By light microscope, epiphyseal plate (EP) height (μm) and chondrocytes hypertrophy (CH) height were measured at six equidistant points, and the values were averaged to obtain a final result for each section. Collected data were analyzed using ANOVA test, and the significant value was set up $p< 0.05$. The mean of EC and CH were lower than control, but mean of ratio EP/CH were higher than control. However, ANOVA showed that there did not differ significantly ($p=0.36$).

1. Introduction

Epiphyseal plate also known as growth plate has been understood as a target organ for longitudinal bone growth which determines a final height. This plate contains organized cartilage structure located between the epiphyseal and metaphyseal bone at the distal ends of the long bones [1,2,3].

Histologically, the epiphyseal plate consists of horizontal zones of chondrocytes at different stages of differentiation. The germinal zone, or resting zone lie toward the epiphysis of the growth plate, contains immature mesenchymal chondrocytes, which are small, round, singly or distributed in pairs within a compact extracellular matrix (ECM). Beneath it is the proliferative zone, chondrocytes assume a flattened, discoid shape and form column-like structures undergo longitudinal cell division and direct the lengthening of the bone in one dimension. And then, proliferating chondrocytes lose their capacity to divide; they differentiate and become prehypertrophic, together with a size increase. Proliferating chondrocytes are located in the transition (maturation or prehypertrophic) zone. In the hypertrophic zone, round cells produce matrix proteins in large amounts. This stage is characterized by abundant in intracellular calcium concentration, which is essential in the production of matrix vesicles. [4,5,6,7,8,9].

Longitudinal bone growth takes place by a process called endochondral ossification, in which a cartilaginous scaffold is replaced by bone in a coordinated fashion. It is regulated by genetic, hormonal growth, and environment factors. The major systemic hormones that control longitudinal bone growth during childhood are GH, insulin-like growth factor 1 (IGF-1), the thyroid hormones triiodothyronine (T3) and thyroxine (T4), and glucocorticoids. During puberty, the sex steroids play the most significant role.Whereas, locally it regulated by factors such as Indian hedgehog (Ihh), parathyroid hormone–related peptide (PTHrP), and fibroblast growth factors (FGFs) [3,4,5,8,9].

Acupuncture has been used to treat various diseases and maintain healthy since thousand years ago. The term acupuncture is derived from the Latin radicals "acus and pungere" meaning needle and puncture. Acupuncture therapy is an act of needles on specific areas of the body to get the desired effect. This area is known as acupuncture points spread across the surface of the body. Selection of points in acupuncture therapy based on the indication and the desired effect. In contemporary western medicine, acupuncture is considered a form of sensory stimulation that affects several classes of afferent nerve fibers and once directed to the spinal cord and the higher centers in the CNS, it triggers the neurophysiological events that generate the therapeutic outcome. The stimuli generated by acupuncture can activate the autonomic centers and Hypothalamic Pituitary- Adrenal (HPA) axis and, thereby, help to maintain homeostasis and improve the efficiency of several systems of the organism [10,11]. Needling acupuncture is believed to increase levels of endorphins and levels of GH in the body.

LASER is the acronym of Light Amplification by Stimulated Emission of Radiation. According to the doses, the laser is divided into the hard laser, medium laser, and low-level laser. The low-level laser which applied to specific acupoint to stimulate body system is known as laserpuncture. Laser is an electromagnetic wave which can inhibit and biostimulate biological tissue, such as improving cellular activities, nervous regeneration in central or peripheral, and also taking roles in hormones and enzymes production. Laserpuncture is one modality of low power laser therapy, non-invasive form of phototherapy which is defined as a traditional acupuncture point stimulation with low intensity and non thermal laser irradiation. Noninvasive and painless nature of the laser is useful to provide comfort and relieve the fear of children. Although the use of low-power, laserpuncture has the same principles of traditional acupuncture and capable of the same biological effects with acupuncture needles [12,13,14,15,16,17]. If acupoint is stimulated by laser with particular dose, it will effect on the body system. The impulse in the form of laser is delivered by the nervous system to higher center and will stimulate hormonal releasing from hypothalamus, and then stimulate hypophysis, and it eventually will release growth factor which influences growth [16,17,18].

Compared to the needle,the laser has more beneficial such as no pain, no termal, precise doses, effective and efficience. Laserpuncture technique combines the advantages of traditional Chinese acupuncture with modern medicine [14].

In this context, the aim of this study was to present histology changes (epiphyseal plate height and chondrocytes hypertrophy height) of epiphyseal plate of adolescent rats stimulated by laserpuncture.

2. Material and methods

2.1. Ethic Statement

Mice kept and bred according to Institutional Guidelines of the Gadjah Mada University Jogjakarta, approved by Laboratorium Farmakologi dan Farmasi Klinik UGM Jogjakarta.

2.2. Animal Model

Forty male Wistar rats which aged three weeks old and weighed more than 40 g randomly divided into group A or B. Each group evenly divided into four subgroups which were a negative control and three subgroups that were stimulated by laser on GV20, ST 36 or combination of GV 20+ST 36 respectively. These acupoints were then applied using the KX Laser GX-2000B which emitted cold red laser (635– 680nm/5mW). Laserpuncture was performed 60 seconds/day for ten days (group A) and 15 days (group B). [Fig. 1]

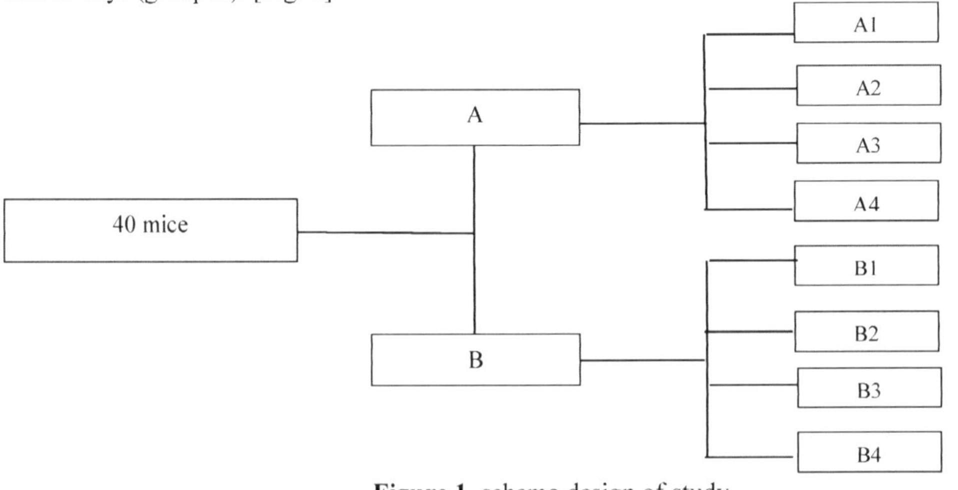

Figure 1. schema design of study

2.3. Laserpuncture

Laserpuncture was performed by laser stimulation using KX Laser GX-2000B (Kangxing), a semiconductor-based low-level laser therapy (LLLT) device emitting a cold red laser (635–680nm/5mW). Before being a laser beam, first, determine the point ST 36 and GV 20. The area around the points was shaved gently with a razor, then laser affixed, a perpendicular laser is turned on for 60 seconds. Control rats were operated with the inactivated probe which was turned off. Acupoint ST 36 is located at two lower limbs, approximately 1mm lateral to the tibial tuberosity, whereas GV 20 located at the peak of head. [Fig.2]

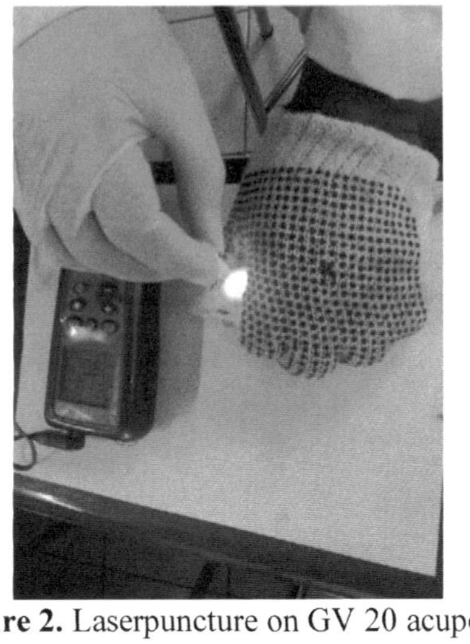

Figure 2. Laserpuncture on GV 20 acupoint

IOP Conf. Series: Journal of Physics: Conf. Series **909** (2017) 012092 doi:10.1088/1742-6596/909/1/012092

2.4. Histology Analysis
Following cervical dislocation, the tibiae were resected for free from other tissues. The specimens were fixed in 10% saline buffered formalin at room temperature for two days, then decalcified in 10% HNO3 for 72 hours. After this, it was prepared for following histological processing: dehydration, diaphanization, paraffin embedding, and microtomy, longitudinal section was performed to a thickness of 3μm sections. The sections were stained with hematoxylin-eosin (HE).

Histology structure was assessed by looking epiphyseal plate height and chondrocytes hypertrophy height. Measurement was made by using a camera attached to the light microscope and an image program. Epiphyseal plate (EP) height (μm) was measured at six equidistant points, and the values were averaged to obtain a final result for each section (Fig. 3). Chondrocytes hypertrophy (CH) height (μm) was measured at six equidistant points, and the values were averaged to obtain a final result for each section. The hypertrophy zone is limited by the first hypertrophic chondrocyte on the upper side and by the first mixed trabeculae on the lower side (Fig. 4). A ratio of the epiphyseal plate height to chondrocyte hypertrophy height (EP/CH) was calculated to assess the histology changes.[Fig.3 and 4]

2.5. Statistical Analysis
Data were expressed as means with standard deviations (SD). All data were analyzed using SPSS version 18.0 for Windows. Statistical significance was determined using ANOVA. Significance was accepted at $P < 0.05$.

3. Result

3.1. Histology Analysis
Measurements of epiphyseal plate zone heights were taken in the central part. The start of the proliferative zone was defined as the point at which the round, individual cells of the resting zone start to become disc shaped and organized into columns. The start of the hypertrophic zone was defined as the point at which proliferative zone chondrocytes round up and become larger. The end of the hypertrophic zone was defined as the vascular invasion front.

The means of EP, CH and ratio EP/CH can be seen in table 1.

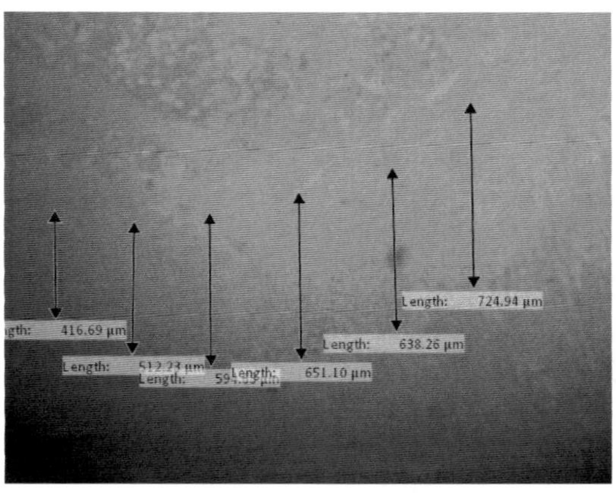

Figure 3. Epiphyseal plate width (μm) and hypertrophic chondrocyte width (μm) were measured at six equidistant points

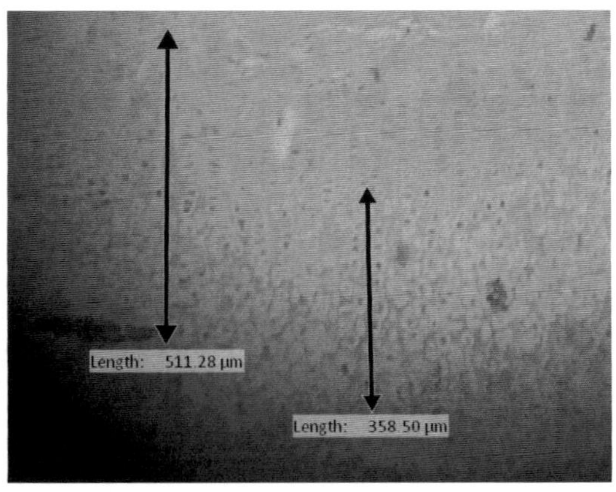

Figure 4. The values were averaged to obtain a final result for each section

Table 1. Height epiphyseal plate (EP) μm, chondrocyte hypertrophy (CH) μm, and ratio EP/CH.

	Mean EP ± SD	Mean CH±SD	Mean EP/CH±SD
ControlA [A1]	482.202 ± 79.189	323.892 ± 73.385	1.550± 0.493
GV20A [A2]	418.802 ± 66.297	283.402 ± 47.228	1.496±0.294
ST36A [A3]	438.812 ± 40.652	291.592 ± 66.983	1.520±0.262
GV20+ST36A [A4]	405.732 ± 35.763	274.332 ± 11.587	1.474±0.121
ControlB [B1]	419.952 ± 65.452	274.462 ± 59.623	1.550±0.141
GV20B [B2]	341.362 ± 90.078	248.642 ± 54.308	1.295±0.215
ST36B [B3]	381.402 ± 74.164	219.302 ±52.985	1.692±0.331
GV20+ST36B [B4}	385.252± 59.365	235.352 ± 59.596	1.692± 0.286

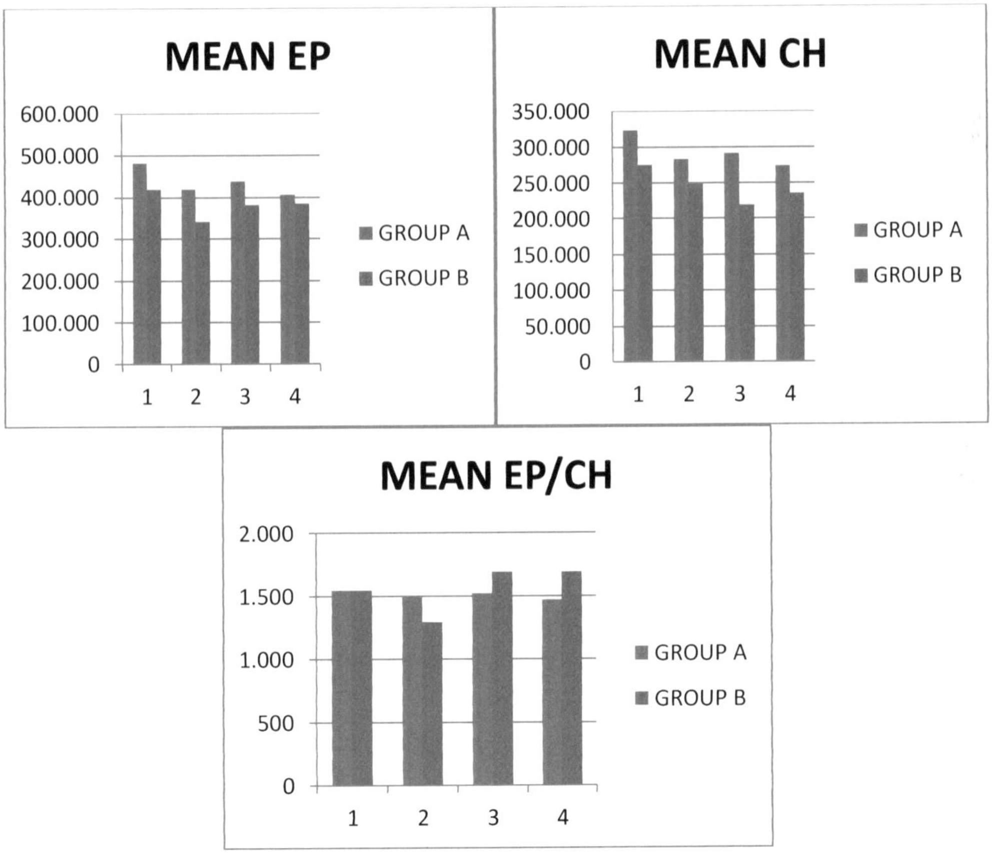

Figure 5. Mean epiphyseal plate (EP) height, mean chondrocyte hypertrophy (CH) height, mean ratio epiphyseal plate/chondrocyte hypertrophy (EP/CH). 1. Control, 2. GV20, 3. ST36 and 4. GV20+ST36

3.2. *Statistical Analysis.*

There are histological changes that can be seen by increasing and decreasing ratio EP/CH depending on a specific acupoint. However, test ANOVA showed that did not differ significantly (mean EP $p=0.083$, mean CH $p=0.123$, and mean EP/CH $p=0.55$, respectively).

4. Discussion

The usage of laserpuncture is widespread with varying indications and varying dosage, but the study on the effect of low-level laser that is used in laserpuncture in epiphyseal plate are few. In this study, we used a low-level laser with 635 nm wavelength, five mW power, and it were given for 60 seconds, so the energy density was 0.3J /cm2. By specific acupoint, treatments were performed in different amounts of time, in which group A received the laser ten times in 10 days, while group B gained laser for 15 days. Here, 2 points were used, namely, point GV 20 that is located at the top of the head and point ST 36 that is located in the two lower limbs (Fig.6). Anatomically, the GV20 point located at the top of the head is adjacent to the brain, and it is associated with the frontal lobes. Whereas ST 36 points were used because it stimulate releasing endorphin, ghrelin, [20] and other neurotransmitters.

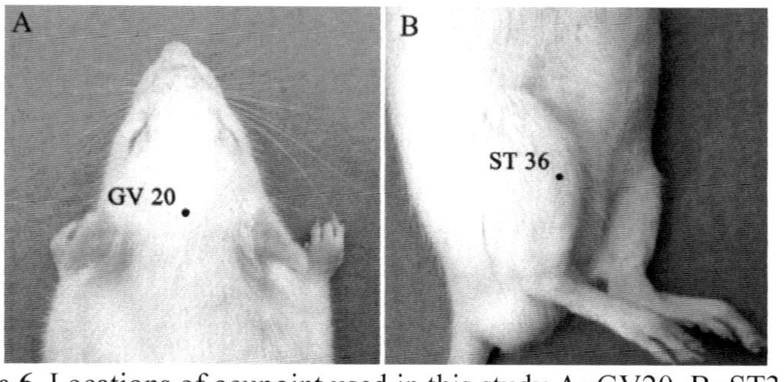

Figure 6. Locations of acupoint used in this study A=GV20, B=ST36 [21]

This study indicates (Fig.5) that histological features are seen from an epiphyseal plate and chondrocyte hypertrophy height differed between subgroups. EP height in each laserpuncture subgroup was reduced, all of the subgroups were lower than control. Also, CH height in each laserpuncture subgroup decreased according to a specific acupoint. It could be seen that the lowest consecutive values were the subgroup of GV20+ST36, GV20 and ST36. Then we look at EP/CH ratio differences. It was seen that laserpuncture effect on EP/CH ratio according to acupoint, where the GV20 was lower than control, whereas the ST36 and GV20+ST36 subgroups were higher than control. However, ANOVA showed that there were not differences between subgroup significantly. Let we look at group A and B differences. When viewed from the duration of stimulation, the mean of ratio EP/CP in group A were smaller than group B, but statistically, there were not differences between A and B significantly.

Unlike this study, Yeom et al. [22] applied laser with a cold red laser (635– 680nm/40mW) on ST36 and SP6 acupoints with energy density 1,2J/cm2. They observed a significant increase in mice tibia length. They report that increasing were on hypertrophy zones and growth factors like IGF-1 and BMP2.

Another study, Cressoni et al. [23] evaluated the epiphyseal plate of rats using histological and radiographic analysis after they had been submitted to the application of low-level laser therapy with a wavelength of 830 nm, and doses of 5 J/ cm2 and 15 J/cm2. They observed that low-level laser therapy have no change in bone length. Similiar to it, de Andrade, A.R et al. [24] concluded that low laser with wavelength 670 nm in doses of 4J/cm2, 8J/cm2 and 16J/cm2 did not cause changes in the zones of epiphyseal cartilage. Compared with this study, they applied the laser on non-acupoint areas, (tibia and mandible only a side), while this study stimulated lasers on acupuncture points.

5. Conclusion

Our results showed that laserpuncture using a 635-nm semiconductor have an effect on histological changes in adolescent rats, but statistically are not significant. The ratio EP/CH on GV 20 is lower than control, whereas the ratio EP/CH in ST 36 and GV20+ST36 are higher than control.

The interesting is that these changes are not significant. This can be explained, that theoretical acupuncture effect to balance the body functions. Maybe, when done under normal conditions, this effect is useful for optimization. It does not get to the extreme point that causes damage to the network. How the laserpuncture can affect the body system is not clearly known, but what can be explained here is that according to the specific points, laserpuncture is saved enough for stimulation body function.

Acknowledgment
Sincere appreciation and thanks to the Laboratorium Patologi Anatomy FK Universitas Sebelas Maret to the support.

References

[1] Buckwalter, J.A., and Sjolund, R.D., 1989. The growth of cornstalks and long bones. *The Iowa orthopedic journal*, *9*, p.25.
[2] Noonan, K.J., Hunziker, E.B., Nessler, J. and Buckwalter, J.A., 1998. Changes in a cell, matrix compartment, and fibrillar collagen volumes between growth- plate zones. *Journal of orthopaedic research*, *16*(4), pp.500-508.
[3] Choi, H., Choi, Y., Kim, J., Bae, J. and Roh, J., 2017. Longitudinal bone growth is impaired by direct involvement of caffeine with chondrocyte differentiation in the growth plate. *Journal of Anatomy*, *230*(1), pp.117-127.
[4] Breur, G.J., Farnum, C.E., Padgett, G.A. and Wilsman, N.J., 1992. Cellular basis of decreased rate of longitudinal growth of bone in pseudoachondroplastic dogs. *J Bone Joint Surg Am*, *74*(4), pp.516-528.
[5] Yeung Tsang, K., Wa Tsang, S., Chan, D. and Cheah, K.S., 2014. The chondrocytic journey in endochondral bone growth and skeletal dysplasia. *Birth Defects Research Part C: Embryo Today: Reviews*, *102*(1), pp.52-73.
[6] Van der Eerden, B.C.J., Karperien, M. and Wit, J.M., 2003. Systemic and local regulation of the growth plate. *Endocrine reviews*, *24*(6), pp.782-801.
[7] Shim, K.S., 2015. Pubertal growth and epiphyseal fusion. *Annals of pediatric endocrinology & metabolism*, *20*(1), pp.8-12.
[8] Sun, M.M.G. and Beier, F., 2014. Chondrocyte hypertrophy in skeletal development, growth, and disease. *Birth Defects Research Part C: Embryo Today: Reviews*, *102*(1), pp.74-82.
[9] Desjardin, C., Charles, C., Benoist-Lasselin, C., Riviere, J., Gilles, M., Chassande, O., Morgenthaler, C., Laloé, D., Lecardonnel, J., Flamant, F. and Legeai-Mallet, L., 2014. Chondrocytes play a major role in the stimulation of bone growth by thyroid hormone. *Endocrinology*, *155*(8), pp.3123-3135.
[10] Cole, H.A., Yuasa, M., Hawley, G., Cates, J.M., Nyman, J.S. and Schoenecker, J.G., 2013. Differential development of the distal and proximal femoral epiphysis and physis in mice. *Bone*, *52*(1), pp.337-346.
[11] Hamvas, S., Havasi, M., Szőke, H., Petrovics, G. and Hegyi, G., 2016. Different Techniques of Acupuncture–Part of the Traditional Chinese Medicine and "Evidence Based Medicine". *JOURNAL OF TRADITIONAL MEDICINE AND CLINICAL NATUROPATHY*, *5*(3), p.1000202.
[12] Villas-Boas, J.D., Dias, D.P.M., Trigo, P.I., Almeida, N.A.D.S., de Almeida, F.Q. and de Medeiros, M.A., 2015. Acupuncture affects autonomic and endocrine but not behavioural responses induced by startle in horses. *Evidence-Based Complementary and Alternative Medicine*, *2015*.
[13] Hamblin, M.R. 2016. Photobiomodulation or low-level laser therapy.*Journal of Biophotonics*. Vol. 9, No. 11-12: 1122-1124
[14] de Oliveira, R.F., da Silva, C.V., Cersosimo, M.C.P., Borsatto, M.C. and de Freitas, P.M., 2015. Laser therapy on points of acupuncture: Are there benefits in dentistry?. *Journal of*

Photochemistry and Photobiology B: Biology, *151*, pp.76-82.

[15] Litscher, G. and Opitz, G., 2012. Technical parameters for laser acupuncture to elicit peripheral and central effects: state-of-the-art and short guidelines based on results from the Medical University of Graz, the German Academy of Acupuncture, and the scientific literature. *Evidence-Based Complementary and Alternative Medicine*, *2012*.

[16] Fuchtenbusch, A., 2014. Laser acupuncture. *AACP ACUPUNCTURE*, *19*, p.20

[17] Khalid, M.Z., 2016. Mechanism of Laser/light beam interaction at cellular and tissue level and study of the influential factors for the application of low-level laser therapy. *arXiv preprint arXiv:1606.04800*.

[18] Soekwanto, L., 2016. The Influence of Laserpuncture Semi Conductor Shooting to Broiller Chicken's Biometry (Doctoral dissertation, Universitas Airlangga).

[19] Mantayborbir, V., Fadjar, M. and Mahendra, A.P., 2013. Exploration laser puntures exposure effect on reproductive point to increasing number of leydig cells catfish (Clarias sp.). *J. Life Sci. Biomed*, *3*(6), pp.444-449.

[20] Nenda, M.M., Lewicki, M. and Mandalunis, P.M., 2016. Histomorphometry of the tibia and mandible of healthy female Wistar rats at different stages of growth. *Experimental Animals*, *65*(2), pp.109-116.

[21] Güçel, F., Bahar, B., Demirtas, C., Mit, S. and Çevik, C., 2012. Influence of acupuncture on leptin, ghrelin, insulin and cholecystokinin in obese women: a randomised, sham-controlled preliminary trial. *Acupuncture in Medicine*, *30*(3), pp.203-207.

[22] Guo, K., Zhu, J., Quan, X., Zhang, D., Chen, X., Yang, C., Li, X., Shi, J., Hu, W. and Zhu, Q., 2015. Comparison of the effects of pretreatment with repeated electroacupuncture at GV20 and ST36 on fatigue in rats. *Acupuncture in Medicine*, *33*(5), pp.406-412.

[23] Yeom, M., Kim, S.H., Lee, B., Zhang, X., Lee, H., Hahm, D.H., Sohn, Y. and Lee, H., 2013. Effects of laser acupuncture on longitudinal bone growth in adolescent rats. *Evidence-Based Complementary and Alternative Medicine*, *2013*.

[24] Cressoni, M.D.C., Giusti, H.H.K.D., Piao, A.C., de Paiva Carvalho, R.L., Anaruma, C.A. and Casarotto, R.A., 2010. Effect of GaAlAs laser irradiation on the epiphyseal cartilage of rats. *Photomedicine and laser surgery*, *28*(4), pp.527-532.

[25] de Andrade, A.R., Meireles, A., Artifon, E.L., Brancalhão, R.M.C., Ferreira, J.R.L. and Bertolini, G.R.F., 2012. The effects of low-level laser therapy, 670 nm, on epiphyseal growth in rats. *The Scientific World Journal*, *2012*.

Institute of Physics
Dirac House, Temple Back
Bristol BS1 6BE UK

ISSN: 1742-6588
ISBN 978-1-5108-5339-3

9 781510 853393

CURRAN ASSOCIATES INC.
proceedings.com

Postgraduate Symposium in Civil and Environmental Engineering (PSCEE 2019)

IOP Conference Series: Materials Science and Engineering
Volume 601

Batu Pahat, Malaysia
31 March 2019

IOP
Institute of Physics